2/10/90

Distance Geometry and Molecular Conformation

CHEMOMETRICS SERIES

Series Editor: **Dr. D. Bawden**
Pfizer Central Research, Sandwich, Kent, England

1. Distance Geometry and Conformational Calculations*
G.M. Crippen

2. Clustering of Large Data Sets*
Jure Zupan

3. Multivariate Data Analysis in Industrial Practice
Paul J. Lewi

4. Correlation Analysis of Organic Reactivity: With particular reference to multiple regression*
John Shorter

5. Information Theoretic Indices for Characterization of Chemical Structures
Danail Bonchev

6. Logical and Combinatorial Algorithms for Drug Design
V.E. Golender *and* **A.B. Rozenblit**

7. Minimum Steric Difference
The MTD method for QSAR studies
Z. Simon, A. Chiriac, S. Holban, D. Ciubotaru *and* **G.I. Mihalas**

8. Analytical Measurement and Information
Advances in the information theoretic approach to chemical analyses
K. Eckschlager *and* **V. Štěpánek**

9. Molecular Connectivity in Structure-Activity Analysis
Lemont B. Kier *and* **Lowell H. Hall**

10. Potential Pattern Recognition in Chemical and Medical Decision Making
D. Coomans *and* **I. Broeckaert**

11. Chemical Pattern Recognition
O. Štrouf

12. Similarity and Clustering in Chemical Information Systems
Peter Willett

13. Multivariate Chemometrics in QSAR: A Dialogue
Peter P. Mager

14. Application of Pattern Recognition to Catalytic Research
I.I. Ioffe

15. Distance Geometry and Molecular Conformation
G.M. Crippen *and* **T.F. Havel**

*Out of print

"My goodness, Toto, I don't think we're in $\mathbb{R}^n$ anymore!"

Distance Geometry and Molecular Conformation

G.M. Crippen

Associate Professor
College of Pharmacy
University of Michigan, U.S.A.

T.F. Havel

Assistant Research Scientist
Molecular Biology Department
Scripps Clinic and Research Foundation, U.S.A.

RESEARCH STUDIES PRESS LTD
Taunton, Somerset, England

JOHN WILEY & SONS INC.
New York · Chichester · Toronto · Brisbane · Singapore

RESEARCH STUDIES PRESS LTD.
24 Belvedere Road, Taunton, Somerset, England TA1 1HD

Marketing and Distribution:
Australia, New Zealand, South-east Asia:
Jacaranda-Wiley Ltd., Jacaranda Press
JOHN WILEY & SONS INC.
GPO Box 859, Brisbane, Queensland 4001, Australia

Canada:
JOHN WILEY & SONS CANADA LIMITED
22 Worcester Road, Rexdale, Ontario, Canada

Europe, Africa:
JOHN WILEY & SONS LIMITED
Baffins Lane, Chichester, West Sussex, England

North and South America and the rest of the world:
JOHN WILEY & SONS INC.
605 Third Avenue, New York, NY 10158, USA

Library of Congress Cataloging-in-Publication Data

Crippen, G.M.
Distance geometry and molecular conformation/G.M. Crippen, T.F. Havel.
p. cm. – (Chemometrics series ; 15)
Bibliography: p
Includes index.
ISBN 0 86380 073 4 ISBN 0 471 92061 4 (Wiley)
1. Stereochemistry. 2. Distance geometry. 3. Conformational analysis. I. Havel, T.F. (Timothy F.), 1953– . II. Title. III. Series.
QD481.C76 1988
541.2'23 – dc19 88-18439
CIP

British Library Cataloguing in Publication Data

Crippen, G.M.
Distance geometry and molecular conformation
1. Stereochemistry. Applications of distance geometry
I. Title II. Havel, T.F. (Timothy Franklin), *1953* – III. Series
541.2'23'01516

ISBN 0 86380 073 4
ISBN 0 471 92061 4 Wiley

ISBN 0 86380 073 4 (Research Studies Press Ltd.)
ISBN 0 471 92061 4 (John Wiley & Sons Inc.)

Printed in Great Britain by Galliard (Printers) Ltd., Great Yarmouth

Table of Contents

PREFACE

This book is an account of its authors' research on the use of distances and other simple geometric invariants as a means of describing the conformation spaces of the complex organic molecules which occur in biology. Since our original book on the subject (G.M. Crippen, "Distance Geometry and Conformational Calculations," Research Studies Press, 1981) we have added so much new material, particularly rigorous mathematical foundations, that we can hardly call the present work a second edition of the old one. The origin of this research is a novel point of view towards the now classical problem of locating the global conformational energy minimum of proteins. Instead of attempting to vary the coordinates of a structure so as to find that combination of interatomic distances which results in the global energy minimum, we have tried to work with the distances directly, and to compute the coordinates from them when we are done. Since the interatomic distances are related to the energy and other experimentally observable parameters of the conformation in a relatively direct and canonical way, the potential advantages of this approach are apparent, and in pursuing it we have discovered a large body of mathematical results of broad applicability in the problems of molecular conformation. Since many of these results were first encountered in the book, "Distance Geometry" by the late Leonard Blumenthal of the University of Missouri, we have given our approach the same name. Its suitability becomes more evident all the time.

While we cannot claim to have solved the global energy minimization problem simply by adopting this point of view, it has allowed us to formulate the problem far more rigorously mathematically than has heretofore been possible. This in turn has allowed us to develop powerful and systematic algorithms for locating low energy conformations. Another benefit of the distance geometry formalism is as a means of summarizing and consolidating all the available experimental information on a conformation which requires essentially no approximations or assumptions other than those which the data can justify. This in turn has proven useful in the direct determination of the conformations of macromolecules in solution by a variety of experimental methods, most notably NMR spectroscopy. It has also proven useful in the rationalization of a variety of structure/activity relations involving the binding of drug and ligand molecules to their receptor proteins. The scope of its applications has in fact lead us to believe that distance geometry may provide the mathematical foundations for the development of a general theory of molecular conformation.

In writing this book, we have had two groups of readers in mind. First, there are many computer-oriented chemists and molecular biologists whose work could benefit from the use of these techniques. For their sake, we have tried to make the book sufficiently self contained that they could, without further searching through the literature, implement any of the distance geometry algorithms whose use is illustrated in the latter sections of the book. Further, we have laced the more mathematical sections with chemical examples, and we have included appendices on relevant areas of mathematics at the end of the book. The appendix on the mathematical notation employed in this book is particularly useful on the first reading.

Second, we hope to make mathematicians aware of the purely mathematical problems which arise in description of *nonrigid* molecular structure, in the expectation that we may see fundamental improvements to our methods for working with such structure in the years to come. The areas of mathematics which would appear to hold the greatest promise are those of combinatorial geometry (geometric graph theory and matroid theory) and semialgebraic geometry (singular varieties over an ordered field defined by inequalities). Given the applied nature of our interest, we hope that we may be forgiven for not presenting the mathematics in the most general possible way, as for example by restricting our field to be the real numbers at all times. We hope nevertheless to convince you that applied mathematics can also be good mathematics.

The book itself consists of two parts. The first part is primarily devoted to the mathematical theory on which our approach is based, and covers both the classical work by Karl Menger and Leonard Blumenthal on distance geometry, as well as more recent results by many other mathematicians. A very considerable debt is owed to the Group in Structural Topology, including Walter Whiteley, Henry Crapo, Neil White and Robert Connelly, for contributions stemming from their research in structural rigidity. The second half of the book deals with algorithmic methods of solving specific chemical problems which the authors have developed, together with numerous examples of their use. The authorship of each of the two parts reflects the responsibility taken by its writer for its content, but in fact both authors have contributed substantially to both parts of the book. A joint introduction attempts to make all readers familiar with the problems, both chemical and the mathematical, with which we shall subsequently deal.

Like all scientific research, our work has not proceeded in isolation but has been influenced by our associations with many scientists and mathematicians over the years. Two individuals, however, have contributed far more than average to our efforts. The first of these, Prof. I.D. Kuntz at the University of California at San Francisco, has been our close friend and collaborator for many years. He contributed greatly to the development of our methods during the years that we were together in San Francisco, and has made several independent applications of distance geometry to the determination of macromolecular structure. The other, Prof. Andreas Dress of the University of Bielefeld, Federal Republic of Germany, is a mathematician who, as a result of his collaboration with Andre Dreiding at the University of Zürich and without any knowledge of our own work on the subject, was led to adopt a formalism for his mathematical studies of nonrigid molecular structure which is essentially identical to our own. His influence on the more mathematical sections of this book has been tremendous. Most of chapter two, the example of cyclohexane at the end of chapter four, and several proofs are his unpublished results, and we thank him for permission to include them here.

In addition, Gordon Crippen thanks the Victorian College of Pharmacy, Melbourne, Australia, and the Fulbright Foundation; and Timothy Havel thanks the Fakultät für Mathematik, University of Bielefeld, Federal Republic of Germany, for their hospitality during the first year of writing this book. Both of us would also like to thank the Department of Pharmaceutical Chemistry, University of California, San Francisco, for its hospitality on many occasions over the years.

T.F. Havel
Scripps Clinic and Research Foundation

G.M. Crippen
University of Michigan

March 1988

1. Introduction and Overview

Molecules are the intellectual property of chemists and geometry is the province of mathematicians. What have they got to do with each other? The answer is hidden in any modern introductory organic chemistry text book, typically a tome in excess of a thousand pages. Although a practicing organic chemist generally is concerned with chemical reactions, easily a quarter of the text is devoted to introducing some hundreds of new vocabulary words for describing the structure of organic molecules, and to presenting a stylized system for drawing pictures of them. As many an after-seminar dinner napkin will attest, organic chemistry is more a written language than a spoken one, simply because words come in a linear stream with time, and molecules are decidedly three-dimensional. The language has evolved over the last 150 years or so, in such a way that it can be taught in a finite time to students, and that it facilitates the thinking, presentation, and discussion of chemical phenomena. However, a certain unity has been lost amidst the welter of "eclipsed", "trans", "gauche", "syn", "R", "stereoisomer", "meso", "para", and the Newman, Fischer, and Haworth projections. For instance, not many chemists realize that *cis/trans* isomerism is simply the two-dimensional analogue of diastereomeric pairs. Taking a fresh look from the more abstract mathe.. .atical viewpoint restores that unity, helps chemists think about yet more complicated problems in molecular structure, enables mathematicians to prove results which have immediate chemical applications, and has actually produced powerful computer algorithms for treating some of the most difficult conformational problems yet addressed.

1.1. The Basic Features of Molecular Structure

This section is intended to introduce nonchemists to some of the terms and concepts commonly used to describe molecular structure, as well as to help chemists see where our approach fits into the wider field of stereochemistry and conformational analysis. For additional background, the reader is referred to standard textbooks, e.g. [Testa1979, Marples1981, March1985].

A *molecule* is a group of atoms which remains spatially intact on the time scale of interest. Its *structure* consists of those features which remain invariant on that time scale. At the quantum mechanical level, a molecule consists of a finite number of nuclei embedded in a cloud of electron density which binds them together. Although many chemists work at this level, the majority of "bench chemists" make do with a much more elementary view of molecules. In this book, we are concerned almost exclusively with *organic* molecules, composed primarily of carbon, hydrogen, nitrogen and oxygen with an occasional sulfur, phosphorus, halogen* or transition metal ion thrown in. As the name implies, these are the molecules of greatest interest in biological systems, and they include the majority of known types of molecules. These relatively few kinds of atoms are bound together almost exclusively by *covalent bonds*, a very strong link which can be broken only by high temperatures or in chemical reactions. Covalent bonds can be single, double or triple, depending on the number of electron pairs involved; this number is also known as the *bond order*. The number of covalent bonds in which an atom participates is called its *valence* (where double bonds count for two and triple for three). Under ordinary circumstances, carbon has valence four, nitrogen three, oxygen two and hydrogen one. The halogens generally have valence one, whereas sulfur may have valence two, three or four, and phosphorus valence three to five, depending on their oxidation states.

The *composition* of a molecule is the number of each type of atom in it. It is specified by the *molecular formula*, in which each type of atom is denoted by the first one or two letters of its name, followed by a subscript which indicates the number of times it occurs in the molecule. The *covalent structure* of a molecule is the way in which its atoms are bonded together. It can be represented mathematically by means of a multigraph whose atoms (vertices) are labelled with their types, and whose couples (edges) indicate the presence of covalent bonds (as shown in Figure 1.1). This graph, known as the *structural formula*, is already an object with interesting mathematical properties, as shown for example

* Meaning one of fluorine, chlorine, bromine or iodine.

by Cayley's and later Pólya's work on the enumeration of *isomers* (molecules with the same composition but different covalent structures; see [Balaban1976]).

molecular formula: C_2H_5BrO

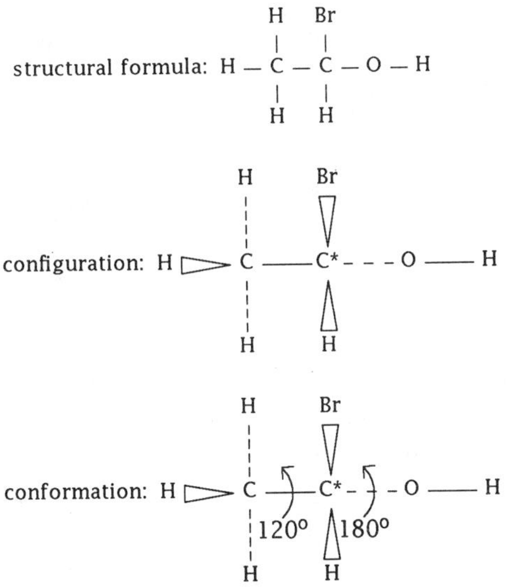

```
                         H    Br
                         |    |
structural formula:  H — C — C — O — H
                         |    |
                         H    H

                     H       Br
                     ¦       ▽
configuration:  H ▷  C ——— C* - - - O ——— H
                     ¦       △
                     H       H

                     H       Br
                     ¦       ▽
conformation:   H ▷  C ——— C* - - - O ——— H
                     ¦ 120°  △ 180°
                     H       H
```

Figure 1.1

The four levels of molecular structure illustrated for the case of the molecule 1-bromoethanol. The molecular formula specifies the composition, i.e. the number of each type of atom in the molecule. The structural formula tells one the covalent structure or the way in which these atoms are bonded together. The configuration describes the stereochemical structure, which is best defined as the spatial arrangement of the atoms within each rigid part of the molecule. The conformation specifies its spatial structure, which is the relative positions of the atoms in space.

Nevertheless, in order to predict how a given molecule will behave a chemist needs to know more than just its structural formula. As first suggested by van't Hoff and Pasteur towards the end of the nineteenth century, molecules are actually three-dimensional entities, and differences in their chemical behavior can often be accounted for by differences in their three-dimensional structure. Fortunately, the spatial structures of molecules exhibit a number of regularities. First, the *bond lengths* between covalently bonded pairs of atoms have very nearly fixed values on the order of 1 to 2 Å, depending on the atoms involved and the order of the bond. Second, the *bond angles* between pairs of bonds which meet at a common atom also have reasonably constant values. Third, there exist torsional "barriers" which may exclude certain *torsion angles.* These are the dihedral angles between the planes spanned by the two consecutive triples in a chain of four covalently bonded atoms (as shown at the bottom of Figure 1.1). Fourth, we have *steric hindrance,* which prevents any two nonbonded atoms from ever coming closer than 1 to 3 Å together in space, which is roughly speaking the sum of their *van der Waals radii.*

As a rule, these geometric conditions are not sufficient to determine the spatial structure of a molecule uniquely. They are sufficient to determine the relative positions of each atom and its nearest covalent neighbors, however, and sometimes larger chunks of the molecule as well. At carbon, nitrogen and oxygen atoms which participate only in single bonds, the bond angles are usually close to the sp^3 value of 109°, so that the covalent neighbors lie near the vertices of a tetrahedron.† Carbons and nitrogens which participate in one double bond have bond angles near to the sp^2 value of 120°, and so lie inside the roughly equilateral triangle which is spanned by their three covalent neighbors. Moreover, when two such carbons are doubly bonded together, they are coplanar with their four nearest neighbors (as in the 2-butene molecule shown in Figure 1.2). Carbons which participate in two double or one triple bond, on the other hand, have bond angles equal to the sp^1 value of 180° and hence are collinear with their neighbors. The *configurations* at each atom or bond in turn can be combined in a variety of ways to yield the *sterochemical structure* of the molecule. The two most important examples of stereoisomerism are the *cis* and *trans* forms of molecules containing carbon-carbon double bonds, and the nonsuperimposable mirror images or *enantiomorphs* of molecules containing *chiral* carbon atoms, which have four chemically distinct substituents. Nonenantiomorphic molecules which differ only in the configurations at their asymmetric carbons are called *diastereomers.*

Even when the stereochemical structure of a molecule is maintained, the relative positions of its atoms may vary with time. The *conformation* of a molecule is its precise spatial structure at some instant in time. In organic molecules, changing from one conformation to another usually involves rotations about single bonds. These rotations are not entirely "free", however, but are hindered by the repulsive forces which exist in the *eclipsed* as opposed to the *staggered* forms. Two conformations which differ by a 120° rotation at one or more single bonds are often called *conformers*; in most small molecules, these

† The symbols sp^n come from quantum mechanics, and their precise meaning is of no real importance here although their geometric implications are important.

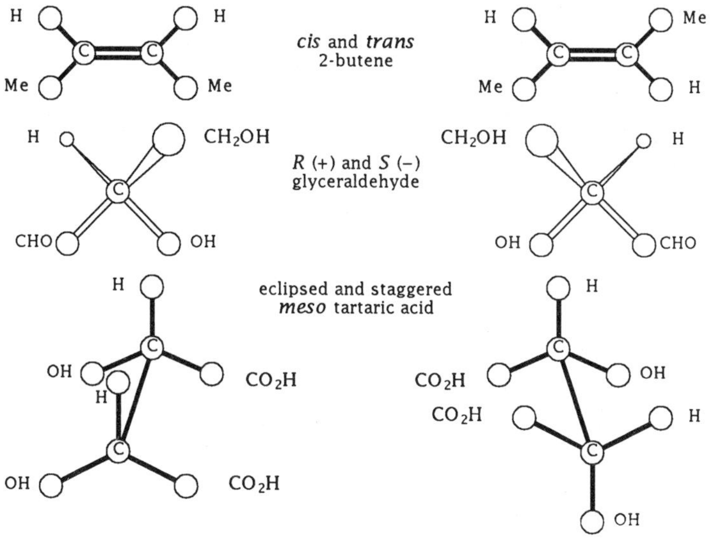

Figure 1.2

Some of the most important stereochemical features of molecules. On the top is shown an example of *cis/trans* isomerism (where Me := CH_3 denotes a methyl group). In the middle, we show a pair of enantiomorphs, each of which is the mirror image of the other. On the bottom, we show the eclipsed and staggered conformers of *meso* tartaric acid. Since the eclipsed conformation has a plane of symmetry, this particular stereoisomer of tartaric acid is achiral even though it contains two asymmetric carbons. Inversion at one of these two carbons leads to a chiral form of tartaric acid, which is a diastereomer of the first.

interconvert rapidly. Other significant rapid intramolecular motions leading to conformational isomerism include the "umbrella" flip motion at singly bonded nitrogens, and the "pseudorotation" at pentavalent phosphorus. Molecules which have more than one such rapidly interconverting conformer are called *mobile*, and we use the term *conformation space* to refer to the set of *all* conformations which are generated by intramolecular motions.

Ultimately speaking, the conformation (space) of a molecule is determined by the Hamiltonian of the system, that is, by the energy associated with each set of possible atomic positions. The differences in the energies of various conformations are due to a variety of *nonbonded* interactions, which even though they are individually too weak to determine any single geometric feature may nevertheless act together in concert to determine the spatial structures of large and complicated molecules such as proteins almost uniquely. The relation between these interactions and the conformational state of a molecule is of great importance in molecular biology, and is one of the main reasons for the authors' interest in the problems of molecular conformation. Over the years, chemists have developed analytic functions which allow them to estimate the energy as a function of the atomic positions reasonably well (see Chapter 9). These *energy functions*, however, tend to be rather complicated, and hence even with the help of a computer it can be difficult to make sound predictions from them.

The traditional approach to describing single conformations favors the least number of parameters that are invariant with respect to rigid translation and rotation of a molecule, namely "internal coordinates" consisting of the bond lengths, the bond angles, and the torsion angles. In the *rigid covalent geometry* approximation, the bond lengths and angles are treated as completely fixed, so that a given spatial structure can be described very compactly indeed by a list of torsion angles alone. On the other hand, the Cartesian coordinates of the atoms are the favored method in computational chemistry, because the algebra is simple and it largely avoids the use of computationally costly trigonometric functions. As will be described presently, whenever possible we have preferred to describe conformations by means of their interatomic distances, because the majority of terms in molecular energy functions, as well as much of the experimental information which is available on the conformational state of molecules, can be expressed most directly in terms of distances.

In recent years, a substantial theory of nonrigid molecular conformation has been developed that enables one to explain the chemical equivalence of atoms on the basis of intramolecular motion, and to enumerate and classify mobile molecules [Brocas1983, Ugi1984]. Although the methods given in [Floersheim1983] to determine the corresponding symmetry groups could easily be applied to the "distance geometry description" (see below), symmetry has not been a central issue in our own work. Instead we have attempted to understand the *geometry* and *kinematics* of mobile molecules, as something distinct from their *physics* and *dynamics*. This is largely just a theoretical version of what chemists do each time they reach for a molecular model. From a physical point of view, a mobile molecule is a cloud of probability density in atomic phase space, but this picture is not something most chemists can use. In contrast, the conformation space of a molecule, as a geometric entity, admits a simple and intuitive description. The ability to work directly with such sets of conformations also has distinct computational advantages, since programs which manipulate such sets directly possess an intrinsic parallelism without the use of any special hardware, and (using the techniques described in this book) one can return to a conventional coordinate based description whenever it becomes necessary to do so. For these reasons, we will now consider how the conformation space of a mobile molecule can be described.

1.2. How to Describe a Mobile Molecule

The following mathematical definition of a molecule, which is due to Dreiding, Dress and Haegi [Dress1983], is useful because it makes perfectly precise our intuitive notion of what a mobile molecule is. It is an explicit description which specifies each and every one of the molecule's possible conformations. Thus suppose that A is the set of atoms in the molecule, and let #A denote the number of elements in A. The composition of the molecule may be specified by a *mapping* (function) $c: A \rightarrow T$, where T is a predefined set of possible atom types (for example, $T := \{H, C, N, O, \cdots\}$). The Cartesian coordinates of a single spatial structure then correspond to another mapping $\mathbf{p}: A \rightarrow \mathbb{R}^3$, where $\mathbb{R}^3$ is regarded as a three-dimensional Euclidean space. We denote the set of all mappings from A into $\mathbb{R}^3$ by $(\mathbb{R}^3)^A$, and when necessary we denote the image $\mathbf{p}(a)$ of such a mapping more explicitly by $[x(a), y(a), z(a)]$. Since no two atoms are ever in the same place at the same time, it is natural to require that the mapping $\mathbf{p}$ be *injective*, meaning $\mathbf{p}(a) = \mathbf{p}(b)$ only if $a = b$, and we refer to an injective mapping $\mathbf{p}$ as an *embedding* of A into $\mathbb{R}^3$. We now take as our geometric definition of a mobile molecule $\mathbf{M}$ the triple $(A, c, \mathbf{P})$ consisting of the set A of its atoms and its composition c together with the set $\mathbf{P} \subset (\mathbb{R}^3)^A$ of *all* embeddings of A which correspond to a possible spatial arrangement of its atoms. We call $\mathbf{P}$ the set of *admissible embeddings* of the molecule. In this definition, it is the chemist whose judgement provides the ultimate test for the membership of each element of $(\mathbb{R}^3)^A$ in $\mathbf{P}$.

With regard to this *Cartesian* description, there are several caveats to be made. First, note that although no explicit mention has been made of either the molecular formula, structural formula or configuration of the molecule, the information necessary to derive any of these features of a molecular structure is clearly contained within it. Second, a great many objects can be described in this way which do not correspond to any possible *physical* molecule; for example, it is possible to specify coordinates with impossible values for bond lengths or angles. The advantage is that this formalism is sufficiently general that just about any possible physical molecule *could* be described within its framework, so that any facts which are true of all such abstract molecules are certainly true of all physical molecules. Finally, even though exchanging any two atoms of the same chemical type results in an equivalent description of the molecule, we assume that the set **P** is defined so that any member of **P** can be obtained from any other member by a rapid intramolecular motion — in the chemist's judgement.

Let $\mathbf{OT} = \mathbf{OT}(3,\mathbb{R}) = \{\mathbf{o}: \mathbb{R}^3 \to \mathbb{R}^3\}$ denote the set of all Euclidean isometries of $\mathbb{R}^3$ (i.e. translations, rotations, reflections and their compositions). Since isometries are necessarily invertible and the composition of isometries is again an isometry, this set becomes a mathematical group under function composition, which is usually called the three-dimensional *Euclidean group* [Birkhoff1977]. Also let $\mathbf{ST} = \mathbf{ST}(3,\mathbb{R})$ denote the set of all *screw translations* of $[x,y,z] \in \mathbb{R}^3$ (i.e. translations, proper rotations and their compositions). Under function composition, the set **ST** becomes a subgroup of **OT** of index two. Any two embeddings $\mathbf{p}, \mathbf{p}'$ such that $\mathbf{p} \circ \mathbf{o} = \mathbf{p}'$ for some $\mathbf{o} \in \mathbf{OT}$ are called *congruent*, whereas if $\mathbf{p} \circ \mathbf{s} = \mathbf{p}'$ for some $\mathbf{s} \in \mathbf{ST}$ they are called *properly congruent*. We shall denote these relations by writing $\mathbf{p} \approx \mathbf{p}'$ and $\mathbf{p} \dot{\approx} \mathbf{p}'$, respectively. It is easily seen that these are equivalence relations, i.e. they are reflexive, symmetric and transitive. Since any two admissible embeddings $\mathbf{p}, \mathbf{p}' \in \mathbf{P}$ with $\mathbf{p} \dot{\approx} \mathbf{p}'$ are geometrically indistinguishable, they necessarily represent the same conformation of the molecule, i.e. the set **P** is *closed* under the screw translations. As a consequence, our description of the molecule is highly redundant. Although this redundancy could be eliminated by selecting a single, canonical set of coordinates (for example center of mass, principal axes coordinates) for each conformation, in the present case it is desirable to accept this redundancy because it allows us to work with the relatively simple ground set, namely $(\mathbb{R}^3)^A$. This sort of thing is actually quite common in mathematics, the best known example probably being the use of homogeneous coordinates in projective geometry.

Within the context of the Cartesian description of a molecule, we are also able to give a precise definition of what is meant by its conformation space. Each $\mathbf{s} \in \mathbf{ST}$ induces a mapping $\mathbf{s}: (\mathbb{R}^3)^A \to (\mathbb{R}^3)^A$ via the definition $\mathbf{p} \mapsto \mathbf{s}(\mathbf{p})$. According to this definition we have $\mathbf{s}_0 \mapsto \mathbf{s}_0(\mathbf{p}) = \mathbf{p}$ for the identity $\mathbf{s}_0 \in \mathbf{ST}$, as well as $\mathbf{s}_i \circ \mathbf{s}_j(\mathbf{p}) \mapsto \mathbf{s}_i(\mathbf{s}_j(\mathbf{p}))$ for all $\mathbf{s}_i, \mathbf{s}_j \in \mathbf{ST}$. In technical terms, this situation is described by saying that the group of screw translations *acts* upon the set $(\mathbb{R}^3)^A$. An *orbit* in $(\mathbb{R}^3)^A$ under this group action is a set of embeddings which differ from a given embedding **p** by the action of a screw translation, i.e. a set $\mathbf{ST}(\mathbf{p})$ of the form $\{\mathbf{p}' \mid \mathbf{p}' \dot{\approx} \mathbf{p}\}$ for some fixed $\mathbf{p} \in (\mathbb{R}^3)^A$. The *conformation space* of the molecule may now be defined as the subset of the *orbit space* $\mathbf{ST}((\mathbb{R}^3)^A)$ which consists of all orbits which are generated by elements of **P**:

$$\mathbf{ST}(\mathbf{P}) \;:=\; \mathbf{P}/\mathbf{ST} \;:=\; \{\mathbf{ST}(\mathbf{p}) \mid \mathbf{p} \in \mathbf{P}\}\,. \tag{1.1}$$

Let the set of atoms in a molecule of water be $A = \{a_1,a_2,a_3\}$, so that $\#A = 3$. Then the various mappings which define its structure might be:

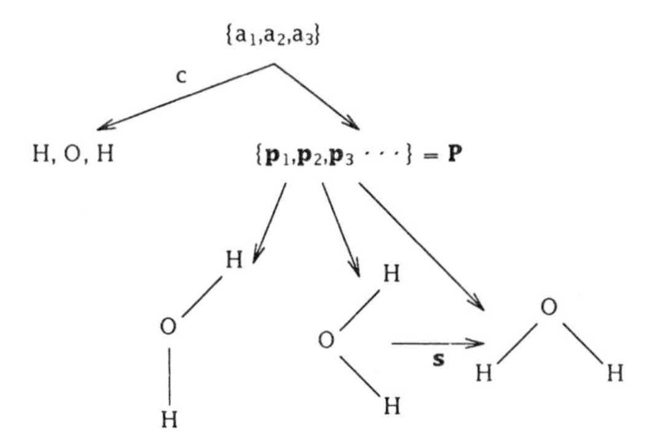

Figure 1.3

Note that the admissible embeddings $\mathbf{p}_2$ and $\mathbf{p}_3$ are related by a screw translation, so their images belong to the same orbit, distinguished by a particular H-H distance, while the image of $\mathbf{p}_1$ is related to that of $\mathbf{p}_2$ by bond angle bending, and is therefore in a different orbit. In any case, all three are elements of the conformation space of water.

In other words, the conformation space of a molecule $\mathbf{M} = (A, c, \mathbf{P})$ is the factor space obtained by *identifying* all members of the set $\mathbf{P}$ which are properly congruent. In contrast to our definition of the molecule as a whole, this definition of its conformation space involves a very complicated ground set, namely the orbit space $\mathbf{ST}((\mathbb{R}^3)^A)$.

The usual way of rendering orbit spaces comprehensible (see §9.5 of [Birkhoff1977] for an excellent introductory overview) is to find a *complete system of invariants*, i.e. a set of functions F on the underlying set $\mathbf{P}$ such that $\mathbf{ST}(\mathbf{p}) = \mathbf{ST}(\mathbf{p}')$ if and only if $f(\mathbf{p}) = f(\mathbf{p}')$ for all $f \in F$. In the case of the Euclidean group $\mathbf{OT}$, the *Fundamental Theorem of Euclidean Geometry* states that the distances between points constitute a complete system of invariants in terms of which all other invariant integral rational functions in the coordinates can be expressed algebraically. As a corollary, it is possible to specify any Euclidean arrangement of points, up to mirror reflection, by means of the distances among these points. This means that one can use the interatomic *distances* as coordinates to uniquely specify an orbit $\mathbf{OT}(\mathbf{p})$, which consists of a single *conformation* $\mathbf{ST}(\mathbf{p}) \in \mathbf{ST}(\mathbf{P})$ together with its mirror image.

The problem with using distances as coordinates is (as we shall soon see) that it is possible to specify sets of distances which do not correspond to any conformation of the molecule. One should keep in mind, however, that other types of inconsistencies can arise even in the Cartesian coordinates of a single conformation. For example, it is often a nontrivial task to find coordinates with reasonable values of the bond lengths and angles, a task which is completely trivial when the distances are used. Although this particular problem is not present when internal coordinates are used, even then it is difficult to avoid sterically unacceptable atomic overlaps, and additional problems arise if the molecule contains flexible rings. Therefore, instead of trying to find a minimal and "independent" set of variables to describe the conformation, we take all of the relevant distance information together and treat it as *equally* important.

Mobile molecules, by definition, contain noninvariant distances as well, but these distances may always be confined between well-defined *lower and upper bounds.* In terms of equations, these are given by

$$l_{\mathbf{P}}(\mathrm{a,b}) \;:=\; inf_{\mathbf{p}\in\mathbf{P}}(\|\mathbf{p}(\mathrm{a})-\mathbf{p}(\mathrm{b})\|)\,, \tag{1.2}$$

$$u_{\mathbf{P}}(\mathrm{a,b}) \;:=\; sup_{\mathbf{p}\in\mathbf{P}}(\|\mathbf{p}(\mathrm{a})-\mathbf{p}(\mathrm{b})\|)\,. \tag{1.3}$$

Clearly, these bounds satisfy $0 \leq l \leq u$ and are finite whenever $\mathbf{P}$ represents the set of admissible embeddings of a molecule. Every pair of functions $l, u : \mathrm{A}\times\mathrm{A}\rightarrow\mathbb{R}$, on the other hand, defines a set of conformations by means of the definition:

$$\mathbf{P}(l,\,u) \;:=\; \{\mathbf{p}\in(\mathbb{R}^3)^{\mathrm{A}} \mid l(\mathrm{a,b}) \;\leq\; \|\mathbf{p}(\mathrm{a})-\mathbf{p}(\mathrm{b})\| \;\leq\; u(\mathrm{a,b}) \;\;\forall\; \mathrm{a,b}\in\mathrm{A}\}\,. \tag{1.4}$$

Note that it is possible to have $\mathbf{P}(l,u) = \{\}$, in which case the distance bounds themselves are inconsistent (e.g. if $l(\mathrm{a,b}) > u(\mathrm{a,b})$ for some pair of atoms $\mathrm{a,b}\in\mathrm{A}$).

Suppose now that we are given a subset $\mathbf{Q}\subset(\mathbb{R}^3)^{\mathrm{A}}$ which is nonempty but otherwise arbitrary. Suppose also that we are given lower and upper distance bounds l, u such that $l = l_{\mathbf{Q}}$ and $u = u_{\mathbf{Q}}$, where $l_{\mathbf{Q}}$ and $u_{\mathbf{Q}}$ are the extrema obtained from Equations (1.2) and (1.3) with $\mathbf{P} := \mathbf{Q}$. Then consistency is assured by the assumption $\mathbf{Q}\neq\{\}$ together with the important general relation

$$\mathbf{Q} \;\subseteq\; \mathbf{P}(l_{\mathbf{Q}},\, u_{\mathbf{Q}}) \;=\; \mathbf{P}(l,\,u)\,, \tag{1.5}$$

where $\mathbf{P}(l,u)$ is obtained via Equation (1.4). It turns out that if we set $\mathbf{Q} := \mathbf{P}$, where $\mathbf{P}$ is the set of admissible embeddings of a molecule which has no distinct diastereomers, we often have $\mathbf{OT}(\mathbf{P}) \approx \mathbf{OT}(\mathbf{P}(l_{\mathbf{Q}},\, u_{\mathbf{Q}}))$ for the corresponding orbit space under the action of the full Euclidean group $\mathbf{OT}$. In this case, the distance bounds $l := l_{\mathbf{Q}} = l_{\mathbf{P}}$ and $u := u_{\mathbf{Q}} = l_{\mathbf{P}}$ can be regarded as a concise description and summary of the conformation space $\mathbf{ST}(\mathbf{P})$ (up to mirror reflection).

This observation constitutes the basis of what is known as the *distance geometry approach* to the problems of molecular conformation. The approach was originally introduced as a means of consolidating all the available experimental data on the solution conformations of biological macromolecules [Crippen1981, Havel1983]. This data may be obtained from a wide variety of experiments, including fluorescence transfer, various kinds of nuclear magnetic resonance spectroscopy, and chemical crosslinking studies, yet it usually can be interpreted in terms of bounds (l,u) on the values of (at least some of) the interatomic distances. If one interprets the data conservatively so that these bounds enclose the *entire range* of values consistent with the experiments, then the conformation space $\mathbf{ST}(\mathbf{P}(l,u))$ which the experimental data define includes the entire range of possible solution conformations $\mathbf{ST}(\mathbf{P})$, and in favorable cases one might hope once again to have $\mathbf{OT}(\mathbf{P}) \approx \mathbf{OT}(\mathbf{P}(l,u))$. The determination of biomolecular structure remains one of the most important applications of distance geometry, but (as we shall see) it has also proven to be a powerful theoretical tool in the study of the actual conformation spaces of more general molecules [Dress1983, Havel1983].

The primary reason for the lack of equality in Equation (1.5) is the fact that the configuration of the molecule cannot be completely determined by bounds on distances alone. If nothing else, distance information alone is insufficient to determine which mirror image or enantiomorph of an asymmetric or chiral molecule is present, and bounds on the distances may also fail to determine the relative handedness of its chiral centers. As a general, *empirical* rule, however, the conformation space of any molecule, *or* our state of limited knowledge of its conformations, can essentially always be adequately delimited by means of lower and upper bounds on the values of its interatomic distances, together with the possible *chiralities* of its asymmetric centers.* As will be shown in Chapter 2, the chiralities of the asymmetric centers in a molecule can be defined by means of an additional function $\tilde{\chi}: A^4 \rightarrow 2^{\{0,\pm 1\}}$, and we let $\mathbf{P}(l,u,\tilde{\chi}) \subseteq (\mathbb{R}^3)^A$ denote the set of all embeddings which are consistent with both the distance bounds and the chirality information. The triple $(l, u, \tilde{\chi})$ of functions is called the *distance geometry description* of conformation space $\mathbf{ST}(\mathbf{P}) \approx \mathbf{ST}(\mathbf{P}(l,u,\tilde{\chi}))$.

* The only severe exceptions occur in molecules which admit multiple "topological" isomers, such as knots and mobius ladders, which are seldom encountered in practice.

1.3. The Fundamental Problem of Distance Geometry

The possibility of inconsistent distance information, which was raised in the last section, is the greatest single difficulty which is encountered in putting the distance geometry approach into practice. At the same time, one of the greatest strengths of the approach is that it defines this problem clearly. For example, if the distance constraints have been determined experimentally, their inconsistency *proves* that a mistake has been made either in the collection of the data or else in its geometric interpretation, which is of course a very useful thing to know. On the other hand, if the constraints are consistent, then we would like to find (explicitly describe) those conformations which satisfy them. We call this the *Fundamental Problem of Distance Geometry*:

> *Given the distance and chirality constraints which define (our state of knowledge of) a mobile molecule, find one or more conformations which satisfy them, or else prove that no such conformations exist.*

Since chirality constraints are not studied until the next chapter, the ensuing discussion will not deal with them. Once the first section of that chapter has been read, however, it should be easy to see how the concepts introduced here can be extended to include them.

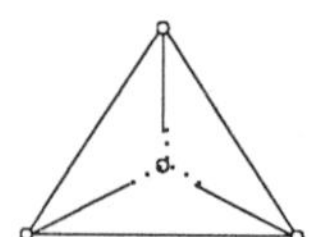
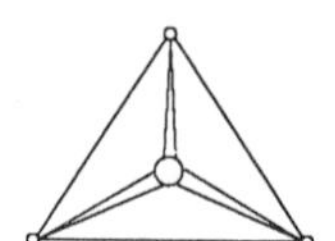
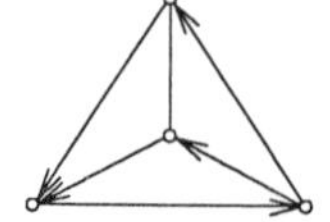

Figure 1.4

Illustration of the various kinds of inconsistencies that can arise for a given set of distance and chirality constraints. On the left, the lengths of the solid lines constitute a set of distances which cannot occur in either the plane or in space. In the middle, we show a set of distances which can occur between points in space, but not in the plane. On the right, we show a set of distances which can occur in the plane, and which must be coplanar in any spatial realization; moreover, in any set of points realizing those distances the handedness (i.e. clockwise or counterclockwise order) of the outermost triangle and the bottom inner triangle must be the same whenever these orders agree on their common exterior edge, as indicated by the arrows.

As it happens, proving inconsistency is presently a relatively difficult task. Basically, one has to show that the given distance (and chirality) constraints in some way fail to satisfy the known mathematical relations which exist between these quantities. Because of the importance of this task, the first part of this book is devoted to a thorough study of these relations. Proving consistency, on the other hand, is at least conceptually straightforward: one has only to exhibit (the coordinates of) a single conformation which satisfies the given distance and chirality constraints. This is known as *coordinatization*, and methods for finding such coordinates, together with their applications, are the main focus of the second half of the book. Our purpose in this section is to show that the importance of the Fundamental Problem extends far beyond the immediate analysis of experimental data, in that if we had a method which reliably solved it, we would in fact be well on our way to solving a great many outstanding problems in molecular conformation.

We have just seen how algorithms which solve the Fundamental Problem can be used to test experimental data for errors. They can also be used to test various hypotheses. For example, suppose we want to know if it is possible for two atoms $a,b \in A$ of a molecule to come into van der Waals contact, where the sum of their van der Waals radii is 3 Å. We take the distance bounds (l,u) which define the conformation space of the molecule, reset $u(a,b) := 3$ Å, and then test the revised bounds for consistency. If they are consistent, we can even use coordinatization to find a conformation in which that contact occurs. It is also possible that the two atoms in question must *always* be in van der Waals contact, even though we have no data which shows this directly. This in turn is true if and only if the bounds obtained by resetting $l(a,b) := 3$ Å are *in*consistent.

Methods of solving the Fundamental Problem can also be used to identify *redundant* constraints, which do not tell us anything about the conformational state of the molecule above and beyond what the rest of the constraints imply. To see how this is done, suppose for the sake of example that we wish to test an upper bound $u(a,b)$ on the distance between two atoms $a,b \in A$ for redundancy. Reset $l(a,b) := u(a,b)+\epsilon$ for some small $\epsilon > 0$, $u(a,b) := \infty$, and leave all other constraints the same. Then $u(a,b)$ was redundant if and only if the revised constraints are inconsistent. In plain English, this means simply that the other constraints do not permit any conformations in which the (a,b)-distance exceeds its upper bound. Obviously, redundancy for lower bounds can be tested by setting $u(a,b) := l(a,b)-\epsilon$ and $l(a,b) := 0$.

There is actually a somewhat finer level of redundancy that can be detected. If instead of setting $u(a,b) := \infty$ as in the previous paragraph, we set $u(a,b) := l(a,b) := u(a,b)+\epsilon$, we may find that these constraints are inconsistent, but that they become consistent when $u(\mathrm{a,b}) := \infty$. In this case we say that $u(\mathrm{a,b})$ was *locally redundant*, while using the term *globally redundant* for those constraints which are redundant but not locally redundant. Another useful concept along these lines are *critical constraints*: an upper bound $u(\mathrm{a,b})$ is said to be critical if setting $l(\mathrm{a,b}) := u(\mathrm{a,b}) - \epsilon$ results in a consistent set of bounds, otherwise it is termed noncritical. Alternatively, a critical constraint is one that cannot be "tightened" by some small amount $\epsilon > 0$ without changing the set of conformations which the geometric constraints as a whole define.

More generally, it may be of interest to determine whether or not two sets of distance constraints (l_1, u_1) and (l_2, u_2) are equivalent, in that they determine the same set of admissible embeddings $\mathbf{P}(l_1, u_1) = \mathbf{P}(l_2, u_2)$. This can be determined by testing the four sets of bounds obtained by resetting $l_1(\mathrm{a,b}) := u_2(\mathrm{a,b})$, $u_1(\mathrm{a,b}) := l_2(\mathrm{a,b})$, $l_2(\mathrm{a,b}) := u_1(\mathrm{a,b})$ and $u_2(\mathrm{a,b}) := l_1(\mathrm{a,b})$ for consistency for each pair $\mathrm{a,b} \in \mathrm{A}$. A rather more elegant approach, however, is as follows. Define the *Euclidean limits* associated with given distance bounds (l, u) as

$$l_E(\mathrm{a,b}) \;:=\; l_{\mathbf{P}}(\mathrm{a,b})\,, \qquad u_E(\mathrm{a,b}) \;:=\; u_{\mathbf{P}}(\mathrm{a,b}) \tag{1.6}$$

for all $\mathrm{a,b} \in \mathrm{A}$, where $\mathbf{P} := \mathbf{P}(l, u)$ (see Equations (1.3) through (1.5)). In other words, the Euclidean limits are the tightest distance bounds that are equivalent to (l, u), and every bound in them is critical, as defined above. Since this definition is clearly canonical, (l_1, u_1) and (l_2, u_2) are equivalent if and only if they determine the same Euclidean limits.* Approaches to the Fundamental Problem that operate by computing various approximations to the Euclidean limits will be studied in Chapter 5.

Another question that is often asked is whether or not two sets of distance bounds (l_1, u_1) and (l_2, u_2) have *any* admissible embeddings in common. This can be done by defining new bounds (l_3, u_3) as

$$l_3(\mathrm{a,b}) \;:=\; \mathit{max}(l_1(\mathrm{a,b}),\, l_2(\mathrm{a,b})) \tag{1.7}$$

and

$$u_3(\mathrm{a,b}) \;:=\; \mathit{min}(u_1(\mathrm{a,b}),\, u_2(\mathrm{a,b})) \tag{1.8}$$

for all $\mathrm{a,b} \in \mathrm{A}$. There exist common conformations if and only if the bounds (l_3, u_3) are consistent, in which case examples of common conformations can be obtained by coordinatization. An analogous question, that has important applications to drug design (i.e. pharmacophore identification; see Chapter 8), is

* In making this statement, we have assumed that the distance bounds (l_1, u_1) and (l_2, u_2) are defined on the same set of atoms A. Should they be defined on distinct sets of atoms A_1 and A_2, then to test for equivalence it is also necessary to determine the correspondence between the two sets. Methods for doing this may be found in [Floersheim1983].

whether or not two or more molecules can assume "similar" conformations. In this case, we are given a one-to-one correspondence between the "active" atoms in each pair of a sequence of homologous molecules, together with the distance geometry descriptions which define these molecules. We then impose very small upper bounds on the distances between corresponding pairs of active atoms in different molecules, which ensure that they are superimposed in any admissible embedding of all the molecules together (note that no lower bounds on the distances between different molecules are imposed). Then the bounds as a whole are consistent if and only if conformations exist in which the relative positions of the active atoms are all similar. Moreover, if enough molecules are simultaneously superimposed in this way, the positions of their active atoms will be nearly unique, so that the spatial structure of the active part of all of the molecules has been determined. This trick was originally proposed by J. Blaney (see also [Sheridan1986]).

Coordinatization algorithms can in fact be used to search for conformations satisfying quite a number of seemingly diverse criteria, or to enumerate "all" possible conformations according to a variety of classification schemes. For example, in complex ring systems it may be quite time consuming to find all possible combinations of the various types of staggered conformations *anti*, *gauche*$^+$ and *gauche*$^-$ by grid searches. By imposing distance bounds characteristic of these different forms and computing coordinates for each (if consistent), we can find all possibilities directly. In the event that the number of possibilities is too large to be enumerated completely, another question that is often asked is how "big" the conformation space of the molecule is. Although methods of solving the Fundamental Problem cannot answer this question directly, coordinatization can still provide us with some idea. This is done by a *Monte Carlo* approach, in which we compute admissible embeddings at random and then compare the conformations with one another to see how different they are. If certain conformations turn up over and over again, we can conclude that they are *rigid*, meaning that they are isolated points in conformation space $\mathbf{ST}(\mathbf{P}(l,u))$. In the extreme case, we may even find that the available constraints permit but a single conformation. This approach in fact enables us to predict with reasonable certainty many features of the molecule's conformation whose existence would be difficult to prove mathematically from the available data, and we shall encounter several examples of its use in the latter chapters of this book.

In conclusion, we remark that the conformation space $\mathbf{ST}(\mathbf{P}(l, u, \tilde{\chi})) \subseteq \mathbf{ST}((\mathbb{R}^3)^A)$ of a molecule which is defined by distance and chirality constraints $(l, u, \tilde{\chi})$ is a mathematical object which is rich in algebraic and topological structure. Although these more abstract properties, e.g. the number of connected components which it has and its fundamental group, are likely to have interesting chemical implications, they are beyond the scope of this book. It is likely, however, that progress in our ability to solve the Fundamental Problem will be closely coupled to improvements in our understanding of the intrinsic nature of conformation space, as defined by distance and chirality constraints.

1.4. An Overview of this Book

This book has been written so that the various chapters are largely independent of one another, and cross-references have been added so that the reader can easily find any missing pieces. In this section we present a chapter by chapter overview of the book,* which should be helpful in discovering the area which is of greatest immediate interest, and we advise the reader to simply start there. At the first reading, we would also recommend skimming the proofs and concentrating instead on the examples.

Chapter 2. In distance geometry, the emphasis is on describing conformations in terms of their interatomic distances, but for completeness we need to know at least the chirality of one asymmetric center (chiral carbon together with its substituents), and we often have some information concerning more than one center. How far can we go on chirality information alone? Granted, there are no experiments currently that yield a great deal of restrictions on many chiral centers, analogous to the way two-dimensional NMR can tell us about many interatomic distances, but a surprising amount of chirality is already built into a molecule at the level of its covalent structure. Chemists are used to looking at tetravalent carbon atoms and deciding whether it is achiral, *R*, or *S*. In §2.1, we explain how this concept can be generalized to any number of dimensions, and in particular to include order along a line and clockwise versus counterclockwise order in the plane. Furthermore, a "chirality" can be sensibly defined for *any* quartet of atoms in the molecule, not just the four substituents of a tetravalent carbon atom. From this, the idea of *cis/trans* isomerism and diastereomers follows very naturally, just by comparing their chiralities.

Not every possible assignment of chiralities in a molecule is consistent, however, and §2.2 gives some of the interrelations they must satisfy. In fact, there is quite a lot of mathematical structure involved, hinging on whether

* In the course of this overview, we shall mention work to which many different people besides ourselves have contributed, but in the interest of brevity no citations are included here. They may be found in the chapters themselves.

certain combinations of atoms are collinear or coplanar, and how much chirality information must be given in order to determine the rest. Here is the mathematical basis for the precise definition and the computer encoding of a number of stereochemical notions. For example, there are many chemical conventions for orienting a molecule in a diagram so as to discuss what substituents are on the same side versus opposite sides, and we show how these are interpreted within the context of the theory. The further conditions the given chiralities must satisfy (§2.3) allow one to define such concepts as the inside and the outside of a molecule in a coordinate independent way. Surprisingly, knowing all the chiralities is equivalent to knowing how many atoms are on the left versus the right side of a number of boundary planes drawn through the molecule. In §2.4 we make some simplifying assumptions, under which we are able to give an algorithm which finds the possible chiralities of the missing quartets from those which are known *a priori*, and then another algorithm produces a possible set of coordinates for the atoms consistent with this information. Thus we can translate abstract chirality information into coordinates as needed for computer graphics and other, more conventional methods of conformational analysis.

Chapter 3. Suppose we are given a full set of exact interatomic distances. They are not necessarily consistent with some arrangement of points in space, even if we go to four or more dimensions. These restrictions in fact provide the basis for a novel approach to Euclidean geometry, which is often more convenient to use in determining the consequences of distance and chirality information than are the usual, coordinate-based approaches. In §3.1 we begin by showing that *any* set of distances can be realized in an appropriate (possibly non-Euclidean) space, how to tell when this space will actually be Euclidean, and what the minimum number of dimensions it must have is. Using this geometric interpretation, we then go on to present a number of different but equivalent consistency checks for Euclidean distances, any one of which might be the most convenient, depending on the circumstances. Much as the methyl group in toluene forces the molecule to be three-dimensional rather than two-dimensional, §3.2 concentrates on the subsets of atoms which determine the dimensionality, and the set-theoretic relations among them. Then §3.3 gives an equivalent consistency check for the given distances that proceeds from a full-dimensional group of atoms and examines the distances to the other atoms. In §3.4 a detailed elimination of the possibilities shows that apart from a single type of six atom exception, to determine the consistency of a set of distances it is sufficient to consider the atoms five at a time. If we do have three-dimensional Euclidean distances, then there are some rather simple relations between them, which are discussed in §3.5. For example, the rigid bond lengths and angles in the neighborhood of a rotatable single bond restrict the range of distance allowed between atoms joined by three bonds to their *cis* and *trans* extreme values. Although a given set of distances for some molecule cannot distinguish between it and its enantiomer, §3.6 shows how the distances discriminate among diastereomers, and indeed how they determine the full set of chiralities discussed in Chapter 2 up to a mirror reflection.

Chapter 4. The results of the previous chapter are all very true, but we usually don't know the exact distance between every pair of atoms in the whole molecule. Suppose now we know a few precise distances, and the others are unknown. The mechanical equivalent of this is to have some sort of universal joint located at each atom and a rigid bar of a fixed length between each pair of atoms having a known distance. Of course, such a molecule can always tumble and translate, but §4.1 tells when enough distances have been fixed between the right pairs of atoms at the right lengths so that no other continuous changes in its conformation are possible. Note, however, that a small deformation of a ball-and-stick model of the rigid chair conformation of cyclohexane can pop it into the smoothly variable boat conformation. The restrictions on sets of forces which the molecule can resist without changing its conformation are covered in §4.2, and likewise what atomic velocities are permitted by a given combination of fixed interatomic distances. This in turn leads to an estimate of the number of degrees of motional freedom it has, which in turn is a measure of the "complexity" of its motions and of conformation space as a whole. In §4.3 we specialize these results to the patterns of fixed distances which are typically found in organic molecules, and show how the corresponding theory can be used to derive simple expressions for the derivatives of functions of the dihedral angles. We also prove that at least six distances are required to fix the relative positions of two rigid molecules.

In §4.4 we then go on to consider the thorny problem of how to find coordinates for conformations which satisfy a set of exact distance constraints. After showing that any general method of doing this requires exponentially increasing amounts of computer time, we consider some simple cases in which it is nevertheless practical. In order for the conformation to be uniquely determined, it is necessary that the constraints hold it rigid, and in order to find out whether or not it is rigid, it is often enough to know which pairs of atoms are involved in fixed distances without knowing the values of those distances. This has considerable bearing on the choice of experiments required to determine a molecule's conformation, and it also enables us to find an ordering for the atoms so that the coordinates of each atom can be computed from those preceding it by triangulation. In the event that the available constraints are not sufficient to hold the molecule rigid, it is much harder to characterize its motions and its conformation space as a whole, but in §4.5 we work out some simple examples, starting with four and five atoms in the plane and proceeding to a full derivation for cyclohexane. Some discussion of more complicated cases is given, but complete characterizations are not yet available.

Chapter 5. Progressing toward more and more complete descriptions of molecules and of the experimental information on their conformation, we consider next the situation where some distances are known exactly, most are unknown, and some now have upper and/or lower bounds. The mechanical analogue of this situation is the tensegrity framework of Buckminster Fuller, with struts that can be compressed only so far, and cables that can be stretched only so far. Things are not as simple nor are generalities as broad as they were with exact distance constraints. First, in an acceptable conformation, a given distance bound may not be active, i.e. that distance may lie somewhere between its upper and lower bounds, as opposed to pressing up against one limit. Second, §5.1 explains why we can no longer predict rigidity in advance of knowing the coordinates. There are, however, clear tests of rigidity of given conformations (or in the alternative sense, tests that the experimentally determined distance bounds prevent continuous changes in the conformation), and interesting ways to combine sets of constraints that hold two overlapping regions of the molecule rigid into a set of constraints that holds the combined region rigid. In fact, §5.2 goes on to show how global rigidity can be detected. That is, not only do the constraints prevent a particular conformation from deforming, but there is only one correct conformation. The "bound smoothing" technique that is used most heavily in practice comes in §5.3, where we explain how to deduce some sort of upper and lower limits on the distances between atom pairs for which no bounds have been determined experimentally. In particular, we present an efficient computer algorithm for inspecting triplets of atoms and using distance bounds on two of the atom pairs to produce limits on the third pair. Given the bounds on the distances between all pairs, we also show how to estimate distances between them in such a way that these distances satisfy the triangle inequality. Then §5.4 goes on to show how the bounds can be further tightened if we look at quartets of atoms, but the algorithms here are not as well developed. The final section, §5.5, gives conditions under which a given set of distance bounds cannot be realized in any number of dimensions, and which arise naturally in the course of penalty function minimization (see below).

Chapter 6. Even if we know the constraints are consistent, we are still left with the problem of finding atomic coordinates which satisfy them. Although the general case of this problem is intractable, the cases we need to treat can be computed in a reasonable amount of time. Some straightforward ways to tackle the problem are outlined in §6.1, and their defects discussed. Triangulation is too sensitive to measurement errors, and minimizing some penalty function tends to converge on some local minimum where not all constraints are satisfied. In §6.2, we present efficient data structures for storing distance and chirality information, along with algorithms for contact checking and bound smoothing which are based on them. Then §6.3 describes the key algorithm, called EMBED, for coming up with good enough trial coordinates to allow minimization of a penalty function to produce completely satisfactory atomic coordinates. This gives us a generally applicable, reasonably reliable method of solving the Fundamental Problem of Distance Geometry. It is explained in enough detail here with references to algorithms in earlier chapters that one could in principle write the corresponding computer program. To understand why we advocate this algorithm, how it really works, and when it runs into trouble, one has to study the earlier chapters. Finally, §6.4 surveys some different approaches proposed by other workers.

Chapter 7. Now that we have a general method for calculating atomic coordinates given upper and lower bounds on some interatomic distances and constraints on some chiral centers, we need to express our chemical knowledge about a molecule in these terms. We begin by showing in §7.1 how to do this for most sorts of chemical experiments, including some clever tricks for making use of data that doesn't look at all like interatomic distances. In the process, we shall touch upon a number of interesting applications of distance geometry to the interpretation of experimental data. Two-dimensional NMR is the single most important source of experimental data for such calculations, and the scope and limitations of this technique, and how one applies the EMBED algorithm to it, are the subject of §7.2. Since computing is easier than doing experiments, §7.3 shows how to calculate in advance whether carrying out certain proposed experiments will be worth the effort in terms of increasing our knowledge about a molecule's conformation. The most challenging applications of distance geometry to experimental data have been in the determination of the conformation of proteins in solution by NMR, and in §7.4 we describe in some detail one case in which this has been done.

Chapter 8. Many of the ideas in the previous chapters can be applied to rather different sorts of chemical problems. In §8.1, we ask the question: can one deduce the structure and the energetics of binding for some drug receptor site (e.g. an enzyme inhibitory site) just from the binding constants for several compounds having known structural formulas? In §8.2 we show how to formulate that in terms of a mathematical model and estimate whether there is any hope at all. Those familiar with the quantitative structure-activity relationships field (QSAR) know this task often involves the search for a pharmacophore. The reader's background in distance geometry by now makes simple the algorithm given for this in §8.3. Then §8.4 gets into the technicalities of two different algorithms for finding all the possible ways to place the (conformationally mobile) drug molecule into the hypothetical binding site. The overall algorithm for building up a site model from the binding data is outlined in §8.5, which includes the placement algorithms, EMBED, and special algorithms for determining the energies of interaction between parts of the site and atoms of the drug molecule. Having developed a site model, §8.6 goes into how to use it to find exceptionally tight binding novel molecules, even if they are chemically unrelated to those used to develop the model. This section includes a useful algorithm for positioning a conformationally mobile molecule in a site model by rigid translations and rotations and adjusting rotatable bonds until specified atoms approach as well as possible certain specified coordinates. In §8.7 we switch to quite a different style of site model, involving Voronoi polyhedra. There are certain advantages, but the methods have not been so extensively developed as of yet. Applications of the original style are covered in §8.8 as illustrations of how to apply the algorithms explained in the earlier sections to real data.

Chapter 9. Chapters 2-7 have concentrated rather strictly on dealing with geometric constraints put on molecules. Now we turn to the problems of finding atomic coordinates that not only obey given distance and chirality constraints, but also have low conformational energy. We begin by discussing the mathematical considerations involved in constrained minimization in §9.1, along with algorithms for actually solving such problems. The emphasis is placed on our preferred method, known as the method of augmented Lagrangians, which is a "local" optimization algorithm where an initial guess at the atomic coordinates is improved by a series of small adjustments. In §9.2 we discuss the problems inherent in locating all physically significant conformations, and show that the amount of computer time required to do this increases exponentially as a function of the size of the molecule, even using the "optimal" algorithm. We then consider "global" optimization algorithms which attempt to search a more limited region of conformation space, with emphasis on a method known as the ellipsoid algorithm, which has been found particularly useful in chemical problems. In §9.3 we consider another global method which operates in a fashion analogous to our coordinate calculation procedure, the EMBED algorithm. The algorithm is rather bizarre in that it searches spaces of dimension higher than three, using the theory presented in Chapter 3. Another alternative is presented in §9.4, where we show that one can also search for global minima in a combinatorial fashion by discretizing the constraints and the energy function.

Appendices. This book demands a lot of mathematical and computer background on the part of the reader, and we felt there were certain topics that were unusual enough to warrant a quick summary. Chapter 3 uses the bilinear algebra covered in Appendix A, as well as the affine geometry in Appendix B. Appendix C provides a painless introduction to projective geometry and Graßmann's theory of extensives, which is used in §4.2 and §4.3. The latter is a superset and ancestor of the ordinary vector analysis used throughout 20th century physics. Chapters 2, 3 and 4 use the matroids described in Appendix D. These are a very general way of handling all kinds of combinatorial problems, from the chirality of molecules to linear programming. The graph theory in Appendix E permeates the whole book, and is perhaps not all that unfamiliar to chemists any more, but it is still hardly a required course in most chemistry curricula. Some of the general numerical computer methods required to implement the algorithms given in this book are covered in Appendix F, and in particular, methods of unconstrained minimization. Finally Appendix G summarizes the notation we have used. Notational conventions are not all that generally recognized in mathematics, but here at least are ours, stated explicitly. Getting through the first few chapters will require frequent reference to the symbols table.

References

Balaban1976.

A.T. Balaban, *Chemical Applications of Graph Theory*, Academic Press, San Francisco, 1976.

Birkhoff1977.

G. Birkhoff and S. MacLane, *A Survey of Modern Algebra, 4th edition*, Macmillan Publishing Co., New York, NY, 1977.

Brocas1983.

J. Brocas, M. Gielen, and R. Willem, *The Permutational Approach to Dynamic Stereochemistry*, McGraw-Hill International Book Co., Cambridge, U.K., 1983.

Crippen1981.

G.M. Crippen, *Distance Geometry and Conformational Calculations*, Chemometrics Research Studies Series, 1, Research Studies Press (Wiley), New York, 1981.

Dress1983.

A.W.M. Dress, A.S. Dreiding, and H.R. Haegi, "Classification of Mobile Molecules by Category Theory," in *Symmetries and Properties of Mobile Molecules: A Comprehensive Survey*, ed. J. Maruani and J. Serre, Studies in Physical and Theoretical Chemistry, vol. 23, Elsevier Scientific, Amsterdam, 1983.

Floersheim1983.

P. Floersheim, K. Wirth, M.K. Huber, D. Pazis, F. Siegerist, H.R. Haegi, and A.S. Dreiding, "From Mobile Molecules to Their Symmetry Groups: a Computer

Implemented Method.," in *Symmetries and Properties of Non-Rigid Molecules*, ed. A. Maruani and J. Serre, Studies in Physical and Theoretical Chemistry, vol. 23, pp. 1-12, Elsevier Scientific Publishing Co., Amsterdam, Holland, 1983.

Havel1983.
T.F. Havel, I.D. Kuntz, and G.M. Crippen, "The Theory and Practice of Distance Geometry," *Bull. Math. Biol.*, *45*, 665-720(1983).

March1985.
J. March, *Advanced Organic Chemistry, 3rd ed.*, Wiley Intersci., Toronto, Canada, 1985.

Marples1981.
B.A. Marples, *Elementary Organic Stereochemistry and Conformational Analysis*, Royal Society of Chemistry, London, U.K., 1981.

Sheridan1986.
R.P. Sheridan, R. Nilakantan, J. Scott Dixon, and R. Venkataraghavan, "The Ensemble Approach to Distance Geometry: Application to the Nicotinic Pharmacophore," *J. Med. Chem.*, *29*, 899-906(1986).

Testa1979.
B. Testa, *Principles of Organic Stereochemistry*, Studies in Organic Chemistry, vol. 6, Marcel Decker, New York, NY, 1979.

Ugi1984.
I. Ugi, J. Dugundji, R. Kopp, and D. Marquarding, *Perspectives in Theoretical Stereochemistry*, Lecture Notes in Chemistry, volume 36, Springer-Verlag, New York, NY, 1984.

PART I

The Principles of Distance Geometry

(T.F. HAVEL)

2. The Mathematics of Chirality

There are several reasons for beginning a book on molecular distance geometry with a chapter on the mathematics of chirality. The first is that, since all the interatomic distances which can occur in a molecule can also occur in its mirror image, it is not possible to specify the absolute configuration of any asymmetric center by means of distance information alone. Of course, if the molecule is rigid and all of the interatomic distances are known exactly, then one need specify the chirality of but a single asymmetric center to uniquely determine the absolute configuration of all others. If the molecule is mobile, however, then certain aspects of its structure, e.g. admissible ranges for the dihedral angles, can only be specified if we include chirality information in our description, and hence this has been done since the earliest applications of distance geometry. Equally important, however, is the fact that the interplay between the two types of information is vital to a complete understanding of the relation between the imposed distance constraints and the conformation space which they define.

The majority of results presented in this chapter are the recent work of several mathematicians working in the field of combinatorial geometry, and a brief history is therefore in order. The mathematical description of chirality, like most ideas whose time has come, was proposed almost simultaneously in several different but equivalent forms by at least five groups of people working more or less independently of one another. The beginnings of these ideas first appear in [Novoa1965] and in [Rockafellar1969]. The first groups to publish detailed treatments were Folkman and Lawrence [Folkman1978], and Bland and Las Vergnas [Bland1977, Bland1978], all of whom settled upon the name "oriented matroids". In particular, Bland *et al.* created considerable excitement in certain circles by showing that the sequence of steps taken by the simplex algorithm when solving a linear program is essentially determined by the oriented matroid of the associated polytope, and not by the details of its geometry.

Many other people subsequently contributed to the theory. For example, Goodman and Pollack [Goodman1983] developed a compact data structure for storing chirality information and an efficient algorithm to compute this data structure from coordinates. Without knowledge of this work, Dreiding and Wirth came up with a different construction (as part of a method for classifying molecular structure in the presence of mobility) which is essentially a representable oriented matroid [Dreiding1980, Dress1983]. Inspired by Dreiding and Wirth's proposals, Dress went on to develop an extensive theory of what he now called "chirotopes", which was later discovered to be fully equivalent to the theory of oriented matroids [Dress1986b] (see also [Lawrence1982]). Again at about the same time, a similar approach was pursued as part of a study of the realizability of PL-spheres by Bokowski and Sturmfels [Bokowski1986], who also developed algorithms for computing the coordinates of certain representable chirotopes.

The present chapter is intended to provide a survey of the above results from the chirotope point of view. The reason we prefer this point of view is that, since it was originally inspired by problems arising in stereochemistry, it is by far the closest to chemical intuition and the easiest to apply to the problems we are interested in. We begin by describing the stereochemical motivation, and then proceed to develop the theory in stages, starting with its simplest aspects and gradually adding on successively more sophisticated features. Considerable time will be spent studying ways of simplifying any given description of molecular chirality while preserving these basic features, i.e. the so-called *restrictions* and *contractions*. A final section discusses some algorithmic aspects of the theory, in particular the Bokowski/Sturmfels algorithm by which one can often compute coordinates which are consistent with a given description of molecular chirality. The above references, especially [Dress1986b], should be consulted for further development of the general theory.

2.1. Chirality and Orientation

Let A be the set of atoms of a molecule, and $\mathbf{p}: A \rightarrow \mathbb{R}^3$ be an admissible embedding of the atoms in space which assigns to each atom $a_i \in A$ a vector of Cartesian coordinates $\mathbf{p}(a_i) = [x(a_i), y(a_i), z(a_i)]$. Also suppose that $a_0 \in A$ is a tetravalent carbon whose four neighbors $a_1, \ldots, a_4$ are of different chemical types and hence distinguishable. We seek some function of their coordinates which allows us to differentiate between the two mirror images of this group of atoms. The simplest such function is the *oriented volume* of the tetrahedron spanned by $a_1, \ldots, a_4$, which may be expressed as a determinant:

$$vol(\mathbf{p}(a_1), \mathbf{p}(a_2), \mathbf{p}(a_3), \mathbf{p}(a_4)) \;:=\; \frac{1}{3!}\, det \begin{bmatrix} 1 & 1 & 1 & 1 \\ x(a_1) & x(a_2) & x(a_3) & x(a_4) \\ y(a_1) & y(a_2) & y(a_3) & y(a_4) \\ z(a_1) & z(a_2) & z(a_3) & z(a_4) \end{bmatrix} . \qquad (2.1)$$

To see that this function actually is capable of making this distinction, we need only note that, if we reflect the molecule in the (x, y)-plane by multiplying the z coordinate by -1, then the sign of the oriented volume also changes. Moreover, it is easily shown that the oriented volume is invariant under rotations and translations, so that this is the only way in which the value of the oriented volume can be changed without actually changing the bond lengths and/or bond angles among the atoms $a_0, \ldots, a_4$. It follows that the absolute value of the oriented volume of a rigid group of atoms is independent of the particular embedding $\mathbf{p}$ used to define it, and that the sign of the oriented volume provides us with an unambiguous criterion by which we can decide whether two optical stereoisomers are different or the same.

If we wish to actually *specify* the absolute configuration, however, things become a little more complicated, because the sign of the determinant also changes whenever any two rows or columns are exchanged. Swapping any two of the last three rows corresponds to changing the orientation of the coordinate system; we can take care of this ambiguity by agreeing that we shall always use a right-handed coordinate system. Swapping any two columns, on the other hand, causes an essentially meaningless change in the sign, since the way in which we order our atoms is of no chemical or geometric significance. Although ways of eliminating this ambiguity exist (see [Dreiding1980]), for our purposes it is better to simply recognize it for what it is and to live with it. For this reason, we define the *chirality* of each quadruple of atoms in $\{a_1, \ldots, a_4\}$ to be the *function* $\chi_{\mathbf{p}}: \{a_1, \ldots, a_4\}^4 \rightarrow \{0, \pm 1\}$, which for any admissible embedding $\mathbf{p}: A \rightarrow \mathbb{R}^3$ is given by

$$\chi_{\mathbf{p}}(b_1, \ldots, b_4) \;:=\; sign(vol_{\mathbf{p}}(b_1, \ldots, b_4)) \;:=\; sign(vol(\mathbf{p}(b_1), \ldots, \mathbf{p}(b_4))) \qquad (2.2)$$

for all $b_1, \ldots, b_4 \in \{a_1, \ldots, a_4\}$. The ambiguity in the sign is now accounted for by the fact that the function is *antisymmetric*, meaning that $\chi_{\mathbf{p}}(\cdots b_i, b_{i+1} \cdots) = -\chi_{\mathbf{p}}(\cdots b_{i+1}, b_i \cdots)$ for $i = 1,2,3$. If the four atoms are coplanar, of course, then the oriented volume and hence the chirality is zero for all quadruples.

In order to define the chirality of the molecule as a whole, we initially assume that *all* of the atoms in A are distinguishable from one another and that the molecule is completely rigid so that all admissible embeddings are related by translation and rotation. Then for any embedding $\mathbf{p}: A \rightarrow \mathbb{R}^3$ of the molecule we define the chirality of the molecule to be the single *chirality function*

$\chi_{\mathbf{p}}: A^4 \rightarrow \{0, \pm 1\}$ whose value on each ordered quadruple of atoms $[b_1, \ldots, b_4] \in A^4$ is given by Equation (2.2) above. In the case that not all of the atoms in the molecule are chemically distinguishable or that the molecule is mobile there may be rigid or internal motions of the molecule which permute the positions of atoms with the same type, or equivalently, which correspond to permutations $\phi: A \rightarrow A$ of the atoms themselves. Such permutations change the function from $\chi_{\mathbf{p}}$ to $\chi_{\mathbf{p} \cdot \phi}$ without changing the chirality of the molecule. As is well-known (see [Longuet-Higgins1963]), these permutations constitute a group, called the *Longuet-Higgins* symmetry group $\mathbf{LH}(A, c, \mathbf{P})$, which consists of exactly those permutations ϕ which leave both the molecule's composition c as well as the set $\mathbf{P}$ of all admissible embeddings invariant, meaning $c = c \circ \phi$ as well as $\mathbf{P} = \mathbf{P} \circ \phi := \{\mathbf{p} \circ \phi \mid \mathbf{p} \in \mathbf{P}\}$. It is also possible that for two different admissible embeddings $\mathbf{p}, \mathbf{p}' \in \mathbf{P}$, we have $\chi_{\mathbf{p}'} \neq \chi_{\mathbf{p} \cdot \phi}$ for all permutations $\phi: A \rightarrow A$. In this case we say that $\chi_{\mathbf{p}}$ and $\chi_{\mathbf{p}'}$ have different *order types*; otherwise, they have the same order type. Our general definition of chirality is as follows:

Definition 2.1. The *chirality* of a molecule (A, c, $\mathbf{P}$) is the *set* of functions:

$$X = \{\chi: A^4 \rightarrow \{0, \pm 1\} \mid \chi = \chi_{\mathbf{p}},\ \mathbf{p} \in \mathbf{P}\}. \tag{2.3}$$

The molecule is called *chiral* if there exists $\chi \in X$ such that $-\chi \notin X$, in which case the set $-X := \{-\chi \mid \chi \in X\}$ is the chirality of its *enantiomorph*.

Note that, if all $\chi \in X$ have the same order type, e.g. if the molecule has but a single conformer, then the chirality is also given by

$$X = \{\chi: A^4 \rightarrow \{0, \pm 1\} \mid \chi = \chi_{\mathbf{p} \cdot \phi} = \chi_{\mathbf{p}} \circ \phi^4,\ \mathbf{P} \circ \phi = \mathbf{P},\ \phi \in \mathbf{LH}(A, c, \mathbf{P})\} \tag{2.4}$$

where $\mathbf{p}$ is some fixed element of $\mathbf{P}$.

Even though the set of functions X is always finite, in practice it is easier and usually also sufficient to use a simpler description of chirality consisting of a single *set-valued* function $\tilde{\chi}: A^4 \rightarrow 2^{\{0, \pm 1\}}$, which is given by

$$\tilde{\chi}(b_1, \ldots, b_4) := \bigcup_{\chi \in X} \{\chi(b_1, \ldots, b_4)\} := \bigcup_{\mathbf{p} \in \mathbf{P}} \{\chi_{\mathbf{p}}(b_1, \ldots, b_4)\} \tag{2.5}$$

for all $b_1, \ldots, b_4 \in A$. We call $\tilde{\chi}$ the *aggregate* chirality. If we define the function $-\tilde{\chi}$ in the obvious way, the molecule as a whole is chiral if $\tilde{\chi} \neq -\tilde{\chi}$. It is possible, however, that the molecule may be chiral in the sense given in Definition 2.1, but that $\tilde{\chi} = -\tilde{\chi}$. No matter which of these descriptions is used, the set of admissible embeddings $\mathbf{P}$ and hence X cannot be defined mathematically but only by the chemist's knowledge of the molecule. Once the set $\mathbf{P}$ has been defined, the stereochemical properties of the molecule follow from the properties of the chirality functions $\chi_{\mathbf{p}}$ obtained from the individual embeddings. For the remainder of this chapter we shall therefore concentrate on the properties of single chirality functions.

As an example of the chirality of a molecule, we consider methanol. If we label the hydroxyl hydrogen H1 and the methyl hydrogens H2, H3, and H4, then $\chi(C,H2,H3,H4)$ is the same in all of its possible conformations. For the first eclipsed conformer, however (here viewed along the C–O axis), we have $\chi(H1,O,C,H2) = 0$, whereas in the staggered conformation all orientations are nonzero. Therefore the eclipsed and staggered conformations have different order types.

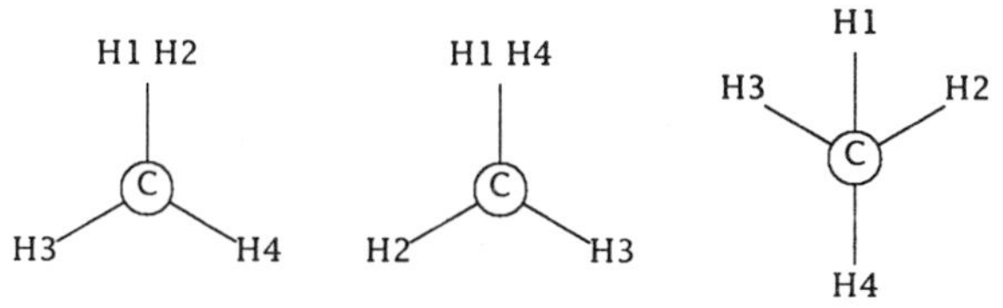

H1 H2   C   H3   H4   H1 H4   C   H2   H3   H1   H3   H2   C   H4

Figure 2.1

On the other hand, going from the first to the second eclipsed conformations can be thought of as a mapping $\phi : H2 \mapsto H4,\ H3 \mapsto H2,\ H4 \mapsto H3$. This produces a different orientation with the same order type, but because this mapping is an admissible permutation (i.e. one which belongs to the mobile symmetry group of methanol), it does not change the chirality of the molecule. This may also be seen from the invariance of the aggragete chirality w.r.t. this permutation, for example

$$\tilde{\chi}(\phi(C),\phi(H2),\phi(H3),\phi(H4)) = \tilde{\chi}(C,H4,H2,H3) = \tilde{\chi}(C,H2,H3,H4) = \{+1\}\ ,$$

while

$$\tilde{\chi}(\phi(H1),\phi(O),\phi(C),\phi(H3)) = \tilde{\chi}(H1,O,C,H4) = \tilde{\chi}(H1,O,C,H3) = \{-1,0,+1\}$$

(since all possible orientations of [H1,O,C,H3] and [H1,O,C,H4] actually occur as the methyl rotates about the C–O bond).

In order to study these chirality functions mathematically, we begin with a simple definition which includes only their most basic properties and proceed to gradually add on successively more complex features. Although the dimension of space is three in all cases of any chemical relevance, in order to identify those features which are characteristic of three dimensions it is useful to formulate our definitions and theorems for the general case of an n-dimensional space. From a mathematical standpoint at least, this is no more difficult to handle than three!

Definition 2.2. Let $\chi : A^{n+1} \to \{0,\pm 1\}$ be a nontrivial, antisymmetric function, i.e. one which satisfies the axioms:

(X0) for some $[b_1, \ldots, b_{n+1}] \in A^{n+1} :\ \ \chi(b_1, \ldots, b_{n+1}) \neq 0\ ;$ (2.6)

and

(X1) for all permutations $\pi: \{1, \ldots, N\} \rightarrow \{1, \ldots, N\}$ (2.7)

and $[b_1, \ldots, b_{n+1}] \in A^{n+1}$:

$$\chi(b_{\pi(1)}, \ldots, b_{\pi(n+1)}) = sign(\pi)\chi(b_1, \ldots, b_{n+1}) .$$

Then χ is called an *orientation* of A of *rank* $n+1$ and *order* $N := \#A$.[†]

To avoid trivial exceptions, it will be useful to regard a map $\chi: A^0 \rightarrow \{\pm 1\}$ as an orientation of rank 0. Note that, as an immediate consequence of antisymmetry,

$$\chi(b_1, \ldots, b_{n+1}) = 0 \quad \text{whenever } b_i = b_j \text{ for some } 1 \leq i < j \leq n+1 . \tag{2.8}$$

Another basic property, which does not hold for all orientations, is:

Definition 2.3. A pair of elements $c_1, c_2 \in A$ which satisfy

$$\chi(b_1, \ldots, b_n, c_1) = \epsilon \cdot \chi(b_1, \ldots, b_n, c_2) \quad \forall \quad [b_1, \ldots, b_n] \in A^n \tag{2.9}$$

for some $\epsilon \in \{\pm 1\}$ are called ϵ-*equivalent*, and we signify this relation by $c_1 \approx c_2$. An orientation is called *proper* if $c_1 \approx c_2 \Rightarrow c_1 = c_2$ for all $c_1, c_2 \in A$.

Let $\underline{A} = [a_1, \ldots, a_N]$ be an indexing of the set of atoms A[‡]. For each such indexing, the orientation $\chi: A \rightarrow \{0, \pm 1\}$ induces a function $\overline{\chi}: \binom{A}{n+1} \rightarrow \{0, \pm 1\}$, which is defined by

$$\overline{\chi}(\{a_{\lambda_1}, \ldots, a_{\lambda_{n+1}}\}) := \chi(a_{\lambda_1}, \ldots, a_{\lambda_{n+1}}) \tag{2.10}$$

for all $\underline{\lambda} \in \Lambda(N, n+1)$. Conversely, every such function $\overline{\chi}$ defines an orientation on A via the same relation together with Equations (2.7) and (2.8). This fact makes it immediately clear that the number of different orientations on A is $3^M - 1$, where $M := \binom{N}{n+1}$ is the number of ordered $(n+1)$-tuples of the integers $\{1, \ldots, N := \#A\}$. We refer to $\overline{\chi}$ as the orientation *versus* $\underline{A}$, whereas χ is called the *antisymmetric* orientation.

In the early days of organic chemistry, it was only possible to determine the *relative* chiralities of asymmetric groups of atoms in chemical compounds. This was done by converting the chemical structure of one molecule into that of another molecule by a sequence of chirality-preserving reactions, and then testing the two molecules for identity via a classical melting point experiment. In our next definition, we attempt to capture the essentials of the concept of

† Here, sign(π) is the signature of the permutation π, which is 1 if the permutation can be written as the composition of an even number of transpositions and -1 if the number is odd.

‡ As explained in Appendix G, an indexing is a one to one numbering of all the elements of a set, which converts the set to a sequence. Such a sequence is indicated by underlining the upper case letter which stands for the set. The individual members of the sequence are specified by taking the corresponding lower case letter and subscripting it by its index. The other symbols appearing below are also defined in Appendix G.

As an example of the orientation associated with an arrangement of points, suppose we have five atoms in $\mathbb{R}^2$:

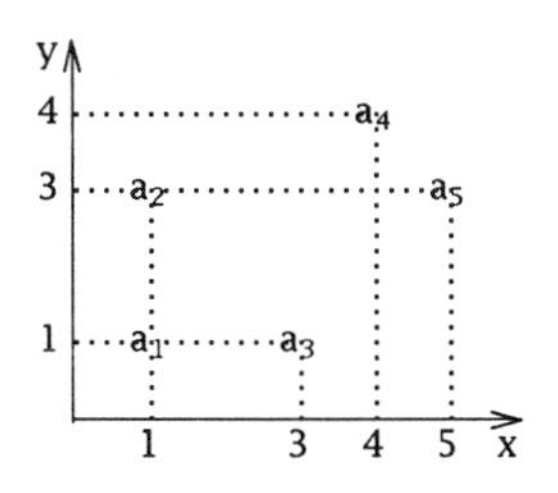

Figure 2.2

If we use a right-handed coordinate system for $\mathbb{R}^2$, an ordered triple of points which occur in counterclockwise order on their circumscribing circle has positive chirality so that, for example:

$$\chi_{\mathbf{p}}(a_3,a_2,a_1) \;=\; sign\ det\begin{bmatrix}1 & 1 & 1\\ 3 & 1 & 1\\ 1 & 3 & 1\end{bmatrix} \;=\; sign(4) \;=\; +1\ .$$

Since χ is not identically zero, the axiom (X0) is obeyed. As an example of the axiom (X1), we let $\pi:\{1,\ldots,5\}\rightarrow\{1,\ldots,5\}$ be the transposition which maps $1\mapsto 2$ and $2\mapsto 1$. Then $sign(\pi) = -1$ by definition and $\chi(a_{\pi(3)}, a_{\pi(2)}, a_{\pi(1)}) = \chi(a_3,a_1,a_2) = -1$. Hence χ is an orientation of the set $\{a_1,\ldots,a_5\}$ of rank 3 and order 5. Observe that, even though $\chi(a_1,a_2,a_4) = \chi(a_1,a_2,a_5) = \chi(a_3,a_1,a_4) = \chi(a_3,a_1,a_5) = \chi(a_3,a_2,a_4) = \chi(a_3,a_2,a_5) = -1$, it is not true that $a_4 \approx a_5$ with $\epsilon = +1$, because $\chi(a_1,a_4,a_5) = -1 \neq \chi(a_1,a_4,a_4) = 0$. In fact, χ is a proper orientation.

relative chirality.

Definition 2.4. The *relative orientation* $\sigma_\chi\colon A^{n+1}\times A^{n+1}\rightarrow\{0,\pm1\}$ associated with an *absolute* orientation $\chi\colon A^{n+1}\rightarrow\{0,\pm1\}$ is given by

$$\sigma_\chi(b_1,\ldots,b_{n+1};\, c_1,\ldots,c_{n+1}) \;:=\; \chi(b_1,\ldots,b_{n+1})\cdot\chi(c_1,\ldots,c_{n+1}) \qquad (2.11)$$

for all $[b_1,\ldots,b_{n+1}], [c_1,\ldots,c_{n+1}] \in A^{n+1}$.

We now show that once we know the relative orientation we also know the absolute orientation up to sign, but that not every function $\sigma\colon A^{n+1}\times A^{n+1}\rightarrow\{0,\pm1\}$ gives rise to an absolute orientation in this way.

LEMMA 2.5. *Given* $\sigma\colon A^{n+1}\times A^{n+1}\rightarrow\{0,\pm1\}$, *there exists an orientation* $\chi\colon A^{n+1}\rightarrow\{0,\pm1\}$ *such that Equation (2.11) holds with* $\sigma = \sigma_\chi$ *if and only if:*

$$(\Sigma 0) \quad \exists\ \underline{B} = [b_1,\ldots,b_{n+1}] \in A^{n+1} \text{ with } \sigma(\underline{B};\underline{B}) \;=\; 1\ ; \qquad (2.12)$$

As an example of a relative orientation, we consider 1,2-dichloroethene in the plane with the unique atom labels:

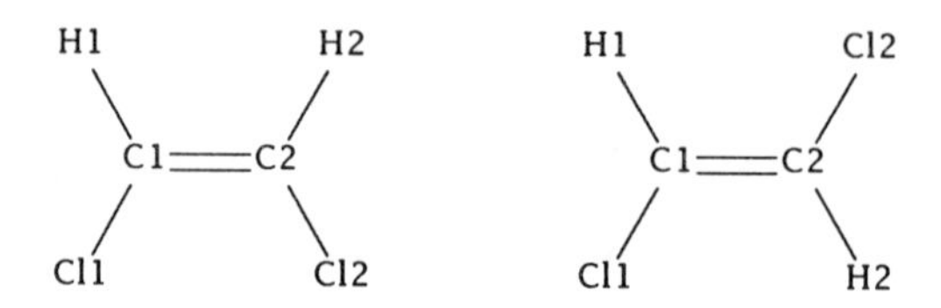

Figure 2.3

Then for the *cis* conformation $\chi(\text{C1,Cl1,H1}) = -1$ while $\chi(\text{C2,Cl2,H2}) = +1$ w.r.t. a right-handed coordinate system, so that $\sigma_\chi(\text{C1,Cl1,H1; C2,Cl2,H2}) = -1$, whereas for the *trans* conformation we have $\chi(\text{C1,Cl1,H1}) = -1$ and $\chi(\text{C2,Cl2,H2}) = -1$ so that $\sigma_\chi(\text{C1,Cl1,H1; C2,Cl2,H2}) = +1$. Note that the relative orientation is the same in both a right- and a left-handed coordinate system.

$$(\Sigma 1) \quad \forall \; [b_1, \ldots, b_{n+1}], [c_1, \ldots, c_{n+1}] \in A^{n+1} \text{ and permutations } \pi: \tag{2.13}$$
$$\sigma(b_1, \ldots, b_{n+1}; c_{\pi(1)}, \ldots, c_{\pi(n+1)}) = sign(\pi)\sigma(b_1, \ldots, b_{n+1}; c_1, \ldots, c_{n+1});$$

$$(\Sigma 2) \quad \forall \; \underline{B} \in A^{n+1} \text{ with } \sigma(\underline{B}; \underline{B}) = 0 : \sigma(\underline{B}; \underline{C}) = \sigma(\underline{C}; \underline{B}) = 0 \quad \forall \; \underline{C} \in A^{n+1}; \tag{2.14}$$

$$(\Sigma 3) \quad \forall \; \underline{B}, \underline{C}, \underline{D} \in A^{n+1} \text{ with } \sigma(\underline{C}; \underline{C}) \neq 0 : \; \sigma(\underline{B}; \underline{C})\cdot\sigma(\underline{C}; \underline{D}) = \sigma(\underline{B}; \underline{D}). \tag{2.15}$$

Furthermore, the orientation χ is unique up to sign.

Proof: The necessity of these conditions is obvious. Hence suppose that for some $\underline{B} \in A^{n+1}$ with $\sigma(\underline{B}; \underline{B}) = 1$, we define $\chi_\sigma: A^{n+1} \to \{0, \pm 1\}$ by

$$\chi_\sigma(\underline{C}) := \sigma(\underline{B}; \underline{C}) \tag{2.16}$$

for all $\underline{C} \in A^{n+1}$. Then by axioms ($\Sigma$0) and ($\Sigma$1), χ_σ is an orientation. Moreover, either $\sigma(\underline{C}; \underline{C}) = 0$ so that both $\sigma(\underline{B}; \underline{C})$ and $\sigma(\underline{C}; \underline{B})$ are zero by axiom (Σ2), or else by axiom (Σ3) with $\underline{D} = \underline{B}$, we have

$$\sigma(\underline{B}; \underline{C})\cdot\sigma(\underline{C}; \underline{B}) = \sigma(\underline{B}; \underline{B}) = 1, \tag{2.17}$$

i.e. the relative orientation satisfies $\sigma(\underline{B}; \underline{C}) = \sigma(\underline{C}; \underline{B})$ for all $\underline{B}, \underline{C} \in A^{n+1}$. Hence

$$\begin{aligned} \chi_\sigma(\underline{C})\chi_\sigma(\underline{D}) &= \sigma(\underline{B}; \underline{C})\sigma(\underline{B}; \underline{D}) \\ &= \sigma(\underline{C}; \underline{B})\sigma(\underline{B}; \underline{D}) = \sigma(\underline{C}; \underline{D}), \end{aligned} \tag{2.18}$$

as desired.

It remains to be shown that χ_σ is unique up to sign. Hence suppose that for some other $\underline{B}' \in A^{n+1}$ with $\sigma(\underline{B}'; \underline{B}') = 1$ we define $\chi'_\sigma: A^{n+1} \to \{0, \pm 1\}$ by Equation

(2.16). Then for $\underline{C} \in A^{n+1}$, if $\sigma(\underline{C};\underline{C}) = 0$, both $\chi_\sigma(\underline{C})$ and $\chi'_\sigma(\underline{C})$ are zero by (Σ2); otherwise, again by (Σ3)

$$\begin{aligned}\chi_\sigma(\underline{C})\chi_\sigma'(\underline{C}) &= \sigma(\underline{B};\underline{C})\sigma(\underline{B}';\underline{C}) \qquad (2.19)\\ &= \sigma(\underline{B};\underline{C})\sigma(\underline{C};\underline{B}') = \sigma(\underline{B};\underline{B}')\end{aligned}$$

which is a constant equal to ± 1, as desired. QED

Suppose we are given the relative orientation $\sigma: A^{n+1}\times A^{n+1}\rightarrow\{0,\pm 1\}$ of four atoms in the plane, arranged as shown:

a_2

a_1 a_3 a_4

Figure 2.4

Then since a_1,a_3,a_4 are collinear we can't choose this triple of atoms as our reference for constructing an absolute orientation $\chi_\sigma: A^{n+1}\rightarrow\{0,\pm 1\}$, but we could use $[a_1,a_2,a_3]$ which, since a_1,a_2,a_3 are arranged in a clockwise fashion, would give us the orientation which the atoms have in a left-handed coordinate system, or we could take $[a_2,a_3,a_4]$ as our reference, which would produce their orientation in a right-handed system.

As a consequence, we may think of a relative orientation (i.e. any function $\sigma: A^{n+1}\times A^{n+1}\rightarrow\{0,\pm 1\}$ which satisfies (Σ0) through (Σ3) above) as being an "enantiomorphic" pair of absolute orientations. It is also possible to define a relative orientation versus a given ordering as $\bar{\sigma}(B;C) := \bar{\chi}(B)\cdot\bar{\chi}(C)$ for all $B, C \subseteq A$ with $\#B = \#C = n+1$.

We now show that each relative orientation on A^{n+1} determines a relative orientation on A^{N-n-1} and vice versa, where $N := \#A$. To do this, for $\underline{\lambda} \in \Lambda(N,k)$ with $1 \leq k \leq N$, we define

$$\Sigma_\lambda := \sum_{i=1}^{k} \lambda_i \, , \qquad (2.20)$$

and recall that $\hat{\underline{\lambda}} \in \Lambda(N,N-k)$ denotes the *complement* of $\underline{\lambda}$, which satisfies $\lambda\cup\hat{\lambda} = \{1,\ldots,N\}$ and $\lambda\cap\hat{\lambda} = \{\}$ (as in Appendix G).

Definition 2.6. Let $\bar{\chi}: \binom{A}{n+1}\rightarrow\{0,\pm 1\}$ be an absolute orientation defined versus an indexing $\underline{A} = [a_1,\ldots,a_N]$ of the set of atoms A, as in Equation (2.10). Then the *dual orientation* $\hat{\bar{\chi}}: \binom{A}{N-n-1}\rightarrow\{0,\pm 1\}$ versus this indexing is given by

$$\hat{\bar{\chi}}(A_\mu) := (-1)^{\Sigma_\mu}\,\bar{\chi}(A_{\hat{\mu}}) \, . \qquad (2.21)$$

for each subset $A_\mu \subseteq A$ with $A_\mu := \{a_{\mu_1}, \ldots, a_{\mu_{N-n-1}}\}$ for some $\mu \in \Lambda(N, N-n-1)$. We shall also denote the corresponding dual antisymmetric orientation defined on A^{N-n-1} via Equation (2.10) by $\hat{\chi}: A^{N-n-1} \to \{0, \pm 1\}$, and the associated dual relative orientation by $\hat{\sigma}: A^{N-n-1} \times A^{N-n-1} \to \{0, \pm 1\}$.

To obtain an explicit example of the dual of an orientation, we consider the same arrangement of atoms as in the previous example (Figure 2.4). The corresponding orientation has order $N = 4$ and rank $n+1 = 3$, so that if $\underline{A} = [a_1, a_2, a_3, a_4]$, then the orientation and its dual are:

$\hat{\underline{\mu}} =$	[123]	[124]	[134]	[234]
$\overline{\chi}(A_{\hat{\mu}}) =$	-1	-1	0	$+1$
$\hat{\overline{\chi}}(A_\mu) =$	-1	$+1$	0	-1
$\Sigma_\mu =$	4	3	2	1
$\mu =$	[4]	[3]	[2]	[1]

Note that, since $\Sigma_\lambda + \Sigma_{\hat{\lambda}} = N(N+1)/2$ for all $\underline{\lambda} \in \Lambda(N, k)$, the dual of the dual satisfies $\hat{\hat{\chi}} = (-1)^{N(N+1)/2} \chi$, while for the associated relative orientation we have $\hat{\hat{\sigma}} = \sigma$. To verify our claim above, it remains to be shown that $\hat{\sigma}$ is well-defined, i.e.

LEMMA 2.7. *If $\hat{\sigma}$ and $\hat{\sigma}_\pi$ are dual relative orientations versus the linear orderings induced by two different indexings $[a_1, \ldots, a_N]$ and $[a_{\pi(1)}, \ldots, a_{\pi(N)}]$ of the set A, where π is a permutation of $\{1, \ldots, N\}$, then $\hat{\sigma} = \hat{\sigma}_\pi$.*

Proof: We show that for the corresponding absolute orientations $\hat{\chi}$ and $\hat{\chi}_\pi$ we have $\hat{\chi} = sign(\pi)\, \hat{\chi}_\pi$. Since any permutation can be written as a composition of transpositions of adjacent integers, it suffices to prove this for a pair of linear orderings which are related by a transposition of the form $i \leftrightarrow i+1$. For $\mu \in \Lambda(N, N-n-1)$, there are three cases to consider:

(i) $a_i, a_{i+1} \in A_{\hat{\mu}}$: in this case, $\overline{\chi}(A_{\hat{\mu}}) = -\overline{\chi}_\pi(A_{\hat{\mu}})$, so $\hat{\overline{\chi}}(A_\mu) = -\hat{\overline{\chi}}_\pi(A_\mu)$.

(ii) $a_i \in A_{\hat{\mu}}, a_{i+1} \in A_\mu$ or $a_i \in A_\mu, a_{i+1} \in A_{\hat{\mu}}$: here $(-1)^{\Sigma_\mu}$ changes sign, while $\overline{\chi}(A_{\hat{\mu}})$ does not, so again $\hat{\overline{\chi}}(A_\mu) = -\hat{\overline{\chi}}_\pi(A_\mu)$.

(iii) $a_i, a_{i+1} \in A_\mu$: then $\hat{\overline{\chi}}(A_\mu) = \hat{\overline{\chi}}_\pi(A_\mu)$, but by antisymmetry

$$\hat{\chi}_\pi(\underline{A}_\mu) = \hat{\chi}(a_{\pi(\mu_1)}, \ldots, a_{\pi(\mu_{N-n-1})}) = sign(\pi) \hat{\chi}(\underline{A}_\mu) = -\hat{\chi}(\underline{A}_\mu), \tag{2.22}$$

since the permutation relating the orders of the sequences $[a_{\mu_1}, \ldots, a_{\mu_{N-n-1}}]$ and $[a_{\pi(\mu_1)}, \ldots, a_{\pi(\mu_{N-n-1})}]$ is a transposition by assumption. QED

Duality is sometimes a useful tool for converting a high dimensional problem into a low dimensional problem which can be more readily visualized.

We close this section by describing two additional ways of reducing the complexity of orientations, albeit with some loss of information.

Definition 2.8. The *restriction* of an orientation $\chi: A^{n+1} \to \{0,\pm 1\}$ to a subset $B \subseteq A$ with $\#B \geq n+1$ is the map $\chi|_B: B^{n+1} \to \{0,\pm 1\}$ obtained by restricting the function χ to B. For $k := \#\underline{C} \leq n+1$, the *contraction* of an orientation $\chi: A^{n+1} \to \{0,\pm 1\}$ w.r.t. an ordered k-tuple $\underline{C} = [c_1, \ldots, c_k] \in A^k$ is the map $\chi|^{\underline{C}}: (A \setminus C)^{n-k+1} \to \{0,\pm 1\}$ defined by:

$$\chi|^{\underline{C}}(b_{k+1}, \ldots, b_{n+1}) \;:=\; \chi(c_1, \ldots, c_k, b_{k+1}, \ldots, b_{n+1}) \tag{2.23}$$

for all $[b_{k+1}, \ldots, b_{n+1}] \in (A \setminus C)^{n-k+1}$.

Contractions are actually used in the standard definition of the absolute configuration of an asymmetric carbon atom.

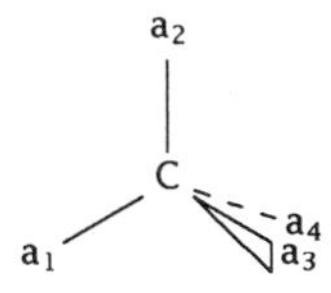

Figure 2.5

Not counting the carbon, the substituents are ordered by increasing atomic number, e.g. $\underline{A} = [a_1,a_2,a_3,a_4]$. Then $\chi|^{[a_1]}(a_2,a_3,a_4)$ is determined by holding a_1 behind the carbon, and using the usual rule of $\chi = 1$ if $[a_2,a_3,a_4]$ are counterclockwise in that ordering. In that case the absolute configuration is *S*, otherwise it is *R*. The counterclockwise rule can be contracted even further to $\chi|^{[a_1,a_2]}(a_3,a_4) = 1$, meaning that if we hold a_2 above the other atoms in the plane of a_1,a_2,a_3, then a_4 is to the right of a_3, as shown.

Note that the restrictions and contractions of orientations need not themselves be orientations, because they may violate the axiom (X0) above. Nevertheless, for $B \subseteq A$ with $\#B \geq n+1$ we shall use the symbol $\sigma|_B$ for the (possibly trivial) map defined by

$$(\underline{C}\,,\,\underline{D}) \;\mapsto\; \chi|_B(\underline{C})\cdot\chi|_B(\underline{D}) \tag{2.24}$$

for all $\underline{C}, \underline{D} \in B^{n+1}$. Similarly, the contraction of an orientation w.r.t. a sequence $\underline{C} \in A^k$ is identically zero unless $\#C = \#\underline{C}$ (i.e. $\underline{C}$ contains no repeated elements), in which case the contraction is determined up to a factor of $\epsilon \in \{\pm 1\}$ by the underlying set C. In this case, the map defined by

$$(\underline{B}\,,\,\underline{D}) \;\mapsto\; \chi|^{\underline{C}}(\underline{B})\cdot\chi^{\underline{C}}(\underline{D}) \tag{2.25}$$

for all $\underline{B}, \underline{D} \in (A \setminus C)^{n-k+1}$ is independent of the ordering $\underline{C}$ given to C, so that we can denote it by $\sigma|^C$ without ambiguity.

Definition 2.9. A sequence $\underline{C}$ in a relative orientation (A, σ) is called *dependent* if $\#\underline{C} > n+1$ or $\chi|^{\underline{C}} = 0$ for one (and hence both) of its orientations; otherwise, it is called *independent.* In particular, a sequence is dependent whenever $\#C < \#\underline{C}$. A subset $C \subseteq A$ is independent whenever the sequence $\underline{C}$ obtained from a given ordering of its elements is independent, in which case $\#C \leq n+1$ and $(A, \sigma|^C)$ is a relative orientation. A subset is *dual independent* if it is independent in the dual relative orientation $(A, \hat{\sigma})$, or equivalently, if its complement contains an independent subset of size $n+1$ so that its restriction to the complement is an orientation.

Geometrically, for a given embedding $\mathbf{p}: A \to \mathbb{R}^n$ the contraction of the corresponding chirality function $\chi_{\mathbf{p}}: A^{n+1} \to \{0, \pm 1\}$ w.r.t. a single point $\mathbf{p}(a) \in \mathbb{R}^n$ produces the chirality of the points obtained by radial projection onto a hyperplane not containing the point $\mathbf{p}(a)$. Similarly, the contraction w.r.t. a longer k-tuple $\underline{C} = [c_1, \ldots, c_k]$ corresponds to the projection by the exclusion of C onto a subspace of dimension $n-k+1$ spanned by $A \setminus C$. The restriction to B, on the other hand, is simply the deletion of the elements of $A \setminus B$ from A. The most interesting thing concerning these operations is:

PROPOSITION 2.10. *If* (A, σ) *is an orientation and* $B, C \subseteq A$ *with* $A \setminus B$ *independent in the dual* $(A, \hat{\sigma})$ *and* C *independent* (A, σ), *then*

$$\hat{\sigma}|_{A \setminus C} = \widehat{(\sigma|^C)} \qquad \text{and} \qquad \hat{\sigma}|^{A \setminus B} = \widehat{(\sigma|_B)} \tag{2.26}$$

Proof: We prove only the first of these equations; the proof of the second is similar. W.l.o.g. we may define both duals versus an indexing $\underline{A} = [a_1, \ldots, a_N]$ such that $a_i \in C$ for $i = 1, \ldots, k =: \#C$, and let $\underline{C} := [a_1, \ldots, a_k]$ and $\underline{A \setminus C} := [a_{k+1}, \ldots, a_N]$. We shall show that

$$\hat{\bar{\chi}}|_{A \setminus C} = \overline{\widehat{(\chi|^{\underline{C}})}}\,. \tag{2.27}$$

For any $\mu \in \Lambda(N, N-n-1)$ with $A_\mu \subseteq A \setminus C$, we have $A_{\hat{\mu}} = A \setminus A_\mu \supseteq C$, and therefore

$$\begin{aligned} \hat{\bar{\chi}}|_{A \setminus C}(A_\mu) &= (-1)^{\Sigma_\mu}\, \bar{\chi}(A_{\hat{\mu}}) \\ &= (-1)^{\Sigma_\mu}\, \chi(a_1, \ldots, a_k, a_{\hat{\mu}_{k+1}}, \ldots, a_{\hat{\mu}_{n+1}})\,. \end{aligned} \tag{2.28}$$

Similarly

$$\begin{aligned} \overline{\widehat{(\chi|^{\underline{C}})}}(A_\mu) &= (-1)^{\Sigma_\mu}\, \overline{\chi|^{\underline{C}}}(\{a_{\hat{\mu}_{k+1}}, \ldots, a_{\hat{\mu}_{n+1}}\}) \\ &= (-1)^{\Sigma_\mu}\, \chi(a_1, \ldots, a_k, a_{\hat{\mu}_{k+1}}, \ldots, a_{\hat{\mu}_{n+1}})\,. \end{aligned} \tag{2.29}$$

QED

To illustrate Proposition 2.10, we once again use the orientation associated with the arrangement of atoms shown in Figure 2.4. If we use the order $\underline{B} := [a_1, a_2, a_3]$ to compute the dual of the contraction w.r.t. $\underline{C} := [a_4]$, we obtain:

$$\begin{array}{rccc} \hat{\mu} = & [12] & [13] & [23] \\ \overline{\chi|^{\underline{C}}}(B_{\hat{\mu}}) = & -1 & 0 & +1 \\ \widehat{\overline{\chi|^{\underline{C}}}}(B_{\mu}) = & +1 & 0 & -1 \\ \mu = & [3] & [2] & [1] \end{array}$$

Comparing this with the dual of the orientation itself (see the example following Definition 2.6), we see that $\widehat{\overline{\chi|^{\underline{C}}}} = \hat{\chi}|_B$.

As we shall see in the coming sections, restrictions and contractions provide us with very useful tools for proving theorems by induction on the order and rank of orientations, respectively. They may also prove useful in devising recursive algorithms on chirotopes, as they have for matroids [Williamson1985].

2.2. Chirotopes and Matroids

We are now ready to consider some additional properties of chirality functions $\chi_{\mathbf{p}} : A^{n+1} \to \{0, \pm 1\}$. Like antisymmetry, these follow from the basic properties which the oriented volume has as a determinant function $det : (\mathbb{R}^{n+1})^{n+1} \to \mathbb{R}$. For this reason, we now prove in detail a characterization of such functions, which is of considerable interest in its own right.

Definition 2.11. Let $\Delta : A^{n+1} \to \mathbb{R}$ be an arbitrary real-valued antisymmetric function (not in general the χ-function of an orientation). We say that Δ satisfies the *Graßmann-Plücker relations* if

$$\sum_{i=1}^{n+2} (-1)^i \cdot \Delta(b_1, \ldots, b_{i-1}, b_{i+1}, \ldots, b_{n+2}) \cdot \Delta(b_i, c_1, \ldots, c_n) = 0 \qquad (2.30)$$

for all $[b_1, \ldots, b_{n+2}] \in A^{n+2}$ and $[c_1, \ldots, c_n] \in A^n$.

These equations play a fundamental role in projective geometry, where they define a manifold embedded in $\mathbb{R}^{\binom{N}{n+1}}$ which is homeomorphic to the manifold of all n-dimensional subspaces of an $(N-1)$-dimensional projective space [Hodge1968] (see also Appendix C). For our purposes, however, the following "dual" interpretation of these equations is of greater immediate significance.†

THEOREM 2.12. *Let $\Delta : A^{n+1} \rightarrow \mathbb{R}$ be an antisymmetric function as above. Then there exists a map* $\mathbf{q} : A \rightarrow \mathbb{R}^{n+1}$ *such that*

$$\Delta(b_1, \ldots, b_{n+1}) \;=\; \mathit{det}(\mathbf{q}(b_1), \ldots, \mathbf{q}(b_{n+1})) \tag{2.31}$$

for all $b_1, \ldots, b_{n+1} \in A$ *if and only if* Δ *satisfies the Graßmann-Plücker relations.*

Suppose we have four points, a, b, c, and d, spaced out at unit intervals along the x-axis, and we take our antisymmetric function Δ to be the oriented areas of triangles formed by each ordered pair of these points and a common origin located at unit distance from the axis.

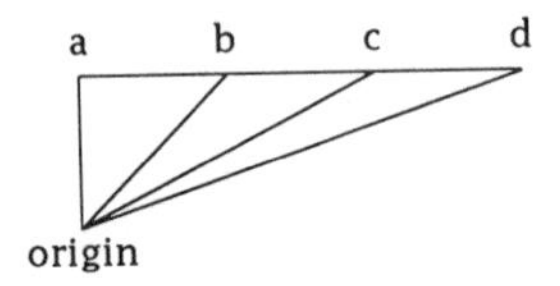

Figure 2.6

Since the areas of these triangles are all equal to *½·base·height*, the three-term Graßmann-Plücker relation is satisfied:

$$\Delta(a,b)\,\Delta(c,d) \;-\; \Delta(a,c)\,\Delta(b,d) \;+\; \Delta(b,c)\,\Delta(a,d)$$
$$=\; \tfrac{1}{2}\Big(1\cdot 1 \;-\; 2\cdot 2 \;+\; 1\cdot 3\Big) \;=\; 0\,.$$

Hence Theorem 2.12 holds with

$$\Delta(e,f) \;=\; \mathit{det}\begin{bmatrix} 1 & 1 \\ \tfrac{1}{2}x_e & \tfrac{1}{2}x_f \end{bmatrix} \;=\; \tfrac{1}{2}\,\mathit{det}\begin{bmatrix} 1 & 1 \\ x_e & x_f \end{bmatrix}$$

for all $e,f \in \{a,b,c,d\}$. This is just our usual formula for the oriented area in terms of the coordinates $[x_a, x_b, x_c, x_d]$.

Proof: (Necessity): Given a map $\mathbf{q} : A \rightarrow \mathbb{R}^{n+1}$, we define $\Delta_{\mathbf{q}} : A^{n+1} \rightarrow \mathbb{R}$ by

$$\Delta_{\mathbf{q}}(b_1, \ldots, b_{n+1}) \;:=\; \mathit{det}(\mathbf{q}(b_1), \ldots, \mathbf{q}(b_{n+1})) \tag{2.32}$$

for all $[b_1, \ldots, b_{n+1}] \in A^{n+1}$. Now, given any $b_1, \ldots, b_{n+2} \in A$, if the dimension $\mathit{dim}(\langle \mathbf{q}(b_1), \ldots, \mathbf{q}(b_{n+2}) \rangle) < n+1$, the Graßmann-Plücker relations are satisfied trivially. Otherwise we may suppose w.l.o.g. that $\mathit{dim}(\langle \mathbf{q}(b_1), \ldots, \mathbf{q}(b_{n+1}) \rangle) = n+1$, so that $\Delta_{\mathbf{q}}(b_1, \ldots, b_{n+1}) \neq 0$. Consider the following system of $n+1$ linear equations in the unknowns $[x_1, \ldots, x_{n+1}]$:

† We thank B. Sturmfels for his help in bringing the following "elementary" proof of this theorem into order.

$$\sum_{i=1}^{n+1} x_i\, \mathbf{q}(b_i) \quad = \quad \mathbf{q}(b_{n+2}) \,. \tag{2.33}$$

Since by Cramer's rule the unknowns x_i are given by

$$x_i \quad = \quad \Delta_{\mathbf{q}}(b_1, \ldots, b_{i-1}, b_{n+2}, b_{i+1}, \ldots, b_{n+1}) \,/\, \Delta_{\mathbf{q}}(b_1, \ldots, b_{n+1}) \,, \tag{2.34}$$

we have

$$\mathbf{0} \quad = \quad \Delta_{\mathbf{q}}(b_1, \ldots, b_{n+1}) \left(\sum_{i=1}^{n+1} x_i\, \mathbf{q}(b_i) - \mathbf{q}(b_{n+2}) \right) \tag{2.35}$$

$$= \quad (-1)^{n+1} \sum_{i=1}^{n+2} (-1)^i\, \Delta_{\mathbf{q}}(b_1, \ldots, b_{i-1}, b_{i+1}, \ldots, b_{n+2}) \cdot \mathbf{q}(b_i) \,.$$

Since for any $\mathbf{q}(c_1), \ldots, \mathbf{q}(c_n) \in \mathbb{R}^{n+1}$, $det(\mathbf{0}, \mathbf{q}(c_1), \ldots, \mathbf{q}(c_n)) = 0$, it follows from the linearity of the determinant in each argument that

$$0 \quad = \quad det\left(\sum_{i=1}^{n+2} (-1)^i \Delta_{\mathbf{q}}(b_1, \ldots, b_{i-1}, b_{i+1}, \ldots, b_{n+2}) \cdot \mathbf{q}(b_i), \quad \mathbf{q}(c_1), \ldots, \mathbf{q}(c_n) \right) \tag{2.36}$$

$$= \quad \sum_{i=1}^{n+2} (-1)^i\, \Delta_{\mathbf{q}}(b_1, \ldots, b_{i-1}, b_{i+1}, \ldots, b_{n+2}) \cdot det(\mathbf{q}(b_i), \mathbf{q}(c_1), \ldots, \mathbf{q}(c_n))$$

which is the desired result.

(Sufficiency): In this case, we are given a function $\Delta : A^{n+1} \rightarrow \mathbb{R}$ and must find a vector $\mathbf{q}(a_i) \in \mathbb{R}^{n+1}$ for every atom $a_i \in A$ $(i = 1, \ldots, N := \#A)$ such that Equation (2.31) holds for all $[b_1, \ldots, b_{n+1}] \in A^{n+1}$. If $\Delta = 0$ any dependent vectors will do; otherwise we can find a sequence $\underline{B} \in A^{n+1}$ such that $\Delta(\underline{B}) \neq 0$, and we assume w.l.o.g. that A has been indexed so that $\underline{B} = [a_1, \ldots, a_{n+1}]$. We let $\mathbf{q}(a_1), \ldots, \mathbf{q}(a_{n+1})$ be any basis of $\mathbb{R}^{n+1}$ scaled so that $det(\mathbf{q}(a_1), \ldots, \mathbf{q}(a_{n+1})) = \Delta(a_1, \ldots, a_{n+1})$. Then we define

$$\mathbf{q}(a_j) \quad := \quad \frac{1}{\Delta(a_1, \ldots, a_{n+1})} \sum_{i=1}^{n+1} \Delta(a_1, \ldots, a_{i-1}, a_j, a_{i+1}, \ldots, a_{n+1}) \cdot \mathbf{q}(a_i) \tag{2.37}$$

for $j = n+2, \ldots, N$. We prove by induction on k that Equation (2.31) holds for all $[b_1, \ldots, b_{n+1}] \in \{a_1, \ldots, a_k\}^{n+1}$, where $n+1 \leq k \leq N$. For $k = n+1$, the result holds by our choice of $\mathbf{q}(a_1), \ldots, \mathbf{q}(a_{n+1})$ above.

For $k > n+1$ it suffices to prove this for a sequence of the form $[b_1, \ldots, b_n, a_k] \in \{a_1, \ldots, a_{k-1}\}^n \times \{a_k\}$. We have

$$det(\mathbf{q}(b_1), \ldots, \mathbf{q}(b_n), \mathbf{q}(a_k)) \tag{2.38}$$

$$= \; det\left(\mathbf{q}(b_1), \ldots, \mathbf{q}(b_n),\; \Delta^{-1}(a_1, \ldots, a_{n+1}) \sum_{i=1}^{n+1} \Delta(a_1, \ldots, a_{i-1}, a_k, a_{i+1}, \ldots, a_{n+1}) \cdot \mathbf{q}(a_i) \right) \tag{2.39}$$

$$= \; \Delta^{-1}(a_1, \ldots, a_{n+1}) \sum_{i=1}^{n+1} \Delta(a_1, \ldots, a_{i-1}, a_k, a_{i+1}, \ldots, a_{n+1}) \cdot det(\mathbf{q}(b_1), \ldots, \mathbf{q}(b_n), \mathbf{q}(a_i)) \tag{2.40}$$

$$= \Delta^{-1}(a_1,...,a_{n+1}) \sum_{i=1}^{n+1} (-1)^{n-i+1}\Delta(a_1,...,a_{i-1},a_{i+1},...,a_{n+1},a_k)\cdot\Delta(b_1,...,b_n,a_i) \tag{2.41}$$

$$= \Delta^{-1}(a_1,...,a_{n+1}) \sum_{i=1}^{n+1} -(-1)^{i}\Delta(a_1,...,a_{i-1},a_{i+1},...,a_{n+1},a_k)\cdot\Delta(a_i,b_1,...,b_n) \tag{2.42}$$

$$= (-1)^{n+2}\Delta(a_k,b_1,\ldots,b_n) \;=\; \Delta(b_1,\ldots,b_n,a_k)\,. \tag{2.43}$$

The step from Equation (2.40) to (2.41) is justified by the induction hypothesis, since $b_1,\ldots,b_n,a_i \in \{a_1,\ldots,a_{k-1}\}^{n+1}$. The last step follows directly from the Graßmann-Plücker relations themselves, which Δ satisfies by hypothesis. QED

This theorem has the following important combinatorial implications:

COROLLARY 2.13. *Let $\chi = \chi_{\mathbf{p}}: A^{n+1} \to \{0,\pm1\}$ be the chirality function given by some embedding $\mathbf{p}: A \to \mathbb{R}^n$ via Equation (2.2), and let $\sigma: A^{n+1}\times A^{n+1} \to \{0,\pm1\}$ be the associated relative orientation. Then for any $\epsilon \in \{\pm1\}$ and $b_1,\ldots,b_{n+2},c_1,\ldots,c_n \in A$:*

$$\text{(GP)}\quad \tau_i \;:=\; \epsilon\cdot(-1)^i\,\sigma(b_1,\ldots,b_{i-1},b_{i+1},\ldots,b_{n+2};\, b_i,c_1,\ldots,c_n) \;\geq\; 0 \tag{2.44}$$

$$\forall\; i = 1,\ldots,n{+}2 \quad\text{implies}\quad \tau_i \;=\; 0 \;\text{ for }\; i = 1,\ldots,n{+}2\,.$$

Proof: This follows immediately from Theorem 2.12, since a sum of nonnegative terms can be zero only if all of the terms are zero. QED

We now incorporate this condition into our abstract study of chirality.

Definition 2.14. A *chirotope* is a relative orientation $\sigma: A^{n+1}\times A^{n+1} \to \{0,\pm1\}$ which satisfies the axiom (GP) (Equation (2.44)). Each of the associated absolute orientations $\pm\chi: A^{n+1} \to \{0,\pm1\}$ is then referred to as an *orientation* of the chirotope.

Note that since transposing any two adjacent elements in either of the sequences $[b_1,\ldots,b_{n+2}]$ or $[c_1,\ldots,c_n]$ changes the sign of every term which occurs in Equation (2.44), to verify that a relative orientation is a chirotope it suffices to check the axiom (GP) for the relative orientation $\bar{\sigma}$ versus an ordering of the set A. Hence the total number of instances which must be checked is not more than $\binom{N}{n}\cdot\binom{N}{n+2}$. As we shall see later, however, these instances are not all independent of one another.

We begin our detailed study of chirotopes by showing that the axiom (GP) is already sufficiently powerful to ensure that the basic projective properties of Euclidean point sets are fulfilled.

PROPOSITION 2.15. *Let $\sigma: A^{n+1}\times A^{n+1} \to \{0,\pm1\}$ be a chirotope, and for $\underline{B} \in A^{n+1}$ define $\sigma(\underline{B}) := \sigma(\underline{B};\, \underline{B})$. Then the collection of subsets $\mathbb{B} \subseteq \binom{A}{n+1}$*

$$\mathbb{B} \;:=\; \{\{b_1,\ldots,b_{n+1}\} \subseteq A \,|\, \sigma(b_1,\ldots,b_{n+1}) = 1\} \tag{2.45}$$

is the set of bases of a matroid on A, which is called the underlying matroid of the chirotope.

As an example of the underlying matroid of a chirotope, we consider the rank 3 chirotope associated with the planar molecule benzene, drawn with unique atom labels:

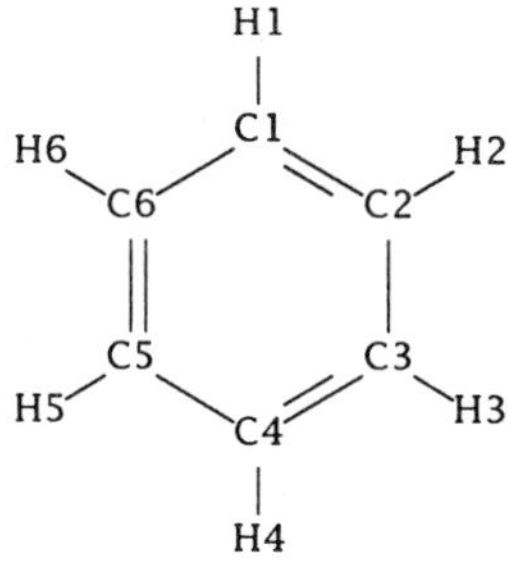

H1
C1
H6 H2
C6 C2
C5 C3
H5 H3
C4
H4

Figure 2.7

Here {H1,C1,C2} and {C2,C3,C4} are bases, whereas {H1,C1,C4} is not. Also, the basis exchange axiom is satisfied, e.g. $\{H1,C1,C2\} \setminus H1 \cup C3 = \{C1,C2,C3\}$ is likewise a base, where $C3 \in \{C2,C3,C4\} \setminus \{H1,C1,C2\}$.

Proof: First, we observe that since $\sigma(b_{\pi(1)}, \ldots, b_{\pi(n+1)}) = \sigma(b_1, \ldots, b_{n+1})$ for all permutations π of $\{1, \ldots, n+1\}$, the set $\mathbb{B}$ is well-defined. Second, by axiom (Σ0) $\sigma \neq 0$, so $\mathbb{B} \neq \{\}$. Third, by antisymmetry $\sigma(b_1, \ldots, b_{n+1}) = 0$ whenever $b_i = b_j$ for distinct $i, j \in \{1, \ldots, n+1\}$, so that $\#B = n+1$ for all $B \in \mathbb{B}$. Thus it suffices to show that the basis axiom holds ((B0) of §D.3). To do this, we let $\chi : A^{n+1} \to \{0, \pm 1\}$ be an orientation of the chirotope, and $\underline{B}$, $\underline{C}$ be two ordered bases in $\mathbb{B}$, so that $\chi^2(\underline{B}) = \chi^2(\underline{C}) = 1$. Then for any $j \in \{1, \ldots, n+1\}$ with $c_j \notin B$ there exists some $\epsilon \in \{\pm 1\}$ with $\epsilon \cdot (-1)^{n+j} \chi(\underline{B}) \chi(\underline{C}) = 1$, and it follows from (the contrapositive of) the axiom (GP) with $b_{n+2} := c_j$ that for some $i \in \{1, \ldots, n+1\}$

$$\epsilon \cdot (-1)^i \chi(b_1, \ldots, b_{i-1}, b_{i+1}, \ldots, b_{n+1}, c_j) \cdot \chi(b_i, c_1, \ldots, c_{j-1}, c_{j+1}, \ldots, c_{n+1}) = -1 \,. \qquad (2.46)$$

From this we get $\sigma(b_i, c_1, \ldots, c_{j-1}, c_{j+1}, \ldots, c_{n+1}) = 1$ as well as $b_i \notin C$, i.e. $\{b_i, c_1, \ldots, c_{j-1}, c_{j+1}, \ldots, c_{n+1}\} \in \mathbb{B}$. QED

As a consequence of this proposition, we have all of the results of matroid theory (see Appendix D) at our disposal to aid us in our study of chirotopes. Those matroids which can be obtained from chirotopes in this way are called *orientable* matroids. In particular, we have a nice correspondence between restriction, contraction and duality in chirotopes and in their underlying orientable matroids.

PROPOSITION 2.16. *Let* (A, σ) *be a chirotope, and let* $(\mathrm{A}, \mathbb{I})$ *be its underlying matroid. Then:*

(i) The relative orientation $(\mathrm{A}, \hat{\sigma})$ obtained by dualization is a chirotope whose underlying matroid is the dual matroid $(\mathrm{A}, \hat{\mathbb{I}})$.

(ii) The relative orientation $(\mathrm{A}, \sigma|_{\mathrm{B}})$ obtained by restriction to any subset $\mathrm{B} \subseteq \mathrm{A}$ whose complement is dual independent is a chirotope whose underlying matroid is the restricted matroid $(\mathrm{B}, \mathbb{I}_{\mathrm{B}})$.

(iii) The relative orientation $(\mathrm{A}, \sigma|^{\mathrm{C}})$ obtained by contraction w.r.t. any independent subset $\mathrm{C} \subseteq \mathrm{A}$ is a chirotope whose underlying matroid is the contracted matroid $(\mathrm{A} \setminus \mathrm{C}, \mathbb{I}^{\mathrm{C}})$.

Proof: Choosing an indexing $\underline{\mathrm{A}}$ of A, we may w.l.o.g. assume that $\#\mathrm{A} =: N = 2n+2$. Then the axiom (GP) for $(\mathrm{A}, \hat{\sigma})$ is exemplified by the statement:

$$0 \leq \epsilon\cdot(-1)^i \hat{\chi}(a_1, \ldots, a_{i-1}, a_{i+1}, \ldots, a_{N-n}) \hat{\chi}(a_i, a_{N-n+1}, \ldots, a_N) \tag{2.47}$$

for some orientation $\hat{\chi}: \mathrm{A}^{N-n-1} \rightarrow \{0, \pm 1\}$ of $(\mathrm{A}, \hat{\sigma})$, $\epsilon \in \{0, \pm 1\}$ and $i = 1, \ldots, n+2 = N-n$ implies equality throughout. Since the r.h.s. of the i-th of these inequalities is equal to

$$\epsilon\cdot(-1)^i(-1)^{(2n+1)(n+1)} \cdot \chi(a_i, a_{n+3}, \ldots, a_{2n+2}) \cdot \chi(a_1, \ldots, a_{i-1}, a_{i+1}, \ldots, a_{n+2}) \tag{2.48}$$

by our definition of the dual, however, this is just an example of the axiom (GP) for (A, σ) times a nonzero constant. That the underlying matroid is the dual matroid is obvious from the definitions, since $\mathrm{A} \setminus \mathrm{B}$ is a base of $\hat{\mathbb{I}}$ iff $0 \neq \hat{\sigma}(\mathrm{A} \setminus \mathrm{B}) = \sigma(\mathrm{B})$ iff B is a base of $\mathbb{I}$. Item (ii) follows simply because the conditions given in Equation (2.44) which involve only the atoms in a subset of A are a subset of those conditions which apply to all of A. Finally, item (iii) follows from (ii) by dualizing (Proposition 2.10). QED

We shall denote the rank function of the underlying matroid by $\rho: 2^{\mathrm{A}} \rightarrow \mathbb{Z}$, and the closure operator by $\mathrm{cl}: 2^{\mathrm{A}} \rightarrow 2^{\mathrm{A}}$.

Referring to the benzene molecule in the previous example (Figure 2.7), we have cl({H1,C1}) = {H1,C1,C4,H4} and ρ({H1,C1}) = 2 = ρ({H1,C1,C4,H4}). The bases of the dual matroid are the set theoretic complements of the bases of the matroid, e.g. $\hat{\mathrm{B}}$ = A \ {H1,C1,C2} = {H2,C3,H3,C4,H4,C5,H5,C6,H6}. The contraction w.r.t. H1 has all bases which contain H1 with H1 deleted as its bases, for example {C1,C2}. Note that the complement of this set in A \ {H1} is the dual base $\hat{\mathrm{B}}$ above, which demonstrates the well-known duality of restriction and contraction in matroids.

In matroid theory, it is possible to take contractions (and restrictions) w.r.t. subsets which are not (complements of dual) independent subsets. We now turn our attention to extending the definitions of restrictions and contractions given above so that all possible matroidal restrictions and contractions are obtained from analogous operations on chirotopes. Among other things, these derivations will demonstrate that the above definitions are reasonably good combinatorial analogues of the corresponding geometric concepts.

LEMMA 2.17. *If* (A, σ) *is a chirotope and* $b_1, \ldots, b_k, c, d_{k+1}, \ldots, d_{n+1}, e_{k+1}, \ldots, e_{n+1} \in A$ *with* $c \in cl(\{b_1, \ldots, b_k\})$, *then*

$$\begin{aligned} &\sigma(b_1, \ldots, b_k, d_{k+1}, \ldots, d_{n+1};\, c, b_2, \ldots, b_k, e_{k+1}, \ldots, e_{n+1}) \\ =\; &\sigma(c, b_2, \ldots, b_k, d_{k+1}, \ldots, d_{n+1};\, b_1, \ldots, b_k, e_{k+1}, \ldots, e_{n+1})\,. \end{aligned} \tag{2.49}$$

Proof: For any $x_1, \ldots, x_{n+2}, y_1, \ldots, y_n \in A$, we define

$$\begin{aligned} &GP_i(x_1, \ldots, x_{n+2};\, y_1, \ldots, y_n) \\ :=\; &(-1)^i \sigma(x_1, \ldots, x_{i-1}, x_{i+1}, \ldots, x_{n+2};\, x_i, y_1, \ldots, y_n)\,. \end{aligned} \tag{2.50}$$

Then we consider the expressions

$$\tau_i \;:=\; GP_i(c, b_1, \ldots, b_k, e_{k+1}, \ldots, e_{n+1};\, b_2, \ldots, b_k, d_{k+1}, \ldots, d_{n+1})\,. \tag{2.51}$$

For $i = 3, \ldots, k+1$, $\tau_i = 0$ since b_{i-1} occurs twice in the second argument, whereas for $i = k+2, \ldots, n+2$ $\tau_i = 0$ because the sequence $[c, b_1, \ldots, b_k]$ is dependent by hypothesis. Thus $\tau_1 = -\tau_2$ by axiom (GP), and this is just Equation (2.49). QED

This lemma immediately imposes severe restrictions on the existence of ϵ-equivalent elements.

COROLLARY 2.18. *For any two independent* $c_1, c_2 \in A$, *we have* $c_1 \approx c_2 \iff \{c_1, c_2\}$ *is dependent. Consequently, all chirality functions* $\chi = \chi_{\mathbf{p}}: A^{n+1} \to \{0, \pm 1\}$ *are proper.*

Proof: If c_1 and c_2 are ϵ-equivalent, then for all $b_2, \ldots, b_n \in A$

$$\chi(c_1, c_2, b_2, \ldots, b_n) \;=\; \epsilon \cdot \chi(c_2, c_2, b_2, \ldots, b_n) \;=\; 0\,, \tag{2.52}$$

meaning $\{c_1, c_2\}$ are dependent.

Now suppose that $\{c_1, c_2\}$ is dependent, but that both $\{c_1\}$ and $\{c_2\}$ are not. By matroid theory there exists $d_2, \ldots, d_{n+1} \in A$ such that $cl(c_i, d_2, \ldots, d_{n+1}) = A$ for $i = 1,2$. Hence for all $e_2, \ldots, e_{n+1} \in A$ it follows from Lemma 2.17 with $k = 1$, $c = c_1$ and $b_1 = c_2$ that

$$\begin{aligned} &\chi(c_1, d_2, \ldots, d_{n+1}) \cdot \chi(c_2, e_2, \ldots, e_{n+1}) \\ =\; &\chi(c_2, d_2, \ldots, d_{n+1}) \cdot \chi(c_1, e_2, \ldots, e_{n+1})\,, \end{aligned} \tag{2.53}$$

i.e. c_1 and c_2 are ϵ-equivalent with $\epsilon = \chi(c_1, d_2, \ldots, d_{n+1}) \cdot \chi(c_2, d_2, \ldots, d_{n+1}) \neq 0$.

Finally given any embedding $\mathbf{p}: A \rightarrow \mathbb{R}^n$ such that

$$\chi(b_1, \ldots, b_{n+1}) = sign(vol(\mathbf{p}(b_1), \ldots, \mathbf{p}(b_{n+1}))) \tag{2.54}$$

for all $b_1, \ldots, b_{n+1} \in A$, suppose that $a_i \approx a_j$ for two distinct $a_i, a_j \in A$. Then by the foregoing, $\{a_i, a_j\}$ is dependent, meaning that the columns $[1, \mathbf{p}(a_i)]^T$ and $[1, \mathbf{p}(a_j)]^T$ are linearly dependent, i.e. one column is a nonzero multiple of the other. Since both columns have a "1" in the first position, the factor relating them must be unity, i.e. $\mathbf{p}(a_i) = \mathbf{p}(a_j)$, which contradicts our definition of an embedding as an injective mapping. QED

For this reason, we have and will continue to ignore a number of minor complications which arise when ϵ-equivalent elements are present, and assume unless otherwise stated that all our chirotopes are proper.

We now determine conditions under which the contractions w.r.t. two independent subsets are essentially the same.

PROPOSITION 2.19. *Given any two independent subsets* $B = \{b_1, \ldots, b_k\} \subseteq A$ *and* $C = \{c_1, \ldots, c_k\} \subseteq A$ *with* $D := A \setminus (B \cup C)$, *we have* $\sigma|^B|_D = \sigma|^C|_D$ *if and only if* $cl(B) = cl(C)$.

Proof: We know from matroid theory that the underlying matroids of the contractions restricted to D are the same (if and) only if $cl(B) = cl(C)$, whence necessity follows by Proposition 2.17. To prove sufficiency we use induction on $K := k - \#(B \cap C)$. The result is trivial for $K = 0 \Leftrightarrow B = C$; otherwise we suppose w.l.o.g. that $b_1 \notin C$. By matroid theory there exists $i \in \{1, \ldots, k\}$ such that $cl((B \setminus b_1) \cup c_i) = cl(C)$ and $c_i \notin B$. Since $k - \#(((B \setminus b_1) \cup c_i) \cap C) = K - 1$, it follows from the induction hypothesis that

$$\begin{aligned} &\sigma(c_i, b_2, \ldots, b_k, e_{k+1}, \ldots, e_{n+1};\, c_i, b_2, \ldots, b_k, f_{k+1}, \ldots, f_{n+1}) \\ = \;&\sigma(c_1, \ldots, c_k, e_{k+1}, \ldots, e_{n+1};\, c_1, \ldots, c_k, f_{k+1}, \ldots, f_{n+1}) \end{aligned} \tag{2.55}$$

for all $e_{k+1}, \ldots, e_{n+1}, f_{k+1}, \ldots, f_{n+1} \in D$. We now choose an independent sequence of atoms $d_{k+1}, \ldots, d_{n+1} \in D$ (versus a suitable indexing of D) with $cl(\{c_i, b_1, \ldots, b_k, d_{k+1}, \ldots, d_{n+1}\}) = A$. Since $c_i \in cl(C) = cl(B)$, by Lemma 2.17 we have

$$\begin{aligned} &\sigma(b_1, \ldots, b_k, d_{k+1}, \ldots, d_{n+1};\, c_i, b_2, \ldots, b_k, e_{k+1}, \ldots, e_{n+1}) \\ = \;&\sigma(c_i, b_2, \ldots, b_k, d_{k+1}, \ldots, d_{n+1};\, b_1, \ldots, b_k, e_{k+1}, \ldots, e_{n+1}) \end{aligned} \tag{2.56}$$

and

$$\begin{aligned} &\sigma(c_i, b_2, \ldots, b_k, f_{k+1}, \ldots, f_{n+1};\, b_1, \ldots, b_k, d_{k+1}, \ldots, d_{n+1}) \\ = \;&\sigma(b_1, \ldots, b_k, f_{k+1}, \ldots, f_{n+1};\, c_i, b_2, \ldots, b_k, d_{k+1}, \ldots, d_{n+1})\,. \end{aligned} \tag{2.57}$$

Since $\sigma(c_i, b_2, \ldots, b_k, d_{k+1}, \ldots, d_{n+1}) = \sigma(b_1, b_2, \ldots, b_k, d_{k+1}, \ldots, d_{n+1}) = 1$, we may multiply the left and right-hand sides of these two equations together and use (Σ3) of Lemma 2.5 to get

$$\begin{aligned}&\sigma(c_i,b_2,\ldots,b_k,e_{k+1},\ldots,e_{n+1};\, c_i,b_2,\ldots,b_k,f_{k+1},\ldots,f_{n+1}) \\ =\;& \sigma(b_1,\ldots,b_k,e_{k+1},\ldots,e_{n+1};\, b_1,\ldots,b_k,f_{k+1},\ldots,f_{n+1})\,.\end{aligned} \tag{2.58}$$

Combining this result with Equation (2.55), we obtain

$$\begin{aligned}&\sigma(b_1,\ldots,b_k,e_{k+1},\ldots,e_{n+1};\, b_1,\ldots,b_k,f_{k+1},\ldots,f_{n+1}) \\ =\;& \sigma(c_1,\ldots,c_k,e_{k+1},\ldots,e_{n+1};\, c_1,\ldots,c_k,f_{k+1},\ldots,f_{n+1})\,,\end{aligned} \tag{2.59}$$

i.e. $\sigma|^B = \sigma|^C$ on D, as desired. QED

Since for all $B \subset A$, if $\underline{C} = [c_1,\ldots,c_k] \in B^k$ is a maximal independent sequence in B, then $cl(\{c_1,\ldots,c_k\}) = cl(B)$, we may therefore define the *contraction* $(A \setminus B, \sigma|^B)$ w.r.t. B as $(A \setminus B, \sigma|^C|_{A\setminus B})$ without ambiguity.

As an example of Proposition 2.19, we consider benzene again (Figure 2.7). Let $B := \{H1,C1\}$ and $C := \{H1,H4\}$, so that $D := A \setminus B \setminus C = \{C2,H2,C3,H3,C4,C5,H5,C6,H6\}$ and $cl(B) = cl(C) = \{H1,C1,C4,H4\}$. Then in the contraction we have, for instance, $\sigma|^B(C2;C6) = \sigma(H1,C1,C2;\,H1,C1,C6) = -1$ and $\sigma|^C(C2;C6) = \sigma(H1,H4,C2;\,H1,H4,C6) = -1$. This shows that it doesn't matter which pair of atoms with the same closure we choose; we will always get the same information about whether any other two atoms are on the same or opposite sides of the line the first pair lies on.

Another consequence of Proposition 2.19 is that for any orientation $\chi : A^{n+1} \to \{0,\pm 1\}$ of a chirotope and any two independent sequences $\underline{B}, \underline{C} \in A^k$ with $cl(B) = cl(C)$ we have $\chi|^{\underline{B}} = \epsilon\cdot\chi|^{\underline{C}}$, where $\epsilon \in \{\pm 1\}$ depends only on the sequences $\underline{B}$, $\underline{C}$. If $\epsilon = +1$ we say that $\underline{B}$ and $\underline{C}$ have the same orientation, whereas if $\epsilon = -1$ we say they have opposite orientation. In this way, each subspace $cl(B) \subseteq A$ inherits a relative orientation from the parent chirotope. More generally, for arbitrary subsets we have:

PROPOSITION 2.20. *For any subset* $B \subseteq A$ *with* $k := \rho(B)$ *and sequence* $\underline{C} = [c_{k+1},\ldots,c_{n+1}] \in A^{n-k+1}$ *with* $cl(B \cup C) = A$, *we have a well-defined chirotope given by* $(B, \sigma|^C|_B)$.

Proof: We must show that for any other sequence $\underline{C}' = [c'_{k+1},\ldots,c'_{n+1}] \in A^{n-k+1}$ with $cl(B \cup C') = A$, we have $\sigma|^C|_B = \sigma|^{C'}|_B$. Given any $d_1,\ldots,d_k,e_1,\ldots,e_k \in B$, either

$$\sigma|^C|_B(d_1,\ldots,d_k;\, e_1,\ldots,e_k) = \sigma|^{C'}|_B(d_1,\ldots,d_k;\, e_1,\ldots,e_k) = 0\,, \tag{2.60}$$

or else $D := \{d_1,\ldots,d_k\}$ and $E := \{e_1,\ldots,e_k\}$ are independent k-tuples in the set

B of rank k and as such satisfy $\mathrm{cl}(D) = \mathrm{cl}(E) = \mathrm{cl}(B)$. Thus for some orientation χ of the chirotope

$$\begin{aligned} & \chi(d_1,\ldots,d_k,c_{k+1},\ldots,c_{n+1})\cdot\chi(d_1,\ldots,d_k,c'_{k+1},\ldots,c'_{n+1}) \\ = \; & \sigma(d_1,\ldots,d_k,c_{k+1},\ldots,c_{n+1};\; d_1,\ldots,d_k,c'_{k+1},\ldots,c'_{n+1}) \qquad (2.61) \\ = \; & \sigma(e_1,\ldots,e_k,c_{k+1},\ldots,c_{n+1};\; e_1,\ldots,e_k,c'_{k+1},\ldots,c'_{n+1}) \\ = \; & \chi(e_1,\ldots,e_k,c_{k+1},\ldots,c_{n+1})\cdot\chi(e_1,\ldots,e_k,c'_{k+1},\ldots,c'_{n+1}) \end{aligned}$$

by Proposition 2.19. Since none of the factors appearing on the far left and right hand sides of this equation is zero, we may multiply through by any two on opposite sides and use the relation $1^2 = -1^2 = 1$ to obtain

$$\begin{aligned} & \sigma(d_1,\ldots,d_k,c_{k+1},\ldots,c_{n+1};\; e_1,\ldots,e_k,c_{k+1},\ldots,c_{n+1}) \qquad (2.62) \\ = \; & \chi(d_1,\ldots,d_k,c_{k+1},\ldots,c_{n+1})\cdot\chi(e_1,\ldots,e_k,c_{k+1},\ldots,c_{n+1}) \\ = \; & \chi(d_1,\ldots,d_k,c'_{k+1},\ldots,c'_{n+1})\cdot\chi(e_1,\ldots,e_k,c'_{k+1},\ldots,c'_{n+1}) \\ = \; & \sigma(d_1,\ldots,d_k,c'_{k+1},\ldots,c'_{n+1};\; e_1,\ldots,e_k,c'_{k+1},\ldots,c'_{n+1}) \end{aligned}$$

as desired. QED

Thus we consider $(B, \sigma|^C|_B)$ to be the *restriction* of (A, σ) w.r.t. an arbitrary (i.e. not necessarily the complement of a dual independent) subset $B \subseteq A$, and we abbreviate it simply by $(B, \sigma|_B)$. It is easily shown that these new, extended definitions of restriction and contraction remain dual to one another, and that they are in complete accord with the usual matroidal definitions.

As an example of Proposition 2.20 we consider once again the benzene molecule of Figure 2.7, and let $B := \{H1, C1, C4\}$. Note that $\rho(B) = 2 < 3 = \rho(A)$, so that B does not contain a base of the chirotope. If $C =: \{C2\}$ then $\mathrm{cl}(B \cup C) = A$, so that $\sigma|^C|_B(H1,C1;\, C4,C1) = \sigma(C2,H1,C1;\, C2,C4,C1) = -1$, meaning that [H1,C1] and [C4,C1] have opposite orientations on the line determined by {H1,C1,C4,H4}, i.e. $\chi|_B(H1,C1) = -\chi|_B(C4,C1)$. Alternatively, we may choose $C = \{H5\}$ and obtain $\sigma|^C|_B(H1,C1;\, C4,C1) = \sigma(H5,H1,C1;\, H5,C4,C1) = -1$, which is the same result.

In closing this section, we briefly discuss the connection between the theory of chirotopes and the other, better-known approach to the study of chirality, which is known as the theory of oriented matroids.

Definition 2.21. A *circuit* in a chirotope is a sequence $\underline{C} = [c_1,\ldots,c_k] \in A^k$ such that $\underline{C}$ is dependent whereas every subsequence $\underline{C}_i := [c_1,\ldots,c_{i-1},c_{i+1},\ldots,c_k]$ is independent (a one element sequence [c] is defined to be a circuit if and only if c is dependent). A *cocircuit* is a circuit of the dual chirotope. If $\chi : A^{n+1} \to \{0, \pm 1\}$ is one of the two orientations of (A, σ) and $\underline{B} = [b_1, \ldots, b_n] \in A^n$ is an independent sequence, the set

$$H(\underline{B}; \epsilon) \;=\; H(b_1, \ldots, b_n; \epsilon) \;:=\; \{c \in A \,|\, \chi(b_1, \ldots, b_n, c) = \epsilon\} \tag{2.63}$$

is called the *positive half-space* determined by $\underline{B}$ if $\epsilon = +1$ and the *negative half-space* if $\epsilon = -1$. Finally, for $\epsilon \in \{\pm 1\}$ the sets

$$\bar{H}(\underline{B}; \epsilon) \;=\; \bar{H}(b_1, \ldots, b_n; \epsilon) \;:=\; H(\underline{B}; \epsilon) \cup \mathrm{cl}(B) \tag{2.64}$$

are called the positive and negative *closed* half-spaces determined by $\underline{B}$.

For consistency, we may also use the symbol $H(\underline{B}; 0)$ to denote the hyperplane $\mathrm{cl}(B)$ spanned by the underlying independent subset $B \subseteq A$.

In the now familiar example of benzene:

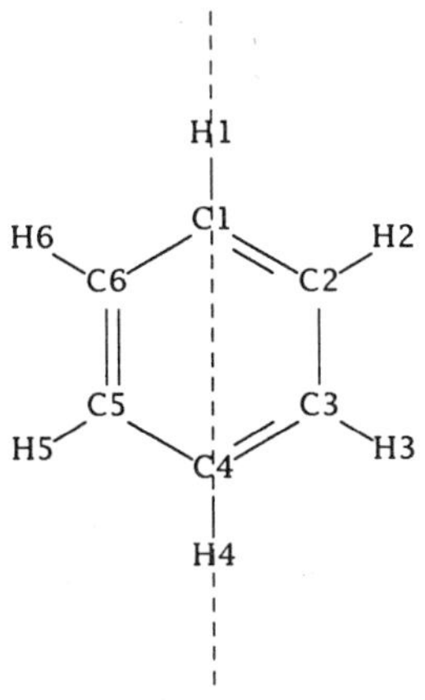

Figure 2.8

both {H2,C2,H3,C3} and {H1,C1,C4} are circuits because they are dependent but the deletion of any one element from these sets results in an independent subset. The set {H1,C1,H4,C4}, on the other hand, is a hyperplane, and its complement $\hat{C}$ = {H2,C2,H3,C3,H5,C5,H6,C6} is a circuit of the dual matroid, or cocircuit. If we let $\underline{B}$:= [H1,C1], then the positive half-space $H(\underline{B}; +1)$ is {C2,H2,C3,H3} while the negative half-space $H(\underline{B}; -1)$ is {C6,H6,C5,H5}, namely all atoms to the right and left of the dividing line shown, respectively. Thus $(\hat{C}, \hat{\eta_X})$ is a signed cocircuit, where $\hat{\eta_X} = 1$ for all atoms in {C2,H2,C3,H3} and $\hat{\eta_X} = -1$ for all atoms in {H5,C5,H6,C6}. The positive closed half-space $\bar{H}(\underline{B}; +1)$ is {H1,C1,C2,H2,C3,H3,C4,H4}, which includes the hyperplane cl(B) itself.

It is easily seen that the underlying sets of circuits and cocircuits are just the circuits and cocircuits of the underlying matroid. Since cocircuits are the complements of hyperplanes, for a given orientation $\chi: A^{n+1} \rightarrow \{0, \pm 1\}$ of a chirotope (A, σ) we can assign to each element "c" of a cocircuit $\underline{C}$ a unique *sign*, which is given by

$$\hat{\eta}_\chi(c) \; := \; \begin{cases} +1 & \text{if } c \in C^+ := H(b_1, \ldots, b_n; +1)\,; \\ -1 & \text{if } c \in C^- := H(b_1, \ldots, b_n; -1)\,. \end{cases} \tag{2.65}$$

where $\underline{B} := [b_1, \ldots, b_n]$ is an ordered basis of the hyperplane $A \setminus C = H(b_1, \ldots, b_n; 0)$ complementary to the cocircuit $\underline{C}$, and $H(b_1, \ldots, b_n; \epsilon)$ $(\epsilon \in \{\pm 1\})$ denotes the half-spaces determined by that basis. Note that since transposing any two elements of the ordered basis $\underline{B}$ or using the enantiomorphic orientation $-\chi$ changes the signs of all the elements in the cocircuit, the function $\hat{\eta}_\chi: C \rightarrow \{\pm 1\}$ is determined only up to sign. Similarly, we can provide the elements of the circuits of a chirotope with a sign by reference to the half-spaces of the dual chirotope, and the signs of the elements in each circuit are determined only up to a common factor of ± 1. Thus with each chirotope is associated a family of *signed circuits* $\{(\underline{C}, \eta)\}$.

THEOREM 2.22. *Let* $\mathbf{C} = \{(\underline{C}, \eta)\}$ *be the collection of signed circuits of a chirotope* (A, σ), *and let* $\hat{\mathbf{C}} = \{(\hat{\underline{C}}, \hat{\eta})\}$ *denote the signed circuits of the dual chirotope. Then*

(O0) $\{C \mid (\underline{C}, \eta) \in \mathbf{C}\}$ and $\{\hat{C} \mid (\hat{\underline{C}}, \hat{\eta}) \in \hat{\mathbf{C}}\}$ (2.66)

are the circuits of a dual pair of matroids ;

(O1) $(\underline{C}, \eta) \in \mathbf{C}\,,\; (\hat{\underline{C}}, \hat{\eta}) \in \hat{\mathbf{C}} \;\Rightarrow\; (\underline{C}, -\eta) \in \mathbf{C}\,,\; (\hat{\underline{C}}, -\hat{\eta}) \in \hat{\mathbf{C}}$; (2.67)

(O2) $(\underline{C}_1, \eta_1),\; (\underline{C}_2, \eta_2) \in \mathbf{C}$ and $\underline{C}_1 = \underline{C}_2 \;\Rightarrow\; \eta_1 = \pm\eta_2$; (2.68)

$(\hat{\underline{C}}_1, \hat{\eta}_1),\; (\hat{\underline{C}}_2, \hat{\eta}_2) \in \hat{\mathbf{C}}$ and $\hat{\underline{C}}_1 = \hat{\underline{C}}_2 \;\Rightarrow\; \hat{\eta}_1 = \pm\hat{\eta}_2$;

(O3) $(\underline{C}, \eta) \in \mathbf{C}\,,\; (\hat{\underline{C}}, \hat{\eta}) \in \hat{\mathbf{C}}$ and $C \cap \hat{C} \neq \{\} \;\Rightarrow$ (2.69)

$(C^+ \cap \hat{C}^-) \cup (C^- \cap \hat{C}^+) \neq \{\}$ and $(C^+ \cap \hat{C}^+) \cup (C^- \cap \hat{C}^-) \neq \{\}$.

Moreover, any collection of signed sets which satisfy these axioms arise in this way from a chirotope.

Definition 2.23. A collection of signed circuits and cocircuits $(A, \{\underline{C}, \eta\}, \{\hat{\underline{C}}, \hat{\eta}\})$ which satisfies axioms (O0) through (O3) is called an *oriented matroid.*

The necessity of axioms (O0) through (O2) follows directly from the foregoing discussion. Axiom (O3), known as the orthogonality property, is essentially a restatement of the axiom (GP). Nevertheless, a complete proof of the equivalence of chirotopes and oriented matroids is nontrivial and will not be given here. The existence of a canonical assignment of signs to the bases of oriented matroids was first demonstrated in [Vergnas1978], and shown to satisfy the Graßmann-Plücker relations by [Lawrence1982], whereas a complete proof of necessity (albeit using a different axiom system of oriented matroids) may be found in [Sturmfels1985] as well as in [Dress1986b]. Mathematicians interested in learning more about these approaches to the theory of oriented matroids, together with extensions to infinite sets and more general coefficient domains, are referred to [Dress1986a]. Most chemists, however, will probably feel that chirotopes provide a much more intuitive approach to the subject than oriented matroids! In addition, chirotopes are a much more compact data structure for algorithmic purposes (see §2.4).

2.3. Affine Chirotopes

We are now ready to consider another property of chirality functions, which is a consequence of the "additivity" of the oriented volumes spanned by each set of $n+1$ points in an n-dimensional affine space. The geometric meaning of this condition in two dimensions is illustrated in the examples that follow.

THEOREM 2.24. *Let* $\Delta : \mathrm{A}^{n+1} \rightarrow \mathbb{R}$ *be an antisymmetric function. Then there exists a function* $\mathbf{p} : \mathrm{A} \rightarrow \mathbb{R}^n$ *such that*

$$\Delta(\mathrm{b}_1, \ldots, \mathrm{b}_{n+1}) \quad = \quad vol(\mathbf{p}(\mathrm{b}_1), \ldots, \mathbf{p}(\mathrm{b}_{n+1})) \qquad (2.70)$$

for all $\mathrm{b}_1, \ldots, \mathrm{b}_{n+1} \in \mathrm{A}$ *if and only if* Δ *satisfies the Graßmann-Plücker relations (Equation (2.30)) together with:*

$$\sum_{i=1}^{n+2} (-1)^i \Delta(\mathrm{b}_1, \ldots, \mathrm{b}_{i-1}, \mathrm{b}_{i+1}, \ldots, \mathrm{b}_{n+2}) \quad = \quad 0 \qquad (2.71)$$

for all $\mathrm{b}_1, \ldots, \mathrm{b}_{n+1} \in \mathrm{A}$.

Proof: We already know from Theorem 2.12 that the Graßmann-Plücker relations are necessary and sufficient for the existence of a map $\mathbf{q} : \mathrm{A} \rightarrow \mathbb{R}^{n+1}$ such that

$$\Delta(\mathrm{b}_1, \ldots, \mathrm{b}_{n+1}) \quad := \quad det(\mathbf{q}(\mathrm{b}_1), \ldots, \mathbf{q}(\mathrm{b}_{n+1})) \qquad (2.72)$$

for all $\mathrm{b}_1, \ldots, \mathrm{b}_{n+1} \in \mathrm{A}$. Thus we must show that these vectors can be written as $\mathbf{q}(\mathrm{a}_i) = \hat{\mathbf{p}}(\mathrm{a}_i) \; := \; [1, p_1(\mathrm{a}_i), \ldots, p_n(\mathrm{a}_i)]$ for some $\mathbf{p}(\mathrm{a}_i) \in \mathbb{R}^n$ and all $\mathrm{a}_i \in \mathrm{A}$ if and only if Equation (2.71) also holds.

The absolute areas spanned by each of the four triangles of a set of four points $\{\mathbf{p}_1, \mathbf{p}_2, \mathbf{p}_3, \mathbf{p}_4\}$ in the plane can always be given signs so that they sum to zero. According to the rule of Mobius, the signs to give the areas can be determined by arbitrarily choosing a triangle, giving it the positive sign and drawing a circle inside of it (or around it) directed in the counterclockwise direction. One then draws similar circles in or around all of the triangles which have a side in common with the first, and directs these circles so that as they go by the common side they pass in the opposite direction from that of the first circle (as shown below). Those triangles whose circles are directed counterclockwise by this rule are likewise given a positive sign, while those whose circles are directed clockwise are given a negative sign.

Let us denote the absolute area of the triangle $[(\mathbf{p}_i, \mathbf{p}_j, \mathbf{p}_k)]$ by $A(i,j,k)$ for all distinct $i,j,k \in \{1,2,3,4\}$. In the figure below, we show the directed circles obtained by the rule of Mobius in the case that the point $\mathbf{p}_4$ is contained in the triangle spanned by the other three points, so that

$$A(2,3,4) + A(1,3,4) + A(1,2,4) - A(1,2,3) = 0 .$$

The signs shown in the figure are *not* those in the above formula, but rather those which the oriented areas $vol(\mathbf{p}_i, \mathbf{p}_j, \mathbf{p}_k)$ have in a right-handed coordinate system. The handy rule for a right-handed coordinate system in $\mathbb{R}^2$ is that $vol(\mathbf{p}_i, \mathbf{p}_j, \mathbf{p}_k)$ is positive if in the course of going around the cycle $\mathbf{p}_i \to \mathbf{p}_j \to \mathbf{p}_k \to \mathbf{p}_i$ we move in a counterclockwise direction, and negative otherwise.

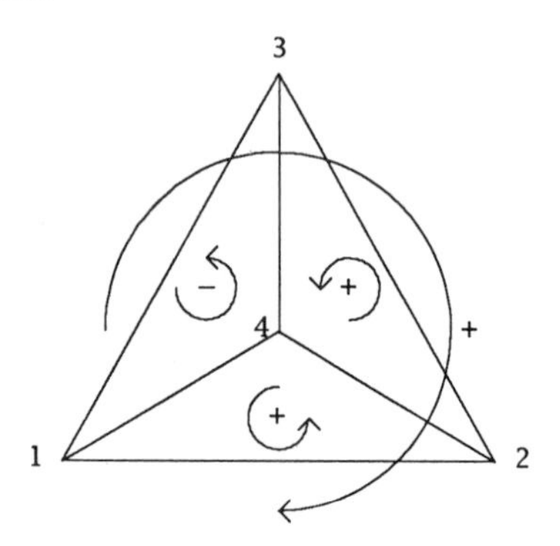

Figure 2.9

It follows that the signed and absolute areas are related by the following formulae:

$$A(2,3,4) = \tfrac{1}{2}\, vol(\mathbf{p}_2, \mathbf{p}_3, \mathbf{p}_4) \; ; \quad A(1,3,4) = -\tfrac{1}{2}\, vol(\mathbf{p}_1, \mathbf{p}_3, \mathbf{p}_4) \; ;$$

$$A(1,2,4) = \tfrac{1}{2}\, vol(\mathbf{p}_1, \mathbf{p}_2, \mathbf{p}_4) \; ; \quad -A(1,2,3) = -\tfrac{1}{2}\, vol(\mathbf{p}_1, \mathbf{p}_2, \mathbf{p}_3) \; ,$$

so that $\sum_i (-1)^i \cdot vol([\mathbf{p}_1, \ldots, \mathbf{p}_4] \setminus \mathbf{p}_i) = 0$, as in Theorem 2.25.

(Necessity): Suppose we have been given the vectors $\mathbf{p}(a_i) \in \mathbb{R}^n$, and define

$$\Delta_{\mathbf{p}}(b_1, \ldots, b_{n+1}) := vol(\mathbf{p}(b_1), \ldots, \mathbf{p}(b_{n+1})) = det\begin{bmatrix} 1 & \cdots & 1 \\ p_1(b_1) & \cdots & p_1(b_{n+1}) \\ \cdot & \cdots & \cdot \\ p_n(b_1) & \cdots & p_n(b_{n+1}) \end{bmatrix} \tag{2.73}$$

for all $b_1, \ldots, b_{n+1} \in A$. Then for any $b_1, \ldots, b_{n+2} \in A$, we consider the following determinant, whose value is obviously zero:

$$0 = det\begin{bmatrix} 1 & 1 & \cdots & 1 \\ 1 & 1 & \cdots & 1 \\ p_1(b_1) & p_1(b_2) & \cdots & p_1(b_{n+2}) \\ \cdot & \cdot & \cdots & \cdot \\ p_n(b_1) & p_n(b_2) & \cdots & p_n(b_{n+2}) \end{bmatrix}. \tag{2.74}$$

Expanding this determinant along the topmost row of ones and applying the definitions in Equation (2.73) gives us

$$\begin{aligned} 0 &= \sum_{i=1}^{n+2} (-1)^i vol(\mathbf{p}(b_1), \ldots, \mathbf{p}(b_{i-1}), \mathbf{p}(b_{i+1}), \ldots, \mathbf{p}(b_{n+2})) \\ &= \sum_{i=1}^{n+2} (-1)^i \Delta_{\mathbf{p}}(b_1, \ldots, b_{i-1}, b_{i+1}, \ldots, b_{n+2}), \end{aligned} \tag{2.75}$$

which is the desired result with $\Delta = \Delta_{\mathbf{p}}$.

(Sufficiency): In this case we can assume that we have already found $\mathbf{q}: A \to \mathbb{R}^{n+1}$ such that Equation (2.72) holds, and we seek $\mathbf{p}: A \to \mathbb{R}^n$ such that Equation (2.70) holds. If the image $\mathbf{q}(A)$ lies in an n-dimensional hyperplane, then $\Delta = 0$ so there is nothing to prove. Otherwise, we choose $b_1, \ldots, b_{n+1} \in A$ such that $\mathbf{q}(b_1), \ldots, \mathbf{q}(b_{n+1})$ are independent, let $c \in A$ be any atom not in $\{b_1, \ldots, b_{n+1}\}$, and index the set A such that $a_i = b_i$ for $i = 1, \ldots, n+1$ and $a_{n+2} = c$. Then the determinant

$$det\begin{bmatrix} 1 & 1 & \cdots & 1 \\ q_0(a_1) & q_0(a_2) & \cdots & q_0(a_{n+2}) \\ q_1(a_1) & q_1(a_2) & \cdots & q_1(a_{n+2}) \\ \cdot & \cdot & \cdots & \cdot \\ q_n(a_1) & q_n(a_2) & \cdots & q_n(a_{n+2}) \end{bmatrix} \tag{2.76}$$

$$= -\sum_{i=1}^{n+2} (-1)^i \; det(\mathbf{q}(a_1), \ldots, \mathbf{q}(a_{i-1}), \mathbf{q}(a_{i+1}), \ldots, \mathbf{q}(a_{n+2})) = 0 \tag{2.77}$$

vanishes by Equation (2.71). This means that the vector $[1, 1, \ldots, 1] \in \mathbb{R}^{n+2}$ lies in the linear span of the (row) vectors

$$\left\{ \begin{array}{c} [\, q_0(a_1),\, q_0(a_2)\,,\, \ldots\,,\, q_0(a_{n+2})\,] \\ [\, q_1(a_1),\, q_1(a_2)\,,\, \ldots\,,\, q_1(a_{n+2})\,] \\ \cdot \\ \cdot \\ \cdot \\ [\, q_n(a_1),\, q_n(a_2)\,,\, \ldots\,,\, q_n(a_{n+2})\,] \end{array} \right\}, \tag{2.78}$$

i.e. there exist scalars $\alpha_0, \ldots, \alpha_n \in \mathbb{R}$, not all zero, such that $\Sigma \alpha_i q_i(a_j) = 1$ for $j = 1, \ldots, n+2$, and w.l.o.g. we may assume that $\alpha_0 \neq 0$. Then

$$\underline{\mathbf{L}} := \begin{bmatrix} \alpha_0 & \alpha_1 & \cdots & \alpha_{n-1} & \alpha_n \\ 0 & 1 & \cdots & 0 & 0 \\ \cdot & \cdot & \cdots & \cdot & \cdot \\ 0 & 0 & \cdots & 1 & 0 \\ 0 & 0 & \cdots & 0 & \alpha_0^{-1} \end{bmatrix} \tag{2.79}$$

is the matrix of a linear transformation $\underline{\mathbf{L}}: \mathbb{R}^{n+1} \rightarrow \mathbb{R}^{n+1}$ such that the zeroth coordinate of $\underline{\mathbf{L}} \cdot \mathbf{q}(a_i)$ is one for all $i = 1, \ldots, n+2$. Moreover, since $det(\underline{\mathbf{L}}) = 1$, we have

$$\begin{aligned} \Delta(c_1, \ldots, c_{n+1}) &= det(\mathbf{q}(c_1), \ldots, \mathbf{q}(c_{n+1})) \\ &= det(\underline{\mathbf{L}} \cdot \mathbf{q}(c_1), \ldots, \underline{\mathbf{L}} \cdot \mathbf{q}(c_{n+1})) \\ &= vol(\mathbf{p}(c_1), \ldots, \mathbf{p}(c_{n+1})) \end{aligned} \tag{2.80}$$

for all $c_1, \ldots, c_{n+1} \in \{a_1, \ldots, a_{n+2}\}$, where $\mathbf{p}$ is defined as the last n components of $\underline{\mathbf{L}} \cdot \mathbf{q}$. Thus $\mathbf{p}: \{a_1, \ldots, a_{n+2}\} \rightarrow \mathbb{R}^n$ is the required mapping on $\{a_1, \ldots, a_{n+2}\}$. Since $\{\mathbf{q}(a_1), \ldots, \mathbf{q}(a_{n+1})\}$ is a linearly independent set of vectors, however, the α_i and hence $\underline{\mathbf{L}}$ is independent of our choice of $c = a_{n+2} \in A \setminus \{a_1, \ldots, a_{n+1}\}$. It follows that $\underline{\mathbf{L}}$ is a determinant preserving transformation which converts $\mathbf{q}(c)$ to a vector of the form $\hat{\mathbf{p}}(c) := [1, \mathbf{p}(c)]$ for all $c \in A$. QED

Together with Theorem 2.12, this theorem provides us with necessary and sufficient conditions for a collection of $\binom{N}{n+1}$ real numbers to be the oriented volumes of the simplices spanned by all $(n+1)$-tuples of a collection of N points in $\mathbb{R}^n$, regarded as an n-dimensional affine space.

The combinatorial implications of Theorem 2.24 are:

COROLLARY 2.25. *Let* $\chi = \chi_{\mathbf{p}}: A^{n+1} \rightarrow \{0, \pm 1\}$ *be a chirality function. Then for any* $\epsilon \in \{\pm 1\}$ *and* $b_1, \ldots, b_{n+2} \in A$*:*

(AD) $$\epsilon \cdot (-1)^i \cdot \chi(b_1, \ldots, b_{i-1}, b_{i+1}, \ldots, b_{n+2}) \geq 0 \quad \text{for all } i = 1, \ldots, n+2 \tag{2.81}$$

$$\text{implies } \chi(b_1, \ldots, b_{i-1}, b_{i+1}, \ldots, b_{n+2}) = 0 \quad \text{for } i = 1, \ldots, n+2\,.$$

To obtain an example of the sufficiency of Theorem 2.24, we consider the alternative case in which we have four points in the plane, none of which is contained in the triangle spanned by the other three points, as shown below (and drawn twice for clarity).

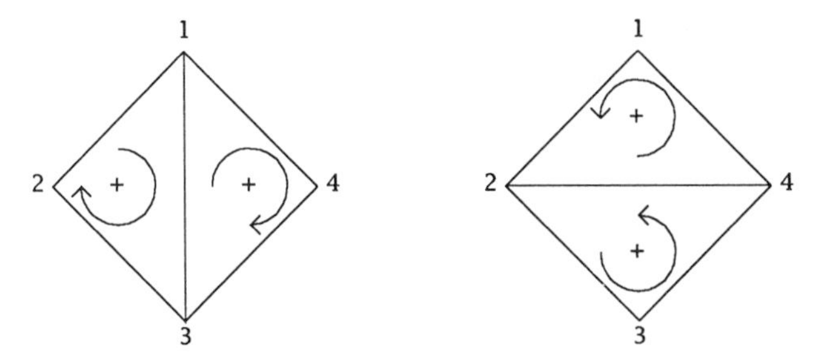

Figure 2.10

If $A(i,j,k)$ denotes the absolute area of $[(\mathbf{p}_i,\mathbf{p}_j,\mathbf{p}_k)]$ as before and we define $\Delta(i,j,k) := A(i,j,k)$, then the rule of Mobius gives us

$$\Delta(2,3,4) - \Delta(1,3,4) + \Delta(1,2,4) - \Delta(1,2,3) = 0 .$$

Since all the ordered triples of points now occur counterclockwise, the determinants $\frac{1}{2}\, vol(\mathbf{p}_i,\mathbf{p}_j,\mathbf{p}_k)$ are all positive and hence equal to the absolute areas $A(i,j,k)$. Thus the mapping $\mathbf{p} :\{1,\ldots,4\}\rightarrow\mathbb{R}^2$ given by $\mathbf{p}(i) := \mathbf{p}_i$ is that guaranteed by the theorem.

Proof: Follows immediately from the Theorem 2.24. QED

This in turn leads us to the next level of refinement in our study of abstract chirality relations.

Definition 2.26. A chirotope (A, σ) is termed *affine* whenever one (and therefore both) of its orientations obeys the affine dependency condition (axiom (AD) above).†

It is worth noting that the axiom (AD) actually characterizes those chirotopes which describe the chirality of one-dimensional point sets.

PROPOSITION 2.27. *Let* $\chi : A^2\rightarrow\{0,\pm1\}$ *be a (not necessarily proper) orientation of rank 2. Then there exists a map* $p : A\rightarrow\mathbb{R}$ *such that*

$$\chi(a,b) = 1 \iff p(a) < p(b) \tag{2.82}$$

if and only if χ *obeys the axiom (AD).*

† In the oriented matroid literature, the equivalent term *totally acyclic* is used.

Proof: Necessity follows at once from Theorem 2.24. To prove sufficiency, for all $b \in A$ we define $P(b) := \{a \in A \mid \chi(a,b) = 1\}$ and $p(b) := \#P(b)$; then the r.h.s. of Equation (2.82) implies

$$0 < p(b) - p(a) \leq \#(P(b) \setminus P(a)) \qquad (2.83)$$
$$= \#\{c \in A \mid \chi(b,c) = -1 \text{ but } \chi(a,c) \neq -1\} .$$

Hence let $c \in P(b) \setminus P(a)$ be some fixed atom in the difference; by axiom (AD), we have

$$\epsilon \cdot \chi(a,b) \geq 0, \quad -\epsilon \cdot \chi(a,c) \geq 0, \quad \epsilon \cdot \chi(b,c) \geq 0 \qquad (2.84)$$
$$\Rightarrow \quad \chi(a,b) = \chi(a,c) = \chi(b,c) = 0$$

for some $\epsilon \in \{\pm 1\}$. Since $\chi(b,c) = -1$ and $\chi(a,c) \neq -1$, this implies $\chi(a,b) = 1$, as desired. Conversely, if $\chi(a,b) = 1$ then $\chi(a,c) = -1$ implies $\chi(b,c) = -1$, so $c \in P(a)$ implies $c \in P(b)$, i.e. $P(a) \subseteq P(b) \iff p(a) \leq p(b)$. Since $a \in P(b)$ and $a \notin P(a)$, it follows that $p(a) < p(b)$. QED

As a consequence, if an affine chirotope of rank 2 is proper its orientations are exactly the one-dimensional chirality functions.

As will be shown in the next section, the axiom (AD) fails to characterize those orientations or even chirotopes which are chirality functions in higher dimensions. Nevertheless, a number of significant additional features of chirality functions follow immediately from this condition. For example, if we define a *loop* in a chirotope to be a dependent singleton, we can now show that:

LEMMA 2.28. *An affine chirotope* (A, σ) *contains no loops.*

Proof: Let $\chi : A \to \{0, \pm 1\}$ be an orientation of the chirotope, and let c be a loop in A. Then for any independent sequence of atoms $[b_1, \ldots, b_{n+1}] \in A^{n+1}$, we have

$$(-1)^i \chi(b_1, \ldots, b_{i-1}, b_{i+1}, \ldots, b_{n+1}, c) = 0 \text{ for } i = 1, \ldots, n+1, \qquad (2.85)$$

whereas $(-1)^{n+2} \cdot \chi(b_1, \ldots, b_{n+1}) \neq 0$ in contradiction with axiom (AD). QED

It is possible that the presence of a few loops is the *only* reason for which a chirotope fails to be affine. If $C := \{c \in A \mid \sigma|^c = 0\}$ and $(A \setminus C, \sigma|_{A \setminus C})$ is affine, we say that (A, σ) is *almost affine.*

The contraction of an affine chirotope need not be affine, as is shown for example by the contraction of

$$A := \{1,2,3\}, \quad \bar{\chi}(\{1,2\}) = \bar{\chi}(\{2,3\}) = \bar{\chi}(\{1,3\}) = 1 \qquad (2.86)$$

w.r.t. $\underline{C} := [2]$, obtaining

$$A \setminus C := \{1,3\}, \quad \chi|^{\underline{C}}(1) = -1, \quad \chi|^{\underline{C}}(3) = 1, \qquad (2.87)$$

which clearly does not satisfy axiom (AD). Restrictions, however, are better

Proposition 2.27 can also be interpreted as a characterization of linear pre-orderings. To see this, we define a *binary relation* to be a collection of ordered pairs $R \subseteq A^2$, and we define a binary relation to be *asymmetric* if $[a,b] \in R \Rightarrow [b,a] \notin R$. A *linear pre-ordering* is a binary relation of the form $R = \{[a,b] \mid p(a) < p(b)\}$ for some placement $p : A \to \mathbb{R}$ of the atoms along the real line; clearly linear pre-orderings are asymmetric. In order to check that a given nonempty asymmetric relation is in fact a linear pre-ordering, we define $\chi(a,b) = +1 = -\chi(b,a)$ if $[a,b] \in R$ and $\chi(a,b) = 0$ if neither $[a,b]$ nor $[b,a] \in R$, then check the function χ to see if it obeys the axiom (AD).

For instance, suppose we have four atoms together with the relation

$$R := \{[a_1,a_2], [a_1,a_4], [a_3,a_4], [a_3,a_1], [a_3,a_2]\} .$$

If we order the atoms by increasing index, we have $\bar{\chi}(\{a_1,a_2\}) = \bar{\chi}(\{a_1,a_4\}) = \bar{\chi}(\{a_3,a_4\}) = +1$ and $\bar{\chi}(\{a_i,a_j\}) = -1$ for all other pairs with $i < j$ except for $\bar{\chi}(\{a_2,a_4\}) = 0$. Since $\bar{\chi}$ clearly satisfies axioms (X0) and (X1), it is an absolute (though improper) orientation w.r.t. the given order of the atoms. Moreover, it is easily seen to satisfy all four instances of the axiom (AD) among the four atoms, e.g. $-\chi(a_2,a_3) = 1$, $+\chi(a_1,a_3) = -1$ and $-\chi(a_1,a_2) = -1$, so that with neither $\epsilon = 1$ nor $\epsilon = -1$ do we get $\chi|_{\{a_1,a_2,a_3\}} \neq 0$ while $\epsilon(-1)^i\chi([a_1,a_2,a_3] \setminus a_i) \geq 0$. Indeed the atoms can be placed on the real line as follows:

$$a_3 \to a_1 \to \begin{matrix} a_2 \\ a_4 \end{matrix} .$$

In terms of the original binary relation, the axiom (AD) is equivalent to a property known as *negative transitivity*, which means $[a,b], [b,c] \notin R \Rightarrow [a,c] \notin R$.

behaved.

PROPOSITION 2.29. *If* (A, σ) *is an affine chirotope, then so is* $(A, \sigma|_B)$ *for all* $B \subseteq A$.

Proof: Let $k := \rho(B)$, $[c_{k+1}, \ldots, c_{n+1}]$ be an independent sequence in $A \setminus B$, and let $b_0, \ldots, b_k \in B$. Then by axiom (AD) for (A, σ),

$$\tau_i := \epsilon\cdot(-1)^i\cdot\chi(b_0,\ldots,b_{i-1},b_{i+1},\ldots,b_k,c_{k+1},\ldots,c_{n+1}) \geq 0 \text{ for } i = 0,\ldots,k ; \quad (2.88)$$

$$\tau_i := \epsilon\cdot(-1)^i\cdot\chi(b_0,\ldots,b_k,c_{k+1},\ldots,c_{i-1},c_{i+1},\ldots,c_{n+1}) \geq 0 \text{ for } i = k+1,\ldots,n+1 ;$$

for some $\epsilon \in \{\pm 1\}$ together would imply $\tau_i = 0$ for $i = 0, \ldots, n+1$. Since $\rho(B) = k$, $[b_0, \ldots, b_k]$ is necessarily dependent so that $\tau_i = 0$ for $i = k+1, \ldots, n+1$. Hence $\tau_i \geq 0$ implies $\tau_i = 0$ for $i = 0, \ldots, k$, which is just the axiom (AD) for $(B, \sigma|_B)$. QED

Since restrictions of affine chirotopes are affine whereas contractions are not, it is immediately clear that the dual of an affine chirotope is not necessarily affine. Nevertheless, it is possible to characterize affine chirotopes in terms of their duals. To this end, we define another type of chirotope.

Definition 2.30. A chirotope (A, σ) of rank $n+1$ is called *spherical* if for all $\epsilon \in \{\pm 1\}$ and independent sequences $[b_1, \ldots, b_n] \in A$, the half-space $H(b_1, \ldots, b_n; \epsilon) \neq \{\}$.

LEMMA 2.31. *A chirotope* (A, σ) *is affine if and only if its dual is spherical.*

Proof: Given distinct $b_1, \ldots, b_{n+2} \in A$, we set $\{c_1, \ldots, c_{N-n-2}\} := A \setminus \{b_1, \ldots, b_{n+2}\}$ ($N := \#A$). Then if we arrange the elements of A in a sequence as $[b_1, \ldots, b_{n+2}, c_1, \ldots, c_{N-n-2}]$, let $\chi : A^{n+1} \to \{0, \pm 1\}$ be an orientation of (A, σ) and use this order to calculate the dual, we obtain

$$\hat{\bar{\chi}}(\{b_i, c_1, \ldots, c_{N-n-2}\}) = (-1)^{i+\Sigma_\mu} \chi(b_1, \ldots, b_{i-1}, b_{i+1}, \ldots, b_{n+2}), \tag{2.89}$$

where $\mu = [n+3, \ldots, N]$. Substitution of this relation into the axiom (AD) gives:

$$\tau_i := \epsilon \cdot (-1)^i \chi(b_1, \ldots, b_{i-1}, b_{i+1}, \ldots, b_{n+2}) \tag{2.90}$$

$$= \epsilon \cdot (-1)^i (-1)^{i+\Sigma_\mu} \hat{\bar{\chi}}(\{b_i, c_1, \ldots, c_{N-n-2}\}) \geq 0 \tag{2.91}$$

for $i = 1, \ldots, n+2$ implies $\tau_i = 0$ for all i. Since $(-1)^i(-1)^{i+\Sigma_\mu} = (-1)^{\Sigma_\mu} \in \{\pm 1\}$ and $\tau_i \neq 0$ for some $i = 1, \ldots, n+2$ whenever $\{c_1, \ldots, c_{N-n-1}\}$ is dual independent, this means simply that $\{\tau_i \mid i = 1, \ldots, n+2\}$ contains both positive and negative terms, i.e. $H(c_1, \ldots, c_{N-n-2}; \epsilon) \neq \{\}$ for all $\epsilon \in \{\pm 1\}$, as desired. Conversely, if $\{c_1, \ldots, c_{N-n-1}\}$ is dual independent, then $H(c_1, \ldots, c_{N-n-2}; \epsilon) \neq \{\}$ for all $\epsilon \in \{\pm 1\}$ implies that $\{\tau_i \mid i = 1, \ldots, n+2\}$ contains both positive and negative terms, which is just the contrapositive of the axiom (AD). QED

As an example of the dual of an affine chirotope, we take the chirotope given in the previous example (following Proposition 2.27) and use the order of the atoms along the real line $[a_3, a_1, a_4, a_2]$ (where we have broken the tie in favor of a_4) to calculate the dual, obtaining:

$\mu =$	[31]	[34]	[32]	[14]	[12]	[42]
$\bar{\chi}(A_\mu) =$	+1	+1	+1	+1	+1	0
$\hat{\bar{\chi}}(A_{\bar{\mu}}) =$	0	+1	−1	−1	+1	−1

for our table of $\hat{\bar{\chi}}$ values, which in turn gives us the following dual half-spaces:

ϵ	$H(1;\epsilon)$	$H(2;\epsilon)$	$H(3;\epsilon)$	$H(4;\epsilon)$
+1	{2}	{3,4}	{4}	{1}
−1	{4}	{1}	{2}	{2,3}

Since none of these is empty the dual is spherical in accord with Lemma 2.31.

We now consider the question: for exactly which subsets $B \subseteq A$ is the contraction $(A \setminus B, \sigma|^B)$ of an affine chirotope (almost) affine?

Definition 2.32. If the contraction of a chirotope (A, σ) by some $B \subseteq A$ is (almost) affine, we say that (A, σ) is *(almost) B-affine.*

This question is easily answered for those subsets $B \subset A$ with $\rho(B) = n$: here for any orientation $\chi: A^{n+1} \to \{0, \pm 1\}$ and ordered basis $\underline{C}$ of cl(B) we have $\chi|^{\underline{C}}$ affine if and only if $\chi|^{\underline{C}}(a) = \epsilon$ for some $\epsilon \in \{\pm 1\}$ and all $a \in A \setminus B$, meaning that $\chi(c_1, \ldots, c_n, a) = \epsilon$, i.e. $(A \setminus B, \sigma|^B)$ is affine if and only if $A \setminus B \subseteq H(c_1, \ldots, c_n; \epsilon)$. Similarly, the contraction is almost affine if and only if $A \setminus B \subseteq \bar{H}(c_1, \ldots, c_n; \epsilon)$. In particular, if (A, σ) is affine of rank 2, then it is almost c-affine if and only if c is a minimal or maximal element in the two linear (pre-)orders associated with (A, σ).

Definition 2.33. A *facet* of a chirotope (A, σ) is a hyperplane $cl(\{b_1, \ldots, b_n\}) \subseteq A$ such that for some $\epsilon \in \{\pm 1\}$ the closed half-space $\bar{H} = \bar{H}(b_1, \ldots, b_n; \epsilon) \supseteq A$. An atom c of an affine chirotope (A, σ) is called an *extreme point* if there exists a closed half-space $\bar{H}$ such that $c = A \setminus \bar{H}$, i.e. $\bar{H} \setminus H$ is a facet of $(A \setminus c, \sigma|_{A \setminus c})$, in which case $\bar{H} \setminus H$ is called a *separating hyperplane* for c.

Let us look for the facets and extreme points of our benzene example.

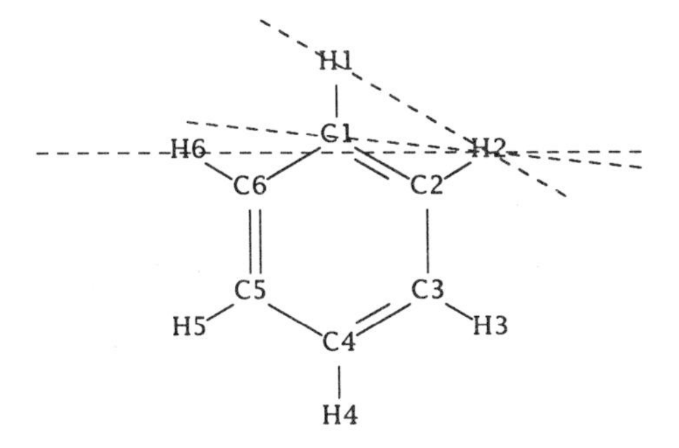

Figure 2.11

The atoms {H1,H2} constitute a facet because $\bar{H}(\{H1,H2\};-1) = A$. An example of an extreme point is provided by H1, because $\bar{H}(\{C1,H2\};-1) = A \setminus \{H1\}$. On the other hand, C1 is not an extreme point, because no matter what pair of atoms we choose and what $\epsilon \in \{\pm 1\}$ we take, $A \setminus \bar{H}$ includes atoms besides C1, e.g. $A \setminus \bar{H}(\{H2,H6\};+1) = \{C1,H1\}$.

Note that, by the above remarks, the contraction w.r.t. a facet is affine, and that a chirotope is spherical if and only if it has no facets.

LEMMA 2.34. *If* c *is an extreme point of an affine chirotope* (A, σ), *then the chirotope is* c-*affine.*

Proof: Let $\chi : A^{n+1} \to \{0, \pm 1\}$ be an orientation of the chirotope and $c = A \setminus \bar{H}$, where $\bar{H} = \bar{H}(b_1, \ldots, b_n; +1)$ w.r.t. this orientation for some independent sequence $[b_1, \ldots, b_n] \subseteq A \setminus c$. We assume that axiom (AD) is violated in the contraction by c and seek a contradiction. Hence suppose we have $d_1, \ldots, d_{n+1} \in \bar{H}$ such that

$$\tau_i \;:=\; \epsilon \cdot (-1)^i \chi(c, d_1, \ldots, d_{i-1}, d_{i+1}, \ldots, d_{n+1}) \;\geq\; 0 \qquad (2.92)$$

for all $i = 1, \ldots, n+1$ and some $\epsilon \in \{\pm 1\}$, but $\tau_i = 1$ for some $i \in \{1, \ldots, n+1\}$. Then by (the contrapositive of) axiom (AD) for (A, σ)

$$\epsilon \cdot \chi(d_1, \ldots, d_{n+1}) \;=\; -1\,. \qquad (2.93)$$

At the same time, however,

$$(-1)^n \chi(d_i, b_1, \ldots, b_n) \;\geq\; 0 \quad \text{for } i = 1, \ldots, n+1 \qquad (2.94)$$

by our assumption $d_i \in \bar{H}$, $i = 1, \ldots, n+1$. Multiplying the inequalities (2.92) and (2.94) together yields

$$\epsilon \cdot (-1)^{n+i} \chi(c, d_1, \ldots, d_{i-1}, d_{i+1}, \ldots, d_{n+1}) \chi(d_i, b_1, \ldots, b_n) \;\geq\; 0 \qquad (2.95)$$

for $i = 1, \ldots, n+1$, which by axiom (GP) implies

$$\epsilon \cdot (-1)^n \chi(d_1, \ldots, d_{n+1}) \chi(c, b_1, \ldots, b_n) \;\leq\; 0 \qquad (2.96)$$

Combining this with Equation (2.93), we conclude that $\chi(b_1, \ldots, b_n, c) \geq 0$, which puts c in $\bar{H}$ contrary to hypothesis. QED

Note that the extreme points of the contraction by an atom are also extreme points in the original chirotope. By induction, contraction of an affine chirotope with respect to a sequence $[c_1, \ldots, c_k]$ such that c_{i+1} is an extreme point of $(A \setminus C_i, \sigma|^{C_i})$ ($\underline{C}_i := [c_1, \ldots, c_i]$) for $i = 1, \ldots, k-1$ is also an affine chirotope, and the closure of such a sequence of extreme points is called a *face* of the chirotope. So far, however, we have said nothing about the existence of such subsets of affine chirotopes.

THEOREM 2.35. *Every affine chirotope* (A, σ) *of rank* $n+1$ *contains at least* $n+1$ *facets. In addition, if the chirotope is proper, it has at least* $n+1$ *extreme points.*

Among other things, this theorem shows immediately that a chirotope is never both spherical and affine.

The proof of Theorem 2.35 is unfortunately too long to be included in this survey (see [Vergnas1980, Dress1986b]). Here we can only say that the theorem marks the beginning of a body of results which show that most of the basic properties of convex polyhedra can be derived using only the axioms of affine chirotopes.

Definition 2.36. Let (A, σ) be an affine chirotope, and let $B \subseteq A$. The *convex hull* of B is defined as

$$cv(B) := \bigcap \left\{ \bar{H}(c_1, \ldots, c_n; \epsilon) \supseteq B \mid [c_1, \ldots, c_n] \in B^n \text{ independent}, \epsilon \in \{\pm 1\} \right\}. \quad (2.97)$$

Clearly, we have $cv(B) \supseteq B$ for all $B \subseteq A$. A set $C \subseteq A$ is called *convex* if $cv(C) = C$; in particular, all closed half-spaces are convex sets.

As an example of the convex hull, we again take benzene and let B = {H1,H2,H3,H4}.

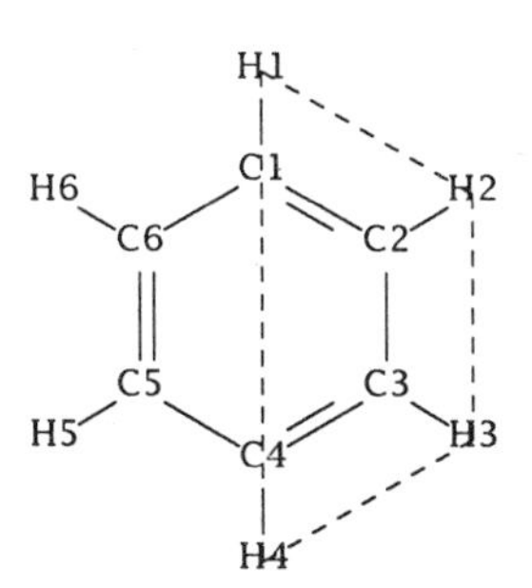

Figure 2.12

Then $\bar{H}(H1,H3;\pm1)$, $\bar{H}(H2,H4;\pm1)$, and $\bar{H}(H1,H4;-1)$ are not among the closed half-spaces $H(c_1, \ldots, c_n; \epsilon)$ used in computing cv(B) (Definition 2.36) because they aren't supersets of B. The closed half-spaces $\bar{H}(H1,H2;-1)$, $\bar{H}(H2,H3;-1)$, and $\bar{H}(H3,H4;-1)$ are used, but without effect since each is equal to A. The only significant closed half-space in the set over which the intersection is taken is $\bar{H}(H1,H4;+1) = \{H1,C1,C2,H2,C3,H3,C4,H4\} \supseteq B$, so that in this case $cv(B) = \bar{H}(H1,H4;+1)$. Note the fact that C2 and C3 are inside cv(B) is evident from the fact that they occur in none of the hyperplanes bounding the closed half-spaces over which the intersection is taken, so that this can be determined without direct reference to the coordinates of the atoms.

Note that, as a result of Proposition 2.29 and Theorem 2.35, the convex hull exists for all $B \subseteq A$. By means of this definition, it becomes possible to use the information contained in chirality functions to identify the "inside" and "outside" of a collection of points without explicit reference to their coordinates. Moreover, can be shown that:

(i) A convex set is equal to the convex hull of its extreme points.

(ii) The faces of the chirotope are the intersections of its facets.

(iii) The collection of all faces, ordered by inclusion, forms a lattice in which every maximal chain (i.e. collection of comparable elements) has the same length (this is known as *Jordan-Dedekind* condition).

We shall not, however, further develop this theory here (see [Vergnas1980, Dress1986b]).

The final feature of affine chirotopes that we shall present is important for algorithmic reasons, because it enables one to encode them in space proportional to N^n, where N is the number of atoms and $n+1$ is the rank. This is a considerable improvement over storing the actual $\bar{\chi}$-function in tabular form, which would require space $O(N^{n+1})$. This compact encoding is known as the λ-function [Goodman1983].

Definition 2.37. Let (A, σ) be an affine chirotope with $N := \#\mathrm{A}$, and $\chi : \mathrm{A}^{n+1} \rightarrow \{0, \pm 1\}$ be an arbitrary orientation thereof. For $\mathrm{b}_1, \ldots, \mathrm{b}_n \in \mathrm{A}$, the *$\lambda$-function* $\lambda : \mathrm{A}^n \rightarrow \{-1, \ldots, N\}$ is given by†

$$\lambda(\mathrm{b}_1, \ldots, \mathrm{b}_n) \;:=\; \begin{cases} -1 & \text{if } \{\mathrm{b}_1, \ldots, \mathrm{b}_n\} \text{ is dependent;} \\ \#\mathrm{H}(\mathrm{b}_1, \ldots, \mathrm{b}_n; +1) & \text{otherwise.} \end{cases} \tag{2.98}$$

Note that as a consequence of the antisymmetry of χ, for all independent sequences $[\mathrm{b}_1, \ldots, \mathrm{b}_n] \in \mathrm{A}^n$

$$\begin{aligned} \#\mathrm{A} \;=\; & \lambda(\mathrm{b}_1, \ldots, \mathrm{b}_i, \mathrm{b}_{i+1}, \ldots, \mathrm{b}_n) + \#\mathrm{cl}(\{\mathrm{b}_1, \ldots, \mathrm{b}_n\}) \\ & + \lambda(\mathrm{b}_1, \ldots, \mathrm{b}_{i+1}, \mathrm{b}_i, \ldots, \mathrm{b}_n)\,. \end{aligned} \tag{2.99}$$

For the λ-function of our benzene example, we have $\lambda(\mathrm{H1,H4}) = \#\{\mathrm{C2,H2,C3,H3}\} = 4$, $\lambda(\mathrm{H4,H1}) = \#\{\mathrm{C5,H5,C6,H6}\} = 4$, and $\#\mathrm{cl}(\mathrm{H4,H1}) = \#\{\mathrm{H1,C1,C4,H4}\} = 4$, which sums to $12 = \#\mathrm{A}$. Observe also that the extreme point H1 can be identified as such by the fact that $\lambda(\mathrm{H1,H2}) = \lambda(\mathrm{H6,H1}) = 0$, where [H6,H2] satisfies $\lambda(\mathrm{H6,H2}) = 1$.

In one dimension, the λ-function simply records, for each point $\mathbf{p}_i$ on the line, the number of points which lie to the left (or, equivalently, right) of it, and (as we saw in the proof of Proposition 2.27) this information is sufficent to enable us to reconstruct the χ-function completely. It is a rather surprising fact that this result holds up in higher dimensions; this is known as the *basic theorem of*

† The λ here has nothing to do with the set $\Lambda(N,n+1)$.

geometric sorting.

THEOREM 2.38. *The* λ*-function of an affine chirotope* (A, σ) *uniquely determines an orientation* $\chi: A^{n+1} \to \{0, \pm 1\}$ *thereof.*

Proof: As usual we shall assume that our chirotope is proper; however the theorem can be shown to hold even if it is not. The proof itself is by double induction, first on $n := \rho(A) - 1$ and then, for fixed n, by induction on $N := \#A$.

Suppose that c is an extreme point of A; these can easily be found from the λ-function, since the extreme points of (A, σ) are just those $c \in A$ for which

$$\lambda(d_1, \ldots, d_{i-1}, c, d_{i+1}, \ldots, d_n) = 0 \quad \forall\ i = \{1, \ldots, n\} \tag{2.100}$$

for some $[d_1, \ldots, d_n] \in A^n$ with $\lambda(d_1, \ldots, d_n) = 1$. Then, whatever the orientation $\chi: A^{n+1} \to \{0, \pm 1\}$ may be, the λ-function $\lambda_n: (A \setminus c)^{n-1} \to \mathbf{Z}$ of the contraction by c will be given by

$$\lambda_n(b_2, \ldots, b_n) := \lambda(c, b_2, \ldots, b_n) \tag{2.101}$$

for all $b_2, \ldots, b_n \in A \setminus c$. Since by Lemma 2.34 the contraction of an affine chirotope w.r.t. an extreme point is again an affine chirotope, it follows from the first induction hypothesis that $\chi|^c$ is uniquely determined by λ_n and hence by λ. We now define the orientation of each sequence $[c, b_1, \ldots, b_n] \in c \times (A \setminus c)^n$ by

$$\chi(c, b_1, \ldots, b_n) := \chi|^c(b_1, \ldots, b_n)\,. \tag{2.102}$$

In terms of this orientation, we can compute the λ-function of $(A \setminus c, \sigma_{A \setminus c})$ as

$$\lambda'(b_1, \ldots, b_n) := \begin{cases} \lambda(b_1, \ldots, b_n) & \text{if } (-1)^n \chi(c, b_1, \ldots, b_n) \leq 0\,; \\ \lambda(b_1, \ldots, b_n) - 1 & \text{if } (-1)^n \chi(c, b_1, \ldots, b_n) = 1\,. \end{cases} \tag{2.103}$$

Hence by the second induction hypothesis, we also know the orientation $\chi(b_1, \ldots, b_{n+1})$ of each $[b_1, \ldots, b_{n+1}] \in (A \setminus c)^{n+1}$, i.e. we know the orientation of all sequences of length $n+1$. QED

For $n = 1$, it is well known that the time required to sort N points on the line is proportional to $N \cdot log(N)$. In higher dimensions, however, it has been found possible to compute the λ-function associated with a collection of points from their coordinates in time proportional to N^n [Edelsbrunner1986]. One important application of this data structure is to obtain a fast method of testing two affine chirotopes to see if they have the same order type, i.e. to find a permutation $\phi: A \to A$ such that $\lambda_1 \circ \phi^n = \lambda_2$, where λ_1 and λ_2 are the λ-functions of the chirotopes involved. In three dimensions, once the λ-functions are known this can actually be done in time proportional to $N := \#A$ [Goodman1983].

To show that Theorem 2.38 does not hold for chirotopes which are not affine, we consider the two arrangements of four vectors in $\mathbb{R}^2$ shown below.

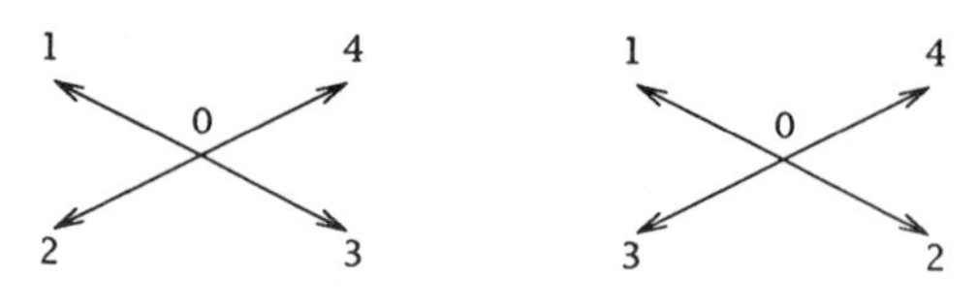

Figure 2.13

Even though the chirotopes associated with these two arrangements are different, in both cases $\lambda(i) = 1$ for $i = 1, \ldots, 4$.

2.4. Enumeration and Coordinatization of Chirotopes

Any algorithm which operates by directly manipulating chirality information must sooner or later deal with the fact that this information may fail to be geometrically consistent. Our first definition makes precise what we mean by this.

Definition 2.39. A chirotope (A, σ) is said to be *representable* if there exists a map $\mathbf{q} : \mathrm{A} \to \mathbb{R}^{n+1}$ such that

$$\chi(\mathrm{b}_1, \ldots, \mathrm{b}_{n+1}) = \chi_{\mathbf{q}}(\mathrm{b}_1, \ldots, \mathrm{b}_{n+1}) := sign(det(\mathbf{q}(\mathrm{b}_1), \ldots, \mathbf{q}(\mathrm{b}_{n+1}))) \qquad (2.104)$$

for all $\mathrm{b}_1, \ldots, \mathrm{b}_{n+1} \in \mathrm{A}$, where $\chi : \mathrm{A}^{n+1} \to \{0, \pm 1\}$ is an orientation of the chirotope. In this case, the function $\mathbf{q}$ is called a *representation* of the chirotope. A chirotope is said to be *affinely representable* if $\mathbf{q}$ can be chosen to be of the form $\hat{\mathbf{p}} = [1, \mathbf{p}]$ for some $\mathbf{p} : \mathrm{A} \to \mathbb{R}^n$.

Given the large number of properties of finite sets of vectors (or points) which can be derived from the relatively simple axioms which define (affine) chirotopes, it is perhaps somewhat surprising to learn that there exist nonrepresentable chirotopes. Basically, there are three things that can go wrong. First, the underlying matroid of the chirotope may fail to be representable. Since the characterization of representable matroids is a long-term outstanding problem in mathematics, this fact alone shows that we can expect the characterization of representable chirotopes to likewise be a very hard problem. Second, the underlying matroid may be representable (so there certainly exists a representable orientation of it), but the given orientation may nevertheless be incompatible with this matroid. A third case we shall distinguish (which is actually an extreme case of the second) occurs when the underlying matroid is uniform, but the

To obtain an example of a chirotope whose underlying matroid is not representable, we consider a hexagon whose vertices [**a**,**e**,**c**,**d**,**b**,**f**] lie alternately on two straight lines in the given order, as shown.

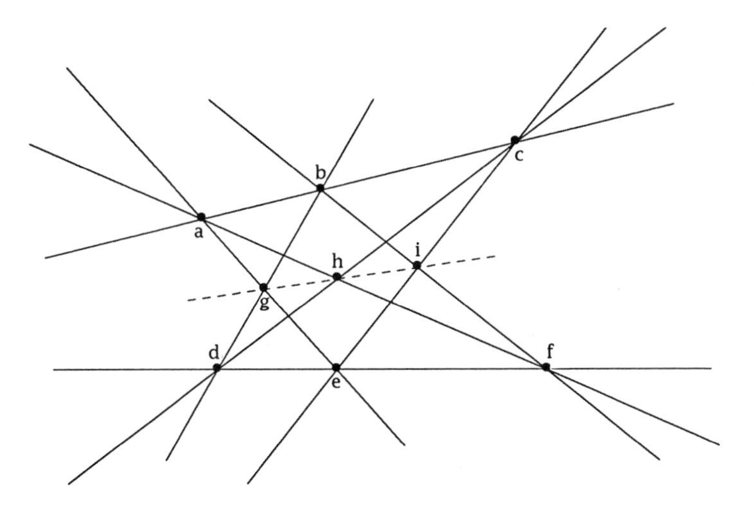

Figure 2.14

Then by the Theorem of Pappus [Coxeter1961], the points **g**,**h**,**i** of intersection of opposite sides of the hexagon must lie on a straight line. Observe that if $\chi : A^3 \to \{0, \pm 1\}$ is the orientation obtained from the figure by the usual counterclockwise rule, then since $\chi(\mathbf{g},\mathbf{h},\mathbf{i}) = 0$ the triple **g**,**h**,**i** plays a neutral role in all possible instances of the axiom (GP). This means that if we alter the chirotope by redefining $\chi(\mathbf{g},\mathbf{h},\mathbf{i}) := 1$, the resulting orientation again satisfies the axiom (GP) and hence is still a chirotope, but any representation of this chirotope would contradict the Theorem of Pappus.

orientation itself is in some way inconsistent. Even this problem is not easy.†

We start out this section by considering the linear representations of affine chirotopes.

LEMMA 2.40. *Let* (A,σ) *be an affine chirotope, and* $\mathbf{q} : A \to \mathbb{R}^{n+1}$ *be a representation thereof. Then for all* $b_1, \ldots, b_{n+1} \in A$ *with* $\sigma(b_1, \ldots, b_{n+1}) = 1$, *we have*

$$(-1)^n \Delta_{\mathbf{q}}(b_1, \ldots, b_{n+1}) \sum_{i=1}^{n+1} (-1)^i \Delta_{\mathbf{q}}(b_1, \ldots, b_{i-1}, b_{i+1}, \ldots, b_{n+2}) \;<\; 0\,, \qquad (2.105)$$

† In fact, it is known that no finite characterization is possible, in the sense that a finite number of chirotopes obtainable by contraction or restriction can be excluded [Bokowski1986].

As an example of a nonrepresentable chirotope whose underlying matroid is representable, we suppose that the points of intersection of opposite sides of the hexagon [**a**,**e**,**c**,**d**,**b**,**f**] are collinear, as in the previous example.

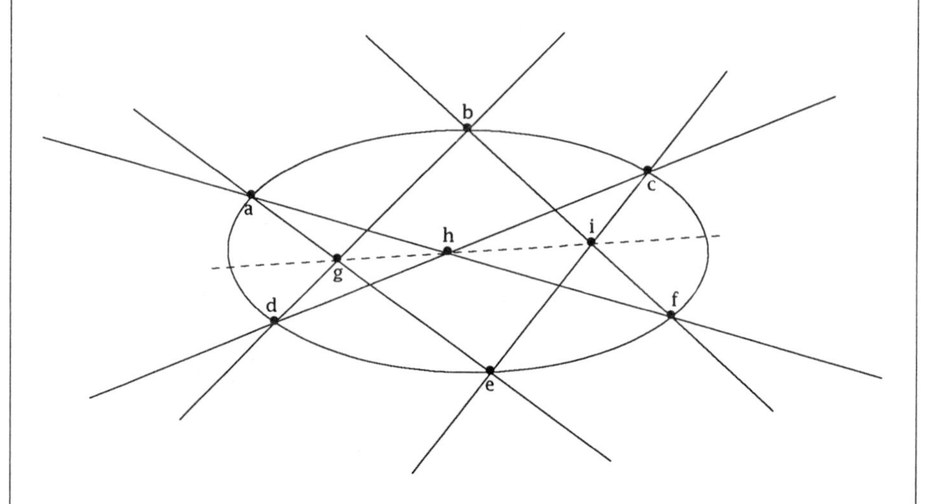

Figure 2.15

Then by the Theorem of Pascal, the vertices of the hexagon must lie on a quadric (in the figure, an ellipse). If (A, σ) is the chirotope associated with the arrangement shown and we redefine $\sigma(\mathbf{a},\mathbf{b},\mathbf{c};\mathbf{d},\mathbf{e},\mathbf{f}) := 1$, we obtain an orientation which still satisfies axiom (GP), because for the only nontrivial instance thereof:

$$\sigma(\mathbf{a},\mathbf{b},\mathbf{c};\mathbf{d},\mathbf{e},\mathbf{f}) = 1, \quad -\sigma(\mathbf{d},\mathbf{b},\mathbf{c};\mathbf{a},\mathbf{e},\mathbf{f}) = 1,$$

$$\sigma(\mathbf{d},\mathbf{a},\mathbf{c};\mathbf{b},\mathbf{e},\mathbf{f}) = -1, \quad -\sigma(\mathbf{d},\mathbf{a},\mathbf{b};\mathbf{c},\mathbf{e},\mathbf{f}) = 1.$$

Nevertheless, any representation of this chirotope would be a violation of the Theorem of Pascal (it is also necessary to check that this chirotope cannot be represented when the vertices lie on a hyperbola).

where $\Delta_{\mathbf{q}}$ is defined as in Theorems 2.12 and 2.24.

Proof: For the sake of simplicity we shall assume that the chirotope is proper; however the result holds even if it is not. Since the restrictions of affine chirotopes are affine, it suffices to prove this for the case $N = n+2$. As an immediate consequence of the axiom (AD), each such affine chirotope has at least one n-element facet, say $\{a_1, \ldots, a_n\}$, so that $\chi(a_1, \ldots, a_n, a_{n+1}) = \chi(a_1, \ldots, a_n, a_{n+2}) =: \epsilon \in \{\pm 1\}$. In addition, either there exists some $i \in \{1, \ldots, n\}$ such that

$\{a_1, \ldots, a_{i-1}, a_{i+1}, \ldots, a_{n+1}\}$ is a separating hyperplane for a_i and a_{n+2}, i.e.

$$\begin{aligned} & -\chi(a_1, \ldots, a_{i-1}, a_{i+1}, \ldots, a_{n+1}, a_{n+2}) \\ = \; & \chi(a_1, \ldots, a_{i-1}, a_{i+1}, \ldots, a_{n+1}, a_i) \; = \; (-1)^{n-i+1} \cdot \epsilon \end{aligned} \tag{2.106}$$

or else

$$(-1)^i \cdot \chi(a_1, \ldots, a_{i-1}, a_{i+1}, \ldots, a_{n+1}, a_{n+2}) \in \{0, (-1)^{n+1} \cdot \epsilon\} \tag{2.107}$$

for $i = 1, \ldots, n$, where by our assumption that the chirotope is proper the orientation occurring in (2.107) is nonzero for some $i =: j \in \{1, \ldots, n\}$. Hence in the latter case we have:

$$\begin{aligned} & (-1)^j \cdot \chi(a_1, \ldots, a_{j-1}, a_{j+1}, \ldots, a_n, a_{n+2}, a_{n+1}) \\ = \; & (-1)^n \cdot \epsilon \; := \; (-1)^n \cdot \chi(a_1, \ldots, a_n, a_{n+2}) \\ = \; & -(-1)^j \cdot \chi(a_1, \ldots, a_{j-1}, a_{j+1}, \ldots, a_n, a_{n+2}, a_j) \, , \end{aligned} \tag{2.108}$$

i.e. $\{a_1, \ldots, a_{j-1}, a_{j+1}, \ldots, a_n, a_{n+2}\}$ is a separating hyperplane for a_j and a_{n+1}.

We conclude that the proper affine chirotope on $n+2$ atoms given by $\chi = \chi_{\mathbf{q}} := sign(\Delta_{\mathbf{q}})$ always has at least one facet together with an extreme point not in the facet, and we assume w.l.o.g. that a_1 is an extreme point and $\{a_2, \ldots, a_{n+1}\}$ is a facet. The remainder of the proof is by induction on n; for $n = 0$ the result is trivial. Otherwise we rewrite the expression in (2.105) as:

$$\begin{aligned} & -(-1)^n \Delta_{\mathbf{q}}(a_1, \ldots, a_{n+1}) \Delta_{\mathbf{q}}(a_2, \ldots, a_{n+2}) \; + \\ & (-1)^{n-1} \Delta_{\mathbf{q}}(a_1, \ldots, a_{n+1}) \sum_{i=1}^{n} (-1)^i \Delta_{\mathbf{q}}(a_1, \ldots, a_i, a_{i+2}, \ldots, a_{n+2}) \, . \end{aligned} \tag{2.109}$$

The first term is negative since $\{a_2, \ldots, a_{n+1}\}$ is a facet, whereas the second is negative by the induction hypothesis, since the contractions of affine chirotopes w.r.t. extreme points are likewise affine (Lemma 2.34). QED

The next result shows that, once we know that a chirotope is representable, it is a trivial matter to decide if it is affinely representable.

THEOREM 2.41. *If an affine chirotope is representable, then it is affinely representable. Furthermore, its orientations are chirality functions whenever the chirotope is proper.*

Proof: Since a representation $\mathbf{q} : A \rightarrow \mathbb{R}^{n+1}$ must be one to one whenever there are no dependent pairs of atoms, the second claim follows at once from Corollary 2.18. To prove the first claim, we let $\underline{A} := [a_1, \ldots, a_N]$ be an ordering such that $\sigma(a_1, \ldots, a_{n+1}) = 1$, and consider an $(n+2)$-tuple $\mu \in \Lambda(k, n+2)$ with $\mu_{n+2} = k$, $n+1 \leq k \leq N$. Proof is by induction on k; the axiom (AD) is vacuous for $k = n+1$. Otherwise, we seek a representation $\mathbf{q} : A \rightarrow \mathbb{R}^{n+1}$ which satisfies

Finally, we give an example which shows that a chirotope can fail to be representable even when its underlying matroid is uniform.

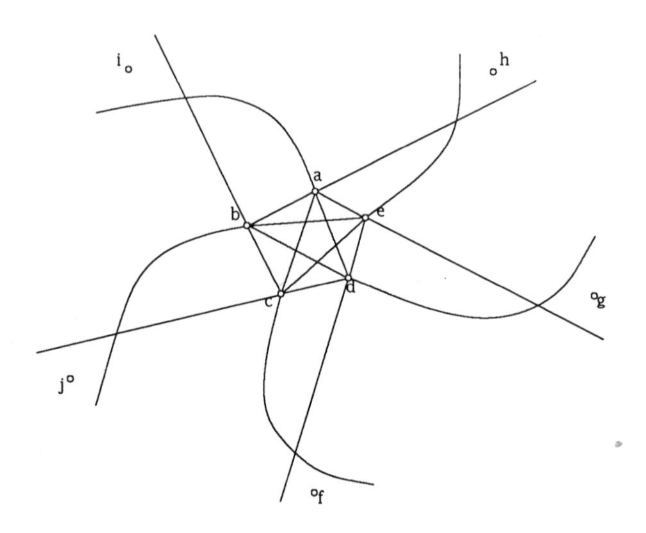

Figure 2.16

The curves in the above figure are meant to show on which side of the 10 hyperplanes spanned by the 10 pairs of points in {**a**,**b**,**c**,**d**,**e**} the points **f**,**g**,**h**,**i**,**j** lie. Thus the orientations of this chirotope are those obtained by the usual counterclockwise rule with the exception of the ordered triples [**a**,**c**,**f**], [**b**,**d**,**g**], [**c**,**e**,**h**], [**d**,**a**,**i**] and [**e**,**b**,**j**], whose orientations are all -1 (as shown). The nonrealizability of this chirotope is a result due to [Bokowski1986].

$$\sum_{i=1}^{n+2}(-1)^i\Delta_{\mathbf{q}}(a_{\mu_1},\ldots,a_{\mu_{i-1}},a_{\mu_{i+1}},\ldots,a_{\mu_{n+2}}) \;=\; 0\,. \tag{2.110}$$

If this relation is not fufilled, we let

$$\alpha(\mu) \;:=\; \frac{-(-1)^n\Delta_{\mathbf{q}}(a_{\mu_1},\ldots,a_{\mu_{n+1}})}{\sum_{i=1}^{n+1}(-1)^i\Delta_{\mathbf{q}}(a_{\mu_1},\ldots,a_{\mu_{i-1}},a_{\mu_{i+1}},\ldots,a_{\mu_{n+2}})}\,, \tag{2.111}$$

and observe that, by Lemma 2.40, $\alpha(\underline{\mu}) > 0$. Thus we may set $\mathbf{q}(a_k) := \alpha(\underline{\mu})\mathbf{q}(a_k)$ to obtain a *new* repesentation of the chirotope which satisfies Equation (2.110).

We claim that this relation in fact now holds for all $\underline{\mu} \in \Lambda(k,n+2)$ with $\mu_{n+2} = k$, i.e. that the function $\alpha(\underline{\mu})$ is a constant independent of our choice of $\underline{\mu}$! To do this, we let $\underline{\nu} \in \Lambda(k,n+2)$ be any other $(n+2)$-tuple with $\nu_{n+2} = k$. Then by the induction hypothesis:

$$\Delta_{\mathbf{q}}(a_{\mu_1}, \ldots, a_{\mu_{n+1}}) = (-1)^{n+1} \sum_{i=1}^{n+1} (-1)^i \Delta_{\mathbf{q}}(a_{\mu_1},\ldots,a_{\mu_{i-1}}, a_{\mu_{i+1}},\ldots,a_{\mu_{n+1}}, a_{\nu_j}) \tag{2.112}$$

for all $j = 1, \ldots, n+1$, and therefore by interchanging the order in the following double summation and applying Theorem 2.12, we get:

$$\begin{aligned} &\sum_{j=1}^{n+1} (-1)^j \Delta_{\mathbf{q}}(a_{\mu_1},\ldots,a_{\mu_{n+1}}) \Delta_{\mathbf{q}}(a_{\nu_1},\ldots,a_{\nu_{j-1}}, a_{\nu_{j+1}},\ldots,a_{\nu_{n+2}}) \\ &= \sum_{j=1}^{n+1} \sum_{i=1}^{n+1} (-1)^{i+j+n+1} \Delta_{\mathbf{q}}(a_{\mu_1},\ldots,a_{\mu_{i-1}}, a_{\mu_{i+1}},\ldots,a_{\mu_{n+1}}, a_{\nu_j}) \Delta_{\mathbf{q}}(a_{\nu_1},\ldots,a_{\nu_{j-1}}, a_{\nu_{j+1}},\ldots,a_{\nu_{n+2}}) \\ &= \sum_{i=1}^{n+2} (-1)^i \Delta_{\mathbf{q}}(a_{\nu_1},\ldots,a_{\nu_{n+1}}) \Delta_{\mathbf{q}}(a_{\mu_1},\ldots,a_{\mu_{i-1}}, a_{\mu_{i+1}},\ldots,a_{\mu_{n+2}}) \,, \end{aligned} \tag{2.113}$$

from which our claim follows. This shows that the new representation $\mathbf{q}$ fulfills the affine dependency condition for all $(n+2)$-tuples in $\Lambda(k,n+1)$, and we conclude that such a representation exists for $k = N$ by induction on k. That the chirotope is affinely representable then follows directly from Theorem 2.24. QED

Because of this theorem, for the remainder of this section we will generally not assume that our chirotopes are affine. Our next theorem allows us to use our knowledge of matroid theory to further simplify the definition of a chirotope.

THEOREM 2.42. *Let* (A, σ) *be a relative orientation such that* $\mathbb{B} = \{B \subseteq A \mid \sigma(B) = 1\}$ *is the set of bases of a matroid on* A. *Then* (A, σ) *is a chirotope if and only if the* *<u>three-term GP-condition</u>* *is satisfied: for all* $b_1, b_2, b_3, c_1, \ldots, c_n \in A$:

$$\begin{aligned} \text{(GP3)} \quad & \epsilon\cdot\sigma(b_1,b_2,c_2,\ldots,c_n;b_3,c_1,\ldots,c_n) \geq 0\,, \\ & -\epsilon\cdot\sigma(b_1,b_3,c_2,\ldots,c_n;b_2,c_1,\ldots,c_n) \geq 0 \quad \text{and} \\ & \epsilon\cdot\sigma(b_2,b_3,c_2,\ldots,c_n;b_1,c_1,\ldots,c_n) \geq 0 \quad \text{for some } \epsilon \in \{\pm 1\} \end{aligned} \tag{2.114}$$

implies equality holds in all three cases .

The necessity of the 3-term GP-condition is obvious, since it is just a special case of the general GP-condition. Because of its length, the proof of the sufficiency of this theorem has not been included here.†

† The only published proof is in [Bland1978], and is phrased in the language of oriented matroids.

An interesting and difficult result, due to [Folkman1978], states that any oriented matroid can be represented by an arrangement of *pseudo-hemispheres* in n-space. In the case of chirotopes of rank three, this says in effect that any such chirotope admits a representation by *pseudolines*. This is a collection of unbounded plane curves, each pair of which cross exactly once, as shown.

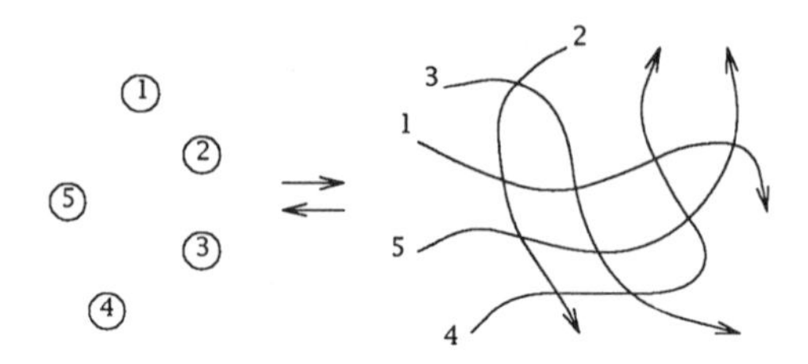

Figure 2.17

The representability of such chirotopes can thus be reduced to the question of whether or not a given arrangement of pseudolines is *stretchable* to an equivalent arrangement of straight lines (see [Mandel1981, Roudneff1987]).

The final simplification which we shall make for the purposes of this section is to restrict ourselves to those chirotopes whose underlying matroids are uniform, so that $\chi(\underline{A}_{\lambda}) \in \{\pm 1\}$ for all $\underline{\lambda} \in \Lambda(N, n+1)$. Such a chirotope is called *simplicial.* Since uniform matroids are always representable, this simplification allows us to avoid the difficult problem of matroid representability. In addition, as a consequence of Theorem 2.42 we need only keep track of the 3-term GP-condition when dealing with simplicial chirotopes, a fact which leads to some drastic simplifications in algorithms for manipulating these chirotopes. Indeed, since any simplicial chirotope on $n+3$ atoms is representable, the 3-term GP-condition is equivalent to the statement "all restrictions to $n+3$ element subsets are representable", and hence can be interpreted as a kind of local representability condition.

As an interesting and useful example, we now present an algorithm for enumerating (an orientation of) all possible simplicial chirotopes on a set A which are consistent with the known values for the χ-function on certain subsets of atoms. This case, of course, is the usual one which occurs in molecules, where only some of the tetrahedra of atoms generally have a fixed, known chirality. To avoid redundancy, we shall enumerate only the orientation $\overline{\chi}$ versus a given ordering $\underline{A} = [a_1, \ldots, a_N]$ of A.

> The benzene chirotope is not simplicial, because for example $\chi(H1,C1,C4) = 0$ due to the collinearity of such triplets of atoms. On the other hand, the chirotope associated with ethylene is simplicial because triplets of atoms are noncollinear.

Algorithm 2.43:

INPUT: A value n, an (ordered) set of atoms $\underline{A}$ and an (incomplete) set of values

$$X := \{X(B_i) \in \{\pm 1\} \mid B_i \subseteq A \text{ and } \#B_i = n+1 \text{ for } i = 1, \ldots, K\} . \quad (2.115)$$

OUTPUT: The set of all X-compatible simplicial chirotopes, i.e. complete assignments of values of $\overline{\chi}$ to $(n+1)$-tuples of atoms such that axiom (GP3) is satisfied and $\overline{\chi}(B_i) = X(B_i)$ for $i = 1, \ldots, K$.

PROCEDURE:

Set $\underline{\mu} := [1, \ldots, n+1]$.

While $\underline{\mu} \neq []$, do:

  If $\underline{A}_\mu$ has been assigned a value $X(\underline{A}_\mu)$ in X, then

    let $S(\underline{\mu}) := \{X(\underline{A}_\mu)\}$;

  otherwise

    let $S(\underline{\mu}) := \{\pm 1\}$.

  For each $(n+1)$-tuple $\underline{\nu} \in \Lambda(N, n+1)$ which is lexicographically smaller than $\underline{\mu}$ such that $\#(\nu \cap \mu) = n-1$, do:

    Let $\{\mu_i, \mu_j\} = \mu \setminus \nu$ $(i < j)$, $\{\nu_k, \nu_l\} = \nu \setminus \mu$ $(k < l)$ and $\underline{\lambda} \in \Lambda(N, n-1)$ be the sequence such that $\lambda = \mu \cap \nu$.

    Then necessarily $\nu_k < \mu_i, \mu_j$, and if in addition $\nu_l < \mu_i, \mu_j$ (so that all of the sequences occurring as arguments to $\overline{\chi}$ in (2.116) below are lexicographically less than $\underline{\mu}$ and hence have already been assigned $\overline{\chi}$-values), then define:

$$\epsilon_1 := \overline{\chi}(a_{\nu_k} \cup a_{\mu_j} \cup A_\lambda)\overline{\chi}(a_{\nu_l} \cup a_{\mu_i} \cup A_\lambda) , \quad (2.116)$$

$$\epsilon_2 := -\overline{\chi}(a_{\nu_k} \cup a_{\mu_i} \cup A_\lambda)\overline{\chi}(a_{\nu_l} \cup a_{\mu_j} \cup A_\lambda) .$$

      If $\epsilon_1 = \epsilon_2$, then

        set $S(\underline{\mu}) := S(\underline{\mu}) \setminus \{\epsilon_1 \cdot \overline{\chi}(A_\nu)\}$.

  If $S(\underline{\mu}) \neq \{\}$, then

    set $\overline{\chi}(A_\mu) := \epsilon \in S(\underline{\mu})$.

If μ is last $(n+1)$-tuple in lexicographic order, then
output chirotope;
otherwise
let μ be the next $(n+1)$-tuple;
otherwise
go backwards in lexicographic order until an $(n+1)$-tuple ν with $S(\nu) = \{\pm 1\}$ is found.
If no such $(n+1)$-tuple exists, then
force exit of main loop by setting $\mu := []$
(all X-compatible chirotopes have been found);
otherwise
let $S(\nu) = S(\nu) \setminus \{\overline{X}(A_\nu)\}$ and μ be the $(n+1)$-tuple following ν in lexicographic order.

This algorithm has been implemented in the MAPLE symbolic programming language [Char1986]. The number of simplicial chirotopes of rank $n+1$ on N atoms is given in Table 2.1 for various values of n and N (observe these numbers are equal when $n+1 = N-n-1$, since such chirotopes come in dual pairs).

Table 2.1.
Numbers of Simplicial Chirotopes of Rank $n+1$ and Order N

$n \backslash N$	3	4	5	6
1	4	24	192	1920
2	1	8	192	11904
3	0	1	16	1920
4	0	0	1	32

For $n = 1$ the general formula appears to be $2^{N-2}(N-1)!$; since the number of affine simplicial rank 2 chirotopes is $\frac{1}{2}N! \approx \frac{1}{2}e^{-N}N^N$, this shows that the fraction of these which are affine goes to zero exponentially as $N \rightarrow \infty$. For $n > 1$ a general formula is not obvious, although the rate of growth remains superexponential. The best-known theoretical upper bound [Goodman1983] is $\frac{1}{2}N^{n(n+1)N}$, which though it may be asymptotically reasonable is rather poor in the absolute sense ($\sim 10^9$ for $n = 2$, $N = 3$).

Rather obviously, for large N the total number of simplicial chirotopes will be more than even a computer can handle. By utilizing one's knowledge of the molecule, however, it may be possible to determine enough of the $\overline{X}$ values to make the enumeration of all X-compatible chirotopes tractable. The above algorithm could also be easily modified to take account of the axiom (AD) as well, in which case the numbers would be much smaller.* This method could then be

* The number of chirotopes to be enumerated can also be reduced by taking account of the internal symmetries of the molecule, and enumerating only those which are not equivalent

used to generate a list of all possible (and probably some impossible) simplicial chiralities for mobile molecules. In this context, it is worth pointing out that chiralities of nonrigid subgroups (e.g. the signs of dihedral angles) have been used as a means of classifying conformation since the earliest computer studies [Hendrickson1967]. More importantly, however, this method provides us with a powerful check on the consistency of a given description of molecular chirality, since if *no* X-compatible orientations consistent with the axioms (GP) and (AD) can be found, the chirality information cannot be realized by *any* conformation of the molecule.

We now present an algorithm which is capable of *coordinizating* simplicial chirotopes, i.e. of finding representations for them [Bokowski1986]. The basic ideas behind this algorithm are as follows. First we assume that the orientation and the ordering have been chosen so that $\chi(a_1, \ldots, a_{n+1}) = 1$. Then in any representation $\mathbf{q}: A \rightarrow \mathbb{R}^{n+1}$ of the chirotope the vectors $\{\mathbf{q}(a_j) \mid j = 1, \ldots, n+1\}$ are independent, and we may use them as a basis for calculating the remaining vectors. The coordinates $\mathbf{q}(a_j)$ of the atoms can be arranged as the columns of an $n+1$ by $N := \#A$ matrix, as follows:

$$\underline{\mathbf{Q}} := \begin{bmatrix} 1 & 0 & \cdots & 0 & q_0(a_{n+2}) & \cdots & q_0(a_N) \\ 0 & 1 & \cdots & 0 & q_1(a_{n+2}) & \cdots & q_1(a_N) \\ \cdot & \cdot & \cdots & \cdot & \cdot & \cdots & \cdot \\ 0 & 0 & \cdots & 1 & q_n(a_{n+2}) & \cdots & q_n(a_N) \end{bmatrix} . \tag{2.117}$$

For each $(n+1)$-tuple $\underline{A}_\mu = [a_{\mu_1}, \ldots, a_{\mu_{n+1}}] \in A^{n+1}$ $(\mu \in \Lambda(N, n+1))$ we let $Q(\underline{A}_\mu)$ be the determinant $det(\underline{\mathbf{Q}}_\mu)$ of the $n+1$ by $n+1$ submatrix $\underline{\mathbf{Q}}_\mu := [\mathbf{q}(a_{\mu_k}) \mid k = 1, \ldots, n+1]$, considered as a function of the unknowns $q_i(a_{\mu_k})$. Then $\mathbf{q}$ (or equivalently $\underline{\mathbf{Q}}$) is a representation of the chirotope if and only if $\chi(\underline{A}_\mu) = sign(Q(\underline{A}_\mu))$ for all $\mu \in \Lambda(N, n+1)$. Observe (by working out a few examples if necessary) that if $\mu_1, \ldots, \mu_k \in \{1, \ldots, n+1\}$ for some $1 \le k \le n$, then $Q(\underline{A}_\mu)$ is really just an $n-k+1$ by $n-k+1$ determinant taken from the last $N-n-1$ columns of $\underline{\mathbf{Q}}$ times a factor of $(-1)^M$, where $M = \sum_{i=1}^{k}(\mu_i - i)$. This determinant is just the complement of the unit matrix in the submatrix $\underline{\mathbf{Q}}_\mu$. In particular, for $j > n+1$ the individual variables $q_i(a_j)$ of the matrix $\underline{\mathbf{Q}}$ themselves are given by

$$\begin{aligned} q_i(a_j) &= (-1)^{n+1-i} Q(a_1, \ldots, a_{i-1}, a_{i+1}, \ldots, a_{n+1}, a_j) \\ &= Q(a_1, \ldots, a_{i-1}, a_j, a_{i+1}, \ldots, a_{n+1}) . \end{aligned} \tag{2.118}$$

We now have a nonlinear (or more precisely, multilinear) system of inequalities to solve in $(n+1)(N-n-1)$ variables, namely

---

under admissible permutations of the atoms.

$$Q(\mathrm{a}_{\mu_1},\ldots,\mathrm{a}_{\mu_{n+1}}) \begin{cases} < 0 & \text{if } \chi(\mathrm{a}_{\mu_1},\ldots,\mathrm{a}_{\mu_{n+1}}) = -1\,; \\ > 0 & \text{if } \chi(\mathrm{a}_{\mu_1},\ldots,\mathrm{a}_{\mu_{n+1}}) = +1\,; \end{cases} \tag{2.119}$$

for all $\binom{N}{n+1}$ $(n+1)$-tuples $\underline{\mu} \in \Lambda(N,n+1)$. These inequalities are not all independent of one another, however. For example, suppose that $\mathbf{q}: \mathrm{A} \rightarrow \mathbb{R}^{n+1}$ is any mapping such that for $\underline{\mu} \in \Lambda(N,n+1)$, $\underline{\nu} \in \Lambda(N,n+1)$ with $\underline{\lambda} = \underline{\mu} \cap \underline{\nu}$, $\#\lambda = n-1$, $\{\mu_i,\mu_j\} = \mu \setminus \nu$ $(i < j)$, and $\{\nu_k,\nu_l\} = \nu \setminus \mu$ $(k < l)$, we have:

$$sign(Q(\mathrm{a}_{\mu_i},\mathrm{a}_{\nu_k},\underline{\mathrm{A}}_\lambda)) = \chi(\mathrm{a}_{\mu_i},\mathrm{a}_{\nu_k},\underline{\mathrm{A}}_\lambda)\,; \quad sign(Q(\mathrm{a}_{\mu_j},\mathrm{a}_{\nu_l},\underline{\mathrm{A}}_\lambda)) = \chi(\mathrm{a}_{\mu_j},\mathrm{a}_{\nu_l},\underline{\mathrm{A}}_\lambda)\,; \tag{2.120}$$

$$sign(Q(\mathrm{a}_{\mu_i},\mathrm{a}_{\nu_l},\underline{\mathrm{A}}_\lambda)) = \chi(\mathrm{a}_{\mu_i},\mathrm{a}_{\nu_l},\underline{\mathrm{A}}_\lambda)\,; \quad sign(Q(\mathrm{a}_{\mu_j},\mathrm{a}_{\nu_k},\underline{\mathrm{A}}_\lambda)) = \chi(\mathrm{a}_{\mu_j},\mathrm{a}_{\nu_k},\underline{\mathrm{A}}_\lambda)\,;$$

and $sign(Q(\underline{\mathrm{A}}_\mu)) = \chi(\underline{\mathrm{A}}_\mu)$, where $Q: \mathrm{A}^{n+1} \rightarrow \mathbb{R}$ is the determinant function induced by $\mathbf{q}$ introduced above. Since the function Q obeys the three-term Graßmann-Plücker relation

$$Q(\underline{\mathrm{A}}_\mu)\cdot Q(\underline{\mathrm{A}}_\nu) - Q(\mathrm{a}_{\mu_i},\mathrm{a}_{\nu_k},\underline{\mathrm{A}}_\lambda)\cdot Q(\mathrm{a}_{\mu_j},\mathrm{a}_{\nu_l},\underline{\mathrm{A}}_\lambda) \tag{2.121}$$

$$+\, Q(\mathrm{a}_{\mu_i},\mathrm{a}_{\nu_l},\underline{\mathrm{A}}_\lambda)\cdot Q(\mathrm{a}_{\mu_j},\mathrm{a}_{\nu_k},\underline{\mathrm{A}}_\lambda) = 0\,,$$

it follows that if $\epsilon \cdot Q(\mathrm{a}_{\mu_i},\mathrm{a}_{\nu_l},\underline{\mathrm{A}}_\lambda) \cdot Q(\mathrm{a}_{\mu_j},\mathrm{a}_{\nu_k},\underline{\mathrm{A}}_\lambda) > 0$ and $-\epsilon \cdot Q(\mathrm{a}_{\mu_i},\mathrm{a}_{\nu_k},\underline{\mathrm{A}}_\lambda) \cdot Q(\mathrm{a}_{\mu_j},\mathrm{a}_{\nu_l},\underline{\mathrm{A}}_\lambda) > 0$ for some $\epsilon \in \{\pm 1\}$, then we necessarily have:

$$\epsilon\cdot Q(\underline{\mathrm{A}}_\mu)\cdot Q(\underline{\mathrm{A}}_\nu) < 0\,. \tag{2.122}$$

Since the orientation χ itself fulfills the 3-term GP-condition by hypothesis it follows that $sign(Q(\underline{\mathrm{A}}_\nu)) = \chi(\underline{\mathrm{A}}_\nu)$ as well, so that we can eliminate the inequality involving $Q(\underline{\mathrm{A}}_\nu)$. In this way, the system of inequalities given in (2.119) can be *reduced* to yield a smaller but equivalent system of inequalities. A system of inequalities is called *minimal* when it cannot be further reduced.

In order to solve any given reduced or minimal system, we must also be able to eliminate individual variables. The following lemma enables us to do this.

LEMMA 2.44. *For $\underline{\mu} \in \Lambda(N,n+1)$, $1 \leq i,k \leq n+1$ and $\mu_k > n+1$*

$$\frac{\partial Q(\mathrm{a}_{\mu_1},\ldots,\mathrm{a}_{\mu_{n+1}})}{\partial q_i(\mathrm{a}_{\mu_k})} = Q(\mathrm{a}_{\mu_1},\ldots,\mathrm{a}_{\mu_{k-1}},\mathrm{a}_i,\mathrm{a}_{\mu_{k+1}},\ldots,\mathrm{a}_{\mu_{n+1}})\,. \tag{2.123}$$

Proof: By Theorem 2.12, the following Graßmann-Plücker relation holds:

$$(-1)^{n+1-k} Q(\mathrm{a}_{\mu_1},\ldots,\mathrm{a}_{\mu_{n+1}}) \tag{2.124}$$

$$= \sum_{j=1}^{n+1} (-1)^j Q(\mathrm{a}_1,\ldots,\mathrm{a}_{j-1},\mathrm{a}_{j+1},\ldots,\mathrm{a}_{n+1},\mathrm{a}_{\mu_k})\, Q(\mathrm{a}_j,\mathrm{a}_{\mu_1},\ldots,\mathrm{a}_{\mu_{k-1}},\mathrm{a}_{\mu_{k+1}},\ldots,\mathrm{a}_{\mu_{n+1}})\,.$$

Since only the i-th term of this summation involves the variable $q_i(\mathrm{a}_{\mu_k})$ (see

Equation (2.118)), the lemma follows at once by differentiating. QED

As a consequence, each determinant can be written as

$$Q(a_{\mu_1}, \ldots, a_{\mu_{n+1}}) \tag{2.125}$$
$$= q_i(a_{\mu_k})\, Q(a_{\mu_1}, \ldots, a_{\mu_{k-1}}, a_i, a_{\mu_{k+1}}, \ldots, a_{\mu_{n+1}}) + Q_{i\mu_k}(a_{\mu_1}, \ldots, a_{\mu_{n+1}}),$$

where $\mu_k > n+1$ and

$$Q_{i\mu_k}(a_{\mu_1}, \ldots, a_{\mu_{n+1}}) := Q(a_{\mu_1}, \ldots, a_{\mu_{n+1}})\Big|_{q_i(a_{\mu_k}) = 0} \tag{2.126}$$

is the function obtained by setting $q_i(a_{\mu_k}) = 0$ in $Q(a_{\mu_1}, \ldots, a_{\mu_{n+1}})$.

We now consider the bounds on the value of any one variable $q_i(a_j)$ which result from the inequalities involving all determinants which contain it. By Equation (2.125), these bounds may be written as:

$$q_i(a_j) < -\frac{Q_{ij}(a_{\mu_1}, \ldots, a_{\mu_{n+1}})}{Q(a_{\mu_1}, \ldots, a_{\mu_{k-1}}, a_i, a_{\mu_{k+1}}, \ldots, a_{\mu_{n+1}})} \quad \text{if} \quad \mu \in U_i(a_j) \tag{2.127}$$

and

$$q_i(a_j) > -\frac{Q_{ij}(a_{\mu_1}, \ldots, a_{\mu_{n+1}})}{Q(a_{\mu_1}, \ldots, a_{\mu_{k-1}}, a_i, a_{\mu_{k+1}}, \ldots, a_{\mu_{n+1}})} \quad \text{if} \quad \mu \in L_i(a_j) \tag{2.128}$$

where $\mu_k = j > n+1$,

$$U_i(a_j) := \{\, \mu \in \Lambda(N, n+1) \mid \mu_k = j \text{ for some } k \in \{1, \ldots, n+1\}, \text{ and} \tag{2.129}$$
$$Q(a_{\mu_1}, \ldots, a_{\mu_{n+1}}) \cdot Q(a_{\mu_1}, \ldots, a_{\mu_{k-1}}, a_i, a_{\mu_{k+1}}, \ldots, a_{\mu_{n+1}}) < 0 \,\};$$

and

$$L_i(a_j) := \{\, \mu \in \Lambda(N, n+1) \mid \mu_k = j \text{ for some } k \in \{1, \ldots, n+1\}, \text{ and} \tag{2.130}$$
$$Q(a_{\mu_1}, \ldots, a_{\mu_{n+1}}) \cdot Q(a_{\mu_1}, \ldots, a_{\mu_{k-1}}, a_i, a_{\mu_{k+1}}, \ldots, a_{\mu_{n+1}}) > 0 \,\}.$$

This system of inequalities is fully equivalent to that given in Equation (2.119). Observe also that if any given variable $q_i(a_j)$ should have $L_i(a_j) = \{\}$ or $U_i(a_j) = \{\}$, then if we can but find values for all remaining variables which satisfy all the inequalities *not* involving the variable $q_i(a_j)$, then it will never be any trouble to finish the job by choosing a value for $q_i(a_j)$ which satisfies all of the inequalities which involve it: we need only make it small enough if it is unbounded from below, or large enough if it is unbounded from above. For all practical intents and purposes, therefore, we can forget about such unbounded variables until we are finished with the rest.

The goal of the algorithm is to find an ordering of the variables so that, if we choose values for the variables one by one in the given order, we can be sure that we will never be trapped in a "blind alley" where a lower bound on a variable exceeds an upper bound, so that no choice is possible for that variable. Such a sequence is called a *solvability sequence.* Clearly, the sequence can end with the unbounded variables in any order, as described above. The entire solvability sequence, in fact, has to be constructed in reverse order (provided it exists!). The procedure is to iteratively eliminate a variable $q_i(\mathrm{a}_j)$ and replace the inequalities involving it by:

$$\frac{Q_{ij}(\mathrm{a}_{\mu_1},\ldots,\mathrm{a}_{\mu_{n+1}})}{Q(\mathrm{a}_{\mu_1},\ldots,\mathrm{a}_{\mu_{k-1}},\mathrm{a}_i,\mathrm{a}_{\mu_{k+1}},\ldots,\mathrm{a}_{\mu_{n+1}})} > \frac{Q_{ij}(\mathrm{a}_{\nu_1},\ldots,\mathrm{a}_{\nu_{n+1}})}{Q(\mathrm{a}_{\nu_1},\ldots,\mathrm{a}_{\nu_{l-1}},\mathrm{a}_i,\mathrm{a}_{\nu_{l+1}},\ldots,\mathrm{a}_{\nu_{n+1}})} \quad (2.131)$$

for all $\underline{\mu} \in \mathrm{L}_i(\mathrm{a}_j)$ $(\mu_k = j)$ and $\underline{\nu} \in \mathrm{U}_i(\mathrm{a}_j)$ $(\nu_l = j)$, or equivalently

$$\begin{aligned} F_{ij}(\mathrm{a}_{\mu_1},\ldots,\mathrm{a}_{\mu_{n+1}};\,\mathrm{a}_{\nu_1},\ldots,\mathrm{a}_{\nu_{n+1}}) &:= \\ Q_{ij}(\mathrm{a}_{\mu_1},\ldots,\mathrm{a}_{\mu_{n+1}})\cdot Q(\mathrm{a}_{\nu_1},\ldots,\mathrm{a}_{\nu_{l-1}},\mathrm{a}_i,\mathrm{a}_{\nu_{l+1}},\ldots,\mathrm{a}_{\nu_{n+1}}) & \quad (2.132) \\ - \; Q_{ij}(\mathrm{a}_{\nu_1},\ldots,\mathrm{a}_{\nu_{n+1}})\cdot Q(\mathrm{a}_{\mu_1},\ldots,\mathrm{a}_{\mu_{k-1}},\mathrm{a}_i,\mathrm{a}_{\mu_{k+1}},\ldots,\mathrm{a}_{\mu_{n+1}}) & \;>/<\; 0\,, \end{aligned}$$

where ">/<" indicates an unspecified strict inequality. As long as these new inequalities are obeyed, it will always be possible to choose a value for the variable $q_i(\mathrm{a}_j)$ so that all inequalities involving it are satisfied. The problem is that the polynomials which occur in these new inequalities can have progressively higher degree, so that the problem gets messier rather than simpler. However, for *certain* choices of variables, at *certain* times in course of the procedure, these polynomials may turn out to be factorizable into products of the determinants we started out with! In this case we have obtained a substantial simplification. For not only can all of the new inequalities be eliminated, but so can all bounds on the remaining variables which involve the eliminated variable $q_i(\mathrm{a}_j)$. Thus, we shall now consider an easy-to-identify situation in which this happens.

PROPOSITION 2.45. *Suppose that* $n+1 < j$, $\underline{\mu} \in \mathrm{L}_i(\mathrm{a}_j)$, $\underline{\nu} \in \mathrm{U}_i(\mathrm{a}_j)$ *and* $\#(\mu \cap \nu) = n$ *with* $\mu_k = j = \nu_l$ *and* $\mu_K \notin \underline{\nu}$, $\nu_L \notin \underline{\mu}$; *then*

$$\begin{aligned} & F_{ij}(\mathrm{a}_{\mu_1},\ldots,\mathrm{a}_{\mu_{n+1}};\,\mathrm{a}_{\nu_1},\ldots,\mathrm{a}_{\nu_{n+1}}) \quad (2.133) \\ = \; & Q(\mathrm{a}_{\mu_1},\ldots,\mathrm{a}_{\mu_{k-1}},\mathrm{a}_{\nu_L},\mathrm{a}_{\mu_{k+1}},\ldots,\mathrm{a}_{\mu_{n+1}})\cdot Q(\mathrm{a}_{\nu_1},\ldots,\mathrm{a}_{\nu_{L-1}},\mathrm{a}_i,\mathrm{a}_{\nu_{L+1}},\ldots,\mathrm{a}_{\nu_{n+1}}) \\ = \; & Q(\mathrm{a}_{\nu_1},\ldots,\mathrm{a}_{\nu_{l-1}},\mathrm{a}_{\mu_K},\mathrm{a}_{\nu_{l+1}},\ldots,\mathrm{a}_{\nu_{n+1}})\cdot Q(\mathrm{a}_{\mu_1},\ldots,\mathrm{a}_{\mu_{K-1}},\mathrm{a}_i,\mathrm{a}_{\mu_{K+1}},\ldots,\mathrm{a}_{\mu_{n+1}})\,. \end{aligned}$$

Proof: Since our indexing of $\underline{\mathrm{A}}$ is arbitrary, we may assume w.l.o.g. that $k = l = 1$ and that $K = L = 2$. Then $\mu_h = \nu_h$ for $h = 1,3,\ldots,n+1$, so our formula for F_{ij} simplifies to:

$$\begin{aligned} & Q(\underline{\mathrm{A}}_\mu)\cdot\partial Q(\underline{\mathrm{A}}_\nu)/\partial q_i(\mathrm{a}_j) \;-\; Q(\underline{\mathrm{A}}_\nu)\cdot\partial Q(\underline{\mathrm{A}}_\mu)/\partial q_i(\mathrm{a}_j) \quad (2.134) \\ = \; & Q(\mathrm{a}_j,\mathrm{a}_{\mu_2},\mathrm{a}_{\mu_3},\ldots,\mathrm{a}_{\mu_{n+1}})\,Q(\mathrm{a}_i,\mathrm{a}_{\nu_2},\mathrm{a}_{\mu_3},\ldots,\mathrm{a}_{\mu_{n+1}}) \end{aligned}$$

$$- Q(a_j,a_{\nu_2},a_{\mu_3},\ldots,a_{\mu_{n+1}})\,Q(a_i,a_{\mu_2},a_{\mu_3},\ldots,a_{\mu_{n+1}})$$

$$= Q(a_i,a_j,a_{\mu_3},\ldots,a_{\mu_{n+1}})\,Q(a_{\nu_2},a_{\mu_2},a_{\mu_3},\ldots,a_{\mu_{n+1}})$$

via Theorem 2.12, i.e. the three-term Graßmann-Plücker relation. Observe that since the r.h.s. of this equation is independent of the variable $q_i(a_j)$ (because the first column of the determinant $Q(a_i,\cdots)$ is the i-th unit vector) the same is true of the l.h.s. Thus we may set $q_i(a_j) := 0$ in the left-hand side to obtain the desired result. QED

The geometric interpretation of the condition $\#(\mu\cap\nu) = n$ may be seen as follows: Assume as above that $k = l = 1$ and $K = L = 2$; then

$$\begin{aligned}
& \nu \in U_i(a_j) \\
\Longleftrightarrow\; & Q(\underline{A}_\nu)\cdot\partial Q(\underline{A}_\nu)/\partial q_i(a_j) \;<\; 0 \qquad (2.135)\\
\Longleftrightarrow\; & Q(a_{\nu_1},\ldots,a_{\nu_{n+1}})\cdot Q(a_i,a_{\nu_2},\ldots,a_{\nu_{n+1}}) \;<\; 0 \\
\Longleftrightarrow\; & \chi(a_{\nu_1},\ldots,a_{\nu_{n+1}})\cdot\chi(a_i,a_{\nu_2},\ldots,a_{\nu_{n+1}}) \;<\; 0\,,
\end{aligned}$$

i.e. $a_j = a_{\nu_1}$ and a_i are on opposite sides of the hyperplane $\{a_{\nu_2},\ldots,a_{\nu_{n+1}}\}$. Similarly, $\underline{A}_\mu \in L_i(a_j)$ if and only if a_j and a_i are on the same side of the hyperplane $\{a_{\mu_2},\ldots,a_{\mu_{n+1}}\}$.

Definition 2.46. A pair of disjoint subsets $L, U \subseteq \Lambda(N,n+1)$ are called *bipartite* if $\#(\mu\cap\nu) = n$ for all $\mu \in L$, $\nu \in U$.

Note in particular that any pair L, U with $L = \{\}$ or $U = \{\}$ is bipartite (since the definition then holds vacuously).

The final point to note before presenting the algorithm is that it is not deterministic, in that there may be more than one possible choice of variable whose sets of lower and upper bounds are bipartite. Nevertheless, since any variable whose bounds are bipartite in some reduced system obviously has bipartite bounds in the same reduced system after the elimination of any other variable, the order in which we eliminate the variables is of no consequence.

Algorithm 2.47:

INPUT: (An orientation of) a simplicial chirotope $(\underline{A}, \chi)$.

OUTPUT: A sequence of pairs of indices $[[i(k), j(k)] \mid k = 1,\ldots,(n+1)(N-n-1)]$ such that $[q_{i(k)}(a_{j(k)}) \mid k = 1,\ldots,(n+1)(N-n-1)]$ is a solvability sequence for the variables.

PROCEDURE:

Do with each $(n+1)$-tuple $\{b_1,\ldots,b_{n+1}\} \subseteq A$ in turn:

  Index the set A so that $\{a_1,\ldots,a_{n+1}\} = \{b_1,\ldots,b_{n+1}\}$

   and $\chi(a_1,\ldots,a_{n+1}) = 1$.

Set $K := M := (n+1)(N-n-1)$, $D_K := \Lambda(N, n+1)$, and $\underline{S}_K := []$.
While $K > 0$ and there exists a minimal system $R \subset \Lambda(N, n+1)$ such that
the bounds $(R \cap D_K \cap L_{ij}, R \cap D_K \cap U_{ij})$ on some variable
$q_i(a_j)$ with $[i, j] \notin S_K$ are bipartite, do:
Put this variable into the solvability sequence by
setting $\underline{S}_K := [i, j]$.
Set $K := K-1$, and $D_K := D_{K+1} \setminus (L_{ij} \cup U_{ij})$.
If $K = 0$, then
a solvability sequence has been found;
otherwise
no solvability sequence exists for this choice of $\{b_1, \ldots, b_{n+1}\} \subseteq A$ as basis.
until a solvability sequence has been found.

The correctness of this algorithm follows more or less immediately from the foregoing discussion. An example illustrating its use is given below.

Although the above algorithm actually succeeds in finding solvability sequences for a surprisingly large fraction of representable simplicial chirotopes, examples are known in which no solvability sequence exists. Those chirotopes which do admit solvability sequences have a number of nice geometric properties. In particular, it is known that the "representation space" defined by the inequalities (2.119) for a such a chirotope has a trivial homotopy group, meaning not only that it consists of a single path-connected component, but that any two paths between a pair of "conformations" are homotopic [Bokowski1986]. A famous conjecture, known as the *isotopy* conjecture [Goodman1983], suggests that this connectedness may hold even for chirotopes without solvability sequences.† The above algorithm has been used to solve a number of interesting outstanding mathematical problems, particularly the realizability of abstract polyhedra [Bokowski1985a]. In combination with Algorithm 2.43 for enumerating chirotopes, it may also prove useful in chemical problems as a systematic method of generating starting points for the numerical refinements commonly used in computational chemistry.

As an example of the use of Algorithm 2.47, taken from [Bokowski1985b], we show how to find values for the nine unknowns in the matrix

$$\underline{Q} = \begin{bmatrix} 1 & 0 & 0 & a & d & g \\ 0 & 1 & 0 & b & e & h \\ 0 & 0 & 1 & c & f & i \end{bmatrix}$$

† This conjecture has recently been shown to be false by a counterexample of rank three and order forty-two, which was constructed by N. White [unpublished].

which constitute a representation of the rank 3, order 6 chirotope:

Table 2.2. Orientations and the Implied Determinant Inequalities

1: $\chi(1,2,3) = +1$ => *nothing*	11: $\chi(2,3,4) = +1$ => $a > 0$
2: $\chi(1,2,4) = +1$ => $c > 0$	12: $\chi(2,3,5) = +1$ => $d > 0$
3: $\chi(1,2,5) = +1$ => $f > 0$	13: $\chi(2,3,6) = +1$ => $g > 0$
4: $\chi(1,2,6) = +1$ => $i > 0$	14: $\chi(2,4,5) = +1$ => $\begin{vmatrix} a & c \\ d & f \end{vmatrix} < 0$
5: $\chi(1,3,4) = +1$ => $b < 0$	15: $\chi(2,4,6) = +1$ => $\begin{vmatrix} a & c \\ g & i \end{vmatrix} < 0$
6: $\chi(1,3,5) = +1$ => $e < 0$	16: $\chi(2,5,6) = -1$ => $\begin{vmatrix} d & f \\ g & i \end{vmatrix} > 0$
7: $\chi(1,3,6) = +1$ => $h < 0$	17: $\chi(3,4,5) = +1$ => $\begin{vmatrix} a & b \\ d & e \end{vmatrix} > 0$
8: $\chi(1,4,5) = +1$ => $\begin{vmatrix} b & c \\ e & f \end{vmatrix} > 0$	18: $\chi(3,4,6) = +1$ => $\begin{vmatrix} a & b \\ g & h \end{vmatrix} > 0$
9: $\chi(1,4,6) = -1$ => $\begin{vmatrix} b & c \\ h & i \end{vmatrix} < 0$	19: $\chi(3,5,6) = +1$ => $\begin{vmatrix} d & e \\ g & h \end{vmatrix} > 0$
10: $\chi(1,5,6) = -1$ => $\begin{vmatrix} e & f \\ h & i \end{vmatrix} < 0$	20: $\chi(4,5,6) = +1$ => $\begin{vmatrix} a & b & c \\ d & e & f \\ g & h & i \end{vmatrix} > 0$

The determinantal inequalities which are implied by this chirotope are also shown. Of these, ten have order 2 or more, one is the determinant of the basis and the remaining nine are variables. If we now consider the variable $g = Q(6,2,3)$ together with the determinant $Q(2,4,6) = cg - ai > 0$, we find that $\partial Q(2,4,6)/\partial g = Q(2,4,1) = c > 0$. Thus $Q(2,4,6) \cdot Q(2,4,1) > 0$, so by Equation (2.130) [2,4,6] belongs to the set of determinantal inequalities L_{34} which give rise to a lower bound on the variable g. The sets L_{ij} and U_{ij} for each variable are shown in Table 2.3.

Table 2.3. Triples whose Determinants Constitute Bounds on Variables		
Variable	*Upper Bounds*	*Lower Bounds*
a	{[2,4,5],[2,4,6],[3,4,5],[3,4,6],[4,5,6]}	{[2,3,4]}
b	{[1,3,4],[1,4,6],[3,4,5],[3,4,6],[4,5,6]}	{[1,4,5]}
c	{[1,4,6]}	{[1,2,4],[1,4,5],[2,4,5],[2,4,6],[4,5,6]}
d	{[3,5,6]}	{[2,3,5],[2,4,5],[2,5,6],[3,4,5],[4,5,6]}
e	{[1,3,5],[1,4,5],[1,5,6],[3,5,6],[4,5,6]}	{[3,4,5]}
f	{[1,4,5],[1,5,6],[2,4,5],[2,5,6],[4,5,6]}	{[1,2,5]}
g	{[2,5,6]}	{[2,3,6],[2,4,6],[3,4,6],[3,5,6],[4,5,6]}
h	{[1,3,6]}	{[1,4,6],[1,5,6],[3,4,6],[3,5,6],[4,5,6]}
i	{[2,4,6]}	{[1,2,6],[1,4,6],[1,5,6],[2,5,6],[4,5,6]}

Since none of these pairs is bipartite, it is necessary to reduce the system of inequalities. In this case, we observe that:

$$\chi(1,4,6)\cdot\chi(3,4,5) = (-1)\cdot(+1) < 0 \text{ and } -\chi(1,4,5)\cdot\chi(3,4,6) = -(+1)\cdot(+1) < 0,$$

so $\chi(1,3,4)\cdot Q(4,5,6) = Q(4,5,6) > 0$ is satisfied whenever the five inequalities involving the remaining triples are all satisfied. Similarly, we can forget about the inequalities $Q(1,5,6) < 0$ and $Q(3,4,6) > 0$.

Eliminating these three inequalities leaves us with the bipartite pair

$$a:\ \mathrm{U}_{14} = \{[2,4,5],[2,4,6],[3,4,5]\}, \text{ and } \mathrm{L}_{14} = \{[2,3,4]\},$$

so that the variable "*a*" can be eliminated along with all inequalities involving it, i.e. all those including "4" but not "1", which leaves us with those shown in Table 2.4.

Table 2.4. Bounds after Eliminating [1,5,6], [3,4,6] and [4,5,6]		
Variable	*Upper Bounds*	*Lower Bounds*
b	{[1,3,4],[1,4,6]}	{[1,4,5]}
c	{[1,4,6]}	{[1,2,4],[1,4,5]}
d	{[3,5,6]}	{[2,3,5],[2,5,6]}
e	{[1,3,5],[1,4,5],[3,5,6]}	{}
f	{[1,4,5],[2,5,6]}	{[1,2,5]}
g	{[2,5,6]}	{[2,3,6],[3,5,6]}
h	{[1,3,6]}	{[1,4,6],[3,5,6]}
i	{}	{[1,2,6],[1,4,6],[2,5,6]}

We now have several variables whose bounds are not only bipartite but which are actually unbounded either above or below. From this point on it is no trouble to finish finding a solvability sequence, e.g. $[i, f, c, h, b, g, d, e, a]$. We leave actually finding an assignment of values to these variables which gives a representation as an exercise.

References

Bland1977.

R.G. Bland, "A Combinatorial Abstraction of Linear Programming," *J. Combinatorial Theory, B23*, 33-57(1977).

Bland1978.

R.G. Bland and M. Las Vergnas, "Orientability of Matroids," *J. Combinatorial Theory, B24*, 94-123(1978).

Bokowski1985a.

J. Bokowski and B. Sturmfels, "Polytopal and Nonpolytopal Spheres - an Algorithmic Approach," Preprint no. 935, Technische Hochschule Darmstadt, 1985.

Bokowski1985b.

J. Bokowski and B. Sturmfels, "Programmsystem zur Realisierung orientierter Matroide," Report PD 85.22 of the Mathematisches Institute der Univ. zu Koln, 1985.

Bokowski1986.

J. Bokowski and B. Sturmfels, "On the Coordinatization of Oriented Matroids," *Discrete Comput. Geom., 1*, 293-306(1986).

Char1986.

B.W. Char, G.J. Fee, K.O. Geddes, G.H. Gonnet, and M.B. Monagan, "A Tutorial Introduction to Maple," *J. Symbolic Computation, 2*, 179-200(1986).

Coxeter1961.

H.S.M. Coxeter, *Introduction to Geometry*, J. Wiley & Sons, New York, NY, 1961.

Dreiding1980.

A.S. Dreiding and K. Wirth, "The Multiplex: A Classification of Finite Ordered Point Sets in oriented *d*-dimensional spaces," *MATCH, 8*, 341-352(1980).

Dress1983.

A.W.M. Dress, A.S. Dreiding, and H.R. Haegi, "Classification of Mobile Molecules by Category Theory," in *Symmetries and Properties of Mobile Molecules: A Comprehensive Survey*, ed. J. Maruani and J. Serre, Studies in Physical and Theoretical Chemistry, vol. 23, Elsevier Scientific, Amsterdam,

1983.

Dress1986a.

A.W.M. Dress, "Duality Theory for Finite and Infinite Matroids with Coefficients," *Advances Math.*, *59*, 97-123(1986).

Dress1986b.

A.W.M. Dress, "Chirotopes and Oriented Matroids," Bayreuth Sommerschule ueber Diskrete Strukturen, algebraische Methoden und Anwendungen, pp. 14-68, 1986.

Edelsbrunner1986.

H. Edelsbrunner, J. O'Rourke, and R. Seidel, "Constructing Arrangements of Lines and Hyperplanes with Applications," *SIAM J. Computing*, *15*, 341-383(1986).

Folkman1978.

J. Folkman and J. Lawrence, "Oriented Matroids," *J. Combinatorial Math.*, *B25*, 199-236(1978).

Goodman1983.

J.E. Goodman and R. Pollack, "Multidimensional Sorting," *SIAM J. Computing*, *12*, 484-507(1983).

Hendrickson1967.

J.B. Hendrickson, "Molecular Geometry. V. Evaluation of Functions and Conformations of Medium Rings," *J. Amer. Chem. Soc.*, *89*, 7042-7061(1967).

Hodge1968.

W.V.D. Hodge and D. Pedoe, *Methods of Algebraic Geometry, volumes I & II*, Cambridge University Press, Cambridge, U.K., 1968.

Lawrence1982.

J. Lawrence, "Oriented Matroids and Multiply Ordered Sets," *Lin. Alg. Appl.*, *48*, 1-12(1982).

Longuet-Higgins1963.

H.C. Longuet-Higgins, "The Symmetry Groups of Non-Rigid Molecules," *Molec. Phys.*, *6*, 445-460(1963).

Mandel1981.

A. Mandel, "Topology of Oriented Matroids," Doctoral dissertation, University of Waterloo, Canada, 1981.

Novoa1965.

L. Gutierrez Novoa, "On *n*-Ordered Sets and Order Completeness," *Pacific J. Math.*, *15*, 1337-1345(1965).

Rockafellar1969.

R.T. Rockafellar, "The Elementary Vectors of a Subspace of R^n," in *Combinatorial Mathematics and its Applications, Proc. of Chapel Hill conf.*, ed. R. C. Bose and T. A. Dowling, pp. 104-127, 1969.

Roudneff1987.

J. Roudneff, "Matroides Orientes et Arrangements de Pseudodroites," Doctoral dissertation, Univ. of Paris, France, 1987.

Sturmfels1985.

B. Sturmfels, "Zur linearen Realisierbarkeit orientierter Matroide," Diplomarbeit, Technische Hochschule Darmstadt, BRD, 1985.

Vergnas1978.

M. Las Vergnas, "Bases in Oriented Matroids," *J. Combinatorial Theory, B25,* 283-289.(1978).

Vergnas1980.

M. Las Vergnas, "Convexity in Oriented Matroids," *J. Combinatorial Theory, B29,* 231-243(1980).

Williamson1985.

S.G. Williamson, *Combinatorics for Computer Scientists,* Computer Science Press, Rockville, MD, 1985.

3. The Fundamental Theory of Distance Geometry

The interatomic distances of a molecule cannot simultaneously assume arbitrary values, but are restricted to very particular combinations of values whose explicit description is essential to the distance geometry approach to the problems of molecular conformation. The general form of these dependencies was first discovered by A. Cayley in 1841 [Cayley1841], but was not systematically studied until 1928, when K. Menger [Menger1928] showed how convexity and many other basic geometric properties could be defined and studied in terms of interpoint distances. Subsequently, Menger proposed a new axiom system for Euclidean geometry which was based upon these relations [Menger1931]. In 1935, I. Schoenberg discovered an equivalent characterization of Euclidean distances [Schoenberg1935], and first realized the connection with bilinear forms. This work was brought together and further clarified by L. Blumenthal, first in the form of a monograph [Blumenthal1938], and later in his classic text "Distance Geometry" [Blumenthal1953], which is still a worthwhile introduction to the subject. In recent years, however, the general theory of bilinear forms defined on affine spaces has been found to provide a simple and elegant framework in which all of these results can be presented in a unified way.

This chapter provides a thorough presentation of the fundamental theory of Euclidean distance geometry at a level which is close to the limits of our current understanding of the subject. Our purpose in giving such a detailed treatment is twofold: First, reference will frequently be made to these results in subsequent chapters of the book, where applications of the theory presented here are our main concern. Second, since more than 30 years have passed since Blumenthal's text was written, it seems desirable to provide a unified modern account of the current state of the art, both for the benefit of mathematicians who might wish to develop it further, and for chemists who wish to learn to apply these results and to keep abreast of future developments. Because our interests here are in the *intrinsic* properties of Euclidean distances, we have adopted a coordinate-free presentation of the subject. The relation between distances and coordinates will be thoroughly studied in Chapter 6 (see also [Gower1982]). Although the distance geometry approach has also been applied to the study of elliptic and hyperbolic geometry, the applied nature of our interests precludes any discussion of this subject (see [Blumenthal1953, Seidel1955]).

In writing this chapter, we have assumed that the reader is familiar with the content of Appendices A and B, which provides the necessary overview of metric affine geometry. Use will also be made of Appendix D on set systems and matroids. As usual in this book, all abstract sets may be assumed to be finite, and we restrict our field to be the real numbers unless otherwise stated. Following the rigorous axiomatic tradition which has become standard in mathematics, we shall begin with the simplest and most general properties of Euclidean distances which hold in any number of dimensions, and successively add on those properties which characterize the three-dimensional case which is our primary concern. Although many of the results presented here were first discovered by the above-mentioned authors, significant parts of the chapter (e.g. the section on metroids) represent more recent work by A. Dress and T. Havel (see [Dress1986]).*

3.1. Premetric Spaces and their Amalgamations

Let A be the set of atoms of a molecule. Each embedding $\mathbf{p}: \mathrm{A} \rightarrow \mathbb{R}^n$ induces a well-defined map $d_{\mathbf{p}}: \mathrm{A} \times \mathrm{A} \rightarrow \mathbb{R}$, which for a,b $\in$ A is given by

$$d_{\mathbf{p}}(\mathrm{a},\mathrm{b}) \ := \ \|\mathbf{p}(\mathrm{a}) - \mathbf{p}(\mathrm{b})\| \ , \tag{3.1}$$

where $\|\mathbf{x}\| := \sqrt{\mathbf{x}\cdot\mathbf{x}}$ is the l^2 or *Euclidean* norm on the vector space $\mathbb{R}^n$.

Definition 3.1. The map $d_{\mathbf{p}}: \mathrm{A}^2 \rightarrow \mathbb{R}$ is called an n-dimensional *Euclidean distance function.*

The following lemma derives some of the simplest properties of these functions.

LEMMA 3.2. *If $d_{\mathbf{p}}$ is a Euclidean distance function and $D_{\mathbf{p}} := d_{\mathbf{p}}^2$, then*

$$\forall\ \mathrm{a,b} \in \mathrm{A}: \quad D_{\mathbf{p}}(\mathrm{a,b}) \ = \ D_{\mathbf{p}}(\mathrm{b,a})\ ; \tag{3.2}$$

$$\forall\ \mathrm{a,b} \in \mathrm{A}: \quad D_{\mathbf{p}}(\mathrm{a,b}) \ = \ 0 \quad \Longleftrightarrow \quad \mathrm{a} = \mathrm{b}\ ; \tag{3.3}$$

$$\forall\ \mathrm{a,b} \in \mathrm{A}: \quad D_{\mathbf{p}}(\mathrm{a,b}) \ \geq \ 0\ ; \tag{3.4}$$

$$\forall\ \mathrm{a,b,c} \in \mathrm{A}: \quad d_{\mathbf{p}}(\mathrm{a,b}) \ \leq \ d_{\mathbf{p}}(\mathrm{a,c}) + d_{\mathbf{p}}(\mathrm{b,c})\ . \tag{3.5}$$

---

* A preliminary draft of this chapter appeared in [Dress1987].

Proof: Equations (3.2), (3.3) and (3.4) are immediate consequences of the definitions and the basic rules of vector algebra. To prove (3.5), we recall the Cauchy-Schwarz inequality: $|\mathbf{u}\cdot\mathbf{v}| \leq \|\mathbf{u}\|\,\|\mathbf{v}\|$ for all $\mathbf{u},\mathbf{v} \in \mathbb{R}^n$. Hence

$$\begin{aligned} D_{\mathbf{p}}(\mathrm{a},\mathrm{b}) &= \|\mathbf{p}(\mathrm{a})-\mathbf{p}(\mathrm{b})\|^2 = \|(\mathbf{p}(\mathrm{a})-\mathbf{p}(\mathrm{c}))+(\mathbf{p}(\mathrm{c})-\mathbf{p}(\mathrm{b}))\|^2 \\ &= \|\mathbf{p}(\mathrm{a})-\mathbf{p}(\mathrm{c})\|^2+\|\mathbf{p}(\mathrm{c})-\mathbf{p}(\mathrm{b})\|^2+2\,(\mathbf{p}(\mathrm{a})-\mathbf{p}(\mathrm{c}))\cdot(\mathbf{p}(\mathrm{c})-\mathbf{p}(\mathrm{b})) \qquad (3.6) \\ &\leq \|\mathbf{p}(\mathrm{a})-\mathbf{p}(\mathrm{c})\|^2+\|\mathbf{p}(\mathrm{c})-\mathbf{p}(\mathrm{b})\|^2+2\,\|\mathbf{p}(\mathrm{a})-\mathbf{p}(\mathrm{c})\|\,\|\mathbf{p}(\mathrm{c})-\mathbf{p}(\mathrm{b})\| \\ &= (d_{\mathbf{p}}(\mathrm{a},\mathrm{c})+d_{\mathbf{p}}(\mathrm{b},\mathrm{c}))^2 \end{aligned}$$

from which the desired result follows by taking positive square roots. QED

Because it says that the length of any side of a triangle does not exceed the sum of the lengths of the other two sides, condition (3.5) is often referred to as the *triangle inequality.*

We now create terms for those real-valued functions D defined on A×A which have just these properties.

Definition 3.3. A *premetric space* $\mathbb{S}$ is a pair (A, D) consisting of a set A together with a function D: A×A→$\mathbb{R}$ which satisfies conditions (3.2) and (3.3) of Lemma 3.2.† If D also satisfies condition (3.4), $\mathbb{S}$ is called a *semimetric space*, and if in addition the function $d := \sqrt{D}$ satisfies the triangle inequality (3.5), then $\mathbb{S}$ is called a *metric space.*

Although it is not the usual mathematical practice, we will persist in referring to the elements of A as *atoms.* We shall refer to the function D itself as a *premetric*, whereas the function d is called a *semimetric* or *metric*, as appropriate. The values d(a,b) of d are called *distances*, while the values D(a,b) are called *squared distances* (even if negative!). Mathematicians will observe that the definition of a premetric space remains valid even when $\mathbb{R}$ is replaced by an arbitrary field $\mathbb{K}$; thus, many of the results of this section can be generalized to that case. Note also that if the size #A of A is N and we index and order the atoms as $\underline{\mathrm{A}} = [\mathrm{a}_1, \ldots, \mathrm{a}_N]$, then the above spaces can be represented by means of a (symmetric) matrix

$$\underline{\mathbf{M}}_D(\underline{\mathrm{A}}) := \begin{bmatrix} 0 & D(\mathrm{a}_1,\mathrm{a}_2) & \cdots & D(\mathrm{a}_1,\mathrm{a}_N) \\ D(\mathrm{a}_1,\mathrm{a}_2) & 0 & \cdots & D(\mathrm{a}_2,\mathrm{a}_N) \\ \cdots & \cdots & \cdots & \cdots \\ D(\mathrm{a}_1,\mathrm{a}_N) & D(\mathrm{a}_2,\mathrm{a}_N) & \cdots & 0 \end{bmatrix}. \qquad (3.7)$$

Thus a (finite) premetric space can be identified with the set of all matrices

† Another possible definition is to relax condition (3.3) by requiring only that $\mathbf{D}_{\mathbf{p}}(\mathrm{a},\mathrm{a}) = 0$ for all a ∈ A. The stronger condition imposed here serves merely to insure that $\mathbf{D}_{\mathbf{p}}$ arises from an injective map $\mathbf{p}$:A→$\mathbb{R}^n$.

For four atoms $\underline{A} = [a_1, \ldots, a_4]$, if

$$\underline{\mathbf{M}}_D(\underline{A}) = \begin{bmatrix} 0 & 1 & 2 & X \\ 1 & 0 & 1 & 2 \\ 2 & 1 & 0 & 1 \\ X & 2 & 1 & 0 \end{bmatrix}$$

and $X = 0$, then the space is premetric if and only if $a_1 = a_4$. If $X < 0$, the space is premetric but not semimetric. The space is semimetric without being metric as long as $0 < X < (\sqrt{2}-1)^2$, because these values violate the triangle inequality $d(a_1,a_2) + d(a_1,a_4) < d(a_2,a_4)$. For $1 > X \geq (\sqrt{2}-1)^2$ the space is metric but fails to be Euclidean, even though all triplets of atoms obey the triangle inequality, because $D(a_1,a_4)$ is shorter than the *cis* value attained when the dihedral angle about the 2–3 bond is 0. For $X = 1$, the space is Euclidean and is embeddable in $\mathbb{R}^2$ as a square configuration, as shown.

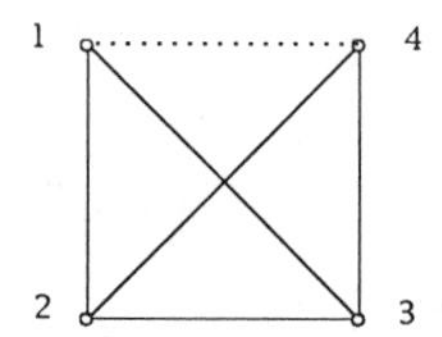

Figure 3.1

Finally, for $1 \leq X \leq \sqrt{5}$, the space is three-dimensional Euclidean.

obtained from $\underline{\mathbf{M}}_D(\underline{A})$ by row/column permutation, and similarly semimetric and metric spaces can be regarded as being essentially just a certain type of matrix.

Definition 3.4. A premetric space $\mathbb{S} = (A, D)$ is *Euclidean* if there exists an embedding $\mathbf{p}: A \rightarrow \mathbb{R}^n$ such that $D = d_{\mathbf{p}}^2$, the square of the corresponding Euclidean distance function. We may also describe this by saying that $\mathbb{S}$ is *embeddable* in an n-dimensional Euclidean space.

Note that a Euclidean premetric space is necessarily a metric space. As we shall see later, however, even though they obey the triangle inequality there exist metric spaces (A, D) which are not Euclidean. It is therefore desirable to seek an analogous algebraic representation which is applicable to arbitrary premetric spaces.

Definition 3.5. A premetric space $\mathbb{S} = (A, D)$ is said to be *embeddable* in a bilinear affine space $\mathbb{A}_B$ with quadratic form Q_B if there exists an embedding $\mathbf{q}: A \rightarrow \mathbb{A}_B$ such that $D(a,b) = Q_B(\mathbf{q}(a) - \mathbf{q}(b))$ for all atoms $a, b \in A$. The embedding $\mathbf{q}$ itself is called an *isometric embedding* of $\mathbb{S}$ in $\mathbb{A}_B$. If such an embedding exists, the *dimension* $dim(\mathbb{S})$ of $\mathbb{S}$ is then the minimum dimension of a bilinear affine space in which it is embeddable.

This brings us to our first:

THEOREM 3.6. *Every premetric space* $\mathbb{S} = (\mathrm{A}, D)$ *can be embedded in a bilinear affine space of dimension* $\#\mathrm{A}-1$ *such that the image of* A *spans the space. This embedding is unique up to isometry.*

Proof: It turns out to be easiest to construct this space using barycentric coordinates. We begin by enlarging the set A to a set $\tilde{\mathbb{R}}[\mathrm{A}]$ consisting of all *formal sums* $\Sigma\alpha_i \mathrm{a}_i$ of the elements in A with $\Sigma\alpha_i = 1$. We then define a vector space $\hat{\mathbb{R}}[\mathrm{A}]$ whose elements are all formal sums such that $\Sigma\beta_i = 0$. Vector addition and scalar multiplication are defined as:

$$\left(\sum_{i=1}^{N} \beta_i \mathrm{a}_i\right) + \left(\sum_{i=1}^{N} \beta'_i \mathrm{a}_i\right) := \sum_{i=1}^{N} (\beta_i + \beta'_i)\mathrm{a}_i \ ; \tag{3.8}$$

$$\gamma \cdot \left(\sum_{i=1}^{N} \beta_i \mathrm{a}_i\right) := \sum_{i=1}^{N} (\gamma\beta_i)\mathrm{a}_i \ , \tag{3.9}$$

where $N := \#\mathrm{A}$. Because of the constraint $\Sigma\beta_i = 0$, the dimension of $\hat{\mathbb{R}}[\mathrm{A}]$ is equal to $\#\mathrm{A}-1$. Furthermore, if we define a map from $\tilde{\mathbb{R}}[\mathrm{A}]\times\hat{\mathbb{R}}[\mathrm{A}]$ to $\tilde{\mathbb{R}}[\mathrm{A}]$ according to the rule:

$$\left(\sum_{i=1}^{N} \alpha_i \mathrm{a}_i, \sum_{i=1}^{N} \beta_i \mathrm{a}_i\right) \mapsto \sum_{i=1}^{N} (\alpha_i + \beta_i)\mathrm{a}_i \ , \tag{3.10}$$

it is easily verified that $\mathbb{A}[\mathrm{A}] := (\tilde{\mathbb{R}}[\mathrm{A}], \hat{\mathbb{R}}[\mathrm{A}])$, together with this map, fulfill all the axioms for an affine space given in Appendix B. The translation which maps $\Sigma\alpha_i \mathrm{a}_i$ to $\Sigma\alpha'_i \mathrm{a}_i$ is obtained simply by subtracting corresponding coefficients to obtain $\Sigma\beta_i \mathrm{a}_i$ with $\Sigma\beta_i = \Sigma(\alpha_i - \alpha'_i) = 0$. We call $\mathbb{A}[\mathrm{A}]$ the *amalgamation space* of A.

We now define a bilinear form on $\hat{\mathbb{R}}[\mathrm{A}]$ according to the equation:

$$B_D(\Sigma\beta_i \mathrm{a}_i; \Sigma\beta'_j \mathrm{a}_j) := -\frac{1}{2} \sum_{i=1}^{N}\sum_{j=1}^{N} D(\mathrm{a}_i, \mathrm{a}_j)\, \beta_i \beta'_j \ . \tag{3.11}$$

The corresponding quadratic form $Q_D\colon \hat{\mathbb{R}}[\mathrm{A}]\to\mathbb{R}$ is known as *Schoenberg's quadratic form*, and the bilinear vector space $(\hat{\mathbb{R}}[\mathrm{A}], B_D)$ will be denoted by $\hat{\mathbb{R}}_D[\mathrm{A}]$. If we define our embedding $\mathbf{q}\colon \mathrm{A}\to\tilde{\mathbb{R}}[\mathrm{A}]$ in the obvious way (i.e. as $\mathbf{q}(\mathrm{a}_j) := \Sigma\delta_{ij}\mathrm{a}_i$ with $\delta_{ij} = 1$ if $i = j$ and 0 otherwise), it is easily verified that

$$Q_D(\mathbf{q}(\mathrm{a}_i) - \mathbf{q}(\mathrm{a}_j)) = D(\mathrm{a}_i, \mathrm{a}_j) \ . \tag{3.12}$$

Thus, $\mathbf{q}$ is the desired isometric embedding of A in the bilinear affine space $\mathbb{A}_D[\mathrm{A}] := (\tilde{\mathbb{R}}[\mathrm{A}], \hat{\mathbb{R}}[\mathrm{A}], B_D)$. We call $\mathbf{q}$ the *canonical embedding* in $\mathbb{A}_D[\mathrm{A}]$.

It is obvious that if $\mathbb{A}' = (\mathbb{R}^{N-1}, \mathbb{V}')$ is any other affine space of dimension $N-1 = \#\mathrm{A}-1$, and if $\mathbf{q}'\colon \mathrm{A}\to\mathbb{A}'$ is an embedding of A into $\mathbb{A}'$ such that $\mathbf{q}'(\mathrm{A})$ spans

$\mathbb{A}'$, then the mappings

$$\tilde{\mathbb{R}}[\mathrm{A}] \to \mathbb{R}^{N-1}:\ \Sigma\alpha_i \mathrm{a}_i \ \mapsto\ \Sigma\alpha_i \mathbf{q}'(\mathrm{a}_i) \tag{3.13}$$

$$\hat{\mathbb{R}}[\mathrm{A}] \to \mathbb{V}':\ \Sigma\beta_i \mathrm{a}_i \ \mapsto\ \Sigma\beta_i(\mathbf{q}'(\mathrm{a}_i) - \mathbf{q}'(\mathrm{a}_1))$$

constitute an isomorphism from $\mathbb{A}[\mathrm{A}]$ onto $\mathbb{A}'$. Hence to verify uniqueness, we need only show that if B' is any other bilinear form on $\hat{\mathbb{R}}[\mathrm{A}]$ whose quadratic form Q' also satisfies $Q'(\mathbf{q}(\mathrm{a}_i) - \mathbf{q}(\mathrm{a}_j)) = D(\mathrm{a}_i, \mathrm{a}_j)$ for all $i, j = 1, \ldots, N$, then $B' = B_D$. If we take $\mathbf{q}(\mathrm{a}_N)$ as our origin, any $\Sigma\beta_i \mathrm{a}_i \in \hat{\mathbb{R}}[\mathrm{A}]$ can be uniquely expressed relative to the basis $[\mathbf{v}_j := \mathbf{q}(\mathrm{a}_j) - \mathbf{q}(\mathrm{a}_N) \mid j = 1, \ldots, N-1]$ as $\Sigma_{j=1}^{N-1} \gamma_j \mathbf{v}_j$. Applying the usual relations between a bilinear form and its corresponding quadratic form, we obtain

$$\begin{aligned} B'(\mathbf{v}_i; \mathbf{v}_j) &= \frac{1}{2}\Big(Q'(\mathbf{q}(\mathrm{a}_i) - \mathbf{q}(\mathrm{a}_N)) + Q'(\mathbf{q}(\mathrm{a}_j) - \mathbf{q}(\mathrm{a}_N)) - Q'(\mathbf{q}(\mathrm{a}_i) - \mathbf{q}(\mathrm{a}_j))\Big) \\ &= \frac{1}{2}\Big(D(\mathrm{a}_i, \mathrm{a}_N) + D(\mathrm{a}_j, \mathrm{a}_N) - D(\mathrm{a}_i, \mathrm{a}_j)\Big) \\ &= \frac{1}{2}\Big(Q_D(\mathbf{q}(\mathrm{a}_i) - \mathbf{q}(\mathrm{a}_N)) + Q_D(\mathbf{q}(\mathrm{a}_j) - \mathbf{q}(\mathrm{a}_N)) - Q_D(\mathbf{q}(\mathrm{a}_i) - \mathbf{q}(\mathrm{a}_j))\Big) \\ &= B_D(\mathbf{v}_i; \mathbf{v}_j). \end{aligned} \tag{3.14}$$

Hence for all $\Sigma\gamma_i \mathbf{v}_i \in \hat{\mathbb{R}}[\mathrm{A}]$

$$\begin{aligned} Q'(\Sigma\gamma_i \mathbf{v}_i) &= \sum_{i=1}^{N-1} \sum_{j=1}^{N-1} B'(\mathbf{v}_i; \mathbf{v}_j)\, \gamma_i \gamma_j \\ &= \sum_{i=1}^{N-1} \sum_{j=1}^{N-1} B_D(\mathbf{v}_i; \mathbf{v}_j)\, \gamma_i \gamma_j \ =\ Q_D(\Sigma\gamma_i \mathbf{v}_i)\,. \end{aligned} \tag{3.15}$$

This shows that $Q' = Q_D$ and $B' = B_D$, as desired. QED

If an orthonormal affine coordinate system is desired, one can simply pick an origin $\mathbf{q}(\mathrm{a}_i) \in \mathbb{A}_D[\mathrm{A}]$ and proceed with a Gram-Schmidt style orthogonalization in $\hat{\mathbb{R}}_D[\mathrm{A}]$. If it should happen that all of the remaining vectors to be orthogonalized are (nearly) isotropic, however, then one may be forced to change origin in the process.

Note that, for a given ordering $\underline{\mathrm{A}} = [\mathrm{a}_1, \ldots, \mathrm{a}_N]$ of the atoms in A, it is possible to regard $\tilde{\mathbb{R}}[\mathrm{A}]$ as an affine hyperplane $\tilde{\mathbb{R}}[\underline{\mathrm{A}}] \approx \mathbb{R}^{N-1}$ contained in $\mathbb{R}^N$, and the associated vector space $\hat{\mathbb{R}}[\mathrm{A}]$ as an $(N-1)$-dimensional subspace $\hat{\mathbb{R}}[\underline{\mathrm{A}}] \approx \mathbb{R}^{N-1}$ of $\mathbb{R}^N$. We shall denote the corresponding affine space $(\tilde{\mathbb{R}}[\underline{\mathrm{A}}], \hat{\mathbb{R}}[\underline{\mathrm{A}}], B_D)$ by $\mathbb{A}_D^{N-1}$. Using these terms, we present another important result:

COROLLARY 3.7. *A premetric space $\mathbb{S} = (\mathrm{A}, D)$ has dimension n if and only if Schoenberg's quadratic form Q_D is of rank n on $\hat{\mathbb{R}}[\underline{\mathrm{A}}]$. In addition, a premetric space is Euclidean if and only if the form is positive semidefinite.*

Proof: A nondegenerate bilinear vector space $(\ll\widehat{\mathbb{R}}[\underline{\mathrm{A}}]\gg,\ \bar{B}_D)$ of dimension equal to the rank of the form can be constructed by factoring the bilinear vector space $\widehat{\mathbb{R}}_D[\underline{\mathrm{A}}] = (\widehat{\mathbb{R}}[\underline{\mathrm{A}}],\ B_D)$ by its radical $\mathrm{rad}(\widehat{\mathbb{R}}_D[\underline{\mathrm{A}}]) := \{\mathbf{v} \in \widehat{\mathbb{R}}[\underline{\mathrm{A}}] \mid B_D(\mathbf{v},\mathbf{w}) = 0 \ \ \forall\ \mathbf{w} \in \widehat{\mathbb{R}}[\underline{\mathrm{A}}]\}$, i.e.

$$\ll\widehat{\mathbb{R}}[\underline{\mathrm{A}}]\gg \ := \ \widehat{\mathbb{R}}[\underline{\mathrm{A}}]/\mathrm{rad}(\widehat{\mathbb{R}}_D[\underline{\mathrm{A}}]) \qquad \text{and} \tag{3.16}$$

$$\bar{B}_D(\mathbf{u}+\mathrm{rad}(\widehat{\mathbb{R}}_D[\underline{\mathrm{A}}]),\ \mathbf{v}+\mathrm{rad}(\widehat{\mathbb{R}}_D[\underline{\mathrm{A}}])) \ := \ B_D(\mathbf{u},\mathbf{v})\ . \tag{3.17}$$

If we now define a partition $<(\tilde{\mathbb{R}}[\underline{\mathrm{A}}])>$ of $\tilde{\mathbb{R}}[\underline{\mathrm{A}}]$ as:

$$<(\tilde{\mathbb{R}}[\underline{\mathrm{A}}])> \ := \ \{\ \{\tau_{\mathbf{v}}(\mathbf{q}) \mid \mathbf{v} \in \mathrm{rad}(\widehat{\mathbb{R}}_D[\underline{\mathrm{A}}])\} \mid \mathbf{q} \in \tilde{\mathbb{R}}[\underline{\mathrm{A}}]\ \} \tag{3.18}$$

where $\tau_{\mathbf{v}}(\mathbf{q})$ denotes the translation of the point $\mathbf{q}$ by $\mathbf{v}$, then $\bar{\mathbb{A}}_D^{N-1} := (<(\tilde{\mathbb{R}}[\underline{\mathrm{A}}])>,\ \ll\widehat{\mathbb{R}}[\underline{\mathrm{A}}]\gg,\ \bar{B}_D)$ is a bilinear affine space of dimension $n := \mathit{rank}(\widehat{\mathbb{R}}[\underline{\mathrm{A}}]) = N-1-\mathit{nul}(\widehat{\mathbb{R}}_D[\underline{\mathrm{A}}])$. The embedding $\mathbf{q}:\ \mathrm{A}\rightarrow\tilde{\mathbb{R}}[\underline{\mathrm{A}}]$ can easily be converted to an isometric mapping into $\bar{\mathbb{A}}_D^{N-1}$ by defining $\bar{\mathbf{q}}(\mathrm{a}) := \{\tau_{\mathbf{v}}(\mathbf{q}(\mathrm{a})) \mid \mathbf{v} \in \mathrm{rad}(\widehat{\mathbb{R}}_D[\underline{\mathrm{A}}])\}$ for each $\mathrm{a} \in \mathrm{A}$.

Because the corresponding quadratic form on $\ll\widehat{\mathbb{R}}[\underline{\mathrm{A}}]\gg$ satisfies $D(\mathrm{a},\mathrm{b}) = \bar{Q}_D(\bar{\mathbf{q}}(\mathrm{a})-\bar{\mathbf{q}}(\mathrm{b})) \neq 0$ for all $\mathrm{a} \neq \mathrm{b}$, the mapping $\bar{\mathbf{q}}$ is again an isometric embedding and hence as in the proof of Theorem 3.6 it can be shown to be unique. Since the rank of the form is invariant under isometry and is never more than the dimension of the space, $\mathit{rank}(\widehat{\mathbb{R}}[\underline{\mathrm{A}}])$ is clearly the minimum dimension for which an isometric embedding of $\mathbb{S}$ can exist. In addition, by Sylvester's law of inertia, all maximal positive definite subspaces of a bilinear vector space have the same dimension. Thus if the space as a whole is positive semidefinite, the maximal positive definite subspaces are also the maximal nondegenerate subspaces which are isometric with the nondegenerate factor space, i.e. the factor space is Euclidean of dimension n. We have therefore shown that the affine factor space $\bar{\mathbb{A}}_D^{N-1}$ is Euclidean of dimension n whenever the form is positive semidefinite of rank n on $\widehat{\mathbb{R}}[\underline{\mathrm{A}}]$. QED

Note that Schoenberg's form Q_D itself is only positive semidefinite on the subspace $\widehat{\mathbb{R}}[\underline{\mathrm{A}}] \subset \mathbb{R}^N$; the matrix $-\tfrac{1}{2}\cdot\mathbf{M}_D(\underline{\mathrm{A}})$ associated with any semimetric space $\mathbb{S} = (\mathrm{A},\ D)$ and ordering $\underline{\mathrm{A}}$ of A in fact always has at least one negative direction, as may be verified by considering the value $Q_D(\mathbf{1}) = \mathbf{1}^T\cdot(-\tfrac{1}{2}\cdot\mathbf{M}_D(\underline{\mathrm{A}}))\cdot\mathbf{1}$, where $\mathbf{1} := [1,1,\ldots,1]$.

In closing, we propose an exercise which may serve to convince the reader that, in the case that Q_D is positive definite on $\widehat{\mathbb{R}}[\underline{\mathrm{A}}]$, B_D really is in accord with the conventional definition of a Euclidean scalar product on $\mathbb{R}^{N-1} \approx \widehat{\mathbb{R}}[\underline{\mathrm{A}}]$. If one sets $D(\mathrm{a}_i,\mathrm{a}_N) = 1$ for all $i = 1,\ldots,N-1$ and $D(\mathrm{a}_i,\mathrm{a}_j) = 2$ for all $1 \leq i < j \leq N-1$, one obtains a Cartesian frame of reference with $\mathbf{q}(\mathrm{a}_N)$ as the origin and the $\mathbf{q}(\mathrm{a}_i)$ located at unit distance along each of the coordinate axes for $i = 1,\ldots,N-1$.

Relative to this frame of reference, the first $N-1$ barycentric coordinates of a point correspond to a set of Cartesian coordinates for it. If one substitutes these values of D into the matrix $-\frac{1}{2}\cdot\mathbf{M}_D(\underline{A})$ and evaluates Q_D on the difference of two points $\mathbf{x} := \sum x_i a_i$ and $\mathbf{y} := \sum y_i a_i$ with $\sum x_i = \sum y_i = 1$, one finds that:

$$Q_D(\mathbf{x}-\mathbf{y}) = \sum_{i=1}^{N-1} (x_i - y_i)^2 , \tag{3.19}$$

as in the usual definition.

A simple case of the proposed exercise is obtained when three atoms a_1, a_2 and a_3 lie at unit distance along the coordinate axes and at the origin in $\mathbb{R}^2$, respectively.

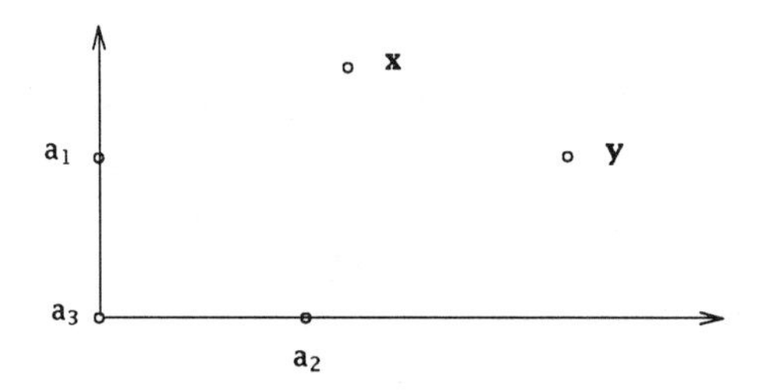

Figure 3.2

Then Schoenberg's quadratic form is:

$$Q_D(\underline{\beta}) = -\frac{1}{2}\cdot [\beta_1,\beta_2,\beta_3] \begin{bmatrix} 0 & 2 & 1 \\ 2 & 0 & 1 \\ 1 & 1 & 0 \end{bmatrix} \begin{bmatrix} \beta_1 \\ \beta_2 \\ \beta_3 \end{bmatrix}$$

$$= -\beta_1\beta_3 - \beta_2\beta_3 - 2\beta_1\beta_2 .$$

Substituting $\beta_i = x_i - y_i$ for $i = 1,2,3$ and applying the relations $\sum x_i = \sum y_i = 1$ then yields $Q_D(\mathbf{x}-\mathbf{y}) = (x_1-y_1)^2+(x_2-y_2)^2$.

3.2. Vector Subspaces and Metroids

Corollary 3.7 provides us with a necessary and sufficient condition for a premetric space to be Euclidean of dimension n. Even though it is possible to determine the signature of Schoenberg's quadratic form by standard numerical techniques, in all but the most trivial cases these require the intervention of a computer. In order to gain direct insight into how the signature of the form depends upon the values of the individual squared distances $D(\mathrm{a},\mathrm{b})$ it is necessary to reduce the problem to a conceptually manageable size, by breaking it down into smaller subproblems whose solutions determine the whole. Hence, the following:

Definition 3.8. A premetric space $\mathbb{S}' = (\mathrm{A}', D') =: \mathbb{S}(\mathrm{A}')$ is termed a *subspace* of a premetric space $\mathbb{S} = (\mathrm{A}, D)$ if $\mathrm{A}' \subseteq \mathrm{A}$ and the restriction $D|_{\mathrm{A}'}$ of D to A′ satisfies $D|_{\mathrm{A}'} = D'$. The premetric space $\mathbb{S}$ in turn is called an *extension* of $\mathbb{S}'$.

Note that if $\mathbb{S}'$ is a subspace of a Euclidean premetric space $\mathbb{S}$ and $\mathbb{A}_D$ and $\mathbb{A}_{D'}$ are their amalgamations, then since $\mathbb{A}_{D'}$ is effectively an affine subspace of $\mathbb{A}_D$, $\mathbb{S}'$ is also a Euclidean premetric space, and $dim(\mathbb{S}') \leq dim(\mathbb{S})$.

As an illustration of Definition 3.8, suppose that $\mathbb{S} = (\mathrm{A}, D)$, $\underline{\mathrm{A}} = [\mathrm{a}_1,\mathrm{a}_2,\mathrm{a}_3]$, and

$$\underline{\mathbf{M}}_D(\underline{\mathrm{A}}) = \begin{bmatrix} 0 & 1 & 2 \\ 1 & 0 & 1 \\ 2 & 1 & 0 \end{bmatrix}.$$

The $\mathbb{S}$ is Euclidean because we could take $\mathbf{p}:\mathrm{a}_1 \mapsto [0,0]$, $\mathbf{p}:\mathrm{a}_2 \mapsto [1,0]$, $\mathbf{p}:\mathrm{a}_3 \mapsto [1,1]$ in $\mathbb{R}^2$. If we consider the subsequence $\underline{\mathrm{A}}' = [\mathrm{a}_1,\mathrm{a}_2]$ of $\underline{\mathrm{A}}$, we obtain the subspace $\mathbb{S}' = (\mathrm{A}', D)$ with matrix

$$\underline{\mathbf{M}}_{D'}(\underline{\mathrm{A}}') = \begin{bmatrix} 0 & 1 \\ 1 & 0 \end{bmatrix}.$$

This subspace is also Euclidean, as may be seen by taking $\mathbf{p}:\mathrm{a}_1 \mapsto [0]$ and $\mathbf{p}:\mathrm{a}_2 \mapsto [1]$ in $\mathbb{R}^1$, which shows that $dim(\mathbb{S}') = 1$.

In the next section we will investigate the relation between the signatures of the various subspaces of the amalgamation space $\mathbb{A}_D$ of (A, D) which are affinely spanned by subsets of the embedded atoms $\mathbf{p}(\mathrm{A})$. In this section, however, we confine ourselves to a simpler problem, namely relations between the *ranks* of the various subspaces $<\mathbf{U}>$ which are linearly spanned by the subsets $\mathbf{U} \subseteq \mathbf{V}$ of a finite set of vectors $\mathbf{V} \subset \mathbb{V}_B$ in a bilinear vector space $\mathbb{V}_B = (\mathbb{V}, B)$. By factoring by a supplement of $<\mathbf{V}>$ if necessary we may assume that $<\mathbf{V}> = \mathbb{V}$, and by factoring the space by its radical we may also assume w.l.o.g. that $\mathbf{V}_B$ is nondegenerate. For two sequences of elements in $\mathbf{V}$, e.g. $\underline{\mathbf{U}} = [\mathbf{u}_1, \ldots, \mathbf{u}_k]$, $\underline{\mathbf{W}} = [\mathbf{w}_1, \ldots, \mathbf{w}_l] \in \underline{2}^{\mathbf{V}}$, we define the $k \times l$ matrix

$$\underline{\mathbf{M}}_B(\underline{\mathbf{U}};\, \underline{\mathbf{W}}) \;:=\; \begin{bmatrix} B(\mathbf{u}_1;\mathbf{w}_1) & \cdots & B(\mathbf{u}_1;\mathbf{w}_l) \\ \cdots & \cdots & \cdots \\ B(\mathbf{u}_k;\mathbf{w}_1) & \cdots & B(\mathbf{u}_k;\mathbf{w}_l) \end{bmatrix}, \tag{3.20}$$

and we abbreviate $\underline{\mathbf{M}}_B(\underline{\mathbf{U}};\, \underline{\mathbf{U}})$ as $\underline{\mathbf{M}}_B(\underline{\mathbf{U}}) = \underline{\mathbf{M}}_B(\mathbf{u}_1, \ldots, \mathbf{u}_k)$. Note that the determinant of $\underline{\mathbf{M}}_B(\underline{\mathbf{U}})$ is actually independent of the order in which we index the elements of $\underline{\mathbf{U}}$, so that if $\#\mathbf{U} = \#\underline{\mathbf{U}}$ (i.e. $\underline{\mathbf{U}}$ contains no repeated elements) this determinant can be considered a function of the underlying set $\mathbf{U}$. Since $det(\underline{\mathbf{M}}_B(\mathbf{u};\mathbf{w})) = B(\mathbf{u};\mathbf{w})$ for all $\mathbf{u},\mathbf{w} \in \mathbf{V}_B$, it is convenient to denote the function mapping $\{\underline{\mathbf{U}}, \underline{\mathbf{W}} \in \underline{2}^{\mathbf{V}} \mid \#\underline{\mathbf{U}} = \#\underline{\mathbf{W}}\} \rightarrow \mathbb{R}$ by $B(\underline{\mathbf{U}};\, \underline{\mathbf{W}}) := det(\underline{\mathbf{M}}_B(\underline{\mathbf{U}};\, \underline{\mathbf{W}}))$, and if $\#\mathbf{U} = \#\underline{\mathbf{U}}$ we abbreviate $B(\mathbf{U}) := B(\underline{\mathbf{U}}) := B(\underline{\mathbf{U}};\, \underline{\mathbf{U}})$.

Definition 3.9. The *bilinear rank* of a subset $\mathbf{U} \subseteq \mathbf{V}$ is defined as

$$\rho_B(\mathbf{U}) \;:=\; dim(<\mathbf{U}>/\mathrm{rad}(<\mathbf{U}>)) \;=\; rank(\underline{\mathbf{M}}_B(\underline{\mathbf{U}})) \tag{3.21}$$

for any ordering $\underline{\mathbf{U}}$ of $\mathbf{U}$. The corresponding function $\rho_B\colon 2^{\mathbf{V}} \rightarrow \mathbb{Z}$ is called a *bilinear rank function.*

The following records a few of the basic properties of bilinear rank functions, which we will subsequently refer to as "axioms".

LEMMA 3.10. *If* $\mathbf{V}$ *is a finite subset of* $\mathbf{V}_B$ *and* $\rho_B\colon 2^{\mathbf{V}} \rightarrow \mathbb{Z}$ *as above:*

(R1) $$\forall\, \mathbf{U} \subseteq \mathbf{V}\colon\;\; 0 \;\le\; \rho_B(\mathbf{U}) \;\le\; \#\mathbf{U}\,; \tag{3.22}$$

(R2) $$\forall\, \mathbf{U} \subseteq \mathbf{V},\; \mathbf{v} \in \mathbf{V}\colon\;\; \rho_B(\mathbf{U}) \;\le\; \rho_B(\mathbf{U} \cup \mathbf{v}) \;\le\; \rho_B(\mathbf{U}) + 2\,. \tag{3.23}$$

Proof: The first assertion follows from $\rho_B(\mathbf{U}) := \mathrm{dim}(<\mathbf{U}>/\,\mathrm{rad}(\mathbf{U})) \le \mathrm{dim}(<\mathbf{U}>) \le \#\mathbf{U}$. To prove the second, we recall that adding one row or column to a matrix can increase its rank by at most 1. Since the metric matrix $\underline{\mathbf{M}}_B(\mathbf{u}_1, \ldots, \mathbf{u}_N, \mathbf{v})$ relative to an ordering $\underline{\mathbf{U} \cup \mathbf{v}} =: [\mathbf{u}_1, \ldots, \mathbf{u}_N, \mathbf{v}]$ of $\mathbf{U} \cup \mathbf{v}$ is obtained from the matrix $\underline{\mathbf{M}}_B(\mathbf{u}_1, \ldots, \mathbf{u}_N)$ by adding both one row and one column, the result follows. QED

The determinant $B(\mathbf{U}) = det(\underline{\mathbf{M}}_B(\underline{\mathbf{U}}))$ is commonly known as the *Gramian* of $\mathbf{U}$. Note that if $\underline{\mathbf{U}}$ is a basis of $<\mathbf{U}>$, then $sign(B(\mathbf{U}))$ is the discriminant of the subspace $<\mathbf{U}>$ by definition. Recall also that a *minor* of a matrix is the determinant of a submatrix thereof, and that a *principal minor* is one whose row and column index sets are equal. A well known result from linear algebra states that the rank of a matrix is equal to the size of its largest nonzero minor. The algebraic content of the following lemma is that in the case of symmetric matrices

As an illustration of Definition 3.9, suppose we are given a bilinear form on $\mathbb{R}^3$ with

$$\underline{\mathbf{M}}_B(\underline{\mathbf{U}}) = \begin{bmatrix} 1 & 2 & -1 \\ 2 & 1 & -2 \\ -1 & -2 & 1 \end{bmatrix},$$

where $\underline{\mathbf{U}} = [\mathbf{u}_1, \mathbf{u}_2, \mathbf{u}_3]$ is the standard basis of $\langle \mathbf{U} \rangle = \mathbb{R}^3$, i.e.

$$\mathbf{u}_1 = \begin{bmatrix} 1 \\ 0 \\ 0 \end{bmatrix}, \quad \mathbf{u}_2 = \begin{bmatrix} 0 \\ 1 \\ 0 \end{bmatrix}, \quad \mathbf{u}_3 = \begin{bmatrix} 0 \\ 0 \\ 1 \end{bmatrix}.$$

Gaussian elimination easily shows that $\underline{\mathbf{M}}_B$ has rank 2, so $2 = \rho_B(\mathbf{U})$. The one-dimensional null space or radical is:

$$\mathrm{rad}(\langle \mathbf{U} \rangle) = \gamma \cdot \begin{bmatrix} 1 \\ 0 \\ 1 \end{bmatrix}, \quad \gamma \in \mathbb{R},$$

since $[1\ 0\ 1] \cdot \underline{\mathbf{M}}_B(\underline{\mathbf{U}}) = [0\ 0\ 0]$. As a B-orthogonal basis, we can take the (unnormalized) vectors

$$\mathbf{v}_1 = \begin{bmatrix} 1 \\ 0 \\ 1 \end{bmatrix}, \quad \mathbf{v}_2 = \begin{bmatrix} 1 \\ -1 \\ 0 \end{bmatrix}, \quad \mathbf{v}_3 = \begin{bmatrix} 2 \\ 1 \\ 1 \end{bmatrix}.$$

Relative to the basis $\underline{\mathbf{V}} = [\mathbf{v}_1, \mathbf{v}_2, \mathbf{v}_3]$, the metric matrix has the diagonal form

$$\underline{\mathbf{M}}_B(\underline{\mathbf{V}}) = \begin{bmatrix} 0 & 0 & 0 \\ 0 & -2 & 0 \\ 0 & 0 & 6 \end{bmatrix}.$$

The elements of the factor space $(\langle \mathbf{U} \rangle / \mathrm{rad}(\langle \mathbf{U} \rangle))$ are the one-dimensional subspaces of the form $\{\gamma \mathbf{v}_1 \mid -\infty < \gamma < \infty\} + \beta \mathbf{v}_2 + \alpha \mathbf{v}_3$, where $\alpha, \beta \in \mathbb{R}$.

the rank is in fact equal to the size of the largest nonzero *principal* minor.

LEMMA 3.11. *Let* $\mathbf{V} \subset \mathbf{V}_B$ *be a finite subset of bilinear vector space; then for* $\mathbf{U} \subseteq \mathbf{V}$

$$\rho_B(\mathbf{U}) = \max_{\mathbf{W} \subseteq \mathbf{U}} \left(\#\mathbf{W} \,\middle|\, B(\mathbf{W}) \neq 0 \right). \tag{3.24}$$

Proof: We must show that $\rho_B(\mathbf{U})$ is equal to the maximum size of a linearly independent subset of $\mathbf{U}$ whose Gramian does not vanish. Let $[\mathbf{v}_1, \ldots, \mathbf{v}_n]$ be a B-orthogonal basis of $\langle \mathbf{U} \rangle$, so that $\underline{\mathbf{M}}_B(\mathbf{v}_1, \ldots, \mathbf{v}_n)$ is the diagonalized metric matrix of the subspace $\langle \mathbf{U} \rangle$. Also suppose w.l.o.g. that the last $n-m$ of these vectors constitute a basis of the radical rad$(\langle \mathbf{U} \rangle, B)$, so that $\rho_B(\mathbf{U}) = m$.

Now let $\mathbf{pr}: \langle \mathbf{U} \rangle \to \langle \mathbf{v}_1, \ldots, \mathbf{v}_m \rangle$ denote the B-orthogonal projection of $\langle \mathbf{U} \rangle$ onto $\langle \mathbf{v}_1, \ldots, \mathbf{v}_m \rangle$. Then $B(\mathbf{u}; \mathbf{u}') = B(\mathbf{pr}(\mathbf{u}); \mathbf{pr}(\mathbf{u}'))$ for all $\mathbf{u}, \mathbf{u}' \in \mathbf{U}$, so that for any subset $\mathbf{W} = \{\mathbf{w}_1, \ldots, \mathbf{w}_k\} \subseteq \mathbf{U}$

$$B(\mathbf{w}_1, \ldots, \mathbf{w}_k) = B(\mathbf{pr}(\mathbf{w}_1), \ldots, \mathbf{pr}(\mathbf{w}_k)) \neq 0 \tag{3.25}$$
$$\Rightarrow \quad dim(\mathbf{pr}(\mathbf{w}_1), \ldots, \mathbf{pr}(\mathbf{w}_k)) = k \, .$$

It follows that $\#\mathbf{W} = k \leq m = \rho_B(\mathbf{U})$ for all $\mathbf{W} \subseteq \mathbf{U}$ with $B(\mathbf{W}) \neq 0$.

On the other hand, since $\mathbf{U}$ spans $< \mathbf{v}_1, \ldots, \mathbf{v}_m >$ we may always find a subset $\mathbf{W} := \{\mathbf{w}_1, \ldots, \mathbf{w}_m\} \subseteq \mathbf{U}$ such that $[\mathbf{pr}(\mathbf{w}_1), \ldots, \mathbf{pr}(\mathbf{w}_m)]$ is a basis of $< \mathbf{v}_1, \ldots, \mathbf{v}_m >$, and since $(<\mathbf{v}_1, \ldots, \mathbf{v}_m>, B)$ is nondegenerate, this m-element subset $\mathbf{W}$ has a nonzero Gramian, i.e. $max(\#\mathbf{W} \mid B(\mathbf{W}) \neq 0) = m = \rho_B(\mathbf{U})$, as desired. QED

For a finite subset $\mathbf{V}$ of $\mathbf{V}_B$, we now define a *bilinear set system* $(\mathbf{V}, \mathbb{F}_B)$ as

$$\mathbb{F}_B := \{\mathbf{F} \subseteq \mathbf{V} \mid B(\mathbf{F}) \neq 0\} \, , \tag{3.26}$$

i.e. as the collection of all subsets of $\mathbf{V}$ whose Gramian does not vanish. The sets $\mathbf{F} \in \mathbb{F}_B$ will be called the *free* subsets of $\mathbf{V}$. As a consequence of Lemma 3.11, the bilinear set system $\mathbb{F}_B$ is given by $\{\mathbf{F} \subseteq \mathbf{V} \mid \rho_B(\mathbf{F}) = \#\mathbf{F}\}$. This set system, in turn, uniquely determines the bilinear rank function via the relation

$$\rho_B(\mathbf{U}) := \max_{\mathbf{F} \subseteq \mathbf{U}} \left(\#\mathbf{F} \;\middle|\; \mathbf{F} \in \mathbb{F}_B \right) . \tag{3.27}$$

Thus ρ_B and $\mathbb{F}_B$ are really just two ways of describing one and the same object.

As an example of a bilinear set system, we consider again the bilinear vector space of the previous example (following Definition 3.9), whose metric matrix versus the standard basis $\underline{\mathbf{U}} = [\mathbf{u}_1, \mathbf{u}_2, \mathbf{u}_3]$ of $\mathbb{R}^3$ is

$$\underline{\mathbf{M}_B(\mathbf{U})} = \begin{bmatrix} 1 & 2 & -1 \\ 2 & 1 & -2 \\ -1 & -2 & 1 \end{bmatrix} .$$

The corresponding bilinear set system is thus

$$\mathbb{F}_B = \{\mathbf{F} \subseteq \mathbf{U} \mid det(\mathbf{M}_B(\mathbf{F}) \neq 0\}$$
$$= \left\{ \{\}, \{\mathbf{u}_1\}, \{\mathbf{u}_2\}, \{\mathbf{u}_3\}, \{\mathbf{u}_1, \mathbf{u}_2\}, \{\mathbf{u}_2, \mathbf{u}_3\} \right\} .$$

Note that free subsets of $\mathbf{V} \subset \mathbf{V}_B$ are necessarily linearly independent. Unless the bilinear form is anisotropic, however, the vanishing of the Gramian does not imply the linear dependence of the corresponding vectors, and hence bilinear set systems are not necessarily stable (i.e. subsets of free sets are not necessarily free), as is shown by the following example:

$$\underline{\mathbf{M}_B(\mathbf{u}, \mathbf{v})} := \begin{bmatrix} 0 & x \\ x & 0 \end{bmatrix} . \tag{3.28}$$

Here we have $B(\mathbf{u},\mathbf{v}) = -x^2 \neq 0$ while $B(\mathbf{u}) = B(\mathbf{v}) = 0$, and hence $\rho_B(\mathbf{u}\cup\mathbf{v}) = \rho_B(\mathbf{u})+2$. As a consequence, bilinear set systems are not necessarily matroids. Recall, however, that the independent subsets of matroids constitute a *balanced* set system (i.e. all maximal independent subsets $\mathbf{I} \subseteq \mathbf{U}$ of a given set $\mathbf{U} \subseteq \mathbf{V}$ have the same size).

PROPOSITION 3.12. *Bilinear set systems are balanced.*

Proof: Suppose $\mathbf{F} \subseteq \mathbf{U} \subseteq \mathbf{V}$ is free; then

$$\{\mathbf{0}\} = \mathrm{rad}(\mathbf{F}) = \langle\mathbf{F}\rangle \cap (\mathbf{F})^\perp \supseteq \langle\mathbf{F}\rangle \cap \langle\mathbf{U}\rangle \cap (\mathbf{U})^\perp = \langle\mathbf{F}\rangle \cap \mathrm{rad}(\mathbf{U}), \quad (3.29)$$

which in turn implies that $\mathbf{F}+\mathrm{rad}(\mathbf{U})$ is an independent subset of the factor space $\langle\!\langle\mathbf{U}\rangle\!\rangle := \langle\mathbf{U}\rangle/\mathrm{rad}(\mathbf{U})$, and so may be augmented to a basis $\mathbf{F}'+\mathrm{rad}(\mathbf{U})$ of $\langle\!\langle\mathbf{U}\rangle\!\rangle$. Since $\langle\mathbf{U}\rangle = \langle\mathbf{F}'\rangle \oplus \mathrm{rad}(\mathbf{U})$, we have

$$\begin{aligned} \langle\mathbf{U}\rangle &= \langle\mathbf{F}'\rangle + \mathrm{rad}(\mathbf{U}) \quad \Rightarrow \qquad (3.30)\\ \mathbf{U}^\perp &= (\mathbf{F}')^\perp \cap \mathrm{rad}(\mathbf{U})^\perp \supseteq (\mathbf{F}')^\perp \cap \langle\mathbf{U}\rangle \ , \end{aligned}$$

as well as

$$\begin{aligned} \{\mathbf{0}\} &= \langle\mathbf{F}'\rangle \cap \mathrm{rad}(\mathbf{U}) \\ &= \langle\mathbf{F}'\rangle \cap \langle\mathbf{U}\rangle \cap (\mathbf{U})^\perp \quad \text{by definition} \qquad (3.31)\\ &\supseteq \langle\mathbf{F}'\rangle \cap (\mathbf{F}')^\perp \cap \langle\mathbf{U}\rangle \quad \text{by (3.30)} \\ &= \mathrm{rad}(\mathbf{F}') \quad \text{since } \mathbf{F}' \subseteq \mathbf{U}\,. \end{aligned}$$

It follows that

$$\rho_B(\mathbf{F}') := \mathit{dim}(\langle\mathbf{F}'\rangle/\mathrm{rad}(\mathbf{F}')) = \mathit{dim}(\langle\mathbf{F}'\rangle) = \#\mathbf{F}' = \rho_B(\mathbf{U})\,, \quad (3.32)$$

as desired. QED

Moreover, it follows easily that:

PROPOSITION 3.13. *The maximal free subsets of a bilinear set system* $(\mathbf{V},\mathbb{F}_B)$ *are the bases of a representable matroid* $(\mathbf{V},\mathbb{I}_B)$*. In addition, if* $\mathbf{I} \in \mathbb{I}_B$ *is a maximal independent subset of an arbitrary subset* $\mathbf{U} \subseteq \mathbf{V}$*, then* $\rho_B(\mathbf{I}) = \rho_B(\mathbf{U})$.

Proof: Since a matroid is a balanced and stable set system, if we define $\mathbb{I}_B$ to be all subsets of sets in $\mathbb{F}_B$ we automatically get a matroid by Proposition 3.12. The bases of this matroid are the maximal subsets in $\mathbb{F}_B$.

To prove the second assertion, we observe that for any maximal independent subset $\mathbf{I}$ of $\mathbf{U}$:

$$\begin{aligned} &\mathbf{I} \text{ is a basis of } \langle\mathbf{U}\rangle \\ \Rightarrow\ &\mathbf{I}+\mathrm{rad}(\mathbf{U}) \text{ generates } \langle\!\langle\mathbf{U}\rangle\!\rangle := \frac{\langle\mathbf{U}\rangle}{\mathrm{rad}(\mathbf{U})} \end{aligned}$$

$$\Rightarrow \quad \mathbf{I}+\text{rad}(\mathbf{U}) \text{ contains a basis } \mathbf{J}+\text{rad}(\mathbf{U}) \text{ of } \ll\mathbf{U}\gg \tag{3.33}$$

$$\Rightarrow \quad \{\mathbf{0}\} = \langle\mathbf{J}\rangle \cap \text{rad}(\mathbf{U}) \supseteq \text{rad}(\mathbf{J}) \quad \Rightarrow \quad \mathbf{J} \in \mathbb{F}_B$$

$$\Rightarrow \quad \rho_B(\mathbf{U}) = \#\mathbf{J} = \rho_B(\mathbf{J}) \leq \rho_B(\mathbf{I}) \leq \rho_B(\mathbf{U})$$

i.e. $\rho_B(\mathbf{I}) = \rho_B(\mathbf{U})$, as claimed. QED

As a consequence, even though the vanishing of the Gramian of a set of vectors does not imply their linear dependence, in a nondegenerate space any linearly independent subset can be augmented to a set of vectors whose Gramian does not vanish.

We now give two axioms for set systems which codify some of the basic properties of bilinear set systems.

Definition 3.14. Let $(V,\mathbb{F})$ be a set system defined on an unspecified finite set V. Such a set system is called a *premetroid* if:

(F1) $\quad F, G \in \mathbb{F},\ \#G < \#F \quad \Rightarrow \quad \exists\ u,v \in F\setminus G \text{ with } G\cup\{u,v\} \in \mathbb{F}\,;$ (3.34)

(F2) $\quad F \in \mathbb{F},\ v \in V \text{ with } F\setminus v \notin \mathbb{F} \quad \Rightarrow \quad \exists\ u \in F\setminus v \text{ with } F\setminus\{u,v\} \in \mathbb{F}\,.$ (3.35)

LEMMA 3.15. *Given a set system* $(V, \mathbb{F})$, *the following are equivalent:*

(i) $\mathbb{F}$ is a premetroid.

(ii) $\mathbb{F}$ is balanced and satisfies axiom (F2).

(iii) $\mathbb{F}$ is balanced and its rank function $\rho: 2^V \rightarrow \mathbf{Z}$: $\rho(U) \mapsto max(\#F \mid F \subseteq U, F \in \mathbb{F})$ satisfies (R2) (Equation (3.23)).

Proof: (*i*) => (*ii*): Suppose we have $F, G \subseteq U \subseteq V$ with $F, G \in \mathbb{F}$ and $\#G < \#F = \rho(U)$. Then by axiom (F1) there exists $u, v \in F\setminus G$ with $G\cup\{u,v\} \in \mathbb{F}$. By setting $G := G\cup\{u,v\}$ and repeating this process, we eventually augment G to a set $G' \in \mathbb{F}$ with $\#G' = \#F$.

(*ii*) => (*iii*): The first inequality of (R2) is obeyed by the rank function of any set system. To prove the second, we take $U \subseteq V$, $v \in V$ and $F \subseteq U\cup v$ with $\#F = \rho(F) = \rho(U\cup v)$. Then if $v \notin F$, $\rho(U\cup v) = \#F = \rho(U)$. Else if $v \in F$ and $F\setminus v \in \mathbb{F}$, then $\rho(U) \geq \#(F\setminus v) = \#F - 1 = \rho(U\cup v) - 1$. Otherwise, $v \in F$ and $F\setminus v \notin \mathbb{F}$, so by axiom (F2) there exists $u \in F\setminus v$ such that $F\setminus\{u,v\}$ is free in U. Hence $\rho(U) \geq \#(F\setminus\{u,v\}) = \#F - 2 = \rho(F\cup v) - 2$.

(*iii*) => (*i*): To prove that axiom (F1) holds, we note that since $\mathbb{F}$ is balanced and $\#G < \#F$, it must be possible to augment G with some $W \subseteq F\setminus G$ with $\#W \neq 0$ so that $G\cup W \in \mathbb{F}$, and we choose W to be minimal with this property. Then for any $w \in W$, if $\rho(G\cup(W\setminus w)) > \rho(G)$, being balanced implies an $H \in \mathbb{F}$ with $G \subseteq H \subseteq G\cup(W\setminus w)$ and $\#H = \rho(G\cup(W\setminus w)) > \rho(G) = \#G$, so $H\setminus G$ is contained in $W\setminus w$ and $G\cup(H\setminus G) \in \mathbb{F}$, which contradicts the minimality of W. Thus $\rho(G\cup(W\setminus w)) = \rho(G)$ for all $w \in W$, and so if $\#W > 2$

$$\rho(G \cup (W \setminus w) \cup w) \;=\; \#G + \#W \;>\; \#G + 2 \;=\; \rho(G \cup (W \setminus w)) + 2\,, \tag{3.36}$$

which contradicts the inequality of axiom (R2).

To prove that axiom (F2) holds, we observe that if $F \setminus v$ is not free, then by the inequalities of axioms (R1) and (R2) we have

$$\#F + 1 \;=\; \#(F \setminus v) + 2 \;>\; \rho(F \setminus v) + 2 \;\geq\; \rho(F) \;=\; \#F\,, \tag{3.37}$$

so that $\rho(F \setminus v) = \#F - 2$. Thus there exists a subset of $\#F \setminus v$ whose rank and size are both equal $\#F - 2$, i.e. an element u of $F \setminus v$ such that $F \setminus \{u,v\} \in \mathbb{F}$, as desired. QED

The following will be useful later in understanding the relations between Euclidean subspaces.

COROLLARY 3.16. *Given a premetroid* $(V, \mathbb{F})$ *in which all subsets size* $n+1$ *or less are stable, the* n*-truncation is a matroid of rank* n.

Proof: Since by hypothesis all subsets of free subsets of size less than or equal to $n+1$ are free, we need only verify that the n-truncation is balanced. Hence let $G, F \subseteq U \subseteq V$ with $G, F \in \mathbb{F}$, $\#G < \#F \leq n$ and $\#F$ maximum subject to these conditions; then by axiom (F1) there exists $f, f' \in F \setminus G$ with $G \cup \{f, f'\} \in \mathbb{F}$. Since $\#(G \cup \{f, f'\}) \leq n+1$, by stability $G \cup f \in \mathbb{F}$, so we can augment G repeatedly until $\#G = \#F$. QED

We now present another property of the rank functions of premetroids which, when combined with axioms (R1) and (R2) above, provides a fully equivalent set of axioms for premetroids in terms of their rank functions.

PROPOSITION 3.17. *Given a finite set* V, *a function* $\rho : 2^V \rightarrow \mathbb{Z}$ *is the rank function of a premetroid if and only if it satisfies axioms (R1), (R2) and:*

(R3) $$\rho(W) < \rho(U) \;\Rightarrow\; \exists\ \{u,v\} \subset U \text{ with } \rho(W \cup \{u,v\}) \;\geq\; \rho(W) + \#\{u,v\}\,. \tag{3.38}$$

Proof: (=>): In light of Lemma 3.15, it suffices to prove that axiom (R3) holds. Given $F \subseteq U$ and $G \subseteq W$ such that $\rho(F) = \#F = \rho(U) > \rho(G) = \#G = \rho(W)$, we have by axiom (F1) elements $u, v \in F \setminus G$ with $G \cup \{u,v\} \in \mathbb{F}$. Thus

$$\begin{aligned} \rho(W \cup \{u,v\}) \;\geq\; \rho(G \cup \{u,v\}) \;&=\; \#(G \cup \{u,v\}) \\ &=\; \#G + \#\{u,v\} \;=\; \rho(W) + \#\{u,v\}\,. \end{aligned} \tag{3.39}$$

(<=): Again in light of the lemma, it suffices to prove that (R3) implies that $\mathbb{F}$ is balanced. Choosing free subsets $F, G \subseteq U \subseteq V$ with $\#G < \#F = \rho(U)$, axiom (R3) gives us a $\{f, f'\} \subset F$ such that

$$\#G + \#\{f,f'\} \;\geq\; \rho(G \cup \{f,f'\}) \;\geq\; \rho(G) + \#\{f,f'\} \;=\; \#G + \#\{f,f'\}\,, \tag{3.40}$$

and hence there exists a $G' = G \cup \{f, f'\} \in \mathbb{F}$. Repeating this procedure with $G := G'$

eventually augments G to a free subset G″ with #G″ = #F. QED

Recall that if $\rho: V \to \mathbb{Z}$ is a matroidal rank function, then $U \subseteq V$, $v, v' \in V$ and $\rho(U \cup v) = \rho(U \cup v') = \rho(U)$ implies $\rho(U \cup v \cup v') = \rho(U)$; for premetroids, the following weaker condition holds:

COROLLARY 3.18. *If* (V, ρ) *is a premetroid and* $V \supseteq W := U \cup \{w_1, w_2, w_3\}$, *then*

$$(R3')\quad \rho(U \cup \{w_i, w_j\}) = \rho(U) \text{ for } i,j = 1,\ldots,3 \quad \Rightarrow \quad \rho(W) = \rho(U)\,. \tag{3.41}$$

Proof: Exercise.

We now return to bilinear set systems for a moment, and derive two more conditions which they satisfy.

PROPOSITION 3.19. *Let* $(\mathbf{V}, \mathbb{F}_B)$ *be a bilinear independence system; then the following are true:*

(F3) If $\mathbf{F} \in \mathbb{F}_B$, $\mathbf{w},\mathbf{x},\mathbf{y},\mathbf{z} \in \mathbf{V}$, $\mathbf{w} \neq \mathbf{x},\mathbf{y}$, and $\mathbf{F} \cup \mathbf{w}$, $\mathbf{F} \cup \mathbf{w} \cup \mathbf{x}$, $\mathbf{F} \cup \mathbf{w} \cup \mathbf{y} \notin \mathbb{F}_B$, then

$$\mathbf{F} \cup \{\mathbf{w},\mathbf{x},\mathbf{y},\mathbf{z}\} \in \mathbb{F}_B \iff \mathbf{F} \cup \{\mathbf{w},\mathbf{z}\} \in \mathbb{F}_B \text{ and } \mathbf{F} \cup \{\mathbf{x},\mathbf{y}\} \in \mathbb{F}_B\,; \tag{3.42}$$

(F4) If $\mathbf{F} \in \mathbb{F}_B$ and $\mathbf{v},\mathbf{w},\mathbf{x},\mathbf{y},\mathbf{z} \in \mathbf{V}$ with $\mathbf{v} \neq \mathbf{w}$, then

$$\begin{gathered}\mathbf{F} \cup \mathbf{v},\ \mathbf{F} \cup \mathbf{w},\ \mathbf{F} \cup \mathbf{v} \cup \mathbf{w} \notin \mathbb{F}_B,\ \mathbf{F} \cup \{\mathbf{v},\mathbf{x}\},\ \mathbf{F} \cup \{\mathbf{v},\mathbf{w},\mathbf{y},\mathbf{z}\} \in \mathbb{F}_B \\ \Rightarrow \quad \mathbf{F} \cup \{\mathbf{v},\mathbf{w},\mathbf{x},\mathbf{y}\} \in \mathbb{F}_B \text{ or } \mathbf{F} \cup \{\mathbf{v},\mathbf{w},\mathbf{x},\mathbf{z}\} \in \mathbb{F}_B\,.\end{gathered} \tag{3.43}$$

Proof: To prove this, we order $\mathbf{V}$ in a sequence $\underline{\mathbf{V}} := [\mathbf{v}_1, \ldots, \mathbf{v}_N]$ such that the $k := \#\mathbf{F}$ vectors in $\mathbf{F}$ come first, and let $\underline{\mathbf{F}} := [\mathbf{v}_1, \ldots, \mathbf{v}_k]$.

(F3): Let $\underline{\mathbf{U}} := [\mathbf{w},\mathbf{x},\mathbf{y},\mathbf{z}]$, and assume without loss of generality that $\mathbf{V} = \mathbf{F} \dot\cup \mathbf{U}$. Let $\mathbf{pr}: \langle\mathbf{V}\rangle \to \langle\mathbf{F}\rangle$ denote the B-orthogonal projection onto $\langle\mathbf{F}\rangle$, $\underline{\mathbf{U}}' := [\mathbf{u}' := \mathbf{pr}(\mathbf{u}) \mid \mathbf{u} \in \underline{\mathbf{U}}]$ and $\underline{\mathbf{U}}'' := [\mathbf{u}'' := \mathbf{u} - \mathbf{pr}(\mathbf{u}) \mid \mathbf{u} \in \underline{\mathbf{U}}]$. Then for $\mathbf{f} \in \underline{\mathbf{F}}$ and $\mathbf{u} \in \underline{\mathbf{U}}$, $B(\mathbf{f};\mathbf{u}) = B(\mathbf{f};\mathbf{u}')$, while for $\mathbf{u},\mathbf{v} \in \underline{\mathbf{U}}$, $B(\mathbf{u};\mathbf{v}) = B(\mathbf{u}';\mathbf{v}') + B(\mathbf{u}'';\mathbf{v}'')$. Because $\mathbf{u}' = \mathbf{pr}(\mathbf{u}) \in \langle\mathbf{F}\rangle$ for all $\mathbf{u} \in \mathbf{U}$, without changing the value of the determinant $det(\underline{\mathbf{M}}_B(\underline{\mathbf{V}}))$ we may add suitable linear combinations of the rows and columns of the matrix $\underline{\mathbf{M}}_B(\underline{\mathbf{F}})$ to the last four rows and columns of $\underline{\mathbf{M}}_B(\underline{\mathbf{V}})$ to reduce all elements of the submatrices $\underline{\mathbf{M}}_B(\underline{\mathbf{F}};\underline{\mathbf{U}})$ and $\underline{\mathbf{M}}_B(\underline{\mathbf{U}};\underline{\mathbf{F}})$ to zero, i.e. to reduce $\underline{\mathbf{M}}_B(\underline{\mathbf{V}})$ to a block diagonal form:

$$\begin{bmatrix} \underline{\mathbf{M}}_B(\underline{\mathbf{F}}) & \underline{\mathbf{M}}_B(\underline{\mathbf{F}};\underline{\mathbf{U}}) \\ \underline{\mathbf{M}}_B(\underline{\mathbf{U}};\underline{\mathbf{F}}) & \underline{\mathbf{M}}_B(\underline{\mathbf{U}}) \end{bmatrix} \mapsto \begin{bmatrix} \underline{\mathbf{M}}_B(\underline{\mathbf{F}}) & \underline{\mathbf{0}} \\ \underline{\mathbf{0}} & \underline{\mathbf{M}}_B(\underline{\mathbf{U}}'') \end{bmatrix} \tag{3.44}$$

with

$$\underline{\mathbf{M}}_B(\underline{\mathbf{U}}'') = \begin{bmatrix} B(\mathbf{w}'';\mathbf{w}'') & B(\mathbf{w}'';\mathbf{x}'') & B(\mathbf{w}'';\mathbf{y}'') & B(\mathbf{w}'';\mathbf{z}'') \\ B(\mathbf{w}'';\mathbf{x}'') & B(\mathbf{x}'';\mathbf{x}'') & B(\mathbf{x}'';\mathbf{y}'') & B(\mathbf{x}'';\mathbf{z}'') \\ B(\mathbf{w}'';\mathbf{y}'') & B(\mathbf{x}'';\mathbf{y}'') & B(\mathbf{y}'';\mathbf{y}'') & B(\mathbf{y}'';\mathbf{z}'') \\ B(\mathbf{w}'';\mathbf{z}'') & B(\mathbf{x}'';\mathbf{z}'') & B(\mathbf{y}'';\mathbf{z}'') & B(\mathbf{z}'';\mathbf{z}'') \end{bmatrix}. \tag{3.45}$$

The hypothesis $\mathrm{F} \cup \mathrm{w} \notin \mathbb{F}_B$ implies $B(\mathbf{w}'';\mathbf{w}'') = 0$, because then

$$0 = det\begin{bmatrix} \underline{\mathbf{M}}_B(\underline{\mathbf{F}}) & \underline{\mathbf{M}}_B(\underline{\mathbf{F}};\mathbf{w}) \\ \underline{\mathbf{M}}_B(\mathbf{w};\underline{\mathbf{F}}) & \underline{\mathbf{M}}_B(\mathbf{w};\mathbf{w}) \end{bmatrix} = det\begin{bmatrix} \underline{\mathbf{M}}_B(\underline{\mathbf{F}}) & \mathbf{0} \\ \mathbf{0}^T & B(\mathbf{w}'';\mathbf{w}'') \end{bmatrix} \tag{3.46}$$
$$= B(\mathbf{F})\cdot B(\mathbf{w}'';\mathbf{w}''),$$

where $B(\mathbf{F}) := det(\underline{\mathbf{M}}_B(\underline{\mathbf{F}}) \neq 0$. Similarly, the assumptions $\mathrm{F}\cup\{\mathrm{w},\mathrm{x}\} \notin \mathbb{F}_B$ and $\mathrm{F}\cup\{\mathrm{w},\mathrm{y}\} \notin \mathbb{F}_B$ imply that $B(\mathbf{w}'',\mathbf{x}'') = -B^2(\mathbf{w}'';\mathbf{x}'') = 0$ and $B(\mathbf{w}'',\mathbf{y}'') = -B^2(\mathbf{w}'';\mathbf{y}'') = 0$. Thus the matrix $\underline{\mathbf{M}}_B(\underline{\mathbf{U}}'')$ assumes the form:

$$\underline{\mathbf{M}}_B(\underline{\mathbf{U}}'') = \begin{bmatrix} 0 & 0 & 0 & * \\ 0 & * & * & * \\ 0 & * & * & * \\ * & * & * & * \end{bmatrix}, \tag{3.47}$$

where "$*$" denotes a nonzero element. Expanding first along the top row and then along the left-hand column, we obtain

$$det(\underline{\mathbf{M}}_B(\underline{\mathbf{U}}'')) = B^2(\mathbf{w}'';\mathbf{z}'')\cdot B(\mathbf{x}'',\mathbf{y}''). \tag{3.48}$$

Hence $B(\mathbf{V}) \neq 0$ if and only if $-B^2(\mathbf{w}'';\mathbf{z}'') = B(\mathbf{w}'',\mathbf{z}'') \neq 0$ and $B(\mathbf{x}'',\mathbf{y}'') \neq 0$, as claimed.

(F4): Let $\underline{\mathbf{U}} := [\mathbf{v},\mathbf{w},\mathbf{x},\mathbf{y},\mathbf{z}]$, assume that $\mathbf{V} := \mathbf{F}\dot{\cup}\mathbf{U}$ and define $\underline{\mathbf{U}}'$, $\underline{\mathbf{U}}''$ as before. As in (F3), $B(\mathbf{V}) = B(\mathbf{F})\cdot det(\underline{\mathbf{M}}_B(\underline{\mathbf{U}}''))$ with $B(\mathbf{F}) \neq 0$, so it suffices to analyze $\underline{\mathbf{M}}_B(\underline{\mathbf{U}}'')$, which has the form shown below:

$$\underline{\mathbf{M}}_B(\underline{\mathbf{U}}'') = \begin{bmatrix} 0 & 0 & * & * & * \\ 0 & 0 & * & * & * \\ * & * & * & * & * \\ * & * & * & * & * \\ * & * & * & * & * \end{bmatrix}. \tag{3.49}$$

This time we expand the submatrix $\underline{\mathbf{M}}_B(\mathbf{v}'',\mathbf{w}'',\mathbf{y}'',\mathbf{z}'')$ along the first two rows, obtaining:

$$B(\mathbf{v}'',\mathbf{w}'',\mathbf{y}'',\mathbf{z}'') = B^2(\mathbf{v}'',\mathbf{w}'';\mathbf{y}'',\mathbf{z}''), \tag{3.50}$$

with similar expressions for $B(\mathbf{v}'',\mathbf{w}'',\mathbf{x}'',\mathbf{y}'')$ and $B(\mathbf{v}'',\mathbf{w}'',\mathbf{x}'',\mathbf{z}'')$. Since $\mathbf{F}\cup\{\mathbf{v},\mathbf{x}\} \in \mathbb{F}_B$ by hypothesis, $B(\mathbf{v}'';\mathbf{x}'') \neq 0$. Thus if we had $B(\mathbf{v}'',\mathbf{w}'';\mathbf{x}'',\mathbf{y}'') = 0$ and $B(\mathbf{v}'',\mathbf{w}'';\mathbf{x}'',\mathbf{z}'') = 0$, the vectors $[B(\mathbf{v}'';\mathbf{y}''), B(\mathbf{w}'';\mathbf{y}'')]$ and $[B(\mathbf{v}'';\mathbf{z}''), B(\mathbf{w}'';\mathbf{z}'')]$ would both depend on $[B(\mathbf{v}'';\mathbf{x}''), B(\mathbf{w}'';\mathbf{x}'')]$ and hence on each other, so that $B(\mathbf{v}'',\mathbf{w}'';\mathbf{y}'',\mathbf{z}'') = 0$, a contradiction. QED

We now use the above to further refine our definition of a premetroid.

Definition 3.20. A *semimetroid* is a premetroid which satisfies axiom (F3) above; if it also satisfies axiom (F4), then it is called a *metroid.*

Those metroids which are bilinear set systems are called *bilinearly representable metroids.* Algebraically, a representable metroid $(V, \mathbb{F})$ is one for which there exists a one to one correspondence between its elements and the row/column indices of a symmetric matrix such that a subset $U \subseteq V$ is in $\mathbb{F}$ if and only if the corresponding principal minor of the matrix is not zero. Of course, these axioms for metroids are not sufficient to guarantee representability. Nevertheless, in addition to all the facts demonstrated above for premetroids, we have the following:

THEOREM 3.21. *Let* $(V, \mathbb{F})$ *be a set system with rank function* ρ*; then*

(i) If $(V, \mathbb{F})$ is a semimetroid, the maximal free subsets are the bases of a matroid $(\mathbf{V}, \mathbb{I})$.

(ii) If $(V, \mathbb{F})$ is a metroid and $(V, \mathbb{I})$ is the associated matroid, the maximal independent subsets $I \subseteq U$ which are contained in any subset $U \subseteq V$ satisfy $\rho(I) = \rho(U)$.

(c.f. Proposition 3.13). These facts show that metroids provide us with a fairly complete combinatorial abstraction of the geometry of bilinear vector spaces, which can be used to better understand them. The reader is referred to [Dress1986] for detailed proofs and further developments of this theory (see also the appendix to [Dress1987]).†

3.3. Affine Metroids and Cayley-Menger Determinants

In this section we shall study the affine version of the metroids introduced in the last section, with particular emphasis on the Euclidean case. As a consequence of our results in §3.1, any premetric space can be regarded as a finite subset of a bilinear affine space and vice versa. In the following, we shall therefore assume that we have been given a premetric space $\mathbb{S} = (A, D)$, and we let $\mathbb{A}_D[A]$ be the associated amalgamation space, $Q_D : A \rightarrow \mathbb{R}$ be Schoenberg's quadratic form on the corresponding bilinear vector space $\mathbf{V}_D$, and $\mathbf{q} : A \rightarrow \tilde{\mathbb{R}}[A]$ be the canonical embedding. We shall also denote the embedded image $\mathbf{q}(A) \subset \tilde{\mathbb{R}}[A]$

† Metroids have recently been found to be completely equivalent to another recently introduced class of set systems, known as Δ-matroids [Bouchet, Dress & Havel, in preparation].

of the atoms A by **bold "A"**, and the image $\mathbf{q}(a) \in \mathbf{A}$ of a single atom by **bold "a"**, so that the translation mapping **a** to $\mathbf{b} \in \tilde{\mathbb{R}}[A]$ can be written directly as "**b**−**a**". The vector subspace spanned by $\{\mathbf{b}-\mathbf{b}' \mid \mathbf{b} \in \mathbf{A}'\}$ for any $\mathbf{b}' \in \mathbf{A}' \subseteq \mathbf{A}$ will be denoted by $<\mathbf{A}'>$ as usual.

Definition 3.22. If (A′, *D*) is a subspace of $\mathbb{S} = (A, D)$, the *affine bilinear rank* of $\mathbf{A}' := \mathbf{q}(A')$ is

$$\rho_D(\mathbf{A}') \;:=\; \begin{cases} 0 & \text{if } \mathbf{A}' = \{\}; \\ \rho_B(<\mathbf{A}'>)+1 & \text{otherwise,} \end{cases} \tag{3.51}$$

where ρ_B is the bilinear rank of the vector subspace $<\mathbf{A}'> \subseteq \hat{\mathbb{R}}[A]$ versus the bilinear form B_D which corresponds to Q_D. The associated set system $(\mathbf{A}, \mathbb{F}_D)$ defined by $\mathbb{F}_D = \{\mathbf{A}' \subseteq \mathbf{A} \mid \rho_D(\mathbf{A}') = \#\mathbf{A}'\}$ is called an *affine bilinear set system.*

As an example of Definition 3.22, let $\mathbb{S} = (A, D)$ with

$$\underline{\mathbf{M}}_D(\underline{A}) \;=\; \begin{bmatrix} 0 & 1 & 2 \\ 1 & 0 & 1 \\ 2 & 1 & 0 \end{bmatrix}$$

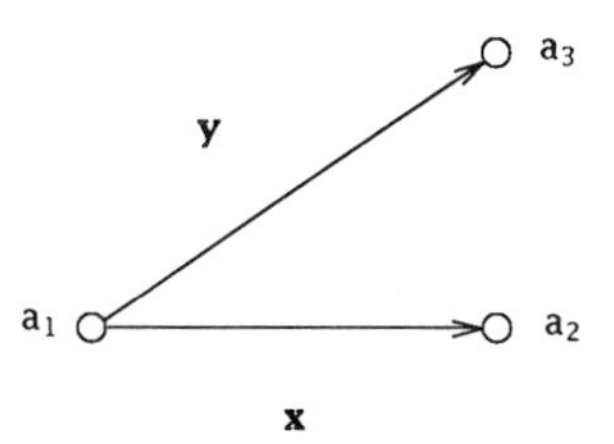

Figure 3.3

If we take $\mathbf{a}_1$ as the origin, then $< \mathbf{A} >$ is the space spanned by $\mathbf{x} := \mathbf{a}_2 - \mathbf{a}_1$ and $\mathbf{y} := \mathbf{a}_3 - \mathbf{a}_1$. Thus if we calculated the corresponding metric matrix, we obtain (see Eq. A.3):

$$\underline{\mathbf{M}}_B(\mathbf{x},\mathbf{y}) \;=\; \begin{bmatrix} B(\mathbf{x};\mathbf{x}) & B(\mathbf{x};\mathbf{y}) \\ B(\mathbf{x};\mathbf{y}) & B(\mathbf{y};\mathbf{y}) \end{bmatrix} \;=\; \begin{bmatrix} 1 & 1 \\ 1 & 2 \end{bmatrix}$$

which has rank 2, so that $\rho_D(A) = 2+1 = 3 = \#A$, i.e. the set of atoms A is free. Note this result holds independent of our choice of origin.

Note that $\rho_D(\mathbf{A}')$ is just one plus the dimension of the premetric subspace $\mathbb{S}(A') := (A', D)$ *induced* by $A' \subseteq A$, and that $\mathbf{A}'$ is free if and only if the corresponding subspace $<\mathbf{A}'>$ of $\mathbf{V}_D$ is nondegenerate.

The reason why we studied metroids so thoroughly in the previous section is:

PROPOSITION 3.23. *Affine bilinear set systems are metroids.*

Proof: We first verify the rank axioms for premetroids. Axiom (R1) (Equation (3.22)) follows immediately from $0 \leq \rho_D(\mathbf{B}) \leq dim(<\mathbf{B}>)+1 \leq \#\mathbf{B}$ for all $\mathbf{B} \subseteq \mathbf{A}$, while the axiom (R2) (Equation (3.23)) follows since for any $\mathbf{a} \in \mathbf{A}$, $\mathbf{b} \in \mathbf{B}$

$$\begin{aligned} \rho_D(\mathbf{B}) \;:=\; 1+\rho_B(<\mathbf{B}>) \;&\leq\; 1+\rho_B(<\mathbf{B}>+<\mathbf{a}-\mathbf{b}>) \;=:\; \rho_D(\mathbf{B}\cup\mathbf{a}) \\ &\leq\; 3+\rho_B(<\mathbf{B}>) \;=:\; 2+\rho_D(\mathbf{B})\,. \end{aligned} \tag{3.52}$$

To prove the axiom (R3) (Equation (3.38)), we must show that for all $\mathbf{B},\mathbf{C} \subseteq \mathbf{A}$ with $\rho_D(\mathbf{C}) < \rho_D(\mathbf{B})$ there exists $\mathbf{a},\mathbf{a}' \in \mathbf{B}$ such that $\rho_D(\mathbf{C}\cup\{\mathbf{a},\mathbf{a}'\}) \geq \rho_D(\mathbf{C})+\#\{\mathbf{a},\mathbf{a}'\}$. If $\mathbf{B}\cap\mathbf{C} \neq \{\}$, we need only take an element in the intersection as origin to obtain this as an immediate consequence of the corresponding result for ρ_B. Otherwise, we take $\mathbf{b} \in \mathbf{B}$, $\mathbf{c} \in \mathbf{C}$ as origins for $<\mathbf{B}>$ and $<\mathbf{C}>$, respectively. Applying the corresponding result for ρ_B, we obtain

$$\rho_B(<\mathbf{C}>+<\mathbf{b}'-\mathbf{b},\mathbf{b}''-\mathbf{b}>) \;\geq\; \rho_B(<\mathbf{C}>)+\#\{\mathbf{b}'-\mathbf{b},\,\mathbf{b}''-\mathbf{b}\} \tag{3.53}$$

for some $\mathbf{b}',\mathbf{b}'' \in \mathbf{B}$. Since $<\mathbf{b},\mathbf{b}',\mathbf{b}''> \;\subseteq\; <\mathbf{c},\mathbf{b},\mathbf{b}',\mathbf{b}''>$, on letting $\mathbf{c}$ be the origin in $<\mathbf{c},\mathbf{b},\mathbf{b}',\mathbf{b}''>$ we find that

$$\begin{aligned} \rho_B(<\mathbf{C}>+<\mathbf{b}-\mathbf{c},\mathbf{b}'-\mathbf{c},\mathbf{b}''-\mathbf{c}>) \;&\geq\; \rho_B(<\mathbf{C}>+<\mathbf{b}'-\mathbf{b},\mathbf{b}''-\mathbf{b}>) \\ &\geq\; \rho_B(<\mathbf{C}>)+\#\{\mathbf{b}'-\mathbf{b},\mathbf{b}''-\mathbf{b}\} \;>\; \rho_B(<\mathbf{C}>)\,. \end{aligned} \tag{3.54}$$

Thus we may use axiom (R3) for ρ_B a second time to obtain:

$$\rho_D(\mathbf{C}\cup\{\mathbf{a},\mathbf{a}'\}) \;=\; 1+\rho_B(<\mathbf{C}>+<\mathbf{a}-\mathbf{c},\mathbf{a}'-\mathbf{c}>) \;\geq\; \rho_D(\mathbf{C})+\#\{\mathbf{a},\mathbf{a}'\} \tag{3.55}$$

for some $\mathbf{a},\mathbf{a}' \in \{\mathbf{b},\mathbf{b}',\mathbf{b}''\}$.

To prove that affine bilinear set systems are true metroids, we need to verify axioms (F3) and (F4) (Equations (3.42) and (3.43)). Given a nonempty free subset $\mathbf{F} \subseteq \mathbf{A}$, this can be done very simply by choosing an origin in $\mathbf{F}$ and applying the corresponding results for the vector case. Since $\rho_D(\mathbf{a}) = 1$ for all $\mathbf{a} \in \mathbf{A}$, however, all one element subsets of $\mathbf{A}$ are free, and hence in the case that $\mathbf{F} = \{\}$ the antecedents of both implications of (F3) and (F4) are always false, i.e. the statements themselves are tautologies. QED

As a consequence of this result, all of the results derived in the previous section for bilinear set systems hold in affine bilinear set systems as well. A metroid is said to be *affinely representable* when there exists a one to one correspondence between its elements and the elements of a premetric space such that free subsets generate nondegenerate affine subspaces. Unlike matroids, however, where any linearly representable matroid whose one element subsets are all independent admits an affine representation, we shall see that there exist bilinearly representable set systems, all of whose singletons are free, which nevertheless do not have an affine representation. It is not known whether affine representability implies bilinear representability or not. At least the example given by the metric space (A, D) with $\mathrm{A} := \{1,2,3,4\}$, $D(i,i+1) := 1$ and $D(i,i+2) := 4$ (modulo 4) shows (by a straightforward albeit lengthy calculation, which we leave to the reader as an exercise) that there might be *no* extension of the bilinear form B_D, defined on $\hat{\mathbb{R}}[\mathrm{A}]$ and used in the definition of the metroidal rank function ρ_D on 2^{A}, to a bilinear form B' defined on the space $\mathbb{R}[\mathrm{A}]$ of all formal linear combinations $\Sigma\gamma_i \mathrm{a}_i$, such that the identity mapping $\mathrm{A} \rightarrow \mathbb{R}[\mathrm{A}]$: $\mathrm{a}_j \rightarrow \Sigma\delta_{ij}\mathrm{a}_i$ induces an isomorphism $(\mathrm{A}, \rho_D) \approx (\mathrm{A}, \rho_{B'})$. Nevertheless, we are able to characterize the free subsets of affinely representable metroids directly in terms of determinants.

Definition 3.24. The *Cayley-Menger determinant*† of a premetric space $(\mathrm{A} = \{\mathrm{a}_1, \ldots, \mathrm{a}_N\}, D)$ is $D(\underline{\mathrm{A}}) := 2(-\tfrac{1}{2})^N det(\tilde{\underline{\mathbf{M}}}_D(\underline{\mathrm{A}}))$, where $\tilde{\underline{\mathbf{M}}}_D(\underline{\mathrm{A}})$ is the bordered matrix:

$$\tilde{\underline{\mathbf{M}}}_D(\underline{\mathrm{A}}) \quad := \quad \begin{bmatrix} 0 & 1 & 1 & \cdots & 1 \\ 1 & 0 & D(\mathrm{a}_1,\mathrm{a}_2) & \cdots & D(\mathrm{a}_1,\mathrm{a}_N) \\ 1 & D(\mathrm{a}_1,\mathrm{a}_2) & 0 & \cdots & D(\mathrm{a}_2,\mathrm{a}_N) \\ \cdot & \cdots & \cdots & \cdots & \cdots \\ 1 & D(\mathrm{a}_1,\mathrm{a}_N) & D(\mathrm{a}_2,\mathrm{a}_N) & \cdots & 0 \end{bmatrix}. \tag{3.56}$$

Note this determinant is once again independent of the order in which we number our atoms, thus enabling us to speak of *the* determinant of the space without ambiguity. If $(\mathrm{A} := \{\mathrm{a}_1,\mathrm{a}_2\}, D)$ is a two atom premetric space, the Cayley-Menger determinant $D(\underline{\mathrm{A}})$ is equal to the single distance squared $D(\mathrm{a}_1,\mathrm{a}_2)$ itself; thus, our use of the same symbol for two seemingly different things is justified. When the premetric is clear, we may also refer to $D(\underline{\mathrm{A}}) = D(\mathrm{A})$ simply as the Cayley-Menger determinant of the atoms A themselves.

LEMMA 3.25. *Let* (A, D) *be a premetric space,* $D(\mathrm{A})$ *be its Cayley-Menger determinant,* $\underline{\mathrm{A}} = [\mathrm{a}_1, \ldots, \mathrm{a}_N]$ *be an ordering of* A, *and* $[\mathbf{v}_j := \mathbf{a}_j - \mathbf{a}_1 \mid j = 2, \ldots, N]$ *be the basis of the amalgamation space* $\mathbb{A}_D$ *obtained by choosing* $\mathbf{a}_1 \in \mathrm{A}$ *as the origin. Then*

$$D(\mathrm{A}) \quad = \quad D(\mathrm{a}_1, \ldots, \mathrm{a}_N) \quad = \quad B_D(\mathbf{v}_2, \ldots, \mathbf{v}_N) \tag{3.57}$$

where $B_D(\mathbf{v}_2, \ldots, \mathbf{v}_N)$ *is the determinant of the metric matrix* $\underline{\mathbf{M}}_B(\mathbf{v}_2, \ldots, \mathbf{v}_N)$.

† This name is due to L. Blumenthal [Blumenthal1953].

Proof: By first bordering the matrix $\underline{\mathbf{M}}_B(\mathbf{v}_2, \ldots, \mathbf{v}_N)$ by two rows and columns each consisting of zeros except for their common diagonal elements, which are ones, and using the relation

$$\begin{aligned} 2\,B_D(\mathbf{a}_i - \mathbf{a}_1;\, \mathbf{a}_j - \mathbf{a}_1) &= Q_D(\mathbf{a}_i - \mathbf{a}_1) + Q_D(\mathbf{a}_j - \mathbf{a}_1) - Q_D(\mathbf{a}_j - \mathbf{a}_i) \\ &= D(\mathrm{a}_1, \mathrm{a}_i) + D(\mathrm{a}_1, \mathrm{a}_j) - D(\mathrm{a}_i, \mathrm{a}_j)\,, \end{aligned} \tag{3.58}$$

we can write this as: $B_D(\mathbf{v}_2, \ldots, \mathbf{v}_N))$

$$= det\begin{bmatrix} 1 & 0 & 0 & \cdots & 0 \\ 0 & 1 & 0 & \cdots & 0 \\ 0 & 0 & B_D(\mathbf{v}_2;\mathbf{v}_2) & \cdots & B_D(\mathbf{v}_2;\mathbf{v}_N) \\ \cdot & \cdot & \cdots & \cdots & \cdots \\ 0 & 0 & B_D(\mathbf{v}_2;\mathbf{v}_N) & \cdots & B_D(\mathbf{v}_N;\mathbf{v}_N) \end{bmatrix}$$

$$= det\begin{bmatrix} 1 & 0 & 1 & \cdots & 1 \\ 0 & 1 & 0 & \cdots & 0 \\ 0 & 1 & B_D(\mathbf{a}_2 - \mathbf{a}_1;\mathbf{a}_2 - \mathbf{a}_1) & \cdots & B_D(\mathbf{a}_2 - \mathbf{a}_1;\mathbf{a}_N - \mathbf{a}_1) \\ \cdot & \cdot & \cdots & \cdots & \cdots \\ 0 & 1 & B_D(\mathbf{a}_2 - \mathbf{a}_1;\mathbf{a}_N - \mathbf{a}_1) & \cdots & B_D(\mathbf{a}_N - \mathbf{a}_1;\mathbf{a}_N - \mathbf{a}_1) \end{bmatrix} \tag{3.59}$$

$$= -det\begin{bmatrix} 0 & 1 & 1 & \cdots & 1 \\ 1 & 0 & 0 & \cdots & 0 \\ 1 & 0 & D(\mathrm{a}_1,\mathrm{a}_2) & \cdots & \tfrac{1}{2}(D(\mathrm{a}_1,\mathrm{a}_2) + D(\mathrm{a}_1,\mathrm{a}_N) - D(\mathrm{a}_2,\mathrm{a}_N)) \\ \cdot & \cdot & \cdots & \cdots & \cdots \\ 1 & 0 & \tfrac{1}{2}(D(\mathrm{a}_1,\mathrm{a}_2) + D(\mathrm{a}_1,\mathrm{a}_N) - D(\mathrm{a}_2,\mathrm{a}_N)) & \cdots & D(\mathrm{a}_1,\mathrm{a}_N) \end{bmatrix}$$

$$= -det\begin{bmatrix} 0 & 1 & 1 & \cdots & 1 \\ 1 & 0 & -\tfrac{1}{2}D(\mathrm{a}_1,\mathrm{a}_2) & \cdots & -\tfrac{1}{2}D(\mathrm{a}_1,\mathrm{a}_N) \\ 1 & -\tfrac{1}{2}D(\mathrm{a}_1,\mathrm{a}_2) & 0 & \cdots & -\tfrac{1}{2}D(\mathrm{a}_2,\mathrm{a}_N) \\ \cdot & \cdots & \cdots & \cdots & \cdots \\ 1 & -\tfrac{1}{2}D(\mathrm{a}_1,\mathrm{a}_N) & -\tfrac{1}{2}D(\mathrm{a}_2,\mathrm{a}_N) & \cdots & 0 \end{bmatrix}$$

$$= -\left(\frac{-1}{2}\right)^N det\begin{bmatrix} 0 & -2 & -2 & \cdots & -2 \\ 1 & 0 & D(\mathrm{a}_1,\mathrm{a}_2) & \cdots & D(\mathrm{a}_1,\mathrm{a}_N) \\ 1 & D(\mathrm{a}_1,\mathrm{a}_2) & 0 & \cdots & D(\mathrm{a}_2,\mathrm{a}_N) \\ \cdot & \cdots & \cdots & \cdots & \cdots \\ 1 & D(\mathrm{a}_1,\mathrm{a}_N) & D(\mathrm{a}_2,\mathrm{a}_N) & \cdots & 0 \end{bmatrix}$$

$$= 2\,(-\tfrac{1}{2})^N\, det(\underline{\tilde{\mathbf{M}}}_D(\mathrm{A}))\,.$$

where we have added multiples of the border of ones to eliminate terms of the form $D(a_1,a_i)$ from the components of the last $N-1$ rows and columns without changing the value of the determinant, but with the effect that these terms now appear in the second row and column. QED

Note the determinant $D(A)$ is independent of our choice of origin $\mathbf{a}_i \in A$.

Continuing the example on bilinear forms in §3.2 (following Definition 3.9), we place an atom a_1 at the origin in $\mathbb{R}^3$ and three more atoms a_2,a_3,a_4 at the ends of the unit vectors $\mathbf{u}_1,\mathbf{u}_2,\mathbf{u}_3$, respectively. Then the bordered matrix of squared distances is:

$$\underline{\tilde{\mathbf{M}}}_D(a_1,\ldots,a_4) = \begin{bmatrix} 0 & 1 & 1 & 1 & 1 \\ 1 & 0 & 1 & 1 & 1 \\ 1 & 1 & 0 & -2 & 4 \\ 1 & 1 & -2 & 0 & 6 \\ 1 & 1 & 4 & 6 & 0 \end{bmatrix}$$

(the "extra" border of ones arises because the metric matrix $\underline{\mathbf{M}}_B$ of the example has ones down the diagonal). The corresponding affine metroid then consists of all subsets of these four atoms except for $\{a_1,a_2,a_4\}$ and $\{a_1,a_2,a_3,a_4\}$.

Proposition 3.26. *If* $\mathbb{S} = (A, D)$ *is a premetric space, then for all* $B \subseteq A$ *the affine bilinear rank function* $\rho_D : 2^{\mathbf{A}} \to \mathbf{Z}$ *is given by:*

$$\rho_D(\mathbf{B}) \quad := \quad \max_{C \subset B} \left(\#C \mid D(C) \neq 0 \right). \tag{3.60}$$

Proof: By definition, $\mathbf{C}$ with $\{\} \neq \mathbf{C} \subseteq \mathbf{B}$ is free whenever $\rho_D(\mathbf{C})-1 := \rho_B(<\mathbf{C}>) = \#\mathbf{C}-1$. For any origin $\mathbf{c}_1 \in \mathbf{C}$, this happens exactly when the Gramian $B(\mathbf{u}_2,\ldots,\mathbf{u}_k)$ of the basis $\mathbf{u}_i := [\mathbf{c}_i - \mathbf{c}_1 \mid i = 2,\ldots,k]$ of $<\mathbf{C}>$ does not vanish, where $k := \#\mathbf{C}$. The result now follows from Lemma 3.25 together with our characterization of ρ_B in terms of Gramians (Lemma 3.11). QED

Thus a subset $\mathbf{B} \subseteq \mathbf{A}$ is free if and only if $D(B) \neq 0$.

By using our previous results on metroids, we can show that given a fixed integer n one can check whether or not the dimension of the premetric space $\mathbb{S}$ equals n in time polynomial in #A.

Corollary 3.27. *A premetric space* $\mathbb{S} = (A, D)$ *has dimension* $\dim(\mathbb{S}) \leq n$ *if and only if* $D(B) = 0$ *for all* $B \subseteq A$ *with* $\#B \in \{n+2, n+3\}$. *More specifically,* $\dim(\mathbb{S}) = n$ *if and only if for one and hence all free subsets* $F \subseteq A$ *with* $\#F = n+1$, $D(F \cup \{a,b\}) = 0$ *for all* $a,b \in A \setminus F$.

Proof: This follows from Corollary 3.18, Proposition 3.23 and Proposition 3.26. QED

We are now ready to turn our attention to the Euclidean case. We start by deriving a simple necessary and sufficient condition for a premetric space to be Euclidean of dimension n.

PROPOSITION 3.28. *A premetric space $\mathbb{S} = (\mathrm{A}, D)$ is Euclidean of dimension $dim(\mathbb{S}) = n > 0$ if and only if for all $\mathrm{B} \subseteq \mathrm{A}$ with $n+1 \geq \#\mathrm{B} > 1$ we have $D(\mathrm{B}) \geq 0$, while $D(\mathrm{B}) = 0$ for all $\mathrm{B} \subseteq \mathrm{A}$ with $\#\mathrm{B} > n+1$, and n is the minimum positive integer with this property.*

Proof: That $D(\mathrm{B}) = 0$ whenever $\#\mathrm{B} > n+1$ follows from Corollary 3.27. Nevertheless, both this result and the nonnegativity of all $D(\mathrm{B})$ with $\#\mathrm{B} \leq n+1$ can be established as follows. For $\mathrm{B} := \{\mathrm{b}_1, \ldots, \mathrm{b}_k\} \subseteq \mathrm{A}$, let $\underline{\mathbf{U}} := [\mathbf{b}_j - \mathbf{b}_1 \mid j = 2, \ldots, k]$, and recall that a bilinear form $B : \mathbb{V} \times \mathbb{V} \rightarrow \mathbb{R}$ is positive (semi)definite if $\forall \mathbf{u} \in \mathbb{V}$, $B(\mathbf{u};\mathbf{u}) \geq 0$, whereas $B(\mathbf{u};\mathbf{u}) = 0 \Rightarrow \mathbf{u} = \mathbf{0}$ $(B(\mathbf{u};\mathbf{v}) = 0\ \forall \mathbf{v} \in \mathbb{V})$. Since $D(\mathrm{B}) = B(\mathbf{U}) := det(\underline{\mathbf{M}}_B(\underline{\mathbf{U}}))$ by Lemma 3.25 and any subspace of a positive semidefinite bilinear vector space is obviously positive semidefinite, it suffices to show that the discriminant of a positive (semi)definite bilinear vector space is $+1$ (or 0). This is easily seen by reducing the metric matrix $\underline{\mathbf{M}}_B(\underline{\mathbf{U}})$ to canonical form, i.e.

$$\underline{\mathbf{M}}_B(\underline{\mathbf{U}}) = \underline{\mathbf{X}}^T \cdot \underline{\Omega} \cdot \underline{\mathbf{X}}, \tag{3.61}$$

where $\underline{\Omega} = \mathbf{diag}(\epsilon_i \in \{0, \pm 1\} \mid i = 2, \ldots, k)$. Since for $\mathbf{Y} = [\mathbf{y}_1, \ldots, \mathbf{y}_k] := \mathbf{X}^{-1}$ we have $\omega_{ii} = \mathbf{y}_i^T \cdot \underline{\mathbf{M}}_B(\underline{\mathbf{U}}) \cdot \mathbf{y}_i \geq 0$ by positive (semi)definiteness, it follows that $\omega_{ii} \in \{0, +1\}$ for $2 \leq i \leq k$ and hence $B(\mathbf{U}) = det^2(\underline{\mathbf{X}}) \cdot \prod_{i=2}^{k} \omega_{ii} \geq 0$.

The converse follows from Corollary 3.7, Lemma 3.25 and the well-known relation between positive (semi)definite quadratic forms and symmetric matrices (cf. Lemma 3.31 below). QED

Even though the above condition is sufficient for a premetric space to be n-dimensional Euclidean, it is much stronger than is actually needed. The rest of this section and much of the next will be devoted to seeing just how it can be weakened.

Definition 3.29. If $\mathbb{S} = (\mathrm{A}, D)$ is a premetric space such that every subspace $(\mathrm{B}, D|_{\mathrm{B}})$ with $\#\mathrm{B} = n+1$ is Euclidean, then $\mathbb{S}$ is called an *n-Euclidean* space. If in addition every subspace of size $n+2$ is Euclidean of dimension not exceeding n, it is called *strongly n-Euclidean*, and if every subspace of size $n+3$ is Euclidean of dimension not exceeding n, it is called *very strongly n-Euclidean.*

Clearly, all semimetric spaces are 1-Euclidean and vice versa. Likewise, 2-

Euclidean spaces are already familiar to us:

LEMMA 3.30. *A semimetric space* $\mathbb{S} = (A, D)$ *is 2-Euclidean if and only if it is a metric space, i.e. if and only if the function* $d = \sqrt{D}$ *obeys the triangle inequality.*

Proof: By the above proposition and the comments following it, a premetric space is 2-Euclidean if and only if all of its two and three atom Cayley-Menger determinants are nonnegative. As is readily verified, such a determinant $D(\mathrm{a,b,c})$ can be factored as:

$$D(\mathrm{a,b,c}) = \tfrac{1}{4}\cdot\left(d(\mathrm{a,b})+d(\mathrm{a,c})+d(\mathrm{b,c})\right)\cdot\left(d(\mathrm{a,b})+d(\mathrm{a,c})-d(\mathrm{b,c})\right)\cdot \left(d(\mathrm{a,b})-d(\mathrm{a,c})+d(\mathrm{b,c})\right)\cdot\left(-d(\mathrm{a,b})+d(\mathrm{a,c})+d(\mathrm{b,c})\right) \tag{3.62}$$

From this it is immediately clear that for all $d: \{\mathrm{a,b,c}\} \to \{0, \mathbb{R}^+\}$, the triangle inequality implies $D(\mathrm{a,b,c}) \geq 0$. On the other hand, if the triangle inequality is violated then at least one of the above factors must be negative, say $d(\mathrm{a,b})+d(\mathrm{a,c})-d(\mathrm{b,c}) < 0$. Thus if $D(\mathrm{a,b,c}) \geq 0$ then at least one other factor must be nonpositive, w.l.o.g. $d(\mathrm{a,b})-d(\mathrm{a,c})+d(\mathrm{b,c}) \leq 0$. Adding these two inequalities together gives $d(\mathrm{a,b}) < 0$, a contradiction. QED

Thus, n-Euclidean spaces can be regarded as a specialization of metric spaces.

Our next lemma is a standard result from linear algebra.

LEMMA 3.31. *An* $N \times N$ *symmetric matrix* $\underline{\mathbf{M}}$ *of rank* n *is positive semidefinite if and only if for some row/column permutation of* $\underline{\mathbf{M}}$ *the principal minors obtained by taking the first* k *rows and columns are positive for* $k = 1, \ldots, n$ *and zero for* $k = n+1, \ldots, N$.

Proof: Let B be the bilinear form on $\mathbb{R}^N$ whose metric matrix is $\underline{\mathbf{M}}$. For a given row/column permutation $\underline{\mathbf{M}}_N$ of $\underline{\mathbf{M}}$, let $\underline{\mathbf{M}}_k$ be the matrix formed from the first k rows and columns of $\underline{\mathbf{M}}_N$.

Since the images $\{\mathbf{u}_i + \mathrm{rad}(\mathbb{R}^N, B)\}$ of the N unit vectors $\mathbf{u}_i \in \mathbb{R}^N$ span the nondegenerate factor space of $\mathbb{R}^N$, we can find a basis composed of them, and the preimages of this basis span a maximal positive definite subspace of $\mathbb{R}^N$. Thus in any row/column permutation such that these preimages come first, $det(\underline{\mathbf{M}}_k) > 0$ for $1 \leq k \leq n$, and since the rank of $\underline{\mathbf{M}}$ is n, all larger determinants are zero.

To prove sufficiency, we first note that since the hypotheses imply that $\mathbb{U}_n := \langle \mathbf{u}_1, \ldots, \mathbf{u}_n \rangle$ is a maximal nondegenerate subspace, for all $\mathbf{v}, \mathbf{w} \in \mathbb{R}^N$ we have $B(\mathbf{v}; \mathbf{w}) = B(\mathbf{pr}_n(\mathbf{v}); \mathbf{pr}_n(\mathbf{w}))$ where $\mathbf{pr}_n: \mathbb{R}^N \to \mathbb{U}_n$ is the B-orthogonal projection. Thus it suffices to prove that $\underline{\mathbf{M}}_n$ is positive definite, which we do by induction on n. Since $det(\underline{\mathbf{M}}_{n-1}) > 0$ by hypothesis, the bilinear subspace $\mathbb{U}_{n-1}$ is nondegenerate and hence there exists a B-orthogonal projection $\mathbf{pr}_{n-1}: \mathbb{R}^n \to \mathbb{U}_{n-1}$.

Any $\mathbf{v} \in \mathbb{R}^n \subseteq \mathbb{R}^N$ can be written as $\mathbf{v} = \mathbf{v}'+\mathbf{v}''$ with $\mathbf{v}' = \mathbf{pr}_{n-1}(\mathbf{v})$, so

$$B_n(\mathbf{v};\mathbf{v}) \;=\; B_n(\mathbf{v}';\mathbf{v}') + B_n(\mathbf{v}'';\mathbf{v}'') \,, \tag{3.63}$$

where B_n is the bilinear form whose metric matrix w.r.t. the basis $[\mathbf{u}_1, \ldots, \mathbf{u}_n]$ is $\underline{\mathbf{M}}_n$. If $\mathbf{v}'' = \mathbf{0}$, i.e. $\mathbf{v} \in \mathbb{U}_{n-1}$, we have $B_n(\mathbf{v};\mathbf{v}) = B_{n-1}(\mathbf{v};\mathbf{v}) > 0$ unless $\mathbf{v} = \mathbf{v}' = \mathbf{0}$ by the induction hypothesis. Otherwise the metric matrix of the form w.r.t. the basis $[\mathbf{u}_1, \ldots, \mathbf{u}_{n-1}, \mathbf{v}'']$ is

$$\underline{\mathbf{M}}'_n \;:=\; \begin{bmatrix} \underline{\mathbf{M}}_{n-1} & \mathbf{0} \\ \mathbf{0}^T & B(\mathbf{v}'';\mathbf{v}'') \end{bmatrix} . \tag{3.64}$$

Since the discriminant $sign(det(\underline{\mathbf{M}}_n)) = +1$ of the form B_n is independent of our choice of basis, $det(\underline{\mathbf{M}}'_n) > 0$, from which it follows that $B_n(\mathbf{v}'';\mathbf{v}'') > 0$. Since $B_n(\mathbf{v}';\mathbf{v}') \geq 0$ by the induction hypothesis, we get $B_n(\mathbf{v};\mathbf{v}) \geq B_n(\mathbf{v}'';\mathbf{v}'') > 0$ unless $\mathbf{v} = \mathbf{0}$, as desired. QED

THEOREM 3.32. *A very strongly n-Euclidean premetric space $\mathbb{S} = (\mathrm{A}, D)$ is Euclidean of dimension n. More specifically, a premetric space is n-dimensional Euclidean if and only if for some indexing* $\{\mathrm{a}_1, \ldots, \mathrm{a}_N\}$ *of the set of atoms* A *we have:*

$$D(\mathrm{a}_1, \ldots, \mathrm{a}_{k+1}) \;>\; 0 \quad \text{for all} \;\; 1 \leq k \leq n \,; \tag{3.65}$$

$$D(\mathrm{a}_1, \ldots, \mathrm{a}_{n+1}, \mathrm{b}, \mathrm{c}) \;=\; 0 \quad \text{for all} \;\; \mathrm{b}, \mathrm{c} \in \{\mathrm{a}_{n+2}, \ldots, \mathrm{a}_N\} \,; \tag{3.66}$$

Proof: The necessity of these conditions is an immediate consequence of Proposition 3.28. To prove sufficiency, we first note that the given conditions together with Corollary 3.27 imply that the dimension of the space is n. Since by Lemma 3.25 the Cayley-Menger determinant $D(\mathrm{a}_1, \ldots, \mathrm{a}_{k+1})$ is numerically equal to the Gramian of the vectors $\{\mathbf{u}_j := \mathbf{a}_j - \mathbf{a}_{k+1} \mid j = 1, \ldots, k\}$, the rest of the theorem then follows from Lemma 3.31. QED

Note that, in the k-dimensional Euclidean case, $B(\mathbf{u}_i;\mathbf{u}_j) = (\mathbf{x}_i - \mathbf{x}_{k+1}) \cdot (\mathbf{x}_j - \mathbf{x}_{k+1})$, where $\mathbf{x}_i = [x_j(\mathrm{a}_i) \mid j = 1, \ldots, k]$ is a vector of Cartesian coordinates for $\mathbf{a}_i$, $i = 1, \ldots, k+1$. Thus we can write the matrix $\underline{\mathbf{M}}_B = \underline{\mathbf{M}}_B(\mathbf{u}_1, \ldots, \mathbf{u}_k)$ as $\underline{\mathbf{X}}^T \cdot \underline{\mathbf{X}}$ where $\underline{\mathbf{X}} = [x_j(\mathrm{a}_i) - x_j(\mathrm{a}_{k+1})]$. Since $det(\underline{\mathbf{X}})$ is $k!$ times the volume of the simplex $[(\mathbf{a}_1, \ldots, \mathbf{a}_{k+1})]$ spanned by the embedded atoms, it follows that $D(\mathrm{A}) = B(\mathrm{A}) := det^2(\underline{\mathbf{X}}))$ is just $(k!)^2$ times the square of this volume. This fact provides us with a nice, simple geometric interpretation of Theorem 3.32.

The theorem itself is, of course, a pleasing result, for it gives us a straightforward, noniterative test for the consistency of a complete set of Euclidean distances. Unfortunately, however, the distances do not appear symmetrically in Equations (3.65) and (3.66), and this can under some circumstances make them more complicated to apply than a brute force check of all Cayley-Menger determinants of $n+3$ atoms or less. Therefore, we shall now consider other ways in which this check can be simplified.

3.4. Pseudo-Euclidean Spaces

Corollary 3.16 shows that the essential projective properties of an n-dimensional Euclidean space are already assured even in a premetric space which is only strongly n-Euclidean. Thus it is a little surprising to find that there exist strongly n-Euclidean spaces which are nevertheless not n-dimensional Euclidean.

Definition 3.33. A strongly n-Euclidean premetric space which is not n-dimensional Euclidean is called an *n-pseudo-Euclidean* space.

In this section we shall prove a characterization of these strange spaces which allows us to strengthen the criteria of §3.3 for isometric embedding in $\mathbb{R}^n$. Our approach uses §44-46 of [Blumenthal1953] as an outline, although the theory of metroids introduced in §3.2 has allowed us to simplify some of the proofs. At the same time, however, some rather specific properties of Euclidean spaces will be needed along the way. The general case is summarized in the following proposition, which is given without proof.

PROPOSITION 3.34. *Let* $\{\mathbf{r}_1, \ldots, \mathbf{r}_{n+3}\}$ *be a set of points in* $\mathbb{R}^n$ *and* $D(\mathbf{r}_i, \mathbf{r}_j) := \|\mathbf{r}_i - \mathbf{r}_j\|^2$*. If for some* $1 \leq m \leq n$ *we have*

$$D(\mathbf{r}_i, \mathbf{r}_j) = D(\mathbf{r}_i, \mathbf{r}_k) \quad \text{for } 1 \leq i \leq n-m+2, \; n-m+3 \leq j < k \leq n+3 \tag{3.67}$$

$$\text{and} \quad D(\mathbf{r}_1, \ldots, \mathbf{r}_{n-m+2}) \neq 0, \quad \text{then } D(\mathbf{r}_{n-m+3}, \ldots, \mathbf{r}_{n+3}) = 0,$$

where $D(\mathbf{r}_i, \ldots, \mathbf{r}_j)$ denotes the Cayley-Menger determinant of the metric space $(\{\mathbf{r}_i, \ldots, \mathbf{r}_j\}, D)$.

Note that, since we are dealing with the Euclidean metric, the Cayley-Menger determinant vanishes if and *only* if the points are affinely dependent. The case $m = 1$ thus states that there exists at most one point with given distances to $n+1$ affinely independent points in $\mathbb{R}^n$, while the case $m = n$ states that there exists at most one point equidistant from any $n+1$ independent points, or equivalently, that any $n+1$ points equidistant from any two distinct points of $\mathbb{R}^n$ are affinely dependent. In fact there always exists exactly one point equidistant from $n+1$ independent points in $\mathbb{R}^n$, which is the center of their circumscribing sphere. Likewise, the case $m = 2$ follows from the (stronger) fact that if there exists a point with given distances from any other n points which span a hyperplane of $\mathbb{R}^n$, then there exist exactly two points with this property which are related by a reflection with respect to the hyperplane.

As an illustration of the case $m = 1$ in Proposition 3.34, we consider an example with $n = 2$. Then if $D(\mathbf{r}_1,\mathbf{r}_4) = D(\mathbf{r}_1,\mathbf{r}_5)$, $D(\mathbf{r}_2,\mathbf{r}_4) = D(\mathbf{r}_2,\mathbf{r}_5)$, and $D(\mathbf{r}_3,\mathbf{r}_4) = D(\mathbf{r}_3,\mathbf{r}_5)$, then having $D(\mathbf{r}_1,\mathbf{r}_2,\mathbf{r}_3) \neq 0$ implies $D(\mathbf{r}_4,\mathbf{r}_5) = 0$, i.e. $\mathbf{r}_4 = \mathbf{r}_5$, as shown on the left-hand-side of the figure below.

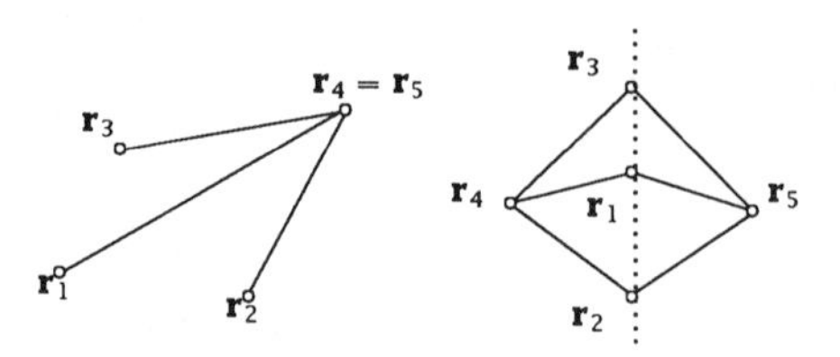

Figure 3.4

On the other hand, if $\mathbf{r}_4 \neq \mathbf{r}_5$ then $D(\mathbf{r}_1,\mathbf{r}_2,\mathbf{r}_3) = 0$, in accord with the elementary geometric fact that the set of all points equidistance from any two distinct points is a line. This is illustrated on the right-hand side of the figure.

Let us first verify that n-pseudo-Euclidean spaces do in fact exist for all $n > 0$. Suppose that $\mathbf{r}_1, \ldots, \mathbf{r}_{n+1}$ are the vertices of an equilateral simplex $S = [(\mathbf{r}_1, \ldots, \mathbf{r}_{n+1})]$ in $\mathbb{R}^n$ of side length $\sqrt{2}\cdot(n+1)$, and let $\mathbf{r}_{n+2}$ be the barycenter of this simplex, so that its distance from each of the other points $\mathbf{r}_1, \ldots, \mathbf{r}_{n+1}$ is $\|\mathbf{r}_{n+2}-\mathbf{r}_i\| = \sqrt{n^2+n}$. For $i = 1, \ldots, n+1$ we let $\mathbf{r}_i'$ be the point obtained by reflecting $\mathbf{r}_{n+2}$ in (the affine span of) the face of the simplex whose vertices are $\{\mathbf{r}_1, \ldots, \mathbf{r}_{i-1}, \mathbf{r}_{i+1}, \ldots, \mathbf{r}_{n+1}\}$, so that $\|\mathbf{r}_{n+2}-\mathbf{r}_i'\| = 2\sqrt{(n+1)/n}$. We then define a premetric space $\mathbb{S} := (\mathrm{A}, D)$ with $\mathrm{A} := \{\mathrm{a}_1, \ldots, \mathrm{a}_{n+3}\}$ according to the rule:

$$D(\mathrm{a}_i,\mathrm{a}_j) = \begin{cases} 2\cdot(n+1)^2 & \text{if } 1 \leq i < j \leq n+1\,; \\ n^2+n & \text{if } 1 \leq i \leq n+1,\ j = n+2, n+3\,; \\ 4(n+1)/n & \text{if } i = n+2,\ j = n+3\,. \end{cases} \tag{3.68}$$

Next, for $i = 1, \ldots, n+1$ we define maps $\mathbf{p}_i\colon \mathrm{A}\setminus\mathrm{a}_i \to \mathbb{R}^n$ by

$$\mathbf{p}_i(\mathrm{a}_j) := \begin{cases} \mathbf{r}_j & \text{for } j = 1, \ldots, i-1, i+1, \ldots, n+2\,; \\ \mathbf{r}_i' & \text{for } j = n+3\,; \end{cases} \tag{3.69}$$

while for $i = n+2, n+3$

$$\mathbf{p}_i(\mathrm{a}_j) := \begin{cases} \mathbf{r}_j & \text{for } j = 1, \ldots, n+1\,; \\ \mathbf{r}_{n+2} & \text{for } i = n+2,\ j = n+3 \text{ or } i = n+3,\ j = n+2\,. \end{cases} \tag{3.70}$$

Alternatively in Proposition 3.34 still with $n=2$, consider the situation for $m=2$ and $D(\mathbf{r}_1,\mathbf{r}_3) = D(\mathbf{r}_1,\mathbf{r}_4) = D(\mathbf{r}_1,\mathbf{r}_5)$ and $D(\mathbf{r}_2,\mathbf{r}_3) = D(\mathbf{r}_2,\mathbf{r}_4) = D(\mathbf{r}_2,\mathbf{r}_5)$. Then if $D(\mathbf{r}_1,\mathbf{r}_2) \neq 0$ we must have $D(\mathbf{r}_3,\mathbf{r}_4,\mathbf{r}_5) = 0$.

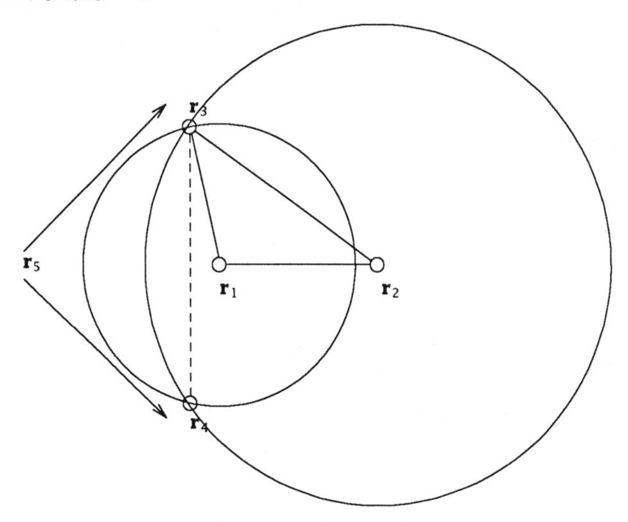

Figure 3.5

Geometrically, this follows from the fact that either $\mathbf{r}_4$ or $\mathbf{r}_5$ or both must be related to $\mathbf{r}_3$ by reflection in the line $[(\mathbf{r}_1,\mathbf{r}_2)]$, so that these three points cannot be distinct.

Then the $\mathbf{p}_i$ are the required isometric embeddings of every $n+2$ atom subspace of $\mathbb{S}$ in $\mathbb{R}^n$. Since by Proposition 3.34 with $m = 1$ there exists at most one point with given distances to any $n+1$ affinely independent points in $\mathbb{R}^n$, however, in any extension $\mathbf{p}: \mathrm{A} \rightarrow \mathbb{R}^n$ of the embeddings $\mathbf{p}_{n+2}$ or $\mathbf{p}_{n+3}$ to all of A both a_{n+2} and a_{n+3} must map to $\mathbf{r}_{n+2}$, i.e. $\|\mathbf{p}(\mathrm{a}_{n+2}) - \mathbf{p}(\mathrm{a}_{n+3})\|^2 = 0$ and not $D(\mathrm{a}_{n+2},\mathrm{a}_{n+3}) = 4n/(n-1)$.

Another surprising fact is that this one example of an n-pseudo-Euclidean space comes rather close to being the only one. Characterizing these spaces, however, will take some doing. Recall that a set of atoms $\mathrm{B} \subseteq \mathrm{A}$ in a premetric space $\mathbb{S} = (\mathrm{A}, D)$ is *free* if $dim(\mathrm{B}, D|_{\mathrm{B}}) = \#\mathrm{B} - 1$.

LEMMA 3.35. *Any n-pseudo-Euclidean space $\mathbb{S} = (\mathrm{A}, D)$ contains at least one free set of $n+1$ atoms $\{\mathrm{b}_1, \ldots, \mathrm{b}_{n+1}\} \subseteq \mathrm{A}$.*

Proof: If all $n+2$ atom sets could be isometrically embedded in $\mathbb{R}^{n-1}$ the space would be very strongly $(n-1)$-Euclidean by definition, and then Theorem 3.32 would imply that it was an $(n-1)$-dimensional Euclidean space contrary to assumption. Hence there exist $n+2$ atoms $b_1, \ldots, b_{n+2} \in A$ which can be isometrically embedded in $\mathbb{R}^n$ but not in $\mathbb{R}^{n-1}$. Since a set of points in $\mathbb{R}^n$ is free if and only if it is affinely independent, it follows that some $n+1$ element subset of $\{b_1, \ldots, b_{n+2}\}$ is free. QED

As defined, n-pseudo-Euclidean spaces are only assumed to be nonembeddable in an n-dimensional Euclidean space; however, a stronger result holds:

PROPOSITION 3.36. *An n-pseudo-Euclidean space $\mathbb{S} = (A, D)$ is not realizable in a Euclidean space of any number of dimensions whatsoever.*

Proof: Since the Gramian of a set of vectors versus a positive semidefinite quadratic form vanishes if and only if the vectors are linearly dependent in the corresponding nondegenerate factor space, the metroids associated with a Euclidean space are necessarily (representable) matroids. In particular, if $\mathbb{S}$ were Euclidean the affine bilinear rank function $\rho_D: 2^A \rightarrow \mathbb{Z}$ would satisfy the matroidal rank axiom:

$$B, C \subseteq A \text{ and } \rho_D(B \cup c) = \rho_D(B) \quad \forall\, c \in C \;\Rightarrow\; \rho_D(B \cup C) = \rho_D(B)\,. \tag{3.71}$$

By the definition of an n-pseudo-Euclidean space, for any $B \subseteq A$ with $\#B = n+1 = \rho_D(B)$ we have $\rho_D(B \cup a) = \rho_D(B)$ for all $a \in A$, so that $\rho_D(A) = n+1$. But then $\mathbb{S}$ is Euclidean of dimension n, a contradiction. QED

We begin our detailed study of n-pseudo-Euclidean spaces $\mathbb{S} = (A, D)$ by restricting ourselves to those with $\#A = n+3$.

LEMMA 3.37. *For every n-pseudo-Euclidean space (A, D) defined on $n+3$ atoms, there exists an ordering $\underline{A} = [a_1, \ldots, a_{n+3}]$ and a map $\mathbf{q}: A \rightarrow \mathbb{R}^n$ such that*

$$D(a_i, a_j) = \|\mathbf{q}(a_i) - \mathbf{q}(a_j)\| \quad \text{for } \{i, j\} \neq \{n+2, n+3\}\,, \tag{3.72}$$

but

$$D(a_{n+2}, a_{n+3}) \neq \|\mathbf{q}(a_{n+2}) - \mathbf{q}(a_{n+3})\|\,. \tag{3.73}$$

Proof: By Lemma 3.35 there exists a free set B of $n+1$ atoms, and we index the atoms so that $B = \{a_1, \ldots, a_{n+1}\}$. By the definition of strongly n-Euclidean there exists isometric embeddings $\mathbf{p}_{n+2}: \{a_1, \ldots, a_{n+1}, a_{n+3}\} \rightarrow \mathbb{R}^n$ and $\mathbf{p}_{n+3}: \{a_1, \ldots, a_{n+1}, a_{n+2}\} \rightarrow \mathbb{R}^n$ such that

$$\|\mathbf{p}_{n+2}(a_i) - \mathbf{p}_{n+2}(a_j)\| = D(a_i, a_j) \quad \text{for all } i, j = 1, \ldots, n+1, n+3\,; \tag{3.74}$$

and

$$\|\mathbf{p}_{n+3}(a_i) - \mathbf{p}_{n+3}(a_j)\| = D(a_i, a_j) \quad \text{for all } i, j = 1, \ldots, n+1, n+2\,. \tag{3.75}$$

Since $\mathbf{B} := \{\mathbf{p}_{n+2}(a_i) \mid i = 1, \ldots, n+1\}$ and $\mathbf{B}' := \{\mathbf{p}_{n+3}(a_i) \mid i = 1, \ldots, n+1\}$ are both affinely independent, there exists a unique isometry $\Psi : \mathbf{B} \to \mathbf{B}'$ between them. Thus if we define $\mathbf{q}$ by:

$$\mathbf{q}(a_i) \;:=\; \begin{cases} \mathbf{p}_{n+3}(a_i) & \text{for } i = 1, \ldots, n+2 \\ \Psi(\mathbf{p}_{n+2}(a_{n+3})) & \text{for } i = n+3\,, \end{cases} \tag{3.76}$$

we see immediately that $D(a_i, a_j) = \|\mathbf{q}(a_i) - \mathbf{q}(a_j)\|$ for all $\{i, j\} \neq \{n+2, n+3\}$, whereas $D(a_{n+2}, a_{n+3}) \neq \|\mathbf{q}(a_{n+2}) - \mathbf{q}(a_{n+3})\|$ since otherwise (A, D) is n-dimensional Euclidean contrary to assumption. QED

We now provide a geometric characterization of this map.

Definition 3.38. Let $\{\mathbf{r}_1, \ldots, \mathbf{r}_{n+1}\}$ be the vertices of a nondegenerate simplex $S = [(\mathbf{r}_1, \ldots, \mathbf{r}_{n+1})]$ in $\mathbb{R}^n$. For any $\mathbf{s} \in \mathbb{R}^n$, let $\mathbf{r}_i'$ be the point obtained by reflecting $\mathbf{s}$ in the face of S opposite to the vertex $\mathbf{r}_i$. Then let $\mathbf{t}$ be the unique point of $\mathbb{R}^n$ which is equidistant from each of the points $\mathbf{r}_i'$, $i = 1, \ldots, n+1$, i.e. we let $\mathbf{t}$ be the center of the circumscribing sphere S of these points. Then the points $\mathbf{s}$ and $\mathbf{t}$ are called *isogonal conjugates* with respect to the simplex S.

This relation is a reciprocal one, so that if $\mathbf{s}$ is the isogonal conjugate of $\mathbf{t}$ with respect to S, then $\mathbf{t}$ is likewise the isogonal conjugate of $\mathbf{s}$ with respect to S.

PROPOSITION 3.39. *Let $\mathbb{S} = (A, D)$ be an n-pseudo-Euclidean space on $n+3$ atoms, assume that A has been indexed such that $B := \{a_1, \ldots, a_{n+1}\}$ is a free $(n+1)$-tuple therein, and let $\mathbf{q} : A \to \mathbb{R}^n$ be the map described in Lemma 3.37. Then $\mathbf{q}(a_{n+2})$ and $\mathbf{q}(a_{n+3})$ are isogonal conjugates with respect to the simplex $S = [(\mathbf{q}(a_1), \ldots, \mathbf{q}(a_{n+1}))]$, neither $\mathbf{q}(a_{n+2})$ nor $\mathbf{q}(a_{n+3})$ lie in any face of S, and $D(a_{n+2}, a_{n+3})$ is the squared radius of the sphere circumscribing the $n+1$ points obtained by reflecting either $\mathbf{q}(a_{n+2})$ or $\mathbf{q}(a_{n+3})$ in the faces of S.*

Proof: For $i = 1, \ldots, n+3$, let $\mathbf{p}_i : A \setminus a_i \to \mathbb{R}^n$ denote the isometric embedding of $A \setminus a_i$ in $\mathbb{R}^n$ which exists by the definition of a n-pseudo-Euclidean space. For each $\mathbf{p}_i$ with $1 \leq i \leq n+1$, as above we may w.l.o.g. assume that $\mathbf{p}_i(a_j) = \mathbf{q}(a_j)$ for all $j \in \{1, \ldots, i-1, i+1, \ldots, n+2\}$. Then necessarily $\mathbf{p}_i(a_{n+3}) \neq \mathbf{q}(a_{n+3})$, since

$$\begin{aligned} D(a_{n+2}, a_{n+3}) \;&=\; \|\mathbf{p}_i(a_{n+2}) - \mathbf{p}_i(a_{n+3})\|^2 \;=\; \|\mathbf{q}(a_{n+2}) - \mathbf{p}_i(a_{n+3})\|^2 \\ &\neq\; \|\mathbf{q}(a_{n+2}) - \mathbf{q}(a_{n+3})\|^2 \end{aligned} \tag{3.77}$$

for $1 \leq i \leq n+1$ by Lemma 3.37. Because

$$\|\mathbf{p}_i(a_{n+3}) - \mathbf{p}_i(a_j)\|^2 \;=\; D(a_{n+3}, a_j) \;=\; \|\mathbf{q}(a_{n+3}) - \mathbf{p}_i(a_j)\|^2 \tag{3.78}$$

for all $j \in \{1, \ldots, i-1, i+1, \ldots, n+1\}$, it follows from Proposition 3.34 with $m = 2$ (and the discussion following it) that $\mathbf{q}(a_{n+3})$ and $\mathbf{p}_i(a_{n+3})$ must be related to each other by reflection with respect to the face of S opposite to $\mathbf{q}(a_i)$, and that $\mathbf{q}(a_{n+3})$ cannot be contained in that face. Furthermore

$$D(\mathrm{a}_{n+3},\mathrm{a}_{n+2}) \;=\; \|\mathbf{p}_i(\mathrm{a}_{n+3})-\mathbf{p}_i(\mathrm{a}_{n+2})\|^2 \;=\; \|\mathbf{p}_i(\mathrm{a}_{n+3})-\mathbf{q}(\mathrm{a}_{n+2})\|^2 \tag{3.79}$$

for all $i = 1, \ldots, n+1$ implies that $\mathbf{q}(\mathrm{a}_{n+2})$ is equidistant to the points $\mathbf{p}_i(\mathrm{a}_{n+3})$ which one gets by reflecting $\mathbf{q}(\mathrm{a}_{n+3})$ with respect to the face opposite to $\mathbf{q}(\mathrm{a}_i)$. So $\mathbf{q}(\mathrm{a}_{n+2})$ and $\mathbf{q}(\mathrm{a}_{n+3})$ are indeed isogonal conjugates with respect to S. QED

Figure 3.6 has been provided for the 2-dimensional case to help make the above argument clear.

The figure below illustrates the proof of Proposition 3.39 for the case $n = 2$.

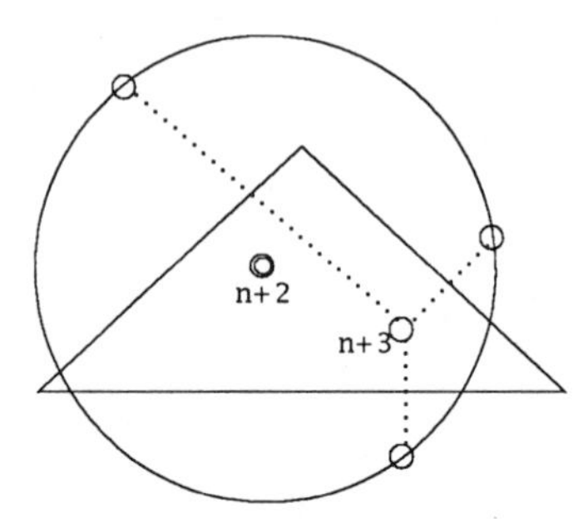

Figure 3.6

The points labelled $n+2$ and $n+3$ are isogonal conjugates of each other with respect to the vertices of the triangle.

COROLLARY 3.40. *All $(n+1)$-tuples of an n-pseudo-Euclidean space* $\mathbb{S} = (\mathrm{A}, D)$ *on $n+3$ atoms are free. In fact, for any two atoms* $\mathrm{a,b} \in \mathrm{A}$ *of an n-pseudo-Euclidean space on $n+3$ atoms there exists a map* $\mathbf{q}: \mathrm{A} \to \mathbb{R}^n$ *such that* $D(\mathrm{x,y}) = \|\mathbf{q}(\mathrm{x})-\mathbf{q}(\mathrm{y})\|^2$ *for all* $\{\mathrm{x,y}\} \neq \{\mathrm{a,b}\}$, *and the two points* $\mathbf{q}(\mathrm{a})$ *and* $\mathbf{q}(\mathrm{b})$ *are isogonal conjugates with respect to the simplex whose vertices are* $\{\mathbf{q}(\mathrm{x}) \mid \mathrm{x} \in \mathrm{A} \setminus \{\mathrm{a,b}\}\}$.

Proof: The argument of Proposition 3.39 shows that all $2(n+1)$ of the $(n+1)$-tuples containing any n atoms from the free $(n+1)$-tuple $\{\mathrm{a}_1, \ldots, \mathrm{a}_{n+1}\}$ are free. Repeating this argument with each of these free $(n+1)$-tuples proves the existence of the map $\mathbf{q}$ for each of them, and shows that the remaining $\frac{1}{2}n(n+1)$ of the $\frac{1}{2}(n+2)(n+3)$ total $(n+1)$-tuples are free as well. Repeating the argument yet again for each of these establishes the Corollary. QED

We now use these facts to establish a yet more surprising result, which shows that there is essentially only one kind of n-pseudo-Euclidean space on

$n+3$ points.

PROPOSITION 3.41. *Let* $\mathbb{S} = (\mathrm{A}, D)$ *be a strongly n-Euclidean space on* $n+2$ *atoms and let* $\mathrm{b} \notin \mathrm{A}$. *Then there exists at most one extension of* $\mathbb{S}$ *to an n-pseudo-Euclidean space on* $\mathrm{A} \cup \mathrm{b}$.

Proof: Let $\mathrm{A} = \{\mathrm{a}_1, \ldots, \mathrm{a}_{n+2}\}$. Suppose we have two extensions $(\mathrm{A}\cup\mathrm{b}, D')$ and $(\mathrm{A}\cup\mathrm{b}, D'')$. Then since $D'|_\mathrm{A} = D = D''|_\mathrm{A}$, the maps $\mathbf{q}', \mathbf{q}'' : \mathrm{A}\cup\mathrm{b} \rightarrow \mathbb{R}^n$ which exist by Lemma 3.37 satisfy

$$D'(\mathrm{a}_i, \mathrm{b}) = \|\mathbf{q}'(\mathrm{a}_i) - \mathbf{q}'(\mathrm{b})\| = \|\mathbf{q}''(\mathrm{a}_i) - \mathbf{q}''(\mathrm{b})\| = D''(\mathrm{a}_i, \mathrm{b}) \tag{3.80}$$

for $i = 1, \ldots, n+1$. Also by Proposition 3.39 $D'(\mathrm{a}_{n+2}, \mathrm{b})$ and $D''(\mathrm{a}_{n+2}, \mathrm{b})$ are both equal to the radius of the sphere circumscribing the points obtained by reflecting the $\mathbf{q}(\mathrm{a}_{n+2})$ with respect to the faces of the simplex spanned by $\{\mathbf{q}(\mathrm{a}_1), \ldots, \mathbf{q}(\mathrm{a}_{n+1})\}$. Thus $D' = D''$. QED

We now consider n-pseudo-Euclidean spaces on $n+4$ atoms.

LEMMA 3.42. *If* $\mathbb{S} = (\mathrm{A}, D)$ *is an n-pseudo-Euclidean space on* $\#\mathrm{A} = n+4$ *atoms, then every* $n+3$ *atom subspace of* $\mathbb{S}$ *is also an n-pseudo-Euclidean space.*

Proof: If $\mathbb{S}$ contained no $n+3$ atom n-pseudo-Euclidean subspaces, then the space would be very strongly n-Euclidean and so Euclidean, a contradiction. Without loss of generality, we may thus suppose that $\{\mathrm{a}_1, \ldots, \mathrm{a}_{n+1}, \mathrm{a}_{n+2}, \mathrm{a}_{n+3}\}$ induces an n-pseudo-Euclidean subspace. If A itself is free in the associated affine bilinear metroid $(\mathrm{A}, \mathbb{F}_D)$, then since no $n+2$ element subset is free it follows from axiom (F2) (Equation (3.35)) that *all* $n+3$ element subsets are free, meaning that they are all n-pseudo-Euclidean, as desired. Otherwise, since metroids are balanced and single element subsets of affinely representable metroids are free, we may augment the remaining $\{\mathrm{a}_{n+4}\}$ to obtain a second n-pseudo-Euclidean subspace, w.l.o.g. $\{\mathrm{a}_1, \ldots, \mathrm{a}_{n+1}, \mathrm{a}_{n+2}, \mathrm{a}_{n+4}\}$. If $\{\mathrm{a}_{n+3}, \mathrm{a}_{n+4}\}$ were not free, i.e. $D(\mathrm{a}_{n+3}, \mathrm{a}_{n+4}) = 0$, no isometric embedding of $\{\mathrm{a}_{n+3}, \mathrm{a}_{n+4}\}$ in $\mathbb{R}^n$ would exist in contradiction to our definition of n-pseudo-Euclidean. Thus we may also augment $\{\mathrm{a}_{n+3}, \mathrm{a}_{n+4}\}$ to an $n+3$ element free subset, w.l.o.g. $\{\mathrm{a}_1, \ldots, \mathrm{a}_{n+1}, \mathrm{a}_{n+3}, \mathrm{a}_{n+4}\}$.

Since all three pairs of these three $n+3$ atom n-pseudo-Euclidean spaces share $n+2$ atoms, by Proposition 3.41 they are pairwise isometric, i.e. we necessarily have:

$$\begin{aligned} D(\mathrm{a}_i, \mathrm{a}_{n+2}) &= D(\mathrm{a}_i, \mathrm{a}_{n+3}) = D(\mathrm{a}_i, \mathrm{a}_{n+4}) \quad \text{for } i = 1, \ldots, n+1 \text{ and} \\ D(\mathrm{a}_{n+2}, \mathrm{a}_{n+3}) &= D(\mathrm{a}_{n+2}, \mathrm{a}_{n+4}) = D(\mathrm{a}_{n+3}, \mathrm{a}_{n+4}) \ . \end{aligned} \tag{3.81}$$

If any other $n+3$ atom subset of A admits an isometric embedding in $\mathbb{R}^n$, say $\{\mathrm{a}_1, \ldots, \mathrm{a}_{i-1}, \mathrm{a}_{i+1}, \ldots, \mathrm{a}_{n+2}, \mathrm{a}_{n+3}, \mathrm{a}_{n+4}\}$, then the n embedded points $\{\mathbf{p}(\mathrm{a}_1), \ldots, \mathbf{p}(\mathrm{a}_{i-1}), \mathbf{p}(\mathrm{a}_{i+1}), \ldots, \mathbf{p}(\mathrm{a}_{n+1})\}$ are equidistant from the vertices $\{\mathbf{p}(\mathrm{a}_{n+2}), \mathbf{p}(\mathrm{a}_{n+3}), \mathbf{p}(\mathrm{a}_{n+4})\}$ of an equilateral triangle, and hence are dependent by

Proposition 3.34 with $m = 2$. At the same time, however, $\{a_1, \ldots, a_{i-1}, a_{i+1}, \ldots, a_{n+1}\}$ is a subset of the $n+3$ atom n-pseudo-Euclidean space induced by $\{a_1, \ldots, a_{n+3}\}$, and so it is free by Corollary 3.40. This contradiction yields the Lemma. QED

However, the most amazing thing of all is the following theorem due to Karl Menger:[†]

THEOREM 3.43. *If* $\mathbb{S} = (A, D)$ *is an* n*-pseudo-Euclidean space, then* $\#A = n+3$.

Proof: As in the proof of Lemma 3.42 the space $\mathbb{S}$ must contain an $n+3$ atom n-pseudo-Euclidean space, w.l.o.g. $(\{a_1, \ldots, a_{n+3}\}, D)$, and from that Lemma it follows that all $n+3$ atom subsets of any $n+4$ atom subset $\{a_1, \ldots, a_{n+4}\}$ containing it will likewise induce an n-pseudo-Euclidean space. We will obtain a contradiction by showing that $D(a_i, a_j)$ is the same for all $1 \leq i < j \leq n+2$ so that $B := \{a_1, \ldots, a_{n+2}\}$ is isometric with an $(n+1)$-dimensional equilateral simplex which is not embeddable in $\mathbb{R}^n$.

Let $i, j, k, l \in \{1, \ldots, n+2\}$ be distinct indices. Since $B \setminus \{a_i\} \cup \{a_{n+3}, a_{n+4}\}$ and $B \setminus \{a_k\} \cup \{a_{n+3}, a_{n+4}\}$ are n-pseudo-Euclidean and share $n+2$ elements, they are isometric by Proposition 3.41 and hence $D(a_i, a_j) = D(a_k, a_j)$. From the n-pseudo-Euclidean spaces induced by $B \setminus \{a_j\} \cup \{a_{n+3}, a_{n+4}\}$ and $B \setminus \{a_l\} \cup \{a_{n+3}, a_{n+4}\}$ we likewise conclude that $D(a_l, a_k) = D(a_j, a_k)$, so that $D(a_l, a_k) = D(a_i, a_j)$ for all $i, j, k, l \in \{1, \ldots, n+2\}$ with $i \neq j$ and $k \neq l$, as desired. QED

It is not known at this time to what extent Menger's theorem holds up in arbitrary bilinear affine spaces. At the very least, some restriction on the characteristic of the field will be necessary, since if $\mathbb{S} = (A, D)$ is a premetric space with $D(a, b) := 2$ for all $a, b \in A$ $(a \neq b)$, then $D(B) = \#B$ for all $B \subseteq A$. Thus if the characteristic of the field is p with $2 < p < \#A$, this space will always be $(p-2)$-pseudo-Euclidean. It should also be observed that Menger's theorem is a characteristically affine result and does not hold for bilinear vector spaces; for example

$$\begin{bmatrix} 1 & -1 & -1 & \cdots & -1 \\ -1 & 1 & -1 & \cdots & -1 \\ -1 & -1 & 1 & \cdots & -1 \\ \cdot & \cdot & \cdot & \cdots & \cdot \\ -1 & -1 & -1 & \cdots & 1 \end{bmatrix} \tag{3.82}$$

is the metric matrix of a bilinear form for which the one dimensional subspace spanned by each single unit vector is positive definite, while the subspace spanned by any two of the unit vectors is degenerate and all larger principal minors are negative.

[†] This is sometimes called *Menger's theorem*, though it should not be confused with the result in graph theory which goes by the same name (and was also proved by Karl Menger).

Menger's theorem has the following important corollary:

COROLLARY 3.44. *A strongly n-Euclidean premetric space* (A, D) *with* $\#\mathrm{A} > n+3$ *is Euclidean of dimension not exceeding n.*

Proof: Immediate from the foregoing material. QED

This theorem can be sharpened slightly, but is sufficient for our purposes. The interested reader is referred to [Blumenthal1953] for details.

3.5. Syzygies between Distances

Algebraic identities between the invariants of groups of transformations are known as *syzygies*. Since, according to Klein's *Erlanger Program*, every geometry is characterized by the group of transformations under which its theorems remain true, syzygies may be thought of as algebraic statements of the theorems of the corresponding geometry. As an example we cite the identities in the oriented volumes which are known as the Graßmann-Plücker relations (§2.2), from which all of the theorems of projective geometry can be derived [Hodge1968]. In this section we shall present some syzygies between Euclidean invariants, e.g. distances, and discuss their geometric interpretations. Unlike the affine and projective syzygies, however, any complete set of Euclidean syzygies must include *in*equalities as well as equations. All of the theorems of Euclidean geometry can in principle be derived from these relations.

In a three-atom semimetric space (A, D), the inequality $D(\mathrm{A}) \geq 0$ is equivalent to

$$4\,D(a_1,a_2)\,D(a_1,a_3) \;\geq\; \Big(D(a_2,a_3)-D(a_1,a_2)-D(a_1,a_3)\Big)^2 , \tag{3.83}$$

which in turn is equivalent to

$$\begin{aligned} d^2(a_1,a_2)+d^2(a_1,a_3)-2\,d(a_1,a_2)\,d(a_1,a_3) \;&\leq\; d^2(a_2,a_3) \\ &\leq\; d^2(a_1,a_2)+d^2(a_1,a_3)+2\,d(a_1,a_2)\,d(a_1,a_3) , \end{aligned} \tag{3.84}$$

where $d := \sqrt{D}$ as usual. Since we have already seen (Lemma 3.30) that $D(\mathrm{A}) \geq 0$ is equivalent to all three triangle inequalities, this means that these inequalities can be succinctly expressed as

$$\begin{aligned} L(a_2,a_3;\,a_1) \;:=\; (d(a_1,a_3)-d(a_1,a_2))^2 \;&\leq\; d^2(a_2,a_3) \\ &\leq\; (d(a_1,a_3)+d(a_1,a_2))^2 \;=:\; U(a_2,a_3;\,a_1) . \end{aligned} \tag{3.85}$$

This shows that in any 2-Euclidean premetric space, each value of the premetric $D(a_i,a_j)$ lies between specific lower and upper limits $L(a_i,a_j;\,a_k)$ and $U(a_i,a_j;\,a_k)$

In other words, Eq. 3.85 expresses the extreme values of $d(a_2,a_3)$ as a_2 moves around the circle centered on a_1 to which it is confined by the fixed value of $d(a_1,a_2)$.

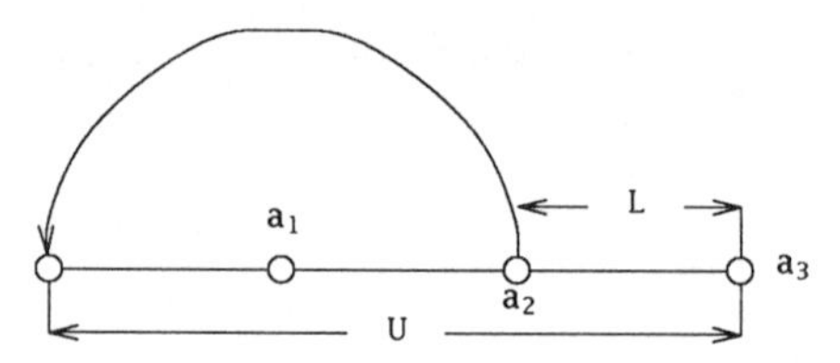

Figure 3.7

which are determined by the premetric among each triple of atoms $a_i,a_j,a_k \in A$. These lower and upper limits are called *triangle inequality limits*, and provide us with our first explicit example of a dependence or syzygy between the distances in a general Euclidean space.

We now consider the Cayley-Menger determinant $D(a_1,a_2,a_3,a_4)$ of a four-atom semimetric space $\mathbb{S} = (A, D)$. By performing a Laplace expansion of this determinant along the last two rows, we obtain

$$D_{34}(a_1, \ldots, a_4;\, X) \;:=\; D(a_1, \ldots, a_4)\Big|_{D(a_3,a_4)\,=\,X} \;=\; A_{34}\, X^2 + B_{34}\, X + C_{34} \tag{3.86}$$

where the coefficients may be found by setting $X := 0$ in $D_{34}(a_1, \ldots, a_4; X)$ and in its first and second derivatives, i.e.

$$A_{34} \;:=\; -\frac{1}{8}\, det\begin{bmatrix} 0 & 1 & 1 \\ 1 & 0 & D(a_1,a_2) \\ 1 & D(a_1,a_2) & 0 \end{bmatrix}; \tag{3.87}$$

$$B_{34} \;:=\; -\frac{1}{4}\, det\begin{bmatrix} 0 & 1 & 1 & 1 \\ 1 & 0 & D(a_1,a_2) & D(a_1,a_4) \\ 1 & D(a_1,a_2) & 0 & D(a_2,a_4) \\ 1 & D(a_1,a_3) & D(a_2,a_3) & 0 \end{bmatrix}; \tag{3.88}$$

$$C_{34} \;:=\; \frac{1}{8}\,det\begin{bmatrix} 0 & 1 & 1 & 1 & 1 \\ 1 & 0 & D(a_1,a_2) & D(a_1,a_3) & D(a_1,a_4) \\ 1 & D(a_1,a_2) & 0 & D(a_2,a_3) & D(a_2,a_4) \\ 1 & D(a_1,a_3) & D(a_2,a_3) & 0 & 0 \\ 1 & D(a_1,a_4) & D(a_2,a_4) & 0 & 0 \end{bmatrix}. \tag{3.89}$$

Since $A_{34} = -1/4\,D(a_1,a_2) < 0$, this quadratic expression can become positive for some $X \in \mathbb{R}$ if and only if

$$B_{34}^2 \;-\; 4A_{34}C_{34} \;=\; D(a_1,a_2,a_3)\cdot D(a_1,a_2,a_4) \;>\; 0\,, \tag{3.90}$$

i.e. if and only if the triangle inequalities among the sets of points $\{a_1,a_2,a_3\}$ and $\{a_1,a_2,a_4\}$ are both strictly obeyed or violated. Expanding $D(a_1, \ldots, a_4)$ as a function of the remaining $D(a_i,a_j)$ shows that in fact $D(a_1, \ldots, a_4) > 0$ implies that the $D(a_i,a_j,a_k)$ are either all positive or all negative for all distinct triples of indices $i,j,k \in \{1,2,3,4\}$. In the general case, we have

LEMMA 3.45. *Let* $(A = \{a_1, \ldots, a_{n+2}\}, D)$ *be an* $(n-1)$*-Euclidean space on* $n+2$ *atoms. If* $D(a_1, \ldots, a_{n+2}) > 0$, *then either* $D(b_1, \ldots, b_{n+1}) \leq 0$ *or* $D(b_1, \ldots, b_{n+1}) > 0$ *for all distinct* $b_1, \ldots, b_{n+1} \in A$.

Proof: If the Cayley-Menger determinant of some $B \subset A$ with $\#B = n+1$ were positive, then since the metroids associated with Euclidean spaces are matroids and hence stable, we have $D(C) > 0$ for all $C \subseteq B$. It follows that the space as a whole is $(n+1)$-dimensional Euclidean by Theorem 3.32, so that since $D(A) > 0$, again by stability, we have $D(B) > 0$ for all $B \subseteq A$. Otherwise, $D(B) \leq 0$ for all $B \subset A$ with $\#B = n+1$. QED

Since this means the discriminants of the quadratic expressions $D_{ij}(a_1,a_2,a_3,a_4;\, X)$ are never negative, this would tend to imply that the multivariate polynomial $D(a_1,a_2,a_3,a_4)$ is factorizable over the reals. To date, however, no such factorization is known, and in fact no factorization over the rationals exists.

Using arguments similar to those given for four atom metric spaces above, it can also be shown that the $n+1$ atom Cayley-Menger determinants must all be strictly negative if any are nonpositive. That it is actually possible to have $D(a_1, \ldots, a_4) > 0$ while all of the triangle inequalities are violated is shown by the following example:

$$\underline{\mathbf{M}}_D(1,2,3,4) \;=\; \begin{bmatrix} 0 & 4 & 16 & 64 \\ 4 & 0 & 1 & 16 \\ 16 & 1 & 0 & 4 \\ 64 & 16 & 4 & 0 \end{bmatrix}. \tag{3.91}$$

Whether the triangle inequalities are all strictly violated or strictly fulfilled, the expression $D_{34}(a_1, \ldots, a_4;\, X)$ will have two real roots, which since $A_{34} < 0$ will both be nonnegative if and only if both $B_{34} > 0$ and $C_{34} \leq 0$. We will show directly by a geometric argument that both roots are indeed nonnegative if the triangle inequalities for a_1, a_2, a_3 and a_1, a_2, a_4 are obeyed.

As noted in §3.3, in the case that the premetric space is Euclidean the value $D(a_1, \ldots, a_4)$ is equal to $(3!)^2$ times the square of the volume of the simplex $[(\mathbf{p}(a_1), \ldots, \mathbf{p}(a_4))]$ spanned by the embedded atoms $\mathbf{p}(a_1), \ldots, \mathbf{p}(a_4) \in \mathbb{R}^3$. Thus in a Euclidean space, $D(a_1, \ldots, a_4) = 0$ if and only if the points $\mathbf{p}(a_1), \ldots, \mathbf{p}(a_4)$ are dependent. On the other hand, in any 2-Euclidean space there exist isometric embeddings $\mathbf{p}_3 : \{a_1, a_2, a_4\} \rightarrow \mathbb{R}^3$ and $\mathbf{p}_4 : \{a_1, a_2, a_3\} \rightarrow \mathbb{R}^3$. For a fixed embedding $\mathbf{p}_4$, we may choose $\mathbf{p}_3$ such that $\mathbf{p}_4(a_1) = \mathbf{p}_3(a_1) =: \mathbf{a}_1$ and $\mathbf{p}_4(a_2) = \mathbf{p}_3(a_2) =: \mathbf{a}_2$. Since a rotation $\underline{\mathbf{R}} : \mathbb{R}^3 \rightarrow \mathbb{R}^3$ of $\mathbf{p}_3(a_4)$ about the line through $\mathbf{a}_1$ and $\mathbf{a}_2$ does not change the distances $\|\mathbf{p}_3(a_4) - \mathbf{a}_1\| = D(a_1, a_4)$ and $\|\mathbf{p}_3(a_4) - \mathbf{a}_2\| = D(a_2, a_4)$, however, the composition $\underline{\mathbf{R}} \circ \mathbf{p}_3$ is again an isometric embedding. Furthermore, it is easily seen that any other isometric embedding $\mathbf{p}'_3 : \{a_1, a_2, a_4\} \rightarrow \mathbb{R}^3$ which satisfies the conditions $\mathbf{p}'_3(a_1) = \mathbf{a}_1$ and $\mathbf{p}'_3(a_2) = \mathbf{a}_2$ must be obtained from $\mathbf{p}_3$ by composing it with such a rotation $\underline{\mathbf{R}}$. That is to say, the set of all embeddings

$$\{\mathbf{p}_3^{\phi} = \mathbf{p}_3 : \{a_1, a_2, a_4\} \rightarrow \mathbb{R}^3 \mid \mathbf{p}_3(a_1) = \mathbf{a}_1, \mathbf{p}_3(a_2) = \mathbf{a}_2\} \tag{3.92}$$

is a one-dimensional family of mappings parametrized by the angle ϕ between the planes spanned by $\mathbf{a}_1, \mathbf{a}_2, \mathbf{a}_3 = \mathbf{p}_4(a_3)$ and $\mathbf{a}_1, \mathbf{a}_2, \mathbf{p}_3^{\phi}(a_4)$ (as shown in Figure 3.8A). Since the volume of the simplex spanned by the points $[(\mathbf{a}_1, \mathbf{a}_2, \mathbf{a}_3, \mathbf{p}_3^{\phi}(a_4))]$ is zero if and only if this angle is 0° or 180°, it follows that the two roots of $D_{34}(a_1, \ldots, a_4; X)$ are

$$\begin{aligned} L(a_3, a_4;\, a_1, a_2) &:= \|\mathbf{p}_3^{0}(a_4) - \mathbf{a}_3\|^2 \geq 0\,; \\ U(a_3, a_4;\, a_1, a_2) &:= \|\mathbf{p}_3^{180}(a_4) - \mathbf{a}_3\|^2 \geq 0\,. \end{aligned} \tag{3.93}$$

Thus in any Euclidean metric space, each value of the squared distance $D(a_3, a_4)$ is bounded above and below as

$$L(a_3, a_4;\, a_1, a_2) \leq D(a_3, a_4) \leq U(a_3, a_4;\, a_1, a_2)\,. \tag{3.94}$$

These lower and upper bounds are known as the *tetrangle inequality limits.* They are analogous to, albeit considerably more complicated than, the triangle inequality limits introduced above. Chemists might think of them as the *cis* and *trans* extreme values of the vicinal distance which occur on rotation about a single bond. The quadratic formula can be used to derive an explicit expression for these limits as a function of the given squared distances $D(a_i, a_j)$ ($\{i, j\} \neq \{3, 4\}$) among the four atoms.

By similar reasoning, it can be shown that in a five-atom 3-Euclidean space $\mathbb{S} = (A, D)$, the Cayley-Menger determinant $D(a_1, \ldots, a_5)$, considered as a function $D_{45}(a_1, \ldots, a_5; X)$ of a single squared distance $X = D(a_4, a_5)$, is a quadratic

The extreme values of one of the distances between a set of four or five atoms when all of the remaining distances are held constant.

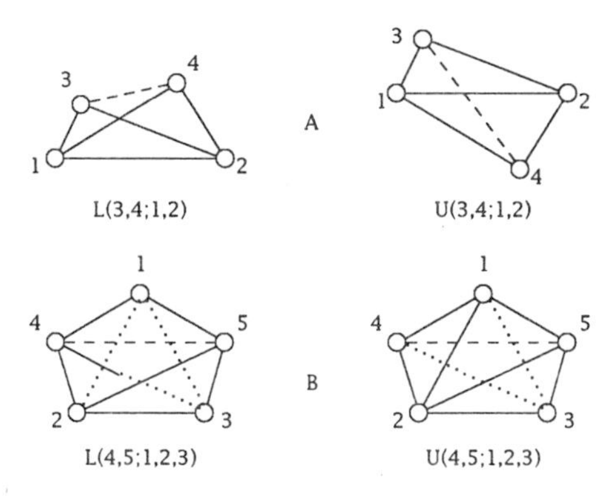

Figure 3.8

The upper pair illustrates the tetrangle case (A), and the lower pair shows the pentangle case (B).

expression with two nonnegative real roots $L(a_4,a_5;\ a_1,a_2,a_3)$ and $U(a_4,a_5;\ a_1,a_2,a_3)$. These roots, called the *pentangle inequality limits*, are shown in Figure 3.8B. Note that if the space is 3-dimensional Euclidean, then $D(a_1,\ldots,a_5) = 0$ so that these two roots are the *only* possible values that $D(a_4,a_5)$ can have. In this case, the values of the premetrics among the five atoms are tightly coupled so that a small change in any single squared distance must be compensated for by suitable changes in the remaining squared distances if the space as a whole is to remain 3-dimensional Euclidean. Similar considerations of course hold for the Cayley-Menger determinants of $n+2$ atoms embedded in n-dimensional Euclidean spaces for all $n > 0$.

In a strongly 3-Euclidean space on a set of six atoms $\{a_1,\ldots,a_6\}$, $D(a_5,a_6)$ must be a common root of the four quadratic equations:

$$\begin{aligned} 0 &= D_{45}(a_1,a_2,a_3,a_5,a_6;\ X) = D_{45}(a_1,a_2,a_4,a_5,a_6;\ X) \\ &= D_{45}(a_1,a_3,a_4,a_5,a_6;\ X) = D_{45}(a_2,a_3,a_4,a_5,a_6;\ X)\ . \end{aligned} \tag{3.95}$$

If these equations have only one root in common, then the space as a whole is necessarily 3-dimensional Euclidean, whereas if these equations all have both their roots in common, then there exists a second solution which corresponds to an n-pseudo-Euclidean space. The polynomial $D_{56}(a_1,\ldots,a_6;X)$ itself factors into the square of a linear function, since the discriminant of this quadratic expression, being the product of five-point Cayley-Menger determinants, is zero. Hence if the space is 3-dimensional Euclidean, i.e. $D(a_1, \ldots, a_6) = 0$, it follows from the quadratic formula that $D(a_5, a_6) = -B_{56}/2A_{56}$, where $A_{56} = -1/4\,D(a_1, \ldots, a_4)$ and

$$B_{56} = \frac{-1}{16}\, det \begin{bmatrix} 0 & 1 & 1 & \cdots & 1 & 1 \\ 1 & 0 & D(a_1,a_2) & \cdots & D(a_1,a_4) & D(a_1,a_5) \\ 1 & D(a_1,a_2) & 0 & \cdots & D(a_2,a_4) & D(a_2,a_5) \\ \cdot & \cdot & \cdot & \cdots & \cdot & \cdot \\ 1 & D(a_1,a_4) & D(a_2,a_4) & \cdots & 0 & D(a_4,a_5) \\ 1 & D(a_1,a_6) & D(a_2,a_6) & \cdots & D(a_4,a_6) & 0 \end{bmatrix}. \tag{3.96}$$

Expanding this determinant along its last row and column shows that in a 3-dimensional Euclidean space $D(a_5, a_6)$ is an inhomogeneous *bilinear* form in the vectors $[D(a_i, a_5) \mid i = 1, \ldots, 4]$ and $[D(a_i, a_6) \mid i = 1, \ldots, 4]$.

In the general case, we observe that for a given indexing of the set of atoms $\underline{A} = [a_1, \ldots, a_N]$, we may regard any (finite) premetric space $\mathbb{S} = (A, D)$ as a vector

$$\mathbf{D}_{\mathbb{S}} := [D(a_1,a_2), D(a_1,a_3), \ldots, D(a_{N-1},a_N)] \in \mathbb{R}^{N(N-1)/2} \tag{3.97}$$

whose components are the values of premetric $D: A \times A \to \mathbb{R}$ in the induced lexicographic order. Conversely, each vector $\mathbf{D} = [D_{12}, \ldots, D_{N-1,N}]$ defines a premetric space $\mathbb{S}_{\mathbf{D}} = \mathbb{S}_{\mathbf{D}}(A) = (A, D)$ with $D(a_i, a_j) := D_{ij}$ for $1 \leq i < j \leq N = \#A$ and $D(a_i, a_i) := 0$ for $1 \leq i \leq N$. Looked at in this way, the Cayley-Menger determinants $D(a_1, \ldots, a_N)$ become homogeneous, multivariate polynomials in the components of these vectors. For $1 \leq n \leq N-1$, we now define a subset of $\mathbb{R}^{N(N-1)/2}$ as

$$\begin{aligned} \mathbb{D}(N,n) &:= \{\mathbf{D} \in \mathbb{R}^{N(N-1)/2} \mid \mathbb{S}_{\mathbf{D}} = (A, D) \text{ is Euclidean of dimension} \leq n\} \\ &= \{\mathbf{D} \in \mathbb{R}^{N(N-1)/2} \mid D(B) \geq 0 \text{ and } D(B) = 0 \text{ if } \#B > n+1\}, \end{aligned} \tag{3.98}$$

where $D(B)$ is the Cayley-Menger determinant of the subspace $\mathbb{S}_{\mathbf{D}}(B) := (B, D|_B)$ of $\mathbb{S}_{\mathbf{D}}(A)$. We will call $\mathbb{D}(N,n)$ a *Cayley-Menger set* of order N and dimension n.

As an explicit example, we consider $\mathbb{D}(4,1)$. Hence suppose that we have a metric space $(\{a_1,a_2,a_3\}, D)$ such that $D(a_1,a_2,a_3) = 0$ with $D(a_1,a_2) = \alpha^2$, $D(a_2,a_3) = \beta^2$ and $D(a_1,a_3) = (\alpha+\beta)^2$ for some $\alpha, \beta > 0$. We now consider the set of all possible one atom extensions of this three atom premetric space, i.e. we take a fourth atom a_4 and we let $D(a_1,a_4) := X$, $D(a_2,a_4) := Y$ and $D(a_3,a_4) = Z$ for arbitrary real numbers $X, Y, Z \in \mathbb{R}$. Since these extensions depend upon three arbitrary parameters, the set of Euclidean extensions can be visualized as a subset of $\mathbb{R}^3$ (see Figure 3.9). Each of the three 3-atom Cayley-Menger determinants involving the new atom a_4 defines a bivariate quadratic expression in the unspecified squared distances X, Y, Z, e.g.

The following picture shows a cross-section of the Cayley-Menger set $\mathbb{D}(4,2)$ which is obtained when the distances among three of the four atoms are held fixed so that these three can be embedded in a Euclidean line.

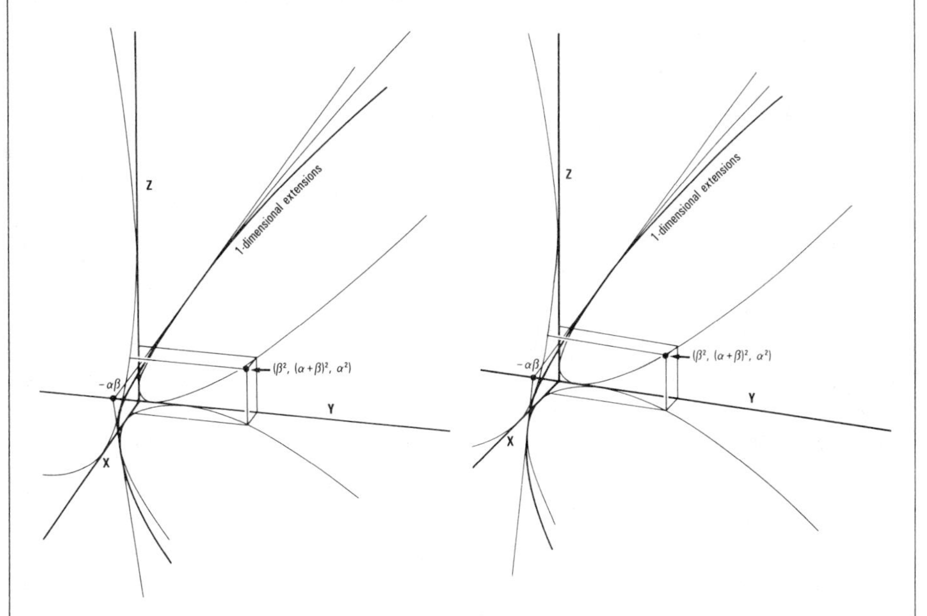

Figure 3.9

The three axes represent the three squared distances involving the fourth atom. The plane in the diagram represents the set of all values consistent with the tetrangle equality. The parabola in this plane represents the set of all values which embed in the Euclidean line; the region inside the parabola represents all values which can be embedded in the Euclidean plane. The isolated point where all three parabolae intersect below the plane is a 1-pseudo-Euclidean space.

$$D(a_1,a_2,a_4) \;=\; -1/4(X-Y)^2 + \alpha^2/2(X+Y) - (\alpha^2/2)^2 \tag{3.99}$$

with similar expressions for the other two determinants. Thus the equation $D(a_1,a_2,a_4) = 0$ defines a parabola in the (X,Y)-coordinate plane. On the other hand, because the triangle equality is satisfied exactly by the values of the premetric among the atoms a_1,a_2,a_3, the four atom Cayley-Menger determinant $D(a_1,\ldots,a_4)$ factors into a square of a linear function, i.e.

$$D(a_1,\ldots,a_4) \;=\; -1/4(\alpha\beta(\alpha+\beta) - \beta X + (\alpha+\beta)Y - \alpha Z)^2 \,. \tag{3.100}$$

Thus the equation $D(a_1,\ldots,a_4) = 0$ defines a plane in the space $[X,Y,Z] \approx \mathbb{R}^3$.

The Cayley-Menger set $\mathbb{D}(4,1)$ is the intersection of the three parabolic surfaces of translation defined by the three triangle equalities with the plane defined by the tetrangle equality, as shown in Figure 3.9. The set of all Euclidean extensions is the part of the plane inside this curve. The single point of intersection of the parabolic cylinders defined by the triangle equalities at $[X,Y,Z] = [\beta^2,(\alpha+\beta)^2,\alpha^2]$ corresponds to the 1-pseudo-Euclidean space.

Cayley-Menger sets play much the same role in Euclidean geometry that the Graßmann manifold plays in projective geometry. Nevertheless, very little is known about the structure of these sets in general. Since $(A, \alpha \cdot D)$ has the same signature as (A, D) does for all $\alpha > 0$, one thing that is immediately clear is that a Cayley-Menger set constitutes a *cone* with apex at the origin in $\mathbb{R}^{N(N-1)/2}$. The set of all semimetrics consistent with the triangle inequality has received some attention in the past [Avis1980]; this is called the *metric cone*. The set $\mathbb{D}(N,N-1)$ of *all* Euclidean premetrics, regardless of dimension, will be called the *Euclidean cone*. All the Cayley-Menger sets $\mathbb{D}(N,n)$ are subsets of this cone; specifically, they are "faces" defined by the vanishing of all Cayley-Menger determinants of $n+2$ atoms. Some other important properties of the Euclidean cone are summarized in the next proposition.

PROPOSITION 3.46. *The Euclidean cone is a convex subset of* $\mathbb{R}^{N(N-1)/2}$. *Its extremal rays are precisely the one-dimensional Euclidean metrics on* A, *and the totality of these rays forms an* $(N-2)$*-dimensional manifold which is canonically homeomorphic to a projective space of this dimension. Moreover, given any two vectors* $\mathbf{D}$, $\mathbf{D}'$ *in the Euclidean cone we have*

$$dim(\mathbb{S}_{\mathbf{D}+\mathbf{D}'}(\{1,\ldots,N\})) \;\leq\; dim(\mathbb{S}_{\mathbf{D}}(\{1,\ldots,N\}) + dim(\mathbb{S}_{\mathbf{D}'}(\{1,\ldots,N\}) \,, \qquad (3.101)$$

where $\mathbb{S}_{\mathbf{D}}(\{1,\ldots,N\})$ *denotes the premetric space* $(\{1,\ldots,N\}, D)$ *whose squared distances are given by the corresponding components of* $\mathbf{D}$, *etc.*

Proof: Letting $n := dim(\mathbb{S}_{\mathbf{D}}(\{1,\ldots,N\}))$ and $n' := dim(\mathbb{S}_{\mathbf{D}'}(\{1,\ldots,N\}))$, by assumption there exist isometric embeddings $\mathbf{p}$ and $\mathbf{p}'$ of $\mathbb{S}_{\mathbf{D}}(\{1,\ldots,N\})$ and $\mathbb{S}_{\mathbf{D}'}(\{1,\ldots,N\})$ in $\mathbb{R}^n$ and $\mathbb{R}^{n'}$. We may regard these as functions mapping $\{1,\ldots,N\}$ into $\mathbb{R}^n \oplus \mathbb{R}^{n'} \approx \mathbb{R}^{n+n'}$, so that for $0 \leq \alpha \leq 1$ the function

$$\mathbf{p}^\alpha := (\sqrt{\alpha}\mathbf{p} + \sqrt{(1-\alpha)}\mathbf{p}') : \{1,\ldots,N\} \rightarrow \mathbb{R}^{n+n'} \qquad (3.102)$$

is an embedding of $\{1,\ldots,N\}$ in $\mathbb{R}^{n+n'}$ such that

$$\begin{aligned} \|\mathbf{p}^\alpha(i) - \mathbf{p}^\alpha(j)\|^2 &= \|\sqrt{\alpha}(\mathbf{p}(i)-\mathbf{p}(j)) + \sqrt{(1-\alpha)}(\mathbf{p}'(i)-\mathbf{p}'(j))\|^2 \\ &= \alpha\,\|(\mathbf{p}(i)-\mathbf{p}(j))\|^2 + (1-\alpha)\,\|(\mathbf{p}'(i)-\mathbf{p}'(j))\|^2 \qquad (3.103) \\ &= \alpha\, D_{ij} + (1-\alpha)\, D'_{ij} \,. \end{aligned}$$

This shows that $\mathbb{D}(N,N-1)$ is a convex set. Also, by setting $\alpha = ½$ in the above we obtain an embedding $\mathbf{p}^{½}$ in $\mathbb{R}^{n+n'}$ such that $\|\mathbf{p}^{½}(i)-\mathbf{p}^{½}(j)\| = ½(\mathbf{D}+\mathbf{D}')$. Multiplying this result by two and using the fact that $\mathbb{D}(N,N-1)$ is a cone then proves Equation (3.101).

To see that the extreme rays are the one-dimensional metrics, we observe that $dim(\mathbb{S}_{\mathbf{D}+\mathbf{D}'}) = 1$ if and only if $\mathbf{D} = \alpha \cdot \mathbf{D}'$ for some $\alpha > 0$. Thus each extremal ray determines a one-dimensional arrangement of points on the line up to an arbitrary nonzero scale factor. The set of all one-dimensional Euclidean arrangements, on the other hand, can be given an $(N-1)$-dimensional global parametrization simply by fixing any one point as the origin, so that the extremal rays are in one-to-one correspondence with lines through the origin in $\mathbb{R}^{N-1}$, i.e. with the points of a projective space of dimension $N-2$. It is easily verified that this correspondence is a canonical homeomorphism. QED

Subsets of $\mathbb{R}^n$ which like $\mathbb{D}(N,n)$ can be defined by polynomial equalities and inequalities are known as *semialgebraic sets*, and their study is an area of mathematics which is known as *semialgebraic geometry*. It would, however, be outside the scope of this book to discuss this difficult subject in any detail (for an introduction, see [Brumfiel1979, Becker1986]).

3.6. Distance and Chirality

Since reflections in $\mathbb{R}^n$ are isometries, the composition of a reflection with an isometric embedding is again an isometric embedding. As a consequence, the absolute configuration or *chirality* of a molecule cannot be established by means of distance information alone. As we saw in Chapter 2, however, the *relative chirality* or oriented matroid structure associated with a given embedding can be specified by means of the function $\sigma_{\mathbf{p}}: \mathrm{A}^{n+1} \times \mathrm{A}^{n+1} \rightarrow \{-1,0,1\}$ given by

$$\sigma_{\mathbf{p}}(b_1, \ldots, b_{n+1}; c_1, \ldots, c_{n+1}) \;:=\; \chi_{\mathbf{p}}(b_1, \ldots, b_{n+1}) \cdot \chi_{\mathbf{p}}(c_1, \ldots, c_{n+1}) \,, \quad (3.104)$$

where $\chi_{\mathbf{p}}: \mathrm{A}^{n+1} \rightarrow \{-1,0,1\}$ is the chirality function given by

$$\chi_{\mathbf{p}}(b_1, \ldots, b_{n+1}) \;:=\; sign(vol(\mathbf{p}(b_1), \ldots, \mathbf{p}(b_{n+1}))) \quad (3.105)$$

for any admissible embedding $\mathbf{p}: \mathrm{A} \rightarrow \mathbb{R}^n$ and all $c_1, \ldots, c_{n+1}, b_1, \ldots, b_{n+1} \in \mathrm{A}$. In this section we will show how the relative chirality function *can* be obtained directly from complete distance information, and present an interesting collection of "cross-syzygies" connecting the affine and purely Euclidean geometric invariants. From the point of view of the last section, these "higher-order" syzygies are relations between the polynomials in the ideals which define the Cayley-Menger sets $\mathbb{D}(N,n)$.

In order to do this, we must first introduce a new kind of determinant.

Definition 3.47. Let $\mathbb{S} = (\mathrm{A}, D)$ be a premetric space and $\underline{\mathrm{B}} = [\mathrm{b}_1, \ldots, \mathrm{b}_{k+1}]$, $\underline{\mathrm{C}} = [\mathrm{c}_1, \ldots, \mathrm{c}_{k+1}]$ be two sequences of atoms in A of length $k+1$. Their *Cayley-Menger bideterminant* is

$$D(\mathrm{b}_1, \ldots, \mathrm{b}_{k+1}; \mathrm{c}_1, \ldots, \mathrm{c}_{k+1}) := 2(-½)^{k+1} det(\tilde{\underline{\mathbf{M}}}_D(\underline{\mathrm{B}}; \underline{\mathrm{C}})) \tag{3.106}$$

$$:= 2\left(\frac{-1}{2}\right)^{k+1} det \begin{bmatrix} 0 & 1 & 1 & \cdots & 1 \\ 1 & D(\mathrm{b}_1,\mathrm{c}_1) & D(\mathrm{b}_1,\mathrm{c}_2) & \cdots & D(\mathrm{b}_1,\mathrm{c}_{k+1}) \\ 1 & D(\mathrm{b}_2,\mathrm{c}_1) & D(\mathrm{b}_2,\mathrm{c}_2) & \cdots & D(\mathrm{b}_2,\mathrm{c}_{k+1}) \\ \cdot & \cdots & \cdots & \cdots & \cdots \\ 1 & D(\mathrm{b}_{k+1},\mathrm{c}_1) & D(\mathrm{b}_{k+1},\mathrm{c}_2) & \cdots & D(\mathrm{b}_{k+1},\mathrm{c}_{k+1}) \end{bmatrix}.$$

Cayley-Menger bideterminants first appeared in the writings of W.K. Clifford and J.J. Sylvester in the mid-nineteenth century [Clifford1864, Sylvester1904-1912] To our knowledge, they next turned up in the writings of J.J. Seidel a century later [Seidel1952], and were subsequently used by L. Blumenthal to prove the completeness of his postulates for the Euclidean plane in a Dover paperback "A Modern View of Geometry" [Blumenthal1961]. Note that this determinant is symmetric in the sequences $\underline{\mathrm{B}}$ and $\underline{\mathrm{C}}$, i.e.

$$D(\mathrm{b}_1, \ldots, \mathrm{b}_{k+1}; \mathrm{c}_1, \ldots, \mathrm{c}_{k+1}) = D(\mathrm{c}_1, \ldots, \mathrm{c}_{k+1}; \mathrm{b}_1, \ldots, \mathrm{b}_{k+1}), \tag{3.107}$$

and vanishes if $\#\mathrm{B} < k+1$ or $\#\mathrm{C} < k+1$ (i.e. if either of the sequences contains repeated elements). At the same time, however, it is antisymmetric with respect to the ordering of the sequences, i.e.

$$D(\mathrm{b}_1, \ldots, \mathrm{b}_i, \mathrm{b}_{i+1}, \ldots, \mathrm{b}_{k+1}; \mathrm{c}_1, \ldots, \mathrm{c}_{k+1}) = \tag{3.108}$$
$$-D(\mathrm{b}_1, \ldots, \mathrm{b}_{i+1}, \mathrm{b}_i, \ldots, \mathrm{b}_{k+1}; \mathrm{c}_1, \ldots, \mathrm{c}_{k+1})$$

for all $i = 1, \ldots, k$. Finally, if $\underline{\mathrm{B}} = \underline{\mathrm{C}}$ and $\#\mathrm{B} = k+1$, then

$$D(\mathrm{b}_1, \ldots, \mathrm{b}_{k+1}; \mathrm{b}_1, \ldots, \mathrm{b}_{k+1}) = D(\mathrm{b}_1, \ldots, \mathrm{b}_{k+1}) = D(\mathrm{B}), \tag{3.109}$$

the Cayley-Menger determinant of the atoms in B.

For any $2 \cdot k$ vectors $\mathbf{u}_1, \ldots, \mathbf{u}_k, \mathbf{v}_1, \ldots, \mathbf{v}_k \in \mathbf{V}_B$ in a bilinear vector space with form B, as in §3.2 we let $\underline{\mathbf{M}}_B(\mathbf{u}_1, \ldots, \mathbf{u}_k; \mathbf{v}_1, \ldots, \mathbf{v}_k) := [B(\mathbf{u}_i; \mathbf{v}_j) \mid i, j = 1, \ldots, k]$, and $B(\mathbf{u}_1, \ldots, \mathbf{u}_k; \mathbf{v}_1, \ldots, \mathbf{v}_k)$ denote the determinant of this matrix.

LEMMA 3.48. *If B_D is the bilinear form in the amalgamation space $\mathbb{A}_D[\mathrm{A}]$ associated with the premetric space* (A, D), *then*

$$D(\mathrm{b}_1,\ldots,\mathrm{b}_{k+1}; \mathrm{c}_1,\ldots,\mathrm{c}_{k+1}) = B_D(\mathbf{b}_1-\mathbf{b}_{k+1},\ldots,\mathbf{b}_k-\mathbf{b}_{k+1}; \mathbf{c}_1-\mathbf{c}_{k+1},\ldots,\mathbf{c}_k-\mathbf{c}_{k+1}). \tag{3.110}$$

The proof is entirely analogous to our derivation of Cayley-Menger determinants from the associated Gramians (Lemma 3.25), and so need not be repeated here.

Instead we will use the lemma to obtain some geometric interpretations of these determinants in the 3-dimensional case. We start with the Cayley-Menger bideterminant of two ordered pairs of atoms $[b_1,b_2]$ and $[c_1,c_2]$ in a premetric space:

$$\frac{1}{2}\,det\begin{bmatrix} 0 & 1 & 1 \\ 1 & D(b_1,c_1) & D(b_1,c_2) \\ 1 & D(b_2,c_1) & D(b_2,c_2) \end{bmatrix} \tag{3.111}$$

$$= \frac{1}{2}\left(D(b_1,c_2) + D(b_2,c_1) - D(b_1,c_1) - D(b_2,c_2)\right)$$

$$= det\left(\underline{\mathbf{M}}_B(\mathbf{b}_1-\mathbf{b}_2;\,\mathbf{c}_1-\mathbf{c}_2)\right) = B_D(\mathbf{b}_1-\mathbf{b}_2;\,\mathbf{c}_1-\mathbf{c}_2)\,.$$

Since $D(a_i,a_j) = Q_D(\mathbf{a}_i-\mathbf{a}_j)$, in the case that $b_2 = c_2$ this formula reduces to the well-known law of cosines for the bilinear form B_D. In the case that the premetric space is Euclidean, we have

$$D(b_1,b_2;\,c_1,c_2) = (\mathbf{b}_1-\mathbf{b}_2)\cdot(\mathbf{c}_1-\mathbf{c}_2) = \|\mathbf{b}_1-\mathbf{b}_2\|\,\|\mathbf{c}_1-\mathbf{c}_2\|\,cos(\Theta_{bc}) \tag{3.112}$$
$$= \sqrt{D(b_1,b_2)\,D(c_1,c_2)}\;cos(\Theta_{bc})\;,$$

where Θ_{bc} is the angle between the lines $((\mathbf{b}_1,\mathbf{b}_2))$ and $((\mathbf{c}_1,\mathbf{c}_2))$. Note that this formula holds even if the lines $((\mathbf{b}_1,\mathbf{b}_2))$ and $((\mathbf{c}_1,\mathbf{c}_2))$ do not intersect.

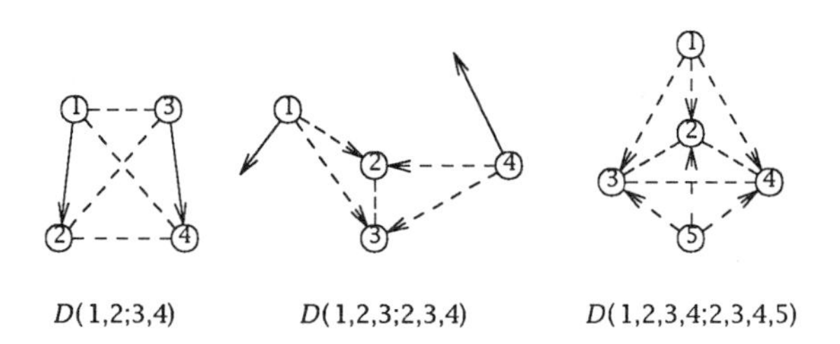

Figure 3.10

In the left-hand drawing, the Cayley-Menger bideterminant is the dot product of the differences vectors between the points labeled 1, 2 and 3, 4, respectively. In the middle drawing, it is the dot product of the cross-products of the vectors drawn with dashed lines. In the right-hand drawing, the Cayley-Menger bideterminant is equal to the product of the triple products of the dashed vectors from 1 and 5, respectively.

We now consider the Cayley-Menger bideterminant of two triples of atoms in a premetric space.

$$\begin{aligned} & D(\mathrm{b}_1,\mathrm{b}_2,\mathrm{b}_3;\ \mathrm{c}_1,\mathrm{c}_2,\mathrm{c}_3) \\ = \ & B_D(\mathbf{b}_1-\mathbf{b}_3,\mathbf{b}_2-\mathbf{b}_3;\ \mathbf{c}_1-\mathbf{c}_3,\mathbf{c}_2-\mathbf{c}_3) \\ = \ & B_D(\mathbf{b}_1-\mathbf{b}_3;\mathbf{c}_1-\mathbf{c}_3)\, B_D(\mathbf{b}_2-\mathbf{b}_3;\mathbf{c}_2-\mathbf{c}_3) - B_D(\mathbf{b}_1-\mathbf{b}_3;\mathbf{c}_2-\mathbf{c}_3)\, B_D(\mathbf{b}_2-\mathbf{b}_3;\mathbf{c}_1-\mathbf{c}_3) \\ = \ & D(\mathrm{b}_1,\mathrm{b}_3;\ \mathrm{c}_1,\mathrm{c}_3)\, D(\mathrm{b}_2,\mathrm{b}_3;\ \mathrm{c}_2,\mathrm{c}_3) - D(\mathrm{b}_1,\mathrm{b}_3;\ \mathrm{c}_2,\mathrm{c}_3)\, D(\mathrm{b}_2,\mathrm{b}_3;\ \mathrm{c}_1,\mathrm{c}_3)\ . \end{aligned} \tag{3.113}$$

In the Euclidean case, we may apply the identity from vector algebra

$$(\mathbf{u}\times\mathbf{v})\cdot(\mathbf{u}'\times\mathbf{v}') \ = \ (\mathbf{u}\cdot\mathbf{u}')(\mathbf{v}\cdot\mathbf{v}') - (\mathbf{u}'\cdot\mathbf{v})(\mathbf{u}\cdot\mathbf{v}') \tag{3.114}$$

(known as *Lagrange's identity*) to conclude that

$$D(\mathrm{b}_1,\mathrm{b}_2,\mathrm{b}_3;\ \mathrm{c}_1,\mathrm{c}_2,\mathrm{c}_3) \ = \ ((\mathbf{b}_1-\mathbf{b}_3)\times(\mathbf{b}_2-\mathbf{b}_3))\cdot((\mathbf{c}_1-\mathbf{c}_3)\times(\mathbf{c}_2-\mathbf{c}_3))\ . \tag{3.115}$$

In the case that the vectors $\mathbf{u}=\mathbf{u}'$ and $\mathbf{v}=\mathbf{v}'$, Lagrange's identity becomes $\|\mathbf{u}\times\mathbf{v}\|^2 = \|\mathbf{u}\|^2\|\mathbf{v}\|^2-(\mathbf{u}\cdot\mathbf{v})^2$, which we can use again to rewrite this last expression as

$$\begin{aligned} & \|(\mathbf{b}_1-\mathbf{b}_3)\times(\mathbf{b}_2-\mathbf{b}_3)\|\ \|(\mathbf{c}_1-\mathbf{c}_3)\times(\mathbf{c}_2-\mathbf{c}_3)\|\ cos(\Phi_{bc}) \ = \\ & \sqrt{D(\mathrm{b}_1,\mathrm{b}_3)D(\mathrm{b}_2,\mathrm{b}_3)-D^2(\mathrm{b}_1,\mathrm{b}_3;\ \mathrm{b}_2,\mathrm{b}_3)}\sqrt{D(\mathrm{c}_2,\mathrm{c}_3)D(\mathrm{c}_1,\mathrm{c}_3)-D^2(\mathrm{c}_2,\mathrm{c}_3;\ \mathrm{c}_1,\mathrm{c}_3)}\ cos(\Phi_{bc}) \end{aligned} \tag{3.116}$$

where Φ_{bc} is now the angle between the planes spanned by the triangles $[(\mathbf{b}_1,\mathbf{b}_2,\mathbf{b}_3)]$ and $[(\mathbf{c}_1,\mathbf{c}_2,\mathbf{c}_3)]$, and the quantities inside the "$\sqrt{\ }$" signs are four times the squared areas of those triangles.

Finally, we consider the Cayley-Menger bideterminant of two quadruples of atoms.

$$\begin{aligned} & D(\mathrm{b}_1,\ldots,\mathrm{b}_4;\ \mathrm{c}_1,\ldots,\mathrm{c}_4) \\ = \ & det\begin{bmatrix} B_D(\mathbf{b}_1-\mathbf{b}_4;\mathbf{c}_1-\mathbf{c}_4) & B_D(\mathbf{b}_2-\mathbf{b}_4;\mathbf{c}_1-\mathbf{c}_4) & B_D(\mathbf{b}_3-\mathbf{b}_4;\mathbf{c}_1-\mathbf{c}_4) \\ B_D(\mathbf{b}_1-\mathbf{b}_4;\mathbf{c}_2-\mathbf{c}_4) & B_D(\mathbf{b}_2-\mathbf{b}_4;\mathbf{c}_2-\mathbf{c}_4) & B_D(\mathbf{b}_3-\mathbf{b}_4;\mathbf{c}_2-\mathbf{c}_4) \\ B_D(\mathbf{b}_1-\mathbf{b}_4;\mathbf{c}_3-\mathbf{c}_4) & B_D(\mathbf{b}_2-\mathbf{b}_4;\mathbf{c}_3-\mathbf{c}_4) & B_D(\mathbf{b}_3-\mathbf{b}_4;\mathbf{c}_3-\mathbf{c}_4) \end{bmatrix}. \end{aligned} \tag{3.117}$$

In the three-dimensional Euclidean case, the bilinear form becomes the scalar product

$$B_D(\mathbf{b}_i-\mathbf{b}_4;\mathbf{c}_j-\mathbf{c}_4) \ = \ \sum_{k=1}^{3}(x_{ik}-x_{4k})(y_{jk}-y_{4k})\ , \tag{3.118}$$

where $[x_{ik}]$, $[y_{jk}]$ are Cartesian coordinates for $\mathbf{b}_i$ and $\mathbf{c}_j$, respectively, so that in terms of matrices we have

$$D(\mathrm{b}_1,\ldots,\mathrm{b}_4;\ \mathrm{c}_1,\ldots,\mathrm{c}_4)$$

$$= \; det\left(\begin{bmatrix} x_{11}-x_{41} & x_{12}-x_{42} & x_{13}-x_{43} \\ x_{21}-x_{41} & x_{22}-x_{42} & x_{23}-x_{43} \\ x_{31}-x_{41} & x_{32}-x_{42} & x_{33}-x_{43} \end{bmatrix} \cdot \begin{bmatrix} y_{11}-y_{41} & y_{21}-y_{41} & y_{31}-y_{41} \\ y_{12}-y_{42} & y_{22}-y_{42} & y_{32}-y_{42} \\ y_{13}-y_{43} & y_{23}-y_{43} & y_{33}-y_{43} \end{bmatrix}\right) \tag{3.119}$$

$$= \; det\begin{bmatrix} 1 & 1 & 1 & 1 \\ x_{11} & x_{21} & x_{31} & x_{41} \\ x_{12} & x_{22} & x_{32} & x_{42} \\ x_{13} & x_{23} & x_{33} & x_{43} \end{bmatrix} det\begin{bmatrix} 1 & 1 & 1 & 1 \\ y_{11} & y_{21} & y_{31} & y_{41} \\ y_{12} & y_{22} & y_{32} & y_{42} \\ y_{13} & y_{23} & y_{33} & y_{43} \end{bmatrix}$$

$$=: \; (3\,!)^2\, vol(\mathbf{b}_1, \ldots, \mathbf{b}_4) \cdot vol(\mathbf{c}_1, \ldots, \mathbf{c}_4)\,.$$

Thus $D(b_1, \ldots, b_4; c_1, \ldots, c_4)$ is equal to the product of the *oriented* volumes of the simplices $[(\mathbf{b}_1, \ldots, \mathbf{b}_4)]$ and $[(\mathbf{c}_1, \ldots, \mathbf{c}_4)]$ spanned by the quadruples of embedded atoms times a factor of $(3\,!)^2$.

More generally, if $B : \mathbb{R}^n \times \mathbb{R}^n \to \mathbb{R}$ is a nondegenerate bilinear form over the reals (or, for that matter, any field $\mathbb{K}$) whose metric matrix with respect to the canonical basis is $\underline{\mathbf{M}}_B$, and if $\mathbf{p} : A \to \mathbb{R}^n$ is an embedding of some set A into $\mathbb{R}^n$ with $D(a,b) := B(\mathbf{p}(a) - \mathbf{p}(b), \mathbf{p}(a) - \mathbf{p}(b))$, it follows from a similar argument that:

$$D(b_1, \ldots, b_{n+1}; c_1, \ldots, c_{n+1}) \tag{3.120}$$

$$= \; (3\,!)^2 \cdot det(\underline{\mathbf{M}}_B) \cdot vol(\mathbf{p}(b_1), \ldots, \mathbf{p}(b_{n+1})) \cdot vol(\mathbf{p}(c_1), \ldots, \mathbf{p}(c_{n+1}))\,.$$

Hence:

PROPOSITION 3.49. *Given an n-dimensional premetric space* (A, D), *the Cayley-Menger bideterminant of two ordered* $(n+1)$*-tuples of atoms* $[b_1, \ldots, b_{n+1}]$ *and* $[c_1, \ldots, c_{n+1}]$ *in* A^{n+1} *is equal to the product of the oriented volumes of the simplices spanned by these two* $(n+1)$*-tuples of atoms in any isometric embedding into an n-dimensional bilinear affine space* $\mathbb{A}_B$ *times a fixed constant factor.*

As an immediate consequence, we have:

COROLLARY 3.50. *For any four* $(n+1)$*-tuples of atoms* $\underline{B}, \underline{C}, \underline{D}, \underline{E} \in A^{n+1}$ *in a premetric space* (A, D) *of dimension n:*

$$D(\underline{B}; \underline{C}) \cdot D(\underline{D}; \underline{E}) \;=\; D(\underline{B}; \underline{E}) \cdot D(\underline{C}; \underline{D})\,. \tag{3.121}$$

We shall soon see that this condition is also sufficient for a Euclidean space to have dimension n. Another, rather interesting result is:

COROLLARY 3.51. *Given an n-dimensional premetric space* (A, D) *and a free* $(n+1)$*-tuple* $\{a_1, \ldots, a_{n+1}\} \in \binom{A}{n+1}$*, the function* $\sigma_D : A^{n+1} \times A^{n+1} \to \{-1, 0, 1\}$ *defined by*

$$\sigma_D(b_1, \ldots, b_{n+1}; c_1, \ldots, c_{n+1}) \tag{3.122}$$

$$= \; sign(D(a_1, \ldots, a_{n+1})) \cdot sign(D(b_1, \ldots, b_{n+1}; c_1, \ldots, c_{n+1}))$$

is the relative orientation of a representable affine chirotope (i.e. totally acyclic oriented matroid) of rank $n+1$.

The example $\mathbf{a}_1 = [0,0]$, $\mathbf{a}_2 = [1,0]$, $\mathbf{a}_3 = [1,1]$ and $\mathbf{a}_4 = [0,1]$ shows that even if the space is Euclidean the signs of the $(k+1)$-point Cayley-Menger bideterminants do not generally yield a chirotope if $k < n$, since

$$\begin{aligned} 1 &= sign((\mathbf{a}_2-\mathbf{a}_1)\cdot(\mathbf{a}_3-\mathbf{a}_1))\cdot sign((\mathbf{a}_3-\mathbf{a}_1)\cdot(\mathbf{a}_4-\mathbf{a}_1)) \\ &= \sigma(\mathbf{a}_1,\mathbf{a}_2;\mathbf{a}_1,\mathbf{a}_3)\cdot\sigma(\mathbf{a}_1,\mathbf{a}_3;\mathbf{a}_1,\mathbf{a}_4) \qquad (3.123) \\ &\neq \sigma(\mathbf{a}_1,\mathbf{a}_2;\mathbf{a}_1,\mathbf{a}_4) \\ &= sign((\mathbf{a}_2-\mathbf{a}_1)\cdot(\mathbf{a}_4-\mathbf{a}_1)) \;=\; 0 \end{aligned}$$

(cf. axiom ($\Sigma 3$) of §2.1).

Using Proposition 3.49, we can also translate the affine dependency relations (i.e. the additivity of oriented volumes) into Cayley-Menger bideterminants as follows:

$$\sum_{i=1}^{n+2} (-1)^i D(\mathrm{b}_1,\ldots,\mathrm{b}_{i-1},\mathrm{b}_{i+1},\ldots,\mathrm{b}_{n+2};\, \mathrm{c}_1,\ldots,\mathrm{c}_{n+1}) \;=\; 0 \qquad (3.124)$$

for all $\mathrm{b}_1,\ldots,\mathrm{b}_{n+2},\mathrm{c}_1,\ldots,\mathrm{c}_{n+1} \in \mathrm{A}$. Similarly, the Graßmann-Plücker relations can be written as:

$$\sum_{i=1}^{n+2} (-1)^i D(\mathrm{b}_1,\ldots,\mathrm{b}_{i-1},\mathrm{b}_{i+1},\ldots,\mathrm{b}_{n+2};\, \mathrm{b}_i,\mathrm{c}_1,\ldots,\mathrm{c}_n) \;=\; 0 \qquad (3.125)$$

for all $\mathrm{b}_1,\ldots,\mathrm{b}_{n+2};\, \mathrm{c}_1,\ldots,\mathrm{c}_n \in \mathrm{A}$. Since Cayley-Menger bideterminants are not proportional to the products of volumes in the case that these sequences have length $k < n$, there is no *a priori* reason to expect that in this case these identities will continue to hold. Nevertheless:

PROPOSITION 3.52. *The affine dependency and Graßmann-Plücker relations are algebraic identities between Cayley-Menger bideterminants, even if n is replaced by $k < n$ in Equations (3.124) and (3.125).*

Proof: To prove the first assertion, we consider the determinant:

$$det \begin{bmatrix} 0 & 1 & 1 & \cdots & 1 \\ 0 & 1 & 1 & \cdots & 1 \\ 1 & D(\mathrm{c}_1,\mathrm{b}_1) & D(\mathrm{c}_1,\mathrm{b}_2) & \cdots & D(\mathrm{c}_1,\mathrm{b}_{k+2}) \\ \cdot & \cdots & \cdots & \cdots & \cdots \\ 1 & D(\mathrm{c}_{k+1},\mathrm{b}_1) & D(\mathrm{c}_{k+1},\mathrm{b}_2) & \cdots & D(\mathrm{c}_{k+1},\mathrm{b}_{k+2}) \end{bmatrix} = 0\,, \qquad (3.126)$$

since the first two rows are equal. Expanding this along the first row yields the affine dependency condition.

The Graßmann-Plücker relation is a little harder, and requires the following equation for the derivatives of Cayley-Menger bideterminants as functions of the values of the premetric:

$$\frac{\partial}{\partial D(b_i,c_j)} D(b_1,\ldots,b_{k+1};\ c_1,\ldots,c_{k+1}) \tag{3.127}$$

$$= \frac{(-1)^{i+j}}{2} D(b_1,\ldots,b_{i-1},b_{i+1},\ldots,b_{k+1};\ c_1,\ldots,c_{j-1},c_{j+1},\ldots,c_{k+1})$$

for all $k > 2$ whenever $b_i \neq c_j\ \forall i,j$, as may easily be verified. We now proceed by induction on k; for $k = 2$ we have

$$-D(b_2,b_3;\ b_1,c_1) + D(b_1,b_3;\ b_2,c_1) - D(b_1,b_2;\ b_3,c_1) = \tag{3.128}$$

$$-\frac{1}{2}\Big(D(b_1,b_3)+D(b_2,c_1)-D(b_1,b_2)-D(b_3,c_1)\Big) +$$

$$\frac{1}{2}\Big(D(b_2,b_3)+D(b_1,c_1)-D(b_1,b_2)-D(b_3,c_1)\Big) +$$

$$-\frac{1}{2}\Big(D(b_2,b_3)+D(b_1,c_1)-D(b_1,b_3)-D(b_2,c_1)\Big) = 0\,.$$

To prove the general case, we first assume that $b_i \neq c_j$ for all $1 \leq i \leq k{+}2$, $1 \leq j \leq k$. Then if we consider the l.h.s. of Equation (3.125) as a function $f: \mathbb{R}^{(k+2)(3k+1)/2} \to \mathbb{R}$ mapping $[D(b,c)\,|\, b = b_i,\ c = b_j \text{ or } c_j]$ into the reals, the gradient of f contains two types of components:

(i) Derivatives with respect to a squared distance of the form $D(b_i,b_j)$;

(ii) Derivatives with respect to a squared distance of the form $D(b_i,c_j)$.

Type (i) is exemplified by

$$\frac{\partial}{\partial D(b_{k+1},b_{k+2})} \sum_{i=1}^{k+2} (-1)^i D(b_1,\ldots,b_{i-1},b_{i+1},\ldots,b_{k+2};\ b_i,c_1,\ldots,c_k) \tag{3.129}$$

$$= \frac{\partial}{\partial D(b_{k+1},b_{k+2})} \Big((-1)^{k+1}\, D(b_1,\ldots,b_k,b_{k+2};\ b_{k+1},c_1,\ldots,c_k) +$$

$$(-1)^{k+2}\, D(b_1,\ldots,b_k,b_{k+1};\ b_{k+2},c_1,\ldots,c_k)\Big)$$

$$= \frac{-1}{2}\Big(D(b_1,\ldots,b_k;\ c_1,\ldots,c_k) - D(b_1,\ldots,b_k;\ c_1,\ldots,c_k)\Big) = 0$$

by Equation (3.127), whereas type (ii) is exemplified by

$$\frac{\partial}{\partial D(b_{k+2},c_k)} \sum_{i=1}^{k+2} (-1)^i D(b_1,\ldots,b_{i-1},b_{i+1},\ldots,b_{k+2};\ b_i,c_1,\ldots,c_k) \tag{3.130}$$

$$= \frac{1}{2} \sum_{i=1}^{k+1} (-1)^i D(b_1,\ldots,b_{i-1},b_{i+1},\ldots,b_{k+1};\ b_i,c_1,\ldots,c_{k-1}) = 0$$

by Equation (3.127) together with the induction hypothesis. Since its gradient is everywhere zero, the function f is constant. Moreover, since f is zero whenever (A, D) is a Euclidean premetric space, this constant is zero. The case $b_i = c_j$ for some values of i and j follows from this result by the continuity of f. QED

The following lemma is due to [Seidel1952]. To present it, for a given sequence $\underline{B} = [b_1, \ldots, b_k]$ we define $\underline{B}_i := \underline{B} \setminus b_i := [b_1, \ldots, b_{i-1}, b_{i+1}, \ldots, b_k]$.

LEMMA 3.53. *If* $\mathbb{S} = (\mathrm{A}, D)$ *is a premetric space, then*

$$D(b_1, \ldots, b_{k-1}) D(b_1, \ldots, b_{k+1}) = D(\underline{B}_k) D(\underline{B}_{k+1}) - D^2(\underline{B}_k; \underline{B}_{k+1}) \tag{3.131}$$

for all $\underline{B} = [b_1, \ldots, b_{k+1}] \in \mathrm{A}^{k+1}$ *with* $k > 1$.

Proof: To prove this, we recall a well-known determinant identity known as *Jacobi's theorem on a minor of the adjugate* (see [Muir1906-1923), which states that a minor of order M in the adjugate matrix is equal to the complementary minor in the original matrix times the determinant of the original matrix to the $(M-1)$-th power. Consider the matrix $\underline{\mathbf{M}}'_D(\underline{B})$ whose determinant is $-D(b_1, \ldots, b_{k+1})$

$$\begin{bmatrix} 0 & 1 & 1 & \cdots & 1 \\ 1 & 0 & -\tfrac{1}{2}D(b_1,b_2) & \cdots & -\tfrac{1}{2}D(b_1,b_{k+1}) \\ 1 & -\tfrac{1}{2}D(b_1,b_2) & 0 & \cdots & -\tfrac{1}{2}D(b_2,b_{k+1}) \\ \cdot & \cdots & \cdots & \cdots & \cdots \\ 1 & -\tfrac{1}{2}D(b_1,b_{k+1}) & -\tfrac{1}{2}D(b_2,b_{k+1}) & \cdots & 0 \end{bmatrix} \tag{3.132}$$

and its adjugate $(\underline{\mathbf{M}}'_D)^*(\underline{B})$

$$\begin{bmatrix} \cdots & \cdots & \cdots & \cdots & \cdots \\ \cdots & -D(\underline{B}_1) & \cdots & -D(\underline{B}_1; \underline{B}_k) & -D(\underline{B}_1; \underline{B}_{k+1}) \\ \cdots & \cdots & \cdots & \cdots & \cdots \\ \cdots & -D(\underline{B}_1; \underline{B}_k) & \cdots & -D(\underline{B}_k) & -D(\underline{B}_{k+1}; \underline{B}_k) \\ \cdots & -D(\underline{B}_1; \underline{B}_{k+1}) & \cdots & -D(\underline{B}_k; \underline{B}_{k+1}) & -D(\underline{B}_{k+1}) \end{bmatrix}. \tag{3.133}$$

The complementary minor in the adjugate of the determinant $D(b_1, \ldots, b_{k-1})$ of the matrix formed from the first k rows and columns of $\underline{\mathbf{M}}'_D(\underline{B})$ is the determinant of the 2×2 matrix shown at the lower right of $(\underline{\mathbf{M}}'_D)^*(\underline{B})$. Thus the proposition follows from Jacobi's theorem with $M = 2$. QED

A variety of interesting syzygies among Cayley-Menger bideterminants can be proved along similar lines. At the moment, however, we are interested in establishing some alternative criteria, expressed in terms of relations among Cayley-Menger bideterminants, for a premetric space to be Euclidean of dimension n. One immediate consequence of Lemma 3.53 is

COROLLARY 3.54. *A premetric space* (A, D) *is Euclidean of dimension* n *if and only if there exists an ordering of the atoms* $\underline{\mathrm{A}} = [a_1, \ldots, a_N]$ *such that*

$$D(a_1, \ldots, a_k) > 0 \quad \text{for } k = 2, \ldots, n+1\,; \tag{3.134}$$

$$D(a_1, \ldots, a_{n+1}, b;\, a_1, \ldots, a_{n+1}, c) = 0 \quad \text{for all } b, c \in A. \tag{3.135}$$

Once again, this criterion singles out an arbitrary $(n+1)$-tuple of atoms, so that in many cases the following criterion is more useful.

PROPOSITION 3.55. *A semimetric space* $\mathbb{S} = (A, D)$ *with* $\#A > n+3$ *is Euclidean of dimension* n *if and only if for* $k = 2, \ldots, n$ *all* $(k+1)$*-tuples of atoms* $b_1, \ldots, b_{k+1} \in A$ *satisfy*

$$D(b_1, \ldots, b_{k-1}, b_k) D(b_1, \ldots, b_{k-1}, b_{k+1}) \geq D^2(b_1, \ldots, b_{k-1}, b_k;\, b_1, \ldots, b_{k-1}, b_{k+1})\,, \tag{3.136}$$

while for all $(n+2)$*-tuples* $b_1, \ldots, b_{n+2}$

$$D(b_1, \ldots, b_n, b_{n+1}) D(b_1, \ldots, b_n, b_{n+2}) = D^2(b_1, \ldots, b_n, b_{n+1};\, b_1, \ldots, b_n, b_{n+2})\,. \tag{3.137}$$

Proof: Necessity is an immediate consequence of Proposition 3.28 and Lemma 3.53. To prove sufficiency, we use induction on k. For $k = 2$, it follows from the hypothesis together with Equation (3.131) that $D(b_1, b_2, b_3) \geq 0$, so that the space is at least 2-Euclidean. Assuming now that the space is $(k-1)$-Euclidean, we consider an arbitrary $(k+1)$-tuple $[b_1, \ldots, b_{k+1}]$ of atoms therein. Since all subsets $\{c_1, \ldots, c_k\} \subset \{b_1, \ldots, b_{k+1}\}$ are Euclidean by our induction hypothesis, $D(c_1, \ldots, c_{i-1}, c_{i+1}, \ldots, c_k) = 0$ for $i = 1, \ldots, k$ implies $D(c_1, \ldots, c_k) = 0$ as well, in which case $\{b_1, \ldots, b_{k+1}\}$ is very strongly $(k-3)$-Euclidean so that $D(b_1, \ldots, b_{k+1}) = 0$. Otherwise the Cayley-Menger determinant of at least one $(k-1)$-tuple is positive, w.l.o.g. $D(b_1, \ldots, b_{k-1}) > 0$. Then again by Equation (3.131)

$$D(b_1, \ldots, b_{k-1}) D(b_1, \ldots, b_{k+1}) = D(b_1, \ldots, b_{k-1}, b_k) D(b_1, \ldots, b_{k-1}, b_{k+1}) - D^2(b_1, \ldots, b_{k-1}, b_k;\, b_1, \ldots, b_{k-1}, b_{k+1}) \geq 0 \tag{3.138}$$

by hypothesis, whence $D(b_1, \ldots, b_{k+1}) \geq 0$, i.e. the space is k-Euclidean. When $k = n+1$, however, it follows similarly from Equation (3.137) that $D(b_1, \ldots, b_{n+2}) = 0$ for all $b_1, \ldots, b_{n+2} \in A$, and hence the space is n-dimensional Euclidean by Corollary 3.44. QED

As a consequence of this proposition, we have

COROLLARY 3.56. *A Euclidean premetric space* (A, D) *which satisfies Equation (3.121) is* n*-dimensional.*

This follows easily by noting that Equation (3.137) is simply a special case of Equation (3.121). Probably, the result also holds for non-Euclidean premetric spaces.

Note the similarity of the Equation (3.136) to the Cauchy-Schwarz inequality; when $k = 2$, it is actually equivalent to it. It is in fact a special case of a more general relation

PROPOSITION 3.57. *Let* (A, D) *be a Euclidean premetric space,* $\underline{\mathrm{A}}$ *be an ordering of* A *and* $\underline{\mathrm{B}}, \underline{\mathrm{C}} \subseteq \underline{\mathrm{A}}$ *be ordered* $(k+1)$*-tuples therein; then*

$$D(\mathrm{B})D(\mathrm{C}) - D^2(\underline{\mathrm{B}};\underline{\mathrm{C}}) \geq 0\,. \tag{3.139}$$

Proof: In order to prove this, we must introduce the concept of the k-th *compound* of an $N\times N$ matrix $\underline{\mathbf{X}}$: This is the matrix $\underline{\mathbf{X}}^{(k)}$ whose elements are all $\binom{N}{k}$ $k\times k$ minors of $\underline{\mathbf{X}}$, arranged in lexicographic order. A well-known result on these matrices states that for any two $N\times N$ matrices $\underline{\mathbf{X}}$ and $\underline{\mathbf{Y}}$, $(\underline{\mathbf{X}}\cdot\underline{\mathbf{Y}})^{(k)} = \underline{\mathbf{X}}^{(k)}\cdot\underline{\mathbf{Y}}^{(k)}$. Since the compound of a diagonal matrix $\underline{\Omega}$ whose components are ± 1 or 0 is clearly a matrix of the same type, it follows that if $\underline{\mathbf{X}}$ is the matrix which reduces the metric matrix $\underline{\mathbf{M}}_B$ of a bilinear form to canonical form, then $\underline{\mathbf{X}}^{(k)}$ reduces the matrix $\underline{\mathbf{M}}_B^{(k)}$ to the corresponding canonical form. From this we see that the k-th compound of a matrix is positive semidefinite if and only if the matrix is. We now consider the $(k-1)$-th compound of the $2(k-1)\times 2(k-1)$ matrix:

$$\begin{bmatrix} B(\mathbf{b}_1-\mathbf{b}_k;\mathbf{b}_1-\mathbf{b}_k) & \cdots & B(\mathbf{b}_1-\mathbf{b}_k;\mathbf{b}_{k-1}-\mathbf{b}_k) & B(\mathbf{b}_1-\mathbf{b}_k;\mathbf{c}_1-\mathbf{c}_k) & \cdots & B(\mathbf{b}_1-\mathbf{b}_k;\mathbf{c}_{k-1}-\mathbf{c}_k) \\ \cdots & \cdots & \cdots & \cdots & \cdots & \cdots \\ B(\mathbf{b}_1-\mathbf{b}_k;\mathbf{c}_{k-1}-\mathbf{c}_k) & \cdots & B(\mathbf{b}_{k-1}-\mathbf{b}_k;\mathbf{c}_{k-1}-\mathbf{c}_k) & B(\mathbf{c}_1-\mathbf{c}_k;\mathbf{c}_{k-1}-\mathbf{c}_k) & \cdots & B(\mathbf{c}_{k-1}-\mathbf{c}_k;\mathbf{c}_{k-1}-\mathbf{c}_k) \end{bmatrix}. \tag{3.140}$$

We form a 2×2 matrix from the four corners of this compound matrix:

$$\begin{bmatrix} B(\mathbf{b}_1-\mathbf{b}_k,\ldots,\mathbf{b}_{k-1}-\mathbf{b}_k) & \begin{matrix} B(\mathbf{b}_1-\mathbf{b}_k,\ldots,\mathbf{b}_{k-1}-\mathbf{b}_k; \\ \mathbf{c}_1-\mathbf{c}_k,\ldots,\mathbf{c}_{k-1}-\mathbf{c}_k) \end{matrix} \\ \begin{matrix} B(\mathbf{b}_1-\mathbf{b}_k,\ldots,\mathbf{b}_{k-1}-\mathbf{b}_k; \\ \mathbf{c}_1-\mathbf{c}_k,\ldots,\mathbf{c}_{k-1}-\mathbf{c}_k) \end{matrix} & B(\mathbf{c}_1-\mathbf{c}_k,\ldots,\mathbf{c}_{k-1}-\mathbf{c}_k) \end{bmatrix}. \tag{3.141}$$

Since principal submatrices of positive semidefinite matrices are again positive semidefinite, this matrix is positive semidefinite, so that its determinant is non-negative. The result now follows from Lemma 3.48. QED

Note that this relation holds as an equality when k is the dimension of the space.

A number of classical inequalities involving the minors of positive definite matrices can similarly be translated into inequalities involving the Cayley-Menger bideterminants of Euclidean metric spaces. As an example, let $\underline{\mathbf{M}}$ be a positive semidefinite $N\times N$ matrix and for a given monotone sequence of integers $\underline{\lambda} \in \Lambda(N,k) := \{\underline{\lambda} \mid 1 \leq \lambda_1 < \cdots < \lambda_k \leq N\}$ let $M(\lambda)$ denote the determinant of the submatrix $\underline{\mathbf{M}}(\lambda)$ whose rows and columns are indexed by λ. According to the well-known Hadamard-Fischer inequality for positive (semi)definite matrices [Muir1906-1923, Carlson1968], for any two $\underline{\mu},\underline{\nu} \in \Lambda(N,k)$ we have

$M(\mu)M(\nu) \geq M(\mu \cap \nu)M(\mu \cup \nu)$. Together with our identities between Cayley-Menger determinants and Gramians (Lemma 3.25), this establishes:

PROPOSITION 3.58. *If* B, C *are any two subsets of the atoms of a premetric space* (A, D) *with* B∩C ≠ {}, *then*

$$D(\mathrm{B})D(\mathrm{C}) \quad \geq \quad D(\mathrm{B}\cap\mathrm{C})D(\mathrm{B}\cup\mathrm{C})\,. \tag{3.142}$$

From our geometric interpretation of these determinants as squared volumes in the nondegenerate factor space induced by B∪C, it is easily seen that if B∩C = {}, then

$$D(\mathrm{B})D(\mathrm{C}) \quad \geq \quad D(\mathrm{B}\cup\mathrm{C})\,. \tag{3.143}$$

References

Avis1980.

D. Avis, "On the Extreme Rays of the Metric Cone," *Canad. J. Math., 32*, 126-144(1980).

Becker1986.

E. Becker, "The Real Spectrum of a Ring and its Application to Semialgebraic Geometry," *Bull. Amer. Math. Soc., 15*, 19-60(1986).

Blumenthal1953.

L. Blumenthal, *Theory and Applications of Distance Geometry,* Cambridge Univ. Press, Cambridge, U.K., 1953. Reprinted by the Chelsea Publishing Co., New York, NY (1970).

Blumenthal1961.

L. Blumenthal, *A Modern View of Geometry,* Dover Publications, Inc., New York, NY, 1961.

Blumenthal1938.

L.M. Blumenthal, *Distance Geometries: A Study of the Development of Abstract Metrics,* The University of Missouri Studies, vol. XIII, no. 2, Univ. of Missouri, Columbia, 1938.

Brumfiel1979.

G.W. Brumfiel, *Partially Ordered Rings and Semi-Algebraic Geometry,* London Math. Soc. Lect. Notes Series no. 37, Cambridge Univ. Press, Cambridge, U.K., 1979.

Carlson1968.

D. Carlson, "On Some Determinantal Inequalities," *Proc. Amer. Math. Soc., 19*, 462-468(1968).

Cayley1841.

A. Cayley, "A Theorem in the Geometry of Position," *Cambridge Math. J., II,* 267-271(1841). See also *Collected Works of Arthur Cayley.*

Clifford1864.

W.K. Clifford, "Analytical Metrics," *Quart. J. Pure Appl. Math., 25,29,30*(1864). See also *Mathematical Papers by W. K. Clifford,* (R. Tucker, ed.), Chelsea Publ. Co. Bronx NY (1968).

Dress1986.

A.W.M. Dress and T.F. Havel, "Some Combinatorial Properties of Discriminants in Metric Vector Spaces," *Advances Math., 62,* 285-312(1986).

Dress1987.

A.W.M. Dress and T.F. Havel, "Fundamentals of the Distance Geometry Approach to the Problems of Molecular Conformation," Proc. Conference on Computer-Aided Geometric Reasoning, INRIA, Sophia Antipolis, FRANCE, 1987.

Gower1982.

J.C. Gower, "Euclidean Distance Geometry," *Math. Scientist, 7,* 1-14(1982).

Hodge1968.

W.V.D. Hodge and D. Pedoe, *Methods of Algebraic Geometry, volumes I & II,* Cambridge University Press, Cambridge, U.K., 1968.

Menger1928.

K. Menger, "Untersuchungen ueber allgemeine Metrik," *Math. Ann., 100,* 75-163(1928).

Menger1931.

K. Menger, "New Foundation of Euclidean Geometry," *Am. J. Math., 53,* 721-745(1931).

Muir1906-1923.

T. Muir, *Theory of Determinants in Historical Order of Development, volumes 1 - 4,* MacMillan Publishing Co., London, U.K., 1906-1923. Reprinted by Dover Publications, New York, NY (1966).

Schoenberg1935.

I.J. Schoenberg, "Remarks to Maurice Frechet's Article 'Sur la Definition Axiomatique d'une Classe d'Espace Distancies Vectoriellement Applicable sur l'Espace de Hilbert'," *Annals Math., 36,* 724-732(1935).

Seidel1955.

J. Seidel, "Angles and Distances in n-Dimensional Euclidean and Non-Euclidean Geometry. I-III," *Indag. Math., 17,* no. 3 & 4(1955). Reprinted *Proc. Koninkl. Nederl. Akademie van Wetenschappen,*

Seidel1952.
J.J. Seidel, "Distance-Geometric Development of 2-Dimensional Euclidean, Hyperbolic and Spherical Geometry. I & II," *Simon Stevin, 29*, 32-50,65-76(1952).

Sylvester1904-1912.
J.J. Sylvester, *Collected Works of J. J. Sylvester, volumes I - IV*, Cambridge Univ. Press, Cambridge, U.K., 1904-1912. Reprinted by Chelsea Publishing Co., New York, NY (1973).

4. Sparse Sets of Exact Distance Constraints

By an "exact distance constraint", we mean a statement to the effect that the distance between two specific atoms of a molecule is exactly equal to some given value "d" in all possible conformations of that molecule. Even though no distance in any real molecule is ever exactly the same in all of its possible conformations, for many distances the variation is small enough to make exact distance constraints a meaningful approximation. In particular, the covalent bond lengths and the geminal distances which are determined by bond angles are "fixed" to within a few percent in most organic molecules, and it often happens that other distances also have very nearly constant values. In a completely "rigid" molecule such as adamantane or dodecahedrane, this can even be true of all the distances. In the general case of mobile molecules, the determination of those conformations which are consistent with exact distance constraints remains an important step towards understanding their conformational behavior, as shown by the success of Dreiding models in conformational analysis [Dreiding1959].

In this chapter we study the kinematics of a collection of points in space which move subject to a sparse (i.e. incomplete) set of exact distance constraints. Such a system of constraints can be visualized as a "bar and joint framework" consisting of a collection of fixed-length bars connected at their endpoints by flexible joints (except that, in our abstract study, there is nothing to prevent two bars from passing through each other). The relations between the stresses, strains and deformations in these bar and joint frameworks is an important area of structural engineering [Smith1983, Pellegrino1986], and as such have been intensively studied for many years. Although the system of forces which exists between the atoms of a conformation of a molecule can be looked upon in this way, there is little advantage in doing so because the density of these interactions and their nonlinear elasticity renders the techniques developed in engineering largely inapplicable. In an attempt to place the engineering problems on a firm theoretical footing, however, an informal society of mathematicians known as the "Group in Structural Topology" has carefully studied the limiting case which results when the individual bars have zero elasticity. The resultant theory, published primarily in the group's own journal, *Structural Topology*, is precisely what is needed for our study of exact distance constraints, and constitutes a major source of material for this chapter.†

4.1. The Motions and Rigidity of Frameworks

If all of the interatomic distances in a molecule have known, fixed values, then it follows from the Fundamental Theorem of Euclidean Geometry that its conformation is uniquely determined except for its overall chirality. In practice, however, only a subset of the distances may have fixed values which are known *a priori*. A natural question to ask is: When are the known distances sufficient to uniquely determine the conformation and hence all of the remaining distances? This turns out to be a rather hard problem, which we prefer to postpone until Chapter 5. Here we shall concentrate on a much simpler version of the same question, namely: When is a given conformation which satisfies the exact distance constraints "locally unique", in the sense that it cannot be approximated arbitrarily closely by a sequence of (noncongruent) conformations which also satisfy those constraints? This question in turn is equivalent to asking: When are these distance constraints sufficient to hold the conformation "rigid", so that there is no way to move any part of it without moving all of it? This specific problem has been the subject of some very elegant papers [Gluck1975, Asimow1978, Asimow1979, Roth1981], and what we present in this section is primarily a summary of that work.

The following definition introduces some of the terminology which we shall use in our study of exact distance constraints.

Definition 4.1. An *abstract framework* $\mathbb{F} := (\mathrm{A}, \mathrm{B}, d)$ consists of graph $\mathbb{G} = (\mathrm{A}, \mathrm{B})$ together with a function $d : \mathrm{B} \rightarrow \mathbb{R}^+$ which assigns to each *bar* $\{\mathrm{a},\mathrm{b}\} \in \mathrm{B}$ its *length* $d(\{\mathrm{a},\mathrm{b}\})$. An *n-dimensional representation** of the abstract framework is a function $\mathbf{p} : \mathrm{A} \rightarrow \mathbb{R}^n$ such that

$$\| \mathbf{p}(\mathrm{a}) - \mathbf{p}(\mathrm{b}) \| \quad = \quad d(\{\mathrm{a}, \mathrm{b}\}) \tag{4.1}$$

for all bars $\{\mathrm{a},\mathrm{b}\} \in \mathrm{B}$. The graph $\mathbb{G}$ alone is called the *underlying graph* of the

† This chapter received the benefit of many useful comments from the Walter Whiteley, Henry Crapo and Robert Connelly during the Seminar in Structural Rigidity at the University of Montreal during the winter and spring of 1987, especially with regard to the numerous examples presented herein. A collective book by the Group in Structural Topology, *The Geometry of Rigid Structures* (Cambridge University Press), is in preparation, and will provide a more complete and general account of rigidity for those who are interested in learning still more.

* Once again, it is just as easy (and more revealing) to study the *n*-dimensional case.

framework.

By a slight abuse of notation, we shall generally abbreviate $d(\mathrm{a,b}) := d(\{\mathrm{a,b}\}) =: d(\mathrm{b,a})$. Note that the composition $\mathbf{t}\circ\mathbf{p}$ of any representation $\mathbf{p}$ of an abstract framework with an isometry $\mathbf{t}: \mathbb{R}^n \to \mathbb{R}^n$ of the ambient space is again a representation, i.e. all mappings $\mathbf{p}'$ which are congruent to a given representation are again representations. Nevertheless, for the purposes of this chapter it will generally be necessary to distinguish between different congruent representations.

Finding a representation of an abstract framework is essentially[‡] just a special case (albeit an important one) of the Fundamental Problem of Distance Geometry which was introduced in §1.2, and which will be further studied in §4.4 and Chapter 6. As is shown by the examples illustrated in Figure 4.1, however, even this special case can be quite complicated. For the present, we shall therefore assume that we have been explicitly given a representation $\mathbf{p}: \mathrm{A} \to \mathbb{R}^n$ which fulfills the imposed exact distance constraints, and proceed to study its geometric properties.

Definition 4.2. An n-dimensional *framework* $\mathbb{F} := (\mathrm{A}, \mathrm{B}, \mathbf{p}) =: \mathbb{G}_{\mathbf{p}}$ consists of a graph $\mathbb{G} = (\mathrm{A}, \mathrm{B})$ together with a map $\mathbf{p}: \mathrm{A} \to \mathbb{R}^n$ which assigns to each atom $\mathrm{a} \in \mathrm{A}$ its *position* $\mathbf{p}(\mathrm{a})$, such that $\mathbf{p}(\mathrm{a}) \neq \mathbf{p}(\mathrm{b})$ for all $\{\mathrm{a,b}\} \in \mathrm{B}$. The framework itself is often called a *realization* of the underlying graph $\mathbb{G}$.

Associated with each such framework is a function $d_{\mathbf{p}}: \mathrm{B} \to \mathbb{R}^+$ which assigns to each couple $\{\mathrm{a,b}\} \in \mathrm{B}$ its *length*

$$d_{\mathbf{p}}(\{\mathrm{a,b}\}) \quad := \quad \|\mathbf{p}(\mathrm{a}) - \mathbf{p}(\mathrm{b})\| \,, \tag{4.2}$$

and thus $\mathbf{p}$ is a representation of the abstract framework $(\mathrm{A}, \mathrm{B}, d_{\mathbf{p}})$. Since frameworks as above are commonly called "bar and joint" frameworks, we shall frequently refer to the atoms $\mathrm{a} \in \mathrm{A}$ or points $\mathbf{p}(\mathrm{a}) \in \mathbb{R}^n$ as the *joints* of the framework, while the term *bar* may refer either to the couple $\{\mathrm{a,b}\} \in \mathrm{B}$ or to the line segment $[(\mathbf{p}(\mathrm{a}), \mathbf{p}(\mathrm{b}))]$.

In the following it is often notationally simpler to work in the vector space $(\mathbb{R}^n)^N \approx \mathbb{R}^{nN}$ of $(n \cdot N)$-tuples of real numbers than it is to work with the vector space $(\mathbb{R}^n)^{\mathrm{A}}$ of n-valued maps over a set A of size N. Thus we choose an arbitrary but fixed order $\underline{\mathrm{A}} = [\mathrm{a}_1, \ldots, \mathrm{a}_N]$ for our set of atoms A, and let $\mathbf{p}(\underline{\mathrm{A}})$ be the nN-dimensional vector:

$$\mathbf{p}(\underline{\mathrm{A}}) \quad := \quad [\mathbf{p}(\mathrm{a}_1) \ldots \mathbf{p}(\mathrm{a}_N)] \tag{4.3}$$

‡ The only real difference is that representations of abstract frameworks are not required to be injective. We have relaxed this requirement because there are no simple conditions that one can impose on the length function d which will insure that all representations $\mathbf{p}: \mathrm{A} \to \mathbb{R}^n$ of an abstract framework have this property.

It follows from our results in Chapter 3 that a given abstract framework may or may not possess a representation; for example, the function d could violate the triangle inequality $d(\mathrm{a,b}) \leq d(\mathrm{a,c}) + d(\mathrm{b,c})$ for some triple of bars $\{\mathrm{a,b}\}, \{\mathrm{a,c}\}, \{\mathrm{b,c}\} \in \mathrm{B}$. The existence of representations also depends upon the dimension; we shall use the symbol $Dim(\mathbb{F}) \in 2^{\mathbf{Z}}$ to denote the set of values of n such that the framework admits an n-dimensional representation, and let $dim(\mathbb{F}) = min(\, n \mid n \in Dim(\mathbb{F}))$. The example shown in Figure 4.1 shows that $Dim(\mathbb{F})$ does not always consist of a consecutive sequence of integers. Representations can be unique, finite in number, or continuously variable.

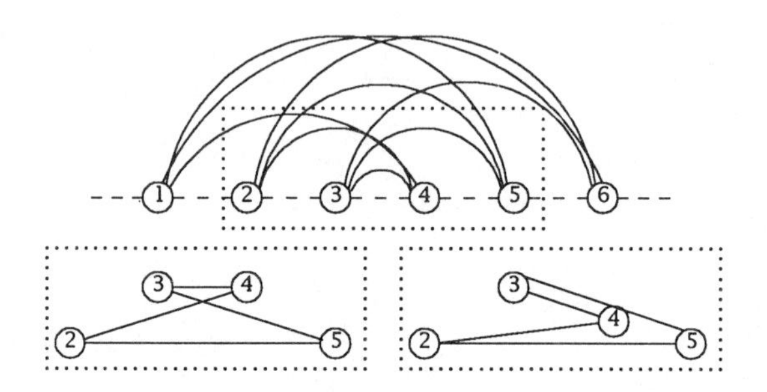

Figure 4.1

A 1-dimensional framework whose underlying graph is $\mathbb{K}_{3,3}$ and whose distances are those shown in the figure cannot be realized in two dimensions with the same bar lengths, although it can be realized in three or more dimensions, e.g. by the coordinates [3,0,0], [0,2,0], [0,0,1], [0,0,0], [-0.61,-0.81,-0.87], [-1,-3/2,-8/3].

which is obtained by concatenation. Other multivectors of this form will be denoted by underlining, e.g. $\underline{\mathbf{x}} \in \mathbb{R}^{nN}$, and their component n-tuples will be denoted by subscripting, so that $\underline{\mathbf{x}} = [\mathbf{x}_1 \ldots \mathbf{x}_N]$. For a given graph $\mathbb{G} = (\mathrm{A, B})$ with $M := \#\mathrm{B}$, we shall also give the set of bars B the corresponding lexicographic order

$$\underline{\mathrm{B}} \;=\; [\{\mathrm{a}_{\mu_1}, \mathrm{a}_{\nu_1}\}, \ldots, \{\mathrm{a}_{\mu_M}, \mathrm{a}_{\nu_M}\}] \tag{4.4}$$

for some $\underline{\mu}, \underline{\nu} \in \{1, \ldots, N\}^M$ with $\mu_{i-1} \leq \mu_i < \nu_i \leq \nu_{i+1}$ for all suitable i. It should be clearly understood, however, that our results are all independent of the order which A and B are given!

Definition 4.3. The *rigidity mapping** $\mathbf{D}_G : \mathbb{R}^{nN} \rightarrow \mathbb{R}^M$ associated with a given graph $\mathbb{G} = (\mathrm{A, B})$ is defined by

$$\mathbf{D}_G(\underline{\mathbf{p}}) \;:=\; [\, \|\mathbf{p}_{\mu_i} - \mathbf{p}_{\nu_i}\|^2 \mid i = 1, \ldots, M] \tag{4.5}$$

for all $\mathbf{p} = [\mathbf{p}_1 \ldots \mathbf{p}_N] \in \mathbb{R}^{nN}$.

Observe that the rigidity mapping $\mathbf{D}_G$ is never injective, since if $\mathbf{s} \in \mathbf{ST}(n,\mathbb{R})$ is any screw translation of $\mathbb{R}^n$, then $\mathbf{D}_G(\underline{\mathbf{s} \circ \mathbf{p}}) = \mathbf{D}_G(\mathbf{p})$ for all $\mathbf{p} \in \mathbb{R}^{nN}$, where $\underline{\mathbf{s} \circ \mathbf{p}}$ denotes the vector $[\mathbf{s}(\mathbf{p}_1) \ldots \mathbf{s}(\mathbf{p}_N)]$. Nevertheless, we shall use the symbol $\mathbf{D}_G^{-1}(\mathbf{D}_G(\mathbf{p}))$ for the set

$$\mathbf{D}_G^{-1}(\mathbf{D}_G(\mathbf{p})) \;:=\; \{\, \mathbf{x} \in \mathbb{R}^{nN} \,|\, \mathbf{D}_G(\mathbf{x}) \;=\; \mathbf{D}_G(\mathbf{p}) \,\} \,. \tag{4.6}$$

We shall also denote by $\mathbf{ST}(\mathbf{p})$ the orbit of $\mathbf{p}$ under the group action given by $(\mathbf{s}, \mathbf{p}) \mapsto \underline{\mathbf{s} \circ \mathbf{p}}$. Then for any framework $(\mathrm{A}, \mathrm{B}, \mathbf{p})$ as above it is clear that $\mathbf{D}_G^{-1}(\mathbf{D}_G(\mathbf{p}(\underline{\mathrm{A}}))) \supseteq \mathbf{ST}(\mathbf{p}(\underline{\mathrm{A}}))$, and it follows from the Fundamental Theorem of Euclidean Geometry that $\mathbf{D}_K^{-1}(\mathbf{D}_K(\mathbf{p}(\underline{\mathrm{A}}))) = \mathbf{ST}(\mathbf{p}(\underline{\mathrm{A}}))$, where $\mathbf{D}_K : \mathbb{R}^{nN} \to \mathbb{R}^{N(N-1)/2}$ is the rigidity mapping associated with the complete graph $\mathbb{K}_N$.

We are now ready to define rigidity as a form of local uniqueness.

Definition 4.4. A framework $\mathbb{F} = (\mathrm{A}, \mathrm{B}, \mathbf{p})$ is called *rigid* if there exists a neighborhood† $\mathbf{U}_\mathbf{p} \subseteq \mathbb{R}^{nN}$ of $\mathbf{p}(\underline{\mathrm{A}})$ such that $\mathbf{U}_\mathbf{p} \cap \mathbf{D}_G^{-1}(\mathbf{D}_G(\mathbf{p}(\underline{\mathrm{A}}))) \;=\; \mathbf{U}_\mathbf{p} \cap \mathbf{ST}(\mathbf{p}(\underline{\mathrm{A}}))$. A *motion* of $\mathbb{F}$ is a continuous mapping $\mathbf{p}^t : [0 \cdots 1] \to (\mathbb{R}^n)^{\mathrm{A}}$ such that $\mathbf{p}^0 = \mathbf{p}$ and $\mathbf{p}^t(\underline{\mathrm{A}}) \in \mathbf{D}_G^{-1}(\mathbf{D}_G(\mathbf{p}(\underline{\mathrm{A}})))$ for all $t \in [0 \cdots 1]$. A *flexing* is a motion which satisfies $\mathbf{p}^t(\underline{\mathrm{A}}) \notin \mathbf{ST}(\mathbf{p}(\underline{\mathrm{A}}))$ for all $t \in (0 \cdots 1]$, and the framework itself is called *flexible* if such a flexing exists.

In Chapter 1 we defined a molecule as "rigid" if its conformation space $\mathbf{ST}(\mathbf{P})$ consists of a finite number of isolated points. The following proposition shows that the above definition is consistent with that given in Chapter 1.

PROPOSITION 4.5. *A framework is rigid if and only if it is not flexible.*

The proof of this proposition and the lemmas which follow involve mathematics beyond the usual prerequisites for this book, and so have been omitted (see [Asimow1979]). Nevertheless, some rather basic definitions from these areas of mathematics are necessary for the purposes of this chapter.‡

Hence let $\mathbf{V} \subseteq \mathbf{V}' \subseteq \mathbb{R}^m$ and $\mathbf{W} \subseteq \mathbb{R}^M$ with $\mathbf{V}'$ open in $\mathbb{R}^m$. A map $\mathbf{D} : \mathbf{V} \to \mathbf{W}$ is called *smooth* if it can be extended to a map $\mathbf{D}' : \mathbf{V}' \to \mathbf{W}$ such that for all $[i_1, \ldots, i_m] \in \{0, \mathbf{Z}^+\}^m$ with $k = \sum i_j$ the partial derivatives $\partial^k \mathbf{D}' / \partial v_1^{i_1} \cdots \partial v_m^{i_m}$ exist, and the set of all smooth functions on $\mathbf{V}$ is denoted by C^∞ (·). If in addition the map $\mathbf{D}$ is bijective and has a smooth inverse $\mathbf{D}^{-1}$, then it is called a *diffeomorphism*, and any two subsets $\mathbf{V}$ and $\mathbf{W}$ which are related in this way are termed *diffeomorphic*. A (smooth) *m-dimensional manifold* $\mathbf{M}$ is a subset of $\mathbb{R}^M$

* We assume that $\mathbf{D}_G$ has been extended by continuity to those points of $\mathbb{R}^{nN}$ which correspond to zero length bars.

† All our neighborhoods are open unless otherwise specified.

‡ An excellent introduction to these concepts may be found in [Guillemin1974].

for which each $\mathbf{m} \in \mathbf{M}$ has a neighborhood $\mathbf{U_m} \subseteq \mathbb{R}^M$ such that $\mathbf{M} \cap \mathbf{U_m}$ is diffeomorphic with an open subset $\mathbf{W_m} \subseteq \mathbb{R}^m$. Intuitively, such a manifold can be thought of as (a piece of) a smooth surface embedded in space, which may bend and twist but does not self-intersect. The diffeomorphisms $\mathbf{D} = \mathbf{D_m}$ then establish a one-to-one correspondence between the points of $\mathbf{W_m} \subseteq \mathbb{R}^m$ and the points in $\mathbf{M} \cap \mathbf{U_m}$, thus providing us with a "local coordinate system" on $\mathbf{M}$ (see Figure 4.2).

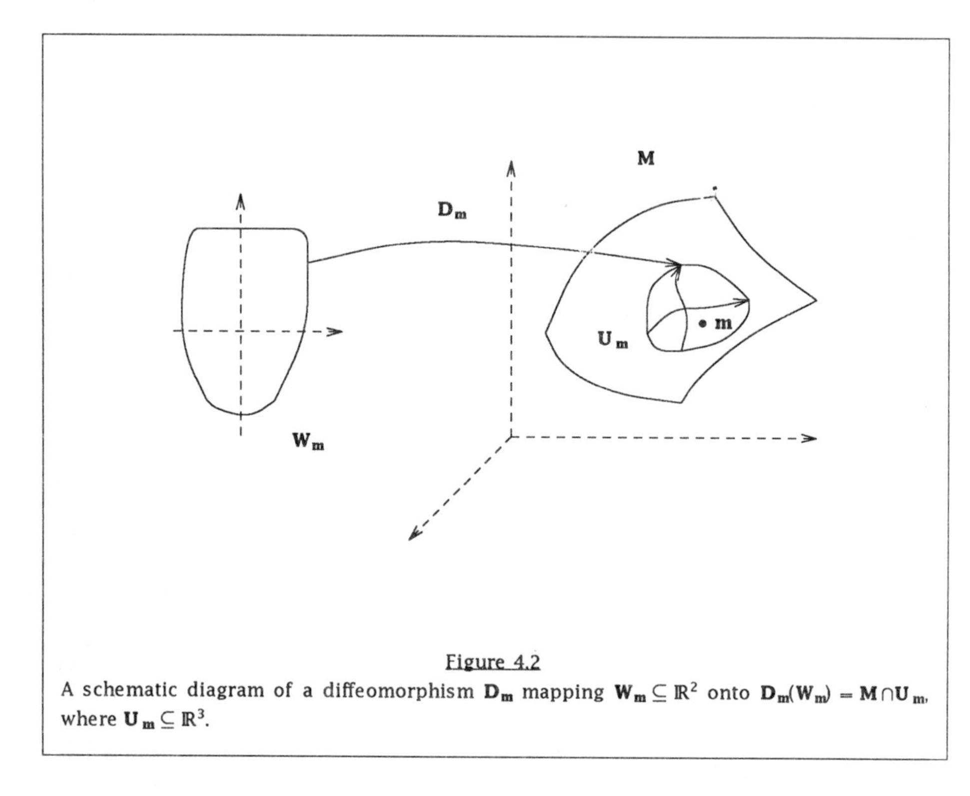

Figure 4.2
A schematic diagram of a diffeomorphism $\mathbf{D_m}$ mapping $\mathbf{W_m} \subseteq \mathbb{R}^2$ onto $\mathbf{D_m}(\mathbf{W_m}) = \mathbf{M} \cap \mathbf{U_m}$, where $\mathbf{U_m} \subseteq \mathbb{R}^3$.

The following lemma establishes sufficient conditions for a subset of $\mathbb{R}^M$, given explicitly as the image of a map, or a subset of $\mathbb{R}^m$, given implicitly as the preimage of a point, to be a manifold.

LEMMA 4.6. *Let* $\mathbf{D} : \mathbb{R}^m \rightarrow \mathbb{R}^M$ *be a smooth map,* $R(\mathbf{D}) := max(rank(\underline{\partial \mathbf{D}}(\mathbf{x})) \mid \mathbf{x} \in \mathbb{R}^m)$, *and* $\mathbf{v} \in \mathbb{R}^m$ *such that* $rank(\underline{\partial \mathbf{D}}(\mathbf{v})) = R(\mathbf{D})$. *Then for some neighborhood* $\mathbf{U_v}$ *of* $\mathbf{v}$, *the image* $\mathbf{D}(\mathbf{U_v}) \subseteq \mathbb{R}^M$ *is a smooth manifold of dimension* $R(\mathbf{D})$, *and the inverse image* $\mathbf{U_v} \cap \mathbf{D}^{-1}(\mathbf{D}(\mathbf{v})) \subseteq \mathbb{R}^m$ *is a smooth manifold of dimension* $m - R(\mathbf{D})$.

A point $\mathbf{v} \in \mathbb{R}^m$ is called a *regular point* of $\mathbf{D}$ if $rank(\underline{\partial \mathbf{D}}(\mathbf{v})) = R(\mathbf{D})$; otherwise, $\mathbf{v}$ is called a *singular point* of $\mathbf{D}$. In the special case that $m = M = R(\mathbf{D})$, the Inverse

Function Theorem implies that $\mathbf{D}$ is locally a diffeomorphism, so that $\mathbf{D}(\mathbf{U_v})$ is open in $\mathbb{R}^M$ and $\mathbf{D}^{-1}(\mathbf{D}(\mathbf{v}))$ is a single point in $\mathbb{R}^m$. Similarly, the special case in which $m > M = R(\mathbf{D})$ follows from the Implicit Function Theorem, and hence the above lemma can be regarded as a straightforward generalization of some well known results from calculus. Our final lemma provides us with an explicit example of a manifold.

LEMMA 4.7. *Suppose that* $[\mathbf{x}_1, \ldots, \mathbf{x}_N]$ *is a sequence of points in* $\mathbb{R}^n$, $\underline{\mathbf{x}} = [\mathbf{x}_1 \ldots \mathbf{x}_N] \in \mathbb{R}^{nN}$ *is their concatenation and that* $k = dim(<\mathbf{x}_1, \ldots, \mathbf{x}_N>)$ *is the dimension of their affine span. Then the orbit*

$$\mathbf{ST}(\underline{\mathbf{x}}) \;:=\; \{\underline{\mathbf{y}} \in \mathbb{R}^{nN} \,|\, \underline{\mathbf{y}} = [\mathbf{s}(\mathbf{x}_1) \ldots \mathbf{s}(\mathbf{x}_N)],\ \mathbf{s} \in \mathbf{ST}(n,\ \mathbb{R})\} \tag{4.7}$$

is a smooth manifold of dimension

$$dim(\mathbf{ST}(\underline{\mathbf{x}})) \;=\; \tfrac{1}{2}(k+1)(2n-k)\,. \tag{4.8}$$

With all that behind us, we can present a simple "predictor" for the rigidity of frameworks.

THEOREM 4.8. *Let* $\mathbb{F} = (\mathrm{A}, \mathrm{B}, \mathbf{p})$ *be an n-dimensional framework with underlying graph* $\mathbb{G} = (\mathrm{A}, \mathrm{B})$, $N := \#\mathrm{A}$ *and* $k := dim(<\mathbf{p}(\mathrm{A})>)$. *Also suppose that* $\mathbf{p}(\underline{\mathrm{A}}) \in \mathbb{R}^{nN}$ *is a regular point of the rigidity mapping* $\mathbf{D}_G : \mathbb{R}^{nN} \to \mathbb{R}^{\#\mathrm{B}}$. *Then* $\mathbb{F}$ *is rigid if and only if*

$$rank(\underline{\partial \mathbf{D}}_G(\mathbf{p}(\underline{\mathrm{A}}))) \;=\; nN \;-\; \tfrac{1}{2}(k+1)(2n-k)\,, \tag{4.9}$$

and $\mathbb{F}$ *is flexible if and only if*

$$rank(\underline{\partial \mathbf{D}}_G(\mathbf{p}(\underline{\mathrm{A}}))) \;<\; nN \;-\; \tfrac{1}{2}(k+1)(2n-k)\,. \tag{4.10}$$

Proof: Let $r := max_{\underline{\mathbf{x}}}(rank(\underline{\partial \mathbf{D}}_G(\underline{\mathbf{x}})))$ and $\underline{\mathbf{p}} := \mathbf{p}(\underline{\mathrm{A}})$. By Lemma 4.6 there exists a neighborhood $\mathbf{U_p} \subseteq \mathbb{R}^{nN}$ of $\underline{\mathbf{p}}$ such that $\mathbf{V_p} := \mathbf{U_p} \cap \mathbf{D}_G^{-1}(\mathbf{D}_G(\underline{\mathbf{p}}))$ is a smooth manifold of dimension $nN - r$, whereas by Lemma 4.7 $\mathbf{W_p} := \mathbf{U_p} \cap \mathbf{ST}(\underline{\mathbf{p}})$ is a smooth submanifold of $\mathbf{V_p}$ of dimension $\tfrac{1}{2}(k+1)(2n-k) \leq nN - r$. Therefore

$$r \;\leq\; nN - \tfrac{1}{2}(k+1)(2n-k)\,, \tag{4.11}$$

and $r = nN - \tfrac{1}{2}(k+1)(2n-k)$ implies $dim(\mathbf{W_p}) = dim(\mathbf{V_p}) \iff \mathbf{W_p} = \mathbf{V_p}$ for some sufficiently small neighborhood $\mathbf{U_p}$, which means that $\mathbb{F}$ is rigid by definition and hence not flexible by Proposition 4.5. Conversely, if $\mathbb{F}$ is flexible then it is not rigid, i.e. $\mathbf{W_p} \subset \mathbf{V_p}$ for all $\mathbf{U_p}$, so $r < nN - \tfrac{1}{2}(k+1)(2n-k)$ as desired. QED

We will call a framework *regular* whenever the corresponding nN-vector $\mathbf{p}(\underline{\mathrm{A}})$ is a regular point of its rigidity mapping.

Since the rank of a matrix can be determined by Gaussian elimination, Theorem 4.8 allows us to establish the rigidity or flexibility of regular realizations of graphs in time polynomial in the number of bars and joints. With this as our motivation, we will now consider a few basic properties which are common to all regular frameworks. We first prove the intuitively appealing fact that a regular framework with too few bars can never be rigid.

COROLLARY 4.9. *If* $\mathbb{F} = (\mathrm{A}, \mathrm{B}, \mathbf{p})$ *is a regular framework,* $N := \#\mathrm{A}$ *and* $M := \#\mathrm{B}$, *then* $\mathbb{F}$ *is flexible if* $M < nN - n(n+1)/2$.

Proof: Let $\underline{\mathbf{p}} := \mathbf{p}(\underline{\mathrm{A}})$ and $k := dim(\langle\mathbf{p}(\mathrm{A})\rangle)$. Since the function $h(x) := nN - ½(x+1)(2n-x)$ satisfies $h(n) = h(n-1) = nN - ½n(n+1)$ and has a unique minimum at $x = n - ½$, we have

$$rank(\underline{\partial \mathbf{D}}(\underline{\mathbf{p}})) \;\leq\; M \;<\; nN - ½n(n+1) \;\leq\; nN - ½(k+1)(2n-k)\,, \tag{4.12}$$

from which it follows that $\mathbb{F}$ is flexible by Theorem 4.8. QED

Next, we show that all regular and rigid realizations of a graph are necessarily "full-dimensional".

COROLLARY 4.10. *If* $(\mathrm{A}, \mathrm{B}, \mathbf{p})$ *is an n-dimensional rigid framework and* $\mathbf{p}(\underline{\mathrm{A}})$ *is a regular point of its rigidity mapping, then* $dim(\langle\mathbf{p}(\mathrm{A})\rangle) = min(\#\mathrm{A}-1, n)$. *Moreover,* $dim(\langle\mathbf{p}(\mathrm{A})\rangle) = \#\mathrm{A}-1$ *only if the underlying graph* $\mathbb{G} = (\mathrm{A}, \mathrm{B})$ *is complete.*

Proof: Let $N := \#\mathrm{A}$. To prove the first statement, we first observe that for any $\mathbf{p} \in (\mathbb{R}^n)^{\mathrm{A}}$ such that $dim(\langle\mathbf{p}(\mathrm{A})\rangle) =: n' \leq n$ there exists a screw-translation $\mathbf{s}$ of $\mathbb{R}^n$ such that $\mathbf{s}{\circ}\mathbf{p}(\mathrm{A}) \subset \tilde{\mathbb{R}}^{n'} \subseteq \mathbb{R}^n$, where $\tilde{\mathbb{R}}^{n'} \approx \mathbb{R}^{n'}$ is the subspace of $\mathbb{R}^n$ given by all vectors whose last $n-n'$ components are zero. At the same time, it is not hard to show that the rank of the Jacobian $\underline{\partial \mathbf{D}}_G$ is invariant under screw-translation, i.e.

$$rank(\underline{\partial \mathbf{D}}_G(\mathbf{p}(\underline{\mathrm{A}}))) \;=\; rank(\underline{\partial \mathbf{D}}_G(\mathbf{t}{\circ}\mathbf{p}(\underline{\mathrm{A}}))) \tag{4.13}$$

for any screw-translation $\mathbf{t} \in \mathbf{ST}(n, \mathbb{R})$. From this it follows that $\mathbf{p}' := \mathbf{s}{\circ}\mathbf{p}$ gives us a regular realization $\mathbb{G}_{\mathbf{p}'}$ in $\mathbb{R}^{n'}$. Also, since any flexing in $\mathbb{R}^{n'}$ can also be viewed as flexing in $\mathbb{R}^n \supseteq \mathbb{R}^{n'}$, the framework $\mathbb{F}' := (\mathrm{A}, \mathrm{B}, \mathbf{p}')$ is necessarily rigid in $\mathbb{R}^{n'}$.

Now define a function $h : \mathbb{R} \to \mathbb{R}$ by

$$h(x) \;:=\; N{\cdot}x - ½(n'+1)(2x-n') \;=\; a{\cdot}x + b \quad \forall\, x \in \mathbb{R}\,, \tag{4.14}$$

where $a = N-(n'+1)$ and $b = n'(n'+1)/2$. If $\mathbf{D}_G$ is the rigidity mapping of the underlying graph of $\mathbb{F}$ and $\mathbb{F}'$, it follows from Theorem 4.8 together with the foregoing considerations that

$$h(n') \;=\; n'N - ½(n'+1)(2n'-n') \;=\; rank(\underline{\partial \mathbf{D}}_G(\mathbf{p}'(\underline{\mathrm{A}}))) \tag{4.15}$$
$$=\; rank(\underline{\partial \mathbf{D}}_G(\mathbf{p}(\underline{\mathrm{A}}))) \;=\; nN - ½(n'+1)(2n-n') \;=\; h(n)\,.$$

Hence either $n = n'$ or $a = N-n'-1 = 0$, which since n' never exceeds $min(N-1, n)$ implies that $n' = min(N-1, n)$, as desired.

To prove the second statement, we observe that if $n \geq N-1$ then $dim(<\mathbf{p}(A)>) = N-1$, so $rank(\partial \mathbf{D}(\mathbf{p}(\underline{A}))) = N(N-1)/2$ by Theorem 4.8 and hence

$$M \;\leq\; N(N-1)/2 \;=\; rank(\partial \mathbf{D}(\mathbf{p}(\underline{A}))) \;\leq\; M\,, \tag{4.16}$$

i.e. $M = N(N-1)/2$ <=> $\mathbb{G}$ is complete. QED

If $\mathbb{G}$ is the complete graph on four atoms minus any one couple, then any realization in $\mathbb{R}^3$ such that no three joints $\mathbf{p}(a_i)$ are collinear is both regular and flexible, so that regular realizations of different dimensions exist. Our final corollary shows that this does not happen when one and hence both of these realizations gives a rigid framework.

COROLLARY 4.11. *Let* $\mathbb{G} = (A, B)$ *be a graph and let* $\mathbb{G}_{\mathbf{p}}$, $\mathbb{G}_{\mathbf{p}'}$ *be two regular realizations thereof in* $\mathbb{R}^n$. *If* $\mathbb{F} := \mathbb{G}_{\mathbf{p}}$ *is rigid, then so is* $\mathbb{F}' := \mathbb{G}_{\mathbf{p}'}$, *and* $dim(<\mathbf{p}(A)>) = dim(<\mathbf{p}'(A)>)$.

Proof: Let $k := dim(<\mathbf{p}(A)>)$, $k' := dim(<\mathbf{p}'(A)>)$, and $\mathbf{D}_G$ be the rigidity mapping of $\mathbb{G}$. As in the proof of Corollary 4.10, we may without loss of generality assume that $k' < k = n$, and as always $k' \leq min(N-1, n)$. Since $\mathbb{F}$ is rigid, by Theorem 4.8

$$rank(\partial \mathbf{D}_G(\mathbf{p}(\underline{A}))) \;=\; kN \;-\; \tfrac{1}{2}k(k+1) \tag{4.17}$$

whereas, likewise by Theorem 4.8

$$rank(\partial \mathbf{D}_G(\mathbf{p}'(\underline{A}))) \;\leq\; k'N \;-\; \tfrac{1}{2}k'(k'+1)\,. \tag{4.18}$$

Since both frameworks are regular the left-hand sides of both of these equations are equal. Thus the function $h(x) = Nx - \tfrac{1}{2}x(x+1)$ satisfies $h(k) \leq h(k')$ and is monotone increasing for all $x \leq N-1$, which means that $k' = k$ as desired. QED

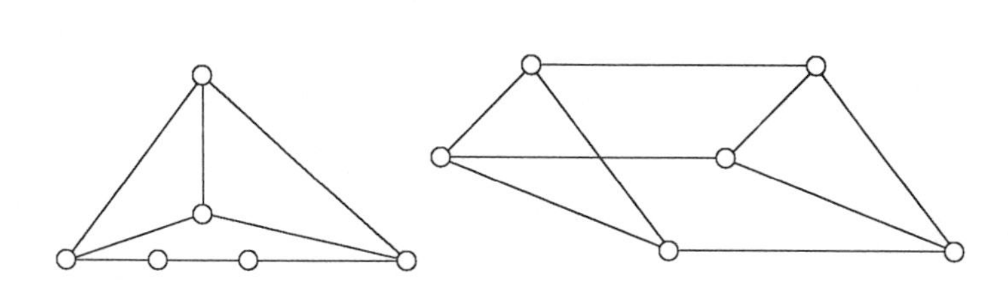

Figure 4.3

Examples in $\mathbb{R}^2$ of a nonrigid realization of a generically rigid graph (right), and of a rigid realization of a generically flexible graph (left).

As shown by the examples given in Figure 4.3 (which are due to B. Grünbaum [Asimow1979]), there exist both nonrigid realizations of graphs whose regular realizations are rigid, and nonflexible realizations of graphs whose regular realizations are flexible. Since all of the above results are valid only for regular realizations and the only way to actually prove that a given framework is regular is to solve the global optimization problem $max(rank(\partial\underline{\mathbf{D}}_G))$, it may at first seem that these results are of little use. Fortunately, nonregular realizations of graphs are the exception rather than the rule!

PROPOSITION 4.12. *Let* $\mathbb{G} = (A, B)$ *be a graph. Then the set of regular realizations of* $\mathbb{G}$ *is open and dense in* $(\mathbb{R}^n)^A$.

Proof: Let $r := max(rank(\partial\underline{\mathbf{D}}_G))$, and consider the polynomial $f : \mathbb{R}^{nN} \rightarrow \mathbb{R}$ which is obtained by taking the sum of the squares of all $r \times r$ subdeterminants of the matrix $\partial\underline{\mathbf{D}}_G$. By our choice of r this polynomial is not uniformly zero and vanishes if and only if its argument is not regular. Since f is continuous, its locus $f^{-1}(0)$ is closed and the set-theoretic complement of the locus, i.e. the set of all regular realizations, is open. Moreover, the polynomial f vanishes on an open set only if its gradient is zero throughout that set. Since each component of the gradient is a polynomial of lower total degree and the result is trivial for constant polynomials, it follows by induction on the total degree that f does not vanish on any open set, i.e. every open set contains regular realizations of $\mathbb{G}$. QED

This topological property of the set of all regular realizations is a very strong one. Via Fubini's theorem [Guillemin1974], one can conclude that versus the "natural" or Lebesgue measure on the space $(\mathbb{R}^n)^A$, the probability that a "randomly" picked realization of a graph will be nonregular is zero!*

The following algorithm allows one to obtain a regular realization of a graph and hence solve the optimization problem $max(rank(\partial\underline{\mathbf{D}}_G))$ with arbitrarily small probability of failure in polynomial time.

Algorithm 4.13:

INPUT: A graph $\mathbb{G} = (A, B)$ and a positive integer $I \in \mathbf{Z}^+$.

OUTPUT: The value of $max(rank(\partial\underline{\mathbf{D}}_G))$ with probability $P(I) \rightarrow 0$ as $I \rightarrow \infty$.

PROCEDURE:

> Choose coordinates for the atoms at random from the interval $[0 \cdots I]$.
> Determine the rank of the resulting Jacobian by Gaussian elimination
> using exact arithmetic.

---

* It should be understood that in actual numerical calculations using finite precision arithmetic, any realization which is even "sufficiently" close to a nonregular one will be effectively nonregular. Under some circumstances, the chances of being sufficiently close may not only be finite, but quite substantial. This can be taken into account by looking at the *condition number* of the Jacobian and its submatrices, rather than its rank.

As $I \rightarrow \infty$, the probability of a random dependence among the columns of $\underline{\partial \mathbf{D}}$ goes to zero. At the same time, however, the time required for Gaussian elimination is exponential in I. The development of algorithms of this type is one approach to obtaining practical methods for solving NP-complete problems [Rabin1976].

Since a framework is either rigid or flexible for "almost all" realizations of the underlying graph, one might expect that this *generic* rigidity or flexibility relates in some way to the structure of the graph. We shall see in §4.4 that this is indeed the case, but before we try to develop such a *combinatorial approach*, it will be useful to consider some alternative ways of looking at the problem.

4.2. The Static and Infinitesimal Properties of Frameworks

Even though the definition of rigidity given in the previous section is easily decided for "almost all" realizations of a given graph $\mathbb{G}$, the admissible embeddings of molecules with internal symmetry are highly exceptional and may therefore be nonregular! An example of such a molecule will be discussed in §4.5. Here we wish to explore some new concepts of rigidity which extend more simply to nonregular frameworks. Because we shall motivate our new definitions by means of physical analogies (which may themselves be of some chemical interest), we restrict ourselves to three dimensions in what follows. These definitions and most of the results in this section, however, can be extended to the n-dimensional case as well.

The first of our definitions is based upon the idea that if the framework is rigid, then any system of forces applied to the joints which has no tendency to translate or rotate the framework as a whole should result in a system of forces within the bars which exactly counteracts the applied system. For this reason, we will now briefly consider an elegant analytic method of describing the effect of a system of forces upon a rigid body which, although originally conceived by Graßmann and Plücker in the latter half of the last century, remained largely forgotten until its recent revival by the Group in Structural Topology [Whiteley1978, Crapo1979, Crapo1982] (see also [Gruenbaum1975]).

A system of forces acting on a rigid body is said to be in *equilibrium* if it produces no translation or rotation. As is well known [Goldstein1950], the effect of a single force $\mathbf{f}_i$ acting upon a point $\mathbf{p}_i$ in a rigid body is completely determined by the inertial tensor of the body together with the components of the force and its torque $\mathbf{p}_i \times \mathbf{f}_i$ about an (arbitrary) origin in space. Since the force and torque which result from the application of a system of forces to a rigid body is given by the vector sum of the individual forces and torques, a system of forces $[\mathbf{f}_i \mid i = 1, \ldots, N]$ is in equilibrium if and only if the sum of the components of the forces and their associated torques about any one origin (and hence all possible origins) vanish, i.e.

$$\sum_{i=1}^{N} \mathbf{f}_i = \mathbf{0} \quad \text{and} \quad \sum_{i=1}^{N} \mathbf{p}_i \times \mathbf{f}_i = \mathbf{0}\,. \tag{4.19}$$

Observe that if we define four-dimensional homogeneous coordinates for the free vector $\hat{\mathbf{f}}_i := [0, \mathbf{f}_i]$ and affine point $\hat{\mathbf{p}}_i := [1, \mathbf{p}_i]$, then (except for some inconsequential signs) the two conditions given in Equation (4.19) can be written as a single equation:

$$\mathbf{F} := \sum_{i=1}^{N} \mathbf{F}_i = \mathbf{0} \quad \text{where} \tag{4.20}$$

$$\mathbf{F}_i := \hat{\mathbf{p}}_i \vee \hat{\mathbf{f}}_i = \begin{bmatrix} 1 \\ p_{1i} \\ p_{2i} \\ p_{3i} \end{bmatrix} \vee \begin{bmatrix} 0 \\ f_{1i} \\ f_{2i} \\ f_{3i} \end{bmatrix} = \begin{bmatrix} f_{1i} \\ f_{2i} \\ f_{3i} \\ p_{1i} f_{2i} - p_{2i} f_{1i} \\ p_{1i} f_{3i} - p_{3i} f_{1i} \\ p_{2i} f_{3i} - p_{3i} f_{2i} \end{bmatrix},$$

and "$\vee$" denotes the outer product of the vectors $\hat{\mathbf{p}}_i$ and $\hat{\mathbf{f}}_i$ (see Appendix C). Thus the effect of a single force $\mathbf{f}_i$ upon a rigid body is completely determined by a *line-bound vector* of length $\|\mathbf{f}_i\|$ on the line parallel to $\mathbf{f}_i$ through the point $\mathbf{p}_i$. Since the line-bound vector $\mathbf{F}_i := \hat{\mathbf{p}}_i \vee \hat{\mathbf{f}}_i$ remains the same when the point $\mathbf{p}_i$ is translated along the line determined by the free vector $\hat{\mathbf{f}}_i$, the effect of the force upon a body remains the same when its point of application is translated along its line of action. Moreover, since the line-bound vector $\mathbf{F}_i$ is simple, its components satisfy the 3-term Graßmann-Plücker relation:

$$\mathbf{F}_i \vee \mathbf{F}_i = 2(F_{1i}F_{4i} - F_{2i}F_{5i} + F_{3i}F_{6i}) = \mathbf{0}\,. \tag{4.21}$$

The 2-vector $\mathbf{F}$ which describes the effect of a system of forces upon a rigid body is obtained by summing the line-bound vectors $\mathbf{F}_i$ of the individual forces, and will not in general be simple. Such an arbitrary composition of forces is called a *wrench.* In Figure 4.4, we give some elementary examples of systems of forces and their associated line-bound vectors.

LEMMA 4.14. *The equilibrium systems of force acting upon the joints of a rigid 3-dimensional framework* $\mathbb{F} = (\mathrm{A}, \mathrm{B}, \mathbf{p})$ *comprise a subspace of* $(\mathbb{R}^3)^{\mathrm{A}}$ *of dimension* $3\#\mathrm{A}-6$ *(*$3\#\mathrm{A}-5$ *if the framework is collinear).*

The three types of systems of forces acting on arbitrary rigid bodies: *simple* forces occur when any two forces act at the same point (or both act along parallel lines); *force couples* occur when two equal but opposite forces act along parallel lines; all other combinations are wrenches, which cannot be reduced to any simple force or force couple acting on the body.

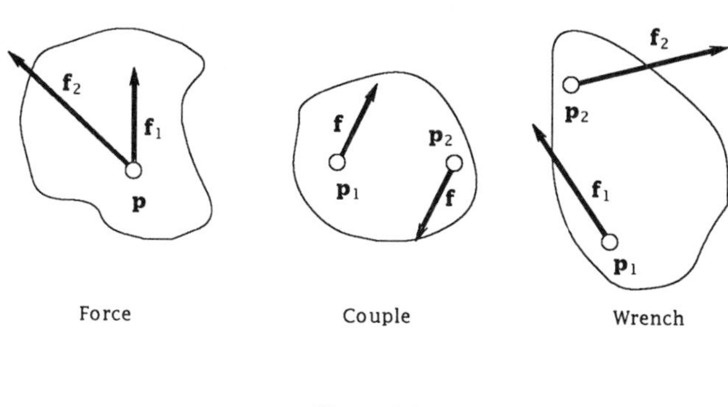

Figure 4.4

Proof: The application of a system of forces $[\mathbf{f}(a_i) \mid i = 1, \ldots, \#A]$ to the joints results in a wrench given by:

$$[\mathbf{f}(a_i) \mid i = 1, \ldots, \#A] \quad \mapsto \quad \mathbf{F} := \sum_{i=1}^{\#A} \hat{\mathbf{p}}(a_i) \vee \hat{\mathbf{f}}(a_i) \, , \tag{4.22}$$

where $\hat{\mathbf{f}} := [0, \mathbf{f}]$. Since the outer product is linear in each argument, this is a linear transformation mapping $(\mathbb{R}^3)^A \rightarrow \mathbb{R}^6$ which is surjective if the framework is noncollinear, and which has the space of all equilibrium systems of forces as its null space or *kernel*. Since the dimension of the range plus the dimension of the kernel of a linear transformation equals the dimension of its domain, the equilibrium systems of forces (assuming that the framework is noncollinear) thus comprise a $(3\#A-6)$-dimensional subspace of $(\mathbb{R}^3)^A$. If, however, the framework is collinear and we choose our origin on the line $((\mathbf{p}(A)))$, then it is clear that the values of $\mathbf{p}(a_i) \times \mathbf{f}(a_i)$ are all in the plane perpendicular to this line so that the image $\mathbf{F} = [\Sigma\, \mathbf{f}(a_i) \,.\, \Sigma\, \mathbf{p}(a_i) \times \mathbf{f}(a_i)]$ is a hyperplane of $\mathbb{R}^6$. QED

Even if the framework is not rigid, we shall continue to say that a system of forces is at equilibrium whenever $\mathbf{F} = \mathbf{0}$ for their result, and the corresponding subspace of $(\mathbb{R}^3)^A \approx \mathbb{R}^{3\#A}$ will be denoted by $\mathbf{U}_{eq} = \{\mathbf{f} \mid \Sigma\, \hat{\mathbf{p}}(a_i) \vee \hat{\mathbf{f}}(a_i) = \mathbf{0}\}$.

For the remainder of this section, we shall set $M := \#\mathrm{B}$, $N := \#\mathrm{A}$, assume a fixed ordering $\underline{\mathrm{A}} = [\mathrm{a}_1, \ldots, \mathrm{a}_N]$ of our atoms, and abbreviate vectors of the form $\mathbf{p}(\mathrm{a}_i) \in \mathbb{R}^3$ by $\mathbf{p}_i$ and those of the form $\mathbf{p}(\underline{\mathrm{A}}) \in \mathbb{R}^{3N}$ by $\underline{\mathbf{p}} = [\mathbf{p}_1 \ldots \mathbf{p}_N]$.

Definition 4.15. Let $\underline{\mathbf{l}}_{ij} \in \mathbb{R}^{3N}$ be the vector whose components are all zero except for the six which correspond to the pair of equal but opposite forces whose vectors are $\mathbf{p}_i - \mathbf{p}_j$ and $\mathbf{p}_j - \mathbf{p}_i$, i.e.

$$\begin{aligned} \underline{\mathbf{l}}_{ij} &:= [\mathbf{0} \ldots \mathbf{0}.\mathbf{p}_i - \mathbf{p}_j.\mathbf{0} \ldots \mathbf{0}.\mathbf{p}_j - \mathbf{p}_i.\mathbf{0} \ldots \mathbf{0}] \\ &= [0, \ldots, 0,\ p_1(\mathrm{a}_i) - p_1(\mathrm{a}_j),\ p_2(\mathrm{a}_i) - p_2(\mathrm{a}_j),\ p_3(\mathrm{a}_i) - p_3(\mathrm{a}_j),\ 0, \cdots \\ &\qquad \cdots, 0,\ p_1(\mathrm{a}_j) - p_1(\mathrm{a}_i),\ p_2(\mathrm{a}_j) - p_2(\mathrm{a}_i),\ p_3(\mathrm{a}_j) - p_3(\mathrm{a}_i),\ 0, \ldots, 0]\,. \end{aligned} \tag{4.23}$$

We call such vectors *elementary loads*, whereas arbitrary compositions of elementary loads $\underline{\mathbf{l}} = \sum r_{ij} \cdot \underline{\mathbf{l}}_{ij}$ will be referred to as *loads*. If $\{\mathrm{a}_i, \mathrm{a}_j\} \in \mathrm{B}$ is a bar in the framework, then the scalar r_{ij} associated with such a load will be referred to as a *stress* on the corresponding bar. A stress may also be referred to as a *tension* (if positive) or a *compression* (if negative)‡.

Observe that a pair of equal but opposite forces $\mathbf{f}_i = -\mathbf{f}_j = r_{ij}(\mathbf{p}_i - \mathbf{p}_j)$ $(r_{ij} \in \mathbb{R})$ applied to the pair of joints $\mathbf{p}_i$, $\mathbf{p}_j$ will always be an equilibrium system of forces

$$\mathbf{F} := \hat{\mathbf{p}}_i \vee \hat{\mathbf{f}}_i + \hat{\mathbf{p}}_j \vee \hat{\mathbf{f}}_j = -r_{ij} \cdot (\hat{\mathbf{p}}_i \vee \hat{\mathbf{p}}_j + \hat{\mathbf{p}}_j \vee \hat{\mathbf{p}}_i) = \mathbf{0}\,, \tag{4.24}$$

as will any composition of such pairs of forces. Thus the subspace $\mathbb{U}_{rf} \subset \mathbb{R}^{3N}$ which is spanned by the elementary loads $\underline{\mathbf{l}}_{ij}$ associated with the bars $\{\mathrm{a}_i, \mathrm{a}_j\} \in \mathrm{B}$ of the framework is contained within the space of equilibrium systems of forces $\mathbb{U}_{eq}$. We can describe this subspace $\mathbb{U}_{rf} \subseteq \mathbb{U}_{eq}$ as follows. If $\underline{\mathrm{B}}$ is the lexicographic ordering of the bars induced by our fixed ordering $\underline{\mathrm{A}}$ of A, we let $\underline{\mathbf{M}}_{\mathbf{p}}$ be the $M \times 3N$ matrix which has the elementary loads associated with the bars of the framework as its rows in this order (see Equation (4.23)). From this it is easily seen that (left) multiplication by the transpose $\underline{\mathbf{M}}_{\mathbf{p}}^T = [\underline{\mathbf{l}}_{ij} \mid \{\mathrm{a}_i, \mathrm{a}_j\} \in \underline{\mathrm{B}}]$ linearly maps each assignment of tensions and compressions to the bars, as an element of $\mathbb{R}^M$, into a composition of elementary loads in $\mathbb{R}^{3N}$, i.e. $\mathbb{U}_{rf}$ is the image of this mapping by definition. We call $\underline{\mathbf{M}}_{\mathbf{p}}$ the *rigidity matrix* of the framework. Note that $2\underline{\mathbf{M}}_{\mathbf{p}} = \underline{\partial \mathbf{D}}_G(\mathbf{p}(\underline{\mathrm{A}}))$, where $\underline{\partial \mathbf{D}}_G$ is the Jacobian of the rigidity mapping associated with $\mathbb{G} = (\mathrm{A}, \mathrm{B})$.

Definition 4.16. A framework is *statically rigid* if $\mathbb{U}_{rf} = \mathbb{U}_{eq}$.

Intuitively, this means that for every equilibrium system of forces that we can apply to the joints of the framework, it is possible to find an assignment of tensions and compressions to the bars such that the resultant system of forces

‡ In order to convert these numbers to the stresses used in engineering, it is necessary to multiply by the length of the bar and divide by its cross-sectional area.

Butane is an example of a statically nonrigid molecule, depicted here with an equilibrium assignment of forces to the terminal atoms which results in bond rotation but no net rotation, translation or stress within the molecule. Hence this molecule admits a nonresolvable system of forces.

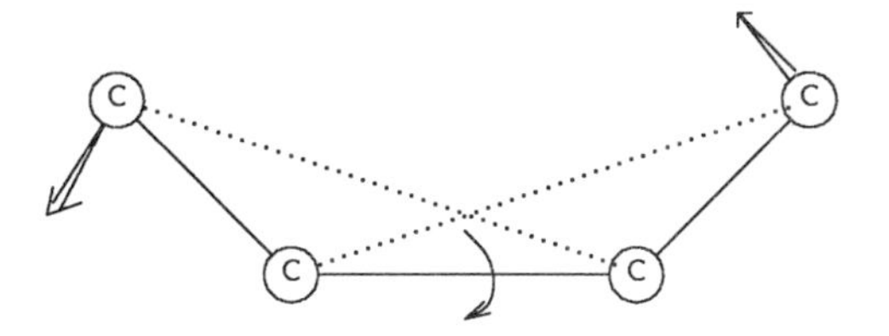

Figure 4.5

exactly balances the applied system. We call such an assignment of scalars to the bars $\mathbf{r} \in \mathbb{R}^M$ a *resolution*† of the equilibrium system of forces $-\underline{\mathbf{M}_\mathbf{p}^T} \cdot \mathbf{r}$, and a system of forces $\mathbf{f} : \mathrm{A} \to \mathbb{R}^3$ *resolvable* whenever a resolution thereof exists.

The kernel of the linear transformation defined by $\underline{\mathbf{M}_\mathbf{p}^T}$ is the subspace of $\mathbb{R}^M$ consisting of all possible resolutions $\mathbf{s} \in \mathbb{R}^M$ of the trivial system of forces $[\mathbf{f}_i \mid i = 1, \ldots, N] = \mathbf{0}$, i.e.

$$\sum_{\{j \mid \{a_i, a_j\} \in \mathrm{B}\}} s_{ij}(\mathbf{p}_i - \mathbf{p}_j) \quad = \quad \mathbf{0} \quad \text{for } i = 1, \ldots, N. \tag{4.25}$$

Such a resolution $\mathbf{s} \in \mathbb{R}^M$ is called an *internal stress* or *self-stress*, and we denote the subspace of all possible self-stresses by $\mathbb{U}_{st} \subseteq \mathbb{R}^M$. Despite the mechanical connotations of our terminology, the reader should be aware that the space $\mathbb{U}_{st}$ of internal stresses is a purely geometric entity which depends only on the framework (and not on the material used to construct it)!

PROPOSITION 4.17. *Let* $\mathbb{F} = (\mathrm{A}, \mathrm{B}, \mathbf{p})$ *be a 3-dimensional framework. Then the following are equivalent:*

(i) *The framework* $\mathbb{F}$ *is statically rigid.*

---

† Henceforth, whenever we write a resolution $\mathbf{r}$ in vector form, we assume that the components r_{ij} have been ordered in the same way as $\underline{\mathrm{B}}$.

Acetylene is rigid, but not statically rigid. Here we indicate a nonresolvable equilibrium system of forces.

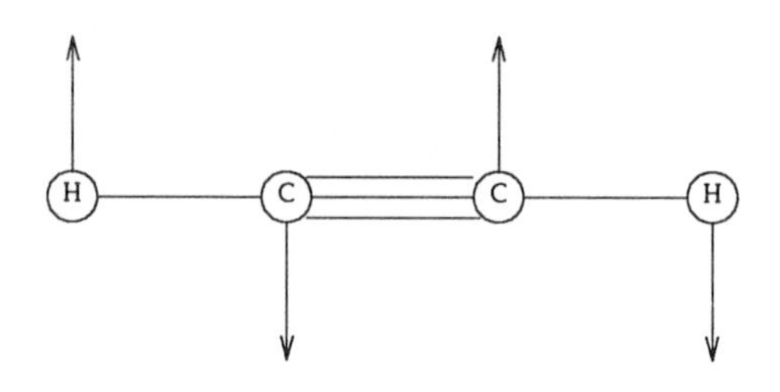

Figure 4.6

(ii) $dim(\mathbf{U}_{rf}) \;=\; 3\#\text{A}-6 \quad (3\#\text{A}-5 \textit{ if } dim(<\mathbf{p}(\text{A})>) = 1).$

(iii) $dim(\mathbf{U}_{st}) \;=\; \#\text{B}-3\#\text{A}+6 \quad (\#\text{B}-3\#\text{A}+5 \textit{ if } dim(<\mathbf{p}(\text{A})>) = 1).$

Proof: Conditions (i) and (ii) are equivalent since $\mathbf{U}_{rf} = \mathbf{U}_{eq} \;<=>\; dim(\mathbf{U}_{rf}) = dim(\mathbf{U}_{eq}) = 3N-6$ ($3N-5$ if $dim(<\mathbf{p}(\text{A})>) = 1$). Since $\mathbf{U}_{rf}$ is the image and $\mathbf{U}_{st}$ the kernel of the transformation defined by multiplication with $\underline{\mathbf{M}}_{\mathbf{p}}^T$, we have $dim(\mathbf{U}_{st})+dim(\mathbf{U}_{rf}) = M$, from which the result (ii) $<=>$ (iii) follows as well. QED

It is easily shown that for planar realizations with $\mathbf{p}: \text{A} \rightarrow \mathbb{R}^2$, the number $3N-6$ in the above must be replaced by $2N-3$ ($2N-2$ if collinear).

We are now ready to consider our second, alternative definition of rigidity [Whiteley1977, Asimow1979, Crapo1979, Crapo1982]. This one is based upon the physically intuitive fact that if a framework is rigid, then any assignment of velocities to the joints that does not change the length of any bar ought to result from a rigid motion (i.e. continuous translation and/or rotation) of the entire framework. Thus suppose that $\mathbf{p}^t: [0 \cdots 1] \rightarrow (\mathbb{R}^3)^{\text{A}}$ is a motion of a framework (A, B, $\mathbf{p}$). Then the instantaneous *strain* or rate of change in the distance $d_{\mathbf{p}}(\text{a}_i, \text{a}_j)$ between the joints $\mathbf{p}_i = \mathbf{p}^0(\text{a}_i)$ and $\mathbf{p}_j = \mathbf{p}^0(\text{a}_j)$ is

$$\dot{d}_{\mathbf{p}}(\text{a}_i, \text{a}_j) \;:=\; \left. \frac{\partial \| \mathbf{p}^t(\text{a}_i) - \mathbf{p}^t(\text{a}_j) \|}{\partial t} \right|_{t=0}$$

$$= \left. \frac{(\mathbf{p}^t(\text{a}_i) - \mathbf{p}^t(\text{a}_j)) \cdot (\partial \mathbf{p}^t(\text{a}_i)/\partial t - \partial \mathbf{p}^t(\text{a}_j)/\partial t)}{\| \mathbf{p}^t(\text{a}_i) - \mathbf{p}^t(\text{a}_j) \|} \right|_{t=0} \tag{4.26}$$

Cyclobutane is a molecule with a nontrival self-stress. The self-stresses are assignments of scalars to the six pairs of atoms such that the sum of the vectors from each atom to the other three, when weighted by the corresponding stresses, is zero, as shown.

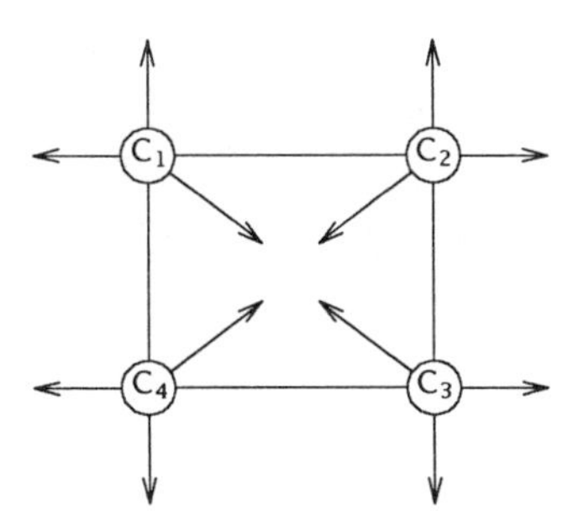

Figure 4.7

Any such assignment must satisfy the $2 \cdot 4 = 8$ linear equations $\Sigma_i s_{ij}(\mathbf{p}_j - \mathbf{p}_i) = \mathbf{0}$ for $j = 1, \ldots, 4$. These equations are not independent of one another, however. Since the resultant system of forces is an equilibrium system it must itself satisfy three additional equations, so the dimension of the space of self-stresses is $6-8+3 = 1$.

$$= \frac{(\mathbf{p}_i - \mathbf{p}_j) \cdot (\mathbf{v}_i - \mathbf{v}_j)}{\|\mathbf{p}_i - \mathbf{p}_j\|}$$

$$= \frac{\mathbf{v}_i \cdot (\mathbf{p}_i - \mathbf{p}_j) + \mathbf{v}_j \cdot (\mathbf{p}_j - \mathbf{p}_i)}{d_{\mathbf{p}}(\mathrm{a}_i, \mathrm{a}_j)},$$

where $\mathbf{v}_i := \partial \mathbf{p}^0(\mathrm{a}_i)/\partial t$ and $\mathbf{v}_j := \partial \mathbf{p}^0(\mathrm{a}_j)/\partial t$.

Definition 4.18. A *producible strain* of a given framework (A, B, **p**) is a vector of the form $[\dot{d}_{\mathbf{p}}(\mathrm{a}_{\mu_1}, \mathrm{a}_{\nu_1}), \ldots, \dot{d}_{\mathbf{p}}(\mathrm{a}_{\mu_M}, \mathrm{a}_{\nu_M})] \in \mathbb{R}^M$ which is obtained from some assignment of velocities $\mathbf{v}: \mathrm{A} \rightarrow \mathbb{R}^3$ to the joints in Equation (4.26), where $\mu, \nu \in \{1, \ldots, N\}^M$ are the indices of the ends of the bars. An *infinitesimal motion* is an assignment of velocities $[\mathbf{v}_i \mid i = 1, \ldots, N]$ to the atoms such that the associated strains $\dot{d}_{\mathbf{p}}(\mathrm{a}_i, \mathrm{a}_j) = 0$ for all $\{\mathrm{a}_i, \mathrm{a}_j\} \in \mathrm{B}$.

Except for the constant $d_{\mathbf{p}}^{-1}(\mathrm{a}_i, \mathrm{a}_j)$, the producible strain $\dot{d}_{\mathbf{p}}(\mathrm{a}_i, \mathrm{a}_j)$ is equal to the ij-th component of the vector $\underline{\mathbf{M}_{\mathbf{p}}} \cdot \underline{\mathbf{v}} \in \mathbb{R}^M$, where $\underline{\mathbf{M}_{\mathbf{p}}}$ is the rigidity matrix as above and $\underline{\mathbf{v}} := [\mathbf{v}_1 \ldots \mathbf{v}_N] \in \mathbb{R}^{3N}$. Thus the infinitesimal motions of the framework are the kernel $\mathbb{U}_{im} \subseteq \mathbb{R}^{3N}$ of the linear transformation defined by multiplication with the matrix $\underline{\mathbf{M}}_{\mathbf{p}}$. Since a matrix and its transpose have the same rank, it follows from Proposition 4.17 that the image of this transformation has dimension at most $3N-6$ ($3N-5$ if collinear), and thus its kernel always has dimension $dim(\mathbb{U}_{im}) \geq 6$. We shall presently show that this six-dimensional part of the kernel in fact corresponds to the infinitesimal motions which are generated by the rigid motions. Such infinitesimal motions are called *trivial.*

Definition 4.19. A framework is *infinitesimally rigid* if it has no nontrivial infinitesimal motions, i.e. if $dim(\mathbb{U}_{im}) = 6$.

Rather than attempting to characterize this new definition of rigidity directly, we will simply determine its relation to our previous definitions.

THEOREM 4.20. *Let* $\mathbb{F} = (\mathrm{A}, \mathrm{B}, \mathbf{p})$ *be a 3-dimensional framework with* $\#\mathrm{A} > 3$. *Then the following are equivalent:*

(i) $\mathbb{F}$ *is statically rigid.*

(ii) $\mathbb{F}$ *is infinitesimally rigid.*

(iii) $\mathbb{F}$ *is regular and rigid.*

Proof: Recall that since the kernel of a linear transformation is the orthogonal complement of the range of its transpose, $dim(\ker(\underline{\mathbf{M}}_{\mathbf{p}})) + dim(\mathrm{rng}(\underline{\mathbf{M}}_{\mathbf{p}}^T)) = 3N$. Since $\mathbb{U}_{im} = \ker(\underline{\mathbf{M}}_{\mathbf{p}})$ and $\mathbb{U}_{rf} = \mathrm{rng}(\underline{\mathbf{M}}_{\mathbf{p}}^T)$ by definition,

$$dim(\mathbb{U}_{im}) = 6 \iff dim(\mathbb{U}_{rf}) = 3N-6 , \tag{4.27}$$

from which the equivalence of (i) and (ii) follows from Proposition 4.17. Since $\partial\mathbf{D}_G(\mathbf{p}(\underline{\mathrm{A}})) = 2\underline{\mathbf{M}}_{\mathbf{p}}$, where $\mathbf{D}_G : \mathbb{R}^{3N} \to \mathbb{R}^M$ is the rigidity mapping associated with $\mathbb{G} = (\mathrm{A}, \mathrm{B})$, to prove that (iii) implies (i) we need only observe that if $\mathbb{F}$ is regular and rigid, then by Theorem 4.8 $rank(\underline{\mathbf{M}}_{\mathbf{p}}) = dim(\mathbb{U}_{rf}) = 3N-6$, whence $\mathbb{F}$ is statically rigid by Proposition 4.17. To prove that (i) implies (iii), we observe that for $N > 3$ the set of all full-dimensional realizations of $\mathbb{G} = (\mathrm{A}, \mathrm{B})$ is open and dense, and since the intersection of any two open dense sets is again open and dense we can find a positioning $\mathbf{q} : \mathrm{A} \to \mathbb{R}^3$ such that $\mathbb{G}_{\mathbf{q}}$ is both full-dimensional and regular, from which it follows that $rank(\mathbf{M}_{\mathbf{q}}) \leq 3N-6$ by Theorem 4.8. If $\mathbb{F}$ is statically rigid, then $rank(\underline{\mathbf{M}}_{\mathbf{p}}) = 3N-6$, and since $\mathbf{M}_{\mathbf{q}}$ is of maximum rank it follows that

$$rank(\underline{\mathbf{M}}_{\mathbf{q}}) \leq 3N-6 = rank(\underline{\mathbf{M}}_{\mathbf{p}}) \leq rank(\underline{\mathbf{M}}_{\mathbf{q}}) , \tag{4.28}$$

which shows that $\mathbb{F} = \mathbb{G}_{\mathbf{p}}$ is a regular and rigid realization of $\mathbb{G}$. QED

Thus static and infinitesimal rigidity are equivalent, and in fact all of the definitions of rigidity introduced so far are "almost always" equivalent! The advantage to working with static/infinitesimal rigidity, however, is that it can be easily determined from the rank of the rigidity matrix even when the framework is not regular.

Even though "almost all" rigid frameworks are statically and infinitesimally rigid, these properties remain dependent upon the geometry of the framework as well as its "topology" (i.e. its underlying graph). To further illustrate this fact, we consider the possible planar realizations of the complete bipartite graph on two sets of three atoms.

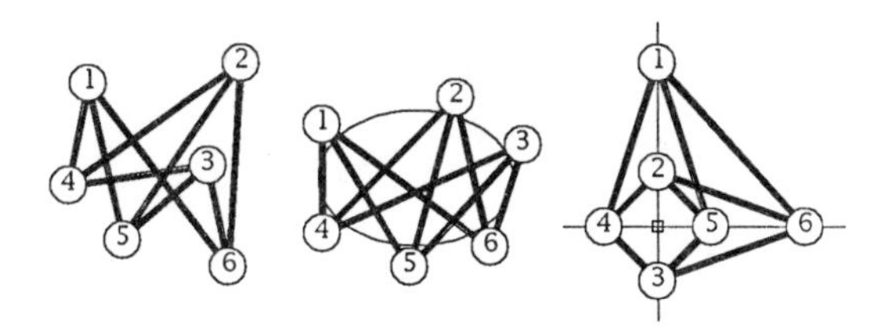

Figure 4.8

Although the generic realizations of this graph are infinitesimally rigid and hence also rigid (left), if the atoms all lie on a quadric it is rigid without being infinitesimally rigid (middle), whereas if the two sets lie on two perpendicular straight lines it is actually flexible (right). Similarly, even though a framework is statically rigid if and only if it is infinitesimally rigid, these two definitions have different things to say about the kinematical properties of the framework. For example, frameworks with an infinite number of joints can be statically rigid without being infinitesimally rigid.

Those with some knowledge of differential geometry will observe that, for regular frameworks, the space of infinitesimal motions $\mathbb{U}_{im}$ is in fact the tangent space $\mathbb{T}_{\mathbf{p}}$ of the manifold $\mathbf{U}_{\mathbf{p}} \cap \mathbf{D}_G^{-1}(\mathbf{D}_G(\mathbf{p}))$ for some sufficiently small neighborhood $\mathbf{U}_{\mathbf{p}}$ of $\mathbf{p}$, whereas the space of resolvable forces $\mathbb{U}_{rf}$ is the orthogonal complement of $\mathbb{T}_{\mathbf{p}}$. Thus every assignment of vectors to the joints of a framework can be uniquely decomposed into the orthogonal sum of a resolvable system of forces and an infinitesimal motion of the framework. Similarly, every assignment of scalars to the bars can be uniquely expressed as the sum of a stress and a producible strain. Of course, this geometric interpretation breaks down in those exceptional cases in which the framework is not regular.

In the event that a framework is not infinitesimally rigid, there is something more that we can say about it.

Definition 4.21. The ***number of degrees of freedom*** $deg(\mathbb{F})$ of a three-dimensional framework is $dim(\mathbf{U}_{im})-6$ ($dim(\mathbf{U}_{im})-5$ if the framework is collinear). More generally, the number of degrees of freedom of an n-dimensional framework is the dimension of its space of infinitesimal motions less the dimension of its space of trivial infinitesimal motions.

Clearly a framework $\mathbb{F}$ is infinitesimally rigid if and only if $deg(\mathbb{F}) = 0$.

The (single nontrivial) infinitesimal motion of cyclobutane corresponds to its lowest vibrational mode.

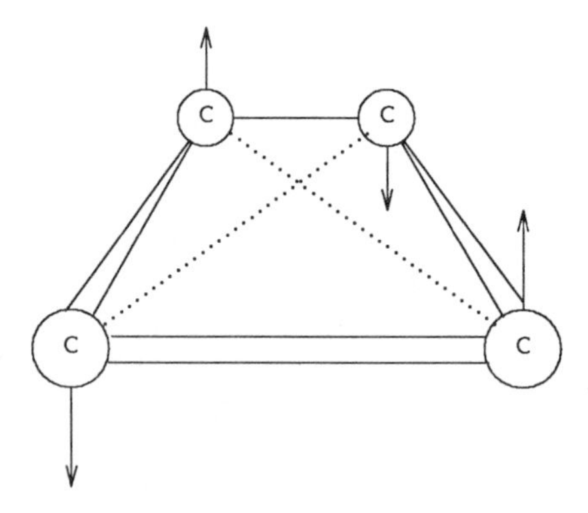

Figure 4.9

Any assignment of vectors to the atoms can be uniquely expressed as a sum of those velocities and those of an equilibrium force system. Similarly, any assignment of scalars to the bars can be written uniquely as the sum of a self-stress and a producible strain.

Chemists might think of this number as follows: If the bars of a framework $\mathbb{F}$ correspond to pairs of atoms in a molecule which are involved in strong interactions (as with covalently bonded or geminal pairs of atoms), then $deg(\mathbb{F})$ will be roughly equal to the number of low-frequency vibrational modes which the molecule has. In order to obtain a simple characterization of this number, we now look a little more closely at the relation between the stresses and producible strains of a framework.

PROPOSITION 4.22. *Let* $\mathbb{F} = (\mathrm{A}, \mathrm{B}, \mathbf{p})$ *be a framework and* $\{a_i, a_j\}$ *be a pair of atoms not in* B. *Then the producible strains on* $\{a_i, a_j\}$ *are trivial, i.e.*

$$(\mathbf{p}_i - \mathbf{p}_j)\cdot(\mathbf{v}_i - \mathbf{v}_j) \quad = \quad \mathbf{0} \tag{4.29}$$

for all infinitesimal motions $\underline{\mathbf{v}} \in \mathbb{U}_{im} \subseteq \mathbb{R}^{3N}$*, if and only if the elementary load* $\underline{\mathbf{l}}_{ij}$ *is resolvable, i.e. if and only if there exists* $\mathbf{r} \in \mathbb{R}^M$ *such that*

$$\sum_{\{k \mid \{a_k, a_l\} \in \mathrm{B}\}} r_{kl}(\mathbf{p}_l - \mathbf{p}_k) \quad = \quad (\delta_{il} - \delta_{jl})(\mathbf{p}_i - \mathbf{p}_j) \quad \text{for } l = 1, \ldots, N\,, \tag{4.30}$$

where δ *is the Kronecker delta.*

Proof: Let $\mathbb{F}' = (\mathrm{A}, \mathrm{B} \cup \{a_i, a_j\}, \mathbf{p})$ be the framework obtained by adding the new bar $\{a_i, a_j\} \notin \mathrm{B}$ and let $\underline{\mathbf{M}}'_{\mathbf{p}}$ be the rigidity matrix of $\mathbb{F}'$. Then Equation (4.29) is equivalent to

$$\underline{\mathbf{M}}'_{\mathbf{p}}\cdot\underline{\mathbf{v}} \quad = \quad \mathbf{0} \quad \forall \;\; \underline{\mathbf{v}} \in \mathbb{U}_{im}\,, \tag{4.31}$$

whereas Equation (4.30) is equivalent to

$$\underline{\mathbf{M}}'^{T}_{\mathbf{p}}\cdot\mathbf{r}' \quad = \quad \mathbf{0} \quad \text{and} \quad r'_{ij} \neq 0 \tag{4.32}$$

for some $\mathbf{r}' \in \mathbb{R}^{M+1}$. If Equation (4.31) is false, we may without loss of generality assume that for some $\underline{\mathbf{v}} \in \mathbb{R}^{3N}$ we have $\underline{\mathbf{M}}'_{\mathbf{p}}\cdot\underline{\mathbf{v}} = \mathbf{u}_{ij}$, the ij-th unit vector in $\mathbb{R}^{M+1}$. Then for any $\mathbf{r}' \in \mathbb{R}^{M+1}$, if $(\mathbf{r}')^T\cdot\underline{\mathbf{M}}'_{\mathbf{p}} = \mathbf{0}^T$ we have $0 = (\mathbf{r}')^T\cdot\underline{\mathbf{M}}'_{\mathbf{p}}\cdot\underline{\mathbf{v}} \; = \; \mathbf{r}'\cdot\mathbf{u}_{ij} = r'_{ij}$ in contradiction with Equation (4.32). Conversely, if Equation (4.32) is false, then for all $\mathbf{r}' \in \mathbb{R}^{M+1}$ with $r'_{ij} \neq 0$ we have $\underline{\mathbf{M}}'^{T}_{\mathbf{p}}\cdot\mathbf{r}' \neq \mathbf{0}$, i.e. $\underline{\mathbf{l}}_{ij}$ is linearly independent of the rows of $\underline{\mathbf{M}}_{\mathbf{p}}$. Hence $\ker(\underline{\mathbf{M}}'_{\mathbf{p}}) \subset \mathbb{U}_{im}$, i.e. there exists a $\underline{\mathbf{v}} \in \mathbb{U}_{im} \setminus \ker(\underline{\mathbf{M}}'_{\mathbf{p}})$, for which $\underline{\mathbf{M}}'_{\mathbf{p}}\cdot\underline{\mathbf{v}} \neq \mathbf{0}$ in contradiction with Equation (4.31). QED

Definition 4.23. A pair of atoms $\{a_i, a_j\}$ of a framework $\mathbb{F} = (\mathrm{A}, \mathrm{B}, \mathbf{p})$ which is not a bar thereof and which satisfies Equations (4.29) and (4.30) is called an *induced bar* of the framework.

Using techniques similar to those used to prove Proposition 4.5, it can be shown that an exact distance constraint involving a bar $\{a_i, a_j\}$ in an abstract framework $(\mathrm{A}, \mathrm{B}, d)$ with $d(a_i, a_j) = \|\mathbf{p}_i - \mathbf{p}_j\|$ for some $\mathbf{p} : \mathrm{A} \to \mathbb{R}^3$ and all $\{a_i, a_j\} \in \mathrm{B}$ is locally redundant (in the sense of §1.2) only if $\{a_i, a_j\}$ is an induced bar of the framework $\mathbb{F} = (\mathrm{A}, \mathrm{B} \setminus \{a_i, a_j\}, \mathbf{p})$ obtained by deleting $\{a_i, a_j\}$ from $\mathbb{F}' := (\mathrm{A}, \mathrm{B}, \mathbf{p})$. The converse of this statement, however, is true only if the framework is regular. We now prove our characterization of the number of degrees of freedom.

THEOREM 4.24. *The number of degrees of freedom of a framework* $\mathbb{F} = (\mathrm{A}, \mathrm{B}, \mathbf{p})$ *is equal to the minimum number of bars which must be added in order to make it statically/infinitesimally rigid.*

Proof: If $deg(\mathbb{F}) > 0$, then there exists a nontrivial infinitesimal motion $\underline{\mathbf{v}} \in \mathbf{U}_{im}$ for which $(\mathbf{v}_i - \mathbf{v}_j)\cdot(\mathbf{p}_i - \mathbf{p}_j) \neq 0$ for some pair $\{a_i, a_j\} \notin B$. This means that the hyperplane given by

$$\mathbf{U}_{ij} \ := \ \{\mathbf{x} \in \mathbb{R}^{3N} \mid (\mathbf{x}_i - \mathbf{x}_j)\cdot(\mathbf{p}_i - \mathbf{p}_j) \ = \ 0\} \tag{4.33}$$

intersects $\mathbf{U}_{im}$ transversely, so that if $\mathbf{U}'_{im}$ is the space of infinitesimal motions of the framework $\mathbb{F}'$ obtained by adding $\{a_i, a_j\}$ to $\mathbb{F}$ we have $\mathbf{U}'_{im} = \mathbf{U}_{im} \cap \mathbf{U}_{ij}$ and $dim(\mathbf{U}'_{im}) = dim(\mathbf{U}_{im}) - 1$, i.e. $deg(\mathbb{F}') = deg(\mathbb{F}) - 1$. The general case follows by induction on $deg(\mathbb{F})$. QED

Note this result holds even if the framework is not regular.

Our final result provides us with some insight into the nature of those exceptional frameworks which are not regular.

PROPOSITION 4.25. *Let* $\mathbb{F} = (A, B, \mathbf{p})$ *be a three-dimensional framework, and* $\mathbf{t}: \mathbb{R}^3 \to \mathbb{R}^3$ *be a projective transformation of* $\mathbb{R}^3$ *which leaves the joints* $\mathbf{p}(A)$ *finite. Then* $\mathbb{F}$ *is statically/infinitesimally rigid if and only if the same is true of* $\mathbb{F}' := (A, B, \mathbf{t}\circ\mathbf{p})$. *More generally,* $\mathbb{F}$ *is regular if and only if* $\mathbb{F}'$ *is.*

Proof: We prove this by showing that the dimension of the image of $\underline{\mathbf{M}}_{\mathbf{p}}^{T}$, i.e. of the subspace $\mathbf{U}_{rf}$, is invariant under projective transformations. Hence let $\hat{\mathbf{t}}: \mathbb{R}^4 \to \mathbb{R}^4$ be a nonsingular linear transformation of the homogeneous coordinates $\hat{\mathbf{p}}_i \in \{1\} \times \mathbb{R}^3$ which induces the projective transformation $\mathbf{t}$ on $\mathbb{R}^3$. The transformation $\hat{\mathbf{t}}$ on $\mathbb{R}^4$ induces a transformation of the Plücker coordinates of the simple 2-vectors which is given by:

$$\mathbf{T}(\mathbf{F}) \ := \ \hat{\mathbf{t}}(\hat{\mathbf{p}}) \vee \hat{\mathbf{t}}(\hat{\mathbf{f}}) \quad \text{for all } \mathbf{F} \ = \ \hat{\mathbf{p}} \vee \hat{\mathbf{f}}\,, \quad \hat{\mathbf{p}}, \hat{\mathbf{f}} \in (\mathbb{R}^4)^A\,, \tag{4.34}$$

and which may be extended to a linear transformation of $\bigvee_2(\mathbb{R}^4) \approx \mathbb{R}^6$ (we leave it to the reader to show that this can always be done in a well-defined way).

For the moment, let $\hat{\mathbf{f}}' := \hat{\mathbf{t}}\circ\hat{\mathbf{f}} = [f_0', \ldots, f_3']$, $\hat{\mathbf{p}}' := \hat{\mathbf{t}}\circ\hat{\mathbf{p}} = [p_0', \ldots, p_3']$ and $\mathbf{F}' := \mathbf{T}(\mathbf{F}) = [F_1', \ldots, F_6']$. If $\hat{\mathbf{f}}'$ is not a free vector (i.e. $f_0' \neq 0$), we may replace $\hat{\mathbf{f}}'$ by $\hat{\mathbf{f}}'' := 1/p_0' \, [0, F_1', F_2', F_3']$ to obtain a free vector such that $\hat{\mathbf{f}}'' \vee \hat{\mathbf{p}}' = \mathbf{F}'$ as before (again, we leave it to the reader to verify this). Hence we may without loss of generality assume that $\hat{\mathbf{f}}' = \hat{\mathbf{t}}\circ\hat{\mathbf{f}}$ is free.

Next, we show that $\hat{\mathbf{t}}$ maps equilibrium systems of forces into systems of forces which are in equilibrium with respect to the positioning $\hat{\mathbf{t}}\circ\hat{\mathbf{p}}$. Defining $\mathbf{F}_i := \mathbf{F}(a_i)$, we have

$$\Sigma \mathbf{F}_i \ = \ \mathbf{0} \ \Rightarrow \tag{4.35}$$

$$\Sigma\, \hat{\mathbf{t}}(\hat{\mathbf{p}}_i) \vee \hat{\mathbf{t}}(\hat{\mathbf{f}}_i) \ = \ \Sigma\, \mathbf{T}\circ\mathbf{F}_i \ = \ \mathbf{T}(\Sigma\, \mathbf{F}_i) \ = \ \mathbf{T}(\mathbf{0}) \ = \ \mathbf{0}\,.$$

Moreover, the induced transformation $\mathbf{T}$ is invertible since $\mathbf{T}^{-1}(\mathbf{F}) = \hat{\mathbf{t}}^{-1}(\mathbf{p}) \vee \hat{\mathbf{t}}^{-1}(\hat{\mathbf{f}})$, so that the equilibrium systems of forces of $\mathbb{F}$ are in one-to-one correspondence with those of $\mathbb{F}'$ under $\mathbf{T}$.

We can also define a linear transformation between the spaces of resolutions of $\mathbb{F}$ and $\mathbb{F}'$ by:

$$\tilde{\mathbf{t}}(\cdots r_{ij}\cdots) \;:=\; [\cdots \alpha_i \alpha_j r_{ij} \cdots], \tag{4.36}$$

where $\hat{\mathbf{t}}(\hat{\mathbf{p}}_i) = [\alpha_i, \mathbf{t}(\mathbf{p}_i)]$. Since $\alpha_i \neq 0 \;\; \forall\; i$ by our assumption that the joints $\mathbf{t}(\mathbf{p}_i)$ are finite, $\tilde{\mathbf{t}}$ is well-defined and the matrix of this transformation $\underline{\mathbf{diag}}[\alpha_i \alpha_j \,|\, \{\mathrm{a}_i, \mathrm{a}_j\} \in \underline{\mathrm{B}}]$ itself is nonsingular. Let $\tilde{t}_{ij}$ be the ij-th component function of $\tilde{\mathbf{t}}$, and observe that the standard homogeneous coordinates of $\mathbf{t} \circ \mathbf{p}_i$ are $\alpha_i^{-1} \hat{\mathbf{t}} \circ \hat{\mathbf{p}}_i$. Under $\tilde{\mathbf{t}}$, the image of the resolution $\mathbf{r} \in \mathbb{R}^M$ of an equilibrium system of forces described by 2-vectors $\{\mathbf{F}_i \,|\, i = 1, \ldots, N\}$ is mapped into a resolution $\tilde{\mathbf{t}}(\mathbf{r})$ of the image of the transformed system $\{\mathbf{T}(\mathbf{F}_i) \,|\, i = 1, \ldots, N\}$, since

$$\begin{aligned}
&\sum_{\{j \,|\, \{\mathrm{a}_i, \mathrm{a}_j\} \in \mathrm{B}\}} \left(\tilde{t}_{ij}(\mathbf{r}) \alpha_i^{-1} \alpha_j^{-1} \cdot (\hat{\mathbf{t}} \circ \hat{\mathbf{p}}_i \vee \hat{\mathbf{t}} \circ \hat{\mathbf{p}}_j) \right) + \mathbf{T}(\mathbf{F}_i) \qquad (4.37) \\
=\; &\sum_{\{j \,|\, \{\mathrm{a}_i, \mathrm{a}_j\} \in \mathrm{B}\}} \left(r_{ij} \cdot \mathbf{T}(\hat{\mathbf{p}}_i \vee \hat{\mathbf{p}}_j) \right) + \mathbf{T}(\mathbf{F}_i) \\
=\; &\mathbf{T}\left(\sum_{\{j \,|\, \{\mathrm{a}_i, \mathrm{a}_j\} \in \mathrm{B}\}} r_{ij} \cdot \hat{\mathbf{p}}_i \vee (\hat{\mathbf{p}}_j - \hat{\mathbf{p}}_i) + \mathbf{F}_i \right) \\
=\; &\mathbf{T}(\mathbf{0}) \;=\; \mathbf{0}\,.
\end{aligned}$$

Hence the image under the vector space isomorphism $\hat{\mathbf{t}}$ of the space $\mathbb{U}_{rf}$ of resolvable systems of forces acting on $\mathbb{F}$ is the space $\mathbb{U}'_{rf}$ of resolvable systems of forces acting on $\mathbb{F}'$, so both spaces have the same dimension and the ranks of the matrices $\underline{\mathbf{M}}_{\mathbf{p}}^T$ and $\underline{\mathbf{M}}_{\mathbf{t}\cdot\mathbf{p}}^T$ are the same, i.e. $\mathbb{F}$ is regular if and only if $\mathbb{F}'$ is. QED

Thus in general, the lack of regularity reflects some projective relation among the joints $\{\mathbf{p}(\mathrm{a}_i) \,|\, i = 1, \ldots, \#\mathrm{A}\}$ of the framework $(\mathrm{A}, \mathrm{B}, \mathbf{p})$. In two dimensions, an elegant characterization of these projectively dependent configurations exists:

THEOREM 4.26. *A two-dimensional framework* $\mathbb{F} = (\mathrm{A}, \mathrm{B}, \mathbf{p})$ *whose graph* (A, B) *is loopless and planar has a nontrivial stress if and only if it is the projection of a spherical polyhedron whose faces are flat in three space.*

This is known as *Maxwell's Theorem*, after the famous physicist James Clerk Maxwell who first proved its necessity. Sufficiency is a later result due to the Group in Structural Topology [Whiteley1982]. Since Maxwell's theorem does not generalize directly to three dimensions, we will not prove it here. Such a characterization would nevertheless be of some interest because, as we shall see later in this chapter, examples of projective dependence are actually found in chemistry and can greatly affect the conformational properties of a molecule.

4.3. The Infinitesimal Kinematics of Organic Molecules

So far we have considered only the rigidity or, a little more generally, the number of degrees of freedom of a molecular model defined by exact distance constraints, or *framework* as we have been calling it. An equally interesting, albeit more difficult, problem is to explicitly describe the motions of which it is capable, or what we call its *kinematics.* Just as the determination of infinitesimal rigidity turned out to be simpler than the finite case, so are the infinitesimal motions of a framework easier to describe than are its finite motions. In fact, we have already shown that the infinitesimal motions of the joints are simply the column dependencies of the rigidity matrix. In most organic molecules, however, the angular velocities about single covalent bonds provide a potentially more concise summary of the internal motions than do the velocity vectors of the individual atoms, firstly because the trivial motions are automatically eliminated from consideration, and secondly because the dihedral angles intrinsically take account of the fact that many covalently bonded groups of atoms move together as single rigid bodies. In this chapter, we present a calculus which describes the *infinitesimal kinematics* of rigid bodies, and use it to obtain an explicit description of the infinitesimal flexings in the space of dihedral angles. This calculus bears a close relation to the calculus of forces acting on a rigid body which was presented in the previous section, and like the calculus of forces it was developed in the latter half of the nineteenth century† only to be largely forgotten until much more recent times [Whiteley1977, Bottema1979, Crapo1982, Whiteley1986a].

Without further ado:

Definition 4.27. A *rigid motion* $\mathbf{m} : [0 \cdots 1] \times \mathbb{R}^n \rightarrow \mathbb{R}^n$ is a transformation of the form

$$\mathbf{p} \mapsto \mathbf{p}^t := \mathbf{m}^t(\mathbf{p}) = \underline{\mathbf{R}}^t \cdot (\mathbf{p} - \mathbf{q}^t) + \mathbf{q}^t + \mathbf{t}^t , \qquad (4.38)$$

where $\underline{\mathbf{R}}^t$ is an $n \times n$ orthogonal matrix, $\mathbf{t}^t$ is the vector of a translation, and $\mathbf{q}^t$ a fixed point of the transformation which results if we hold t constant and replace $\mathbf{t}^t$ by $\mathbf{0} \in \mathbb{R}^n$, all of which are smooth functions of time. An *infinitesimal motion* of $\mathbb{R}^n$ is the vector field $(\mathbf{p}^t, \dot{\mathbf{p}}^t)$ which is obtained by assigning to each point

† The interested reader interested in learning something about the origins of the subject is referred to the book, "The Theory of Screws" by R. S. Ball [Ball1900]. A brief account may also be found in Klein's treatise on Geometry [Klein1939].

$\mathbf{p}^t \in \mathbb{R}^n$ its velocity $\dot{\mathbf{p}}^t := \partial \mathbf{p}^t / \partial t$ due to a rigid motion.

LEMMA 4.28. *The infinitesimal motions of $\mathbb{R}^n$ can be written in the form:*

$$\mathbf{p} \mapsto \mathbf{v} := \dot{\underline{\mathbf{R}}} \cdot \mathbf{p} + \mathbf{s} \tag{4.39}$$

where $\dot{\underline{\mathbf{R}}}$ is a skew-symmetric *$n \times n$ matrix and $\mathbf{s}$ is an arbitrary vector in $\mathbb{R}^n$. If $\dot{\underline{\mathbf{R}}}_1$, $\mathbf{s}_1$ and $\dot{\underline{\mathbf{R}}}_2$, $\mathbf{s}_2$ are the infinitesimal motions of $\mathbb{R}^n$ which result from two rigid motions $\mathbf{m}_1^t$ and $\mathbf{m}_2^t$ at time t, respectively, then the infinitesimal motion which results from their composition $\mathbf{m}_1^t \circ \mathbf{m}_2^t$ is given by $\dot{\underline{\mathbf{R}}}_1 + \dot{\underline{\mathbf{R}}}_2$, $\mathbf{s}_1 + \mathbf{s}_2$. In particular, if $\mathbf{m}_2^t$ is the inverse of $\mathbf{m}_1^t$, then $\dot{\underline{\mathbf{R}}}_2 = -\dot{\underline{\mathbf{R}}}_1$ and $\mathbf{s}_2 = -\mathbf{s}_1$.*

Proof: Without loss of generality, we may assume that $\underline{\mathbf{R}}^0 = \underline{\mathbf{I}}$ and $\mathbf{t}^0 = \mathbf{0}$, so that $\mathbf{p}^0 = \mathbf{p}$, and compute its velocity at $t = 0$. If we write $\mathbf{x} := \mathbf{x}^0$ and $\dot{\mathbf{x}} := \partial \mathbf{x} / \partial t|_{t=0}$ for the derivative of a vector-valued function $\mathbf{x}^t$ as usual, differentiation of the right-hand side of (4.38) yields

$$\begin{aligned} \dot{\mathbf{p}} &= \dot{\underline{\mathbf{R}}} \cdot (\mathbf{p}^0 - \mathbf{q}^0) - \underline{\mathbf{R}}^0 \cdot \dot{\mathbf{q}} + \dot{\mathbf{q}} + \dot{\mathbf{t}} \\ &=: \dot{\underline{\mathbf{R}}} \cdot (\mathbf{p} - \mathbf{q}) + \dot{\mathbf{t}}, \end{aligned} \tag{4.40}$$

which is just Equation (4.39) with $\mathbf{s} := -\dot{\underline{\mathbf{R}}} \cdot \mathbf{q} + \dot{\mathbf{t}}$. Showing that the composition of infinitesimal motions is obtained by summation of their corresponding matrix / vector pairs is proved similarly, and the fact that the inverse infinitesimal motions are obtained by negation follows immediately from this. To see that $\dot{\underline{\mathbf{R}}}$ is skew-symmetric, we observe that if $\underline{\mathbf{R}}^t$ is the orthogonal matrix of a rigid motion, the matrix of its inverse $(\underline{\mathbf{R}}^t)^T$ is obtained by transposition, from which it follows that $-\dot{\underline{\mathbf{R}}} = \dot{\underline{\mathbf{R}}}^T$. QED

In three-dimensions, multiplication by a skew-symmetric matrix is really just a fancy way of writing the cross product, since

$$\begin{bmatrix} 0 & -r_{12} & r_{13} \\ r_{12} & 0 & -r_{23} \\ -r_{13} & r_{23} & 0 \end{bmatrix} \cdot \begin{bmatrix} p_1 \\ p_2 \\ p_3 \end{bmatrix} = \begin{bmatrix} p_3 r_{13} - p_2 r_{12} \\ p_1 r_{12} - p_3 r_{23} \\ p_2 r_{23} - p_1 r_{13} \end{bmatrix} = \mathbf{r} \times \mathbf{p}, \tag{4.41}$$

where $\mathbf{r} := [r_{23}, r_{13}, r_{12}]$. Thus the infinitesimal motions of $\mathbb{R}^3$ can be written as $\mathbf{p} \mapsto \mathbf{r} \times \mathbf{p} + \mathbf{s}$. Alternatively, we can write them in homogeneous form as:

$$\underline{\mathbf{M}} \cdot \hat{\mathbf{p}} := \begin{bmatrix} 0 & -s_1 & -s_2 & -s_3 \\ s_1 & 0 & -r_{12} & r_{13} \\ s_2 & r_{12} & 0 & -r_{23} \\ s_3 & -r_{13} & r_{23} & 0 \end{bmatrix} \cdot \begin{bmatrix} 1 \\ p_1 \\ p_2 \\ p_3 \end{bmatrix} = \begin{bmatrix} -\mathbf{s} \cdot \mathbf{p} \\ \mathbf{r} \times \mathbf{p} + \mathbf{s} \end{bmatrix}, \tag{4.42}$$

i.e. as multiplication of the standard homogeneous coordinates $\hat{\mathbf{p}} = [1, \mathbf{p}]$ of $\mathbf{p}$ by a single 4×4 skew-symmetric matrix. Geometrically, this means that the infinitesimal rotations about the origin in $\mathbb{R}^4$ acting on the affine hyperplane

$\{1\}\times\mathbb{R}^3 \subseteq \mathbb{R}^4$ generate the infinitesimal motions of $\mathbb{R}^3$.

The length of the vector $\mathbf{r}$ which represents an infinitesimal rotation about the origin in $\mathbb{R}^3$ has an interesting interpretation, which may be found by considering the derivative of a smooth *elementary rotation*

$$\frac{\partial}{\partial t}\begin{bmatrix} 1 & 0 & 0 \\ 0 & cos(\omega t) & -sin(\omega t) \\ 0 & sin(\omega t) & cos(\omega t) \end{bmatrix}_{t=0} = \begin{bmatrix} 0 & 0 & 0 \\ 0 & 0 & -\omega \\ 0 & \omega & 0 \end{bmatrix}, \tag{4.43}$$

for which $\|\mathbf{r}\| = |\omega|$, the angular velocity of the rotation. Also, because $\mathbf{r}\times\mathbf{r} = \mathbf{0}$, this vector necessarily lies on the axis of the (infinitesimal) rotation, and its direction is obtained by the usual right-hand rule for cross products. On the other hand, since $\mathbf{r}\cdot\mathbf{s} := \mathbf{r}\cdot(-\mathbf{r}\times\mathbf{q}+\mathbf{t}) = \mathbf{r}\cdot\mathbf{t}$, the quantity $(\mathbf{r}\cdot\mathbf{s})/\|\mathbf{r}\|$ is the component of $\mathbf{t}$ on the axis of rotation, and $\sigma := (\mathbf{r}\cdot\mathbf{s})/\|\mathbf{r}\|^2$ is the ratio of the rate of translation along the axis to the rate of turning about it, which is called the *pitch* of the motion. Observe that

$$\begin{aligned} \mathbf{s} \; := \; -\mathbf{r}\times\mathbf{q} + \mathbf{t} \; &= \; -\mathbf{r}\times(\mathbf{q} - \|\mathbf{r}\|^{-2}\mathbf{r}\times\mathbf{t}) + \sigma\mathbf{r} \\ &= \; -\mathbf{r}\times\mathbf{q}' + \mathbf{t}', \end{aligned} \tag{4.44}$$

where $\mathbf{q}' := \mathbf{q} - \|\mathbf{r}\|^{-2}\mathbf{r}\times\mathbf{t}$ and $\mathbf{t}' := \sigma\mathbf{r}$. Hence $\mathbf{r}\times\mathbf{q}'+\mathbf{s} = \mathbf{t}' = \mathbf{0}$ in the case of a pure infinitesimal rotation, and any infinitesimal motion of $\mathbb{R}^3$ can be represented by a pure infinitesimal rotation about an axis $\mathbf{q}'+\alpha\mathbf{r}$ ($\alpha \in \mathbb{R}$) in space together with an infinitesimal translation $\mathbf{t}'$ parallel to that axis.

This representation of the infinitesimal motions of $\mathbb{R}^3$ by pairs of vectors has a special name.

Definition 4.29. The concatenation $[\mathbf{r}.\mathbf{s}] \in \mathbb{R}^6$ of the vectors $\mathbf{r}$ and $\mathbf{s}$ which represent an infinitesimal motion is called a *screw*. A *rotor* is a screw such that $\mathbf{r}\cdot\mathbf{s} = 0$.

Hence a rotor either has pitch $\sigma = 0$ or else $\|\mathbf{r}\| = 0$, and so represents either a pure infinitesimal rotation or translation. Rotors themselves admit an interesting representation. If we expand the determinant of the skew-symmetric matrix given in Equation (4.42) above, we get

Any infinitesimal motion can be uniquely expressed as a sum of an infinitesimal rotation and translation parallel to the axis of rotation.

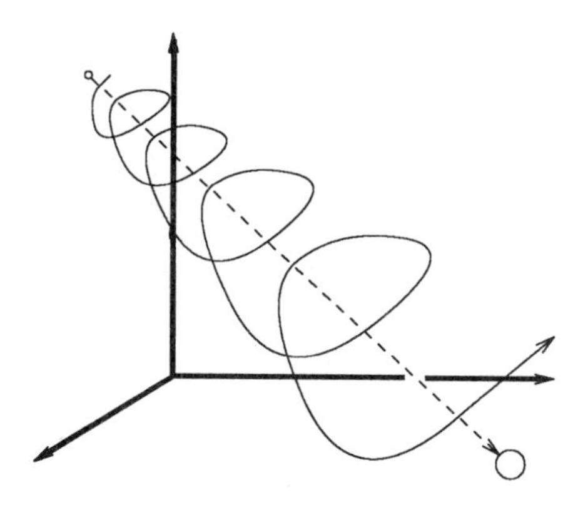

Figure 4.10

Thus a screw is a rotor if and only if the infinitesimal motion it describes is either a pure rotation or else a pure translation.

$$det(\underline{\mathbf{M}}) \;=:\; det\begin{bmatrix} 0 & -M_{12} & M_{13} & -M_{14} \\ M_{12} & 0 & -M_{23} & M_{24} \\ -M_{13} & M_{23} & 0 & -M_{34} \\ M_{14} & -M_{24} & M_{34} & 0 \end{bmatrix} \tag{4.45}$$

$$= \; (M_{12}M_{34} - M_{13}M_{24} + M_{14}M_{23})^2 \;=\; (\mathbf{r}\boldsymbol{\cdot}\mathbf{s})^2 \,.$$

Hence if $\mathbf{r}\boldsymbol{\cdot}\mathbf{s} = 0$, the six components of the screw satisfy a three-term Graßmann-Plücker relation, so that rotors can be regarded as *simple* 2-vectors in $\bigvee_2(\mathbb{R}^4)$ (see Appendix C). It follows from Theorem 2.12 that there exist vectors $\{\mathbf{w}_i = [x_i, y_i] \in \mathbb{R}^2 \,|\, i = 1, \ldots, 4\}$ such that

$$M_{\hat{\mu}} \;=\; \mathbf{w}_{\mu_1} \vee \mathbf{w}_{\mu_2} \;=\; det\begin{bmatrix} x_{\mu_1} & x_{\mu_2} \\ y_{\mu_1} & y_{\mu_2} \end{bmatrix} \tag{4.46}$$

for all six of the $\mu \in \Lambda(4,2)$ (note that, for reasons that will soon become clear, we have mapped the $\hat{\mu}$-th component of the matrix onto the outer product of the complementary pair of vectors $\mathbf{w}_{\mu_1} \vee \mathbf{w}_{\mu_2}$). Equivalently, we can say that there exist vectors $\mathbf{x} = [x_1, \ldots, x_4]$, $\mathbf{y} = [y_1, \ldots, y_4]$ in $\mathbb{R}^4$ such that

$$\begin{bmatrix} M_{34} \\ M_{24} \\ M_{23} \\ M_{14} \\ M_{13} \\ M_{12} \end{bmatrix} = \begin{bmatrix} det\begin{bmatrix} x_1 & x_2 \\ y_1 & y_2 \end{bmatrix} \\ \cdot \\ \cdot \\ \cdot \\ det\begin{bmatrix} x_3 & x_4 \\ y_3 & y_4 \end{bmatrix} \end{bmatrix} = \begin{bmatrix} x_1 \\ x_2 \\ x_3 \\ x_4 \end{bmatrix} \vee \begin{bmatrix} y_1 \\ y_2 \\ y_3 \\ y_4 \end{bmatrix}. \tag{4.47}$$

That is, we can identify each rotor of $\mathbb{R}^4$ with the outer product of a pair of vectors in $\mathbb{R}^4$. The rotor represents a pure infinitesimal translation if both these vectors are the homogeneous coordinates of free vectors in $\mathbb{R}^3$ (i.e. their first components are zero), and otherwise it represents a pure infinitesimal rotation. Of course, the components of arbitrary antisymmetric matrices do not satisfy the Graßmann-Plücker relations, so that arbitrary screws are essentially compound 2-vectors.

LEMMA 4.30. *Multiplying a vector in $\mathbb{R}^4$ by an antisymmetric matrix results in the dual of the outer product of the vector with the (possibly compound) 2-vector whose components are those of the matrix, i.e.*

$$\begin{bmatrix} 0 & -M_{12} & M_{13} & -M_{14} \\ M_{12} & 0 & -M_{23} & M_{24} \\ -M_{13} & M_{23} & 0 & -M_{34} \\ M_{14} & -M_{24} & M_{34} & 0 \end{bmatrix} \begin{bmatrix} z_1 \\ z_2 \\ z_3 \\ z_4 \end{bmatrix} = \left(\begin{bmatrix} M_{34} \\ \cdot \\ \cdot \\ \cdot \\ M_{12} \end{bmatrix} \vee \begin{bmatrix} z_1 \\ z_2 \\ z_3 \\ z_4 \end{bmatrix} \right)^* . \tag{4.48}$$

Proof: Because both sides of Equation (4.48) clearly constitute linear transformations of the indeterminate vector $\mathbf{z} = [z_1, z_2, z_3, z_4]$ and compound 2-vectors can always be expressed as a sum of simple ones, it suffices to verify this equation for simple 2-vectors, say $[M_{\hat{\mu}} = (\mathbf{x} \vee \mathbf{y})_\mu \mid \mu \in \Lambda(4,2)]$. Moreover, since a linear transformation is uniquely determined by the image of a basis, we need only verify that both sides are the same for each of the four unit vectors in $\mathbb{R}^4$. For example, the image of $\mathbf{u}_1 = [1,0,0,0]$ is

$$\begin{bmatrix} 0 & -M_{12} & M_{13} & -M_{14} \\ M_{12} & 0 & -M_{23} & M_{24} \\ -M_{13} & M_{23} & 0 & -M_{34} \\ M_{14} & -M_{24} & M_{34} & 0 \end{bmatrix} \begin{bmatrix} 1 \\ 0 \\ 0 \\ 0 \end{bmatrix} = \begin{bmatrix} 0 \\ M_{12} \\ -M_{13} \\ M_{14} \end{bmatrix} = \begin{bmatrix} -det\begin{bmatrix} x_2 & y_2 & 0 \\ x_3 & y_3 & 0 \\ x_4 & y_4 & 0 \end{bmatrix} \\ det\begin{bmatrix} x_1 & y_1 & 1 \\ x_3 & y_3 & 0 \\ x_4 & y_4 & 0 \end{bmatrix} \\ -det\begin{bmatrix} x_1 & y_1 & 1 \\ x_2 & y_2 & 0 \\ x_4 & y_4 & 0 \end{bmatrix} \\ det\begin{bmatrix} x_1 & y_1 & 1 \\ x_2 & y_2 & 0 \\ x_3 & y_3 & 0 \end{bmatrix} \end{bmatrix}, \tag{4.49}$$

i.e. $(\mathbf{x}\vee\mathbf{y}\vee\mathbf{u}_1)^*$, as desired. The proof for the unit vectors $\mathbf{u}_2, \mathbf{u}_3, \mathbf{u}_4$ is similar.† QED

Since $\mathbf{x}\vee\mathbf{y}\vee(a\mathbf{x}+b\mathbf{y}) = \mathbf{0}$ for all $a, b \in \mathbb{R}$ and $\mathbf{x}, \mathbf{y} \in \mathbb{R}^4$, one immediate consequence of the lemma is that the two vectors whose outer product equals a given rotor span the null space of the corresponding 4×4 antisymmetric matrix. Geometrically, this means that $\underline{\mathbf{M}}$ is the antisymmetric matrix which arises from an elementary rotation in $\mathbb{R}^4$. Another consequence of the above is the fact that the space of trivial infinitesimal motions of a full-dimensional framework is six-dimensional, in accord with our claim in §4.2.

As our first application of this theory, we show how it can be used to derive the derivative of the square of an interatomic distance with respect to a dihedral angle about a covalent single bond connecting two atoms. This calculation is important in computational chemistry, because most conformational energy functions are defined directly in terms of the (squared) interatomic distances, while many conformational energy minimization procedures use these *torsion* angles as their variables. Using the chain rule and the fact that the mapping between squared distances and dihedral angles is smooth, the derivative of such an energy function $E:(\mathbb{R}/2\pi)^k \to \mathbb{R}$ with respect to a dihedral angle ϕ can be written as:

$$\frac{\partial E}{\partial \phi} = \frac{\partial E}{\partial \mathbf{D}_K} \cdot \frac{\partial \mathbf{D}_K}{\partial \phi}, \tag{4.50}$$

where $\mathbf{D}_K$ denotes the vector of all squared distances. Since E is usually a simple

† The index swapping and sign changes here are admittedly a bit hard to keep track of. Note in particular that in $\mathbb{R}^4$ the dual is not quite its own inverse, since $(\mathbf{x}^*)^* = -\mathbf{x}$.

function of the squared distances, the hard part of this calculation is getting the derivatives of the squared distances with respect to the dihedral angles. Anyone who has had to do this using vector algebra will appreciate the simplicity of the following derivation.

LEMMA 4.31. *If* a–b–c–d *is a covalently bonded chain of atoms and* $\mathbf{p}(a), \ldots, \mathbf{p}(d) \in \mathbb{R}^3$ *are their position vectors, the derivative of* $\|\mathbf{p}(a)-\mathbf{p}(d)\|^2$ *with respect to the dihedral angle* ϕ *about the* b–c *bond is:*

$$\frac{\partial\|\mathbf{p}(a)-\mathbf{p}(d)\|^2}{\partial\phi} = 2\,\frac{(\mathbf{p}(b)-\mathbf{p}(a))\times(\mathbf{p}(c)-\mathbf{p}(b))\cdot(\mathbf{p}(d)-\mathbf{p}(c))}{\|\mathbf{p}(b)-\mathbf{p}(c)\|}, \tag{4.51}$$

Proof: Applying the chain rule again, we get:

$$\begin{aligned}\frac{\partial\|\mathbf{p}(a)-\mathbf{p}(d)\|^2}{\partial\phi} &= \frac{\partial\|\mathbf{p}(a)-\mathbf{p}(d)\|^2}{\partial\underline{\mathbf{p}}}\cdot\frac{\partial\underline{\mathbf{p}}(a,b,c,d)}{\partial\phi} \\ &= \frac{\partial\|\mathbf{p}(a)-\mathbf{p}(d)\|^2}{\partial\underline{\mathbf{p}}}\cdot\mathbf{v}(a,b,c,d)\,,\end{aligned} \tag{4.52}$$

where $\underline{\mathbf{p}} = \mathbf{p}(a,b,c,d)$ is a 12-dimensional vector of atom positions and $\mathbf{v}(a,b,c,d)$ is a 12-vector of the velocities of the atoms as a result of a unit angular velocity rotation about b–c. Since the *relative* positions of the atoms a,b,c are invariant under the rotation, we may consider their absolute positions $\mathbf{p}(a)$, $\mathbf{p}(b)$ and $\mathbf{p}(c)$ to be fixed so that $\mathbf{v}(a) = \mathbf{v}(b) = \mathbf{v}(c) = \mathbf{0}$. If we now define $\hat{\mathbf{v}}(d) := [-\mathbf{p}(d)\cdot\mathbf{v}(d), \mathbf{v}(d)]$, $\hat{\mathbf{p}}(a) := [1, \mathbf{p}(a)]$ and abbreviate $\hat{\mathbf{p}}(a)$ to $\mathbf{a} = [1, a_1, a_2, a_3]$ (with similar expressions for $\mathbf{b}, \mathbf{c}, \mathbf{d} \in \mathbb{R}^4$), the above derivative simplifies to:

$$\begin{aligned}\frac{\partial\|\mathbf{p}(a)-\mathbf{p}(d)\|^2}{\partial\phi} &= \frac{\partial\|\mathbf{a}-\mathbf{d}\|^2}{\partial\phi} \\ &= 2(\mathbf{d}-\mathbf{a})\cdot\hat{\mathbf{v}}(d) \;=\; 2(\mathbf{d}-\mathbf{a})\vee\hat{\mathbf{v}}^*(d)\end{aligned} \tag{4.53}$$

since, by the definition of the dual given in Appendix C, $\mathbf{x}\cdot\mathbf{y} = \mathbf{x}\vee\mathbf{y}^*$ for all $\mathbf{x}, \mathbf{y} \in \mathbb{R}^4$. By Proposition 4.30 and the foregoing discussion of rotors, the dual of the velocity vector is obtained by applying the rotor $\mathbf{b}\vee\mathbf{c}$ to $\mathbf{d}$ and dualizing, i.e.

$$\hat{\mathbf{v}}^*(d) = \frac{-1}{\|\mathbf{b}-\mathbf{c}\|}\cdot\begin{bmatrix}1\\ b_1\\ b_2\\ b_3\end{bmatrix}\vee\begin{bmatrix}1\\ c_1\\ c_2\\ c_3\end{bmatrix}\vee\begin{bmatrix}1\\ d_1\\ d_2\\ d_3\end{bmatrix}, \tag{4.54}$$

and its outer product with the difference vector is obtained simply by completing the determinant:

$$\frac{\partial\|\mathbf{a}-\mathbf{d}\|^2}{\partial\phi} = \frac{-2}{\|\mathbf{b}-\mathbf{c}\|}\cdot det\begin{bmatrix}0 & 1 & 1 & 1\\ d_1-a_1 & b_1 & c_1 & d_1\\ d_2-a_2 & b_2 & c_2 & d_2\\ d_3-a_3 & b_3 & c_3 & d_3\end{bmatrix} \tag{4.55}$$

$$= \frac{2}{\|\mathbf{b}-\mathbf{c}\|}\cdot det\begin{bmatrix}1 & 1 & 1 & 1\\ a_1 & b_1 & c_1 & d_1\\ a_2 & b_2 & c_2 & d_2\\ a_3 & b_3 & c_3 & d_3\end{bmatrix}.$$

Equation (4.51) follows readily from this together with some basic vector algebra. QED

As an exercise, the reader may want to show that the second derivative of $\|\mathbf{a}-\mathbf{f}\|^2$ with respect to a second angle ϕ' about the bond d−e in a chain of atoms a−b−c−d−e−f is given by:

$$\frac{\|\mathbf{b}-\mathbf{c}\|\,\|\mathbf{d}-\mathbf{e}\|}{2}\cdot\frac{\partial^2\|\mathbf{a}-\mathbf{f}\|^2}{\partial\phi\partial\phi'} = det\begin{bmatrix}1 & 1 & 1\\ a_1 & b_1 & c_1\\ a_2 & b_2 & c_2\end{bmatrix} det\begin{bmatrix}1 & 1 & 1\\ d_1 & e_1 & f_1\\ d_2 & e_2 & f_2\end{bmatrix} \tag{4.56}$$

$$+\, det\begin{bmatrix}1 & 1 & 1\\ a_1 & b_1 & c_1\\ a_3 & b_3 & c_3\end{bmatrix} det\begin{bmatrix}1 & 1 & 1\\ d_1 & e_1 & f_1\\ d_3 & e_3 & f_3\end{bmatrix} + det\begin{bmatrix}1 & 1 & 1\\ a_2 & b_2 & c_2\\ a_3 & b_3 & c_3\end{bmatrix} det\begin{bmatrix}1 & 1 & 1\\ d_2 & e_2 & f_2\\ d_3 & e_3 & f_3\end{bmatrix}$$

(note this formula with d = b, e = c and f = d also gives us the second derivative with respect to a single dihedral angle).

This lemma enables us to extend our previous criterion for when an infinitesimal motion preserves a distance to the case of rigid bodies.

PROPOSITION 4.32. *Suppose two rigid bodies in* $\mathbb{R}^3$ *are connected by a rigid bar* $[(\mathbf{p},\mathbf{q})]$, *so that the distance* $\|\mathbf{p}-\mathbf{q}\|$ *between the endpoints of the bar is fixed. Then if the body containing the point* $\mathbf{p}$ *is subjected to (the infinitesimal motion described by) the screw* $\mathbf{S}\in\bigvee_2(\mathbb{R}^4)$ *and the body containing the point* $\mathbf{q}$ *is subjected to the screw* $\mathbf{T}\in\bigvee_2(\mathbb{R}^4)$, *we have:*

$$\mathbf{S}\vee\hat{\mathbf{p}}\vee\hat{\mathbf{q}} + \mathbf{T}\vee\hat{\mathbf{q}}\vee\hat{\mathbf{p}} = \mathbf{0}\,. \tag{4.57}$$

Proof: By linearity, it sufficies to prove this for the case in which $\mathbf{S}=\mathbf{s}\vee\mathbf{s}'$, $\mathbf{T}=\mathbf{t}\vee\mathbf{t}'$ are simple rotors. Since the angular velocity $\dot{\phi}$ associated with the rotor $\mathbf{S}$ is equal to the length $\|\mathbf{s}-\mathbf{s}'\|$, and by Lemma 4.31 the strain on $[(\mathbf{p},\mathbf{q})]$ due to a unit angular velocity rotation about the line $((\mathbf{s},\mathbf{s}'))$ is $2\|\mathbf{s}-\mathbf{s}'\|^{-1}\cdot\hat{\mathbf{q}}\vee\mathbf{s}\vee\mathbf{s}'\vee\hat{\mathbf{p}}$ (with similar expressions for $\mathbf{T}$ and its angular velocity $\dot{\phi}'$), if the composition of these rotors preserves the squared distance $\|\mathbf{p}-\mathbf{q}\|^2$, we have

$$0 = \frac{\partial\|\mathbf{p}-\mathbf{q}\|^2}{\partial t} = \frac{\partial\|\mathbf{p}-\mathbf{q}\|^2}{\partial\phi}\cdot\dot{\phi} + \frac{\partial\|\mathbf{p}-\mathbf{q}\|^2}{\partial\phi'}\cdot\dot{\phi}' \tag{4.58}$$

$$= \quad 2\cdot\hat{\mathbf{q}}\vee\mathbf{s}\vee\mathbf{s}'\vee\hat{\mathbf{p}} + 2\cdot\hat{\mathbf{p}}\vee\mathbf{t}\vee\mathbf{t}'\vee\hat{\mathbf{q}}\,,$$

which is the desired result. QED

In the case that one body is at rest and we subject the other to a screw $\mathbf{S} \in \bigvee_2(\mathbb{R}^4)$, Equation (4.57) simplifies to the vanishing of a symmetric bilinear form $B(\mathbf{S};\mathbf{C}) := \mathbf{S}\vee\mathbf{C}$, where $\mathbf{C} := \hat{\mathbf{p}}\vee\hat{\mathbf{q}}$ is the line-bound vector or *constraint* generated by the ends of the bar. Note in particular that $B(\mathbf{C};\mathbf{C}) = 0$ is just the Graßmann-Plücker relation for the constraint $\mathbf{C}$.

If we define an assembly of two rigid bodies linked by bars $\{[(\mathbf{p}_i,\mathbf{q}_i)] \mid i = 1,\ldots,l\}$ to be *infinitesimally rigid* whenever the bar and joint framework obtained by adding all possible bars within each of the two indexed sets of points $\{\mathbf{p}_i \mid i = 1,\ldots,l\}$ and $\{\mathbf{q}_i \mid i = 1,\ldots,l\}$ is infinitesimally rigid, we obtain a rather interesting:

COROLLARY 4.33. *A set of bars* $\{[(\mathbf{p}_i,\mathbf{q}_i)] \mid i = 1,\ldots,l\}$ *connecting two rigid bodies in* $\mathbb{R}^3$ *will result in an infinitesimally rigid assembly if and only if* $<\mathbf{C}_1,\ldots,\mathbf{C}_l> \;=\; \bigvee_2(\mathbb{R}^4)$, *where* $\mathbf{C}_i = \hat{\mathbf{p}}_i\vee\hat{\mathbf{q}}_i$ *for* $i = 1,\ldots,l$. *In particular, a minimum of six bars are required to hold one rigid body fixed with respect to another.*

Proof: Since the form $B:(\mathbf{S},\mathbf{C}) \mapsto \mathbf{S}\vee\mathbf{C}$ defined above is easily seen to be nondegenerate, any screw which is B-orthogonal to a set of constraints $\{\mathbf{C}_i \mid i = 1,\ldots,l\}$ which spans $\bigvee_2(\mathbb{R}^4)$ is necessarily trivial. This means that any infinitesimal motion of one body with respect to the other which produces no strain on all of the bars is trivial, i.e. the assembly is infinitesimally rigid by definition. Because infinitesimal rigidity implies rigidity and $dim(\bigvee_2(\mathbb{R}^4)) = 6$, the second claim is a trivial consequence of the first. QED

Since $dim(\mathbf{C}_1,\ldots,\mathbf{C}_6) = 6$ in the generic case, any six bars between two bodies will "almost always" be sufficient to hold them rigid relative to each other. In those special positionings for which $dim(\mathbf{C}_1,\ldots,\mathbf{C}_6) < 6$, the lines determined by $\mathbf{C}_1,\ldots,\mathbf{C}_6$ are members of a projectively dependent family of lines which is known as a *line complex* [Jessop1903] (see Figure 4.11 below).

Using this corollary, it is possible to develop a theory of *bar and body* frameworks which is in many respects simpler than the theory of bar and joint frameworks given in the previous sections of this chapter [Tay1984a, White1987]. Unfortunately, this theory appears to be of relatively little chemical relevance. It has been pointed out, however, that a certain closely related type of framework appears to be ideally suited for the analysis of the infinitesimal kinematics of organic molecules in terms of their dihedral angles [Tay1984b].

Definition 4.34. A *hinge and body framework*† is a triple (A, $\mathbb{B}$, $\mathbf{p}$) consisting of a set A (whose elements we call atoms as before), a subset $\mathbb{B} \subseteq 2^A$ of *bodies* which satisfies $\#(B \cap B') \in \{0,2\}$ for all $B, B' \in \mathbb{B}$, and a function $\mathbf{p}: A \to \mathbb{R}^3$ which assigns to each atom its *position*. The nonempty intersections $H := \{B \cap B' \neq \{\} \mid B, B' \in \mathbb{B}\}$ of pairs of bodies are called *hinges*, and we further assume that $\mathbf{p}(a) \neq \mathbf{p}(b)$ for all $\{a,b\} \in H$.

The connection with organic molecules is easy to see: The hinges are flexible connections between rigid groups of atoms in the molecule, which correspond to the chemical bonds between their common pair of atoms. Situations in which $\#(B \cap B') = 1$ appear to be at best uncommon in chemistry, and hence this case has not been included. If any pair of rigid bodies have three or more points in common, of course, then the entire pair is a single rigid body unless these points are collinear. In this case (which is also rare in chemistry), we can consider the bodies to be connected by multiple collinear hinges.

The analysis of hinge and body frameworks is based upon the following result.

LEMMA 4.35. *If two bodies* $\mathbf{p}(B_1)$, $\mathbf{p}(B_2)$ *in space which are joined together by a hinge* $B_1 \cap B_2 \neq \{\}$ *are subjected to an infinitesimal motion with screws* $\mathbf{S}_1$, $\mathbf{S}_2$, *respectively, then*

$$\mathbf{S}_2 - \mathbf{S}_1 \;=\; \omega \cdot \hat{\mathbf{p}}(a) \vee \hat{\mathbf{p}}(b)\,. \tag{4.59}$$

for some scalar $\omega \in \mathbb{R}$.

Proof: Since the velocity vectors at $\mathbf{p}(a)$ due to the screws $\mathbf{S}_1$ and $\mathbf{S}_2$ are equal to the last three components of $(\mathbf{S}_1 \vee \hat{\mathbf{p}}(a))^*$ and $(\mathbf{S}_2 \vee \hat{\mathbf{p}}(a))^*$ with similar equations for $\mathbf{p}(b)$, the fact that $\mathbf{p}(B_1)$ and $\mathbf{p}(B_2)$ are tied together at $\mathbf{p}(a)$ and $\mathbf{p}(b)$ implies that

$$\mathbf{S}_1 \vee \hat{\mathbf{p}}(a) \;=\; \mathbf{S}_2 \vee \hat{\mathbf{p}}(a) \quad \text{and} \quad \mathbf{S}_1 \vee \hat{\mathbf{p}}(b) \;=\; \mathbf{S}_2 \vee \hat{\mathbf{p}}(b)\,. \tag{4.60}$$

Choose any two atoms $c \in B_1 \setminus \{a,b\}$, $d \in B_2 \setminus \{a,b\}$ such that $< \hat{\mathbf{p}}(a), \ldots, \hat{\mathbf{p}}(d) > \; = \mathbb{R}^4$ (if $dim(\hat{\mathbf{p}}(B_1 \cup B_2)) < 4$ we can simply add a noncoplaner point to $\hat{\mathbf{p}}(B_1)$ without changing its relation to $\hat{\mathbf{p}}(B_2)$). Then by Equation (4.58) of Proposition 4.32, the rate of change in $\|\mathbf{p}(c) - \mathbf{p}(d)\|^2$ is given by

$$\frac{\partial \|\mathbf{p}(c) - \mathbf{p}(d)\|^2}{\partial \phi} \cdot \dot{\phi} \;=\; (\mathbf{S}_2 - \mathbf{S}_1) \vee \hat{\mathbf{p}}(c) \vee \hat{\mathbf{p}}(d)\,. \tag{4.61}$$

If this change is due to a rotation about the line $((\mathbf{p}(a), \mathbf{p}(b)))$, then by Lemma 4.31

$$\frac{\partial \|\mathbf{p}(c) - \mathbf{p}(d)\|^2}{\partial \phi} \cdot \dot{\phi} \;=\; \frac{2 \cdot \dot{\phi}}{\|\mathbf{p}(a) - \mathbf{p}(b)\|} \cdot \hat{\mathbf{p}}(a) \vee \hat{\mathbf{p}}(b) \vee \hat{\mathbf{p}}(c) \vee \hat{\mathbf{p}}(d)\,, \tag{4.62}$$

† This definition is functionally identical with what is usually called a *panel and hinge framework*.

which implies we should choose:

$$\omega \;=\; \frac{2\dot{\phi}}{\|\hat{\mathbf{p}}(\mathrm{a})-\hat{\mathbf{p}}(\mathrm{b})\|} \;=\; \frac{(\mathbf{S}_2-\mathbf{S}_1)\vee\hat{\mathbf{p}}(\mathrm{c})\vee\hat{\mathbf{p}}(\mathrm{d})}{\hat{\mathbf{p}}(\mathrm{a})\vee\hat{\mathbf{p}}(\mathrm{b})\vee\hat{\mathbf{p}}(\mathrm{c})\vee\hat{\mathbf{p}}(\mathrm{d})}\,. \tag{4.63}$$

Since $\hat{\mathbf{p}}(\{\mathrm{a,b,c,d}\})$ is a basis of $\mathbb{R}^4$, $[\mathbf{x}\vee\mathbf{y}\,|\,\mathbf{x},\mathbf{y}\in\hat{\mathbf{p}}(\{\mathrm{a,b,c,d}\})]$ is likewise a basis of $\bigvee_2(\mathbb{R}^4)$. Hence to verify that this choice of ω actually does result in Equation (4.59), it suffices to show that for all $\mathbf{x},\mathbf{y}\in\hat{\mathbf{p}}(\{\mathrm{a,b,c,d}\})$, we have

$$B(\mathbf{S}_2-\mathbf{S}_1;\,\mathbf{x}\vee\mathbf{y}) \;=\; \omega\cdot B(\hat{\mathbf{p}}(\mathrm{a})\vee\hat{\mathbf{p}}(\mathrm{b});\,\mathbf{x}\vee\mathbf{y}) \tag{4.64}$$

where B is the nondegenerate bilinear form of Corollary 4.33. For $\{\mathbf{x},\mathbf{y}\}=\hat{\mathbf{p}}(\{\mathrm{c,d}\})$, this holds as an identity by our choice of ω. On the other hand, for any $\mathbf{x}\in\hat{\mathbf{p}}(\{\mathrm{a,b}\})$ we have

$$(\mathbf{S}_2-\mathbf{S}_1)\vee\mathbf{x} \;=\; \mathbf{0} \;=\; \omega\cdot\hat{\mathbf{p}}(\mathrm{a})\vee\hat{\mathbf{p}}(\mathrm{b})\vee\mathbf{x}\,, \tag{4.65}$$

where the left-hand-side is zero by Equation (4.60) and the right by the fact that "$\vee$" is alternating. Thus $B(\mathbf{S}_1-\mathbf{S}_2;\,\mathbf{x}\vee\mathbf{y})=0=\omega\cdot B(\hat{\mathbf{p}}(\mathrm{a})\vee\hat{\mathbf{p}}(\mathrm{b});\,\mathbf{x}\vee\mathbf{y})$ for all $\mathbf{y}\in\hat{\mathbf{p}}(\{\mathrm{a,b,c,d}\})$, as desired. QED

Definition 4.36. An infinitesimal flexing of a hinge and body framework $\mathbb{F}=(\mathrm{A},\mathbb{B},\mathbf{p})$ is an infinitesimal flexing of the bar and joint framework $\mathbb{F}'=(\mathrm{A},\mathrm{B},\mathbf{p})$ which is obtained by adding all possible bars $\mathrm{B}=\{\{\mathrm{a,b}\}\,|\,\mathrm{a,b}\in\mathrm{B},\,\mathrm{B}\in\mathbb{B}\}$ between the atoms in each body.

Associated with each hinge and body framework is a graph $\mathbb{G}=(\mathbb{B},\mathbb{H})$, where

$$\mathbb{H} \;:=\; \{\{\mathrm{B},\mathrm{B}'\}\,|\,\mathrm{B},\mathrm{B}'\in\mathbb{B},\,\#(\mathrm{B}\cap\mathrm{B}')=2\}\,. \tag{4.66}$$

This graph will now be used to obtain a characterization of the infinitesimal flexings of hinge and body frameworks in terms of their dihedral angles.

THEOREM 4.37. *Let* $\mathbb{F}=(\mathrm{A},\mathbb{B},\mathbf{p})$ *be a hinge and body framework, and let* H *be the set of hinges thereof. Then an assignment* $\dot{\phi}:\mathrm{H}\to\mathbb{R}$ *of scalars to the hinges are the angular velocities of an infinitesimal flexing of the framework if and only if for every cycle* $[\{\mathrm{B}_1,\mathrm{B}_1'\},\ldots,\{\mathrm{B}_K,\mathrm{B}_K'\}]$ $(\mathrm{B}_i'=\mathrm{B}_{i+1(mod\ K)})$ *of hinged bodies in the graph* $\mathbb{G}=(\mathbb{B},\mathbb{H})$ *we have*

$$\sum_{i=1}^{K}\omega(\mathrm{b}_i,\mathrm{b}_i')\cdot\hat{\mathbf{p}}(\mathrm{b}_i)\vee\hat{\mathbf{p}}(\mathrm{b}_i') \;=\; \mathbf{0} \tag{4.67}$$

where $\{\mathrm{b}_i,\mathrm{b}_i'\}=\mathrm{B}_i\cap\mathrm{B}_i'$ *for* $i=1,\ldots,K$, *and*

$$\omega(\mathrm{b}_i,\mathrm{b}_i') \;:=\; \frac{2\dot{\phi}(\mathrm{b}_i,\mathrm{b}_i')}{\|\mathbf{p}(\mathrm{b}_i)-\mathbf{p}(\mathrm{b}_i')\|}\,,\quad 1\le i<j\le L:=\#\mathbb{H}\,.$$

Proof: For any infinitesimal motion of the bodies $\hat{\mathbf{p}}(\mathbb{B}) = \{\hat{\mathbf{p}}(B_1), \ldots, \hat{\mathbf{p}}(B_K)\}$, described by screws $\mathbf{S}_1, \ldots, \mathbf{S}_K$ respectively, which preserves the constraints $\mathbf{p}(b_i) \vee \mathbf{p}(b_i')$ imposed by the hinges $B_i \cap B_i'$), we have

$$\sum_{i=1}^{K} \omega(b_i, b_i') \cdot \hat{\mathbf{p}}(b_i) \vee \hat{\mathbf{p}}(b_i') \quad = \quad \sum_{i=1}^{K} (\mathbf{S}_i - \mathbf{S}_{i+1(mod\, K)}) \quad = \quad \mathbf{0} \tag{4.68}$$

by Lemma 4.35, thus proving necessity. On the other hand, if an assignment of scalars $\omega : H \rightarrow \mathbb{R}$ satisfies this equation, the six components of the screws $\mathbf{S}_i = [s_{1i}, \ldots, s_{6i}]$ ($i = 1, \ldots, M := \#\mathbb{B}$) we seek are the solutions of

$$\underline{\mathbf{M}}_G \cdot \mathbf{s}_j \quad = \quad \mathbf{h}_j \tag{4.69}$$

for $j = 1, \ldots, 6$, where $\underline{\mathbf{M}}_G$ is the L by M node-arc incidence matrix of an orientation of the graph $\mathbb{G}$, $\mathbf{s}_j \in \mathbb{R}^M$ is a vector of the j-th components of the $\mathbf{S}_i$, and $\mathbf{h}_j \in \mathbb{R}^L$ is a vector of the j-th components of the rotors $\omega(b_i, b_i') \cdot \hat{\mathbf{p}}(b_i) \vee \hat{\mathbf{p}}(b_i')$ associated with the (directed) bonds $[b_i, b_i']$ ($\{b_i, b_i'\} \in H$). We must show that these six systems of equations are feasible, i.e. that

$$\mathbf{t}_j^T \cdot \underline{\mathbf{M}}_G \quad = \quad \mathbf{0} \quad \Rightarrow \quad \mathbf{t}_j \cdot \mathbf{h}_j \quad = \quad \mathbf{0}\,. \tag{4.70}$$

The solutions $\mathbf{t}_j$ of the left-hand system of these equations are simply the *circulations* of the graph $\mathbb{G}$ [Papadimitriou1982], and since any circulation is a linear combination of the arc-indicator vectors $\mathbf{c}$ of the cycles of the graph which satisfy $\mathbf{c} \cdot \mathbf{h}_j = \mathbf{0}$ by hypothesis, we have $\mathbf{t}_j \cdot \mathbf{h}_j = 0$ as desired. QED

This theorem was first proved (in this century at least!) by Crapo and Whiteley [Crapo1982]. Shortly thereafter, Braun [Braun1987] independently proved the same result by an ingeneous variational argument based on Go's recursive equations for the derivatives of squared distances with respect to the dihedral angles [Abe1984]. Using this technique, Braun was further able to derive equations which assure the vanishing of the second derivatives of the squared distances with respect to the dihedral angles as well,† and has gone on to develop algorithms which use these equations to efficiently restore ring closure in cyclic molecules after small changes in their dihedral angles. These may be of considerable use in energy minimization and molecular dynamics simulations of such molecules. The second-order conditions have also been derived for the general case of bar and joint frameworks [Connelly1980] (see also §5.2).

In closing, we remark that the relation of the arc-indicator vectors of cycles (Theorem 4.37) in the (oriented) graph $\mathbb{G}$ to the rigidity of the framework $\mathbb{F}$

---

† The complexity of these equations when expressed in the notation of vector algebra is, however, somewhat daunting. If Equation (4.56) for the second derivatives of the squared distances could be used to derive Braun's equations directly, it could lead to some substantial simplifications and hence further insight into the meaning of the equations.

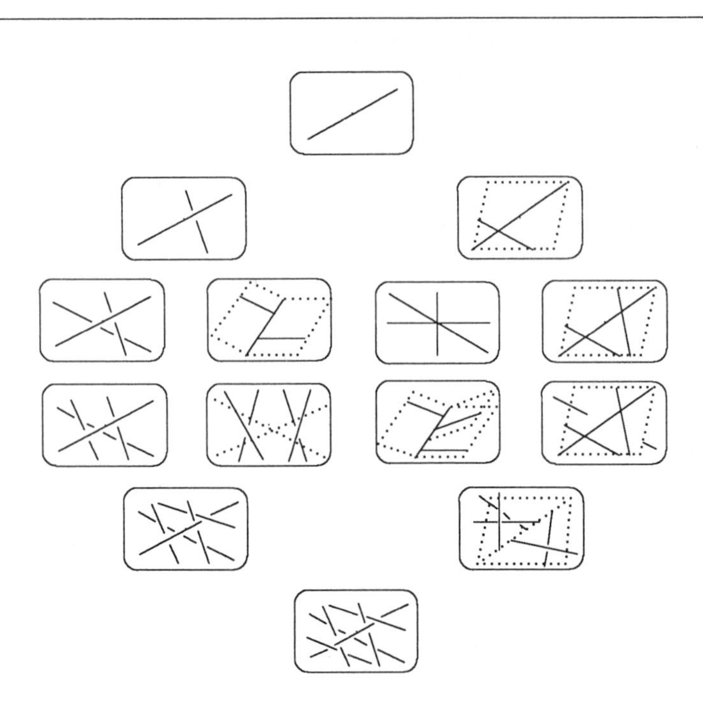

Figure 4.11

The number of degrees of freedom in a hinge and body framework is greater than the generic count of $6(\#\mathbb{B}-1)-5\#\mathbb{H}$ if and only if the rotors defined by the hinges are linearly dependent. The set of lines whose 2-vectors lie in the linear span of a given set of 2-vectors is called a *line variety*. Line varieties in turn can be classified combinatorially in terms of the incidence structure of their bases. Note in particular that there are two types of line complex (5 dimensional line varieties), whose bases are shown on the second row up.

provides a combinatorial approach to the generic rigidity of hinge and body frameworks similar to that presented in the next section for bar and joint frameworks. We present this result without proof [Tay1984b, Whiteley1986b].

THEOREM 4.38. *Let* $\mathbb{F} = (\mathrm{A}, \mathbb{B}, \mathbf{p})$ *be an* n*-dimensional hinge and body framework,* $\mathbb{H} \subseteq \binom{\mathbb{B}}{2}$ *be the pairs of bodies joined by hinges, and set* $m := n(n+1)/2$. *Then the following are equivalent:*

(i) $\mathbb{F}$ *is generically* n*-isostatic;*

(ii) *The multigraph* $\mathbb{G}^m = (\mathbb{B}, \mathbb{H}^m)$ *obtained by replacing each couple* $\{B, B'\} \in \mathbb{H}$ *by m copies of the same pair can be covered by exactly* $m-1$ *couple-disjoint spanning trees;*

(iii) $(m-1)\#\mathbb{H} = m(\#\mathbb{B}-1)$, *while for all partitions*‡ $\mathbb{G}^* = (\mathbb{B}^*, \mathbb{H}^*)$ *of* $\mathbb{G}$ *we have* $(m-1)\#\mathbb{H}^* \geq m(\#\mathbb{B}^*-1)$.

It should be noted that any two "hinges" in a molecule (i.e. rotatable chemical bonds) are always coincident and hence are not generic whenever they have an atom in common. Nevertheless, it has been conjectured that the above theorem remains valid even when this sort of degeneracy is present [Tay1984b].

4.4. The Coordinatization of Abstract Frameworks

In this section we study the problem of coordinatizing abstract frameworks (A, B, d), i.e. of computing representations $\mathbf{p}: A \to \mathbb{R}^n$ such that $\|\mathbf{p}(a)-\mathbf{p}(b)\| = d(a,b)$ for all $\{a,b\} \in B$. As previously observed, this is just a special case of the Fundamental Problem of Distance Geometry. Although algorithms are described in Chapter 6 by which the Fundamental Problem can often be solved, considerable room for improvement remains in terms of their efficiency, reliability and the accuracy of the solutions they find. One reason for studying the simpler problem of coordinatizing frameworks is that we can prove directly that it will be difficult to find an algorithm which meets all these criteria for abstract frameworks. Since these difficulties will also have to be dealt with in devising an improved algorithm for the general case of the Fundamental Problem, it behooves us to learn how to deal with them. One rather interesting discovery we shall make is that there exists a fairly large class of abstract frameworks for which exact representations can be found within reasonable amounts of computer time. These frameworks are necessarily generically rigid, and the coordinatization algorithms which work on them rely heavily on the graph-theoretic structure of generically rigid frameworks. Hence we shall study this structure in detail, and make some suggestions as to how these algorithms might be extended to deal with larger classes of abstract frameworks.

Let us begin by considering some very general computational aspects of the problem of coordinatizing abstract bar and joint frameworks, or what we will

‡ A *partition* of a graph $\mathbf{G} = (A, B)$ is the graph $\mathbf{G}^* = (A^*, B^*)$ obtained from a partition $\bar{A}$ of A by defining the elements of A^* to be the members $a^* \in \bar{A}$ of the partition, and connecting two distinct atoms a^*, b^* in $\mathbf{G}^*$ by a couple whenever any two atoms $a \in a^*$, $b \in b^*$, one from each of the corresponding members of the partition, were adjacent in $\mathbf{G}$.

henceforth call the Framework Coordinatization Problem. In order to achieve this generality, we shall ignore the amount of time it takes to solve any given instance of the problem by any specific algorithm on any particular computer, and consider only the *rate* at which the amount of time required to solve the problem must increase with its size in the worst possible case. This rate is called the *complexity* of the problem. In order to place the problem in its proper perspective, we provide a brief description of the relevant complexity classes into which such computational problems are generally placed on the adjacent page. A more complete account of the theory of computational complexity may be found in [Garey1979].

In their most general form, it is not at all obvious that the Fundamental Problem of Distance Geometry or even the Framework Coordinatization Problem alone is decidable. As observed earlier, however (§3.5), these problems belong to an area of mathematics known as *semialgebraic geometry*, which deals, roughly speaking, with the solutions of systems of algebraic equalities and inequalities over the real numbers. Thanks to some very high-level and general results due to the logician A. Tarski [Tarski1951], an algorithm is known by which a solution to any such system can be found within a finite amount of time. This shows in particular that the Fundamental Problem of Distance Geometry is indeed decidable. Unfortunately, although Tarski's algorithm is finite, it is of no practical use, since the amount of time it requires can exceed the estimated age of the universe even on rather small problems! Nevertheless, the mere existence of this algorithm is important because it proves that the Fundamental Problem at least *can* always be solved exactly, and what remains is to find a way of doing so practically in those cases of greatest interest.

Since the number of squared distances is bounded by the square of the number of atoms and the time required to compute these quantities from Cartesian coordinates is bounded by a polynomial in the number of digits in the coordinates, it is clear that the amount of time to verify that any given set of coordinates is a solution to an instance of Framework Coordinatization Problem is polynomial in the number of bits necessary to store the coordinates. Unfortunately, it is *not* at all clear that the number of bits necessary to store the coordinates can be bounded by a polynomial in the number of bits used to encode the input distance constraints. Even an object so simple as an equilateral triangle cannot be embedded in the plane with *rational* coordinates, which means a fixed-point encoding is not adequate. Hence the presence of the Framework Coordinatization Problem in NP is at the time of writing an open question. It has been shown, however, that this problem is NP-hard [Saxe1979], and we will now give the simplest available proof of this result.

The most important complexity classes are:

Undecidable Problems: For the problems in this class, it can be rigorously proven that there exists no algorithm by which they can be reliably solved. This does not mean that any particular instance of the problem cannot be solved, but rather that the set of possibilities which must be examined to do so can be infinite. For example, there exists no provably finite algorithm for finding integer solutions of multivariate integral polynomial equations, because there exists no *a priori* bound on the size of the integers in the solution.

Polynomial Problems: This class is usually designated by the letter P. For the problems in P, algorithms are known whose running time is bounded from above by some polynomial in the size of their instances (which is, roughly speaking, the minimum number of binary digits necessary to encode them). This distinction is important, because any other type of algorithm will generally require an exponentially increasing amount of time to solve some sequence of instances of the problem, and hence even going to a faster computer will allow only a very modest increase in the size of the instances which can be solved in a given amount of time.

Nondeterministic Polynomial: These problems are distinguished by the fact that, even if no polynomial or possibly even finite algorithm for solving them is known, algorithms do exist by which at least one solution to every instance of any problem in the class can be *verified* in polynomial time. Another way of looking at them is as the class of problems which can be solved in polynomial time on a computer with an infinite number of parallel processors. This class is usually designated by the letters NP.

NP-complete: As the name implies, this class is a subclass of NP. Its problems have two characteristics in common. The first is that no polynomial-time algorithm for solving them is known. The second is that if anyone ever finds a polynomial-time algorithm for solving one of them, they will have succeeded in finding a polynomial-time algorithm for *every* member of this class! This is because polynomial-time algorithms are known by which any instance of any problem in this class can be translated into an instance of any other problem in this class, so that a solution to the latter provides a solution to the former. We describe this by saying that the former problem can be *reduced* to the latter in polynomial time.

Strongly NP-complete: It has been found that some NP-complete problems can be solved in polynomial time if the number of digits in the numbers involved in their input is kept below certain limits. No polynomial time algorithm is known for the problems of this class even when the size of their numbers is restricted.

(Strongly) NP-hard: Strictly speaking, the term NP-complete applies only to *decision problems*, which require only "yes" or "no" answers. If a known NP-complete decision problem can be reduced to any other problem, even one not itself in NP, then a polynomial-time algorithm for the latter would provide a polynomial-time algorithm for the former, so the latter can be regarded as being at least as hard. In particular, if a strongly NP-complete problem can be reduced to some other problem, the latter is called strongly NP-hard.

We first establish a basic fact that is necessary for us to apply the theory of NP-completeness to the this problem.

LEMMA 4.39. *The problem of computing a 1-dimensional representation of an abstract framework* $\mathbb{F} = (\mathrm{A}, \mathrm{B}, d)$ *with an integer-valued length function* $d : \mathrm{B} \rightarrow \mathbf{Z}$ *is in NP.*

Proof: If $S(d(\mathrm{B}))$ is the size of the largest integer occurring in $d(\mathrm{B})$, then by placing one of the atoms of each connected component of $\mathbb{G} = (\mathrm{A}, \mathrm{B})$ at the origin we can ensure that the largest integer in any representation has size less than $S(d(\mathrm{B}))\cdot diam(\mathrm{A}, \mathrm{B})$, where $diam(\mathrm{A}, \mathrm{B})$ is the largest diameter of any of the connected components of $\mathbb{G}$. Thus the total number of bits needed to store the coordinates is bounded from above by $log_2(\#\mathrm{A}^2\cdot S(d(\mathrm{B})))$, which is of the same order as the number of bits necessary to encode the framework itself. QED

To prove that the 1-representability of frameworks is NP-complete, we must show that some known NP-complete problem can be reduced to it in polynomial time. The easiest to use is the *Partition Problem*:

> Given a list of positive integers $\underline{Z} \in \mathbf{Z}^N$, does there exist a partition of $\{1, \ldots, N\}$ into subsets I and J such that the sums over these subsets satisfy $\sum_{i \in \mathrm{I}} z_i = \sum_{j \in \mathrm{J}} z_j$?

This is one of the earliest known NP-complete problems [Karp1975].

PROPOSITION 4.40. *The 1-representability of abstract bar and joint frameworks with integral bar lengths is NP-complete.*

To reduce the partition problem on the list $\underline{Z} = [2,3,1,2,4]$ to a one-dimensional representability problem, we let $\mathrm{A} := \{1, \ldots, 6\}$, $\mathrm{B} := \{\{1,2\}, \ldots, \{5,6\}, \{1,6\}\}$, $d(i, i+1) := z_i$ and $d(1,6) := 0$.

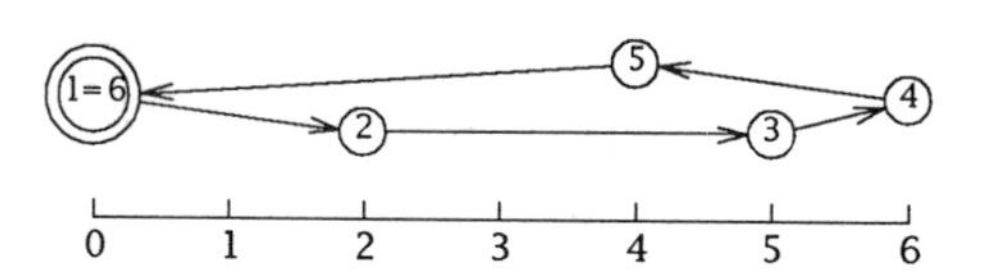

Figure 4.12

The one-dimensional representation shown above gives us a partition of $\{1, \ldots, 5\}$ such that the sum of the lengths of the right arrows equals the sum of the lengths of the left arrows, or 6.

Proof: Given a list $\underline{Z} \in \mathbb{Z}^N$, let

$$A := \{1, \ldots, N+1\}, \quad B := \{\{i, i+1\} \mid i = 1, \ldots, N\} \cup \{1, N+1\}, \tag{4.71}$$

$$d(i, i+1) := z_i \text{ for } i = 1, \ldots, N \text{ and } d(1, N+1) := 0 .$$

Then any map $p = \mathbf{p} : B \rightarrow \mathbb{R}^1$ which is a 1-dimensional representation of the abstract framework (A, B, d) gives us a partition $I \dot{\cup} J = \{1, \ldots, N\}$ according to the rule:

$$I := \{k \mid z_k \in Z \mid p(k) < p(k+1)\}, \quad J := \{k \mid z_k \in Z \mid p(k) > p(k+1)\} \tag{4.72}$$

(see Figure 4.12). QED

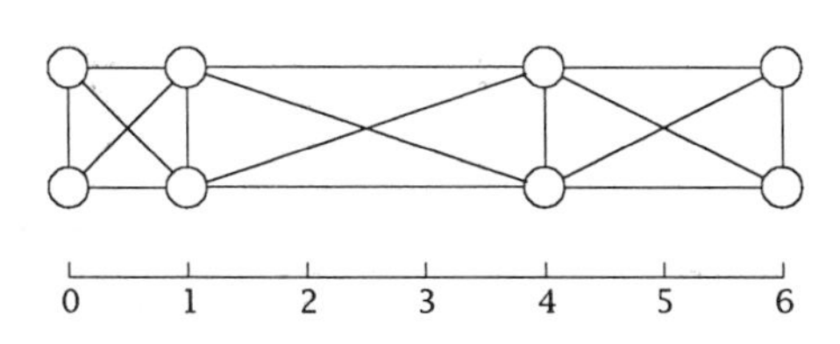

Figure 4.13

The use of trusses to reduce the 1-representability of integral abstract frameworks to two (and higher)-dimensional representability.

It is easy to extend this result to higher dimensions by means of suitable "trusses", as illustrated in Figure 4.13. Thus we have:

COROLLARY 4.41. *The representability of integral abstract frameworks is* NP-*hard in all dimensions.*

Since the representability of abstract frameworks is a special case of the Fundamental Problem of Distance Geometry, we then have the following very important result.

COROLLARY 4.42. *The Fundamental Problem of Distance Geometry is* NP-*hard.*

Because the Partition Problem is not *strongly* NP-complete, it might still be reasonable to hope that if the size of the integers is limited (as is the case when our distance measurements are limited in precision and a fixed-point encoding is used), a polynomial-time algorithm for the Framework Coordinatization Problem or even the Fundamental Problem of Distance Geometry might be found. Unfortunately, Saxe has shown that the former (and hence the latter) problem is actually *strongly* NP-hard. In addition, they have shown that the problem of finding a representation of an abstract framework which satisfies the distance constraints to within certain predefined tolerances is also strongly NP-hard, and that even when one representation of the framework is already known, the problem of finding a second noncongruent representation remains NP-hard [Saxe1979].

While these negative results are not desirable, they are far from being a complete disaster. As we shall see in Chapter 6, reasonably efficient algorithms by which approximate solutions to chemically relevant instances of the Fundamental Problem of Distance Geometry can usually be found are available, and in certain specific cases, e.g. when all of the distances in the molecule are known exactly, polynomial time algorithms for Framework Coordinatization Problems actually do exist. A more general case in which the coordinatization of an abstract framework is often (though not always) relatively easy occurs when the underlying graph of the framework is generically rigid (cf. § 4.1). Although the above results show that the coordinatization of such frameworks is still an NP-hard problem, the number of possible representations which must be considered is at least (generically) finite so that exact, combinatorial algorithms are possible. In order to see how this is done we must present some basic results on generically rigid graphs (see [Tay1985] for a recent review). These results are also of some intrinsic chemical interest, because they show how one can usually predict the number of conformational degrees of freedom which a molecule has directly from its covalent structure (structural formula).

We begin by considering some of the combinatorial relations between infinitesimally rigid frameworks with a common realization, generic or otherwise. Since the infinitesimal rigidity of a framework is determined by the rank of its rigidity matrix, the graph-theoretic structure characteristic of these frameworks must follow from the linear dependencies which exist between the rows of their corresponding rigidity matrices. As usual, the language in which these "patterns" of dependencies can best be described is that of matroid theory. Hence let $\mathbb{K} = \mathbb{K}(A) := (A, \binom{A}{2})$ be the complete graph on the set of atoms A, and let $\mathbf{p} : A \to \mathbb{R}^n$ be an arbitrary positioning thereof. If $\underline{\mathbf{M}}_\mathbf{p}$ is the rigidity matrix of the framework $(A, \binom{A}{2}, \mathbf{p})$, we denote by $\underline{\mathbf{M}}_\mathbf{p}(B)$ the submatrix of $\underline{\mathbf{M}}_\mathbf{p}$ whose rows correspond to the elements of $B \subseteq \binom{A}{2}$.

Definition 4.43. An n-dimensional *infinitesimal rigidity matroid* $\mathbf{M}_\mathbf{p} = \mathbf{M}_\mathbf{p}(A) = (\binom{A}{2}, \mathbb{I}_\mathbf{p})$ on $\binom{A}{2}$ is the matroid whose independent sets $\mathbb{I}_\mathbf{p}$ consist of those $B \subseteq \binom{A}{2}$ for which the submatrix $\underline{\mathbf{M}}_\mathbf{p}(B)$ has $rank(\underline{\mathbf{M}}_\mathbf{p}(B)) = \#B$. Given a graph $\mathbb{G} = (A, B)$ on the set of atoms A, we say that $\mathbb{G}$ is $\mathbf{p}$-*independent* whenever its set of couples B is independent in $\mathbf{M}_\mathbf{p}$, and $\mathbf{p}$-*isostatic* if B is a basis of this matroid. Extending our previous way of speaking slightly, we'll also say that $\mathbb{G}$ is $\mathbf{p}$-*rigid* if B contains a basis of $\mathbf{M}_\mathbf{p}$. Thus a graph is $\mathbf{p}$-isostatic if and only if it is both $\mathbf{p}$-rigid and $\mathbf{p}$-independent.

Let us first see what some common matroid-theoretic concepts have to say about $\mathbf{p}$-rigidity. By Theorem 4.8 the rank of $\mathbf{M_p}$ is $n\#\mathrm{A} - \frac{1}{2}(k+1)(2n-k)$ where $k := dim(<\mathbf{p}(\mathrm{A})>)$, and hence *all* $\mathbf{p}$-isostatic graphs contain exactly this many couples. From the point of view of statics, a graph is $\mathbf{p}$-independent if and only if the corresponding framework admits no internal stresses. From the point of view of kinematics, a graph is $\mathbf{p}$-rigid if and only if the corresponding framework admits no nontrivial infinitesimal motions. A pair of atoms $\{a,b\} \notin \mathrm{B} \subseteq \binom{\mathrm{A}}{2}$ is in the closure cl(B) in the matroid $\mathbf{M_p}(\mathrm{A})$ if and only if {a,b} is an induced bar of the framework $\mathbb{F} = (\mathrm{A}, \mathrm{B}, \mathbf{p})$, and thus $\mathbb{G} = (\mathrm{A}, \mathrm{B})$ is $\mathbf{p}$-rigid if and only if $\mathrm{cl}(\mathrm{B}) = \binom{\mathrm{A}}{2}$. The circuits of $\mathbf{M_p}(\mathrm{A})$ are those $\mathrm{C} \subseteq \binom{\mathrm{A}}{2}$ such that $\mathrm{C} \setminus \{a,b\}$ is $\mathbf{p}$-independent and $\{a,b\} \in \mathrm{cl}(\mathrm{C} \setminus \{a,b\})$ for all $\{a,b\} \in \mathrm{C}$.

To illustrate the utility of the matroidal view-point, let us show how a couple of results that we need later follow immediately from it.

PROPOSITION 4.44. *For a given mapping* $\mathbf{p}: \mathrm{A} \to \mathbb{R}^n$ *and graph* $\mathbb{G} = (\mathrm{A}, \mathrm{B})$ *with* $b, b', c, c' \in \mathrm{A}$ *and* $\{b,b'\}, \{c, c'\} \notin \mathrm{B}$, *suppose that* $\{c,c'\}$ *is not an induced bar of* $\mathbb{F} = (\mathrm{A}, \mathrm{B}, \mathbf{p})$ *but that* $\{c,c'\}$ *is an induced bar of* $\mathbb{F}_1 := (\mathrm{A}, \mathrm{B} \cup \{b,b'\}, \mathbf{p})$. *Then* $\{b,b'\}$ *is an induced bar of* $\mathbb{F}_2 := (\mathrm{A}, \mathrm{B} \cup \{c,c'\}, \mathbf{p})$, *and the underlying graph of* $\mathbb{F}_1$ *is* $\mathbf{p}$-*isostatic (or rigid or independent) if and only if the same is true of the underlying graph of* $\mathbb{F}_2$.

Proof: This is essentially just a translation of the matroidal closure axiom (K4) of Appendix D into the setting of infinitesimal rigidity matroids. QED

We also have the following rather broad substitution principle [Whiteley1984].

PROPOSITION 4.45. *For a given mapping* $\mathbf{p}: \mathrm{A} \to \mathbb{R}^n$, *suppose that* $\mathbb{G} = (\mathrm{A}, \mathrm{B})$ *is* $\mathbf{p}$-*isostatic (independent or rigid). For* $\mathrm{A}' \subseteq \mathrm{A}$ *with* $dim(<\mathbf{p}(\mathrm{A}')>) = k$, *let* $\mathbf{s}: \mathbb{R}^n \to \mathbb{R}^k$ *be an isometry mapping* $\mathbf{p}(\mathrm{A}')$ *into* $\mathbb{R}^k$, *and define* $\mathbf{p}': \mathrm{A}' \to \mathbb{R}^k$ *by* $\mathbf{p}' := \mathbf{s} \circ \mathbf{p}|_{\mathrm{A}'}$. *If the induced subgraph* $\mathbb{G}(\mathrm{A}') := (\mathrm{A}', \mathrm{B}(\mathrm{A}'))$ *is* $\mathbf{p}'$-*isostatic, then for any* $\mathrm{C} \subseteq \binom{\mathrm{A}'}{2}$ *such that* $(\mathrm{A}', \mathrm{C})$ *is likewise* $\mathbf{p}'$-*isostatic the graph* $\hat{\mathbb{G}} := (\mathrm{A}, (\mathrm{B} \setminus \mathrm{B}(\mathrm{A}')) \cup \mathrm{C})$ *obtained by replacing the couples* $\mathrm{B}(\mathrm{A}')$ *by* C *is again* $\mathbf{p}$-*isostatic (independent or rigid).*

Proof: For any matroid $(\mathrm{X}, \mathbb{I})$ with rank function ρ and closure operator cl we have $\rho(\mathrm{Y} \cup \mathrm{Z}) = \rho(\mathrm{Y} \cup \mathrm{Z}')$ for all $\mathrm{Y}, \mathrm{Z}, \mathrm{Z}' \subseteq \mathrm{X}$ with $\mathrm{cl}(\mathrm{Z}) = \mathrm{cl}(\mathrm{Z}')$. The conditions of the Proposition imply that $\mathrm{cl}(\mathrm{B}(\mathrm{A}')) = \binom{\mathrm{A}'}{2} = \mathrm{cl}(\mathrm{C})$, so that (letting $\mathrm{Y} = \mathrm{B} \setminus \mathrm{B}(\mathrm{A}')$, $\mathrm{Z} = \mathrm{B}(\mathrm{A}')$ and $\mathrm{Z}' = \mathrm{C}$) we have:

$$\rho(\mathrm{B}) \;=\; \rho((\mathrm{B} \setminus \mathrm{B}(\mathrm{A}')) \cup \mathrm{B}(\mathrm{A}')) \;=\; \rho((\mathrm{B} \setminus \mathrm{B}(\mathrm{A}')) \cup \mathrm{C})\,. \tag{4.73}$$

This shows that $\mathbb{G}_1$ is $\mathbf{p}$-rigid if and only if $\mathbb{G}_2$ is. Moreover since $\#\mathrm{B}(\mathrm{A}') = \#\mathrm{C} = k\#\mathrm{A}' - \binom{k+1}{2}$, $\mathbb{G}_1$ and $\mathbb{G}_2$ have equal numbers of couples in them, and hence the rank equals the number of couples in $\mathbb{G}_1$ if and only if the same is true of $\mathbb{G}_2$, as desired. QED

We now turn our attention to the generic case.

Definition 4.46. Let $\mathbf{p}: \mathrm{A} \to \mathbb{R}^n$ be a *generic positioning*, i.e. one such that the frameworks (A, B, $\mathbf{p}$) are regular for all $\mathrm{B} \subseteq \binom{\mathrm{A}}{2}$. The *unique* infinitesimal rigidity matroid which is obtained from such a mapping is called the n-dimensional *generic rigidity matroid* on A, and is denoted by $\mathbf{M}_n = \mathbf{M}_n(\mathrm{A}) = (\binom{\mathrm{A}}{2}, \mathbb{I}_n)$. A graph (A, B) is called *generically n-independent* if its set of couples B is an independent subset of $\mathbf{M}_n$, and *generically n-isostatic* if B is a basis of $\mathbf{M}_n$.

Clearly a graph is generically n-rigid if and only if it contains a spanning n-isostatic subgraph. Since generic realizations are full-dimensional, it follows from Theorem 4.8 that the rank of this matroid or number of couples in an n-isostatic graph is just $n \cdot \#\mathrm{A} - n(n+1)/2$. All subsequent discussion will be centered on the generic case, and hence we shall usually drop the qualifier "generic" in what follows.

We begin with a result which shows that the structure of (generically) rigid graphs has much to say about the structure of generic rigidity matroids [Asimow1979, Graver1984].

THEOREM 4.47. *Let* $\mathrm{cl}: 2^{\binom{\mathrm{A}}{2}} \to 2^{\binom{\mathrm{A}}{2}}$ *be the closure operator associated with the generic rigidity matroid* $\mathbf{M}_n(\mathrm{A})$. *Then for all* $\mathrm{B}, \mathrm{B}' \subseteq \binom{\mathrm{A}}{2}$ *with* $\mathrm{cl}(\mathrm{B}) = \binom{\mathrm{A}(\mathrm{B})}{2}$ *and* $\mathrm{cl}(\mathrm{B}') = \binom{\mathrm{A}(\mathrm{B}')}{2}$, *we have:*

$$\mathrm{cl}(\mathrm{B} \cup \mathrm{B}') = \begin{cases} \binom{\mathrm{A}(\mathrm{B})}{2} \cup \binom{\mathrm{A}(\mathrm{B}')}{2} & \text{if } \#(\mathrm{A}(\mathrm{B}) \cap \mathrm{A}(\mathrm{B}')) < n; \\ \binom{\mathrm{A}(\mathrm{B} \cup \mathrm{B}')}{2} & \text{if } \#(\mathrm{A}(\mathrm{B}) \cap \mathrm{A}(\mathrm{B}')) \geq n. \end{cases} \tag{4.74}$$

A proof of this (intuitively obvious) result may be obtained by using the fact that since for any $\mathrm{a} \notin \mathrm{A}(\mathrm{B}) \cap \mathrm{A}(\mathrm{B}')$ the vectors from $\mathbf{p}(\mathrm{a})$ to $\mathbf{p}(\mathrm{A}(\mathrm{B}) \cap \mathrm{A}(\mathrm{B}'))$ are linearly independent in any generic realization $\mathbf{p}: \mathrm{A} \to \mathbb{R}^n$, the framework can resolve any elementary load of the form $\mathbf{p}(\mathrm{b}) - \mathbf{p}(\mathrm{b}')$ with $\mathrm{b} \in \mathrm{A}(\mathrm{B})$ and $\mathrm{b}' \in \mathrm{A}(\mathrm{B}')$.

Since $\mathbb{K}_{3,3}$ is generically rigid but has no generically rigid proper subgraphs, it is clear that this theorem is not sufficient to characterize generic rigidity matroids. Nevertheless, the theorem suggests a means by which generically rigid representations for two abstract frameworks, say $\mathbb{F}_1 = (\mathrm{A}_1, \mathrm{B}_1, \mathbf{p}_1)$ and $\mathbb{F}_2 = (\mathrm{A}_2, \mathrm{B}_2, \mathbf{p}_2)$, can be combined to yield a generic representation of a larger framework $\mathbb{F} = (\mathrm{A}_1 \cup \mathrm{A}_2, \mathrm{B}_1 \cup \mathrm{B}_2, \mathbf{p})$ whenever $\#(\mathrm{A}_1 \cap \mathrm{A}_2) \geq n$: we simply let $\mathbf{p}$ be the extension of both $\mathbf{p}_1$ and $\mathbf{s} \circ \mathbf{p}_2$ to $\mathrm{A}_1 \cup \mathrm{A}_2$ for some isometry $\mathbf{s}$ of $\mathbb{R}^n$ such that

$$\mathbf{s} \circ \mathbf{p}_2(\mathrm{A}_1 \cap \mathrm{A}_2) = \mathbf{p}_1(\mathrm{A}_1 \cap \mathrm{A}_2)\,. \tag{4.75}$$

Note that this extension is unique up to isometry if $\#(\mathrm{A}_1 \cap \mathrm{A}_2) > n$, while if

$\#(A_1 \cap A_2) = n$ there are two congruence classes depending on whether $\mathbf{s}$ is proper or improper.

An alternative approach to characterizing generic rigidity matroids is to obtain a graph-theoretic description of their independent sets, and in particular of their bases. Our approach to characterizing the bases of $\mathbf{M}_n$ or n-isostatic graphs is to determine conditions under which an atom can be added to or deleted from an n-isostatic graph while preserving its rigidity and independence [Tay1985].

LEMMA 4.48. *Let* $\mathbb{G} = (A, B)$ *be a graph and* $a \in A$ *be an n-valent atom thereof with incident couples* $B_a := \{\{a, b_i\} \mid i = 1, \ldots, n\}$. *Then the graph* $\mathbb{G}' = (A', B') := (A \setminus a, B \setminus B_a)$ *is n-isostatic (n-rigid or n-independent) if and only the graph* $\mathbb{G}$ *is.*

Proof: We will first prove that $\mathbb{G}'$ is independent if and only if $\mathbb{G}$ is, which we do by showing that $\mathbb{G}'$ has a nontrivial self-stress in a generic realization if and only if $\mathbb{G}$ does.

Let $\mathbf{p}' : A' \rightarrow \mathbb{R}^n$ be a generic positioning of the atoms of $\mathbb{G}'$, and define a positioning $\mathbf{p} : A \rightarrow \mathbb{R}^n$ of A by defining $\mathbf{p}|_{A'} := \mathbf{p}'$ and $\mathbf{p}(a)$ to be any point of $\mathbb{R}^n$ such that $dim(\mathbf{p}(a \cup A_a)) = dim(\mathbf{p}(a \cup b_1 \cup \cdots \cup b_n)) = n$. Then any self-stress in $\mathbb{F} := (A, B, \mathbf{p})$ must be zero on the bars of B_a because

$$\sum_{i=1}^{n} s_i(\mathbf{p}(a) - \mathbf{p}(b_i)) = 0 \quad \Rightarrow \quad s_i = 0 \quad \text{for } i = 1, \ldots, n \tag{4.76}$$

by the generic linear independence of vectors $\{\mathbf{p}(a) - \mathbf{p}(b_i) \mid i = 1, \ldots, n\}$. If $\mathbb{G}'$ is isostatic the framework $\mathbb{F}' := (A', B', \mathbf{p}')$ admits no nontrivial self-stresses by definition, and this together with Equation (4.76) implies that $\mathbb{F}$ admits no nontrivial self-stresses either. Conversely, if $\mathbf{p} : A \rightarrow \mathbb{R}^n$ is a generic positioning of the atoms of $\mathbb{G}$ then $dim(\mathbf{p}(a \cup A_a)) = n$, so any self-stress of $\mathbb{F}$ must be zero on all bars of B_a and hence any self-stress of $\mathbb{F}$ gives a self-stress of $\mathbb{F}'$ by restriction to $B \setminus B_a$. It follows that if $\mathbb{F}$ has no nontrivial self-stresses then neither does $\mathbb{F}'$, as desired.

To prove that $\mathbb{G}'$ is isostatic and only if $\mathbb{G}$ is, we need only observe that since $val(a) = n$, $\#B = n\#A - n(n+1)/2$ if and only if $\#B' = n\#A' - n(n+1)/2$. To prove that $\mathbb{G}$ is rigid if $\mathbb{G}'$ is, we observe that if $\mathbb{G}'$ is rigid it contains an isostatic subgraph $\hat{\mathbb{G}}$ and the subgraph obtained by adding a and the couples incident it in $\mathbb{G}$ to $\hat{\mathbb{G}}$ is an isostatic subgraph of $\mathbb{G}$. On the other hand, if $\mathbb{G}$ is rigid then the independent set of couples B_a can be augmented to the set of couples of an isostatic subgraph $\hat{\mathbb{G}}$, so that $\hat{\mathbb{G}} \setminus a$ is an isostatic subgraph of $\mathbb{G}'$. QED

This one condition is already strong enough to enable us to characterize 1-

isostatic graphs [Tay1985].

THEOREM 4.49. *A graph is 1-isostatic if and only if it is a spanning tree.*

Proof: The theorem is obvious for any graph $\mathbb{G} = (A, B)$ with $\#A = 2$. On the other hand, any 1-isostatic graph $\mathbb{G} = (A, B)$ has $\#B = \#A - 1$, whereas $\Sigma_i \, val(a_i) = 2\#B$ for any graph, implying that any 1-isostatic graph always has at least two 1-valent atoms. Hence if $\mathbb{G}$ is a 1-isostatic graph, $a \in A$ is a 1-valent atom thereof and $\{a,b\} \in B$ is the unique couple connecting it to the rest of $\mathbb{G}$, the induction hypothesis together with Lemma 4.48 implies that $\mathbb{G}' = (A \setminus a, B \setminus \{a,b\})$ is a tree. Since any graph which is obtained from a tree by connecting a new atom to it by a single couple is again a tree, the result follows. QED

The matroids whose bases are the spanning trees of a graph $\mathbb{G}$ are one of the most important examples of matroids, and are called *graphic matroids.* Thus we can say that the 1-dimensional generic rigidity matroids are the graphic matroids associated with the complete graphs, that a graph is 1-rigid if and only if it is connected, and that a graph is 1-independent if and only if it contains no cycles. These conditions can all be tested in polynomial time [Williamson1985].

More importantly from our point of view, the above characterization immediately suggests a combinatorial algorithm for finding one-dimensional representations of abstract frameworks whose underlying graphs are also 1-rigid: One simply finds a spanning tree in the graph, and then proceeds to traverse the tree laying each new atom "a" down on the real line at a distance $d(a,b)$ to the left of the previous atom "b". This determines all the distances from "a" to all the preceding atoms, and if these agree with those given in the specification of the abstract framework the traversal can proceed; otherwise the search must backtrack and try to place some atom to the right of its predecessor in the tree. We shall soon see that these ideas can be extended to higher dimensions as well.

We now derive a second condition under which the deletion and addition of atoms preserves independence and rigidity [Tay1985].

LEMMA 4.50. *Let* $\mathbb{G} = (A, B)$ *be a graph.*

(i) *For any* $\{b,b'\} \in B$ *it is possible to delete* $\{b,b'\}$ *and add a new atom "a" connected to the rest of* $\mathbb{G}$ *by* $n+1$ *couples* B_a *so that the resultant graph* $\mathbb{G}' = (A \cup a, (B \cup B_a) \setminus \{b,b'\})$ *is n-isostatic (n-rigid or n-independent) if* $\mathbb{G}$ *is.*

(ii) *If* $a \in A$ *is an* $(n+1)$*-valent atom of* $\mathbb{G}$, *then for some* $\{b,b'\} \notin B$ *with* $b,b' \in A_a$ *the graph* $\mathbb{G}' := (A \setminus a, (B \setminus B_a) \cup \{b,b'\})$ *is n-isostatic (n-rigid or n-independent) if* $\mathbb{G}$ *is.*

Proof: (i): Let A_a be any $n+1$ atom subset of A such that $b, b' \in A_a$, define $B_a := \{\{a,c\} \mid c \in A_a\}$, and let $\mathbf{p}: A \to \mathbb{R}^n$ be a generic positioning of A. Then $dim(\mathbf{p}(A_a)) = n$, and we extend $\mathbf{p}$ to a positioning $\mathbf{p}': A \cup a \to \mathbb{R}^n$ of $(A \cup a, (B \cup B_a) \setminus \{a,b\})$ by defining $\mathbf{p}'|_A := \mathbf{p}$ and $\mathbf{p}'(a)$ to be any point on the line $((\mathbf{p}(b), \mathbf{p}(b')))$ other than $\mathbf{p}(b)$ and $\mathbf{p}(b')$. Since $\mathbf{p}'(a) \neq \mathbf{p}(b') = \mathbf{p}'(b')$ it cannot be in the hyperplane spanned by the n atoms $A_a \setminus b$, from which it follows that the framework $\mathbb{F}'' := (A \cup a, (B \cup B_a) \setminus \{a,b\}, \mathbf{p}')$ is infinitesimally independent (as in the proof of Lemma 4.48). Because $(\{a,b,b'\}, \{\{a,b'\}, \{b,b'\}\}, \mathbf{p}'|_{\{a,b,b'\}})$ is a tree hence 1-isostatic framework on the line $((\mathbf{p}'(b), \mathbf{p}'(b')))$ by Theorem 4.49, Proposition 4.45 now implies that the framework $\mathbb{F}' := (A \cup a, (B \cup B_a) \setminus \{b,b'\}, \mathbf{p}')$ is also infinitesimally independent. Since $\mathbb{G}'$ has an infinitesimally independent, i.e. maximum rank (albeit not generic) realization, $\mathbb{G}'$ is generically n-independent, as desired. From this together with $\#B' = \#B + n$, it follows that $\mathbb{G}'$ is n-isostatic if $\mathbb{G}$ is; likewise, the fact that n-rigid graphs contain n-isostatic subgraphs allows us to conclude that $\mathbb{G}'$ is also n-rigid if $\mathbb{G}$ is.

(ii): If $\mathbb{G}$ is independent, then in any generic positioning $\mathbf{p}: A \to \mathbb{R}^n$ we have $dim(\mathbf{p}(A_a)) = n$. We now remove one of the bars incident "a", say $\{a,b\}$, to obtain a new framework $\mathbb{F}''$. If any of the pairs in $\binom{A_a}{2}$ is neither a bar nor an induced bar of $\mathbb{F}''$, then by Lemma 4.48 we may add it to B and delete the n-valent atom "a" to obtain an independent framework $\mathbb{F}'$ with underlying graph $\mathbb{G}'$. If they are all (induced) bars of $\mathbb{F}''$, however, then $\{a,b\}$ is an induced bar of $\mathbb{F}''$ so that there exists a nonzero stress on that bar in $\mathbb{F} = (A, B, \mathbf{p})$, a contradiction. That $\mathbb{G}'$ is n-isostatic or n-rigid if $\mathbb{G}$ is then follows from the usual counting arguments. QED

It turns out that Lemmas 4.48 and 4.50 together are sufficient to characterize the 2-isostatic graphs [Tay1985].

Definition 4.51. Let $\mathbb{G} = (A, B)$ be a graph. A *Henneberg n-sequence* for $\mathbb{G}$ is an ordering $\underline{A} = [a_1, \ldots, a_N]$ of A such that the graphs $\mathbb{G}_i = (A_i, B(A_i))$ with $A_i := \{a_1, \ldots, a_i\}$ satisfy $\mathbb{G}_n := (A_n, \binom{A_n}{2})$, $\mathbb{G}_N := \mathbb{G}$ and for $i = n+1, \ldots, N$ each of the graphs $\mathbb{G}_i$ is obtained from $\mathbb{G}_{i-1}$ by one of the following operations:

(1) The addition of an n-valent atom which is connected to any n atoms of $\mathbb{G}_{i-1}$.

(2) The deletion of a bar and addition of an $(n+1)$-valent atom which is connected to the atoms of the deleted couple and $n-1$ other atoms of $\mathbb{G}_{i-1}$.

A graph is called *n-simple* when it can be built-up by means of operation (1) alone.

Note that we could just as well have defined Henneberg n-sequences by saying that $\mathbb{G}_i$ was obtained from $\mathbb{G}_{i+1}$ by the deletion of an n-valent atom or the

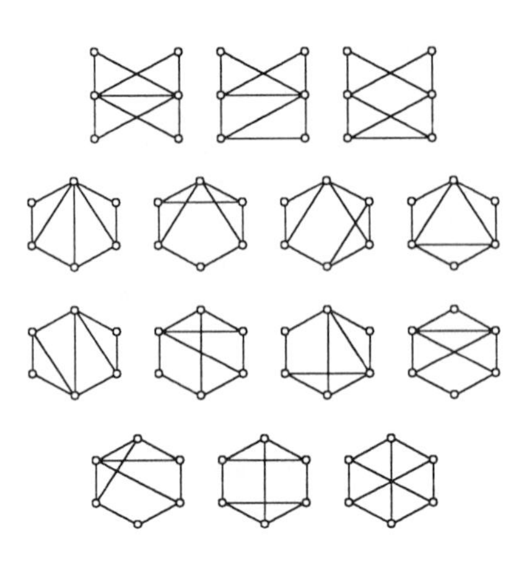

Figure 4.14

All nonisomorphic 2-isostatic graphs on six atoms. Only the last two on the bottom right are not simple.

deletion of an $(n+1)$-valent atom together with the addition of couple among the atoms to which it was adjacent. The name comes from the German engineer Henneberg, who first used such sequences to analyze rigidity [Henneberg1911].

THEOREM 4.52. *A graph* $\mathbb{G} = (\mathrm{A}, \mathrm{B})$ *is 2-isostatic if and only if there exists a Henneberg 2-sequence for its atoms.*

Proof: To prove this, we must show: (a) that any graph which can be built-up from a single couple by a Henneberg 2-sequence is 2-isostatic; (b) that any 2-isostatic graph $\mathbb{G}$ can be reduced to a single couple by a Henneberg 2-sequence. To prove (a), we need only note that since a single couple is 2-isostatic, Lemmas 4.48 and 4.50 assure us that every graph $\mathbb{G}_i$ up to and including $\mathbb{G}_N = \mathbb{G}$ will be 2-isostatic.

To prove (b), we first observe that any 2-isostatic graph satisfies $\#\mathrm{B} = 2\#\mathrm{A}-3$ as well as $min(val(\mathrm{a}) \mid \mathrm{a} \in \mathrm{A}) \geq 2$ for all $\mathrm{a} \in \mathrm{A}$. At the same time, $\Sigma_{\mathrm{a} \in \mathrm{A}} val(\mathrm{a}) = 2\#\mathrm{B}$ as always, and hence any 2-isostatic graph must possess either 2 or 3-valent atoms. In the event that it contains 2-valent atoms we can just delete one of them to obtain a 2-isostatic graph on one less atom. If it contains only 3-valent atoms, however, then none of these atoms, together with the three atoms adjacent to it, can induce a complete graph on four atoms since such a graph is not 2-independent. Hence it is possible to delete any one of these 3-

valent atoms and to add a new couple among its neighbors to obtain a graph on one less atom which is 2-isostatic by Lemma 4.50. Iterating on this procedure, we must eventually reach a 2-isostatic graph on two atoms, i.e. a single couple, and so have found a Henneberg 2-sequence for $\mathbb{G}$. QED

Note that the sufficiency of this condition holds in all dimensions, i.e. a graph which has a Henneberg n-sequence is n-isostatic.

Theorem 4.52 is only one of several possible characterizations of 2-isostatic graphs. For completness, the remaining characterizations are given in the following theorem, which we state without proof (see [Tay1985], and references therein).

THEOREM 4.53. *Let* $\mathbb{G} = (\mathrm{A}, \mathrm{B})$ *be a graph. Then the following are equivalent:*

(i) $\mathbb{G}$ *is 2-isostatic, i.e.* B *is a base of* $\mathbf{M}_2$.

(ii) $\#\mathrm{B} = 2\#\mathrm{A} - 3$, *and for every* $\mathrm{A}' \subseteq \mathrm{A}$ *the corresponding induced subgraph* $\mathbb{G}(\mathrm{A}') := (\mathrm{A}', \mathrm{B}(\mathrm{A}'))$ *satisfies*

$$\#\mathrm{B}(\mathrm{A}') \;\leq\; 2\#\mathrm{A}' - 3\,. \tag{4.77}$$

(iii) *For each and every couple* $\{\mathrm{b},\mathrm{b}'\} \in \mathrm{B}$ *of* $\mathbb{G}$ *there exists a pair of spanning trees* $\mathrm{T}_1, \mathrm{T}_2 \subseteq \mathrm{B}$ *in* $\mathbb{G}$ *such that* $\mathrm{T}_1 \cup \mathrm{T}_2 = \mathrm{B}$ *and* $\mathrm{T}_1 \cap \mathrm{T}_2 = \{\mathrm{b},\mathrm{b}'\}$.

Condition (ii) is known as *Laman's condition*, after the mathematician who first proved it [Laman1970]. Intuitively, this condition states that if a graph has only the theoretical minimum number of couples necessary for it to be rigid, then these couples must be "evenly distributed" if it is to actually be rigid.

Theorem 4.52 allows us to give an exact algorithm which can decide whether or not a graph is generically 2-isostatic, by finding a Henneberg 2-sequence for it.

Algorithm 4.54:

INPUT: A graph $\mathbb{G} = (\mathrm{A}, \mathrm{B})$.

OUTPUT: A Henneberg 2-sequence $\underline{\mathrm{A}} = [\mathrm{a}_1, \ldots, \mathrm{a}_N]$ for $\mathbb{G}$ if one exists.

PROCEDURE:

- Set $i := \#\mathrm{A}$ and $\mathbb{G}_i = (\mathrm{A}_i, \mathrm{B}_i) := \mathbb{G}$.
- While not done, do:
  - If $\mathbb{G}_i$ has a 1-valent atom, then
    - *done* := *true* ($\mathbb{G}$ is not 2-isostatic);
  - else if $i = 2$ and $\mathbb{G}_i$ consists of a single couple, then
    - *done* := *true* ($\mathbb{G}$ is isostatic);
  - else if there exists a 2 or 3-valent atom in A_i, then
    - Let $\mathrm{a}_i \in \mathrm{A}_i$ be such an atom.
    - If $val(\mathrm{a}_i) = 2$, then:

Delete it together with the couples incident to it
to obtain a new graph $\mathbb{G}_{i-1}$.
Set $k(i) := 0$ and $i := i-1$.
else if $val(\mathrm{a}_i) = 3$, then:
Let $K(i) \leq 3$ be the number of pairs of atoms in $\mathrm{A}_{\mathrm{a}_i}$ which
are not in B_i, and let $\{\mathrm{b}_{i1}, \mathrm{b}'_{i1}\}, \ldots, \{\mathrm{b}_{iK(i)}, \mathrm{b}'_{iK(i)}\}$
be the members of this set.
Set $k(i) := 1$.
Set $\mathrm{A}_{i-1} := \mathrm{A} \setminus \mathrm{a}_i$ and $\mathrm{B}_{i-1} := (\mathrm{B}_i \setminus \mathrm{B}_{\mathrm{a}_i}) \cup \{\mathrm{b}_{ik(i)}, \mathrm{b}'_{ik(i)}\}$
to obtain the graph $\mathbb{G}_{i-1}$ (new couple).
Set $i := i-1$.
else if there exists an index $j \in \{i, \ldots, \#\mathrm{A}\}$ such that
$K(j) > k(j) > 0$, then:
Let j be the minimum such index,
set $k(j) := k(j)+1$ and $i := j$.
Set $\mathrm{A}_{i-1} := \mathrm{A} \setminus \mathrm{a}_i$ and $\mathrm{B}_{i-1} := (\mathrm{B}_i \setminus \mathrm{B}_{\mathrm{a}_i}) \cup \{\mathrm{b}_{ik(i)}, \mathrm{b}'_{ik(i)}\}$
to obtain the graph $\mathbb{G}_{i-1}$ (new couple).
Set $i := i-1$.
else
done := *true* ($\mathbb{G}$ is not 2-isostatic).

The reader is invited to find Henneberg 2-sequences for the graphs shown in Figure 4.14.

We note that the above algorithm can easily be modified to determine whether or not a graph is 2-independent, rather than merely 2-isostatic, by deleting all 1-valent atoms as soon as any are formed by the deletion of a 2 or a 3-valent atom, and then proceeding as before with the remainder of the graph. In conjunction with the (maximum cardinality) greedy algorithm (see Appendix D) this method could be used to determine the number of degrees of freedom which the graph has in any generic realization. It does not give us a polynomial time algorithm, however, because it may be necessary to check as many as three possible couples to add back after the deletion of each 3-valent atom. It is nevertheless possible to determine whether or not a graph is 2-independent in polynomial time, although the algorithms for doing so are fairly complicated [Imai1981, Lovasz1982]. Since the probabilistic algorithm given in §4.1 is so simple, efficient and very nearly reliable, in most cases it would not be worthwhile to use them. The Henneberg 2-sequence, however, provides a good deal more information about the graph than the mere knowledge of its rigidity or independence, and hence the additional effort needed to find it can often be justified.†

† A probabilistic or polynomial rigidity prediction algorithm could be used to speed up Algorithm 4.54, by allowing one to decide which couples one could add back in the steps labelled

In particular, if the underlying graph of an abstract framework has a spanning subgraph which is n-simple, then a Henneberg 2-sequence permits us to find a two-dimensional representation for it. To see how such subgraphs can be found, we first note that since the above algorithm deletes 2-valent atoms in preference to 3-valent, it will always find a simple Henneberg 2-sequence if one exists. By the same token, once one has a test for independence it is easy to enumerate *all* the bases of a matroid (exercise!), so we can also find all 2-isostatic spanning subgraphs and test each in turn to see if it is also 2-simple.

Algorithm 4.55:

INPUT: An abstract framework $\mathbb{F} = (A, B, \mathbf{p})$.

OUTPUT: A representation $\mathbf{p}: A \rightarrow \mathbb{R}^2$ thereof if one exists and the underlying graph $\mathbb{G} = (A, B)$ contains a 2-simple spanning subgraph.

PROCEDURE:

```
For each 2-isostatic spanning subgraph H of G:
  If there exists a simple Henneberg 2-sequence for H:
    Index the atoms accordingly as [a_1, . . . ,a_N].
    Set p(a_1) := 0 and p(a_2) := [d(a_1,a_2), 0].
    Set i := 3 and found := false.
    While not found do
      Compute p(a_i) by triangulation w.r.t.
       its neighbors in H, and set mark(i) = true.
      If ||p(a_i)−p(a_j)|| ≠ d(a_i,a_j) for some j < i, then:
        If there exists a k ≤ i with mark(k) = true, then:
          Let k be the maximum such index, and reflect p(a_k)
           in the line spanned by its two neighbors.
          Set mark(k) := false and i := k+1.
        Else halt    (G is not representable).
      Else if i = N then
        Set found := true    (a representation has been found).
      Else set i := i+1.
    Print representation and halt.
```

Although the above brute force approach to finding a 2-simple spanning subgraph is rather inefficient, it is not hard to improve upon it by standard backtracking techniques.

The above algorithm could fail if the only existing representation of the graph was nongeneric, i.e. if $\mathbf{p}(a) = \mathbf{p}(b)$ for some $a \neq b$, but this is very unlikely to happen if the distribution of bar lengths in the framework is reasonably

"new couple" (above) and still expect to find a Henneberg 2-sequence.

random. A greater limitation is imposed by the fact that it is only applicable to frameworks whose underlying graphs have a 2-simple spanning subgraph. For frameworks which lack this property, it is sometimes possible to divide the graph into 2-simple subgraphs which each have two or more atoms in common, to find representations for each subgraph as above, and then to merge these representations as described following Theorem 4.47. Even then, however, the algorithm fails to be applicable to all frameworks (as happens, for example, if the underlying graph is $\mathbb{K}_{3,3}$). Extending Algorithm 4.55 so that it can find a 2-dimensional representation of any abstract framework whose underlying graph is 2-isostatic (and hence any 2-rigid framework whatsoever) is at the time of writing an open problem which deserves further attention.

A picture proof that the molecule adamantane is generically rigid in three-space:

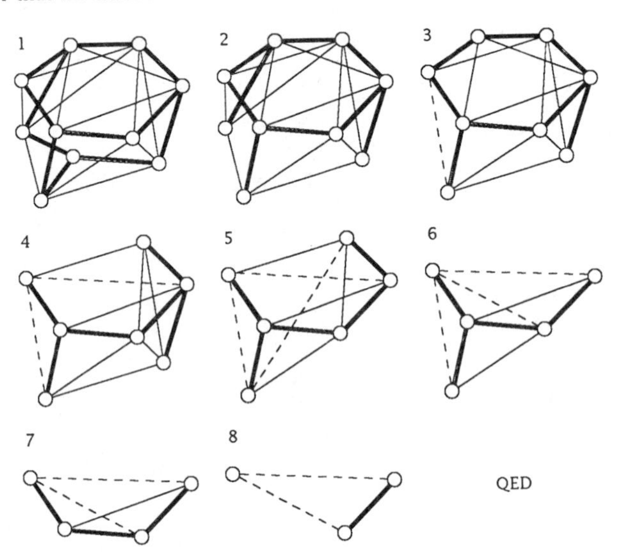

Figure 4.15

The proof proceeds by showing that the graph of the corresponding framework has a spanning subgraph which admits a Henneberg 3-sequence and so is 3-isostatic.

Algorithm 4.54 could easily be extended to find Henneberg 3-sequences for graphs as well, thereby permiting us to determine the 3-rigidity and/or 3-independence of graphs, as well as the number of conformational degrees of freedom of the framework. An example of the use of this extended algorithm on a chemically relevant problem is shown in Figure 4.15. In contrast to the two dimensional case, this algorithm is not applicable to all frameworks, because it is possible for a 3-isostatic graph with $\#B = 3\#A-6$ to have no atoms of valence less than $n+2 = 5$, so that a Henneberg 3-sequence for it does not exist (e.g. as in the edge graph of the icosahedron). In the event that the underlying graph was 3-simple, however, Algorithm 4.55 could be extended in the obvious way to find a representation for it.

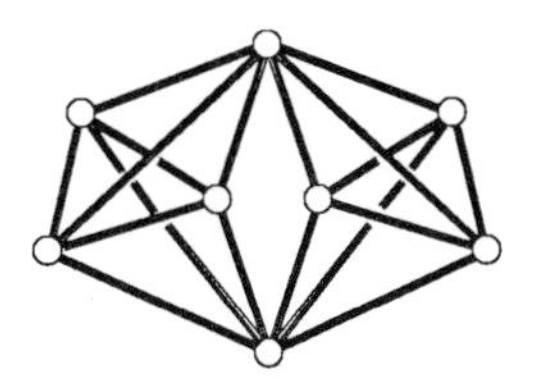

Figure 4.16

A flexible circuit in the three-dimensional generic rigidity matroid. None of the subgraphs obtained by deletion of any one couple from this graph are 3-isostatic.

A general characterization of the 3-rigid graphs, however, has yet to be found, and all of the conditions listed in Theorem 4.53 for the 2-dimensional case fail to generalize. For example, the 3-dimensional analogue of Laman's condition

$$\#B' \leq 3\#A(B') - 6 \quad \forall \; B' \subseteq B \tag{4.78}$$

remains necessary for a graph to be 3-isostatic, but it is no longer sufficient, as shown by the example in Figure 4.16. Several interesting conjectures have been made [Tay1985], but as of the time of writing none have been extablished with certainty. The prevailing opinion among experts in the field, however, is that the problem is not far from being solved. If this is so, then it may also not be long before an exact algorithm for the coordinatization of generically 3-rigid abstract frameworks is found.

4.5. Some Examples of Conformation Spaces

In the preceding sections, we have shown how one can determine if a set of exact distance constraints is sufficient to hold a given conformation of a molecule "rigid," and what its infinitesimal motions are in the event that they fail to hold it rigid. In this section, we wish to study the finite (and generally large) internal motions of a molecule which respect its stereochemical structure, or what we call molecular *kinematics*. In addition to the intrinsic interest of large amplitude molecular motions in chemistry [Orville-Thomas1974, Lowery1981, Serre1983], the mathematical "structure" of the set of all conformations of a molecule which arise through such motions provides us with an extension of the usual concept of molecular "structure" which includes those cases in which the molecule is mobile. In studying this new type of structure, it is a little more convenient to use abstract frameworks (see §4.1).

Definition 4.56. The *n-dimensional conformation space* $\mathbf{ST}(\mathbf{P})$ of an abstract framework $\mathbb{F} = (\mathrm{A}, \mathrm{B}, d : \mathrm{B} \to \mathbb{R}^{+})$ is the set $\mathbf{P} = \mathbf{P}_n(\mathbb{F}) \subseteq (\mathbb{R}^n)^{\mathrm{A}}$ of all of its possible representations, modulo the proper isometries $\mathbf{ST} = \mathbf{ST}(n, \mathbb{R})$.

Even though the framework $\mathbb{F}$ itself constitutes a precise description of its associated conformation space, just as the Cartesian coordinates of each of the possible conformations of a rigid molecule provide a description which is for many purposes more useful than a list of its invariant distances, so is it often desirable to have an explicit *global parametrization* of the conformation space of a molecule for which the available exact distance constraints are not sufficient to hold it rigid. For example, if we consider the abstract framework defined by the exact distance constraints which follow from the constitution of an acyclic saturated hydrocarbon, then the dihedral angles about carbon-carbon single bonds constitute a global parametrization of its conformation space. This parametrization falls short of being a truly global description of the conformation space of such molecules, simply because it does not take account of the fact that certain conformations are forbidden by long-range* interactions, in particular the van der Waals repulsions between the atoms (this exclusion of possible conformations is sometimes called the *excluded volume effect*). Even if one restricts oneself to the corresponding abstract frameworks, however, the dihedral angles fail to provide a global parametrization of the conformation space when the framework contains flexible rings, because only certain combinations of angles are consistent with the ring closure conditions.

* The term "long-range" here does not refer to distance, but rather to the number of covalent bonds on the shortest path connecting the atoms in the chemical graph of the molecule.

In such cases, global parametrizations of the conformation spaces of abstract frameworks do not exist. The mathematical description of these complex structures is in fact a difficult and incompletely solved problem, which we shall not attempt to undertake in its full generality. Instead, we shall show how one might go about determining some of the simpler and more basic features of the conformation space of a framework, or what is known as its *topological structure*. Although no general methods of deriving the global topological structure of the conformation space of an abstract framework are known, in the special case that the space is a smooth manifold (see §4.1) powerful mathematical tools are available which relate the more easily obtained local properties of the space to its global properties. What follows makes no pretense at being a rigorous or complete treatment of this extensive field of mathematics, but only an attempt to illustrate the basic concepts and some of the more useful techniques on some simple examples which bear an analogy to chemical problems. We begin with a definition which summarizes some of the basic terminology of topology.†

Definition 4.57. The *topological structure* of a subset $\mathbf{X} \subseteq \mathbb{R}^M$ are those features thereof which are preserved under homeomorphisms.‡ A *homeomorphism* is a bijective and continuous mapping $\mathbf{h}: \mathbf{X} \rightarrow \mathbf{Y} \subseteq \mathbb{R}^m$ whose inverse $\mathbf{h}^{-1}$ is also continuous. The continuity of a function $\mathbf{g}: \mathbf{X} \rightarrow \mathbf{Y}$ can be defined in two equivalent ways: First, we have the usual definition that for all $\mathbf{x}, \mathbf{x}' \in \mathbf{X}$ and $\epsilon > 0$ there exists $\delta > 0$ such that $\|\mathbf{x} - \mathbf{x}'\| < \delta$ implies $\|\mathbf{g}(\mathbf{x}) - \mathbf{g}(\mathbf{x}')\| < \epsilon$. Second, we have the purely topological definition, which states that for all open subsets $\mathbf{O} \subseteq \mathbf{Y}$ the pre-image $\mathbf{g}^{-1}(\mathbf{O})$ is open in $\mathbf{X}$. Since the inverse of a homeomorphism is also continuous, this latter definition shows that homeomorphisms induce a one-to-one correspondence between the open sets in $\mathbf{X}$ and those in $\mathbf{Y}$. A collection of open sets $\mathbb{O} = \{\mathbf{O} \subseteq \mathbf{X}\}$ whose union is all of $\mathbf{X}$ is called an *open covering* of the *topological space* $\mathbf{X}$. In the special case that each of the sets in $\mathbb{O}$ is homeomorphic to a subset of $\mathbb{R}^m$, the topological space $\mathbf{X}$ is called an *m*-dimensional *topological manifold*, and the open sets of the covering are called *coordinate patches*. If $\{\mathbf{h}: \mathbf{O} \rightarrow \mathbb{R}^m \mid \mathbf{O} \in \mathbb{O}\}$ is a collection of homeomorphisms, one for each of the coordinate patches, then the set $\{(\mathbf{O}, \mathbf{h}) \mid \mathbf{O} \in \mathbb{O}\}$ is called an *atlas* for the manifold $\mathbf{X}$. In contrast to a global parametrization, each such homeomorphism provides us with a purely *local parametrization*, valid only on the coordinate patch $\mathbf{O}$ in question (see Figure 4.2). Note that the smooth manifolds introduced in §4.1 are also topological manifolds; an example which shows that the converse is not true is shown in Figure 4.17. Those points of a

† The reader is referred to the excellent texts by Munkres [Munkres1975, Munkres1984] for more detailed introductions to topology.

‡ The intuitive definition of topology as "rubber sheet" geometry is probably already familiar to the reader.

The circle (left) is a smooth 1-dimensional manifold. The "diamond" (middle) is a 1-dimensional manifold which is homeomorphic (topologically equivalent) to a circle, but since it has four differential singularities (cusps), it is not a smooth manifold. One the other hand, the "clover-leaf" (right) is not a manifold at all, since it has five topological singularities (bifurcations).

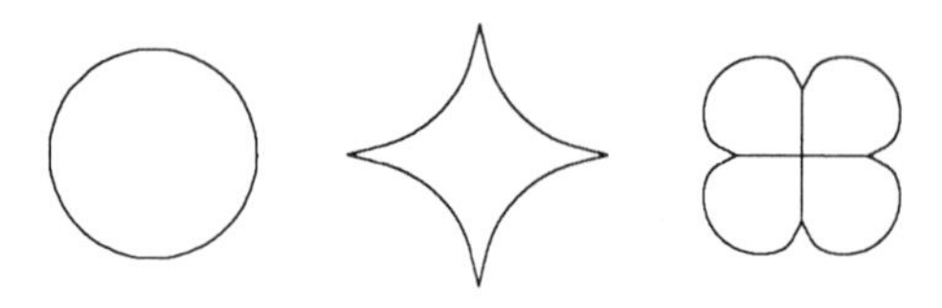

Figure 4.17

topological space $\mathbf{X} \subseteq \mathbb{R}^M$ which are contained in no open subset $\mathbf{O} \subseteq \mathbf{X}$ which is homeomorphic to a subset of $\mathbb{R}^m$ are called *topological singularities*; those points which are not topological singularities but which are nevertheless contained in no open subset diffeomorphic to a subset of $\mathbb{R}^m$ are called *differential singularities.*

We will take our first examples of the conformation spaces of "cyclic" frameworks from the relatively simple two-dimensional case, beginning with a triangle with fixed length sides:

$$\mathbb{F} \;=\; (\mathrm{A} = \{1,2,3\},\ \mathrm{B} = \{\{1,2\},\ \{1,3\},\ \{2,3\}\},\ d : \mathrm{B} \rightarrow \mathbb{R}^+)\ . \tag{4.79}$$

By our results in Chapter 3, the conformation space of this abstract framework is nonempty if and only if the function "d" obeys the triangle inequality. If this is the case, then since the underlying graph is the complete graph $\mathbb{K}_3$ it follows from the Fundamental Theorem of Euclidean Geometry that the two-dimensional conformation space $\mathbf{ST}(\mathbf{P}_2(\mathbb{F}))$ consists of but a single pair of elements, one for each of the two mirror images of the triangle (assuming that the triangle is non-degenerate).

A rather less trivial example is provided by a quadrilateral with fixed length sides, i.e.

$$\mathbb{F} \;=\; \left(\mathrm{A} = \{1,2,3,4\},\ \ \mathrm{B} = \{\{1,2\},\ \{2,3\},\ \{3,4\},\ \{1,4\}\},\ \ d : \mathrm{B} \rightarrow \mathbb{R}^+\right). \tag{4.80}$$

It is easily seen that this has a nontrivial conformation space $\mathbf{ST}(\mathbf{P}_2(\mathbb{F}))$ if and only if

$$d(\pi(1),\pi(4)) \le d(\pi(1),\pi(2)) + d(\pi(2),\pi(3)) + d(\pi(3),\pi(4)) \tag{4.81}$$

for all cyclic permutations $\pi:\{1,2,3,4\}\rightarrow\{1,2,3,4\}$. If this is the case, the topological structure of the conformation space can be determined by the following method [Thurston1984]. Without loss of generality we assume that $d(2,3) = max(d(i,i+1)\,|\,1 \le i \le 4)$ (where $i+1$ is computed modulo 4), and remove the irrelevant translational and rotational degrees of freedom by fixing the position and orientation of the bar $\{2,3\}$, e.g. by setting $\mathbf{p}(2) := [0,0]$ and $\mathbf{p}(3) := [d(2,3), 0]$. If we break the cycle at the atom labelled "1", the end of the bar $[(\mathbf{p}(1),\mathbf{p}(2))]$ must lie on the circle with center $\mathbf{p}(2)$ and radius $d(1,2)$. Similarly, the end of the bar $[(\mathbf{p}(1),\mathbf{p}(4))]$ must lie on an annulus centered on atom $\mathbf{p}(3)$ with inner radius $|\,d(1,4) - d(3,4)\,|$ and outer radius $d(1,4) + d(3,4)$ (see Figure 4.18 for details).

If the distances $d(1,2)$, $d(2,3)$, $d(3,4)$ and $d(1,4)$ are known, the possible positions of atom 1 (double circle) w.r.t. atom 2 constitute a circle centered on 2 of radius $d(1,2)$, whereas the possible positions of atom 1 w.r.t. atom 3 form an annulus centered on 3 of inner and outer radius $|d(3,4) - d(1,4)|$ and $d(3,4) + d(1,4)$, respectively. Thus atom 1 must lie on the segment(s) of the circle contained within the annulus.

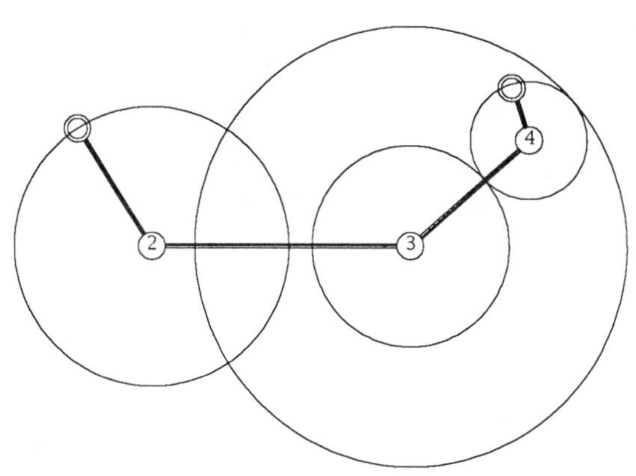

Figure 4.18

The position of the atom "1" in the intact framework must therefore lie on the intersection of the circle with the annulus. By our assumption that $d(2,3)$ is maximum, this intersection will be a full circle whenever

(P0) $d(1,2) \leq d(1,4) + d(3,4) - d(2,3)$, (4.82)

a single arc whenever

(P1) $d(1,4) + d(3,4) - d(2,3) < d(1,2) \leq d(2,3) - |d(1,4) - d(3,4)|$, (4.83)

and consists of two disjoint arcs symmetrically placed above and below the line segment [(2,3)] whenever

(P2) $d(2,3) - |d(1,4) - d(3,4)| < d(1,2)$. (4.84)

Similar conditions, henceforth referred to as (Q0), (Q1) and (Q2), hold for the possible positions of atom "4".

LEMMA 4.58. *Let* $\mathbb{F}$ *be a quadrilateral as in Equation (4.80) with* $d(2,3) = max(d(i,i+1) \mid 1 \leq i \leq 4)$, *and suppose that the lengths of the bars preclude the framework becoming collinear, so that the inequalities of Equation (4.81) are strict and*

$$\begin{aligned} d(1,2) + d(2,3) &\neq d(1,4) + d(3,4) ; \\ d(1,2) + d(1,4) &\neq d(2,3) + d(3,4) ; \\ d(1,2) + d(3,4) &\neq d(2,3) + d(1,4) . \end{aligned} \tag{4.85}$$

Then (P0) => (Q2), (P2) <= (Q0) and (P1) <=> (Q1). In addition, if both (P2) and (Q2) hold, then in any representation of $\mathbb{F}$ *the joints* $\mathbf{p}(1)$ *and* $\mathbf{p}(4)$ *are on the same side of the line* $((\mathbf{p}(2), \mathbf{p}(3)))$.

Proof: The proof is straightforward but long, and hence is left as an exercise. QED

It is interesting to observe that since by the triangle inequality

$$\begin{aligned} max(|d(2,3)-d(1,2)|, |d(1,4)-d(3,4)|) &\leq \|\mathbf{p}(1)-\mathbf{p}(3)\| \\ &\leq min(d(2,3)+d(1,2), d(1,4)+d(3,4)) \end{aligned} \tag{4.86}$$

and

$$\begin{aligned} max(|d(2,3)-d(3,4)|, |d(1,4)-d(1,2)|) &\leq \|\mathbf{p}(2)-\mathbf{p}(4)\| \\ &\leq min(d(2,3)+d(3,4), d(1,4)+d(1,2)) , \end{aligned} \tag{4.87}$$

for all representations $\mathbf{p}: A \to \mathbb{R}^2$ of $\mathbb{F}$, Lemma 4.58 can be interpreted as a statement about which combinations of triangle inequality limits can be attained by the indeterminate distances $d_{13}(\mathbf{p}) := \|\mathbf{p}(1)-\mathbf{p}(3)\|$ and $d_{24}(\mathbf{p}) := \|\mathbf{p}(2)-\mathbf{p}(4)\|$ in some representation of the framework $\mathbb{F}$ (assuming that it cannot become collinear). These three possibilities are illustrated in Figure 4.19.

The exact position of joint $\mathbf{p}(1)$ is uniquely determined by the value of the oriented angle Θ_{123} between the half-lines $[(\mathbf{p}(2), \mathbf{p}(1)))$ and $[(\mathbf{p}(2), \mathbf{p}(3)))$, or

The three possible types of noncollinear quadrilaterals with fixed length sides are shown below, together with the sequences of orientations of the four 3-tuples of their joints in the order that these occur as the quadrilaterals go through their cycle (or cycles) of motion.

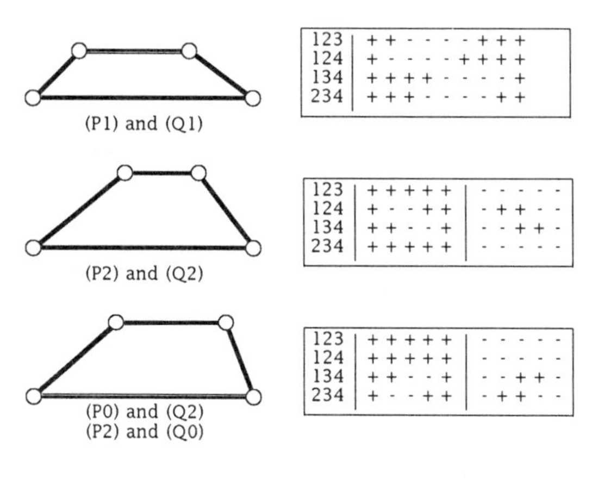

Figure 4.19

equivalently of the indeterminate distance $d_{13} = d_{13}(\mathbf{p})$ together with the orientation $\chi_{123} = \chi_{123}(\mathbf{p}) := sign(vol_{\mathbf{p}}(1,2,3))$. Thus for fixed values of these parameters, there are at most two possible conformations of the framework which differ in the sign of the angle between the half-lines $[(\mathbf{p}(4), \mathbf{p}(1)))$ and $[(\mathbf{p}(4), \mathbf{p}(3)))$, i.e. in the orientation $\chi_{134} = \chi_{134}(\mathbf{p})$. These observations allow us to prove:

PROPOSITION 4.59. *If $\mathbb{F}$ is a quadrilateral defined as in Equation (4.80) which cannot become collinear (as above), and $d_1 < d_2 < d_3 < d_4 = d(2,3)$ are the values of the fixed distances around the cycle sorted in increasing order, then the conformation space of $\mathbb{F}$ is a smooth one-dimensional manifold consisting of either one (if $d_2+d_3 < d_1+d_4$) or two (if $d_2+d_3 > d_1+d_4$) disjoint circles.*

Proof: It is easily seen from Lemma 4.58 that the set

$$\mathbf{P} \;:=\; \{\mathbf{p} \in (\mathbb{R}^2)^A \,|\, \mathbf{p} \text{ is a representation of } \mathbb{F} \text{ with } \mathbf{p}(2), \mathbf{p}(3) \text{ fixed}\} \qquad (4.88)$$

is compact (i.e. closed and bounded in $(\mathbb{R}^2)^A \approx \mathbb{R}^8$), and that it consists of one connected component if the conditions ((P1), (Q1)) prevail, whereas it consists of two components related by mirror reflection if conditions ((P0), (Q2)), ((P2), (Q0)) or ((P2), (Q2)) prevail. It is straightforward to show that this is the case if and only if $d_2+d_3 < d_1+d_4$ and $d_2+d_3 > d_1+d_4$, respectively. Since the only compact connected one-dimensional manifolds are circles and $\mathbf{ST}(\mathbf{P}) \approx \mathbf{P}$, the theorem will follow if we can but show that $\mathbf{P}$ is a smooth manifold.

To do this, we cover our set $\mathbf{P}$ with the open sets given by

$$\mathbf{P}_{13}(\chi_{123},\chi_{134}) \;:=\; \{\mathbf{p} \in (\mathbb{R}^2)^{\mathrm{A}} \mid \chi_{123}(\mathbf{p}) = \chi_{123},\ \chi_{134}(\mathbf{p}) = \chi_{134}\} \tag{4.89}$$

for all $\chi_{123}, \chi_{134} \in \{\pm 1\}$, and

$$\mathbf{P}_{24}(\chi_{234},\chi_{124}) \;:=\; \{\mathbf{p} \in (\mathbb{R}^2)^{\mathrm{A}} \mid \chi_{234}(\mathbf{p}) = \chi_{234},\ \chi_{124}(\mathbf{p}) = \chi_{124}\} \tag{4.90}$$

for all $\chi_{234}, \chi_{124} \in \{\pm 1\}$. It follows from the foregoing considerations that the squared distances D_{13} and D_{24} constitute local coordinates on the intersections of these open sets with $\mathbf{P}$. Verifying that these functions are smooth and hence that the pairs $(\mathbf{P} \cap \mathbf{P}_{ij}, D_{ij})$ constitute a covering of $\mathbf{P}$ by smooth coordinate patches is left to the reader as an exercise. QED

Further details concerning this example are given in the text surrounding Figure 4.19.

Although it is somewhat more difficult to do so, with a little patience and some models† one can show that in the event that Equation (4.85) is not fulfilled the topological structure of $\mathbf{ST}(\mathbf{P})$ will generally consist of two circles which intersect in two distinct points. These points of intersection are topological singularities, so that the space is no longer a manifold. The one exception is the symmetric or *equilateral* quadrilateral with $d(i,i+1) = d(j,j+1)$ $(mod\ 4)$ for all $i,j \in \{1,\ldots,4\}$; in this case $\mathbf{ST}(\mathbf{P})$ consists of three circles, each of which intersects each of the other two in exactly one point (see Figure 4.20).

While this example is of little chemical relevance, the thorough analysis given above is well justified because it illustrates several interesting properties of conformation spaces which seem to generalize to much more complicated and higher-dimensional abstract frameworks $\mathbb{F} = (\mathrm{A}, \mathrm{B}, d)$. These are:

(1) The structure of the space $\mathbf{ST}(\mathbf{P}_n(\mathbb{F}))$ becomes much more complex when internal symmetries are present in the abstract framework, i.e. when there exist nontrivial permutations $\phi: \mathrm{A} \rightarrow \mathrm{A}$ such that $d(\mathrm{a},\mathrm{b}) = d(\phi(\mathrm{a}),\phi(\mathrm{b}))$ for all $\{\mathrm{a},\mathrm{b}\} \in \mathrm{B}$. Observe that for a "random" length function d, such symmetries have zero probability of occurring, and hence we conjecture that in the *generic* case the conformation space will be a manifold.

(2) The topological singularities of $\mathbf{ST}(\mathbf{P}_n(\mathbb{F}))$ often correspond to positions in which triangle, tetrangle and higher-order limits are attained (see §3.5). These in turn correspond to the presence of projective dependencies between the joints of the framework $(\mathrm{A}, \mathrm{B}, \mathbf{p})$, i.e. the collinearity of joints,

† Plastic soda straws and pins are suitable material for making such models.

Topological structure of the conformation space of the equilateral quadrilateral consists of three circles each of which intersects the other two in exactly one point. A representative from each of these circles is indicated beside it.

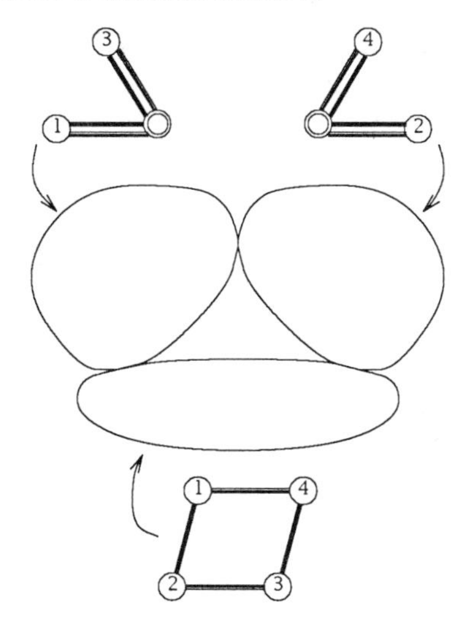

Figure 4.20

parallelism of bars, etc. More generally, they are confined to the nonregular realizations of the underlying graph $\mathbb{G} = (\mathrm{A}, \mathrm{B})$ (although not all nonregular realizations are singularities).

(3) In the event that the conformation space $\mathbf{ST}(\mathbf{P}_n(\mathbb{F}))$ is a smooth manifold, one can often obtain an atlas for it by fixing the orientations of selected $(n{+}1)$-tuples of atoms to obtain an open covering $\{\mathbf{O}\}$ of $\mathbf{P}_n(\mathbb{F})$, and then for each open set $\mathbf{O} \subseteq \mathbf{P}_n(\mathbb{F})$ fixing the values of a set of indeterminate distances $\{d(i,j) \mid \{i,j\} \in \mathrm{B}' \subseteq \binom{\mathrm{A}}{2}\}$ which is sufficient to render the graph $(\mathrm{A}, \mathrm{B} \cup \mathrm{B}')$ both n-isostatic and n-simple (see Definition 4.51), so as to obtain a well-defined smooth parametrization of the open subset $\mathbf{ST}(\mathbf{O}) \subseteq \mathbf{ST}(\mathbf{P}_n(\mathbb{F}))$ being considered.

Let us consider briefly one more example from the plane, the pentagon with fixed length sides:

$$\mathbb{F} \;=\; (\mathrm{A} = \{1,2,3,4,5\},\ \mathrm{B} = \{\{1,2\},\{2,3\},\{3,4\},\{4,5\},\{1,5\}\},\ d: \mathrm{B} \rightarrow \mathbb{R}^+)\,. \tag{4.91}$$

Since the number of indeterminate distances whose values must be fixed to render the framework n-simple and hence n-rigid is clearly two, the number of generic degrees of freedom of this framework and hence the generic local dimension of its conformation space $\mathbf{ST}(\mathbf{P}) = \mathbf{ST}(\mathbf{P}_2(\mathbb{F}))$ is likewise two. Using techniques similar to those used for the quadrilateral above, one can show that the topological structure of the conformation space is either a single sphere or else a pair of disjoint spheres in the generic case. Since this argument introduces no new principles, we will not pursue it here. Instead, we consider the special case of an equilateral pentagon, in which all of the side-lengths are equal. Unlike the equilateral quadrilateral above, it is easily shown that $\mathbf{ST}(\mathbf{P})$ remains a manifold in the symmetric case.

We do this by defining an open cover of $\mathbf{P}$:

$$\mathbb{O} \;=\; \{\mathbf{P}_i(\eta) \mid i \in \{1,\ldots,5\},\, \eta \in \{\pm 1\}^3\}\,, \tag{4.92}$$

where each member of this cover is given by:

$$\mathbf{P}_i(\eta) \;:=\; \{\mathbf{p} \in (\mathbb{R}^2)^{\mathrm{A}} \mid sign(vol_{\mathbf{p}}(i,i+j,i+j+1)) = \eta_j,\ j = 1,2,3\} \tag{4.93}$$

(the indices $i+j$ and $i+j+1$ are computed modulo 5). On each member $\mathbf{P}_i(\eta)$ of this covering we now choose the two indeterminate squared distances $D_{i,i+2}$ and $D_{i,i+3}$ as our parameters. The corresponding squared distance functions are obviously smooth and invertible on each of the open sets $\mathbf{P}_i(\eta)$ (provided we fix the positions of any two consecutive atoms around the cycle), and as in the proof of Proposition 4.59 their inverses can likewise be shown to be smooth. Thus we have covered the conformation space $\mathbf{ST}(\mathbf{P})$ by a smooth atlas consisting of 40 coordinate patches (see Figure 4.21). It is easily seen from this covering that the space consists of a single connected component.

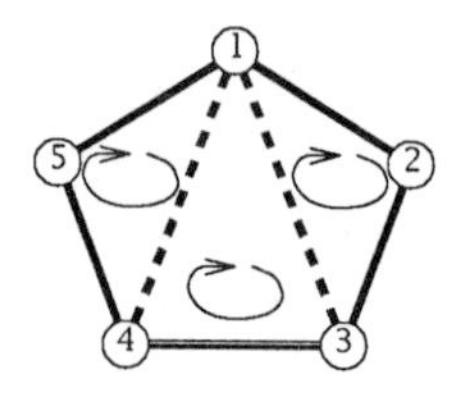

Figure 4.21

Equilateral pentagon together with distances (dashed) and orientations (arrows) of a coordinate patch.

Despite its relative simplicity, the size of this covering makes it a bit painful to determine the topological structure of **ST(P)** by elementary techniques when the pentagon is equilateral. Using an important technique, called *Morse theory* [Morse1969], it is nevertheless straightforward to prove the following result [Havel1988].

PROPOSITION 4.60. *The topological structure of the conformation space* $\mathbf{ST}(\mathbf{P}) = \mathbf{ST}(\mathbf{P}_2(\mathbb{F}))$ *of the pentagon is the compact, connected and orientable 2-dimensional manifold of genus* $G(\mathbf{ST}(\mathbf{P})) = 4$.

The genus of a two-dimensional manifold can be thought of as the number of "holes" in the closed surface of the manifold as a subset of $\mathbb{R}^3$. Thus a sphere has genus zero, a torus genus one, and so on. It is a basic result in topology that the value of this single integer invariant is sufficient to determine the structure of connected, compact and orientable two-dimensional manifolds completely up to homeomorphism [Munkres1984].

We now turn our attention to the three-dimensional case, beginning with cyclopentane:

$$\mathbb{F} \;=\; (\mathrm{A},\, \mathrm{B},\, d : \mathrm{B} \rightarrow \mathbb{R}^+)\,, \tag{4.94}$$

$$\mathrm{A} \;=\; \{1, 2, 3, 4, 5\}\,; \quad \mathrm{B} \;=\; \binom{\mathrm{A}}{2}\,;$$

$$d(i,j) \;\mapsto\; \begin{cases} 1 & i = j+1 \quad (mod\ 5)\,; \\ \sqrt{X} & i = j+2 \quad (mod\ 5)\,. \end{cases}$$

PROPOSITION 4.61. *The three-dimensional conformation space* $\mathbf{ST}(\mathbf{P}) = \mathbf{ST}(\mathbf{P}_3(\mathbb{F}))$ *of the cyclopentane framework is nontrivial if and only if* $X = ½(\sqrt{5}+3)$ *or* $X = ½(\sqrt{5}-3)$, *in which cases the only conformations are the regular convex and inverted planar pentagons, respectively.*

Proof: In proving this proposition we adopt a purely algebraic approach based on our distance-theoretic results from Chapter 3. Since the framework $\mathbb{F}$ has $\mathbb{K}_5$ as its underlying graph, it is representable in $\mathbb{R}^3$ if and only if all Cayley-Menger determinants on four atoms or less are nonnegative while the one and only five-atom Cayley-Menger determinant vanishes. The three-atom Cayley-Menger determinants are nonnegative if and only if the distances all satisfy the triangle inequality, i.e. $0 \le \sqrt{X} \le 2$, whereas the four-atom Cayley-Menger determinants are equal to:

$$D(1,2,3,4) \;=\; -1/4 \cdot (X^3 - 2X^2 - 2X + 1)\,. \tag{4.95}$$

This cubic is easily seen to be nonnegative if and only if

$$X \;\le\; -1 \qquad \text{or} \qquad \sqrt{5}/2 - 3/2 \;\le\; X \;\le\; \sqrt{5}/2 + 3/2\,, \tag{4.96}$$

and since $X \ge 0$ the latter possibility is the only one we have to consider. The five-atom Cayley-Menger determinant is:

$$D(1,2,3,4,5) \;=\; 1/16\cdot\left(5X^4 - 30X^3 + 55X^2 - 30X + 5\right) \;=\; 0\,. \qquad (4.97)$$

This last polynomial has $\sqrt{5}/2 \pm 3/2$ as its roots (where each root is two-fold degenerate). Both of these values satisfy Equation (4.96), so that the i to $i+2$ distances $\sqrt{X}$ are either $\sqrt{5}/2+1/2$ or $\sqrt{5}/2-1/2$. These distances are those found in the convex and inverted regular planar pentagons, respectively (see Figure 4.20). The uniqueness of these two representations (up to screw-translation) now follows from the Fundamental Theorem of Euclidean Geometry. QED

The above proof is much simpler than the trigonometric proof which is usually given (see, for example, [Dunitz1979]).

We now consider the case of cyclohexane (assuming tetrahedral bond angles):

$$\mathbb{F} \;=\; (\mathrm{A},\,\mathrm{B},\,d:\mathrm{B}\rightarrow\mathbb{R}^+)\,; \qquad (4.98)$$

$$\mathrm{A} \;:=\; \{1,\ldots,6\}\,; \qquad \mathrm{B} \;:=\; \tbinom{\mathrm{A}}{2}\setminus\{\{1,4\},\{2,5\},\{3,6\}\}\,;$$

$$d(i,j) \;:=\; \begin{cases} 1 & \text{if } j = i+1 \pmod 6\,; \\ \sqrt{8/3} & \text{if } j = i+2 \pmod 6\,. \end{cases}$$

In this case, it is easy to find a Henneberg 3-sequence for the underlying graph $\mathbb{G} = (\mathrm{A},\mathrm{B})$, so that the representations of $\mathbb{G}$ are generically 3-rigid. Thus we would expect the conformation space of cyclohexane to consist of a finite number of isolated points. It is something of a surprise, therefore, to discover that cyclohexane has a nonrigid "boat" form as well as a rigid "chair" form (see Figure 4.22). The reason for this is that the boat form consists of a one-parameter family $\mathbf{p}^\alpha:\mathrm{A}\rightarrow\mathbb{R}^3$ of nonregular conformations. One way to see this is to consider the triangles $[(\mathbf{p}^\alpha(1),\mathbf{p}^\alpha(2),\mathbf{p}^\alpha(3))]$ and $[(\mathbf{p}^\alpha(4),\mathbf{p}^\alpha(5),\mathbf{p}^\alpha(6))]$ to be two rigid bodies joined by the six bars $[(\mathbf{p}^\alpha(1),\mathbf{p}^\alpha(5))]$, $[(\mathbf{p}^\alpha(1),\mathbf{p}^\alpha(6))]$, $[(\mathbf{p}^\alpha(2),\mathbf{p}^\alpha(4))]$, $[(\mathbf{p}^\alpha(2),\mathbf{p}^\alpha(6))]$, $[(\mathbf{p}^\alpha(3),\mathbf{p}^\alpha(4))]$, $[(\mathbf{p}^\alpha(3),\mathbf{p}^\alpha(5))]$. Then by Corollary 4.33, these bars fail to hold the bodies infinitesimally rigid w.r.t. one another only if the Plücker coordinates $\mathbf{p}^\alpha(i)\vee\mathbf{p}^\alpha(j)$ of the bars are linearly dependent and so lie on a line complex for all α. On the other hand, by Theorem 4.37 an infinitesimal flexing exists if and only if the Plücker coordinates of the bonds $\mathbf{p}(i)\vee\mathbf{p}(i+1)$ are dependent. By means of standard identities in projective geometry [Hodge1968], this in turn can be shown to happen if and only if the planes

$$((\mathbf{p}^\alpha(1),\mathbf{p}^\alpha(2),\mathbf{p}^\alpha(3)))\,,\quad ((\mathbf{p}^\alpha(1),\mathbf{p}^\alpha(5),\mathbf{p}^\alpha(6)))\,, \qquad (4.99)$$

$$((\mathbf{p}^\alpha(2),\mathbf{p}^\alpha(4),\mathbf{p}^\alpha(5)))\,,\quad ((\mathbf{p}^\alpha(3),\mathbf{p}^\alpha(4),\mathbf{p}^\alpha(6)))$$

are copunctual (i.e. all meet in a common point), in which case the same is also true of the planes

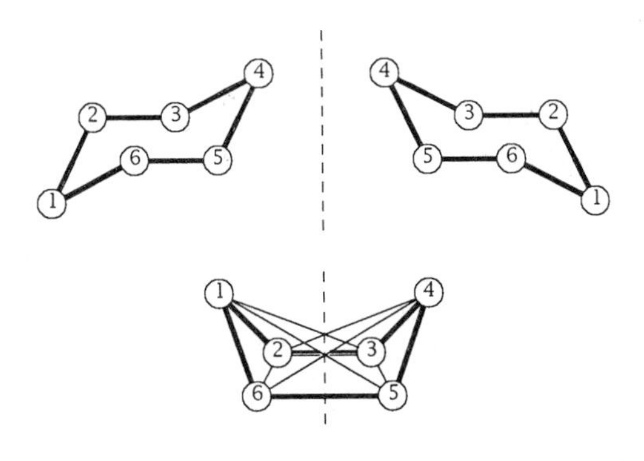

Figure 4.22

The two mirror images of the chair form of the cyclohexane framework, together with a representation of the boat form. The thin lines are those necessary to complete the Bricard octahedron, and the dashed line is its axis of symmetry.

$$((\mathbf{p}^\alpha(1), \mathbf{p}^\alpha(2), \mathbf{p}^\alpha(5))) , \quad ((\mathbf{p}^\alpha(1), \mathbf{p}^\alpha(3), \mathbf{p}^\alpha(6))) , \tag{4.100}$$
$$((\mathbf{p}^\alpha(2), \mathbf{p}^\alpha(3), \mathbf{p}^\alpha(4))) , \quad ((\mathbf{p}^\alpha(4), \mathbf{p}^\alpha(5), \mathbf{p}^\alpha(6))) .$$

All of these conditions, however, are only sufficient to guarantee that the cyclohexane framework admits an infinitesimal flexing, and hence we shall not bother to verify them here. Of perhaps greater interest is the fact that, if we consider the triangles listed in (4.99) and (4.100) to be the faces of an octahedron, then the boat form of cyclohexane can be regarded as an example of one of the famous flexible Bricard octahedra [Bricard1987, Connelly1979].

The conformation space of the cyclohexane framework was first characterized mathematically in the last century by [Sachse1892], who resorted to rather hair-raising trigonometric methods. A somewhat more elegant but incomplete derivation was offered much later by [Hazebroek1951]. The following algebraic *tour de force*, which is a direct solution of the system of Cayley-Menger determinants in the indeterminate squared distances, is unpublished work due to [Dress1982].†

† This approach should not be confused with the empirical description of ring puckering which has been proposed by [Cremer1975], which does not attempt to maintain fixed bond

To begin with, the indeterminate squared distances D_{14}, D_{25}, D_{36} must lie within the triangle and tetrangle inequality limits which are imposed by the remaining known distances. Assuming unit bond lengths, the triangle inequality limits are easily seen to be:

$$0.633 \approx \sqrt{8/3}-1 \leq D_{ij} \leq \sqrt{8/3}+1 \approx 2.633 \quad (i = j+3 \; mod \; 6) . \qquad (4.101)$$

The tetrangle limits, on the other hand, are found by examining the roots of those four-point Cayley-Menger determinants which involve only one indeterminate squared distance, e.g.

$$216 \cdot D_{14}(1,2,3,4; D_{14}) = -54 D_{14}^2 + 492 D_{14} - 950 , \qquad (4.102)$$

and

$$216 \cdot D_{13}(1,2,4,6; D_{14}) = -144 D_{14}^2 + 672 D_{14} - 400 . \qquad (4.103)$$

These polynomials have roots $[25/9, 19/3]$ and $[7/3-2\sqrt{6}/3, 7/3+2\sqrt{6}/3]$, respectively, so that the squared tetrangle limits are simply the tighter of these two pairs of bounds, or $[25/9, 7/3+2\sqrt{6}/3]$. Since the square roots of these numbers $[1.667, 1.991]$ are tighter than the triangle inequality limits, the latter need not be considered.

The five-atom Cayley-Menger determinants, of course, must all be zero, and since

$$\begin{aligned} D(1,2,3,4,5) &= D(1,2,4,5,6) ; \\ D(1,2,3,4,6) &= D(1,3,4,5,6) ; \qquad (4.104) \\ D(1,2,3,5,6) &= D(2,3,4,5,6) ; \end{aligned}$$

(as the reader can easily verify), we have exactly three bivariate quadratic equations in the indeterminates to deal with, for example

$$\begin{aligned} 0 &= 1296 \cdot D_{14,25}(1,2,3,4,5; D_{14}, D_{25}) \qquad (4.105) \\ &= 81 \cdot D_{14}^2 D_{25}^2 - 594(D_{14} D_{25}^2 + D_{14}^2 D_{25}) + 225(D_{14}^2 + D_{25}^2) \\ &\quad + 3492(D_{14} D_{25}) + 750(D_{14} + D_{15}) - 14575 . \end{aligned}$$

By making the substitution $D_{14} := S(x)$, $D_{25} := S(y)$ and $D_{36} := S(z)$, where

$$S(x) := (11-2x)/3 \qquad (4.106)$$

etc., we obtain the following rather drastic simplification of these equations:

$$E(x, y) := 81 \cdot D_{14,25}(1,2,3,4,5; S(x), S(y)) \qquad (4.107)$$

lengths and angles. An early example of the use of the distance geometry approach to analyze the kinematics of frameworks may be found in [Schoenberg1969].

$$= x^2y^2 - 24(x^2+y^2+xy) + 32(x+y) = 0;$$

$$E(x, z) := 81\cdot D_{14,25}(6,1,2,3,4;\ S(x), S(z)) \tag{4.108}$$

$$= x^2z^2 - 24(x^2+z^2+xz) + 32(x+z) = 0;$$

$$E(y, z) := 81\cdot D_{14,25}(5,6,1,2,3;\ S(y), S(z)) \tag{4.109}$$

$$= y^2z^2 - 24(y^2+z^2+yz) + 32(y+z) = 0.$$

LEMMA 4.62. *For any solution* $[x, y, z]$ *of the system of equations given in (4.107) through (4.109), we have either*

$$x = y = z \quad \text{or else} \quad xy+xz+yz = 0. \tag{4.110}$$

Proof: Suppose that $x \neq y$; then

$$0 = E(x, z) - E(y, z) \tag{4.111}$$

$$= (x-y)(x+y)z^2 - 24((x-y)(x+y) + (x-y)z) + 32(x-y)$$

so that, on dividing through by $x-y \neq 0$, we obtain

$$F(x,y,z) := (x+y)z^2 - 24(x+y+z) + 32 = 0. \tag{4.112}$$

Similarly, if $x \neq z$, we obtain

$$F(x,z,y) := (x+z)y^2 - 24(x+y+z) + 32 = 0, \tag{4.113}$$

and hence

$$F(x,y,z) - F(x,z,y) = x(z^2-y^2) + yz(z-y) \tag{4.114}$$

$$= (z-y)(xy+xz+yz) = 0.$$

It follows that if $y \neq z$ we have

$$xy + xz + yz = 0, \tag{4.115}$$

as desired.

On the other hand, if $y = z$ it follows from

$$E(y,z) = E(y,y) = y^4 - 72y^2 + 64y = 0 \tag{4.116}$$

and

$$F(x,y,z) = F(x,y,y) = y^3 + x(y^2 - 24) - 48y + 32 = 0 \tag{4.117}$$

that

$$(y^2+2yx)(y^2-24) = y^4 - 24y^2 + 2yx(y^2-24) \tag{4.118}$$

$$= y^4 - 24y^2 + 2y(-y^3+48y-32) \quad \text{by (4.117)}$$

$$= -y^4 + 72y^2 - 64y = 0.$$

Since it is impossible to have $y^2 = 24$ while $x \neq y = z \neq x$ (for otherwise we get the contradiction $E(y,y) = 24^2 - 72 \cdot 24 \pm 64\sqrt{24} \neq 0$), we obtain

$$y^2 + 2xy \;=\; xy + xz + yz \;=\; 0\,, \tag{4.119}$$

as before.

The only remaining possibility is $x = y = z$. QED

With the aid of this lemma, we can completely characterize all possible solutions of Equations (4.107) through (4.109) in terms of a simpler system of equations.

COROLLARY 4.63. *The simultaneous solutions of the equations* $E(x, y) = E(x, z) = E(y, z) = 0$ *consist exactly of those triples* $[x, x, x] \in \mathbb{R}^3$ *such that*

$$E(x, x) \;=\; x^4 - 72x^2 + 64 \;=\; 0\,, \tag{4.120}$$

together with the triples $[x, y, z] \in \mathbb{R}^3$ *which satisfy*

$$G(x, y, z) \;:=\; xy + xz + yz \;=\; 0 \tag{4.121}$$

and

$$H(x, y, z) \;:=\; -xyz - 24(x+y+z) + 32 \;=\; 0\,. \tag{4.122}$$

Proof: Since $-xyz - 24(x+y+z) + 32 \;=\; F(x,y,z) - z(xy+xz+yz)$, the necessity of Equations (4.120) through (4.122) follows immediately from Lemma 4.62. To prove the converse we observe that since $E(x,x) = x^4 - 72x^2 + 64$, we have $E(x,y) = E(x,z) = E(y,z) = 0$ for all solutions of Equation (4.120) with $x = y = z$. When Equations (4.121) and (4.122) hold, on the other hand, it follows from Equations (4.112) and (4.113) that

$$0 \;=\; (x-y)F(x,y,z) \;=\; E(x,z) - E(y,z) \tag{4.123}$$

and

$$0 \;=\; (x-z)F(x,z,y) \;=\; E(x,y) - E(y,z)\,, \tag{4.124}$$

i.e. $E(x,y) = E(y,z) = E(x,z)$, whence

$$\begin{aligned} 3E(x, y) \;&=\; E(x, y) + E(x, z) + E(y, z) \qquad (4.125)\\ &=\; x^2y^2 + x^2z^2 + y^2z^2 - 24(2x^2 + 2y^2 + 2z^2 + xy + xz + yz)\\ &\qquad + 64(x+y+z)\\ &=\; (xy+xz+yz)^2 - 2(x^2yz + xy^2z + xyz^2) - 24(2(x+y+z)^2\\ &\qquad - 3(xy+xz+yz)) + 64(x+y+z)\\ &=\; 2(x+y+z)(-xyz - 24(x+y+z) + 32)\\ &\qquad + (xy+xz+yz)^2 + 72(xy+xz+yz)\,, \end{aligned}$$

so that $E(x,y) = E(x,z) = E(y,z) = 0$ by Equations (4.121) and (4.122), as desired. QED

We now consider the hexangle equality which, when we express it in terms of the variables x, y, z:

$$I(x,y,z) \;:=\; 243/32 \cdot D_{14,25,36}(1,\ldots,6;\, S(x), S(y), S(z)) \;=\; 0\,. \qquad (4.126)$$

COROLLARY 4.64. *The solutions* $[x,y,z]$ *of the equations* $E(x,y) = E(x,z) = E(y,z) = I(x,y,z) = 0$ *consist exactly of the triples* $[0,0,0]$ *and* $[8,8,8]$, *together with the simultaneous solutions of Equations (4.121) and (4.122).*

Proof: Factorizing, we obtain

$$E(x,x) \;=\; x(x-8)(x^2+8x-8) \qquad (4.127)$$

(with similar expressions for $E(y,y)$ and $E(z,z)$), together with

$$\begin{aligned} I(x,y,z) \;=\; & (xy+xz+xz)(64-48(x+y+z) \\ & + 11(xy+xz+yz) - 2xyz) \;=\; 0\,. \end{aligned} \qquad (4.128)$$

In particular, the relation

$$\begin{aligned} I(x,x,x) &= 3x^2(-2x^3+32x^2-144x+64) \\ &= 3x^2(x-8)^2(1-2x) \end{aligned} \qquad (4.129)$$

shows that the triples $[0,0,0]$ and $[8,8,8]$ satisfy the equations $E(x,y) = E(x,z) = E(y,z) = I(x,y,z) = 0$. Furthermore, since $G(x,y,z) = 0$ occurs as a factor in Equation (4.128), the solutions of (4.121) and (4.122) also fulfill these equations.

Conversely, if $x = y = z$ does *not* hold, it follows at once from Corollary 4.63 that the simultaneous solutions of the equations $E(x,y) = E(x,z) = E(y,z) = I(x,y,z) = 0$ also satisfy Equations (4.121) and (4.122). Otherwise $x = y = z$ is necessarily a solution of the equation $E(x,x) = E(y,y) = E(z,z) = 0$. From the factorizations shown in Equations (4.127) and (4.129), one sees at once that the only roots of these polynomials which are also roots of $I(x,x,x)$ are 0 and 8. QED

The remaining two solutions of Equation (4.127) are $-4-2\sqrt{6}$ and $-4+2\sqrt{6}$. When $x = y = z = -4-2\sqrt{6}$, we obtain $I(x,y,z) \approx 1.68\times10^5$, whereas for $x = y = z = -4+2\sqrt{6}$ we obtain $I(x,y,z) \approx -12.847$. The former solution gives squared distances of $D_{ij} = (19+4\sqrt{6})/3 \approx 9.600 > 3^2$ for all $i = j+3 \bmod 6$, so it is not even 2-Euclidean, whereas the latter solution can be shown to satisfy the tetrangle inequalities and hence corresponds to a 3-pseudo-Euclidean space.

Having narrowed down the range of possibilities this far, we are now ready to prove our distance-theoretic characterization of the conformation space of the cyclohexane framework:

THEOREM 4.65. *The conformation space of the cyclohexane framework* $\mathbb{F}$ *defined in Equation (4.98) consists exactly of those mappings* $\mathbf{p}: A \to \mathbb{R}^3$ *such that the function* $D_{\mathbf{p}}: A^2 \to \mathbb{R}^+$: $D_{\mathbf{p}}(i,j) := \| \mathbf{p}(i) - \mathbf{p}(j) \|^2$ *satisfies*

$$D_{\mathbf{p}}(i,j) \;=\; \begin{cases} 0 & \text{if } i = j\,; \\ 1 & \text{if } i = j+1 \ (mod\ 6)\,; \\ 8/3 & \text{if } i = j+2 \ (mod\ 6)\,; \end{cases} \tag{4.130}$$

and

$$D_{\mathbf{p}}(i,j) \;=\; 11/3 \qquad \text{if } i = j+3 \ (mod\ 6)\,, \tag{4.131}$$

along with those $\mathbf{p}: A \to \mathbb{R}^3$ *for which Equation (4.130) together with the equations*

$$\begin{aligned} D_{\mathbf{p}}(1,4) &= (11 - 2s_\theta(1 + 2\cos(\theta))) \\ D_{\mathbf{p}}(2,5) &= (11 - 2s_\theta(1 + 2\cos(\theta + 2\pi/3))) \\ D_{\mathbf{p}}(3,6) &= (11 - 2s_\theta(1 + 2\cos(\theta + 4\pi/3))) \end{aligned} \tag{4.132}$$

are fulfilled for some value of θ, *where* s_θ *is the unique solution of the equation* $2(1 - \cos(3\theta)s^3) - 72s + 32 = 0$ *which lies in the interval* $[0 \cdots 1]$.

Proof: By Corollary 4.64, the only solutions to the five and six-point inequalities (when expressed in terms of the variables $x = S^{-1}(D_{14})$, $y = S^{-1}(D_{25})$ and $z = S^{-1}(D_{36})$) are the triples [0,0,0] and [8,8,8] together with the simultaneous solutions of Equations (4.121) and (4.122). The solution $x = y = z = 8$ gives squared distances of $D_{ij} = S(8) = -5/3 < 0$ for all $i = j+3$ *mod* 6, which is obviously invalid, whereas the solution $x = y = z = 0$ gives $D_{ij} = S(0) = 11/3$, which lies between both the tetrangle limits (see Equations (4.101) through (4.103)). Hence in order to verify that this is actually a *bona fide* three-dimensional solution, we need only verify the one remaining type of tetrangle inequality, e.g.

$$\begin{aligned} 0 &\le D_{13,24}(1,2,4,5;\, D_{14},\, D_{25}) \\ &= -2/27\left(27(D_{14}D_{25}^2 + D_{14}^2 D_{25}) - 198\, D_{14}D_{25} - 75(D_{14} + D_{25}) + 550\right). \end{aligned} \tag{4.133}$$

It is easily seen that $D_{14} = D_{25} = 11/3$ is actually a *root* of this polynomial, so that such quadruples of atoms are actually coplanar in this solution. The solution itself corresponds to the chair form of cyclohexane (see Figure 4.22).

To prove the rest of the theorem, let us first consider the solutions to Equation (4.121). Since $G(x,y,z)$ is a homogeneous polynomial of degree two, its nonnegative zeros form a cone whose apex is situated at the origin in $\mathbb{R}^3$. Furthermore, since

$$(x + y + z)^2 \;=\; x^2 + y^2 + z^2 \,+\, 2G(x,\,y,\,z)\,, \tag{4.134}$$

the intersection of this cone with the plane defined by $x+y+z = r > 0$ is a circle of radius r whose center is located at the point $[r/3, r/3, r/3]$ (see Figure 4.23).

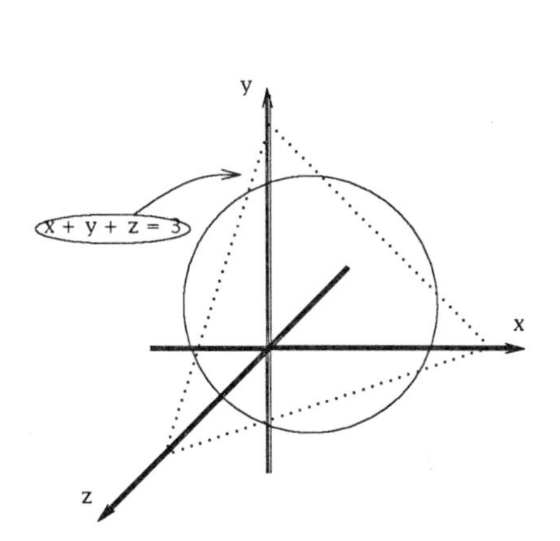

Figure 4.23

The solutions to Equation 4.121 which lie in the plane $x+y+z = r$ form a circle in the plane centered on the point $[r/3, r/3, r/3]$.

If we take the center as the origin for this plane, the vector $[2r/3, -r/3, -r/3]$ from this origin to the point $[r, 0, 0]$ on the circle as our first coordinate axis, and the vector $[r/3, r(1-\sqrt{3})/3, r(1+\sqrt{3})/3]$ perpendicular to the first as our second coordinate axis, then we can parametrize this circle as:

$$\begin{aligned} \begin{bmatrix} x(\theta) \\ y(\theta) \\ z(\theta) \end{bmatrix} &= r/3 \begin{bmatrix} 1 \\ 1 \\ 1 \end{bmatrix} + r/3\,cos(\theta) \begin{bmatrix} 2 \\ -1 \\ -1 \end{bmatrix} + r/3\,sin(\theta) \begin{bmatrix} 1 \\ 1-\sqrt{3} \\ 1+\sqrt{3} \end{bmatrix} \\ &= r/3 \begin{bmatrix} 1+2\,cos(\theta) \\ 1+2\,cos(\theta+2\pi/3) \\ 1+2\,cos(\theta+4\pi/3) \end{bmatrix} \end{aligned} \tag{4.135}$$

where $\theta \in [0 \cdots 2\pi]$. It follows that the solutions $[x, y, z]$ of Equation (4.121) can all be expressed in the form:

$$\begin{bmatrix} x \\ y \\ z \end{bmatrix} = \begin{bmatrix} x(\theta,s) \\ y(\theta,s) \\ z(\theta,s) \end{bmatrix} = \begin{bmatrix} s(1+2\,cos(\theta)) \\ s(1+2\,cos(\theta+2\pi/3)) \\ s(1+2\,cos(\theta+4\pi/3)) \end{bmatrix}, \tag{4.136}$$

where $s := r/3$. It remains to be shown that for each $\theta \in [0 \cdots 2\pi]$ there exists exactly one value of $s =: s_\theta$ such that $x = x(\theta, s_\theta)$, $y = y(\theta, s_\theta)$, and $z = z(\theta, s_\theta)$ satisfy Equation (4.122) (and hence Equation (4.128)), together with the tetrangle inequalities.

Let us begin by transforming Equation (4.122):

$$\begin{aligned} 0 &= H'(\theta, s) := H(x(\theta,s), y(\theta,s), z(\theta,s)) \\ &= -s^3(1+2\cos(\theta))(1+2\cos(\theta+2\pi/3))(1+2\cos(\theta+4\pi/3)) \\ &\quad - 72s + 32 . \end{aligned} \tag{4.137}$$

By means of standard trigonometric identities (best seen by resorting to complex arithmetic) this can be reduced to

$$H'(\theta, s) = 2(1-\cos(3\theta))s^3 - 72s + 32 . \tag{4.138}$$

A graph of this function versus s for $\theta = 0, 2\pi/30, 4\pi/30, 6\pi/30, 8\pi/30, \pi/3$ is shown in Figure 4.24. From this graph, it is immediately clear that for any fixed value of $\theta \in (0 \cdots 2\pi]$ the equation $H'(\theta, s) = 0$ has three real roots s_θ, s'_θ, s''_θ, one of which, say s_θ, always lies in the interval $[0 \cdots 1]$, while another satisfies $s'_\theta \geq 4 > 1$ and the last one obeys $s''_\theta \leq -4.45 < 0$.

To finish the proof we need only show that $x = x(\theta, s)$, $y = y(\theta, s)$ and $z = z(\theta, s)$ always fulfill the tetrangle inequalities for $s = s_\theta$ and always violate them for $s = s'_\theta, s''_\theta$. We begin by considering tetrangle inequality limits (Equations (4.102) – (4.103)). Observe that since

$$x + y + z = 3s \tag{4.139}$$

while the tetrangle limits themselves imply

$$S(7/3+2\sqrt{6}/3) = 2 - \sqrt{6} \leq x, y, z \leq 4/3 = S(25/9) , \tag{4.140}$$

the roots of $H'(\theta, s)$ consistent with these limits necessarily satisfy

$$-4.45 < 2 - \sqrt{6} \leq s \leq 4/3 < 4 . \tag{4.141}$$

This eliminates the roots s'_θ and s''_θ. The proof that $x = x(\theta, s_\theta)$, $y = y(\theta, s_\theta)$ and $z = z(\theta, s_\theta)$ actually fulfill

$$\begin{aligned} 0 &\leq D_{13,24}(1,2,4,5; S(x), S(y)) \\ &= 2/27\left(8(xy^2+x^2y) - 44(x^2+y^2) - 88xy + 192(x+y)\right) \end{aligned} \tag{4.142}$$

etc. is left as a (tedious but straightforward) exercise for the reader. QED

This distance-theoretic characterization, of course, does not enable us to distinguish between enantiomorphs (mirror images) of cyclohexane. Since the boat form of cyclohexane can assume a symmetric conformation which has a

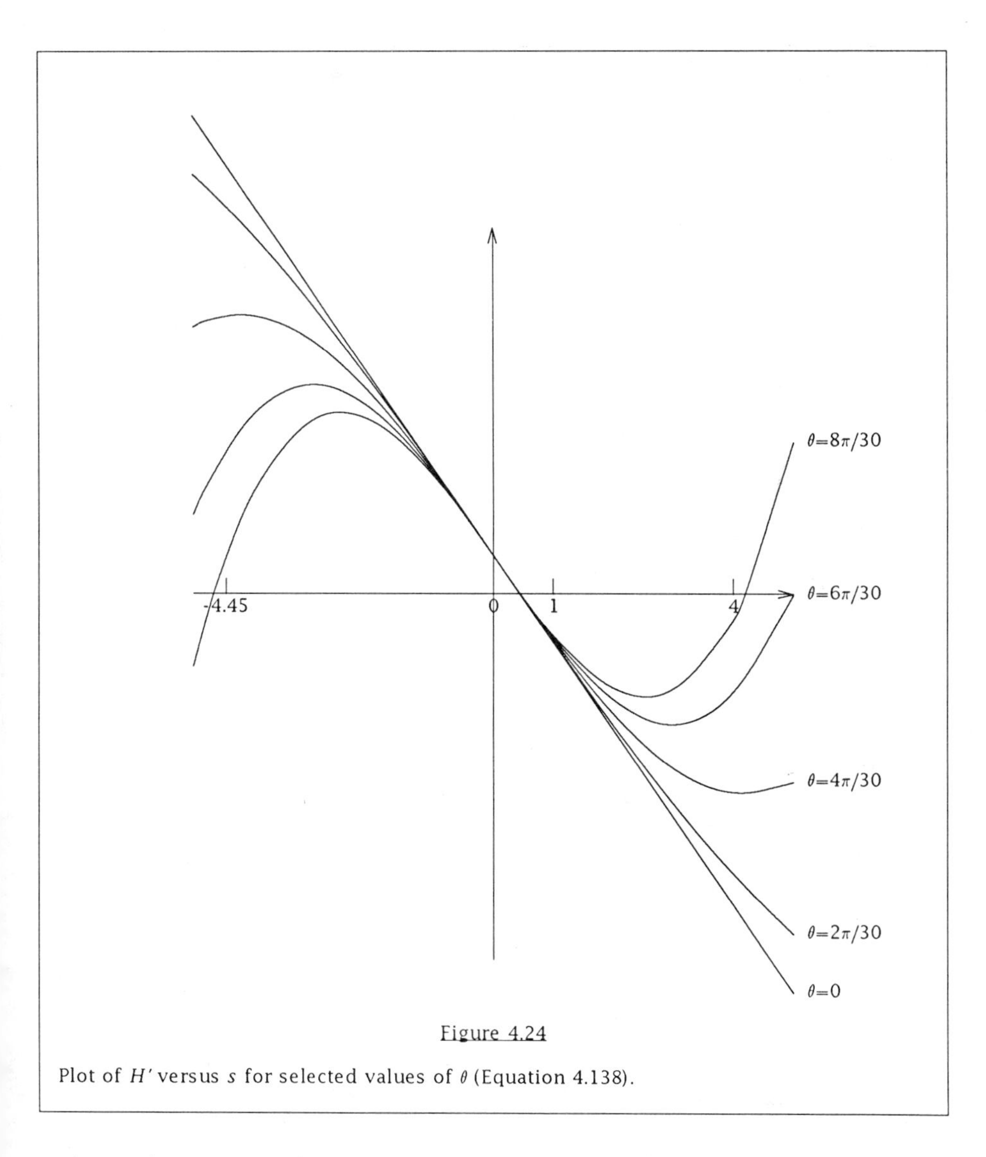

Figure 4.24

Plot of H' versus s for selected values of θ (Equation 4.138).

plane of symmetry, however, the mirror images of any two conformations thereof can interconvert by an internal motion so that the boat form is achiral. Also, since the root s_θ of the polynomial in (4.139) is locally a smooth function of its coefficients, the single "circle" of conformations which constitutes the boat form is smooth. We summarize these observations in the following:

COROLLARY 4.66. *The topological structure of the conformation space of the cyclohexane framework consists of two isolated points, one for each of the two mirror images of the chair form*†*, together with a single smooth circle which corresponds to the possible conformations of the boat form.*

Since the sequence of conformations which are obtained by traversing the circle characteristic of the boat form includes all conformations which would be obtained by a cyclic permutation of the atoms $1 \mapsto 2 \cdots 6 \mapsto 1$ (up to mirror reflection), the corresponding motion is known to chemists as a pseudorotation [Strauss1983]. Equivalently, we can say that the Longuette-Higgins symmetry group of this conformer is the group of rotations of the symmetric plane hexagon.

The next logical step in our series of examples is the framework obtained from the seven membered ring cycloheptane. In this case it has been empirically observed by means of Dreiding models [Dreiding1959] that the conformation space consists of two disjoint smooth circles, one of which corresponds to a "chair" conformation and the other of which corresponds to a "boat" (Figure 4.25).

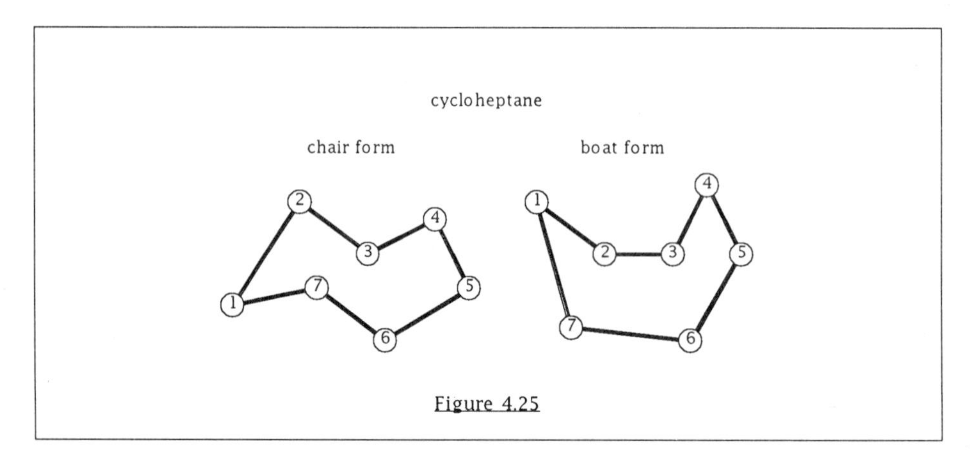

Figure 4.25

Although these observations are fully in accord with the easily verified fact that this framework has one generic degree of freedom, due to the sheer algebraic complexity of the problem a complete mathematical proof of this fact has not

† The chair form is chiral because we have assumed unique labellings of the atoms for the purposes of mathematical analysis. These two mirror images are of course chemically indistinguishable.

yet been achieved. In the case of the eight membered ring cyclooctane, even with the use of models it has not yet been possible to predict the topological structure! Instead of pursuing these difficult problems here, we will present a heuristic derivation (which, to the best of our knowledge, is unpublished work due to Dreiding, Dress and Haegi) which uses our empirical knowledge of the structure of cycloheptane to predict what the conformation space of the fused ring frameworks of various bicycloheptanes look like (see Figure 4.26), thereby gleaning some insight into what can occur when "higher-order" correlations in the motions of frameworks are present.

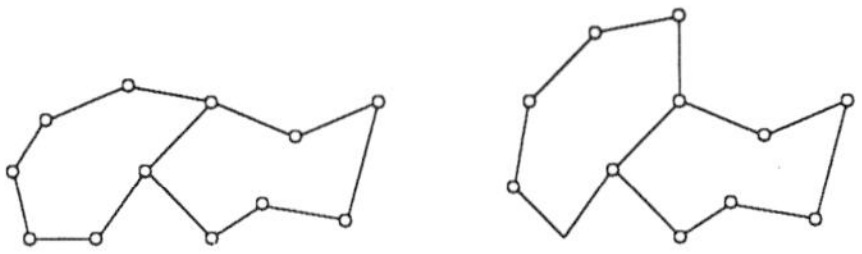

Figure 4.26

Bicycloheptane ring systems can be fused either *cis* (left) or *trans* to their common bond.

To do this, one observes using Dreiding models that as the boat form of cycloheptane goes through its pseudorotation cycle, there are at most two different conformations with a given value of any single dihedral angle, and there are in fact exactly two conformations unless this angle attains its minimum or maximum value (as shown in the graph of Figure 4.27). When two boat-form frameworks are fused together across a bond in the *cis* configuration, the two dihedral angles about the common bond in each of the two rings will be equal. Hence for a given value of this angle there will be at most four different conformations, and when this angle reaches one of its extreme values there will be exactly one. The topology of the conformation space thus consists of a pair of circles which have two points of intersection in common, where the "traveler" through conformation space has four choices of a direction to go (as shown in Figure 4.28). On the other hand, if the two rings are fused in the *trans* configuration, the two dihedral angles about the common bond differ by 120° so that these bifurcations cannot occur. In this case, the conformation space consists topologically of two disjoint circles, one for each of the two mirror images. Although these circles are smooth, an examination of the Dreiding models reveals an amazingly contorted cycle of motion.

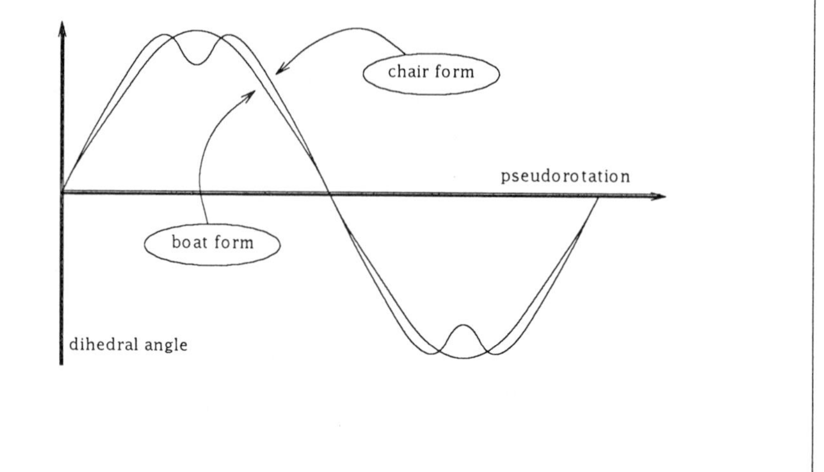

Figure 4.27
Variation of any dihedral angle of cycloheptane in its chair and boat conformations as a function of their pseudorotation parameters.

As the chair form of bicycloheptane goes through its pseudorotation, the graph of any one dihedral angle describes a pair of "camel's backs" (Figure 4.27). Thus if a pair of rings, one in the boat form and one in the chair form, are fused together in the *cis* configuration, the resultant conformation space will have four topological singularities of the "four-way intersection" type† (as is also shown in Figure 4.28). More interesting, however, is the fact that there are also four points at which the relative directions of pseudorotation in the two rings must change abruptly, giving rise to purely differential singularities. In the case of two chairs fused *cis*, the space will have ten four-way intersections, along with eight abrupt changes in direction, and the plot of one pseudorotation against the other assumes the symmetrical form seen in Figure 4.28. Although it is difficult to establish with certainty from this sort of semiempirical analysis, we believe that when two boat/chair or chair/chair cycloheptane rings are fused *trans* the result

† Since the extremal dihedral angles of the chair and boat frameworks are not exactly equal, this is not strictly true of the frameworks. In the actual molecules, however, the deformations of the bond lengths and angles from their equilibrium values which are necessary to make them equal are insignificant, and hence this analysis is expected to be valid (neglecting van der Waals repulsions, of course).

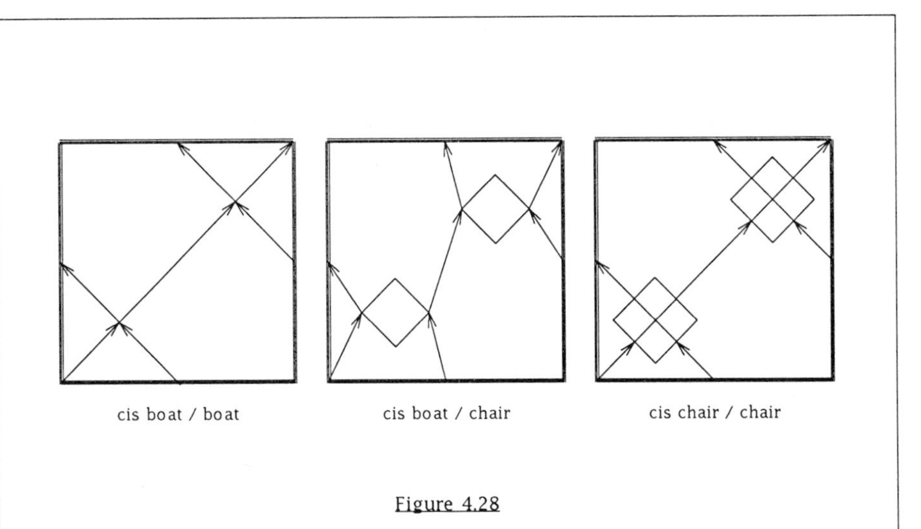

Figure 4.28

Conformation spaces of each of the possible stereoisomers of *cis* bicycloheptane.

will once again be a pair of smooth circles (albeit yet more convoluted than before).

References

Abe1984.

H. Abe, W. Braun, T. Noguti, and N. Go, "Rapid Calculation of First and Second Derivatives of Conformational Energy Functions with Respect to Dihedral Angles for Proteins," *Computers and Chemistry, 8*, 239-247(1984).

Asimow1978.

L. Asimow and B. Roth, "The Rigidity of Graphs," *Trans. Amer. Math. Soc., 245*, 279-289(1978).

Asimow1979.

L. Asimow and B. Roth, "The Rigidity of Graphs. II.," *J. Math. Anal. Appl., 68*, no. 1, 171-190(1979).

Ball1900.

R.S. Ball, *A Treatise on the Theory of Screws*, Cambridge Univ. Press, Cambridge, U.K., 1900.

Bottema1979.

O. Bottema and B. Roth, *Theoretical Kinematics*, North-Holland Publishing Co., Amsterdam-New York, 1979. North-Holland Series in Applied

Mathematics and Mechanics, 24.

Braun1987.

W. Braun, "Local Deformation Studies of Chain Molecules: Differential Conditions for Changes of Dihedral Angles," *Biopolymers, 26,* 1691-1704(1987).

Bricard1987.

R. Bricard, "Mémoire sur la Théorie de l'Octaèdre Articulé," *J. Math. Pures Appl. (5), 3,* 279-289(1987).

Connelly1979.

R. Connelly, "The Rigidity of Polyhedral Surfaces," *Math. Mag., 52,* no. 5, 275-283(1979).

Connelly1980.

R. Connelly, "The Rigidity of Certain Cabled Frameworks and the Second-Order Rigidity of Arbitrarily Triangulated Convex Surfaces," *Advances in Math., 37,* no. 3, 272-299(1980).

Crapo1979.

H. Crapo, "Structural Rigidity," *Structural Topology, 1979,* no. 1, 26-45(1979).

Crapo1982.

H. Crapo and W. Whiteley, "Statics of Frameworks and Motions of Panel Structures, a Projective Geometric Introduction," *Structural Topology, 1982,* no. 6, 43-82(1982). With a French translation.

Cremer1975.

D. Cremer and J.A. Pople, "A General Definition of Ring Puckering Coordinates," *J. Am. Chem. Soc., 97,* 1354-1358(1975).

Dreiding1959.

A.S. Dreiding, "Einfache Molekularmodelle," *Helv. Chim. Acta, 42,* 1339-1344(1959).

Dress1982.

A.W.M. Dress, *Vorlesungen ueber kombinatorische Geometrie,* 1982. Unpublished notes, Univ. Bielefeld, W. Germany (in German).

Dunitz1979.

J.D. Dunitz, *X-ray Analysis and the Structure of Organic Molecules,* Cornell University Press, Ithaca, NY, 1979.

Garey1979.

M.R. Garey and D.S. Johnson, *Computers and Intractability: A Guide to the Theory of NP-Completeness,* Freeman, San Francisco, 1979.

Gluck1975.

H. Gluck, "Almost All Simply Connected Closed Surfaces are Rigid," in

Geometric Topology, Lecture Notes in Mathematics, vol. 438, Springer-Verlag, Berlin, FRG, 1975.

Goldstein1950.
H. Goldstein, *Classical Mechanics*, Addison-Wesley, Reading, MA, 1950.

Graver1984.
J.E. Graver, *A Combinatorial Approach to Infinitesimal Rigidity*, 1984. Unpublished manuscript.

Gruenbaum1975.
B. Gruenbaum and G. Shephard, *Lectures in Lost Mathematics*, Seattle, WA, 1975. Unpublished manuscript.

Guillemin1974.
V. Guillemin and A. Pollack, *Differential Topology*, Prentice-Hall, Inc., Englewood Cliffs, NJ, 1974.

Havel1988.
T.F. Havel, "The Use of Distances as Coordinates in Computer-Aided Proofs of Theorems in Euclidean Geometry," Symbolic Computations in Geometry, IMA preprint #389, Institute for Mathematics and its Applications, Univ. of Minnesota, Minneapolis, U.S.A., 1988.

Hazebroek1951.
P. Hazebroek and L.J. Oosterhoff, *Disc. Faraday Soc.*, *10*, 87(1951).

Henneberg1911.
L. Henneberg, *Die graphische Statik der starren Systeme*, Liepzig, 1911. Johnson reprint, 1968.

Hodge1968.
W.V.D. Hodge and D. Pedoe, *Methods of Algebraic Geometry, volumes I & II*, Cambridge University Press, Cambridge, U.K., 1968.

Imai1981.
H. Imai, *A note on Sugihara's μ-Nonnegative Recognition Algorithm*, 1981. Unpublished manuscript.

Jessop1903.
C.M. Jessop, *A Treatise on the Line Complex*, Chelsea Publishing Co., Bronx, New York, 1903. Reprinted in 1969.

Karp1975.
R.M. Karp, "On the Complexity of Computational Problems," *Networks*, *5*, 45-68(1975).

Klein1939.
F. Klein, *Elementary Mathematics from an Advanced Standpoint. II.*

Geometry, Dover Publications, London, U.K., 1939. Translation of the third edition (1925) from German to English.

Laman1970.

G. Laman, "On Graphs and Rigidity of Plane Skeletal Structures," *J. Engineering Math., 4,* 331-340(1970).

Lovasz1982.

L. Lovasz and Y. Yemini, "On Generic Rigidity in the Plane," *SIAM J. Alg. Discrete Meth., 3,* 91-98(1982).

Lowery1981.

A.H. Lowery, "Experimental Investigation of Large Amplitude Motions: illustrations from current research," in *Diffraction Studies of Non-Crystalline Substances,* ed. I. Hargittai and W. J. Orville-Thomas, Studies in Physical and Theoretical Chemistry, vol. 13, pp. 199-241, Elsevier North-Holland, New York, NY, 1981.

Morse1969.

M. Morse and S.S. Cairns, *Critical Point Theory in Global Analysis and Differential Topology,* Pure and Applied Mathematics, 33, Academic Press, New York, NY, 1969.

Munkres1975.

J.R. Munkres, *Topology: A First Course,* Prentice-Hall, Inc., Englewood Cliffs, NJ, 1975.

Munkres1984.

J.R. Munkres, *Elements of Algebraic Topology,* Benjamin/Cummings, Menlo Park, CA, 1984.

Orville-Thomas1974.

W.J. Orville-Thomas, editor, *Internal Rotation in Molecules,* Monographs in Chemical Physics, Wiley Interscience, London, U.K., 1974.

Papadimitriou1982.

C.H. Papadimitriou and K. Steiglitz, *Combinatorial Optimization: Algorithms and Complexity,* Prentice-Hall, Englewood Cliffs, NJ, 1982.

Pellegrino1986.

S. Pellegrino and C.R. Calladine, "Matrix Analysis of Statically and Kinematically Indeterminate Frameworks," *Int. J. Solids Structures, 22,* no. 4, 409-428(1986).

Rabin1976.

M.O. Rabin, "Probabilistic Algorithms," in *Algorithms and Complexity: New Directions and Recent Results,* ed. J. F. Traub, Academic Press, New York, 1976.

Roth1981.

B. Roth, "Rigid and Flexible Frameworks," *Amer. Math. Monthly, 88,* no. 1, 6-21(1981).

Sachse1892.

H. Sachse, *Z. physik. Chemie, 10,* 203(1892).

Saxe1979.

J.B. Saxe, "Embeddability of Weighted Graphs in *k*-Space is Strongly NP-Hard," Carnegie-Mellon Computer Science Department Report, 1979.

Schoenberg1969.

I.J. Schoenberg, "Linkages and Distance Geometry, I & II," *Nederl. Akad. Wetensch. Proc. Ser. A (= Indag. Math.), 72 (=31),* 43-63(1969).

Serre1983.

J. Serre, "Historical Survey and Recent Progress in the Theory of Non-Rigid Molecules," in *Symmetries and Properties of Non-Rigid Molecules,* ed. A. Maruani and J. Serre, Studies in Physical and Theoretical Chemistry, vol. 23, pp. 1-12, Elsevier Scientific Publishing Co., Amsterdam, Holland, 1983.

Smith1983.

T.R. Graves Smith, *Linear Analysis of Frameworks,* Ellis Horwood, Limited, Chichester, U.K., 1983.

Strauss1983.

H.L. Strauss, "Pseudorotation: A Large Amplitude Molecular Motion," *Ann. Rev. Phys. Chem., 34,* 301-328(1983).

Tarski1951.

A. Tarski, *A Decision Method for Elementary Algebra and Geometry,* Univ. of California Press, Berkeley, CA, 1951. (second edition).

Tay1984a.

T.S. Tay, "Rigidity of Multi-Graphs. I. Linking Rigid Bodies in *n*-Space," *J. Comb. Theory B, 36,* 95-112(1984).

Tay1984b.

T.S. Tay and W. Whiteley, "Recent Advances in the Generic Rigidity of Structures," *Structural Topology, 9,* 31-38(1984).

Tay1985.

T.S. Tay and W. Whiteley, *Generating Isostatic Frameworks,* 1985, pp. 21-68, 1985.

Thurston1984.

W.P. Thurston and J.R. Weeks, "The Mathematics of Three-Dimensional Manifolds," *Sci. Amer., 251,* 108-120(1984).

Whiteley1977.

W. Whiteley, *Introduction to Structural Geometry. I. Infinitesimal Motions and Infinitesimal Rigidity*, 1977. Unpublished manuscript.

Whiteley1978.

W. Whiteley, *Introduction to Structural Geometry. II. Statics and Stresses*, 1978. Unpublished manuscript.

Whiteley1982.

W. Whiteley, "Motions and Stresses of Projected Polyhedra," *Structural Topology, 1982*, no. 7, 13-38(1982).

Whiteley1984.

W. Whiteley, "Infinitesimally Rigid Polyhedra. 1. Statics of Frameworks," *Trans. Amer. Math. Soc., 285*, 269-295(1984).

Whiteley1986a.

W. Whiteley, *Infinitesimally Rigid Polyhedra. 4. Motions of Hinged-Panel Manifolds*, 1986. Unpublished manuscript.

Whiteley1986b.

W. Whiteley, *The Union of Matroids and the Rigidity of Frameworks*, 1986. To appear, SIAM J. Alg. Discrete Meth.

White1987.

N.L. White and W. Whiteley, "The Algebraic Geometry of Motions in Bar and Body Frameworks," *SIAM J. Alg. Discrete Meth., 8*, 1-32(1987).

Williamson1985.

S.G. Williamson, *Combinatorics for Computer Scientists*, Computer Science Press, Rockville, MD, 1985.

5. Bounds on Distances

Any reasonably complete geometric description of the structure of a mobile molecule must include bounds on the values of its interatomic distances. For example, the van der Waals repulsion between pairs of atoms increases so rapidly as the atoms come close together that the interatomic distance between any pair of atoms always exceeds some nontrival lower bound, which is roughly the sum of their van der Waals radii. Although attractive noncovalent forces between pairs of atoms which are by themselves sufficiently strong to insure that an interatomic distance is always below some nontrivial upper bound are relatively uncommon, molecules are by definition limited in their spatial extent, so that some upper bound on their interatomic distances always exists. In addition, the covalent bond lengths and the geminal distances which determine bond angles can vary slightly, which can be accounted for only if distinct lower and upper bounds on their distances are used, and various kinds of experimental data may imply the presence of interatomic proximities in the molecule which are not due to direct interaction between any one pair of atoms (see Chapter 7).

In this chapter we shall extend the results presented in the previous chapter for exact distance constraints to the more general case in which the distance between each pair of atoms {a,b} in the molecule is confined within given lower and upper bounds $l(a,b)$, $u(a,b)$. The organizing theme is to start with the relatively simple local properties of the conformations consistent with such a system of *distance constraints*, and progress to the more sophisticated global properties. We shall see that conformations subject to distance bounds can be modeled by a new type of framework, called a *tensegrity framework*, which includes "cables" and "struts" in addition to bars. By means of standard methods in analysis and convexity theory we will be able to develop an astonishingly rich theory of tensegrity frameworks.* We then show how this theory can be used to determine those combinations of distance bounds attained in a spatial arrangement of

* In writing this material, we have made use of preliminary versions of those chapters in the forthcoming book, *The Geometry of Rigid Structures* by the Group in Structural Topology (Cambridge University Press), which deal with tensegrity frameworks.

atoms which minimizes or maximizes a given distance, and hence to use the fundamental relations of distance geometry to compute approximations to the Euclidean limits (see §1.3), via a process known as "bound smoothing". Finally, we shall show how the theory enables us to classify those patterns of distance bounds which can be violated at local and global minima of "error functions", which in turn play a vital role in our numerical methods of solving the Fundamental Problem of Distance Geometry (see § 6.1).

5.1. The Local Theory of Tensegrity Frameworks

We begin our study of the relations between distance bounds and the conformation space which they define by considering some of its local properties, i.e. the motions and stresses which can occur in a single conformation consistent with the distance bounds. In this context, the only bounds of interest are those which are actually attained by the given conformation, so that all "slack" bounds can be ignored. To model this situation, we introduce a generalization of bar and joint frameworks which is called a "tensegrity framework" after Buckminster Fuller, who popularized the architectural applications of these structures [Fuller1975]. The first reasonably quantitative accounts of tensegrity frameworks are found in [Pugh1976, Calladine1978], where simple rules for constructing a number of types of rigid tensegrity frameworks are given. Inspired by the numerous examples and conjectures in [Gruenbaum1975], this was followed by several considerably more detailed mathematical studies by members of the Group in Structural Topology [Rosenberg1980, Roth1981], from which most of the results of this section were obtained. We shall find that many, though by no means all, of the properties that we found in bar and joint frameworks generalize to these new frameworks. As in §4.2 we shall state and prove most of our results in the three-dimensional case, while using the plane for most of our examples. Unless otherwise stated, however, all of these results can easily be extended to the n-dimensional case.

Definition 5.1. A *tensegrity framework* $\mathbb{F} = (A, L, U, \mathbf{p})$ is an object consisting of a finite set A of *joints*, a set $L \subseteq \binom{A}{2}$ of *struts*, a set $U \subseteq \binom{A}{2}$ of *cables* and a *positioning* $\mathbf{p} : A \rightarrow \mathbb{R}^3$.

The graphs (A, L) and (A, U) are called the *strut* and *cable* graphs of $\mathbb{F}$, while the graph $\mathbb{G} = (A, L \cup U)$ is called the *underlying graph* of $\mathbb{F}$. The triple $\mathbb{H} = (A, L, U)$ will often be referred to as the *underlying tensegrity*, and we will use the notation $\mathbb{H}_{\mathbf{p}} = \mathbb{F}$ to denote the *realization* of $\mathbb{H}$ by $\mathbf{p}$. The set $B := L \cap U$ will be referred to as the bars of the framework; a tensegrity framework is called *pure* if $B = \{\}$. Clearly, if $B = L = U$ then $\mathbb{F}$ is just a bar and joint framework as before. At the same time, as illustrated in Figure 5.1†, any tensegrity framework can be regarded as being essentially a type of bar and joint framework.

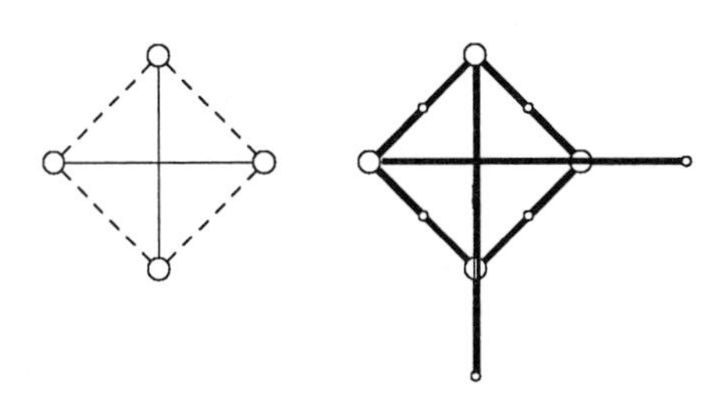

Figure 5.1

A tensegrity framework and its bar and joint equivalent. It is obvious that one is rigid if and only if the other is. The same does not hold for their infinitesimal properties, however, since the bar and joint framework obtained in this way is never infinitesimally rigid.

For the purposes of this chapter, it is convenient to rephrase our previous definition of the (finite) rigidity of bar and joint frameworks slightly (Definition 4.4). To do this, we introduce the notion of the equivalence of bar and joint frameworks: Two realizations $\mathbb{F} = \mathbb{G}_{\mathbf{p}}$ and $\mathbb{F}' = \mathbb{G}_{\mathbf{p}'}$ of a given graph $\mathbb{G} = (A, B)$ are *equivalent* (denoted as $\mathbb{F} \equiv \mathbb{F}'$) if $\mathbf{D}_G(\mathbf{p}) = \mathbf{D}_G(\mathbf{p}')$, where $\mathbf{D}_G$ is the rigidity mapping of $\mathbb{G}$ (Definition 4.3). Note that if the positionings $\mathbf{p}$ and $\mathbf{p}'$ themselves are congruent, i.e. $\mathbf{p} \approx \mathbf{p}'$, we certainly have $\mathbb{F} \equiv \mathbb{F}'$, though not conversely. However, it is easily seen that $\mathbb{F}$ is rigid if and only if for all positionings $\mathbf{p}'$ in some sufficiently small neighborhood $\mathbf{U}_{\mathbf{p}} \subseteq (\mathbb{R}^3)^A$ of $\mathbf{p}$ with $\mathbb{F} \equiv \mathbb{F}'$ we have $\mathbf{p} \approx \mathbf{p}'$ (or equivalently, $\mathbf{p} \dot{\approx} \mathbf{p}'$).

The analogue of this equivalence relation for tensegrity frameworks goes as follows:

Definition 5.2. Given two realizations $\mathbb{F} = (A, L, U, \mathbf{p})$ and $\mathbb{F}' = (A, L, U, \mathbf{p}')$ of the same tensegrity, we say that $\mathbb{F}$ *dominates* $\mathbb{F}'$ if

$$\|\mathbf{p}(a) - \mathbf{p}(b)\| \;\leq\; \|\mathbf{p}'(a) - \mathbf{p}'(b)\| \qquad (5.1)$$

for all $\{a, b\} \in L$, and

† In the drawings of this chapter, thin solid lines are used to represent struts, dashed lines to represent cables and heavy solid lines to represent bars.

$$\|\mathbf{p}(a)-\mathbf{p}(b)\| \geq \|\mathbf{p}'(a)-\mathbf{p}'(b)\| \tag{5.2}$$

for all $\{a, b\} \in U$. We may also say that $\mathbb{F}$ is a *dominant* of $\mathbb{F}'$, and we denote this by writing $\mathbb{F} \succ \mathbb{F}'$ and $\mathbb{F}' \prec \mathbb{F}$.

This relation enables us to extend our definition of rigidity to tensegrity frameworks.

Definition 5.3. A tensegrity framework $\mathbb{F} = (A, L, U, \mathbf{p})$ is *rigid* if there exists a neighborhood $\mathbf{U_p} \subseteq (\mathbb{R}^3)^A$ of $\mathbf{p}$ such that for all frameworks $\mathbb{F}' = (A, L, U, \mathbf{p}')$ with $\mathbf{p}' \in \mathbf{U_p}$ the relation $\mathbb{F} \succ \mathbb{F}'$ implies $\mathbf{p} \approx \mathbf{p}'$.

In analogy to bar and joint frameworks, the tuple (A, L, l, U, u) obtained from a tensegrity framework by defining maps $l: L \rightarrow \mathbb{R}$ and $u: U \rightarrow \mathbb{R}$ by $l(a,b) := \|\mathbf{p}(a)-\mathbf{p}(b)\|$ for all $\{a,b\} \in L$ and $u(a,b) := \|\mathbf{p}(a)-\mathbf{p}(b)\|$ for all $\{a,b\} \in U$ is called an *abstract* tensegrity framework, in which case any realization $\mathbb{F}' = \mathbb{H}_{\mathbf{p}'}$ of $\mathbb{H} = (A, L, U)$ which satisfies $\mathbb{F}' \prec \mathbb{F}$ is called a *representation* of the abstract framework. Thus, we can also say that a tensegrity framework $\mathbb{F}$ is rigid if all representations $\mathbf{p}' \in (\mathbb{R}^3)^A$ of the same abstract framework in some neighborhood $\mathbf{U_p}$ of $\mathbf{p}$ satisfy $\mathbf{p}' \approx \mathbf{p}$. Similarly, we can define motions and flexings of tensegrity frameworks.

Definition 5.4. A *motion* of a tensegrity framework $\mathbb{F} = (A, L, U, \mathbf{p})$ is a mapping $\mathbf{p}^t: [0 \cdots 1] \rightarrow (\mathbb{R}^3)^A$ such that $\mathbf{p}^0 = \mathbf{p}$ while for all $t \in [0 \cdots 1]$ the tensegrity frameworks $\mathbb{F}^t = (A, L, U, \mathbf{p}^t)$ satisfy $\mathbb{F}^t \prec \mathbb{F}$. A motion is called a *flexing* if $\mathbf{p}^t \not\approx \mathbf{p}$ for all $t \in (0 \cdots 1]$, and a tensegrity framework is *flexible* if there exists a flexing of it.

The proof the following proposition is analogous to the proof of Proposition 4.5, and is omitted for the same reasons (see [Roth1981]).

PROPOSITION 5.5. *A tensegrity framework is rigid if and only if it is not flexible.*

Just as with bar and joint frameworks, a much simpler theory can be developed by linearizing the polynomials which occur in the inequalities (5.1) and (5.2), i.e. by studying the infinitesimal and static rigidity of tensegrity frameworks. We shall indicate briefly how the corresponding definitions for bar and joint frameworks must be extended. As in §4.2, we assume a fixed ordering $\underline{A}$ of our atoms and abbreviate $\mathbf{p}(a_i)$ by $\mathbf{p}_i$.

Definition 5.6. A *resolution* of an equilibrium system of forces $\mathbf{f}: A \rightarrow \mathbb{R}^3$ in a tensegrity framework $\mathbb{F} = (A, L, U, \mathbf{p})$ is an assignment of scalars r_{ij} to the couples in $L \cup U$ such that

$$\sum_{\{j \mid \{a_i, a_j\} \in L \cup U\}} r_{ij}(\mathbf{p}_i - \mathbf{p}_j) = \mathbf{f}_i \tag{5.3}$$

for $i = 1, \ldots, \#A$, while

$$r_{ij} \geq 0 \qquad \text{for all } \{a_i, a_j\} \in U \setminus L \tag{5.4}$$

and

$$r_{ij} \leq 0 \qquad \text{for all } \{a_i, a_j\} \in L \setminus U \tag{5.5}$$

(no restriction on the sign of those r_{ij} for which $\{a_i, a_j\} \in L \cap U$). The tensegrity framework $\mathbb{F}$ is *statically rigid* if every equilibrium system of forces is resolvable.

Thus in any resolution, the stress on every cable must be a tension while the stress on every strut must be a compression.

Similarly, we may define the infinitesimal rigidity of tensegrity frameworks as follows:

Definition 5.7. An *infinitesimal motion* of a tensegrity framework $\mathbb{F} = (A, L, U, \mathbf{p})$ is a mapping $\mathbf{v} : A \rightarrow \mathbb{R}^3$ such that

$$(\mathbf{p}_i - \mathbf{p}_j)\cdot(\mathbf{v}_i - \mathbf{v}_j) \begin{cases} \geq 0 & \text{if } \{a_i, a_j\} \in L \setminus U\,; \\ \leq 0 & \text{if } \{a_i, a_j\} \in U \setminus L\,; \\ = 0 & \text{if } \{a_i, a_j\} \in L \cap U\,. \end{cases} \tag{5.6}$$

The tensegrity framework $\mathbb{F}$ is *infinitesimally rigid* if every infinitesimal motion $\mathbf{v} \in (\mathbb{R}^3)^A$ thereof lies in the tangent space $\mathbb{T}_\mathbf{p}$ of $\mathbf{ST}(\mathbf{p})$, which (assuming $dim(\mathbf{p}(A)) = 3$) is given by

$$\mathbb{T}_\mathbf{p} := \{\mathbf{x} \in (\mathbb{R}^3)^A \,|\, (\mathbf{p}_i - \mathbf{p}_j)\cdot(\mathbf{x}_i - \mathbf{x}_j) = 0 \ \text{ for } i, j = 1, \ldots, \#A\}\,. \tag{5.7}$$

Thus in any infinitesimal motion, the rate of change in the length of each couple is nonnegative for all struts, nonpositive for all cables and zero for all bars.

One trivial consequence of the above definitions worth noting is the fact that if a tensegrity framework $\mathbb{F} = (A, L, U, \mathbf{p})$ is statically or infinitesimally rigid, then the *reverse* framework $\tilde{\mathbb{F}} = (A, U, L, \mathbf{p})$ obtained by swapping the cables and struts is likewise statically or infinitesimally rigid, respectively. A somewhat less obvious but not less interesting fact is that, just as for bar and joint frameworks, static and infinitesimal rigidity are equivalent. In order to prove this result we shall need some standard results from convexity theory.

Definition 5.8. Let $\mathbf{K} = \{\sum_{i=1}^{M} \alpha_i \mathbf{x}_i \,|\, \alpha_i \geq 0,\ i = 1, \ldots, M\}$ be the closed convex cone generated by M vectors $\mathbf{x}_i \in \mathbb{R}^m$. The *polar* of $\mathbf{K}$ is the subset of $\mathbb{R}^m$ defined by $\mathbf{K}^+ := \{\mathbf{y} \in \mathbb{R}^m \,|\, \mathbf{x}\cdot\mathbf{y} \geq 0 \ \ \forall \mathbf{x} \in \mathbf{K}\}$.

The following lemma simply collects some corollaries to the Hahn-Banach theorem, whose proofs may be found in standard texts, e.g. [Tiel1984, Rockafel-

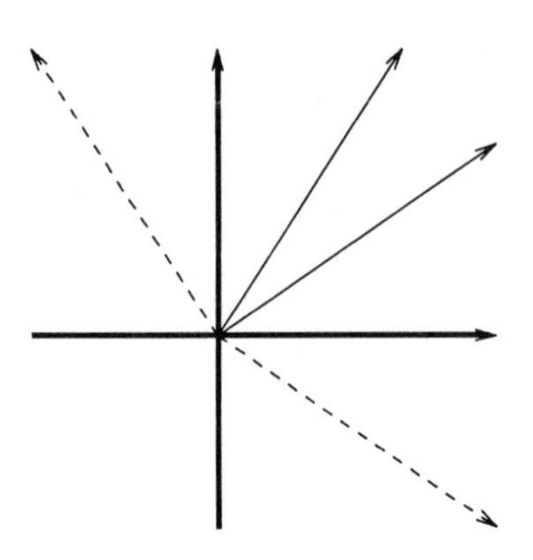

Figure 5.2
The two-dimensional convex cones generated by the dashed and solid vectors are polars of each other. Observe that this relation is a reciprocal one.

lar1970] (see Figure 5.2 for an illustration).

LEMMA 5.9.

(i) *Given two closed convex cones* $\mathbf{K}_1, \mathbf{K}_2 \subseteq \mathbb{R}^m$ *with* $\mathbf{K}_1 \subseteq \mathbf{K}_2$, *we have* $\mathbf{K}_1^+ \supseteq \mathbf{K}_2^+$.

(ii) *If* $\mathbf{K} \subseteq \mathbb{R}^m$ *is a closed convex cone, then* $\mathbf{K}^{++} = \mathbf{K}$.

We are now ready for our first theorem [Roth1981].

THEOREM 5.10. *A tensegrity framework* $\mathbb{F} = (\mathrm{A}, \mathrm{L}, \mathrm{U}, \mathbf{p})$ *is infinitesimally rigid if and only if it is statically rigid.*

Proof: Let $\underline{\mathbf{l}}_{ij} \in \mathbb{R}^{3\#\mathrm{A}}$ be the load associated with each couple $\{\mathrm{a}_i, \mathrm{a}_j\} \in \mathrm{L} \cup \mathrm{U}$, and let $\underline{\mathbf{M}}_{\mathbf{p}}$ be the $(\#\mathrm{L}+\#\mathrm{U})$ by $3\#\mathrm{A}$ matrix

$$\underline{\mathbf{M}}_{\mathbf{p}}^T \;=\; [\underline{\mathbf{l}}_{ij} \,|\, \{\mathrm{a}_i, \mathrm{a}_j\} \in \mathrm{U}].[-\underline{\mathbf{l}}_{ij} \,|\, \{\mathrm{a}_i, \mathrm{a}_j\} \in \mathrm{L}]\,, \tag{5.8}$$

where the "." denotes the concatenation as usual. Then by definition the set of resolvable forces $\mathbf{U}_{rf} \subseteq \mathbb{R}^{3\#\mathrm{A}}$ is the convex cone $\mathbf{K}_{\mathbf{p}}$ generated by the rows of $\underline{\mathbf{M}}_{\mathbf{p}}$, while the set of infinitesimal motions $\mathbf{U}_{im} \subseteq \mathbb{R}^{3\#\mathrm{A}}$ is polar $\mathbf{K}_{\mathbf{p}}^+$ of this cone. Recall that $\mathbb{F}$ is infinitesimally rigid if and only if $\mathbf{U}_{im} \subseteq \mathbb{T}_{\mathbf{p}}$, the tangent space of $\mathbf{ST}(\mathbf{p})$, while the space of equilibrium systems of forces, being the resolvable systems of any infinitesimally rigid full-dimensional bar and joint framework, is simply $\mathbb{U}_{eq} = \mathbb{T}_{\mathbf{p}}^{\perp}$.

By part (i) of Lemma 5.9, we have $\mathbf{K}_{\mathbf{p}}^{+} \subseteq \mathbb{T}_{\mathbf{p}}$ if and only if $\mathbb{T}_{\mathbf{p}}^{+} \subseteq \mathbf{K}_{\mathbf{p}}^{++}$. But $\mathbb{T}_{\mathbf{p}}^{+} = \mathbb{T}_{\mathbf{p}}^{\perp}$ since $\mathbb{T}_{\mathbf{p}}$ is a subspace of $\mathbb{R}^3$, while $\mathbf{K}_{\mathbf{p}}^{++} = \mathbf{K}_{\mathbf{p}} = \mathbf{U}_{rf}$ by part (ii) of the lemma. Hence the infinitesimal motions are all in $\mathbb{T}_{\mathbf{p}}$ if and only if the equilibrium forces are all resolvable. QED

Thus in contrast to bar and joint frameworks, where $\mathbf{U}_{im}$ and $\mathbf{U}_{rf}$ were the orthogonal complements of each other, in tensegrity frameworks these sets are the polar cones of each other. The matrix $\underline{\mathbf{M}}_{\mathbf{p}}$ above is called the *rigidity matrix* of the tensegrity framework.

It can also be shown that when $\mathbb{H}_{\mathbf{p}}$ is a generic realization of the underlying tensegrity $\mathbb{H} = (A, L, U)$, then infinitesimal, static and finite rigidity coincide [Roth1981]. We shall content ourselves with a proof that flexibility implies infinitesimal flexibility, which also allows us to introduce a very interesting method of generating nonregular realizations of a given graph or tensegrity, which is called the *averaging method* [Whiteley1987].

PROPOSITION 5.11. *Let $\mathbb{F} = (A, L, U, \mathbf{p})$ be a tensegrity framework.*

(i) *If $\mathbf{v}: A \to \mathbb{R}^3$ is an infinitesimal flexing of $\mathbb{F}$ and we define $\mathbb{F}' := (A, L, U, \mathbf{p}+\mathbf{v})$ as well as $\mathbb{F}'' := (A, L, U, \mathbf{p}-\mathbf{v})$, then $\mathbb{F}' \prec \mathbb{F}''$ and $\mathbf{p}' \not\approx \mathbf{p}''$, i.e. $\mathbb{F}'$ is strictly dominated by $\mathbb{F}''$.*

(ii) *If $\mathbb{F}' = (A, L, U, \mathbf{p}')$ is strictly dominated by $\mathbb{F}$, then $\mathbf{p}'-\mathbf{p}$ is an infinitesimal flexing of $\mathbb{F}'' := (A, L, U, \tfrac{1}{2}(\mathbf{p}+\mathbf{p}'))$.*

Proof: (i) Assume that $\mathbf{v} \in (\mathbb{R}^3)^A$ is an infinitesimal flexing of $\mathbb{F}$. Then for any couple $\{a_i, a_j\} \in L \cup U$:

$$\begin{aligned} d_{F'}^2(a_i, a_j) &:= \|(\mathbf{p}_i+\mathbf{v}_i) - (\mathbf{p}_j+\mathbf{v}_j)\|^2 \\ &= \|(\mathbf{p}_i-\mathbf{p}_j) + (\mathbf{v}_i-\mathbf{v}_j)\|^2 \\ &= \|\mathbf{p}_i-\mathbf{p}_j\|^2 + 2(\mathbf{p}_i-\mathbf{p}_j)\cdot(\mathbf{v}_i-\mathbf{v}_j) + \|\mathbf{v}_i-\mathbf{v}_j\|^2 . \end{aligned} \tag{5.9}$$

On the other hand,

$$\begin{aligned} d_{F''}^2(a_i, a_j) &:= \|(\mathbf{p}_i-\mathbf{v}_i) - (\mathbf{p}_j-\mathbf{v}_j)\|^2 \\ &= \|(\mathbf{p}_i-\mathbf{p}_j) - (\mathbf{v}_i-\mathbf{v}_j)\|^2 \\ &= \|\mathbf{p}_i-\mathbf{p}_j\|^2 - 2(\mathbf{p}_i-\mathbf{p}_j)\cdot(\mathbf{v}_i-\mathbf{v}_j) + \|\mathbf{v}_i-\mathbf{v}_j\|^2 . \end{aligned} \tag{5.10}$$

Hence by Equation (5.6), we have

$$d^2_{F'}(a_i,a_j) - d^2_{F''}(a_i,a_j) \;=\; 4(\mathbf{p}_i-\mathbf{p}_j)\cdot(\mathbf{v}_i-\mathbf{v}_j) \qquad (5.11)$$

$$\begin{cases} \geq 0 & \text{if } \{a_i,a_j\} \in L \setminus U\,; \\ \leq 0 & \text{if } \{a_i,a_j\} \in U \setminus L\,; \\ = 0 & \text{if } \{a_i,a_j\} \in L \cap U\,; \end{cases}$$

and it follows that $\mathbb{F}'$ is dominated by $\mathbb{F}''$. If the infinitesimal motion is nontrivial, then this dominance is strict.

(ii) Assume that $\mathbb{F}'$ is strictly dominated by $\mathbb{F}$, and let $\{a_i,a_j\} \in L \cup U$. Then

$$\tfrac{1}{2}\big((\mathbf{p}'_i+\mathbf{p}_i)-(\mathbf{p}'_j+\mathbf{p}_j)\big)\cdot\big((\mathbf{p}'_i-\mathbf{p}_i)-(\mathbf{p}'_j-\mathbf{p}_j)\big)$$

$$= \tfrac{1}{2}\big((\mathbf{p}'_i-\mathbf{p}'_j)+(\mathbf{p}_i-\mathbf{p}_j)\big)\cdot\big((\mathbf{p}'_i-\mathbf{p}'_j)-(\mathbf{p}_i-\mathbf{p}_j)\big) \qquad (5.12)$$

$$= \tfrac{1}{2}\big(\|\mathbf{p}'_i-\mathbf{p}'_j\|^2-\|\mathbf{p}_i-\mathbf{p}_j\|^2\big) \begin{cases} \geq 0 & \text{if } \{a_i,a_j\} \in L \setminus U\,; \\ \leq 0 & \text{if } \{a_i,a_j\} \in U \setminus L\,; \\ = 0 & \text{if } \{a_i,a_j\} \in L \cap U\,, \end{cases}$$

as desired. QED

From this, we obtain our claim:

COROLLARY 5.12. *If a framework is flexible then it is infinitesimally flexible.*

Proof: If the framework $\mathbb{F} = (A, L, U, \mathbf{p})$ is flexible then there exists a sequence $[\mathbf{p}^k \mid k = 1,\ldots,\infty]$ whose limit is $\mathbf{p}$ and such that each of the frameworks $\mathbb{F}^k = (A, L, U, \mathbf{p}^k)$ satisfies $\mathbb{F}^k \prec \mathbb{F}$. Without loss of generality we assume that $\Sigma_i \mathbf{p}_i = \mathbf{0}$. By Proposition 5.11, each of the frameworks $(A, L, U, \tfrac{1}{2}(\mathbf{p}+\mathbf{p}^k))$ admits a nontrivial infinitesimal flexing $\mathbf{v}^k := \mathbf{p}^k - \mathbf{p}$. We can assure that there is no net translation, i.e. $\Sigma_i \mathbf{v}_i^k = \mathbf{0}$, by setting

$$\mathbf{v}_i^k \;:=\; \mathbf{v}_i^k - \#A^{-1}\Sigma_i \mathbf{v}_i^k\,. \qquad (5.13)$$

Then, we can assure that the framework $\mathbb{F}$ has no net angular momentum w.r.t. $\mathbf{v}^k$, i.e. $\Sigma_i \mathbf{v}_i^k \times \mathbf{p}_i = 0$, by setting

$$\mathbf{v}_i^k \;:=\; \mathbf{v}_i^k - \mathbf{p}_i \times \mathbf{r}^k\,, \qquad (5.14)$$

for a suitable choice of $\mathbf{r}^k \in \mathbb{R}^3$ (which may be found by solving a nonsingular 3×3 system of equations, as the reader can easily verify). Finally, we normalize the result: $\mathbf{v}^k := \mathbf{v}^k / \sqrt{\Sigma_i \|\mathbf{v}_i^k\|^2}$. This gives us an infinite sequence $[\mathbf{v}^k]$ on the unit sphere, and since the sphere is a compact set this sequence has a limit point. Any subsequence converging to this limit point respects the nonstrict inequalities which define an infinitesimal motion, so that the limit $\mathbf{v}$ is an infinitesimal motion of $\mathbb{F}$. This motion is nontrivial, since each element of the sequence satisfies $\Sigma_i \mathbf{v}_i^k = 0$ and $\Sigma_i \mathbf{v}_i^k \times \mathbf{p}_i = 0$ by construction. QED

The first real difference that we encounter between bar and joint frameworks and tensegrity frameworks stems from the fact that, as illustrated in Figure 5.3, the infinitesimal/static rigidity of tensegrity frameworks is not a generic property which holds for "almost all" realizations.

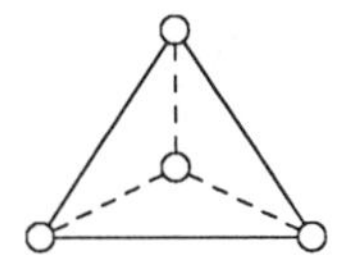
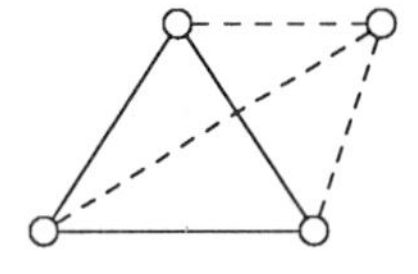

Figure 5.3
Two different generic realizations of the same tensegrity, one of which is infinitesimally rigid (left) and the other of which is flexible (right).

Understanding why requires us to consider the self-stresses of a tensegrity framework, and hence we shall concentrate on static rigidity in what follows.

Definition 5.13. A *self-stress* of a tensegrity framework $\mathbb{F} = (A, L, U, \mathbf{p})$ is an assignment of scalars $s_{ij} \in \mathbb{R}$ to the couples $L \cup U$ such that

$$s_{ij} \begin{cases} \leq 0 & \text{for all } \{a_i, a_j\} \in L \setminus U\,; \\ \geq 0 & \text{for all } \{a_i, a_j\} \in U \setminus L\,; \end{cases} \tag{5.15}$$

and

$$\sum_{\{j \mid \{a_i, a_j\} \in L \cup U\}} s_{ij}(\mathbf{p}_i - \mathbf{p}_j) \;=\; \mathbf{0}\,. \tag{5.16}$$

A self-stress is *proper* if it is nonzero on some couple in $L \wedge U = (L \setminus U) \cup (U \setminus L)$, *spanning* if for each $a_i \in A$ there exists $\{a_i, a_j\} \in L \cup U$ with $s_{ij} \neq 0$, and *strict* if $s_{ij} \neq 0$ for all $\{a_i, a_j\} \in L \wedge U$.

As before, when we write a self-stress in vector form $\mathbf{s} \in \mathbb{R}^{\#(L \cup U)}$, we have ordered the components s_{ij} lexicographically relative to our fixed ordering $\underline{A}$ of A.

It is obvious that a framework $\mathbb{F}$ can be statically rigid only if the *associated bar and joint framework* $\bar{\mathbb{F}} := (A, L \cup U, \mathbf{p})$ is statically rigid. We now show that each self-stress of a statically rigid bar and joint framework determines a statically rigid tensegrity framework [Roth1981].

THEOREM 5.14. *Let* $\mathbb{F} = (A, L, U, \mathbf{p})$ *be a tensegrity framework. Then the following are equivalent:*

(i) $\mathbb{F}$ is statically rigid.

(ii) $\mathbb{F}$ has a strict self-stress and the associated bar and joint framework $\bar{\mathbb{F}}$ is statically rigid.

Proof: If $\mathbb{F}$ is statically rigid, then $\bar{\mathbb{F}}$ is likewise statically rigid, and for each $\{a_i, a_j\} \in L \setminus U$ the elementary load $\underline{\mathbf{l}}_{ij}$ is resolvable by $\mathbb{F}$, while for each $\{a_i, a_j\} \in U \setminus L$ the elementary load $-\underline{\mathbf{l}}_{ij}$ is likewise resolvable. This defines a proper self-stress $\mathbf{s} \in \mathbb{R}^{\#(L \cup U)}$ of $\mathbb{F}$ which has the value -1 or $+1$ as the couple $\{a_i, a_j\}$ is a strut or cable, respectively, and (since the sign on $\{a_i, a_j\}$ must be either zero or the same as that of s_{ij} in any other self-stress of $\mathbb{F}$) the sum of these self-stresses over all couples in $L \cup U$ is a strict self-stress of $\mathbb{F}$.

Conversely, if $\mathbb{F}$ has a strict self-stress $\mathbf{s} \in \mathbb{R}^{\#(L \cup U)}$ and the associated bar and joint framework $\bar{\mathbb{F}}$ is statically rigid, then any equilibrium system of forces $\mathbf{f} \in (\mathbb{R}^3)^A$ has a resolution $\mathbf{r} \in \mathbb{R}^{\#(L \cup U)}$ by $\bar{\mathbb{F}}$. For all $\alpha > 0$, the sum $\mathbf{r} + \alpha\mathbf{s}$ is again a resolution of the force system $\mathbf{f}$, and for sufficiently large α, its components will have the same signs as $\mathbf{s}$ and hence be the required resolution of $\mathbf{f}$ by $\mathbb{F}$. QED

It can also be shown that any *finitely* rigid tensegrity framework admits at least a *proper* self-stress [Connelly1982].

Generally speaking, there are two kinds of regions of conformation space in which a tensegrity (A, L, U) with statically rigid realizations fails to be statically rigid. The first occur where the associated bar and joint framework $\bar{\mathbb{F}}$ fails to be statically rigid, and correspond to nonregular points of the rigidity mapping $\mathbf{D}_G$ of the underlying graph $\mathbb{G} = (A, L \cup U)$. The second occur when the sign patterns of the possible self-stresses of the underlying graph are incompatible with our choice of cables U and struts L. These sign patterns have not been fully characterized, but have received some study [White1983]. Another interesting consequence of the above characterization is the fact that, since a statically rigid tensegrity framework has to have a nontrivial self-stress, the total number of cables and struts present must be at least one more than the minimum number of couples which are necessary to make the associated bar and joint framework statically rigid, i.e. at least $3\#A - 5$ in three dimensions. Unlike bar and joint frameworks, however, there exist statically rigid tensegrity frameworks which are minimal in the sense that the deletion of any cable or strut results in a flexible framework, but which have more than this minimum number of couples (see Figure 5.4 for examples). The characterization of these minimal frameworks is also an open problem, but we can show:

PROPOSITION 5.15. *In any minimal statically/infinitesimally rigid tensegrity framework* $\mathbb{F} = (A, L, U, \mathbf{p})$*, we have:*

$$3\#A - 6 \ < \ \#L + \#U \ \leq \ 6\#A - 12 \,. \tag{5.17}$$

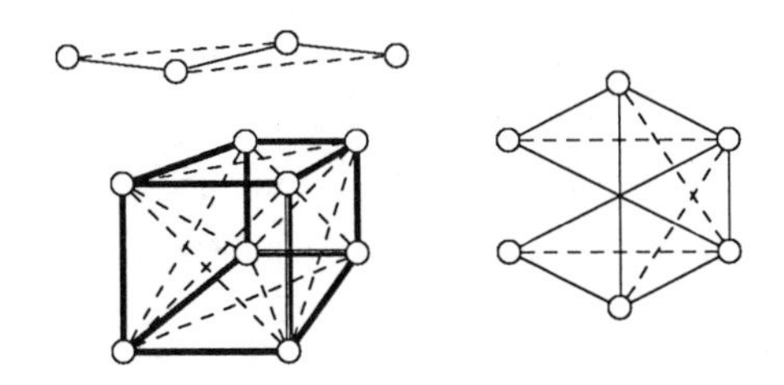

Figure 5.4

Some minimal infinitesimally rigid tensegrity frameworks on the line, in the plane and in space.

Proof: The framework $\mathbb{F}$ is infinitesimally rigid if and only if $\mathbf{U}_{im} = \mathbb{T}_{\mathbf{p}}$, the zero vector of the $3\#A-6$ dimensional factor space $(\mathbb{R}^3)^{A}/\mathbb{T}_{\mathbf{p}}$. Each cable or strut defines a half-space in the factor space according to the relation $(\mathbf{p}(a)-\mathbf{p}(b))\cdot(\mathbf{v}(a)-\mathbf{v}(b)) \le 0$ or ≥ 0, respectively, and the intersection of these half-spaces is $\mathbf{U}_{im}$ by definition. The maximum size of a collection of half-spaces of an m-dimensional vector space whose intersection is $\{\mathbf{0}\}$ and which is minimal with respect to this property obviously does not exceed $2m$, while a simple inductive argument shows that the size of such a collection of half-spaces is always greater than m. QED

We now show how the above can be used to obtain a concrete test for the static rigidity of tensegrity frameworks [Rosenberg1980].

PROPOSITION 5.16. *Let* $\mathbb{F} = (A, L, U, \mathbf{p})$ *be a tensegrity framework,* $\underline{\mathbf{M}}_{\mathbf{p}}$ *be its rigidity matrix, and suppose w.l.o.g. that the rank of the matrix* $\underline{\mathbf{M}}'_{\mathbf{p}}$ *consisting of the first* $3\#A-6$ *rows of* $\underline{\mathbf{M}}_{\mathbf{p}}$ *is* $3\#A-6$. *If* $\underline{\mathbf{M}}''_{\mathbf{p}}$ *is the matrix consisting of the remaining* $k := \#L+\#U-3\#A+6$ *rows of* $\underline{\mathbf{M}}_{\mathbf{p}}$ *and* $\underline{\mathbf{N}}_{\mathbf{p}}$ *is the* $(3\#A-6)\times k$ *matrix such that* $-\underline{\mathbf{M}}''_{\mathbf{p}} = \underline{\mathbf{N}}_{\mathbf{p}}^{T}\cdot\underline{\mathbf{M}}'_{\mathbf{p}}$, *then* $\mathbb{F}$ *is infinitesimally/statically rigid if and only if the inhomogeneous system of linear inequalities*

$$\underline{\mathbf{N}}_{\mathbf{p}}\cdot\mathbf{x} \ge \mathbf{1}\,, \quad \mathbf{x} \ge \mathbf{0} \tag{5.18}$$

has a solution, where $\mathbf{1} = [1, 1, \ldots, 1]$.

Proof: If $\mathbb{F}$ is statically rigid, then there exists a vector $\mathbf{s} \in \mathbb{R}^{\#L+\#U}$ such that $\mathbf{s} > \mathbf{0}$ and $\mathbf{s}^T\cdot\underline{\mathbf{M}}_{\mathbf{p}} = \mathbf{0}^T$. Define $\mathbf{s}'$, $\mathbf{s}''$ to be the vectors consisting of the first $3\#A-6$ and last $k = \#L+\#U-3\#A+6$ elements of $\mathbf{s}$, respectively. Then by the definition of the matrix $\underline{\mathbf{N}}_{\mathbf{p}}$ we have

$$(\underline{\mathbf{M}}'_{\mathbf{p}})^T\cdot(\mathbf{s}' - \underline{\mathbf{N}}_{\mathbf{p}}\cdot\mathbf{s}'') = \mathbf{0}\,. \tag{5.19}$$

Because the rows of $\underline{\mathbf{M}}'_\mathbf{p}$ are independent, this implies that $\mathbf{s}' - \underline{\mathbf{N}}_\mathbf{p}\cdot\mathbf{s}'' = \mathbf{0}$, and since $\mathbf{s} > \mathbf{0}$ we conclude that the inequation $\underline{\mathbf{N}}_\mathbf{p}\cdot\mathbf{x} > \mathbf{0}$ has a solution $\mathbf{s}'' > \mathbf{0}$. Dividing through by the smallest element in $\mathbf{s}'$ then gives us a solution to $\underline{\mathbf{N}}_\mathbf{p}\cdot\mathbf{x} \geq \mathbf{1}$, $\mathbf{x} \geq \mathbf{0}$. Conversely, given an $\mathbf{x}$ which satisfies these inequations, the vector $\mathbf{y} := \mathbf{x} + \epsilon\mathbf{1}$ for sufficiently small $\epsilon > 0$ will satisfy $\underline{\mathbf{N}}_\mathbf{p}\cdot\mathbf{y} > \mathbf{0}$ and $\mathbf{y} > \mathbf{0}$. Thus setting $\mathbf{s}'' := \mathbf{y}$ and $\mathbf{s}' := \underline{\mathbf{N}}_\mathbf{p}\cdot\mathbf{y}$ yields a positive vector $\mathbf{s} := [\mathbf{s}'.\mathbf{s}'']$ such that $\mathbf{s}^T\cdot\underline{\mathbf{M}}_\mathbf{p} = \mathbf{0}$ by Equation (5.19), which can easily be converted into a strict self-stress of $\mathbb{F}$. QED

Although the feasibility of the system of inequalities in (5.18) can be tested by standard linear programming algorithms [Luenberger1973], it is probable that more efficient algorithms, employing the special structure of the matrix $\underline{\mathbf{N}}_\mathbf{p}$, could be devised [Rosenberg1980].

In closing, we consider some ways of building new statically and infinitesimally rigid tensegrity frameworks out of old ones. Recall that by Proposition 4.25 a bar and joint framework $\mathbb{F} = (\mathrm{A}, \mathrm{B}, \mathbf{p})$ is infinitesimally/statically rigid if and only if all other realizations of (A, B) obtained from $\mathbf{p}$ by composition with a projective transformation $\mathbf{t}$ such that the joints $\mathbf{t}\circ\mathbf{p}(\mathrm{A})$ remain finite are likewise infinitesimally/statically rigid. Although nonsingular *affine* transformations of tensegrity frameworks preserve their static and infinitesimal rigidity, the example of Figure 5.3 shows already that this does not hold for general projective transformations. Nevertheless, the following theorem gives us a general and extremely useful way of getting new infinitesimally/statically rigid tensegrity frameworks out of the projective images of old ones (providing of course that one knows enough about projective geometry to use it; see for example [Pedoe1970]).

THEOREM 5.17. *Let* $\mathbf{t}: \mathbb{R}^3 \to \mathbb{R}^3$ *be a projective transformation which leaves the joints* $\mathbf{p}(\mathrm{A})$ *of a tensegrity framework* $\mathbb{F} = (\mathrm{A}, \mathrm{L}, \mathrm{U}, \mathbf{p})$ *finite, and let* $\mathrm{L}' \subseteq \mathrm{L}$, $\mathrm{U}' \subseteq \mathrm{U}$ *be the sets of struts and cables* $\{a_i, a_j\}$, *respectively, such that the line segments* $[(\mathbf{p}_i, \mathbf{p}_j)]$ *cross the plane* $\mathbf{H} \subset \mathbb{R}^3$ *which is mapped to infinity by* $\mathbf{t}$. *Then the tensegrity framework*

$$\mathbb{F}_\mathbf{t} \;:=\; (\mathrm{A}, (\mathrm{L}\setminus\mathrm{L}')\cup\mathrm{U}', (\mathrm{U}\setminus\mathrm{U}')\cup\mathrm{L}', \mathbf{t}\circ\mathbf{p}) \tag{5.20}$$

obtained by swapping the cables and struts of L′ *and* U′ *in the projective image of* $\mathbb{F}$ *is infinitesimally/statically rigid if and only if the original framework* $\mathbb{F}$ *is.*

Because no direct use will be made of this result in the sequel and its proof is fairly involved, we shall not prove it here (see [Roth1981]). Figure 5.5, however, has been provided to show how one uses the planar analogue of this result.

Another way of generating new rigid tensegrity frameworks from old ones (at least in the plane) which *will* be very important to us later on is by an

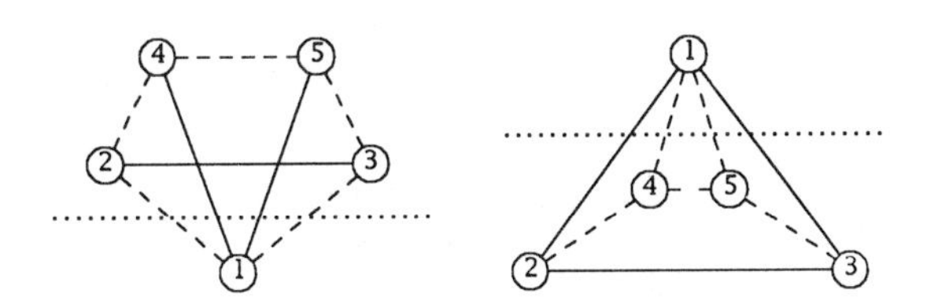

Figure 5.5

A statically/infinitesimally rigid planar tensegrity framework and its projective transform. The line which is mapped to (from) infinity by the projective transformation is dotted.

operation known as "tensegrity exchange". Let $\mathbb{F}' = (A', L', U', \mathbf{p}')$ and $\mathbb{F}'' = (A'', L'', U'', \mathbf{p}'')$ be two planar tensegrity frameworks such that $A' \cap A'' = \{b, c\} \in (L' \setminus U') \cap (U'' \setminus L'')$ with $\mathbf{p}'(b) = \mathbf{p}''(b)$ and $\mathbf{p}'(c) = \mathbf{p}''(c)$. The *tensegrity exchange* $\mathbb{F} = (A, L, U, \mathbf{p})$ of $\mathbb{F}'$ and $\mathbb{F}''$ is obtained by setting $A := A' \cup A''$, $L := L' \cup L'' \setminus \{b, c\}$, $U := U' \cup U'' \setminus \{b, c\}$ and $\mathbf{p}(a') := \mathbf{p}'(a')$, $\mathbf{p}(a'') := \mathbf{p}''(a'')$ for all $a' \in A'$, $a'' \in A''$. Figure 5.6 shows an example of this operation.

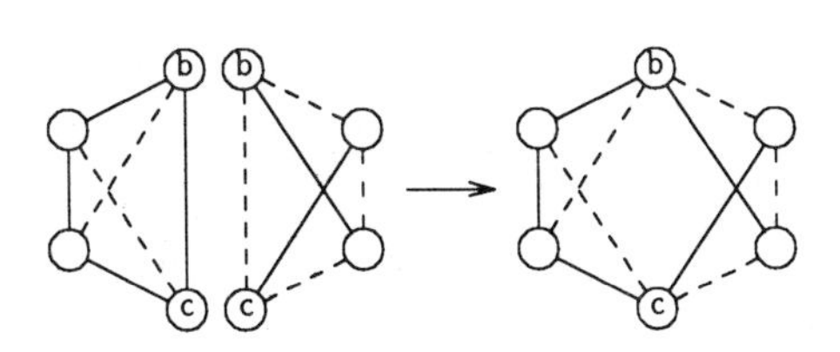

Figure 5.6

Illustration of the tensegrity exchange of two infinitesimally/statically rigid tensegrity frameworks (see text for complete explanation). On the left are the frameworks $\mathbb{F}'$ and $\mathbb{F}''$ with common joints b and c; on the right is their tensegrity exchange $\mathbb{F}$.

PROPOSITION 5.18. *If $\mathbb{F}'$, $\mathbb{F}''$ are two planar infinitesimally/statically rigid tensegrity frameworks and $\mathbb{F}$ is their tensegrity exchange as above, then $\mathbb{F}$ is likewise infinitesimally/statically rigid.*

Proof: Since $\mathbb{F}'$ and $\mathbb{F}''$ are statically rigid, each has a strict self-stress which is negative in $\mathbb{F}'$ and positive in $\mathbb{F}''$ on their common couple $\{b,c\} \in L' \cap U''$. Thus adding appropriate positive multiples of these self-stresses results in a strict self-stress of $\mathbb{F}$. Since the associated bar and joint frameworks $\bar{\mathbb{F}}'$ and $\bar{\mathbb{F}}''$ are infinitesimally/statically rigid and have a self-stress which is nonzero on $\{b,c\}$, the corresponding bar and joint frameworks with that bar deleted are likewise infinitesimally/statically rigid. From this it follows (cf. Theorem 4.47) that the associated bar and joint framework $\bar{\mathbb{F}}$ is infinitesimally/statically rigid, so that $\mathbb{F}$ is infinitesimally/statically rigid by Theorem 5.14. QED

A tensegrity framework will be called *basic* when it cannot be constructed from two smaller tensegrity frameworks by a tensegrity exchange.

5.2. The Global Theory of Tensegrity Frameworks

Most rigid tensegrity frameworks which are constructed for architectural purposes as well as those which we shall have occasion to study later on have fewer than $3\#A-5$ couples in them and so cannot be infinitesimally or statically rigid. Such tensegrity frameworks are made rigid by shortening a cable or lengthening a strut until it can go no further, given the restrictions on the lengths of the remaining couples, at which point a special (nongeneric) positioning is attained in which a proper self-stress exists. Thus we are motivated to seek new methods which are applicable even when the joints are in special position. One possible approach is to take into account the fact that in any smooth flexing $\mathbf{p}^t$ not only must the velocities $\dot{\mathbf{p}} = \partial \mathbf{p}^t / \partial t \,|_{t=0}$ of the joints satisfy the usual conditions for an infinitesimal flexing, but their accelerations of $\ddot{\mathbf{p}} = \partial^2 \mathbf{p}^t / \partial t^2 \,|_{t=0}$ must also satisfy the relations obtained by taking second derivatives of the constraint (in)equalities. For example, if $\{a,b\} \in L \cap U$ is a bar of the framework then the second derivative of its length must vanish as well as its first, giving:

$$
\begin{aligned}
0 &= \tfrac{1}{2}\, \partial^2/\partial t^2 \,\|\mathbf{p}^t(a)-\mathbf{p}^t(b)\|^2 \Big|_{t=0} \\
&= \|\dot{\mathbf{p}}(a)-\dot{\mathbf{p}}(b)\|^2 + (\mathbf{p}(a)-\mathbf{p}(b))\cdot(\ddot{\mathbf{p}}(a)-\ddot{\mathbf{p}}(b))\,.
\end{aligned}
\tag{5.21}
$$

These considerations lead us to the definition [Connelly1980]:

Definition 5.19. A *second-order flexing* of an n-dimensional tensegrity framework $\mathbb{F} = (A, L, U, \mathbf{p})$ is a pair of mappings $\mathbf{p}', \mathbf{p}'' \in (\mathbb{R}^n)^A$ which satisfies the relations:

(1) For all $\{a,b\} \in L \cap U$, we have both $(\mathbf{p}(a)-\mathbf{p}(b))\cdot(\mathbf{p}'(a)-\mathbf{p}'(b)) = 0$ as well as $\|\mathbf{p}'(a)-\mathbf{p}'(b)\|^2 + (\mathbf{p}(a)-\mathbf{p}(b))\cdot(\mathbf{p}''(a)-\mathbf{p}''(b)) = 0$.

(2) For all $\{a,b\} \in U \setminus L$, either:

(a) $(\mathbf{p}(a)-\mathbf{p}(b))\cdot(\mathbf{p}'(a)-\mathbf{p}'(b)) < 0$,
or else

(b) $(\mathbf{p}(a)-\mathbf{p}(b))\cdot(\mathbf{p}'(a)-\mathbf{p}'(b)) = 0$ and
$\|\mathbf{p}'(a)-\mathbf{p}'(b)\|^2 + (\mathbf{p}(a)-\mathbf{p}(b))\cdot(\mathbf{p}''(a)-\mathbf{p}''(b)) \leq 0$.

(3) For all $\{a,b\} \in L \setminus U$, either:

(a) $(\mathbf{p}(a)-\mathbf{p}(b))\cdot(\mathbf{p}'(a)-\mathbf{p}'(b)) > 0$,
or else

(b) $(\mathbf{p}(a)-\mathbf{p}(b))\cdot(\mathbf{p}'(a)-\mathbf{p}'(b)) = 0$ and
$\|\mathbf{p}'(a)-\mathbf{p}'(b)\|^2 + (\mathbf{p}(a)-\mathbf{p}(b))\cdot(\mathbf{p}''(a)-\mathbf{p}''(b)) \geq 0$.

The framework is *second-order rigid* if for every second-order flexing $(\mathbf{p}', \mathbf{p}'')$ we have $\mathbf{p}' \in \mathbb{T}_{\mathbf{p}}$.

Connelly has shown that second-order rigidity implies finite rigidity [Connelly1980], thereby demonstrating that this is a valid approach to proving the rigidity of nonregular tensegrity frameworks. For our immediate purposes, however, the above definition is of interest primarily as a conceptual tool. Instead, we shall pursue a rather different approach which offers significantly deeper insight into the properties of conformation space as a whole. This is based upon a new definition:

Definition 5.20. An n-dimensional tensegrity framework $\mathbb{F} = (A, L, U, \mathbf{p})$ is said to be *globally rigid* if all realizations $\mathbb{H}_{\mathbf{p}'}$ of the same tensegrity $\mathbb{H} = (A, L, U)$ satisfy $\mathbf{p} \approx \mathbf{p}'$. In this case, we will also say that the corresponding abstract framework (A, L, l, U, u) is *uniquely representable.*

Clearly a globally rigid tensegrity framework is rigid, in the local sense of the word used up to now.

We shall see that globally rigid tensegrity frameworks play a fundamental role in both the theory and in the applications of distance geometry to molecular problems. For example, if a conformation consistent with the available distance constraints can be found and the tensegrity framework obtained by taking the active constraints is globally rigid, then the molecule has at most two conformations, related to one another by mirror reflection, and hence an essentially trivial conformation space. Although establishing such global properties is usually a difficult if not impossible task, it turns out that there exist powerful relations between the global and local properties of tensegrity frameworks which in many (though by no means all) cases enable us to do exactly that! This sort of situation is very rare in mathematics, and the possibilities it offers are by themselves sufficiently exciting to justify a purely geometric approach to the problems of molecular conformation.

Paradoxically, the method which is used to obtain these results is based on the so-called "energy" forms, which were first used in this context by Robert Connelly [Connelly1982]. To see how this works, let $\mathbb{F} = (\mathrm{A, L, U}, \mathbf{p})$ be an n-dimensional tensegrity framework, and consider a function $E: \mathbb{R}^{nN} \to \mathbb{R}$ of the form

$$E(\mathbf{x}) = \Sigma_{\mathrm{L} \cup \mathrm{U}} e_{ij}(\|\mathbf{x}(\mathrm{a}_i) - \mathbf{x}(\mathrm{a}_j)\|^2), \tag{5.22}$$

where $e_{ij}: \mathbb{R} \to \mathbb{R}$ are any functions which:

(a) for $\{\mathrm{a}_i, \mathrm{a}_j\} \in \mathrm{L} \cap \mathrm{U}$: have a unique minimum with value $e_{ij}(x)$ at $x = \|\mathbf{p}(\mathrm{a}_i) - \mathbf{p}(\mathrm{a}_j)\|^2$;

(b) for $\{\mathrm{a}_i, \mathrm{a}_j\} \in \mathrm{L} \setminus \mathrm{U}$: are strictly decreasing for all arguments $x < \|\mathbf{p}(\mathrm{a}_i) - \mathbf{p}(\mathrm{a}_j)\|^2$ and satisfy $e_{ij}(x) = \mathit{constant}$ otherwise;

(c) for $\{\mathrm{a}_i, \mathrm{a}_j\} \in \mathrm{U} \setminus \mathrm{L}$: are strictly increasing for all arguments $x > \|\mathbf{p}(\mathrm{a}_i) - \mathbf{p}(\mathrm{a}_j)\|^2$ and satisfy $e_{ij}(x) = \mathit{constant}$ otherwise.

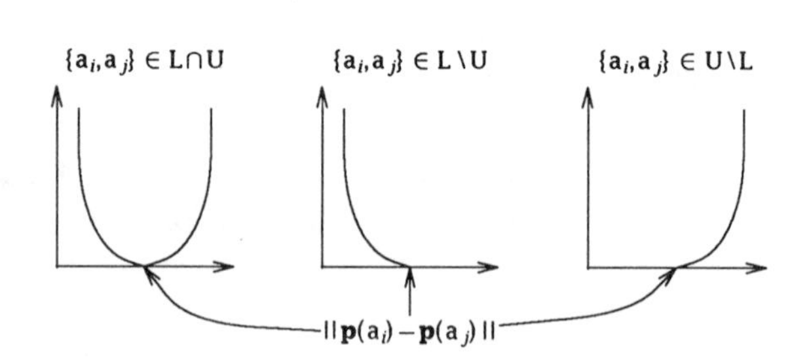

Figure 5.7

Plots of the individual terms of a function which has a given positioning $\mathbf{p}: \mathrm{A} \to \mathbb{R}^n$ of the tensegrity $\mathbb{H} = (\mathrm{A, L, U})$ as a global minimum if and only if the framework $\mathbb{H}_{\mathbf{p}}$ is globally rigid.

Clearly $\mathbb{F}$ is globally rigid if and only if it is the global minimum of such a function*, and hence one approach to proving global rigidity is to choose the e_{ij} to be sufficiently simple, subject to the conditions (a) through (c), that there is some hope of showing that the given realization is the global minimum of their sum up to congruence. Although these conditions are a little too restrictive to allow us

* These functions are essentially error functions, as defined in Chapter 6 (see also §5.5).

to do this, it turns out that we can weaken them slightly and weight the individual terms e_{ij} so as to achieve almost the same result.

Definition 5.21. Let $\mathbb{F} = (\mathrm{A, L, U}, \mathbf{p})$ be a tensegrity framework, $\underline{\mathbf{p}} := \mathbf{p}(\underline{\mathrm{A}})$, $N := \#\mathrm{A}$, $M := \#(\mathrm{L} \cup \mathrm{U})$ and $\mathbf{s} \in \mathbb{R}^M$ be a proper spanning self-stress of $\mathbb{F}$. The *energy form* of $\mathbb{F}$ with respect to $\mathbf{s}$ is the quadratic function:

$$E_{\mathbf{s}} : \mathbb{R}^{nN} \rightarrow \mathbb{R} : \quad \underline{\mathbf{p}} \mapsto \sum_{\{\mathrm{a}_i, \mathrm{a}_j\} \in \mathrm{L} \cup \mathrm{U}} s_{ij} \| \mathbf{p}_i - \mathbf{p}_j \|^2 . \tag{5.23}$$

The name, of course, stems from the fact that its value can be thought of as the energy of the framework due to the compression of its struts and the stretching of its cables from an equilibrium length of zero versus a harmonic potential energy function. Note that $\mathbb{F} \succcurlyeq \mathbb{F}'$ implies $E_{\mathbf{s}}(\underline{\mathbf{p}}) \geq E_{\mathbf{s}}(\underline{\mathbf{p}}')$ for any two positionings $\mathbf{p}, \mathbf{p}'$ of the same tensegrity, although the converse of this statement does not hold. Some other extremely useful properties which these forms have are summarized below [Connelly1982]:

LEMMA 5.22. *Let* $\mathbb{H} = (\mathrm{A, L, U})$, $\mathbb{F} = \mathbb{H}_{\mathbf{p}}$ *for some positioning* $\mathbf{p} \in (\mathbb{R}^n)^{\mathrm{A}}$, $\mathbf{s} \in \mathbb{R}^M$ *be a proper spanning self-stress of* $\mathbb{F}$ *and* $E_{\mathbf{s}} : \mathbb{R}^{nN} \rightarrow \mathbb{R}$ *be the corresponding energy form. Then:*

(i) $\underline{\mathbf{p}}$ *is a critical point of* $E_{\mathbf{s}}$, *i.e.* $\nabla E_{\mathbf{s}}(\underline{\mathbf{p}}) = \mathbf{0}$.

(ii) *For any critical point* $\underline{\mathbf{q}} \in \mathbb{R}^{nN}$ *of* $E_{\mathbf{s}}$ *the tensegrity framework* $\mathbb{H}_{\mathbf{q}}$ *has* $\mathbf{s}$ *as a self-stress, and for every other realization* $\mathbb{H}_{\mathbf{q}}$ *with* $\mathbf{s}$ *as a self-stress* $\underline{\mathbf{q}}$ *is a critical point of* $E_{\mathbf{s}}$.

(iii) $E_{\mathbf{s}}(\underline{\mathbf{q}}) = 0$ *for every critical point* $\underline{\mathbf{q}} \in \mathbb{R}^{nN}$.

(iv) *The set of critical points of* $E_{\mathbf{s}}$ *is invariant under affine transformation.*

Proof:

$$\text{(i)} \quad \nabla E_{\mathbf{s}}(\underline{\mathbf{p}}) \;=\; 2\Big[\sum_{\{j \,|\, \{\mathrm{a}_i, \mathrm{a}_j\} \in \mathrm{L} \cup \mathrm{U}\}} s_{ij}(\mathbf{p}_i - \mathbf{p}_j) \;|\; i = 1, \ldots, N \Big] \;=\; \mathbf{0} \tag{5.24}$$

by the definition of a self-stress of $\mathbb{H}_{\mathbf{p}}$.

(ii) The same argument holds for any other realization $\mathbf{q}$ of $\mathbb{F}$ with the same self-stress, while any time $\mathbf{q}$ is a critical point of $E_{\mathbf{s}}$ Equation 5.24 shows that $\mathbf{s}$ is a self-stress by definition.

(iii) Since $2 E_{\mathbf{s}}(\underline{\mathbf{q}}) = \nabla E_{\mathbf{s}}(\underline{\mathbf{q}}) \cdot \underline{\mathbf{q}}$, if $\underline{\mathbf{q}}$ is a critical point of $E_{\mathbf{s}}$ then $E_{\mathbf{s}}(\underline{\mathbf{q}}) = 0$.

(iv) Let $\mathbf{t}(\mathbf{x}) := \underline{\mathbf{M}} \cdot \mathbf{x} + \mathbf{m}$ be an affine transformation.

$$\begin{aligned} \nabla^T E_{\mathbf{s}}(\underline{\mathbf{t}}(\mathbf{q})) &:= 2\big[\cdots, \textstyle\sum_{\{j \,|\, \{\mathrm{a}_i, \mathrm{a}_j\} \in \mathrm{L} \cup \mathrm{U}\}} s_{ij}(\mathbf{t}(\mathbf{q}_i) - \mathbf{t}(\mathbf{q}_j)), \cdots \big] \\ &= 2\big[\cdots, \textstyle\sum_{\{j \,|\, \{\mathrm{a}_i, \mathrm{a}_j\} \in \mathrm{L} \cup \mathrm{U}\}} s_{ij}(\underline{\mathbf{M}} \cdot \mathbf{q}_i - \underline{\mathbf{M}} \cdot \mathbf{q}_i), \cdots \big] \\ &= 2\big[\cdots, \underline{\mathbf{M}} \cdot \textstyle\sum_{\{j \,|\, \{\mathrm{a}_i, \mathrm{a}_j\} \in \mathrm{L} \cup \mathrm{U}\}} s_{ij}(\mathbf{q}_i - \mathbf{q}_i), \cdots \big] \end{aligned} \tag{5.25}$$

$$= 2\left[\cdots,\underline{\mathbf{M}}\cdot\mathbf{0},\cdots\right] = \mathbf{0}\,.$$

QED

If for $\mathbf{x} \in (\mathbb{R}^n)^A$ we let $\underline{\mathbf{X}} = \underline{\mathbf{X}}(\underline{A})$ be the matrix whose columns are $[\mathbf{x}(a_i) \mid i = 1, \ldots, N]$, and $\mathbf{x}^k \in \mathbb{R}^N$ be the rows thereof transposed, then it is easily shown that any energy form $E_\mathbf{s}$ can be written in terms of matrices as

$$E_\mathbf{s}(\mathbf{x}) = \sum_{k=1}^{n} (\mathbf{x}^k)^T \cdot \underline{\mathbf{S}} \cdot \mathbf{x}^k = \mathbf{1}^T \cdot \underline{\mathbf{X}} \cdot \underline{\mathbf{S}} \cdot \underline{\mathbf{X}}^T \cdot \mathbf{1}\,, \tag{5.26}$$

where $\mathbf{1} = [1, \ldots, 1]$ and $\underline{\mathbf{S}}$ is an $N \times N$ symmetric matrix whose off-diagonal entries are

$$S_{ij} := \begin{cases} -s_{ij} & \text{if } \{a_i, a_j\} \in L \cup U\,; \\ 0 & \text{otherwise}\,; \end{cases} \tag{5.27}$$

and whose diagonal entries are

$$S_{ii} := -\sum_{j \neq i}^{N} S_{ij}\,, \tag{5.28}$$

The matrix $\underline{\mathbf{S}}$ is called a *stress matrix* of the framework $\mathbb{F} = (A, L, U, \mathbf{p})$. Observe that if $\underline{\mathbf{S}}$ were positive definite, the energy form $E_\mathbf{s}$ as a whole would also be positive definite and hence would have a unique minimum at $\mathbf{p}$. Since the row/column sums of any stress matrix $\underline{\mathbf{S}}$ are clearly zero, however, the vector $\mathbf{1} = [1, 1, \ldots, 1]$ is always in the null space and hence an energy form can never be definite. This is not surprising, since if $\mathbf{x}^k = \mathbf{1} \in \mathbb{R}^N$ and $\mathbf{x}^l = \mathbf{0} \in \mathbb{R}^N$ for $l \neq k$ then $\mathbf{x} \in (\mathbb{R}^n)^A$ is in fact the vector field resulting from a unit velocity translation along the k-th coordinate axis, which does not change the interjoint distances or, as a consequence, the energy. More generally, we expect the null space of $\underline{\mathbf{S}}$ to contain at least the (component vectors of those) vector fields which result from the trivial infinitesimal motions. The following proposition provides us with a complete characterization of the null space.

PROPOSITION 5.23. *Let $\underline{\mathbf{S}}$ be the stress matrix associated with a proper spanning self-stress $\mathbf{s}$ of an n-dimensional tensegrity framework $\mathbb{H}_\mathbf{p} = (A, L, U, \mathbf{p})$ such that $dim(\mathbf{p}(A)) = m \leq n$, and suppose (adding superfluous coordinates if necessary) that $n+1 \geq nul(\underline{\mathbf{S}})$; then:*

(i) *The nullity $nul(\underline{\mathbf{S}}) \geq m+1$.*

(ii) *If $nul(\underline{\mathbf{S}}) > m+1$, there is another realization $\mathbb{H}_\mathbf{q}$ with $nul(\underline{\mathbf{S}}) = dim(\mathbf{q}(A))+1$ which also has $\underline{\mathbf{S}}$ as a stress matrix, and such that $\mathbf{p}(A)$ is an orthogonal projection of $\mathbf{q}(A)$.*

(iii) *Assuming $m = n$, for all $\mathbf{q} \in (\mathbb{R}^n)^A$ with $\underline{\mathbf{S}} \cdot \mathbf{q}^k = \mathbf{0}$ for $k = 1, \ldots, n$ there exists a (possibly singular) affine transformation $\mathbf{t} : \mathbb{R}^n \to \mathbb{R}^n$ with $\mathbf{t} \circ \mathbf{p} = \mathbf{q}$.*

Proof: Let $\hat{\mathbf{p}} := [1, \mathbf{p}]$ be projective coordinates for the joints and $\underline{\hat{\mathbf{P}}} := [\hat{\mathbf{p}}(a_i) \mid i = 1, \ldots, \#A]$. Then by the definition of a self-stress, $\underline{\mathbf{S}}$ is a stress matrix for $\mathbb{H}_\mathbf{p}$ if and only if $\underline{\hat{\mathbf{P}}} \cdot \underline{\mathbf{S}} = \mathbf{0}$. Since $\underline{\hat{\mathbf{P}}}$ has $m+1$ independent columns by assumption, it follows that $nul(\underline{\mathbf{S}}) \geq m+1$.

To prove (ii), we may assume that $nul(\underline{\mathbf{S}}) = n+1$, and that $\mathbf{p}$ has zeros for its last $n-m$ coordinates. Then we may replace the zero rows of $\underline{\hat{\mathbf{P}}}$ by a basis for the orthogonal complement of the nonzero rows in the null space of $\underline{\mathbf{S}}$ to obtain a matrix $\underline{\hat{\mathbf{Q}}}$ with $\underline{\hat{\mathbf{Q}}} \cdot \underline{\mathbf{S}} = \mathbf{0}$ as before. Defining $\hat{\mathbf{q}}(a_i)$ to be the columns of this matrix, the desired mapping $\mathbf{q}$ is obtained simply by stripping off the leading ones from these columns.

To prove (iii), we choose $\mathbf{p} \in (\mathbb{R}^n)^A$ as in the proof of part (ii), so that $dim(\mathbf{p}(A)) = n$. Let $\mathbb{U} := \{\mathbf{t} \circ \mathbf{p} \mid \mathbf{t} \text{ affine}\}$ be the subspace of $(\mathbb{R}^n)^A$ generated by the $n(n+1)$-dimensional vector space of all (possibly singular) affine transformations. By Lemma 5.22 the null space of $\underline{\mathbf{S}}$ contains the linear subspace $\mathbb{U}^k := \{\mathbf{u}^k \mid \mathbf{u} \in \mathbb{U}\}$, and by symmetry with respect to coordinate permutation these are equal to the same subspace of $\mathbb{R}^{\#A}$ for all k. In particular, their direct sum is $\mathbb{U}$ and, since they are n in number, their dimensions are all equal to $n+1 = nul(\underline{\mathbf{S}})$, from which it follows that $\mathbb{U}^k$ is the null space of $\underline{\mathbf{S}}$ for all k, as claimed. QED

Observe that the above proposition provides us with an upper bound on the dimension of tensegrity frameworks having a given stress matrix. Of still greater interest is the fact that if a tensegrity framework has a stress matrix which is positive semidefinite of the maximum possible rank, those frameworks which it dominates are greatly restricted.

THEOREM 5.24. *Let $\mathbb{H} = (A, L, U)$ be a tensegrity, $\mathbf{p}: A \rightarrow \mathbb{R}^n$ be a positioning thereof with $dim(\mathbf{p}(A)) = n$, and $\underline{\mathbf{S}}$ be a stress matrix of $\mathbb{H}_\mathbf{p}$ which is positive semidefinite of nullity $n+1$. Then all positionings $\mathbf{q}: A \rightarrow \mathbb{R}^n$ such that $\mathbb{H}_\mathbf{p}$ dominates $\mathbb{H}_\mathbf{q}$ are related to $\mathbf{p}$ by an affine transformation $\mathbf{t}: \mathbb{R}^n \rightarrow \mathbb{R}^n$, i.e. $\mathbf{q} = \mathbf{t} \circ \mathbf{p}$.*

Proof: If $\mathbb{H}_\mathbf{q} \prec \mathbb{H}_\mathbf{p}$, then $E_\mathbf{s}(\mathbf{q}) \leq E_\mathbf{s}(\mathbf{p})$, where by Lemma 5.22 $\mathbf{p}$ is a critical point of $E_\mathbf{s}$ with $E_\mathbf{s}(\mathbf{p}) = 0$. But since $E_\mathbf{s}$ is a positive semidefinite quadratic form it can never be negative, from which it follows that $\mathbf{p}$ and hence also $\mathbf{q}$ are global minima of $E_\mathbf{s}$ with value zero. This in turn implies that $\mathbf{q}^k$ is in the null space of $\underline{\mathbf{S}}$ for $k = 1, \ldots, n$, and the result follows at once from Proposition 5.23 (iii). QED

COROLLARY 5.25. *Given $\mathbb{H}$, $\mathbf{p}$ and $\mathbf{q}$ as in Theorem 5.24, there exists an affine flexing $\mathbf{t}^\alpha: \mathbb{R}^{n+1} \rightarrow \mathbb{R}^n$ with $\mathbf{t}^0 = \mathbf{i}$ (the identity) and $\mathbf{t}^1 \circ \mathbf{p} = \mathbf{q}$. In particular, if $\mathbb{H}_\mathbf{p}$ is rigid, then it is globally rigid.*

Proof: Letting $\mathbf{t}$ be the affine mapping of Theorem 5.24, we have $\mathbf{t}^\alpha := (1-\alpha)\mathbf{i}+\alpha\mathbf{t}$. QED

Since it is relatively easy to check for the existence of an affine flexing, this provides us with the desired method of proving the global rigidity of tensegrity frameworks. Note, however, that it is quite possible for a globally rigid framework to have stress matrices which are not positive semidefinite, and thus the hard part is usually finding a positive semidefinite stress matrix (if one exists!). Figure 5.8 illustrates the use of this method in the case of a one-dimensional self-stress space, where this problem does not arise.

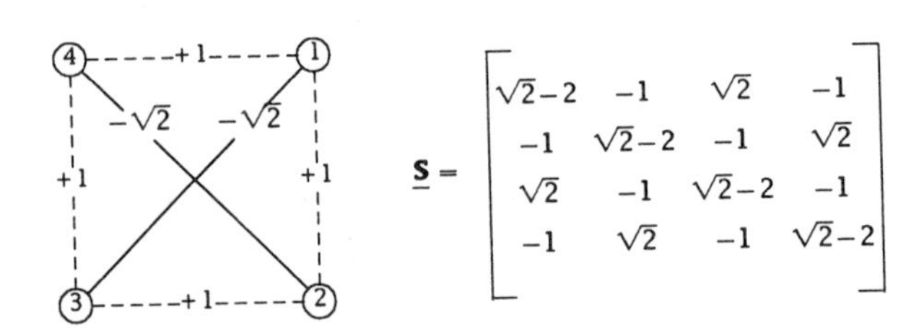

Figure 5.8

The planar tensegrity framework shown on the left has a one-dimensional self-stress space generated by the stresses indicated on its struts and cables. The corresponding positive semidefinite rank 1 stress matrix is shown on the right. Since this framework is infinitesimally rigid, it is rigid and hence the definiteness of its stress matrix proves that it is globally rigid.

Our next result provides us with some insight into the negative directions of energy forms [Connelly1987]. We begin with a lemma which will simplify our subsequent arguments substantially.

LEMMA 5.26. *If a tensegrity framework $\mathbb{F} = (A, L, U, \mathbf{p})$ is second-order flexible, then there exists a second-order flexing $(\mathbf{q}', \mathbf{q}'')$ of it such that*

$$\|\mathbf{q}'(a)-\mathbf{q}'(b)\|^2 + (\mathbf{q}(a)-\mathbf{q}(b))\cdot(\mathbf{q}''(a)-\mathbf{q}''(b)) \begin{cases} \geq 0 & \text{if } \{a,b\} \in L \setminus U\,; \\ = 0 & \text{if } \{a,b\} \in L \cap U\,; \\ \leq 0 & \text{if } \{a,b\} \in U \setminus L\,. \end{cases} \tag{5.29}$$

Proof: If $(\mathbf{p}', \mathbf{p}'')$ is a second-order flex as in Definition 5.19, simply set $(\mathbf{q}', \mathbf{q}'') := (\mathbf{p}', \mathbf{p}''+\alpha\mathbf{p}')$ for some sufficiently large $\alpha > 0$. QED

Under these conditions, we can rewrite the equations defining flexings in matrix form as:

$$\underline{\mathbf{M}}^{\circ}_{\mathbf{p}}\cdot\underline{\mathbf{p}}' = \mathbf{0} \qquad \text{and} \qquad \underline{\mathbf{M}}^{*}_{\mathbf{p}}\cdot\underline{\mathbf{p}}' \leq \mathbf{0} \tag{5.30}$$

for the first-order, together with:

$$\underline{\mathbf{M}}^{\circ}_{\mathbf{p}'}\cdot\underline{\mathbf{p}}' + \underline{\mathbf{M}}^{\circ}_{\mathbf{p}}\cdot\underline{\mathbf{p}}'' = \mathbf{0} \qquad \text{and} \qquad \underline{\mathbf{M}}^{*}_{\mathbf{p}'}\cdot\underline{\mathbf{p}}' + \underline{\mathbf{M}}^{*}_{\mathbf{p}}\cdot\underline{\mathbf{p}}'' \leq \mathbf{0} \tag{5.31}$$

for the second-order, where $\underline{\mathbf{M}}^{\circ}_{\mathbf{p}}$ denotes the usual rigidity matrix of the bar and joint framework (A, L∩U, $\mathbf{p}$), and $\underline{\mathbf{M}}^{*}_{\mathbf{p}}$ is the rigidity matrix of the pure tensegrity framework (A, L \ U, U \ L, $\mathbf{p}$). In addition, if we partition our self-stress $\mathbf{s} \in \mathbb{R}^{\#(L\cup U)}$ into two parts:

$$\mathbf{s}^{\circ} := [s_{ij} \mid \{a_i, a_j\} \in \underline{L\cap U}] \tag{5.32}$$

and

$$\mathbf{s}^{*} := [s_{ij} \mid \{a_i, a_j\} \in \underline{U\setminus L}].[-s_{ij} \mid \{a_i, a_j\} \in \underline{L\setminus U}] \tag{5.33}$$

(in the same order as the rows of the matrices), it is easily seen that for $\mathbf{x} \in \mathbb{R}^{nN}$ the energy form itself can be written as

$$E_{\mathbf{s}}(\mathbf{x}) = (\mathbf{s}^{\circ})^{T}\cdot\underline{\mathbf{M}}^{\circ}_{\mathbf{x}}\cdot\mathbf{x} + (\mathbf{s}^{*})^{T}\cdot\underline{\mathbf{M}}^{*}_{\mathbf{x}}\cdot\mathbf{x}\,, \tag{5.34}$$

PROPOSITION 5.27. *An infinitesimal motion* $\mathbf{p}'$ *of a tensegrity framework* $\mathbb{F} = (A, L, U, \mathbf{p})$ *extends to a second-order flexing* $(\mathbf{p}', \mathbf{p}'')$ *if and only if* $E_{\mathbf{s}}(\mathbf{p}') \leq 0$ *for all proper self-stresses* $\mathbf{s} \in \mathbb{R}^{\#(L\cup U)}$.

Proof: Given a solution $(\mathbf{p}', \mathbf{p}'')$ to Equation 5.31, we take any proper self-stress $\mathbf{s}$ of $\mathbb{F}$ and partition it into $\mathbf{s}^{\circ}$ and $\mathbf{s}^{*}$ as above. Then by its definition $\mathbf{s}^{*} \geq \mathbf{0}$ and

$$(\mathbf{s}^{\circ})^{T}\cdot\underline{\mathbf{M}}^{\circ}_{\mathbf{p}} + (\mathbf{s}^{*})^{T}\cdot\underline{\mathbf{M}}^{*}_{\mathbf{p}} = \mathbf{0}^{T}\,, \tag{5.35}$$

so that by Equation (5.31)

$$\begin{aligned} 0 &\geq (\mathbf{s}^{\circ})^{T}\cdot\underline{\mathbf{M}}^{\circ}_{\mathbf{p}'}\cdot\underline{\mathbf{p}}' + (\mathbf{s}^{\circ})^{T}\cdot\underline{\mathbf{M}}^{\circ}_{\mathbf{p}}\cdot\underline{\mathbf{p}}'' + (\mathbf{s}^{*})^{T}\cdot\underline{\mathbf{M}}^{*}_{\mathbf{p}'}\cdot\underline{\mathbf{p}}' + (\mathbf{s}^{*})^{T}\cdot\underline{\mathbf{M}}^{*}_{\mathbf{p}}\cdot\underline{\mathbf{p}}'' \\ &= (\mathbf{s}^{\circ})^{T}\cdot\underline{\mathbf{M}}^{\circ}_{\mathbf{p}'}\cdot\underline{\mathbf{p}}' + (\mathbf{s}^{*})^{T}\cdot\underline{\mathbf{M}}^{*}_{\mathbf{p}'}\cdot\underline{\mathbf{p}}' =: E_{\mathbf{s}}(\mathbf{p}')\,, \end{aligned} \tag{5.36}$$

as desired.

Conversely, suppose that for all proper self-stresses we have $E_{\mathbf{s}}(\mathbf{p}') \leq 0$, and for $\underline{\mathbf{p}}' := \mathbf{p}'(\underline{A})$ consider the matrices

$$\underline{\mathbf{N}}^{\circ} := [\underline{\mathbf{M}}^{\circ}_{\mathbf{p}'}\cdot\underline{\mathbf{p}}',\ \underline{\mathbf{M}}^{\circ}_{\mathbf{p}}] \tag{5.37}$$

and

$$\underline{\mathbf{N}}^{*} := [\underline{\mathbf{M}}^{*}_{\mathbf{p}'}\cdot\underline{\mathbf{p}}',\ \underline{\mathbf{M}}^{*}_{\mathbf{p}}]\,. \tag{5.38}$$

Together the rows of these matrices generate a convex cone $\mathbf{K} := \{\mathbf{x}^T \cdot \underline{\mathbf{N}}^o + \mathbf{y}^T \cdot \underline{\mathbf{N}}^* \mid \mathbf{y} \geq \mathbf{0}\}$, and we claim that $[1, \mathbf{0}] \notin \mathbf{K}$. For if it were we would have some $\mathbf{s} \in \mathbb{R}^{\#(L \cup U)}$ with

$$(\mathbf{s}^o)^T \cdot \underline{\mathbf{M}}^o_{\mathbf{p}'} \cdot \mathbf{p}' + (\mathbf{s}^*)^T \cdot \underline{\mathbf{M}}^*_{\mathbf{p}'} \cdot \mathbf{p}' = 1 > 0 \quad (5.39)$$

as well as

$$(\mathbf{s}^o)^T \cdot \underline{\mathbf{M}}^o_{\mathbf{p}} + (\mathbf{s}^*)^T \cdot \underline{\mathbf{M}}^*_{\mathbf{p}} = \mathbf{0}^T , \quad (5.40)$$

i.e. $\mathbf{s}$ would be a self-stress with $E_{\mathbf{s}}(\mathbf{p}') = 1 > 0$ contrary to our hypothesis. It follows from Lemma 5.9 (ii) that there exists a vector $\hat{\mathbf{q}} := [\alpha, \mathbf{q}] \in -\mathbf{K}^+$ for which $\alpha = [1, \mathbf{0}] \cdot [\alpha, \mathbf{q}] > 0$ as well as

$$\underline{\mathbf{N}}^o \cdot \hat{\mathbf{q}} = \mathbf{0} \qquad \text{and} \qquad \underline{\mathbf{N}}^* \cdot \hat{\mathbf{q}} \leq \mathbf{0} \quad (5.41)$$

(in this form, Lemma 5.9 (ii) is known as Farkas' Lemma). Thus the desired extension exists with $\mathbf{p}'' := \alpha^{-1}\mathbf{q}$. QED

Although we shall not do so here, it is in fact possible to show that the projective transform (as defined in §5.1) of any tensegrity framework which has a positive semidefinite stress matrix also has a positive semidefinite stress matrix. Similarly, given the tensegrity exchange of any two such tensegrity frameworks we can likewise find a positive semidefinite stress matrix for it. Thus, although these operations do not necessarily preserve global rigidity, they can result in at most affine flexings. These facts allow us to construct a rather large variety of globally rigid tensegrity frameworks (see Figure 5.9). Another approach to constructing globally rigid tensegrity frameworks is based upon a standard geometric construction known as the Gale transform [McMullen1978].

Definition 5.28. Let $\mathbf{P} = \{\mathbf{p}_1, \ldots, \mathbf{p}_N\}$ be a set of points in $\mathbb{R}^n$ ($N > n$) such that $dim(\mathbf{P}) = n$, and let $\underline{\hat{\mathbf{P}}}$ be a matrix whose columns are their standard projective coordinates $\hat{\mathbf{p}}_i := [1, \mathbf{p}_i]$. The *Gale transform* of $\mathbf{P}$ is any set $\mathbf{X} = \{\mathbf{x}_1, \ldots, \mathbf{x}_N\}$ of vectors in $\mathbb{R}^{N-n-1}$ such that the *rows* of the matrix $\underline{\mathbf{X}} = [\mathbf{x}_1, \ldots, \mathbf{x}_N]$ span the orthogonal complement of those of $\underline{\hat{\mathbf{P}}}$ in $\mathbb{R}^N$.

Hence the rows of $\underline{\mathbf{X}}$ constitute a basis for the space of affine dependencies of the points $\mathbf{P}$. Note this definition is valid for any vectors in $\mathbb{R}^{n+1}$, even if they are not of the form $\hat{\mathbf{p}}$ for $\mathbf{p} \in \mathbb{R}^n$.

To see what all this has to do with global rigidity, suppose $\mathbb{H}_{\mathbf{p}} = (A, L, U, \mathbf{p})$ is an n-dimensional tensegrity framework which has a positive semidefinite stress matrix $\underline{\mathbf{S}}$ of nullity $n+1$. Like any positive semidefinite matrix, $\underline{\mathbf{S}}$ has a square root $\underline{\mathbf{Y}}$ which is $(N-n-1) \times N$ ($N = \#A$) and which satisfies $\underline{\mathbf{Y}}^T \cdot \underline{\mathbf{Y}} = \underline{\mathbf{S}}$. Hence

$$\underline{\mathbf{0}} = \underline{\mathbf{S}} \cdot \underline{\hat{\mathbf{P}}}^T = \underline{\mathbf{Y}}^T \cdot \underline{\mathbf{Y}} \cdot \underline{\hat{\mathbf{P}}}^T \quad (5.42)$$

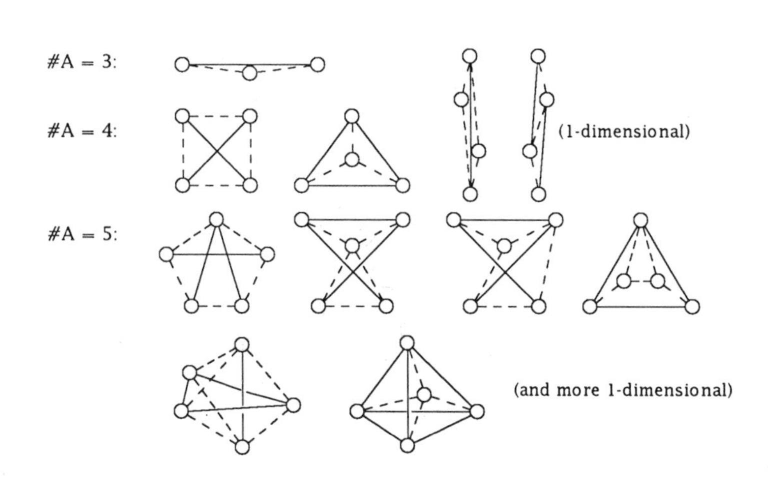

Figure 5.9

All known globally rigid tensegrity frameworks on the line, in the plane and in space for up to five atoms.

which, since $rank(\underline{\mathbf{Y}}) = N-n-1$, implies that $\underline{\mathbf{Y}} \cdot \underline{\hat{\mathbf{P}}}^T = \underline{\mathbf{0}}$, i.e. the rows of $\underline{\mathbf{Y}}$ are dependencies of $\hat{\mathbf{P}}$. Because these are $N-n-1$ in number, they are in fact a basis of the space of dependencies, so that the columns of $\underline{\mathbf{Y}}$ constitute a Gale transform of $\mathbf{P}$. Conversely, given any full rank $(N-n-1) \times N$ matrix $\underline{\mathbf{Y}}$ whose row sums vanish and some Gale transform $\hat{\mathbf{P}} = [\mathbf{1}, \mathbf{P}^T]^T$ of its columns, we can interpret the matrix $\underline{\mathbf{S}} := \underline{\mathbf{Y}}^T \cdot \underline{\mathbf{Y}}$ as a stress matrix of the tensegrity framework $\mathbb{F} = (\mathrm{A}, \mathrm{L}, \mathrm{U}, \mathbf{p})$ where $\mathrm{A} := \{1, \ldots, N\}$, $\mathrm{L} := \{\{i,j\} \mid S_{ij} < 0\}$, $\mathrm{U} := \{\{i,j\} \mid S_{ij} > 0\}$ and $\mathbf{p}(i) := \mathbf{p}_i$. In fact, all possible tensegrity frameworks with a positive semidefinite stress matrix $\underline{\mathbf{S}}$ of nullity $n+1$ can be obtained in this way!

We now demonstrate the utility of this construction by proving a simple result which we shall need later from it.

PROPOSITION 5.29. *Let* $\mathbb{K} = (\mathrm{A}, \binom{\mathrm{A}}{2})$ *be the complete graph on* $N = \#\mathrm{A}$ *atoms and let* $\mathbf{p} : \mathrm{A} \to \mathbb{R}^{N-2}$ *be a positioning thereof. If we define* $\underline{\hat{\mathbf{P}}}_i := \hat{\mathbf{p}}(\mathrm{a}_1, \ldots, \mathrm{a}_{i-1}, \mathrm{a}_{i+1}, \ldots, \mathrm{a}_N)$ *and* $\hat{P}_i := det(\underline{\hat{\mathbf{P}}}_i)$ *for* $i = 1, \ldots, N$, *then all positive semidefinite stress matrices of* $\mathbb{K}_{\mathbf{p}}$ *are multiples of* $\underline{\mathbf{S}} = [(-1)^{i+j} \hat{P}_i \hat{P}_j \mid i, j = 1, \ldots, N]$.

Proof: By Cramer's rule, the one-dimensional space of linear dependencies of $\hat{\mathbf{P}}$ is easily seen to be generated by $[(-1)^i \hat{P}_i \mid i = 1, \ldots, N]$. Squaring this matrix gives the matrix $\underline{\mathbf{S}}$. QED

Another example of this construction is shown in the box below.

To find all tensegrity frameworks with positive semidefinite nullity four stress matrices whose associated bar and joint framework is a realization of the graph of the octahedron, we proceed as follows:

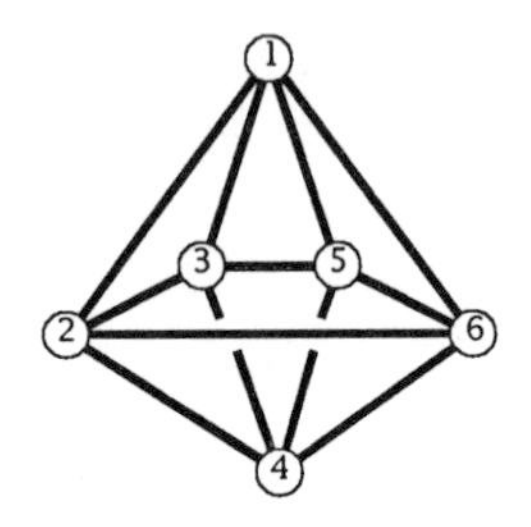

Figure 5.10

Since there are three nonadjacent pairs of joints, the Gale transform $\mathbf{Z}$ must consist of three pairs of orthogonal 2-dimensional vectors. Together with the condition that the row sums of the corresponding matrix $\underline{\mathbf{Z}}$ must vanish, we find the set of all Gale transforms depends on four parameters x_1, x_2, y_1, y_2, as follows:

$$\underline{\mathbf{Z}} = \begin{bmatrix} x_1 & x_2 & -x_1-x_2 & -y_1 & -y_2 & y_1+y_2 \\ y_1 & y_2 & -y_1-y_2 & x_1 & x_2 & -x_1-x_2 \end{bmatrix}.$$

The set of all positive semidefinite stress matrices of rank two or less whose nonzero elements are those of the adjacency matrix of (a subgraph of) the octahedral graph is $\underline{\mathbf{Z}}^T \cdot \underline{\mathbf{Z}}$. For example, if $x_1 = -2$, $x_2 = 1$, $y_1 = -1$ and $y_2 = 2$, we get:

$$\underline{\mathbf{Z}}^T \cdot \underline{\mathbf{Z}} = \begin{bmatrix} 5 & -4 & -1 & 0 & 3 & -3 \\ -4 & 5 & -1 & -3 & 0 & 3 \\ -1 & -1 & 2 & 3 & -3 & 0 \\ 0 & -3 & 3 & 5 & -4 & -1 \\ 3 & 0 & -3 & -4 & 5 & -1 \\ -3 & 3 & 0 & -1 & -1 & 2 \end{bmatrix}.$$

This gives the globally rigid tensegrity framework shown in the next box below.

We point out that the above proposition is an example of a general rule which states that the possible stresses of any given framework can always be factored into polynomials in the oriented volumes spanned by its joints [White1983]. This in turn shows that the rigid tensegrity frameworks which can exist for a given positioning $\mathbf{p} \in (\mathbb{R}^n)^A$ of their joints are intimately tied to the chirality of the set of points $\mathbf{p}(A)$. Although these relations appear to be quite complicated, they are worthy of further study because (as we shall see in §5.5) they show how the pattern of attractive and repulsive forces which can exist between the atoms of a molecule at a minimum of any conformational energy function is connected to its geometric structure.

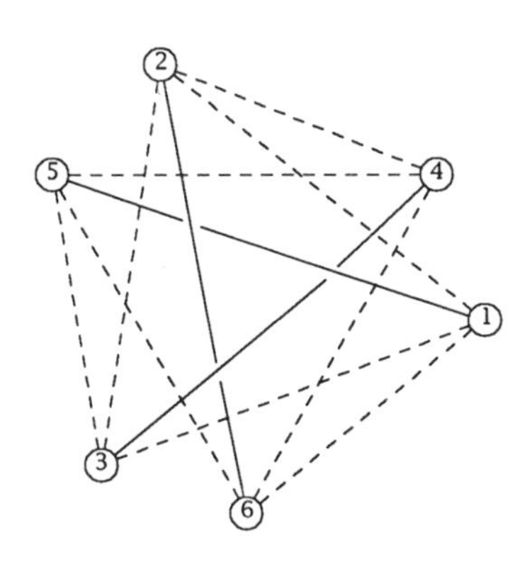

Figure 5.11

A center of mass positioning for the joints of the globally rigid octahedral tensegrity discovered in the previous example can be obtained from its Gale transform $\underline{\mathbf{Z}}$ by finding a basis for the orthogonal complement of the columns of $[\mathbf{1}, \underline{\mathbf{Z}}^T]$. The reader is invited to search for other globally rigid octahedral tensegrity frameworks which are not equivalent to the above under renumbering of the joints.

5.3. Triangle Inequality Bound Smoothing

We now turn our attention from direct consideration of the rigidity, statics and kinematics of individual conformations to those properties which are common to all possible conformations of the molecule which are consistent with the available distance information. Specifically, for a given set of lower and upper bounds

$$l,\ u \in F(A) := \{\, f : A \times A \rightarrow [0 \cdots \infty) \mid f(a,b) = f(b,a) \text{ and } f(a,a) = 0 \} \quad (5.43)$$

on *all* of the interatomic distances in the molecule, we shall consider the problem of computing the corresponding *triangle inequality limits.* This problem was first studied in [Havel1983], where it was shown that it could be solved in polynomial time, and later by [Dress1987], where the complete characterization given below was first proved.

The triangle inequality limits may be regarded as an approximation to the complete Euclidean limits introduced in §1.3, which are given by:

$$l_E(\mathrm{a,b}) \;:=\; inf(\,\|\mathbf{p}(\mathrm{a})-\mathbf{p}(\mathrm{b})\|\mid \mathbf{p}\in\mathbf{P}(l,u)) \tag{5.44}$$

and

$$u_E(\mathrm{a,b}) \;:=\; sup(\,\|\mathbf{p}(\mathrm{a})-\mathbf{p}(\mathrm{b})\|\mid \mathbf{p}\in\mathbf{P}(l,u))\,, \tag{5.45}$$

where

$$\mathbf{P}(l,\,u) \;:=\; \{\mathbf{p}\in(\mathbb{R}^3)^{\mathrm{A}}\mid l(\mathrm{a,b})\leq\|\mathbf{p}(\mathrm{a})-\mathbf{p}(\mathrm{b})\|\leq u(\mathrm{a,b})\;\;\forall\;\mathrm{a,b}\in\mathrm{A}\} \tag{5.46}$$

is the set of all (l,u)-admissible Cartesian coordinates.* In order to define the triangle inequality limits, we shall call a function $f\in\mathrm{F(A)}$ $(l,\,u)$*-admissible* if the condition $l(\mathrm{a,b})\leq f(\mathrm{a,b})\leq u(\mathrm{a,b})$ holds for all $\mathrm{a,b}\in\mathrm{A}$.

Definition 5.30. The *triangle inequality limits* associated with a set of distance bounds $l,\,u\in\mathrm{F(A)}$ are given by:

$$l_3(\mathrm{a,b}) \;=\; l_3^{(l,u)}(\mathrm{a,b}) \;:=\; inf(\,f(\mathrm{a,b})\mid f\in\mathrm{F_3(A)}\text{ is }(l,u)\text{-admissible}) \tag{5.47}$$

and

$$u_3(\mathrm{a,b}) \;=\; u_3^{(l,u)}(\mathrm{a,b}) \;:=\; sup(\,f(\mathrm{a,b})\mid f\in\mathrm{F_3(A)}\text{ is }(l,u)\text{-admissible})\,, \tag{5.48}$$

where

$$\mathrm{F_3(A)} \;:=\; \{\,f\in\mathrm{F(A)}\mid 0\leq f(\mathrm{a,b})\leq f(\mathrm{a,c})+f(\mathrm{b,c})\;\;\forall\;\mathrm{a}\neq\mathrm{b}\neq\mathrm{c}\neq\mathrm{a}\text{ in A}\} \tag{5.49}$$

is the set of all *metrics* on A, i.e. functions in F(A) which are consistent with the triangle inequality.† We shall formally define $l_3(\mathrm{a,b}):=\infty$ and $u_3(\mathrm{a,b}):=-\infty$ for all $\mathrm{a}\neq\mathrm{b}$ when there exists no (l,u)-admissible metric in $\mathrm{F_3(A)}$.

Since the triangle inequality is a necessary but not sufficient condition for f to be a Euclidean distance function $d_{\mathbf{p}}(\mathrm{a,b}):=\|\mathbf{p}(\mathrm{a})-\mathbf{p}(\mathrm{b})\|$ (cf. Chapter 3), we have $l\leq l_3\leq l_E$ and $u\geq u_3\geq u_E$ as functions in F(A).

The above definition, although mathematically sound, does not suggest any means of actually computing the triangle inequality limits save possibly by the

* The fact that we are using the set $\mathbf{P}(l,\,u)$ of all (l,u)-admissible positionings instead of embeddings (as in Chapter 1) does not affect the definition, since embeddings are a dense subset of $\mathbf{P}$ and, in any event, $l>0$ in most cases of chemical interest.

† Since we have not stipulated $f(\mathrm{a,b})=0 \iff \mathrm{a}=\mathrm{b}$, the functions we are dealing with here are more commonly referred to as *pseudometrics*. For the purposes of the present discussion, this distinction is of no consequence.

horribly inefficient expedient of solving $O(\#A^2)$ linear programs in $O(\#A^2)$ variables and $O(\#A^3)$ constraints. To obtain an equivalent characterization which is better suited to computational purposes, observe that l_3 and u_3 (if they are finite) are both in F(A) and that if we set $l := l_3$ and $u := u_3$ in Equations (5.47) and (5.48), we obtain the same functions l_3, u_3 back again. Thus the triangle inequality limits are in a certain sense *self-consistent*, and indeed this condition can be used to characterize them via:

$$l_3^{(l,u)}(a,b) \quad := \quad \inf_{l_3', u_3' \in F(A)} \left(l_3'(a,b) \;\middle|\; \begin{array}{l} l_3',\, u_3' \text{ are self-consistent} \\ \text{and } (l,u)\text{-admissible} \end{array} \right). \tag{5.50}$$

and

$$u_3^{(l,u)}(a,b) \quad := \quad \sup_{l_3', u_3' \in F(A)} \left(u_3'(a,b) \;\middle|\; \begin{array}{l} l_3',\, u_3' \text{ are self-consistent} \\ \text{and } (l,u)\text{-admissible} \end{array} \right). \tag{5.51}$$

In order to characterize self-consistent pairs of functions, we make use of a result which was first derived in the course of a study on the approximation of metric spaces by weighted trees, as is commonly done in the construction of evolutionary trees from measures of the dissimilarity between protein and nucleotide sequences [Dress1984].

LEMMA 5.31. *Given* $f \in F_3(A)$, *let* x *be any element not in* A *and let* $l_x, u_x : A \to \mathbb{R}$ *be two real-valued functions defined on* A *with* $0 \leq l_x \leq u_x$. *Then there exists an extension of* f *to* $f' \in F_3(A \cup x)$ *with* $l_x(a) \leq f(a,x) \leq u_x(a)$ *for all* $a \in A$ *if and only if for all* $a, b \in A$:

$$f(a,b) \;\leq\; u_x(a) + u_x(b)\,; \tag{5.52}$$

$$l_x(a) \;\leq\; f(a,b) + u_x(b)\,. \tag{5.53}$$

Proof: Necessity is trivial; to prove the converse, we define

$$P_A := \{ v : A \to \mathbb{R} \mid v(a) + v(b) \geq f(a,b) \geq l_x(b) - v(a) \;\; \forall\, a,b \in A \} \tag{5.54}$$

and

$$T_A := \{ v : A \to \mathbb{R} \mid v(a) = \mathit{sup}(\, f(a,b) - v(b),\ l_x(b) - f(a,b) \mid b \in A) \;\forall a \in A \} \tag{5.55}$$

Thus T_A consists of those members of P_A which are minimal, and in particular $v \in P_A$, $v' \in T_A$ and $v \leq v'$ implies $v = v'$. We claim that for all $v \in P_A$ there exists some $v' \in T_A$ with $v' \leq v$. If for any $v \in P_A$ and $a \in A$ we define $p_a(v) : A \to \mathbb{R}$ by

$$p_a(v)(b) \;:=\; v(b) \qquad \forall\; b \neq a \tag{5.56}$$

and

$$p_a(v)(a) \;:=\; \mathit{sup}(\, f(a,b) - v(b),\ l_x(b) - f(a,b) \mid b \in A)\,, \tag{5.57}$$

then it follows from the above definitions, together with $l_x \geq 0$, that $p_a(\nu) \leq \nu$ as a function on A, $p_a(\nu) \in P_A$ and therefore:

$$p_a(\nu)(a) \;=\; sup(\, f(a,b) - p_a(\nu)(b),\;\; l_x(b) - f(a,b) \mid b \in A) \,. \tag{5.58}$$

Also by definition, for any $\nu \in P_A$ we have $p_a(\nu) = \nu\ \forall a \in A$ if and only if $\nu \in T_A$. Since $l_x \leq p_a(\nu)$ for all $a \in A$ and $\nu \in P_A$, it follows from Zorn's Lemma (a standard result in the theory of partially ordered sets, which is in fact equivalent to the Axiom of Choice) that there does indeed exist a $\nu' \in T_A$ with $\nu' \leq \nu$.

In particular, since $u_x \in P_A$ by (5.52) and (5.53), there exists $u_x' \leq u_x$ with $u_x' \in T_A$. Therefore, if we now define the extension $f' : (A \cup x) \times (A \cup x) \to \mathbb{R}$ by

$$f'(x,a) \;:=\; f'(a,x) \;:=\; u_x'(a) \quad \forall\, a \in A \tag{5.59}$$

together with $f'(x,x) := 0$, we see at once that $l_x(a) \leq f'(a,x) \leq u_x(a)$, as desired. To verify that $f' \in F_3(A \cup x)$ we must check two types of inequalities, namely $f'(a,b) \leq f'(a,x) + f(x,b)$ and $f'(a,x) \leq f'(a,b) + f(b,x)$. To prove the first of these, we simply observe that since $u_x' \in T_A \subseteq P_A$, for all $a,b \in A$ we have:

$$f'(a,b) \;=\; f(a,b) \;\leq\; u_x'(a) + u_x'(b) \;=\; f'(a,x) + f'(x,b) \,. \tag{5.60}$$

To prove the second, we observe that for all $a,b,c \in A$:

$$\begin{aligned} u_x'(a) &= sup(\, f(a,c) - u_x'(c),\;\; l_x(c) - f(a,c) \mid c \in A) \\ &\leq sup(\, f(a,b) + f(b,c) - u_x'(c),\;\; l_x(c) + f(a,b) - f(b,c) \mid c \in A) \\ &= f(a,b) + sup(\, f(b,c) - u_x'(c),\;\; l_x(c) - f(b,c) \mid c \in A) \\ &= f(a,b) + u_x'(b) \,. \end{aligned} \tag{5.61}$$

QED

An equivalent statement of this lemma which the reader might find more palatable says that whenever a metric space (A, f) has the property that for all $a,b \in A$ there exists an extension of the subspace ({a,b}, f) to {a,b,x} which lies between the lower and upper bounds $l_x, u_x : A \to \mathbb{R}$, then there exists an extension of the entire space (A, f) to ($A \cup x$, f) which lies within the bounds. The lemma can then be generalized without difficulty to show that if (A, f) and (A′, f') are any two metric spaces with $A \cap A' = \{\}$, and $l_3, u_3 : A \times A' \to \mathbb{R}$ are lower and upper bounds such that for any $a,b \in A$ and $a',b' \in A'$ there exist extensions of ({a,b}, f) to both {a,b,a′} and {a,b,b′} as well as extensions of ({a′,b′}, f') to both {a,a′,b′} and {b,a′,b′} which lie within these bounds, then there exists an extension of both spaces to $A \cup A'$ which lies within the bounds.

We are now ready to present our promised characterization of the triangle inequality limits.

THEOREM 5.32. *For any pair of functions l, $u \in F(A)$ the following two conditions are equivalent:*

(i) *The functions l and u are self-consistent, i.e. $l = l_3^{(l,u)}$ and $u = u_3^{(l,u)}$.*

(ii) *The function u satisfies the triangle inequality, i.e.*

$$u(a,b) \leq u(a,c) + u(b,c) \tag{5.62}$$

for all $a,b,c \in A$, *while l and u together satisfy*

$$l(a,b) \leq l(a,c) + u(b,c) \tag{5.63}$$

for all $a,b,c \in A$.

Proof: ((i) => (ii)): Since the set of (l,u)-admissible functions in $F_3(A)$ is compact (i.e. closed and bounded), self-consistency is equivalent to the assertion that for every $a,b \in A$ there exist two (l,u)-admissible functions $f_l = f_l^{(a,b)}$, $f_u = f_u^{(a,b)}$ in $F_3(A)$ which satisfy:

$$f_l(a,b) = l(a,b) \quad \text{and} \quad f_u(a,b) = u(a,b), \tag{5.64}$$

respectively. Hence for all $a,b,c \in A$, we have

$$u(a,b) = f_u(a,b) \leq f_u(a,c) + f_u(b,c) \leq u(a,c) + u(b,c) \tag{5.65}$$

as well as

$$\begin{aligned} l(a,c) \leq f_l(a,c) \leq f_l(a,b) + f_l(b,c) &= l(a,b) + f_l(b,c) \\ &\leq l(a,b) + u(b,c). \end{aligned} \tag{5.66}$$

((ii) => (i)): For every $a,b \in A$ we must demonstrate the existence of the functions $f_l = f_l^{(a,b)}$ and $f_u = f_u^{(a,b)}$ defined above. Obtaining $f_u^{(a,b)}$ is easy: simply set $f_u^{(a,b)} := u$ for all $a,b \in A$. To establish the existence of the function $f_l^{(a,b)}$ we use induction on $N := \#A$; for a set $A = \{a,b\}$ of size two we simply set $f_l^{(a,b)}(a,b) := l(a,b)$. We now assume that for some $N > 2$ the theorem is true for all functions defined on sets of cardinality $N-1$. Thus the desired function $f_l = f_l^{(a,b)}$ exists on any $C := \{a,b,c_3,\ldots,c_{N-1}\} \subseteq A$ with $\#C = N-1$. For $x \in A \setminus C$, we define:

$$l_x(c) := l(c,x) \quad \text{and} \quad u_x(c) := u(c,x) \tag{5.67}$$

for all $c \in C$. Since these functions together with the restriction $f|_C$ of f to C satisfy the hypothesis of Lemma 5.31, it follows that the desired function $f_l = f_l^{(a,b)}$ also exists on the N-element subset $A = C \cup x$. QED

The following corollary highlights those consequences of the theorem which are of greatest immediate interest.

COROLLARY 5.33. *If $l, u \in F_A$ are a pair of not necessarily self-consistent functions which allow for at least one (l,u)-admissible $f \in F_3(A)$, then the corresponding upper triangle limits $u_3 = u_3^{(l,u)}$ satisfy (5.62) with $u = u_3$, while u_3 together with the lower triangle limits satisfy (5.63) with $l = l_3$. Conversely, if a pair of functions $l, u \in F_A$ satisfy (5.62) and (5.63), then there necessarily exists an (l,u)-admissible $f \in F_3(A)$.*

Proof: Immediate from the Theorem 5.32. QED

Thus, the theorem provides us with necessary conditions for the geometric consistency of any given distance bounds l, u.

In order to see how the above characterization is used to compute the triangle inequality limits, we first consider the upper triangle inequality limits. By virtue of Theorem 5.32 together with Equations (5.50) and (5.51), these (assuming that they exist) are given by

$$u_3(a,b) \;:=\; \sup_{v \in F(A)} \left(v(a,b) \mid v(x,y) \leq \min(u(x,y),\, v(x,z)+v(y,z)) \;\; \forall\; x,y,z \in A \right) \quad (5.68)$$

for all $a,b \in A$, which shows, in particular, that they are independent of the lower bounds l.

Let $\mathbb{G}(u) := (A,\, B := \binom{A}{2},\, w{:}B \to \mathbb{R})$ be the complete graph on A each of whose couples $C = \{a,b\}$ is given the *weight* $w(C) := u(a,b)$. As is well-known (see, for example, [Lawler1976]), a family of paths in $\mathbb{G}$ from any one atom $a \in A$ to all other atoms $b \in A$ whose lengths are given by $s(a,b)$ is a family of shortest paths with respect to the given weights w if and only if each is at least as short as the shortest "short-cut" consisting of the shortest path from the atom a to some other atom $x \in A$, followed by the direct path from x to b of length $u(a,b)$, i.e.

$$s(a,b) \;=\; \inf(s(a,x) + u(x,b) \mid x \in A)\,. \quad (5.69)$$

These are known as *Bellman's equations.*

LEMMA 5.34. *The upper triangle inequality limits are equal to the shortest paths in the graph $\mathbb{G}(u)$.*

Proof: For any function $v \in F(A)$ and $a,b \in A$ we have $v(a,b) \geq \inf(v(a,x)+u_3(x,b) \mid x \in A)$, so all we need to show is the opposite inequality. Since $u_3 \leq u$ and u_3 obeys the triangle inequality, $u_3(a,b) \leq u_3(a,x)+u(x,b)$ for all $x \in A$, so that $u_3(a,b) \leq \inf(u_3(a,x)+u(x,b) \mid x \in A)$, as desired. QED

Thus, the triangle inequality limits can be computed simply by applying any all-pairs shortest paths algorithm (see, for example, [Lawler1976, Aho1983]) to the graph $\mathbb{G}(u)$.

Similarly, it follows from Theorem 5.32 that the lower triangle inequality limits are given by:

$$l_3(\mathrm{a,b}) \;:=\; \inf_{v \in F(A)} \Big\{ v(\mathrm{a,b}) \mid l(\mathrm{x,y}) \le v(\mathrm{x,y}) \le v(\mathrm{x,z}) + u_3(\mathrm{y,z}) \;\; \forall\, \mathrm{x,y,z} \in \mathrm{A} \Big\}, \tag{5.70}$$

where u_3 are the upper triangle inequality limits as above. In [Havel1983] we presented a somewhat complicated algorithm for computing the lower triangle inequality limits from the upper triangle inequality limits and the given lower bounds. It turns out to be more efficient, however, to compute both the lower and upper limits together, which can be done by reducing the problem to an all-pairs shortest paths problem involving negative weights on some couples (see also [Havel1984a]). To do this, we rewrite Equation (5.70) as

$$-l_3(\mathrm{a,b}) \;:=\; \sup_{-v \in F(A)} \Big\{ v(\mathrm{a,b}) \mid v(\mathrm{x,y}) \le \min(-l(\mathrm{x,y}),\, v(\mathrm{x,z}) + u_3(\mathrm{y,z})) \;\; \forall\, \mathrm{x,y,z} \in \mathrm{A} \Big\}. \tag{5.71}$$

We then have the following characterization of this supremum, which was first demonstrated in [Havel1983].

LEMMA 5.35. *For all* a,b $\in$ A*:*

$$-l_3(\mathrm{a,b}) \;=\; \inf(u_3(\mathrm{a,x}) + u_3(\mathrm{b,y}) - l(\mathrm{x,y}) \mid \mathrm{x,\,y} \in \mathrm{A})\,. \tag{5.72}$$

Proof: Let $v(\mathrm{a,b}) := \inf(u_3(\mathrm{a,x}) + u_3(\mathrm{b,y}) - l(\mathrm{x,y}) \mid \mathrm{x,y} \in \mathrm{A})$. By Equation (5.63) we have $-l_3(\mathrm{a,b}) \le -l_3(\mathrm{b,x}) + u_3(\mathrm{a,x})$ as well as $-l_3(\mathrm{b,x}) \le -l_3(\mathrm{x,y}) + u_3(\mathrm{b,y}) \le -l(\mathrm{x,y}) + u_3(\mathrm{b,y})$, so that $-l_3(\mathrm{a,b}) \le v(\mathrm{a,b})$. Since trivially $v(\mathrm{a,b}) \le -l(\mathrm{a,b})$, to prove the opposite inequality it suffices to show that $v(\mathrm{b,c}) \le v(\mathrm{a,b}) + u_3(\mathrm{a,c})$ (cf. (5.71)). Because the upper bounds obey the triangle inequality (5.62), we have

$$\begin{aligned} u_3(\mathrm{a,c}) + v(\mathrm{a,b}) &= \inf(u_3(\mathrm{a,c}) + u_3(\mathrm{a,x}) + u_3(\mathrm{b,y}) - l(\mathrm{x,y}) \mid \mathrm{x,y} \in \mathrm{A}) \\ &\ge \inf(u_3(\mathrm{c,x}) + u_3(\mathrm{b,y}) - l(\mathrm{x,y}) \mid \mathrm{x,y} \in \mathrm{A}) \\ &= v(\mathrm{b,c})\,, \end{aligned} \tag{5.73}$$

as desired. QED

We now define a weighted *di*graph $\mathbb{G}(l,\, u) = (\mathrm{A,\, C},\, w)$, where $\mathrm{C} = (\mathrm{A} \times \{0,1\}) \times (\mathrm{A} \times \{0,1\})$ and for all $\{\mathrm{a,b}\} \in \mathrm{C}$ the weights $w : \mathrm{C} \to \mathbb{R}$ are given by $w(\mathrm{a},0;\, \mathrm{b},0) := w(\mathrm{b},0;\, \mathrm{a},0) := u(\mathrm{a,b})$, $w(\mathrm{a},1;\, \mathrm{b},1) := w(\mathrm{b},1;\, \mathrm{a},1) := u(\mathrm{a,b})$, $w(\mathrm{a},0;\, \mathrm{b},1) := w(\mathrm{b},0;\, \mathrm{a},1) := -l(\mathrm{a,b})$ and $w(\mathrm{a},1;\, \mathrm{b},0) := w(\mathrm{b},1;\, \mathrm{a},0) := \infty$. Clearly the shortest paths from each (a,0) to (b,0) and from (a,1) to (b,1) are the upper triangle inequality limits $u_3(\mathrm{a,b})$ as before. The shortest paths from each (a,0) to (b,1), however, go over exactly one arc of weight $-l(\mathrm{x,y})$, and hence their lengths are given by $u_3(\mathrm{a,x}) + u_3(\mathrm{b,y}) - l(\mathrm{x,y})$. (Note that the arcs going from (a,0) to (b,1) are essentially "one-way", which insures the absence of negative cycles and hence the existence of finite shortest paths.) Vice versa, since for each combination of the form $u_3(a,x)+u_3(b,y)-l(x,y)$ there is a path from (a,0) to (b,1) of this length, the shortest such path is the infimum given in Lemma 5.35.

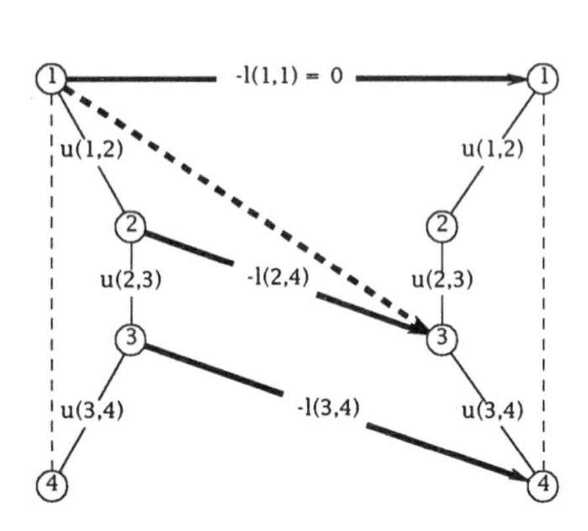

Figure 5.12

Diagram showing the weighted digraph on the set of nodes $A = \{1,2,3,4\}$ whose shortest-path lengths yield both upper and lower triangle inequality limits. The circles in the left-hand half of the digraph represent nodes of the form $(i,0)$ $(i \in A)$, while the circles in the right-hand side represent nodes of the form $(i,1)$ $(i \in A)$. The arrows represent one-way arcs, e.g. $[(1,0),(3,1)]$, while the thin solid lines represent symmetric pairs of arcs, e.g. $[(1,0),(2,0)]$ and $[(2,0),(1,0)]$, both of which are given the indicated weight. The corresponding dashed arrows and lines indicate lower and upper limits whose corresponding paths in the digraph are present in the figure.

We now present a modification of Floyd's well-known shortest paths algorithm, which computes the triangle inequality limits in time of order $O(\#A^3)$, a process we call *triangle inequality bound smoothing.*

Algorithm 5.36 (TRIANGLE):

INPUT: Variables: *NoofAtoms* together with the symmetric matrices *LowerBound* and *UpperBound.*

OUTPUT: The lower and upper triangle inequality limits *LowerLimit, UpperLimit.*

PROCEDURE:

Set *LowerLimit* := *LowerBound*, *UpperLimit* := *UpperBound*;

For k := 1 to *NoofAtoms* do:

For i := 1 to *NoofAtoms*-1 do:

For j := $i+1$ to *NoofAtoms* do:

If $UpperLimit[i,j] > UpperLimit[i,k] + UpperLimit[k,j]$ then

UpperLimit[*i*, *j*] := *UpperLimit*[*i*, *k*] + *UpperLimit*[*k*, *j*].
If *LowerLimit*[*i*, *j*] < *LowerLimit*[*i*, *k*] − *UpperLimit*[*k*, *j*] then
LowerLimit[*i*, *j*] := *LowerLimit*[*i*, *k*] − *UpperLimit*[*k*, *j*];
else
if *LowerLimit*[*i*, *j*] < *LowerLimit*[*j*, *k*] − *UpperLimit*[*k*, *i*] then
LowerLimit[*i*, *j*] := *LowerLimit*[*j*, *k*] − *UpperLimit*[*k*, *i*].
If *LowerLimit*[*i*, *j*] > *UpperLimit*[*i*, *j*] then
print('Erroneous Bounds.') and halt.

PROPOSITION 5.37. *The above procedure correctly computes the triangle inequality limits corresponding to any nonerroneous distance bounds.*

Proof: Given our reduction of the triangle inequality limits to shortest paths in the weighted digraph $\mathbb{G}(l, u)$, the proof is completely analogous to the well-known proof of Floyd's algorithm (see, for example, [Aho1983]). One verifies that, after each iteration of the outermost loop over k, for all $m \leq k$ the path length between each pair $(a_i, 0)$ and $(a_j, 0)$ is at least as short as the length of any path passing through $(a_m, 0)$, the path length between each pair $(a_i, 1)$ and $(a_j, 1)$ is at least as short as the length of any path passing through $(a_m, 1)$, while the path length from each $(a_i, 0)$ to $(a_j, 1)$ is at least as short as the length of any path passing through either $(a_m, 0)$ or $(a_m, 1)$. QED

It should be mentioned that, in the event that the distance bounds are found to be erroneous by reaching the contradiction *LowerLimit*[*i*, *j*] > *UpperLimit*[*i*, *j*] above, the incorrect bound(s) must be either on the current shortest path from $(a_i, 0)$ to $(a_j, 0)$ or else on the shortest path from $(a_i, 0)$ to $(a_j, 1)$. These can be recovered by storing the paths in an integer matrix, as described for example in [Aho1983]. Although $O(\#A^3)$ is close to the best-known complexity for the all-pairs shortest paths problem in general, in problems of chemical interest the matrix *UpperBound* is generally quite sparse, while most of the entries in *LowerBound* are too small to have any effect, so that more efficient algorithms which exploit these facts are possible (see §6.2).

As an interesting application of the above algorithm, we present the following algorithm for generating "quasirandom" metric spaces within any given set of self-consistent bounds on the individual distances.

Algorithm 5.38 (METRIZE):

INPUT: Variables: *NoofAtoms* together with the symmetric matrices *LowerBound* and *UpperBound*.

OUTPUT: A metric space whose distances *Distance* all lie within their respective bounds.

PROCEDURE:

For i := 1 to *NoofAtoms* − 1 do:

```
For j := i+1 to NoofAtoms do:
  TRIANGLE (NoofAtoms, LowerBound, UpperBound, LowerLimit, UpperLimit);
  LowerBound := LowerLimit, UpperBound := UpperLimit;
  Distance[i,j] := RANDOM(LowerBound[i,j], UpperBound[i,j]);
  LowerBound[i,j] := UpperBound[i,j] := Distance[i,j];
```

where RANDOM is a function returning a random number within the interval defined by its two real arguments, and TRIANGLE computes the triangle inequality limits as above.

COROLLARY 5.39. *The METRIZE procedure correctly computes a metric space lying within any self-consistent distance bounds.*

Proof: The correctness of this procedure is a consequence of the following two facts: (i) By Theorem 5.32, for each pair of atoms $a,b \in A$ there exist functions $f_l, f_u \in F_3(A)$ which are admissible with respect to the lower and upper limits l_3 and u_3 returned by TRIANGLE and which satisfy $f_l(a,b) = l_3(a,b)$ and $f_u(a,b) = u_3(a,b)$, respectively. (ii) The subset of $F_3(A)$ consisting of all (l_3,u_3)-admissible functions, being the intersection of the hyperrectanguloid defined by the distance bounds and the polyhedron defined by the triangle inequalities, is a convex set. Thus for any one pair of atoms $a,b \in A$, all values between the corresponding lower and upper triangle inequality limits are attainable for some (l_3,u_3)-admissible $f \in F_3(A)$. QED

Since the complexity of TRIANGLE is $O(\#A^3)$, the complexity of the simple version of METRIZE presented here is $O(\#A^5)$, which is polynomial but not otherwise impressive. By means of suitable updating procedures, however, it is possible to reduce this complexity to $O(\#A^3)$ [Havel1984a] (see also §6.2). We have used the term "quasirandom" because, even if the order in which the atoms are considered is random, the algorithm definitely does not produce a sampling which is uniform with respect to the Lebesque measure on the set of all metric spaces which are consistent with the bounds.

5.4. Tetrangle Inequality Bound Smoothing

Although the triangle inequality limits can be efficiently computed, they are not a very good approximation to the actual Euclidean limits in general. This appears to be particularly true of the lower triangle inequality limits. The next obvious step is to take account of the tetrangle and higher-order inequalities which were introduced and discussed in depth in Chapter 3. To this end, we define

$$F_4(A) \;:=\; \{\, f \in F_3(A) \mid D_f(w,x,y,z) \geq 0 \;\; \forall w,x,y,z \in A \,\}\,, \tag{5.74}$$

where $D_f(w,x,y,z)$ denotes the Cayley-Menger determinant of the four atom metric space $(\{w,x,y,z\}, f^2)$. Since these functions are a special type of metric, we may sometimes call any pair (A, f^2) with $f \in F_4(A)$ a *tetric space*. Clearly this is essentially just a 3-Euclidean space, as defined in Chapter 3.

Definition 5.40. The *tetrangle inequality limits* associated with a given set of lower and upper bounds l, $u \in F(A)$ are given by

$$l_4(a,b) \;:=\; \inf_{f \in F_4(A)} \left(f(a,b) \mid f \text{ is } (l,u)\text{-admissible} \right) \tag{5.75}$$

and

$$u_4(a,b) \;:=\; \sup_{f \in F_4(A)} \left(f(a,b) \mid f \text{ is } (l,u)\text{-admissible} \right). \tag{5.76}$$

Even though they are perfectly well-defined, no general method of computing the tetrangle inequality limits is known. The primary reason for this is that no analogue of Theorem 5.32 is known for the tetrangle inequality limits, and without having such an algebraic system of inequalities which characterizes those bounds which are self-consistent with respect to Equations (5.75) and (5.76) we do not properly know what it is we are looking for. What we are able to do, in analogy with what we do in order to compute the triangle limits, is to solve for the tetrangle limits between one pair in each subset of four atoms and to iteratively go through all such pairs and quadruples until no further changes occur [Havel1983]. Although the limits $l_{(4)}$, $u_{(4)}$ which result from this procedure are certainly valid and improved approximations to the Euclidean limits, in that the relations $l_3 \leq l_{(4)} \leq l_4 \leq l_E$ and $u_3 \geq u_{(4)} \geq u_4 \geq u_E$ hold, as we shall see they are not the same in general. In addition, we have neither a proof that this procedure will always converge in a finite number of iterations, nor that the limits $l_{(4)}$ and $u_{(4)}$ which result are independent of the order in which the quadruples are considered.

In this section, we first show how one can efficiently and reliably compute the tetrangle inequality limits between one pair of a quadruple of atoms, by showing that only a finite number of possible solutions need be considered. This enables us to give a noniterative method of calculating the actual tetrangle limits among any quadruple of atoms, which in turn allows us to easily calculate the iterative approximations $l_{(4)}$, $u_{(4)}$ among all the atoms. We shall also demonstrate a remarkable correspondence between these iterative limits and the set of tensegrity frameworks on four atoms which may be obtained by repeated tensegrity exchanges among a small set of globally rigid planar tensegrity frameworks. Then we shall use this tensegrity interpretation to derive a specific example in which these iterative limits are not equal to the tetrangle inequality limits. Finally, we shall make some conjectures regarding a complete characterization of the tetrangle inequality limits and the computational complexity of computing them, and discuss briefly how the tensegrity interpretation may be used to compute improved approximations to the Euclidean limits.

The lower and upper tetrangle inequality limits on any one of the six distances (henceforth, the (3,4)-distance) among four atoms $\mathrm{B} := \{\mathrm{b}_1, \ldots, \mathrm{b}_4\} \in \binom{\mathrm{A}}{4}$, if not equal to the given bounds $l(\mathrm{b}_3,\mathrm{b}_4)$ or $u(\mathrm{b}_3,\mathrm{b}_4)$, respectively, are by their definition equal to the solutions of the following minimization and maximization problems:

$$\text{minimize/maximize}: \quad f^2(\mathrm{b}_3,\mathrm{b}_4) \tag{5.77}$$

$$\text{subject to}: \quad l^2(\mathrm{b}_i,\mathrm{b}_j) \leq f^2(\mathrm{b}_i,\mathrm{b}_j) \leq u^2(\mathrm{b}_i,\mathrm{b}_j) \quad (\{i,j\} \neq \{3,4\}) \tag{5.78}$$

$$\text{and}: \quad f \in \mathrm{F}_4(\mathrm{B})\,. \tag{5.79}$$

Note that the symbols $f^2(\mathrm{b}_i,\mathrm{b}_j)$ which appear above are variables which depend upon each other implicitly through the constraints (5.78) and (5.79). To emphasize this fact, we shall often abbreviate them by $F_{ij} = f^2(\mathrm{b}_i,\mathrm{b}_j)$. We call any $[F_{12}, \ldots, F_{34}] \in \mathbb{R}^6$, or equivalently $f \in \mathrm{F}(\mathrm{B})$, which satisfies the conditions (5.78) and (5.79) of these optimization problems a *feasible* solution. Before attempting to solve the general problem, it is helpful to first solve the special case thereof which occurs when $l(\mathrm{b}_i,\mathrm{b}_j) = u(\mathrm{b}_i,\mathrm{b}_j) := d_{ij}$ for some fixed constants $d_{ij} = d_{ji}$ and all $\{i,j\} \neq \{3,4\}$. By our results in §3.5, the minimum and maximum values of F_{34} over all feasible solutions to this equality constrained problem are equal to the *cis* and *trans* limits shown in Figure 3.8, and may be found analytically by solving for the roots of the quadratic polynomial

$$D_{34}(\mathrm{b}_1, \ldots, \mathrm{b}_4; X) := \frac{1}{8} \det \begin{bmatrix} 0 & 1 & \cdots & 1 & 1 \\ 1 & 0 & \cdots & d_{13}^2 & d_{14}^2 \\ \cdot & \cdot & \cdots & \cdot & \cdot \\ \cdot & \cdot & \cdots & \cdot & \cdot \\ \cdot & \cdot & \cdots & \cdot & \cdot \\ 1 & d_{31}^2 & \cdots & 0 & X \\ 1 & d_{41}^2 & \cdots & X & 0 \end{bmatrix} \tag{5.80}$$

$$=: A_{34}X^2 + B_{34}X + C_{34}$$

which is obtained by Laplace expansion of the above Cayley-Menger determinant along its last two rows or columns. For future reference, we denote these roots by

$$L(d_{12}, d_{13}, d_{14}, d_{23}, d_{24}) := \frac{-B_{34} + \sqrt{B_{34}^2 - 4A_{34}C_{34}}}{2A_{34}} \tag{5.81}$$

and

$$U(d_{12}, d_{13}, d_{14}, d_{23}, d_{24}) \; := \; \frac{-B_{34} - \sqrt{B_{34}^2 - 4A_{34}C_{34}}}{2A_{34}} \tag{5.82}$$

if they exist, and otherwise we formally define $L := \infty$ and $U := -\infty$. Since $A_{34} = -d_{12}^2/4 < 0$, we have $L \leq U$ as functions on $\mathbb{R}^5$.

In [Dress1987], we showed how the optimization problem given in Equations (5.77) through (5.79) could be solved by utilizing the fact that when $\#B = 4$ the condition $f \in F_4(B)$ is necessary and sufficient for the existence of an isometric positioning $\mathbf{p}: B \rightarrow \mathbb{R}^3$, so that this condition can be eliminated by changing the variables of the optimization from the squared distances $F_{ij} = f^2(b_i, b_j)$ to Cartesian coordinates $\mathbf{p}_i = \mathbf{p}(b_i)$. Here, however, we prefer to give a purely distance-geometric proof which generalizes more easily to the analogous problems obtained by considering pentuples and larger numbers of atoms. Both the Cartesian and distance-geometric proofs rely on the well-known Kuhn-Tucker first-order optimality conditions [Luenberger1973] (see also §9.1):

LEMMA 5.41. *If $\mathbf{x} \in \mathbb{R}^m$ is a local maximum of the constrained optimization problem:*

$$\text{maximize: } g(\mathbf{x}) \; ; \text{ subject to: } h_i(\mathbf{x}) \geq 0 \, , \; i = 1, \ldots, M \; ; \tag{5.83}$$

and the gradients $\nabla h_j(\mathbf{x})$ of the active constraints $h_j(\mathbf{x}) = 0$ are linearly independent, then there exist scalars $\lambda_j \geq 0$, known as Lagrange multipliers, *with $\nabla g(\mathbf{x}) + \Sigma_j \lambda_j \nabla h_j(\mathbf{x}) = \mathbf{0}$.*

Geometrically, this means simply that $\nabla g(\mathbf{x})$ lies in the convex cone generated by the gradients of the active constraints $h_j(\mathbf{x}) = 0$.

PROPOSITION 5.42. *Unless some triple of atoms is collinear in the maximizing f^+ or minimizing f^- solution to Equations (5.78) and (5.79) i.e. $D_{f^\pm}(c_1, c_2, c_3) = 0$ for distinct atoms $c_1, c_2, c_3 \in B$, the optimum values are:*

$$\begin{aligned} F_{34}^+ \;=\; & max(U(l(b_1,b_2), u(b_1,b_3), u(b_1,b_4), u(b_2,b_3), u(b_2,b_4)), \\ & U(u(b_1,b_2), l(b_1,b_3), l(b_1,b_4), u(b_2,b_3), u(b_2,b_4)), \\ & U(u(b_1,b_2), u(b_1,b_3), u(b_1,b_4), l(b_2,b_3), l(b_2,b_4))) \end{aligned} \tag{5.84}$$

for the maximization, and

$$\begin{aligned} F_{34}^- \;=\; & min(L(u(b_1,b_2), u(b_1,b_3), l(b_1,b_4), l(b_2,b_3), u(b_2,b_4)), \\ & L(u(b_1,b_2), l(b_1,b_3), u(b_1,b_4), u(b_2,b_3), l(b_2,b_4)), \\ & L(l(b_1,b_2), l(b_1,b_3), u(b_1,b_4), l(b_2,b_3), u(b_2,b_4)), \\ & L(l(b_1,b_2), u(b_1,b_3), l(b_1,b_4), u(b_2,b_3), l(b_2,b_4))) \end{aligned} \tag{5.85}$$

for the minimization.

Proof: We have shown above that the maximizing and minimizing solutions of the equality constrained problem with $F_{ij} = d_{ij}^2$ for all $\{i,j\} \neq \{3,4\}$ occur when the constraint $D_f(\mathrm{B}) \geq 0$ is active, i.e. $D_f(\mathrm{B}) = 0$. Whatever the maximizing or minimizing solution $f^{\pm} \in \mathrm{F}_4(\mathrm{B})$ of the inequality constrained problem may be, the equality constrained problem which is obtained by setting $d_{ij}^2 := F_{ij}^{\pm}$ will have exactly the same maximizing or minimizing solution $f^{\pm}$. It follows that in any maximizing or minimizing solution $f^{\pm}$ to Equations (5.78) and (5.79) we likewise have $D_{f^{\pm}}(\mathrm{B}) = 0$. Furthermore, by our assumption that no triples of atoms $\mathrm{C} \in \binom{\mathrm{B}}{3}$ are collinear in the optimum solutions, the constraints $D_f(\mathrm{C}) \geq 0$ will never be active. The only remaining constraints are the inequalities $l^2(\mathrm{b}_i,\mathrm{b}_j) \leq F_{ij}$ and $F_{ij} \leq u^2(\mathrm{b}_i,\mathrm{b}_j)$, which by definition are active if and only if $l^2(\mathrm{b}_i,\mathrm{b}_j) = F_{ij}$ and $F_{ij} = u^2(\mathrm{b}_i,\mathrm{b}_j)$, respectively.

The Kuhn-Tucker conditions, of course, can be applied to minimization problems as well as upper bound constraints by replacing the objective or constraint function by its negative. In the case of Equations (5.77) through (5.79), the gradients of the objective and constraint functions $\pm F_{ij}$ are simply the appropriately signed unit vectors $\{\pm\mathbf{u}_{ij} \mid 1 \leq i < j \leq 4\}$ in our six-dimensional space of squared distances. The gradient of the function in the active constraint $D_f(\mathrm{B}) = 0$, on the other hand, is

$$\nabla D_f(\mathrm{B}) \quad = \quad \begin{bmatrix} \dfrac{\partial D_f(\mathrm{B})}{\partial F_{12}} \\ \cdot \\ \cdot \\ \cdot \\ \dfrac{\partial D_f(\mathrm{B})}{\partial F_{34}} \end{bmatrix} \quad = \quad \begin{bmatrix} D_f(\mathrm{b}_1,\mathrm{b}_3,\mathrm{b}_4;\ \mathrm{b}_2,\mathrm{b}_3,\mathrm{b}_4) \\ -D_f(\mathrm{b}_1,\mathrm{b}_2,\mathrm{b}_4;\ \mathrm{b}_2,\mathrm{b}_3,\mathrm{b}_4) \\ D_f(\mathrm{b}_1,\mathrm{b}_2,\mathrm{b}_3;\ \mathrm{b}_2,\mathrm{b}_3,\mathrm{b}_4) \\ D_f(\mathrm{b}_1,\mathrm{b}_2,\mathrm{b}_4;\ \mathrm{b}_1,\mathrm{b}_3,\mathrm{b}_4) \\ -D_f(\mathrm{b}_1,\mathrm{b}_2,\mathrm{b}_3;\ \mathrm{b}_1,\mathrm{b}_3,\mathrm{b}_4) \\ D_f(\mathrm{b}_1,\mathrm{b}_2,\mathrm{b}_3;\ \mathrm{b}_1,\mathrm{b}_2,\mathrm{b}_4) \end{bmatrix} \tag{5.86}$$

(cf. proof of Proposition 3.52), where we have ordered the components of the gradient lexicographically by index pair and $D_f(\cdots;\ \cdots)$ denotes the Cayley-Menger bideterminant of two triples of atoms in the premetric space $(\mathrm{B},\ f^2)$, as defined in §3.6. Since the space is 3-Euclidean and $D_f(\mathrm{B}) = 0$, it follows from the results of that section that these bideterminants are equal to $(2!)^2$ times the products of the oriented areas of the triangles spanned by the triples in the associated (two-dimensional Euclidean) amalgamation space.

The gradients of the constraints are clearly linearly independent unless $D_{f^{\pm}}(\mathrm{b}_1,\mathrm{b}_2,\mathrm{b}_3;\ \mathrm{b}_1,\mathrm{b}_2,\mathrm{b}_4) = 0$, in which case at least one of the two triples $\{\mathrm{b}_1,\mathrm{b}_2,\mathrm{b}_3\}$ or $\{\mathrm{b}_1,\mathrm{b}_2,\mathrm{b}_4\}$ is collinear contrary to assumption. Thus, the Kuhn-Tucker conditions apply. If we now fix the multiplier of the gradient of the constraint function $D_f(\mathrm{B})$ at -1 by dividing through by the negative of its Lagrange multiplier, the Kuhn-Tucker optimality conditions imply that

$$\nabla D_f(\mathrm{B}) \;=\; \sum_{i<j} \alpha_{ij} \mathbf{u}_{ij} \,, \tag{5.87}$$

where the multipliers α_{ij} satisfy:

$$\alpha_{ij} \begin{cases} \geq 0 & \text{if } F_{ij} = u^2(\mathrm{b}_i,\mathrm{b}_j)\,; \\ \leq 0 & \text{if } F_{ij} = l^2(\mathrm{b}_i,\mathrm{b}_j)\,; \\ = 0 & \text{otherwise}\,; \end{cases} \tag{5.88}$$

for all $\{i,j\} \neq \{3,4\}$, and

$$\alpha_{34} \begin{cases} > 0 & \text{if we are minimizing } F_{34}\,; \\ < 0 & \text{if we are maximizing } F_{34}\,. \end{cases} \tag{5.89}$$

Since in any dependence among these seven vectors in $\mathbb{R}^6$ we obviously must have $\alpha_{ij} \neq 0$ for all $1 \leq i < j \leq 4$, this shows us at once that in any constrained optimum all five of the squared distances $\{F_{ij} \,|\, \{i,j\} \neq \{3,4\}\}$ are equal to either their lower or to their upper bounds. Furthermore, $F_{ij} = l^2(\mathrm{b}_i,\mathrm{b}_j)$ if and only if the ij-th term of $\nabla D_f(\mathrm{B})$ is negative, while $F_{ij} = u^2(\mathrm{b}_i,\mathrm{b}_j)$ if and only if the ij-th term of $\nabla D_f(\mathrm{B})$ is positive. Since Cayley-Menger bideterminants are directly proportional to the products of the oriented areas spanned by the triples, however, their signs constitute the orientations of an affine simplicial chirotope (see §2.3), and up to relabelling there are only two rank three affine simplicial chirotopes on four atoms: one in which one atom is in the convex span of the other three, and one in which no atom is in the convex span of the others. Demonstrating that these combinations of signs give the possible combinations of squared distances at their lower and upper bounds listed in Equations (5.84) and (5.85) for the maximization and minimization problem, respectively, is left to the reader as an exercise. QED

In Figure 5.13, we provide a pictorial presentation of the combinations of bounds which occur in Equations (5.84) and (5.85), which most readers will probably find easier to digest.

The really interesting thing about this proposition is that the multipliers α_{ij} in the dependence between the gradients of the objective function and the constraints are, by Proposition 5.29 and our geometric interpretation of Cayley-Menger bideterminants as the products of oriented volumes, equal to a strict self-stress of a realization of the complete graph $\mathbb{K}_4$ in the plane which gives rise to a positive semidefinite stress matrix. Thus, given an optimum solution $f^{\pm}$ to Equations (5.77) through (5.79), if we take those pairs of atoms whose distances are at their upper bounds to be cables, those whose distances are at their lower bounds to be struts, and $\{\mathrm{b}_3,\mathrm{b}_4\}$ to be a cable if we are minimizing this (squared) distance and to be a strut if we are maximizing it, we obtain a *globally* rigid

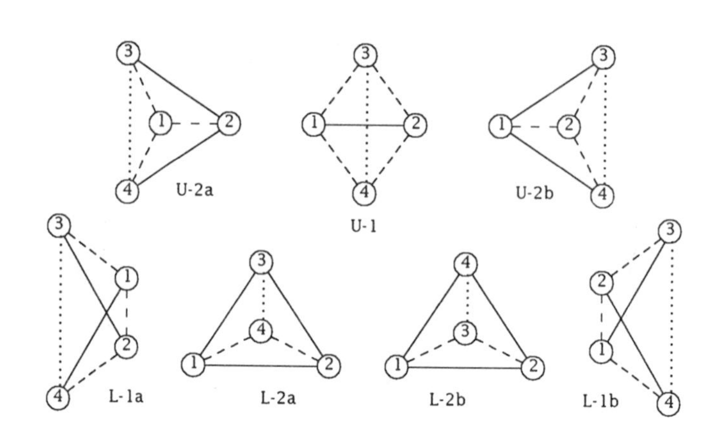

Figure 5.13

The combinations of bounds which can become active in an upper (above) or lower (below) tetrangle inequality limit on the (3,4)-distance among a set of four points, given that no three points are collinear, as in Proposition 5.42. As usual, thin solid lines indicate distances at their lower bounds, dashed lines indicate distances at their upper bounds and the dotted lines indicated the distances whose tetrangle inequality limits are being computed.

tensegrity framework. The combinations of active constraints depicted in Figure 5.13 are essentially the seven possible noncongruent, labelled, globally rigid and fully two-dimensional tensegrity frameworks. We shall soon see that this interpretation can be turned around, allowing us to obtain limits on distances from a given set of globally rigid tensegrity frameworks.

We now consider the "degenerate" solutions of Equations (5.78) and (5.79) which occur when one or more triples of atoms become collinear. We are particularly interested in those limits $l_4(\mathrm{b}_3,\mathrm{b}_4)$ and $u_4(\mathrm{b}_3,\mathrm{b}_4)$ which are not equal to the triangle inequality limits $l_3(\mathrm{b}_3,\mathrm{b}_4)$ and $u_3(\mathrm{b}_3,\mathrm{b}_4)$. Given that the condition $f \in \mathrm{F}_4(\mathrm{B})$ implies the existence of an isometric positioning $\mathbf{p}: \mathrm{B} \to \mathbb{R}^3$, the following lemma is geometrically obvious. Its proof, however, requires a tedious case by case analysis which must be left to the few readers who would be interested.

LEMMA 5.43. *Let* $\mathrm{B} = \{\mathrm{b}_1, \ldots, \mathrm{b}_4\}$ *be a quadruple of atoms,* $l, u \in \mathrm{F}(\mathrm{B})$ *be bounds on the distances thereof and* $f \in \mathrm{F}(\mathrm{B})$ *be a feasible solution of Equations (5.78) and (5.79) with* $l^2(\mathrm{b}_3,\mathrm{b}_4) \le F_{34} \le u^2(\mathrm{b}_3,\mathrm{b}_4)$.

(i) If $D_f(\mathrm{c}_1,\mathrm{c}_2,\mathrm{c}_3) = 0$ *for all triples of atoms* $\mathrm{c}_1,\mathrm{c}_2,\mathrm{c}_3 \in \mathrm{B}$, *then* f *is both a maximizing and a minimizing solution of Equations (5.78) and (5.79) if*

$$u(b_1,b_3)+u(b_2,b_3) \;=\; l(b_1,b_2) \;=\; u(b_1,b_4)+u(b_2,b_4) \tag{5.90}$$

or

$$u(b_1,b_2)+u(b_1,b_3) \;=\; l(b_2,b_3) \quad \text{and} \quad u(b_1,b_2)+u(b_2,b_4) \;=\; l(b_1,b_4) \tag{5.91}$$

or

$$u(b_1,b_2)+u(b_2,b_3) \;=\; l(b_1,b_3) \quad \text{and} \quad u(b_1,b_2)+u(b_1,b_4) \;=\; l(b_2,b_4)\,. \tag{5.92}$$

Otherwise, it is a maximizing solution if and only if $f(b_3,b_4) = u_3(b_3,b_4)$, *and a minimizing solution if and only if* $f(b_3,b_4) = l_3(b_3,b_4)$.

(ii) If all triples are not collinear, then at most one triple is. In the case that $\{b_1,b_2,b_3\}$ *is the only collinear triple, then either*

$$l(b_1,b_2) \;=\; u(b_1,b_3)+u(b_2,b_3) \quad or \quad l(b_1,b_3) \;=\; u(b_1,b_2)+u(b_2,b_3) \tag{5.93}$$
$$or \quad l(b_2,b_3) \;=\; u(b_1,b_2)+u(b_1,b_3)\,,$$

or else f *is neither a maximizing nor a minimizing solution of Equations (5.78) and (5.79). A similar statement holds with* b_4 *replacing* b_3 *if* $\{b_1,b_2,b_4\}$ *is the only collinear triple.*

(iii) If the case that either $\{b_1,b_3,b_4\}$ *or* $\{b_2,b_3,b_4\}$ *is collinear,* f *is a maximizing solution to Equations (5.78) and (5.79) if and only if* $f(a_3,a_4) = u_3(b_3,b_4)$, *and* f *is a minimizing solution to Equations (5.78) and (5.79) if and only if* $f(a_3,a_4) = l_3(b_3,b_4)$.

Part (i) of the lemma says that optimum solutions in which all triples are collinear and which are not triangle inequality limits can occur only if all triples are collinear in all feasible solutions. Part (ii) says that if exactly one of the two triples involving both atoms b_1 and b_2 are collinear in an optimum solution to Equations (5.77) through (5.79), then that triple must be collinear in all feasible solutions. Part (iii) of the lemma says that the only optima which can occur when exactly one of the two triples involving both atoms b_3 and b_4 is collinear are triangle inequality limits. These cases are illustrated in Figure 5.14.

The degenerate tetrangle limits in parts (i) and (ii) of the lemma turn out to be covered by Equations (5.84) and (5.85). Before we can apply these equations, however, we must decide if any triangle inequality limits on the (3,4)-distance can be attained by some solution f of Equations (5.78) and (5.79). This requires yet one more:

LEMMA 5.44. *Given a nondegenerate triangle* $[(\mathbf{p},\mathbf{q},\mathbf{s})]$ *in the plane, the squared distance from its apex* $\mathbf{s}$ *to a point* $\mathbf{r}$ *on its base* $[(\mathbf{p},\mathbf{q})]$ *is given by:*

$$d_{rs}^2 \;=\; \frac{d_{pr}(d_{qs}^2-d_{qr}^2) + d_{qr}(d_{ps}^2-d_{pr}^2)}{d_{pr}+d_{qr}}\,, \tag{5.94}$$

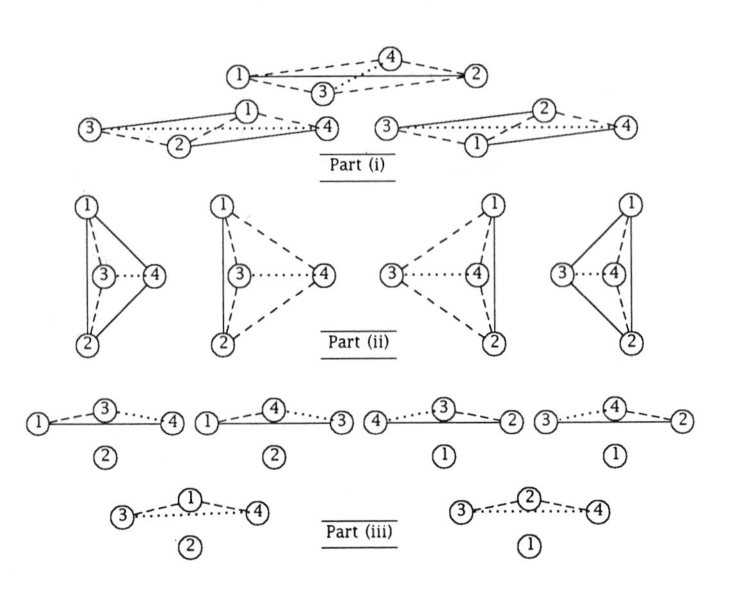

Figure 5.14

The degenerate tetrangle inequality limits discussed in Lemma 5.43. In part (i), the lower and upper tetrangle inequality limits are equal; in part (ii), the upper are shown in the center and the lower on the outside; in part (iii), the only limits which need be considered are triangle limits, as shown.

where $d_{rs} := \|\mathbf{r} - \mathbf{s}\|$, *etc.*

Proof: This is actually a special case of Equations (5.81) and (5.82) which occurs when the discriminant of the quadratic vanishes, so that the roots L and U are equal. However, it is both easier and more informative to derive it directly. By the law of cosines for the angle θ between the sides of the triangle at the vertex $\mathbf{p}$:

$$\frac{d_{ps}^2 + d_{pr}^2 - d_{rs}^2}{d_{ps}\, d_{pr}} \quad = \quad 2\cos(\theta) \quad = \quad \frac{d_{ps}^2 + d_{pq}^2 - d_{qs}^2}{d_{ps}\, d_{pq}} \,. \tag{5.95}$$

The result follows from this together with the relation $d_{pr} + d_{qr} = d_{pq}$. QED

Let us denote the function which returns the apex to base squared distance d_{rs}^2 from the remaining distances by $T(d_{pr}, d_{qr}, d_{ps}, d_{qs})$. Note that this function will fail if the distances given to it violate the triangle inequality, i.e.

$d_{pr}+d_{qr} > d_{ps}+d_{qs}$ or $d_{pr}+d_{qr} < |d_{ps}-d_{qs}|$.

COROLLARY 5.45. *Given three distinct collinear points* $\mathbf{p},\mathbf{q},\mathbf{r}$ *in the plane whose mutual distances satisfy* $d_{pr}+d_{qr} = d_{pq}$, *together with six real numbers* $0 \leq l_{xs} \leq u_{xs}$ $(x \in \{p,q,r\})$, *there exists a point* $\mathbf{s}$ *whose distances* d_{xs} *to* $\mathbf{p}$, $\mathbf{q}$ *and* $\mathbf{r}$ *satisfy*

$$l_{ps} \leq d_{ps} \leq u_{ps}, \quad l_{qs} \leq d_{qs} \leq u_{qs} \quad \text{and} \quad l_{rs} \leq d_{rs} \leq u_{rs} \tag{5.96}$$

if and only if $d_{pr}+d_{qr} \leq u_{ps}+u_{qs}$ *as well as*

$$u_{rs} \geq L_{rs} \quad \text{and} \quad l_{rs} \leq U_{rs} \tag{5.97}$$

where

$$U_{rs} := \begin{cases} (d_{qr}+u_{qs})^2 & \text{if } d_{pr}+d_{qr} < u_{ps}-u_{qs} \\ (d_{pr}+u_{ps})^2 & \text{if } d_{pr}+d_{qr} < u_{qs}-u_{ps} \\ T(d_{pr}, d_{qr}, u_{ps}, u_{qs}) & \text{otherwise}, \end{cases} \tag{5.98}$$

and

$$L_{rs} := \begin{cases} max^2\begin{pmatrix} 0, \; d_{qr}-u_{qs}, \; d_{pr}-u_{ps}, \\ l_{qs}-d_{qr}, \; l_{ps}-d_{pr} \end{pmatrix} & \text{if } d_{pr}+d_{qr} > l_{ps}+l_{qs} \\ (l_{ps}-d_{pr})^2 & \text{if } d_{pr}+d_{qr} < l_{ps}-l_{qs} \\ (l_{qs}-d_{qr})^2 & \text{if } d_{pr}+d_{qr} < l_{qs}-l_{ps} \\ T(d_{pr}, d_{qr}, l_{ps}, l_{qs}) & \text{otherwise}. \end{cases} \tag{5.99}$$

Proof: Since

$$\partial/\partial d_{ps} \; T(d_{pr}, d_{qr}, d_{ps}, d_{qs}) = 2\,d_{ps}d_{qr} > 0 \tag{5.100}$$

and

$$\partial/\partial d_{qs} \; T(d_{pr}, d_{qr}, d_{ps}, d_{qs}) = 2\,d_{qs}d_{pr} > 0, \tag{5.101}$$

the function T is monotone increasing in these arguments and hence the maximum and minimum values which d_{rs}^2 can have for fixed values of d_{pr} and d_{qr} may be calculated simply by putting in the maximum and minimum possible values of d_{ps} and d_{qs} – provided that these combinations of bounds obey the triangle inequality. Otherwise, depending on the triangle inequality violation which occurs, the triangle inequality limits enumerated above must be used. The point $\mathbf{s}$ exists if and only if these limits are compatible with the given bounds. QED

We shall let COLLINEAR($d_{pr}, d_{qr}, l_{ps}, l_{qs}, l_{rs}, u_{ps}, u_{qs}, u_{rs}$) be the function which returns the boolean value "*true*" if the conditions of the corollary are satisfied. This can then be used to check if any triangle inequality limits on the

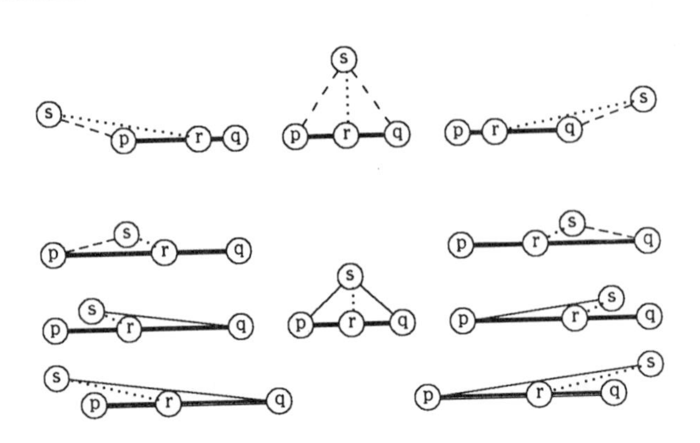

Figure 5.15

Cases to check when computing COLLINEAR. The upper limit U_{rs} on the rs-distance squared is obtained from one of the three cases shown in the topmost row of the figure; the lower limit L_{rs} is obtained from one of the seven cases shown below. The quadruple can be made collinear if and only if the interval defined by these squared limits overlaps with the interval defined by the given squared bounds on the rs-distance.

(3,4) distance are attainable, as follows:

Algorithm 5.46 (TRI_CHECK):
INPUT: A quadruple of atoms $B = \{b_1, \ldots, b_4\}$ together with lower and upper bounds $l, u \in F(B)$ on their mutual distances.
OUTPUT: The triangle inequality limits on the (3,4)-distance, together with two boolean flags, *l_col* and *u_col* which are *true* whenever the triangle inequality limits on the (3,4)-distance are attainable without violating the given bounds or the tetrangle inequality.
PROCEDURE:

Set *l_col* := *false* and *u_col* := *false*.
Let $u_3(b_3,b_4)$ be the minimum of the limits:
$U[3,1,4] := u(b_1,b_3)+u(b_1,b_4)$;
$U[3,2,4] := u(b_2,b_3)+u(b_2,b_4)$;
$U[3,1,2,4] := u(b_1,b_3)+u(b_1,b_2)+u(b_2,b_4)$;
$U[4,1,2,3] := u(b_1,b_4)+u(b_1,b_2)+u(b_2,b_3)$.
Let $l_3(b_3,b_4)$ be the maximum of the limits:
$L[0] := 0$;
$L[1,3,4] := l(b_1,b_4)-u(b_1,b_3)$;
$L[2,3,4] := l(b_2,b_4)-u(b_2,b_3)$;

$L[1,4,3] := l(b_1,b_3) - u(b_1,b_4)$;
$L[2,4,3] := l(b_2,b_3) - u(b_2,b_4)$;
$L[1,3,4,2] := l(b_1,b_2) - u(b_1,b_3) - u(b_2,b_4)$;
$L[1,4,3,2] := l(b_1,b_2) - u(b_2,b_3) - u(b_1,b_4)$;
$L[1,2,3,4] := l(b_1,b_4) - u(b_1,b_2) - u(b_2,b_3)$;
$L[2,1,3,4] := l(b_2,b_4) - u(b_1,b_2) - u(b_1,b_3)$.
$L[1,2,4,3] := l(b_1,b_3) - u(b_1,b_2) - u(b_2,b_4)$;
$L[2,1,4,3] := l(b_2,b_3) - u(b_1,b_2) - u(b_1,b_4)$;
For each 3-atom upper limit with $U[3,i,4] = u_3(b_3,b_4)$:
  Let $j \in \{1, \ldots ,4\} \setminus \{i,3,4\}$.
  If COLLINEAR($u(b_i,b_3), u(b_i,b_4), l(b_j,b_3), l(b_j,b_4), l(b_i,b_j), u(b_j,b_3), u(b_j,b_4), u(b_i,b_j)$):
    Set *u_col* := *true*.
For each 4-atom upper limit with $U[k,i,j,l] = u_3(b_3,b_4)$:
  If $u(b_j,b_k) \geq u(b_i,b_k) + u(b_i,b_j) \geq l(b_j,b_k)$
  and $u(b_i,b_l) \geq u(b_j,b_l) + u(b_i,b_j) \geq l(b_i,b_l)$:
    Set *u_col* := *true*.
If $l_3(b_3,b_4) = L[0] = 0$:
  Choose $i, j \in \{1, \ldots ,4\} \setminus \{3,4\}$ with $i \neq j$.
  Set $l'(c_1,c_2) := l(b_1,b_2)$, $u'(c_1,c_2) := u(b_1,b_2)$,
    $l'(c_1,c_3) = max(l(b_1,b_3), l(b_1,b_4))$, $u'(c_1,c_3) = min(u(b_1,b_3), u(b_1,b_4))$,
    $l'(c_2,c_3) = max(l(b_2,b_3), l(b_2,b_4))$, $u'(c_2,c_3) = min(u(b_2,b_3), u(b_2,b_4))$.
  If TRIANGLE($\{c_1,c_2,c_3\}$, l', u'): *l_col* := *true*.
For each 3-atom lower limit with $L[i,k,l] = l_3(b_3,b_4)$:
  Let $j \in \{1, \ldots ,4\} \setminus \{i,3,4\}$.
  If COLLINEAR($u(b_i,b_k), l_3(b_k,b_l), l(b_i,b_j), l(b_j,b_l), l(b_j,b_k), u(b_i,b_j), u(b_j,b_l), u(b_j,b_k)$):
    Set *l_col* := *true*.
For each 4-atom lower limit with $L[i,k,l,j] = l_3(b_3,b_4)$ and $\{k,l\} = \{3,4\}$:
  If $u(b_i,b_k) \geq l(b_i,b_j) - u(b_i,b_k) \geq l(b_j,b_k)$
  and $u(b_i,b_l) \geq l(b_i,b_j) - u(b_j,b_l) \geq l(b_i,b_l)$:
    Set *l_col* := *true*.
For each 4-atom lower limit with $L[i,j,k,l] = l_3(b_3,b_4)$ and $\{k,l\} = \{3,4\}$:
  If $u(b_j,b_l) \geq l(b_i,b_l) - u(b_i,b_j) \geq l(b_j,b_l)$
  and $u(b_i,b_k) \geq u(b_i,b_j) + u(b_j,b_k) \geq l(b_i,b_k)$:
    Set *l_col* := *true*.

Four atom triangle inequality limits are checked simply by placing the atoms down on a line with the distances equal to the corresponding active bounds for that limit, while three atom collinearities may be checked with the COLLINEAR function above. In addition, one must check to see if the atoms b_3 and b_4 can be superimposed, i.e. if their triangle inequality lower limit is zero, by means of a function "TRIANGLE" which returns *false* whenever the three atoms and the bounds among them passed to it contain triangle inequality violations. In the interest of space, the proof of the correctness of the above is omitted.

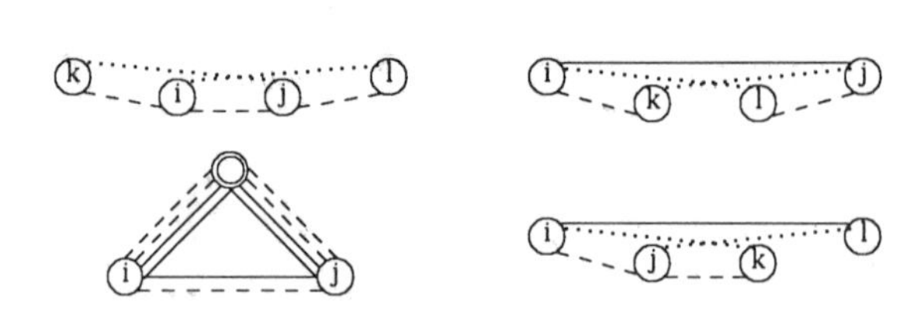

Figure 5.16

The four point collinearities which TRI_CHECK must examine to decide if any are attainable. The four point upper collinearities (upper left) and lower collinearities (right) can occur when the dotted distances are within their bounds. The points indexed k and l can be superimposed whenever the lengths of the sides of the triangle (lower left) are all within the bounds indicated.

It is now a relatively simple matter to present our algorithm for computing the tetrangle inequality limits among each quadruple of atoms.

Algorithm 5.47 (T4_LIMITS):

INPUT: A quadruple of atoms $B = \{b_1, \ldots, b_4\}$, together with lower and upper bounds $l, u \in F(B)$ on their mutual distances.

OUTPUT: The tetrangle inequality limits $l_4(b_3,b_4)$ and $u_4(b_3,b_4)$ implied by the bounds l, u.

PROCEDURE:

Set $ll := 0$ and $ul := \infty$.

TRI_CHECK(l_col, u_col, $l_3(b_3,b_4)$, $u_3(b_3,b_4)$);

If not u_col then:

Set

$$ul := max(U(l(b_1,b_2), u(b_1,b_3), u(b_1,b_4), u(b_2,b_3), u(b_2,b_4)), \quad U(u(b_1,b_2), l(b_1,b_3), l(b_1,b_4), u(b_2,b_3), u(b_2,b_4)), \quad U(u(b_1,b_2), u(b_1,b_3), u(b_1,b_4), l(b_2,b_3), l(b_2,b_4))) \; ;$$

If not l_col then:

Set

$$ll := min(L(u(b_1,b_2), u(b_1,b_3), l(b_1,b_4), l(b_2,b_3), u(b_2,b_4)), \quad L(u(b_1,b_2), l(b_1,b_3), u(b_1,b_4), u(b_2,b_3), l(b_2,b_4)),$$

$$L(l(b_1,b_2),\ l(b_1,b_3),\ u(b_1,b_4),\ l(b_2,b_3),\ u(b_2,b_4)),$$
$$L(l(b_1,b_2),\ u(b_1,b_3),\ l(b_1,b_4),\ u(b_2,b_3),\ l(b_2,b_4)))\ ;$$

Set

$$u_4(b_3,b_4)\ :=\ min(\sqrt{ul},\ u_3(b_3,b_4))\ .$$

and

$$l_4(b_3,b_4)\ :=\ max(\sqrt{ll},\ l_3(b_3,b_4))\ .$$

Finally, for the sake of completeness we sketch an algorithm by which the iterative approximations $l_{(4)}$, $u_{(4)}$ to the tetrangle inequality limits themselves may be computed.

Algorithm 5.48 (TETRANGULAR):

INPUT: A pair of symmetric matrices *LowerBound* and *UpperBound*, representing lower and upper bounds on the distances among an indexed set of atoms $\underline{A}$ of size *NoofAtoms*.

OUTPUT: A pair of symmetric matrices *LowerLimit* and *UpperLimit* with the property that their restriction to the indices of any quadruple of atoms are the tetrangle inequality limits among those atoms.

PROCEDURE:

```
Set LowerLimit := LowerBound, UpperLimit := UpperBound.
Repeat:
  For i1 := 1 to NoofAtoms-1 do:
  For i2 := i1+1 to NoofAtoms do:
  For i3 := 1 to NoofAtoms-1 do:
  For i4 := i3+1 to NoofAtoms do:
    Replace LowerLimit[i3, i4] and UpperLimit[i3, i4]
     by their tetrangle inequality limits,
     as computed by the T4_LIMITS algorithm above.
    If LowerLimit[i3, i4] > UpperLimit[i3, i4] then
      print('Erroneous bounds.') and halt.
Until no changes in LowerLimit or UpperLimit have been made
 on the last cycle through.
```

As previously mentioned, we currently have no proof that this algorithm converges after a finite number of iterations, or that the limits it computes are independent of the ordering of the atoms $\underline{A}$.* It should also be mentioned that, just as with the TRIANGLE routine described in the last section, it is possible to

* Infinite cycles have been encountered in the execution of the above algorithm, but they seem to be rather rare in practice. In no case have we found a dependence upon the order in which the quadruples are considered.

record the history of the bound smoothing process so that if the above routine halts because of errors in the distance bounds it is possible to execute traceback which finds the offending bounds. An account of how this is done may be found in [Havel1982, Easthope1988].

Just as for the tetrangle inequality limits among four atoms, it is possible to interpret many of the iterative limits $l_{(4)}$, $u_{(4)}$ in terms of tensegrity frameworks with a positive semidefinite stress matrix.

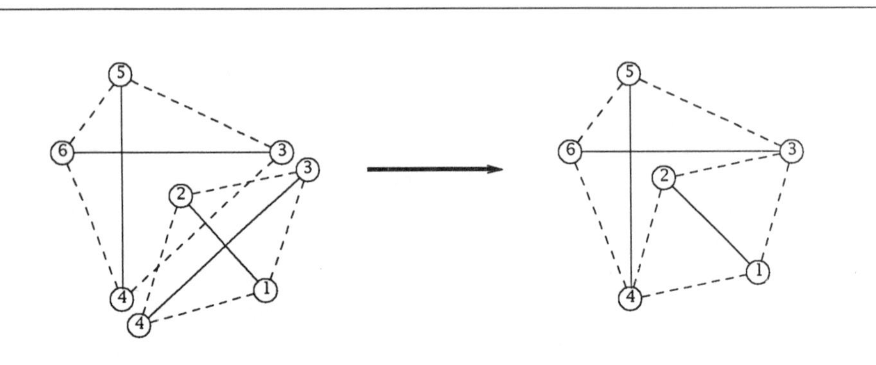

Figure 5.17

Figure illustrating one way in which the basic tensegrity frameworks of the tetrangle inequality limits can be combined to yield a new limit on the distances among the atoms (see text for a detailed explanation).

To see how this is done, suppose that at some point in the execution of the iterative algorithm we lower the current (3,4) upper limit to the tetrangle inequality limit which corresponds to case U-1 in Figure 5.13, setting

$$UpperLimit[3,4] \;:=\; U^{½}(l(a_1,a_2),\, u(a_1,a_3),\, u(a_1,a_4),\, u(a_2,a_3),\, u(a_2,a_4)) \;. \qquad (5.102)$$

Then suppose that at a later point in the execution of the algorithm we use this improved limit *UpperLimit*[3,4], together with the bounds among the quadruple $\{a_3,a_4,a_5,a_6\}$, to lower the (4,5) upper limit in an analogous way, setting

$$UpperLimit[4,5] \;:=\; U^{½}(l(a_3,a_6),\, UpperLimit[3,4],\, u(a_3,a_5),\, u(a_4,a_6),\, u(a_5,a_6)) \;. \qquad (5.103)$$

This results in a five-atom tetrangle limit (see Figure 5.17). Consider the abstract tensegrity framework obtained from this limit by regarding the couples whose distances are at their upper bounds to be cables, those whose distances are at their lower bounds to be struts, and the couple $\{a_4,a_5\}$ to be a strut which pushes this distance to its upper limit, i.e.

$$\mathbb{F} \;:=\; (\{a_1,\,.\,.\,.\,,a_6\},\, L,\, l,\, U,\, u) \qquad (5.104)$$

with

$$L := \{\{a_1,a_2\},\{a_3,a_6\},\{a_4,a_5\}\} \tag{5.105}$$

and

$$U := \{\{a_1,a_3\},\{a_1,a_4\},\{a_2,a_3\},\{a_2,a_4\},\{a_3,a_5\},\{a_4,a_6\},\{a_5,a_6\}\} . \tag{5.106}$$

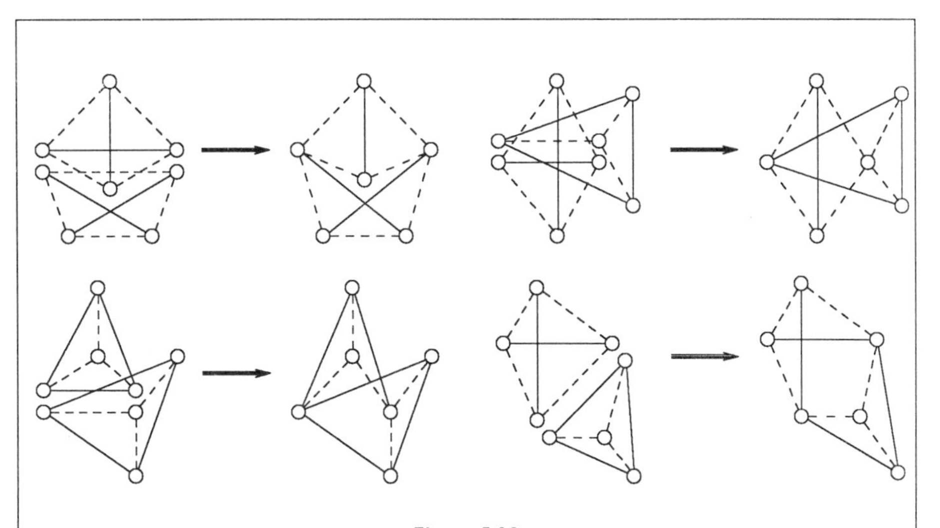

Figure 5.18

Figure illustrating some other ways in which the basic tensegrity frameworks of the tetrangle inequality limits can be combined.

Observe that this framework is obtained by performing a tensegrity exchange (see §5.1) with respect to the couple $\{a_3,a_4\}$ on the (globally rigid) tensegrity frameworks which correspond to the U-1 limits on the (3,4) and (4,5) distances, as depicted in Figure 5.17. It turns out that most of the combinations of distance bounds which are active in any upper or lower iterative limit may be obtained by repeated tensegrity exchanges, starting from the three types of globally rigid tensegrity frameworks which exist on four atoms or less.† This is further illustrated and clarified by the additional examples shown in Figure 5.18 and 5.19. Although the frameworks which are obtained in this way are not globally rigid in general, as noted in §5.2 they do have a positive semidefinite stress matrix, so

† The exception is the case in which one four-atom tetrangle inequality limit is used to "tighten" another in a quadruple which contains three atoms in common with the first. This can actually happen, but it remains to be seen how the tensegrity interpretation can be extended to this case.

that their motions (in the case of 𝔽 above, rotation about the (3,4)-axis) are always affine motions which do not change the (3,4)-distance.

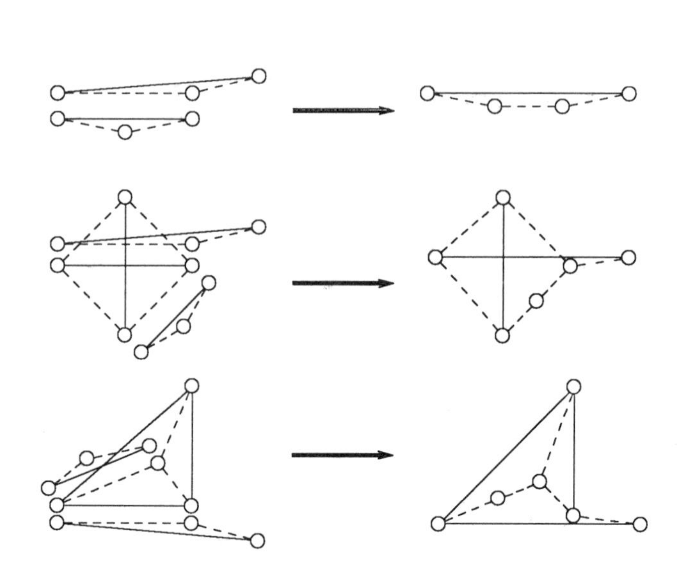

Figure 5.19

Figure illustrating how the basic tensegrity framework of the triangle inequality limits can be combined with those of the triangle (top) and tetrangle (center and bottom) to yield new limits on the distances among the atoms.

In the case of the triangle inequality limits, all one-dimensional globally rigid tensegrity frameworks and hence all combinations of active distance bounds could be obtained by repeated tensegrity exchanges on the single type of globally rigid one-dimensional tensegrity framework on three atoms, which is in fact what made the efficient, noniterative algorithm TRIANGLE possible.

We now prove that the iterative limits $l_{(4)}$ and $u_{(4)}$ computed by the above algorithm are not in general equal to the tetrangle inequality limits l_4 and u_4‡. We do this by showing that a globally rigid tensegrity framework on five atoms $B = \{b_1, \ldots, b_5\}$ which *cannot* be obtained by tensegrity exchange from any smaller globally rigid tensegrity frameworks gives rise to the combination of active distance bounds in a five-atom tetrangle inequality lower limit on a pair of

‡ Thereby showing that an earlier attempt on the part of the authors to prove that they are equal to the tetrangle limits is in fact incorrect (see Algorithm 2.5 of [Havel1983]).

atoms $\{b_4,b_5\}$ which are connected by a cable. This framework is called the *Cauchy pentagon* (see Figure 5.20), and its global rigidity is a result due to [Connelly1982].

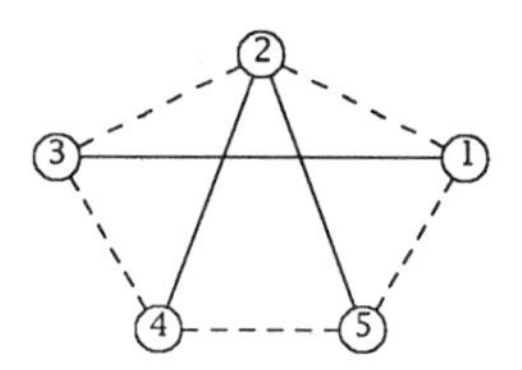

Figure 5.20

The cables and struts of the globally rigid basic tensegrity framework known as the Cauchy pentagon (excluding the cable between joints 4 & 5) give the active upper and lower bounds of a tetrangle inequality lower limit on the (4,5)-distance.

PROPOSITION 5.49. *The lower and upper bounds corresponding to the cables and struts of the Cauchy pentagon minus the couple* $\{b_4,b_5\}$ *(see Figure 5.20) are the active distance bounds in a tetrangle inequality lower limit on the* (4,5)*-distance which is not found by the iterative algorithm.*

Proof: Showing that this limit is not found by the iterative algorithm is left as an exercise. To prove that the (4,5)-distance in the Cauchy pentagon is in fact a tetrangle inequality lower limit for this combination of bounds, we first show that the Kuhn-Tucker conditions are satisfied for the problem of minimizing F_{45} subject to $f \in F_4(B)$ and F_{ij} within the bounds implied by the cables and struts of this tensegrity framework with $\{i,j\} \neq \{4,5\}$. The first thing to notice is that, in such a planar structure, all five of the tetrangle inequality constraints are in fact active. Nevertheless, in order to satisfy the Kuhn-Tucker conditions we shall only need two of them: $D_f(b_1,b_2,b_3,b_4) \geq 0$ and $D_f(b_1,b_2,b_4,b_5) \geq 0$. The gradients of these Cayley-Menger determinants are

$$\nabla D_f(b_1,b_2,b_3,b_4) = \begin{bmatrix} D_f(b_1,b_3,b_4;\ b_2,b_3,b_4) \\ -D_f(b_1,b_2,b_4;\ b_2,b_3,b_4) \\ D_f(b_1,b_2,b_3;\ b_2,b_3,b_4) \\ 0 \\ D_f(b_1,b_2,b_4;\ b_1,b_3,b_4) \\ -D_f(b_1,b_2,b_3;\ b_1,b_3,b_4) \\ 0 \\ D_f(b_1,b_3,b_4;\ b_1,b_2,b_4) \\ 0 \\ 0 \end{bmatrix} \tag{5.107}$$

and

$$\nabla D_f(b_1,b_2,b_4,b_5) = \begin{bmatrix} D_f(b_1,b_4,b_5;\ b_2,b_4,b_5) \\ 0 \\ -D_f(b_1,b_2,b_5;\ b_2,b_4,b_5) \\ D_f(b_1,b_2,b_4;\ b_2,b_4,b_5) \\ 0 \\ D_f(b_1,b_2,b_5;\ b_1,b_4,b_5) \\ -D_f(b_1,b_2,b_4;\ b_1,b_4,b_5) \\ 0 \\ 0 \\ D_f(b_1,b_2,b_4;\ b_1,b_2,b_5) \end{bmatrix}, \tag{5.108}$$

where we have ordered the components lexicographically by index pair. Multiplying the first of these by $D_f(b_1,b_2,b_5;\ b_2,b_4,b_5)$, the second by $D_f(b_1,b_2,b_3;\ b_2,b_3,b_4)$ and adding gives

$$\begin{bmatrix} D_f(1,2,3;2,3,4)\,D_f(1,4,5;2,4,5) + D_f(1,3,4;2,3,4)\,D_f(1,2,3;2,3,4) \\ -D_f(1,2,4;2,3,4)\,D_f(1,2,5;2,4,5) \\ 0 \\ D_f(1,2,3;2,3,4)\,D_f(1,2,4;2,4,5) \\ D_f(1,2,4;1,3,4)\,D_f(1,2,5;2,4,5) \\ D_f(1,2,3;2,3,4)\,D_f(1,2,5;1,4,5) - D_f(1,2,3;1,3,4)\,D_f(1,2,5;2,4,5) \\ -D_f(1,2,3;2,3,4)\,D_f(1,2,4;1,4,5) \\ D_f(1,2,3;1,2,4)\,D_f(1,2,5;2,4,5) \\ 0 \\ D_f(1,2,3;2,3,4)\,D_f(1,2,4;1,2,5) \end{bmatrix} \tag{5.109}$$

where $D_f(1,2,3;2,3,4)$ is just an abbreviated notation for $D_f(b_1,b_2,b_3;b_2,b_3,b_4)$, and so on. The gradients of the active distance constraint functions are of course equal to the appropriate unit vectors $\mathbf{u}_{ij} \in \mathbb{R}^{10}$ for those at their upper bounds and to their negatives for those at their lower bounds. Since the Cayley-Menger bideterminants are the products of oriented areas and the pentagon is convex, all of the Cayley-Menger bideterminants which occur in $\nabla D_f(b_1,b_2,b_3,b_4)$ and $\nabla D_f(b_1,b_2,b_4,b_5)$ above are positive. Thus the nonzero elements in the above linear combination are all clearly either positive or negative as the corresponding couple is a cable or strut, respectively, and hence can be canceled by adding an appropriate positive multiple of the corresponding signed unit vector to it, with the possible exception of the (2,4)-component, which contains both positive and negative terms. In this case, however, we can use Corollary 3.50 to swap triples of atoms between the bideterminants, obtaining

$$\begin{aligned} & D_f(b_1,b_2,b_3;\, b_2,b_3,b_4)\, D_f(b_1,b_2,b_5;\, b_1,b_4,b_5) \\ & \quad - D_f(b_1,b_2,b_3;\, b_1,b_3,b_4)\, D_f(b_1,b_2,b_5;\, b_2,b_4,b_5) \\ = \; & D_f(b_1,b_2,b_3;\, b_1,b_2,b_5)\big(D_f(b_1,b_4,b_5;\, b_2,b_3,b_4) - D_f(b_1,b_3,b_4;\, b_2,b_4,b_5)\big)\,, \end{aligned} \tag{5.110}$$

followed by the three-term Graßmann-Plücker relation to show that

$$\begin{aligned} & D_f(b_1,b_4,b_5;\, b_2,b_3,b_4) - D_f(b_1,b_3,b_4;\, b_2,b_4,b_5) \\ = \; & -D_f(b_1,b_4,b_5;\, b_3,b_4,b_5)\,. \end{aligned} \tag{5.111}$$

Hence the (2,4)-component is actually equal to $-D_f(b_1,b_2,b_3;\, b_1,b_2,b_5)$ $D_f(b_1,b_4,b_5;\, b_3,b_4,b_5) < 0$, as it should be.

Of course, the Kuhn-Tucker conditions are only necessary conditions for optimality, and moreover the variable F_{35} is not included in either of the two constraint inequalities $D_f(b_1,b_2,b_3,b_4) \geq 0$ and $D_f(b_1,b_2,b_4,b_5) \geq 0$ which we have considered thus far. Nevertheless, they tell us that F_{45} is, in the space $F_4(B)$ and in the neighborhood of the $f^* \in F_4(B)$ whose values $f^*(b_i,b_j)$ are the distances in the Cauchy pentagon, a locally increasing function of all F_{ij} at their lower bounds and a locally decreasing function of all F_{ij} at their upper bounds. So, these variables cannot be changed so as to further reduce F_{45} unless some change in the unconstrained squared distances F^*_{14} and/or F^*_{35} is simultaneously made. It is easily seen, however, that these distances are at their lower tetrangle inequality limits, so that they cannot be decreased without violating either the distance bounds or the tetrangle inequality. In addition, the signs of both the (1,4) and (3,5)-components of the gradient of the constraint function $D_f(b_1,b_3,b_4,b_5)$

$$\nabla D_f(b_1,b_3,b_4,b_5) = \begin{bmatrix} 0 \\ D_f(b_1,b_4,b_5;\ b_3,b_4,b_5) \\ -D_f(b_1,b_3,b_5;\ b_3,b_4,b_5) \\ D_f(b_1,b_3,b_4;\ b_3,b_4,b_5) \\ 0 \\ 0 \\ 0 \\ D_f(b_1,b_3,b_5;\ b_1,b_4,b_5) \\ -D_f(b_1,b_3,b_4;\ b_1,b_3,b_4) \\ D_f(b_1,b_3,b_4;\ b_1,b_3,b_5) \end{bmatrix} \tag{5.112}$$

are both negative, whereas the sign of the (4,5)-component is positive, so that in some neighborhood of f^* the objective function F_{45} is locally an increasing function of both F_{14} and F_{35} on the surface consistent with the constraint $D_f(b_1,b_3,b_4,b_5) = 0$, and the Cauchy pentagon represents at least a strict constrained local minimum of F_{45} in the space $F_4(B)$. Moreover, because the Cauchy pentagon is globally rigid, this is necessarily a global minimum in Euclidean space. Since the above implies that any second solution in $F_4(B)$ must be separated from the Euclidean one by some finite amount, such non-Euclidean solutions (if they exist) can always be eliminated by making the lower and upper bounds very nearly equal on all couples in the framework. QED

We *conjecture* that the cables and struts in all globally rigid planar tensegrity frameworks in fact correspond to possible tetrangle inequality limits, and that moreover all the combinations of active distance constraints which determine the tetrangle limits may be obtained by tensegrity exchanges among them. Although the globally rigid planar tensegrity frameworks have not been characterized, it is possible to construct sequences of them (for example, the Cauchy polygons; see [Connelly1982]) on N atoms for all $N = 5, \ldots, \infty$. Thus this conjecture, if true, implies that there exists no polynomial algorithm for computing the tetrangle inequality limits, and hence we further conjecture that tetrangle inequality bound smoothing will be found to belong to that class of intrinsically exponential problems which is known as *NP*-complete [Garey1979] (see also §4.4). If this is true, then it would appear we have been trying to solve the wrong problem. Rather than searching for an efficient algorithm to compute the tetrangle and higher-order inequality limits, we should be searching for those limits which are implied by a set of tensegrities with globally rigid realizations and their tensegrity exchanges whose combinatorial properties enable us to derive an efficient noniterative algorithm. Since this is probably where future progress lies, we feel that the following result, though it is but a trivial consequence of the definition of global rigidity, is important enough to be stated as an independent theorem.

THEOREM 5.50. *Suppose we are given lower and upper bounds* $l, u \in F(A)$ *among a set of atoms in an n-dimensional Euclidean space together with a distinguished pair of atoms* a,b ∈ A, *and suppose there exists a subset* B ⊆ A *with* a,b ∈ B *and a positioning* $\mathbf{p}: B \to \mathbb{R}^n$ *of these atoms in n-space such that the tensegrity framework* $\mathbb{F}_{ab} = \mathbb{F}_{ab}(l, u, \mathbf{p}) := (B, L \cup \{a,b\}, U, \mathbf{p})$ *is globally rigid, where*

$$L := \{\{x,y\} \in \tbinom{B}{2} \mid l(x,y) = \|\mathbf{p}(x) - \mathbf{p}(y)\|\} \tag{5.113}$$

and

$$U := \{\{x,y\} \in \tbinom{B}{2} \mid u(x,y) = \|\mathbf{p}(x) - \mathbf{p}(y)\|\} . \tag{5.114}$$

Then all positionings $\mathbf{q}: A \to \mathbb{R}^n$ *such that* $l(x,y) \leq \|\mathbf{q}(x) - \mathbf{q}(y)\| \leq u(x,y)$ *for all* x,y ∈ A *further satisfy*

$$\|\mathbf{p}(a) - \mathbf{p}(b)\| \geq \|\mathbf{q}(a) - \mathbf{q}(b)\| . \tag{5.115}$$

Similarly, if the tensegrity framework $\mathbb{F}^{ab} = \mathbb{F}^{ab}(l, u, \mathbf{p}) := (B, L, U \cup \{a,b\}, \mathbf{p})$ *is globally rigid, where* L *and* U *are defined by Equations (5.113) and (5.114), then all positionings* $\mathbf{q}: A \to \mathbb{R}^n$ *such that* $l(x,y) \leq \|\mathbf{q}(x) - \mathbf{q}(y)\| \leq u(x,y)$ *further satisfy*

$$\|\mathbf{p}(a) - \mathbf{p}(b)\| \leq \|\mathbf{q}(a) - \mathbf{q}(b)\| . \tag{5.116}$$

Thus, the (a,b)-distances which occur in the tensegrity frameworks $\mathbb{F}_{ab}$ and $\mathbb{F}^{ab}$ constitute inviolate upper and lower limits on the (a, b) distance in any positioning of the atoms which satisfies the given bounds.

5.5. Local Minima and Zeros of Error Functions

The next chapter is devoted to the numerical methods which have proved the most useful in solving chemically relevant instances of the Fundamental Problem of Distance Geometry. These methods all involve the minimization of an "error function" which consists of a sum of functions of the squared distances, one for each distance bound, each of which is nonnegative and vanishes if and only if the bound is satisfied (see §6.1 for further details). The purpose of this section is to characterize the "patterns" of bound violations which can occur at a local minimum of an error function in terms of a certain type of tensegrity framework, and to show how the global theory of tensegrity frameworks allows us to unambiguously identify combinations of violations of the distance bounds which can *never* be completely eliminated, no matter how drastically the conformation is changed. In order to derive these results, we shall need the following decomposition of the Hessian of any function $F:\mathbb{R}^{3N}\rightarrow\mathbb{R}$ which can be expressed as a sum of functions of the squared distances, i.e.

$$F(\underline{\mathbf{p}}) \;=\; \sum_{i<j} f_{ij}(\|\mathbf{p}_i-\mathbf{p}_i\|^2)\,, \tag{5.117}$$

which seems to have first been observed by [Connelly1983]. In order to present this decomposition, we let $\underline{\mathbf{S}}_{\mathbf{p}} \in \mathbb{R}^{N\times N}$ be the matrix whose elements are

$$S_{ij} \;:=\; -2\, f'_{ij}(\|\mathbf{p}_i-\mathbf{p}_j\|^2) \;:=\; -2\left.\frac{\partial f_{ij}(D)}{\partial D}\right|_{D\,=\,\|\mathbf{p}_i-\mathbf{p}_j\|^2} \tag{5.118}$$

for $i\neq j$, and $S_{ii} := -\sum_{j\neq i} S_{ij}$. Also let $\underline{\mathbf{S}}_{\mathbf{p}}^{ij} \in \mathbb{R}^{3\times 3}$ be the matrix $\underline{\mathbf{diag}}[S_{ij}, S_{ij}, S_{ij}]$, and $\underline{\mathbf{S}}_{\mathbf{p}}^{(3)} \in \mathbb{R}^{3N\times 3N}$ be the matrix whose 3×3 submatrices are the $\underline{\mathbf{S}}_{\mathbf{p}}^{ij}$:

$$\underline{\mathbf{S}}_{\mathbf{p}}^{(3)} \;:=\; \begin{bmatrix} \underline{\mathbf{S}}_{\mathbf{p}}^{11} & \underline{\mathbf{S}}_{\mathbf{p}}^{12} & \cdots & \underline{\mathbf{S}}_{\mathbf{p}}^{1N} \\ \underline{\mathbf{S}}_{\mathbf{p}}^{12} & \underline{\mathbf{S}}_{\mathbf{p}}^{22} & \cdots & \underline{\mathbf{S}}_{\mathbf{p}}^{2N} \\ \cdot & \cdot & \cdots & \cdot \\ \underline{\mathbf{S}}_{\mathbf{p}}^{1N} & \underline{\mathbf{S}}_{\mathbf{p}}^{2N} & \cdots & \underline{\mathbf{S}}_{\mathbf{p}}^{NN} \end{bmatrix}. \tag{5.119}$$

Finally, let $\underline{\mathbf{C}}_{\mathbf{p}}$ be the diagonal matrix whose nonzero elements are the second derivatives of the f_{ij} in lexicographic order, i.e.

$$\underline{\mathbf{C}}_{\mathbf{p}} \;:=\; \underline{\mathbf{diag}}\Big[4\, f''_{ij}(\|\mathbf{p}_i-\mathbf{p}_j\|^2) \,\big|\, 1\le i<j\le N\Big]\,, \tag{5.120}$$

and $\underline{\mathbf{M}}_{\mathbf{p}}$ be the $\binom{N}{2}\times 3N$ rigidity matrix of the complete bar and joint framework on $\{1,\ldots,N\}$:

$$\underline{\mathbf{M}}_{\mathbf{p}} \;:=\; \big[\,[\,(\delta_{ik}-\delta_{jk})(\mathbf{p}_i-\mathbf{p}_j) \,|\, 1\le i<j\le N\,] \;|\; k=1,\ldots,N\,\big] \tag{5.121}$$

(where δ_{ik} is the Kronecker delta).

LEMMA 5.51. *Let $F:\mathbb{R}^{3N}\rightarrow\mathbb{R}$, $\underline{\mathbf{S}}_{\mathbf{p}}^{(3)}$, $\underline{\mathbf{C}}_{\mathbf{p}}$ and $\underline{\mathbf{M}}_{\mathbf{p}}$ all be defined as in Equations (5.117) through (5.121). Then the second derivative of F in the direction $\underline{\mathbf{v}} \in \mathbb{R}^{3N}$ can be written as:*

$$\underline{\mathbf{v}}^T\cdot\underline{\nabla^2 F(\mathbf{p})}\cdot\underline{\mathbf{v}} \;=\; \underline{\mathbf{v}}^T\cdot(\underline{\mathbf{S}}_{\mathbf{p}}^{(3)} + \underline{\mathbf{M}}_{\mathbf{p}}^T\cdot\underline{\mathbf{C}}_{\mathbf{p}}\cdot\underline{\mathbf{M}}_{\mathbf{p}})\cdot\underline{\mathbf{v}}\,, \tag{5.122}$$

Proof: To obtain the first derivative of F at $\underline{\mathbf{p}}$ in the direction given by $\underline{\mathbf{v}}$, we differentiate $F(\underline{\mathbf{p}}+\alpha\underline{\mathbf{v}})$ with respect to α and set $\alpha=0$, obtaining:

$$\begin{aligned} \nabla F(\underline{\mathbf{p}})\cdot\underline{\mathbf{v}} &= \left.\frac{\partial F(D)}{\partial D}\right|_{D\,=\,\|\mathbf{p}_i-\mathbf{p}_j\|^2}\cdot\left.\frac{\partial\|(\mathbf{p}_i+\alpha\mathbf{v}_i)-(\mathbf{p}_j+\alpha\mathbf{v}_j)\|^2}{\partial\alpha}\right|_{\alpha=0} \\ &= \frac{1}{2}\sum_{i=1}^{N}\sum_{j=1}^{N} f'_{ij}(\|\mathbf{p}_i-\mathbf{p}_j\|^2)\cdot(2(\mathbf{p}_i-\mathbf{p}_j)\cdot(\mathbf{v}_i-\mathbf{v}_j)) \qquad (5.123) \\ &= 2\sum_{i=1}^{N}\left((\mathbf{p}_i\cdot\mathbf{v}_i)\sum_{j=1}^{N} f'_{ij}(\|\mathbf{p}_i-\mathbf{p}_j\|^2)\right) - 2\sum_{i=1}^{N}\sum_{j=1}^{N}\Big((\mathbf{p}_i\cdot\mathbf{v}_j)\, f'_{ij}(\|\mathbf{p}_i-\mathbf{p}_j\|^2)\Big) \end{aligned}$$

$$= 2 \sum_{i=1}^{N} \mathbf{p}_i^T \cdot \underline{\mathbf{S}}_{\mathbf{p}}^{ii} \cdot \mathbf{v}_i - \sum_{i=1}^{N} \sum_{j=1}^{N} \mathbf{p}_i^T \cdot \underline{\mathbf{S}}_{\mathbf{p}}^{ij} \cdot \mathbf{v}_j$$

$$= \underline{\mathbf{p}}^T \cdot \underline{\mathbf{S}}_{\mathbf{p}}^{(3)} \cdot \underline{\mathbf{v}} .$$

Similarly, the second derivative of F at $\underline{\mathbf{p}}$ in the direction $\underline{\mathbf{v}}$ is obtained by differentiating $F(\underline{\mathbf{p}}+\alpha\underline{\mathbf{v}})$ twice, obtaining:

$$\underline{\mathbf{v}}^T \cdot \nabla^2 F(\underline{\mathbf{p}}) \cdot \underline{\mathbf{v}} = \sum_{i<j} \Big(f'_{ij}(\|\mathbf{p}_i - \mathbf{p}_j\|^2) \cdot 2 \|\mathbf{v}_i - \mathbf{v}_j\|^2 \tag{5.124}$$

$$+ f''_{ij}(\|\mathbf{p}_i - \mathbf{p}_j\|^2) \cdot 4((\mathbf{p}_i - \mathbf{p}_j) \cdot (\mathbf{v}_i - \mathbf{v}_j))^2 \Big) .$$

Translated into matrix terms, this is just Equation (5.122). QED

Observe that if $\underline{\mathbf{p}}$ is a minimum (or other critical point) of the function F, then (as suggested by our notation) the off-diagonal elements of the matrix $\underline{\mathbf{S}}_{\mathbf{p}}$ form a self-stress $\mathbf{s} \in \mathbb{R}^M$ ($M := \binom{N}{2}$) of the complete bar and joint framework, since

$$\mathbf{0} = \nabla F(\underline{\mathbf{p}}) = \left[2 \Sigma_j f'_{ij}(\|\mathbf{p}_i - \mathbf{p}_j\|^2) \cdot (\mathbf{p}_i - \mathbf{p}_j) \,|\, i = 1, \ldots, N \right] . \tag{5.125}$$

Thus under these conditions $\underline{\mathbf{S}}_{\mathbf{p}}$ is a stress matrix of the complete framework on $\{1, \ldots, N\}$, and the first term $\underline{\mathbf{v}}^T \cdot \underline{\mathbf{S}}_{\mathbf{p}}^{(3)} \cdot \underline{\mathbf{v}}$ of the decomposition is the energy form $\mathbf{s}^T \cdot \underline{\mathbf{M}}_{\mathbf{v}} \cdot \underline{\mathbf{v}}$ defined by this self-stress (cf. Definition 5.21). The matrix $\underline{\mathbf{M}}_{\mathbf{p}}^T \cdot \underline{\mathbf{C}}_{\mathbf{p}} \cdot \underline{\mathbf{M}}_{\mathbf{p}}$ which occurs in the second term is called the *stiffness matrix* of the framework; the diagonal elements of the matrix $\underline{\mathbf{C}}_{\mathbf{p}}$ can be thought of as a measure of the extent to which the elasticities of the bars of the framework deviate from the ideal of Hook's law $f_{ij}(D) = k_{ij} D$.

In the case that $\underline{\mathbf{p}} = \mathbf{p}(\underline{\mathrm{A}})$ for some $\mathbf{p} \in (\mathbb{R}^3)^{\mathrm{A}}$ and $F: \mathbb{R}^{3\#\mathrm{A}} \rightarrow \mathbb{R}$ is an error function versus lower and upper distance bounds $l, u \in \mathrm{F}(\mathrm{A})$, the individual functions f_{ij} are of course locally constant and equal to zero whenever

$$l(\mathrm{a}_i, \mathrm{a}_j) < \|\mathbf{p}_i - \mathbf{p}_j\| < u(\mathrm{a}_i, \mathrm{a}_j) . \tag{5.126}$$

Hence these components of $\underline{\mathbf{C}}_{\mathbf{p}}$ may be dropped along with the corresponding rows of the matrix $\underline{\mathbf{M}}_{\mathbf{p}}$ to yield the rigidity matrix of a framework whose bars correspond to those pairs of atoms whose lower or upper bounds are violated by the distances from $\mathbf{p}$. Of greater interest, however, is the pure tensegrity framework to which this bar and joint framework is associated.

Definition 5.52. The tensegrity $\mathbb{H}(l, u) = (\mathrm{B}, \mathrm{L}, \mathrm{U})$ whose struts, cables and joints are given by

$$\mathrm{L} := \{\{\mathrm{a},\mathrm{a}'\} \in \tbinom{\mathrm{A}}{2} \,|\, \|\mathbf{p}(\mathrm{a}) - \mathbf{p}(\mathrm{a}')\| < l(\mathrm{a},\mathrm{a}')\} , \tag{5.127}$$

$$\mathrm{U} := \{\{\mathrm{a},\mathrm{a}'\} \in \tbinom{\mathrm{A}}{2} \,|\, \|\mathbf{p}(\mathrm{a}) - \mathbf{p}(\mathrm{a}')\| > u(\mathrm{a},\mathrm{a}')\} , \tag{5.128}$$

and

$$B := \{b \in A \mid \exists\, a \in A: \{a,b\} \in L \text{ or } \{a,b\} \in U\}\,, \tag{5.129}$$

respectively, is called the tensegrity *induced* by $\mathbf{p}$ with respect to the bounds $l, u \in F(A)$.

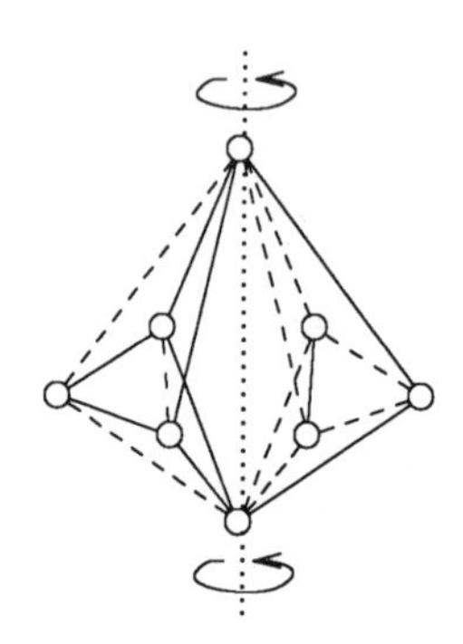

Figure 5.21

The induced tensegrity of a minimum of an error function which is not rigid, because the underlying bar and joint framework is not rigid.

The example shown in Figure 5.21 shows that the tensegrity frameworks induced by the minima of error functions need not be infinitesimally rigid or even rigid. Nevertheless, it helps.

PROPOSITION 5.53. *If* $\mathbb{F} = (A, L, U, \mathbf{p})$ *is an infinitesimally rigid tensegrity framework and* $\mathbf{s} \in \mathbb{R}^{\#(L \cup U)}$ *is a self-stress thereof, then there exist distance bounds* $l, u \in F(A)$ *together with an error function* $F: \mathbb{R}^{3\,\#A} \to \mathbb{R}$ *for those bounds such that* $\underline{\mathbf{p}} = \mathbf{p}(\underline{A})$ *is a strict local minimum of* F *and* $\nabla F(\underline{\mathbf{p}}) = \mathbf{s}$.

Proof: To prove this, for some small $\epsilon > 0$ we define $l(a_i, a_j) := \|\mathbf{p}_i - \mathbf{p}_j\| + \epsilon$ for all $\{a_i, a_j\} \in L$ and $u(a_i, a_j) := \|\mathbf{p}_i - \mathbf{p}_j\| - \epsilon$ for all $\{a_i, a_j\} \in U$. The result then follows from the fact that the value of an unspecified function and its first and second derivatives at any point can be chosen independently. Hence for $F = \sum_{i<j} f_{ij}$ we choose each $f_{ij}(\|\mathbf{p}_i - \mathbf{p}_j\|^2) := 1$, $f'_{ij}(\|\mathbf{p}_i - \mathbf{p}_j\|^2) := s_{ij}$ and then make the $f''_{ij}(\|\mathbf{p}_i - \mathbf{p}_j\|^2)$ sufficiently large and positive that the Hessian matrix of F at $\underline{\mathbf{p}}$,

i.e. $\underline{\mathbf{S}_\mathbf{p}}+\underline{\mathbf{M}_\mathbf{p}}^T\cdot\underline{\mathbf{C}_\mathbf{p}}\cdot\underline{\mathbf{M}_\mathbf{p}}$, is positive semidefinite of nullity six. The details are left to the reader. QED

It should be noted that, if the induced tensegrity framework is not regular at a minimum of an error function F, then it is quite possible that it is also not statically / infinitesimally rigid there, even though the associated bar and joint framework is generically rigid. Furthermore, since these minima are by their nature nongeneric realizations of the underlying induced tensegrity, the probability that a given minimum will be nonregular is distinctly positive. Fortunately, there also exists a necessary criterion which the minima of F all fulfill [Connelly1988]. In order to prove this result, we make use of the following analogue of Proposition 4.22:

LEMMA 5.54. *Let* $\mathbb{F} = (\mathrm{A}, \mathrm{L}, \mathrm{U}, \mathbf{p})$ *be a tensegrity framework. Then there exists a self-stress* $\mathbf{s} \in \mathbb{R}^{\#(\mathrm{L}\cup\mathrm{U})}$ *such that* $s_{ij} > 0$ *(* < 0*) if and only if there exists* *__no__* *infinitesimal motion* $\mathbf{v} \in (\mathbb{R}^3)^\mathrm{A}$ *with* $(\mathbf{p}(\mathrm{a}_i)-\mathbf{p}(\mathrm{a}_j))\cdot(\mathbf{v}(\mathrm{a}_i)-\mathbf{v}(\mathrm{a}_j)) < 0$ *(* > 0*, resp.).*

The proof of this (intuitively obvious) result will be left as a (not altogether trivial) exercise. Its significance here stems from the fact that, if we assume the individual f'_{ij} of an error function $F = \sum_{i<j} f_{ij}$ to be strictly increasing (as is the case for all the error functions used in practice), then the values of these derivatives f'_{ij} at a minimum will have the right signs and hence constitute a strict self-stress of the induced tensegrity by its very definition. Together with the above lemma, this implies that all infinitesimal motions of the induced tensegrity framework are also infinitesimal motions of the associated bar and joint framework.

THEOREM 5.55. *Let* $\mathbf{p} = \mathbf{p}(\underline{\mathrm{A}})$ *be a strict nonzero minimum of an error function* $F{:}\,\mathbb{R}^{3\,\#\mathrm{A}}\to\mathbb{R}$ *for the distance bounds* $l, u \in \mathrm{F}(\mathrm{A})$ *which is of the form* $F(\mathbf{p}) = \sum_{i<j} f_{ij}(\|\mathbf{p}_i-\mathbf{p}_j\|^2)$, *where each* $f'_{ij}{:}\,\mathbb{R}\to\mathbb{R}$ *is strictly increasing. Then the induced tensegrity framework* $\mathbb{H}_\mathbf{p}(l, u)$ *is second-order rigid.*

Proof: If $\mathbb{H}_\mathbf{p}(l,u)$ is infinitesimally rigid, then it is also second-order rigid and we are done. Otherwise, let $\mathbf{p}' \in \mathbb{R}^{3\,\#\mathrm{A}}$ be an infinitesimal flexing of it, and

$$\mathbf{s} \;:=\; [\, f'_{ij}(\|\mathbf{p}_i-\mathbf{p}_j\|^2) \mid \{\mathrm{a}_i,\mathrm{a}_j\} \in \underline{\mathrm{L}\cup\mathrm{U}}] \tag{5.130}$$

be the self-stress defined by the derivatives of the f_{ij} in lexicographic order. By Proposition 5.27, $\mathbf{p}'$ extends to a second-order flexing with acceleration $\mathbf{p}''$ only if $E_\mathbf{s}(\mathbf{p}') \le 0$, where $E_\mathbf{s}$ is the energy form induced by $\mathbf{s}$. Since by Lemma 5.54 and the remarks following it $\underline{\mathbf{M}_\mathbf{p}}\cdot\mathbf{p}' = \mathbf{0}$, it follows from Lemma 5.51 that for the second derivative we have

$$\mathbf{p}'^T\cdot\underline{\nabla^2 F}(\mathbf{p})\cdot\mathbf{p}' \;=\; \mathbf{p}'^T\cdot\underline{\mathbf{S}_\mathbf{p}}\cdot\mathbf{p}' \;=\; E_\mathbf{s}(\mathbf{p}') \;\le\; 0\,, \tag{5.131}$$

contradicting our hypothesis that $\mathbf{p}$ is a strict minimum of F. QED

The particular kind of second-order rigidity which is obtained by the minimization of an error function is sometimes called *prestress stability* by structural engineers. This reflects the fact that physical frameworks built with elastic bars which fail to be rigid because the nullity of their stiffness matrices exceeds six can sometimes be made rigid by imposing a self-stress upon them, such that the sum of their stiffness and stress matrices becomes positive semidefinite of nullity exactly six.

COROLLARY 5.56. *At a strict minimum of an error function satisfying the conditions of Theorem 5.55, the induced tensegrity is rigid.*

Proof: This follows from Connelly's proof that second-order rigidity implies rigidity [Connelly1980]. QED

Since noncongruent realizations of rigid frameworks are often (i.e. when they are simple) related by reflections of one part of the framework with respect to another, we expect that the local minima of error functions will often be related in this way. This has indeed been observed in practice [Havel1984b, Wagner1987]

Our final, and most important, result shows that the first term of Connelly's decomposition of the Hessian can actually provide us with global information about the existence of solutions to the Fundamental Problem of Distance Geometry.

THEOREM 5.57. *Let* $\mathbf{p} = \mathbf{p}(\underline{A})$ *be a nonzero strict minimum of an error function* $F = \sum_{i<j} f_{ij}$ *for the distance bounds* $l, u \in \mathrm{F}(\mathrm{A})$*, where each* f_{ij} *is strictly monotone, and for* $M := \binom{\#\mathrm{A}}{2}$ *let* $\mathbf{s} \in \mathbb{R}^M$ *be the self-stress of the induced tensegrity framework whose components are the* $f'_{ij}(\|\mathbf{p}_i - \mathbf{p}_j\|^2)$*. If the corresponding stress matrix* $\underline{\mathbf{S}}_\mathbf{p}$ *is positive semidefinite, then* $F(\mathbf{q}) > 0$ *for all* $\mathbf{q} \in \mathbb{R}^{3\,\#\mathrm{A}}$*, i.e. the bounds* l, u *are geometrically inconsistent.*

Proof: Since the stress matrix is positive semidefinite, all other realizations of the induced tensegrity $\mathbb{H}(l, u) = (\mathrm{B}, \mathrm{L}, \mathrm{U})$ are necessarily related to $\mathbf{p}$ by an affine flexing (Corollary 5.25). Since $\mathbf{p}$ is a strict minimum of F, Corollary 5.56 shows that the tensegrity framework $\mathbb{H}_\mathbf{p}(l, u)$ is rigid, meaning that it has no flexings and hence no other representations at all. This in turn means that in any other list of Cartesian coordinates for the atoms $\mathbf{q} = \mathbf{q}(\underline{A}) \in \mathbb{R}^{3\,\#\mathrm{A}}$, either

$$\|\mathbf{q}_i - \mathbf{q}_j\| < \|\mathbf{p}_i - \mathbf{p}_j\| < l(\mathrm{a}_i, \mathrm{a}_j) \tag{5.132}$$

for some pair of atoms $\{\mathrm{a}_i, \mathrm{a}_j\} \in \mathrm{L}$, or else

$$\|\mathbf{q}_i - \mathbf{q}_j\| > \|\mathbf{p}_i - \mathbf{p}_j\| > u(\mathrm{a}_i, \mathrm{a}_j) \tag{5.133}$$

for some pair of atoms $\{\mathrm{a}_i, \mathrm{a}_j\} \in \mathrm{U}$, i.e. some bound violation has grown worse. Since F is by definition zero if and only if all of the bounds are completely fulfilled, it follows that $F(\mathbf{q}) > 0$, as claimed. QED

Suppose we are given the four-atom error function for the nontrivial distance bounds $u(a_1,a_i) = 2$ $(i = 2,3,4)$ and $l(a_j,a_k) = 5$ $(2 \le j < k \le 4)$:

$$
\begin{aligned}
& E(\mathbf{p}_1,\ldots,\mathbf{p}_4) \\
= \; & max^2(0, \|\mathbf{p}_1-\mathbf{p}_2\|^2-4) + max^2(0, \|\mathbf{p}_1-\mathbf{p}_2\|^2-4) + max^2(0, \|\mathbf{p}_1-\mathbf{p}_2\|^2-4) + \\
& min^2(0, \|\mathbf{p}_2-\mathbf{p}_4\|^2-25) + min^2(0, \|\mathbf{p}_2-\mathbf{p}_3\|^2-25) + min^2(0, \|\mathbf{p}_3-\mathbf{p}_4\|^2-25) \; .
\end{aligned}
$$

This has a strict minimum at $\mathbf{p}_1 = [0,0]$, $\mathbf{p}_2 = [0,2.811]$, $\mathbf{p}_3 = [-2.434,-1.405]$ and $\mathbf{p}_4 = [2.434,-1.405]$, whose induced tensegrity framework is shown below.

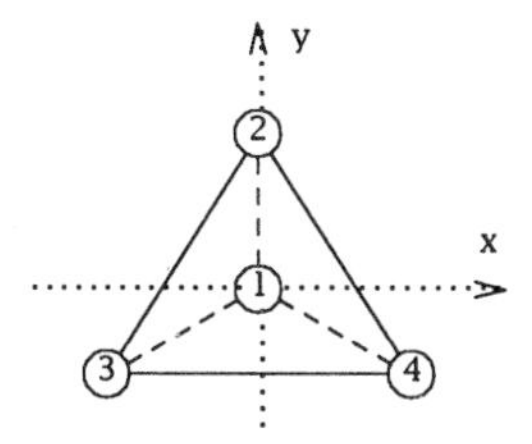

The corresponding stress matrix is given by

$$
\underline{\mathbf{S}}_{\mathbf{p}} = \begin{bmatrix} 11.7 & -3.9 & -3.9 & -3.9 \\ -3.9 & 1.3 & 1.3 & 1.3 \\ -3.9 & 1.3 & 1.3 & 1.3 \\ -3.9 & 1.3 & 1.3 & 1.3 \end{bmatrix},
$$

and has only one nonzero eigenvalue $15.6 > 0$. Since the induced tensegrity framework is infinitesimally rigid, it is rigid and hence globally rigid, which proves that the error function E has no zeros.

It should be noted that this test for the inconsistency of a given set of distance bounds is not infallible, in that minima $\mathbf{p} \in \mathbb{R}^{3\,\#A}$ of an error function F may exist in which the corresponding stress matrix $\underline{\mathbf{S}}_{\mathbf{p}}$ is not positive semidefinite, even though the function F has no zeros. The choice of error function is to a large extent arbitrary, however, and in many cases it may be possible to multiply its terms f_{ij} by *weights* $w_{ij} \ge 0$ in such a way that the given coordinates $\mathbf{p}$ remain a minimum of the new error function $F_w := \sum_{i<j} w_{ij} f_{ij}$ (so that the

derivatives $w_{ij} f'_{ij}$ remain a stress of the induced tensegrity), but the corresponding stress matrix becomes positive semidefinite. Then, the above theorem shows once again that the weighted error function F_w has no zeros, from which it follows that the original function F has no zeros either. More generally, in the event that one can find such an error function F_w, one would like to be able to make the weight and hence the stress zero on as many of the terms as possible, subject to the condition that the corresponding stress matrix $\underline{\mathbf{S}}_w$ stays positive semidefinite. Because the incorrect distance bound(s) will always correspond to couples having a nonzero stress, by this means one may be able to isolate the mistakes in the data.*

Since Connelly's decomposition of the Hessian depends only upon the fact the function F can be expressed as a sum of functions f_{ij} of the individual (squared) distances, it is clearly applicable to at least the most important terms of the usual conformational potential energy functions. Unfortunately, even when F is an error function, the positive semidefiniteness of the stress matrix $\underline{\mathbf{S}}_\mathbf{p}$ obtained from the derivatives f'_{ij} at a critical point $\underline{\mathbf{p}} \in \mathbb{R}^{3\#A}$ does not guarantee that one has found a global minimum of F! All it says is that in any other conformation, a distance between a pair of atoms involved in an attractive interaction must get larger, or else a distance between a pair of atoms involved in a repulsive interaction must get smaller. The error or energy gained this way, however, may be more than compensated for by what happens to the remaining distances. Nevertheless, the directional information contained in the stress matrix does seem to be of a less local character than that which is obtained from the Hessian as a whole, and it is possible that improved methods of escaping from local minima in both error and energy functions could be developed by utilizing it properly.

References

Aho1983.

A. Aho, J. Hopcroft, and J. Ullman, *Data Structures and Algorithms*, Addison-Wesley, Reading, Mass., 1983.

Calladine1978.

C.R. Calladine, "Buckminster Fuller's "Tensegrity" Structures and Clerk Maxwell's Rules for the Construction of Stiff Frames," *Internat. J. Solids and Structures, 14*, 161-172(1978).

* A computer graphics program to enable one to do this interactively is currently under development.

Connelly1980.

R. Connelly, "The Rigidity of Certain Cabled Frameworks and the Second-Order Rigidity of Arbitrarily Triangulated Convex Surfaces," *Advances in Math., 37*, no. 3, 272-299(1980).

Connelly1982.

R. Connelly, "Rigidity and Energy," *Invent. Math., 66*, no. 1, 11-33(1982).

Connelly1983.

R. Connelly, 1983. Personal communication.

Connelly1987.

R. Connelly and W. Whiteley, 1987. Personal communication.

Connelly1988.

R. Connelly and W. Whiteley, 1988. In preparation.

Dress1984.

A.W.M. Dress, "Trees, Tight Extensions of Metric Spaces, and the Cohomological Dimension of Certain Groups: A Note on the Combinatorial Properties of Metric Spaces," *Advances Math., 53*, 321-402(1984).

Dress1987.

A.W.M. Dress and T.F. Havel, "Shortest Path Problems and Molecular Conformation," *Discrete Applied Math.*(1987). In press.

Easthope1988.

P.L. Easthope and T.F. Havel, "Computational Experience with an Algorithm for Tetrangle Inequality Bound Smoothing," *Bull. Math. Biol.*(1988). In preparation.

Fuller1975.

R.B. Fuller, *Synergetics: Explorations in the Geometry of Thinking,* MacMillian, New York, NY, 1975.

Garey1979.

M.R. Garey and D.S. Johnson, *Computers and Intractability: A Guide to the Theory of NP-Completeness,* Freeman, San Francisco, 1979.

Gruenbaum1975.

B. Gruenbaum and G. Shephard, *Lectures in Lost Mathematics,* Seattle, WA, 1975. Unpublished manuscript.

Havel1982.

T.F. Havel, *The Combinatorial Distance Geometry Approach to the Calculation of Molecular Conformation,* University of California, Berkeley, California, 1982. Ph.D. Dissertation

Havel1983.

T.F. Havel, I.D. Kuntz, and G.M. Crippen, "The Theory and Practice of Distance Geometry," *Bull. Math. Biol., 45,* 665-720(1983).

Havel1984a.

T.F. Havel and K. Wuthrich, "A Distance Geometry Program for Determining the Structures of Small Proteins and Other Macromolecules from Nuclear Magnetic Resonance Measurements of 1H-1H Proximities in Solution," *Bull. Math. Biol., 46,* 281-294(1984).

Havel1984b.

T.F. Havel, *Conformational Enantiomorphism in Proteins,* 1984. Unpublished manuscript

Lawler1976.

E.L. Lawler, *Combinatorial Optimization, Networks and Matroids,* Holt, Rinehart and Winston, New York, NY, 1976.

Luenberger1973.

D.G. Luenberger, *Introduction to Linear and Nonlinear Programming,* Addison-Wesley, Menlo Park, CA, 1973.

McMullen1978.

P. McMullen, "Transforms, Diagrams and Representations," in *Contributions to Geometry (Proc. Geometry Symposium, Siegen, W. Germany, June 1978),* ed. J. Tolke, J. M. Wills, Birkhauser, Basel, Switzerland, 1978.

Pedoe1970.

D. Pedoe, *A Course of Geometry for Colleges and Universities,* Cambridge University Press, Cambridge, U.K., 1970.

Pugh1976.

A. Pugh, *Introduction to Tensegrity,* Univ. of California Press, Berkeley, CA, 1976.

Rockafellar1970.

R.T. Rockafellar, *Convex Analysis,* Princeton Univ. Press, Princeton, NJ, 1970.

Rosenberg1980.

I.G. Rosenberg, "Structural Rigidity I: Foundations and Rigidity Criteria," *Annals Discrete Math., 8,* 143-161(1980).

Roth1981.

B. Roth and W. Whiteley, "Tensegrity Frameworks," *Trans. Amer. Math. Soc., 265,* no. 2, 419-446(1981).

Tiel1984.

J. van Tiel, *Convex Analysis: An Introductory Text,* J. Wiley and Sons, New York, NY, 1984.

Wagner1987.

G. Wagner, W. Braun, T.F. Havel, T. Schauman, N. Go, and K. Wuthrich, "Solution Structure of the Basic Pancreatic Trypsin Inhibitor Determined by Two Different Distance Geometry Algorithms from the Same Experimental Data," *J. Mol. Biol.*, *196*, 611-639(1987).

Whiteley1987.

W. Whiteley, 1987. Personal communication.

White1983.

N.L. White and W. Whiteley, "The Algebraic-Geometry of Stresses in Frameworks," *SIAM J. Alg. Discrete Meth.*, *4*, 481-511(1983).

PART II

The Applications of Distance Geometry

(GORDON M. CRIPPEN)

6. The Computation of Coordinates

This chapter is devoted to practical computational methods of solving the *Fundamental Problem of Distance Geometry*, which was introduced in Chapter 1. Given some collection of distance and chirality constraints, this is the problem of calculating coordinates in ordinary three-dimensional Euclidian space for the atoms of a molecule. These coordinates are necessary to prove the given constraints were consistent (if indeed they were!), to provide the starting point for molecular energy minimization, to display conformations consistent with the experimentally derived distance constraints by means of computer graphics, and to compare them with conformations derived from other approaches. Keep in mind that the original distance constraints are usually insufficient to restrict the resulting coordinates to a single conformation, and it is important to learn what range of conformations is permitted. In reality the molecule might be continually changing some features of its conformation, in which case this range will necessarily include all possible conformations. It is also possible that the distance constraints which could be derived from the available experimental data were insufficient, in which case the range may suggest further experiments directed toward examining those indeterminate features.

The fact that the Fundamental Problem of Distance Geometry is NP-hard (see §4.4), though sobering, does not mean that practical algorithms by which it can be solved do not exist. What it does mean is that the design of such algorithms will probably always remain something of an art, and that we cannot reasonably expect to find a single algorithm which is the "best" one to use on all possible instances of the problem. A number of different approaches to developing usable (though not provably polynomial) algorithms for NP-hard problems are available. The two most important of these are:

(1) Efficient approximation algorithms which produce results which are reasonably "close" to exact solutions in polynomial time.

(2) Algorithms which produce exact solutions in time which are polynomial in the "average" case.

Although no algorithm is known at this time for the Fundamental Problem which has been mathematically *proved* to satisfy either of these criteria, algorithms have been developed and shown to come reasonably close to satisfying both criteria by means of numerous applications to practical, every-day chemical problems.

Most of the rest of this chapter is concerned with EMBED, our practical but somewhat involved algorithm for solving the Fundamental Problem. The natural question at this point is: why go to all that trouble? After all, the problem of computing an 3-dimensional representation of a complete abstract framework $\mathbb{F} = (\mathrm{A}, \binom{\mathrm{A}}{2}, d)$ can be solved in linear time. Just choose the first three atoms so that their Cayley-Menger determinant satisfies $D(\mathrm{a}_1,\mathrm{a}_2,\mathrm{a}_3) > 0$. Then our linear time algorithm calculates the Cartesian coordinates $\mathbf{p}_i = [x_i, y_i, z_i]$ by means of simple triangulation:

$$
\begin{aligned}
\mathbf{p}_1 &:= [0, 0, 0]\,; \\
\mathbf{p}_2 &:= [d(\mathrm{a}_1,\mathrm{a}_2), 0, 0]\,; \qquad (6.1)\\
\mathbf{p}_3 &:= \left[\frac{D(\mathrm{a}_1,\mathrm{a}_2;\, \mathrm{a}_1,\mathrm{a}_3)}{d(\mathrm{a}_1,\mathrm{a}_2)},\, \sqrt{d^2(\mathrm{a}_1,\mathrm{a}_3) - x_3^2},\, 0\right];
\end{aligned}
$$

and

$$
\begin{aligned}
\mathbf{p}_i := \Bigg[&\frac{D(\mathrm{a}_1,\mathrm{a}_i;\, \mathrm{a}_2,\mathrm{a}_i)}{d(\mathrm{a}_1,\mathrm{a}_2)}, \qquad (6.2)\\
&\frac{D(\mathrm{a}_1,\mathrm{a}_i;\, \mathrm{a}_2,\mathrm{a}_3) - x_i(x_2 - x_3)}{y_3},\, \sqrt{d^2(\mathrm{a}_1,\mathrm{a}_i) - x_i^2 - y_i^2}\Bigg],
\end{aligned}
$$

for $i = 4, \ldots, \#\mathrm{A}$ (here, the D's denote Cayley-Menger bideterminants with respect to the premetric d^2). If we avoid the actual computation of square roots by computing symbolically over the field obtained by adjoining y_3 and the z_i to the rationals whenever these quantities are irrational, the time required to compute each triple of coordinates is thus constant.

One problem with this triangulation procedure is that it assumes we can find a rigid tetrahedron of four atoms in the molecule such that the distances to all of the remaining atoms are known exactly. Although the triangulation method based on Henneberg sequences which was introduced in §4.4 is considerably more flexible, it still fails to be applicable to most of the problems which occur in chemistry, simply because most distances are not known exactly. Of course, if the bounds are fairly "tight", one could just make a reasonable guess at the values of the distances needed for triangulation. Unfortunately, triangulation is numerically extremely unstable, so that a small mistake in the value of any one distance can make a big change in the coordinates, with the effect that the coordinates no longer even come close to fitting the remaining distances (for example, when $d(a_1,a_3) \approx d(a_2,a_3) \gg d(a_1,a_2)$). The result is that small errors in the positions of the quadruple of atoms being used as the "base" for the triangulation can result in large errors in the relative positions of all the remaining atoms. We have therefore sought out methods that treat all points the same and produce coordinates that are representative in some sense of the overall set of distance constraints.

Our own approach to solving the Fundamental Problem breaks the task down into three successive stages. The first stage takes the input distance bounds, which are usually quite sparse and imprecise, and attempts to estimate the actual range of values each distance can assume for some conformation of the molecule consistent with this input data. This is done by the bound smoothing techniques described in the preceding chapter. In the second step, a list of Cartesian coordinates for the atoms is computed such that the distances, as derived from these coordinates, lie largely within the ranges from the previous step. The method by which this is done is known as the *EMBED* algorithm. The fit to the data which the EMBED algorithm produces is generally not perfectly accurate, however, and in addition does not take account of the chirality constraints. Hence the third stage uses standard numerical optimization methods to minimize an *error function* whose magnitude measures the deviations of the coordinates from the input distance and chirality constraints. Because the coordinates produced by the EMBED algorithm usually come close to fitting the data, these minimizations often converge quite well even in problems with thousands of variables.

Thus the organization of this chapter is as follows: In the first section we consider various possible error functions which can be used to define the sets of atomic coordinates which are admissible with respect to the input distance and chirality constraints. We also provide a brief discussion of the various methods which can be used to minimize them, as well as the problems which are inherent in this task. In the second section we shall introduce practical data structures for storing distance and chirality constraints, along with efficient algorithms for bound smoothing and for the evaluation of error functions. The third section presents a detailed derivation and statement of the EMBED algorithm itself. A final section discusses some of the many alternative methods which have been used to compute coordinates from distance information.

6.1. Analytic Formulations of the Problem

In order to present the numerical methods which we have found most useful in solving chemically relevant instances of the Fundamental Problem, it now becomes necessary for us to formulate the Fundamental Problem *analytically*, rather than *synthetically* as we have thus far. Since the squared distances and chiralities are polynomial functions of the Cartesian coordinates, from the analytic point of view an admissible embedding of a molecule is simply a solution to a system of polynomial equations and inequalities. For algorithmic purposes, however, it is expedient to combine all these polynomials together into a single function, which in a sense measures the deviation of any given embedding of the atoms from admissibility.

Definition 6.1. For a given distance geometry description $(l, u, \tilde{\chi})$ of a molecule, an *error function* is any continuous function $E:(\mathbb{R}^3)^A \rightarrow \mathbb{R}$ such that for each $\mathbf{p} \in (\mathbb{R}^3)^A$ we have:

(i) $E(\mathbf{p}) \geq 0$;

(ii) $E(\mathbf{s} \circ \mathbf{p}) = E(\mathbf{p})$ for all $\mathbf{s} \in \mathbf{ST}(3, \mathbb{R})$;

(iii) $E(\mathbf{p}) = 0 \iff \mathbf{p} \in \mathbf{P}(l, u, \tilde{\chi})$;

where $\mathbf{ST}(3,\mathbb{R})$ denotes the direct isometries of $\mathbb{R}^3$, and $\mathbf{P}(l, u, \tilde{\chi})$ denotes the set of all admissible embeddings, as defined in Chapter 1.

We shall generally consider error functions to be functions on $\mathbb{R}^{3\#A}$ relative to a predefined ordering of the atoms $\underline{A}$, i.e. as functions of the Cartesian coordinates $\underline{\mathbf{p}} = \mathbf{p}(\underline{A})$. In addition, for practical reasons we shall usually require our error functions to have continuous first and second derivatives, and that they be monotone in the sense that any change in the conformation which increases all of the constraint violations also increases the error.

Since $E(\underline{\mathbf{p}}) \geq 0$ and $E(\underline{\mathbf{p}}) = 0$ if and only if the coordinates $\underline{\mathbf{p}}$ satisfy *all* the given distance and chirality constraints, the minimization of such a function, if it converges to zero error, will produce the coordinates of an admissible conformation of the molecule. The major difficulty which is encountered in putting this approach into practice is that the error function used will generally have *local minima* in it. At such points $\underline{\mathbf{p}}$ the error $E(\underline{\mathbf{p}})$ exceeds zero, but at all points $\underline{\mathbf{q}}$ in some sufficiently small neighborhood of $\underline{\mathbf{p}}$ we have $E(\underline{\mathbf{p}}) < E(\underline{\mathbf{q}})$, so that it is not possible to further improve the fit to the geometric constraints by means of small changes to the coordinates $\underline{\mathbf{p}}$. This so-called *local minimum problem* is also encountered in the minimization of conformational energy functions (see Chapter 9), and despite years of research there exists no really good way of overcoming it. Fortunately, the error functions we usually use appear to have far fewer local minima than the usual conformational energy functions, so that with a reasonably good choice of starting coordinates it is often possible to reduce the error almost to zero by means of local minimization alone (see Appendix F). For the time being, it will be enough to know that finding a conformation with near-zero error is our primary goal.

There is an important point in the philosophy of the use of such an error function. The distance constraints are viewed as correct, rather than subject to unknown (possibly random) errors. Therefore acceptable coordinates must have error function values of zero or only very slightly greater than zero due to the minimization converging on the edge of the feasible region. This is very different from the "least squares statistical" approach, where the constraints are thought to possibly contain errors, so that having the error function substantially greater than zero corresponds to a best fit to erroneous input data and is probably an improvement on it. The difficulty with the "least squares" view is that it becomes hard to say that a best fit conformation is *wrong* and one must search for a rather different one that will be right, or one must check the given constraints for inconsistencies. On the other hand, if many conformations with error near zero can be found, it is clear that the available geometric information is insufficient to determine a unique conformation. This may in fact be due to the presence of intramolecular motion, so that the molecule does not have a unique conformation. The least-squares approach, however, attempts to obtain a single "best" fit to the data whether a unique conformation exists or not. If one does not exist, the fit to the data will be poor, and the range of conformations which are obtained will be determined by the (usually *ad hoc*) weights which are assigned to the individual data, rather than by the actual range of conformations which are present. In reality, the "average" solution conformation of a molecule may well be present in essentially zero concentration!

If we leave out the chirality information for the time being (e.g., by setting $\tilde{\chi} := \{0, \pm 1\}$), the simplest example of an error function is given by

$$F(\mathbf{p}) := \sum_{i<j} \left(min^2\Big(0, \frac{\|\mathbf{p}_i - \mathbf{p}_j\|^2}{l^2(\mathrm{a}_i, \mathrm{a}_j)} - 1\Big) + max^2\Big(0, \frac{\|\mathbf{p}_i - \mathbf{p}_j\|^2}{u^2(\mathrm{a}_i, \mathrm{a}_j)} - 1\Big) \right). \qquad (6.3)$$

It is easily seen that this function has all of the properties specified in Definition 6.1. Note that the contributions to the error from each term of this function is proportional to the fractional violation of the constraint, rather than to the absolute violation. Even though the terms could be weighted in any way without changing the zeros of the error function, in the presence of residual errors which most nonlinear minimization algorithms leave behind, one generally wants to give short distance constraints such as covalent bond lengths and van der Waals overlaps a high weight to ensure that the near-zeros of the function do not contain any features which violate elementary chemical common sense. This particular weighting scheme is simple and produces adequate results without any juggling.

Although good results have been obtained with this function, somewhat better results are usually achieved by using a slight modification of it. In order to motivate this modification, we introduce a particular type of function which plays an important role in optimization theory. Recall that a set $\mathbf{X} \subseteq \mathbb{R}^M$ is *convex* if the line segment $[(\mathbf{x}, \mathbf{x}')]$ between any two points $\mathbf{x}, \mathbf{x}' \in \mathbf{X}$ is contained in $\mathbf{X}$, i.e. $\alpha\mathbf{x}+(1-\alpha)\mathbf{x}' \in \mathbf{X}$ for all $\alpha \in [0 \cdots 1]$.

Definition 6.2. For any $\mathbf{X} \subseteq \mathbb{R}^M$, a function $f: \mathbf{X} \rightarrow \mathbb{R}$ is called *convex* if

$$f(\alpha\mathbf{x}+(1-\alpha)\mathbf{x}') \leq \alpha f(\mathbf{x})+(1-\alpha) f(\mathbf{x}') \tag{6.4}$$

for all $\mathbf{x}, \mathbf{x}' \in \mathbf{X}$ and $\alpha \in [0 \cdots 1]$.

By considering the quadratic approximation which is obtained from the Taylor expansion of f to second order it is easily shown that a function f is convex if and only if its Hessian is everywhere positive semidefinite. The reason convex functions are of interest to us here is:

PROPOSITION 6.3. *If the domain $\mathbf{X} \subseteq \mathbb{R}^M$ of a convex function $f: \mathbf{X} \rightarrow \mathbb{R}$ is convex, then the subset $\mathbf{Y} \subseteq \mathbf{X}$ where f achieves its minimum value is convex, and f has no local minima in $\mathbf{X}$.*

A proof of this well-known result may be found in [Luenberger1973].

Unfortunately, it is not possible to design a convex function which meets all the criteria for an error function given in Definition 6.1. This follows immediately from the existence of generically rigid abstract bar and joint frameworks which admit multiple, locally unique representations, each of which must be a distinct minimum of any error function. Thus the nonexistence of a convex error function is, in a sense, a consequence of the NP-hardness of the Fundamental Problem of Distance Geometry. Nevertheless, it is enlightening to determine just why the error function F fails to be convex.

LEMMA 6.4. *The squared distance $D(\mathbf{p}_1, \mathbf{p}_2) := \|\mathbf{p}_1-\mathbf{p}_2\|^2$ is a convex function of the coordinates $\mathbf{p}_1, \mathbf{p}_2 \in \mathbb{R}^n$ for all $n > 0$.*

Proof: The Hessian of D is easily seen to consist of 2×2 diagonal blocks of the form:

$$\begin{bmatrix} 2 & -2 \\ -2 & 2 \end{bmatrix}. \tag{6.5}$$

Since these blocks are positive semidefinite, the Hessian as a whole is also positive semidefinite, and it follows from the remark following Definition 6.2 that D

is convex. QED

From this it is easily shown that:

COROLLARY 6.5. *The contribution to the error function F from terms involving the upper bounds is convex with respect to the coordinates.*

Proof: These terms $max^2(0, (D/u^2 - 1))$ of F are obtained by multiplying the squared distance by the positive constant u^{-2}, subtracting the convex (constant) function -1, composing the result with the nondecreasing convex functions $x \mapsto max(0,x)$ and $x \mapsto x^2$ $(x \geq 0)$. Since these operations all preserve convexity, the corollary is established. QED

This shows that the subset of $(\mathbb{R}^3)^A$ which is defined by the upper bounds on the distances is a convex set, which might be described as a sort of "curvilinear" polyhedron obtained by the intersection of the convex sets which are defined by the individual upper bound constraints.

Thus we see that the lack of convexity of F is caused by those terms involving the lower bounds. These terms fail to be convex because the function $x \mapsto min(0, x)$ is not convex. In one-dimension, however, these terms may be modified so as to become "locally" convex.

LEMMA 6.6. *In one-dimension, the inverse squared distance between two points $x, y \in \mathbb{R}$, when restricted to the subset of $\mathbb{R}^2$ given by $\{[x, y] \mid x < y\}$, is a convex function of the coordinates x and y.*

Proof: The Hessian of the function $(x-y)^{-2}$ is:

$$(x-y)^{-4} \cdot \begin{bmatrix} 6 & -6 \\ -6 & 6 \end{bmatrix}, \tag{6.6}$$

which is clearly positive semidefinite. QED

Our new error function will be given by:

$$G(\mathbf{p}) := \sum_{i<j} \left(max^2(0, \frac{l^2(a_i, a_j)}{\|\mathbf{p}_i - \mathbf{p}_j\|^2} - 1) + max^2(0, \frac{\|\mathbf{p}_i - \mathbf{p}_j\|^2}{u^2(a_i, a_j)} - 1) \right), \tag{6.7}$$

COROLLARY 6.7. *In one-dimension, the error function G is convex on the subsets of its domain $\mathbb{R}^N$ ($N := \#A$) given by:*

$$\{\mathbf{x} \in \mathbb{R}^N \mid x_{\pi(1)} < \cdots < x_{\pi(N)}\} \tag{6.8}$$

for each permutation π of $\{1, \ldots, N\}$.

Proof: Follows at once from Lemma 6.6 using arguments similar to those used to prove Corollary 6.5. QED

Although G fails to be even locally convex in higher dimensions, it has been found computationally that it is lacking at least some of the local minima of F.

Two of these are shown in Figure 6.1.

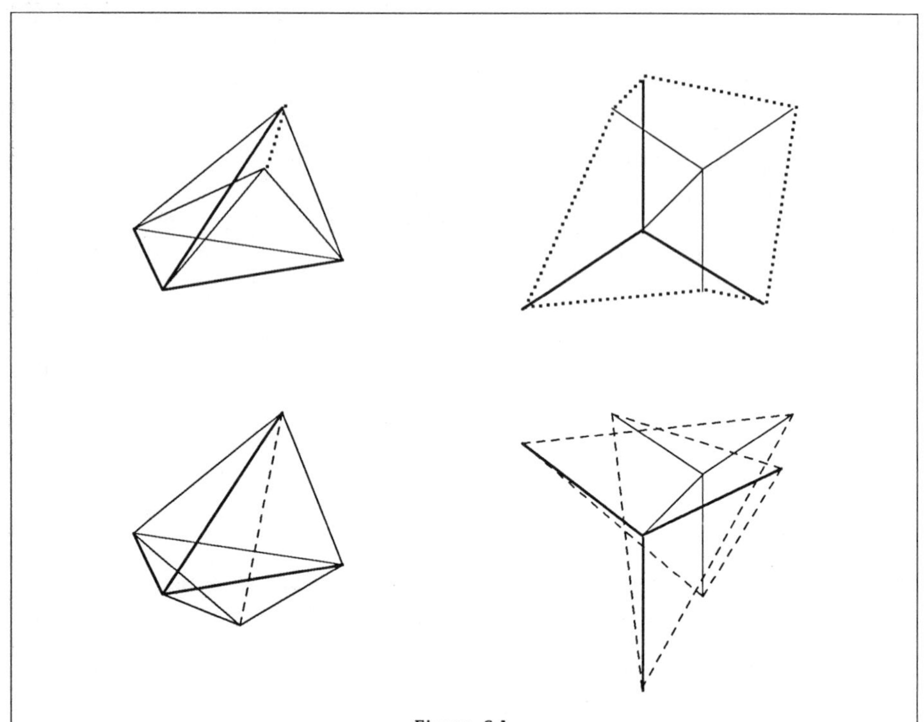

Figure 6.1

Some local minima of the error function F which are not minima of the function G. Inverted trigonal bipyramid (left), twisted propellane (right). All distance constraints are satisfied in the lower row, as indicated by the dashed and solid lines. In the upper row, dotted lines indicate violated constraints.

A more thorough evaluation of possible error functions for distance constraints may still be a highly worthwhile enterprise.

We are now ready to present an error function which, when added to the distance error function, takes account of the chirality constraints:

$$C(\mathbf{p}) \ := \sum_{\underline{\lambda} \in \Lambda(N,4)} \left(max^2(0,\, l(\underline{A}_\lambda) - vol_{\mathbf{p}}(\underline{A}_\lambda)) + max^2(0,\, vol_{\mathbf{p}}(\underline{A}_\lambda) - u(\underline{A}_\lambda)) \right). \qquad (6.9)$$

In this definition of C, $\underline{A}_\lambda$ is a sequence of four distinct atoms $[a_{\lambda_1}, \ldots, a_{\lambda_4}]$ in the order inherited from a given ordering $\underline{A}$ of the set of atoms, $vol_{\mathbf{p}}(\underline{A}_\lambda) = 1/6\, det(\hat{\mathbf{p}}_\lambda)$ is the oriented volume spanned by the sequence of points $[\mathbf{p}_{\lambda_1}, \ldots, \mathbf{p}_{\lambda_4}]$ and $l, u : A^4 \rightarrow \mathbb{R}$ are lower and upper bounds on the value of this oriented volume. In terms of vector algebra, the oriented volume spanned by these four points is given by one sixth the triple product of the vectors from any one of the points to the other three, i.e.

$$vol_{\mathbf{p}}(\underline{A}_\lambda) \;=\; \frac{1}{6}(\mathbf{p}_{\lambda_1}-\mathbf{p}_{\lambda_4})\cdot(\mathbf{p}_{\lambda_2}-\mathbf{p}_{\lambda_4})\times(\mathbf{p}_{\lambda_3}-\mathbf{p}_{\lambda_4}) \;. \qquad (6.10)$$

Although this function makes no explicit mention of the values of $\tilde{\chi}$, it is clear that all possible values with the exception of $\{\pm 1\}$ can be modeled in these terms. For example, if for some quadruple of atoms $\underline{B} \in A^4$ we have $\tilde{\chi}(\underline{B}) = \{0\}$, then we set $l(\underline{B}) = u(\underline{B}) = 0$, while if $\tilde{\chi}(\underline{B}) = \{0,+1\}$ (or $\{+1\}$), we set $l(\underline{B}) = 0$ and $u(\underline{B}) = \infty$. The one omitted value $\{\pm 1\}$ is rarely of any interest in chemical problems, and can usually be accounted for by means of distance constraints even when it does occur. We call C the *chirality error function.*

The additional flexibility obtained by using general lower and upper bounds on the oriented volumes rather than simple chirality constraints is not of much use in *defining* the conformation of a molecule, but it turns out to be quite handy for the purposes of actually *computing* it via the numerical algorithms that we usually use (see Appendix F). One reason for this is that most methods of minimizing error functions leave at least a slight residual error, and hence it is necessary to penalize them for even being too close to the wrong chirality if one wants it to actually come out right. Another place where bounds on the oriented volumes have proven useful is in imposing bounds upon the torsion angles about single bonds. Although bounds on torsion angles can be specified by means of distance and chirality constraints alone, the quality of numerical convergence is markedly improved when lower and upper bounds on the oriented volume equal to the extremes which occur as the angle varies between its bounds are included in the chirality error function C (see §7.1).

Complete descriptions of the many different methods of nonlinear optimization which can be used to minimize error functions and thereby solve the Fundamental Problem of Distance Geometry are given in Appendix F and the references therein. Here we will only indicate briefly what some of their advantages and disadvantages are. The method of steepest descents is relatively easy to implement and requires very little memory, but often fails to reduce the error to near-zero values where other methods succeed. The Newton-Raphson method and its variants, on the other hand, are very efficient at locating minima exactly, but require memory proportional to the square of the number of atoms. Since the precise location of minima is not of great interest in these sorts of calculations, we have found the conjugate gradient method and its various extensions to be among the best methods for error function optimization. Another method which has proven effective is the projected Newton method used, for example, in the MM2 program [Burkert1982]. Although the ellipsoid algorithm is too inefficient for use on large problems, on problems with up to about fifty dihedral angles it has proven to be a particularly powerful method of eliminating distance constraint violations. [Billeter1986a, Billeter1986b]. Other, more specialized methods have also been developed, such as the quartic line search algorithm described in [Havel1983].

We have seen that any error function which properly accounts for the geometric information characteristic of molecules must admit local minima. Thus when the minimization of an error function fails to yield coordinates which satisfy the geometric constraints, it is important to decide if it has become trapped in a local minimum, or whether the data are geometrically inconsistent so that there are no minima with value zero. As we saw earlier (cf. §5.3 & 5.4), bound smoothing gives us one means of proving geometric inconsistency, but geometrically inconsistent sets of distance constraints can pass this test. Another way to decide is simply to repeat the minimization from a variety of different sets of starting coordinates, since the correctness of the data will be proved if the minimization converges to zero error even once. Unfortunately, in large problems there are usually so many local minima around that convergence to zero error is almost never achieved even when the data are correct, and it is quite difficult to decide whether the error levels attained reflect incorrect data or not. Furthermore, when the data are not correct one would really like to be able to find the reason(s) *why* it is not, so that the mistake(s) can be corrected. If a single incorrect constraint is always violated after minimization from a variety of starting coordinates, of course, it will be obvious where the problem is. More commonly, however, these minimizations will result in a number of different compromises between the correct and the incorrect data, so that the mistakes are very hard to pinpoint. We hope that this difficulty will be alleviated soon by applying the global theory of tensegrity frameworks (see §5.5) to unambiguously identify combinations of constraint violations which can *never* be completely eliminated, no matter how drastically the conformation is changed.

6.2. Efficient Data Structures for EMBED

In this section, we shall describe the data structures which are necessary to implement efficient algorithms for bound smoothing and the evaluation of error functions. With a modern computer, it is of course a simple matter to evaluate these error functions together with their derivatives even for molecules with thousands of atoms. Since these evaluations must be repeated many times in the course of a minimization, however, efficiency is not a trivial matter. Thus we begin by describing some techniques by which these evaluations can be implemented efficiently.

Although the total number of quadruples in a molecule is of order $\#A^4$, in most experimentally available definitions of a molecule the total number of quadruples $\underline{B} \in A^4$ involved in nontrivial chirality constraints (i.e. those with $\tilde{\chi}(\underline{B}) \neq \{0, \pm 1\}$) generally grows only linearly in #A. As a consequence, only the nontrivial oriented volume constraints are stored, usually as a simple list, and all those not stored are implicitly assumed to be trivial. In a structured language such as PASCAL, this list usually consists of an array of records containing the indices of the quadruples with respect to some ordering $\underline{A}$ of the atoms, together with the lower and upper bounds on the oriented volume spanned by the embedded quadruple $\mathbf{p}_\lambda$ ($\underline{\lambda} \in \Lambda(\#A, 4)$) in the order inherited from $\underline{A}$. In a nonstructured language such as FORTRAN, four separate arrays would have to be used for the indices, together with two further arrays for the lower and upper bounds (although hopefully these would all be placed together in their own named common block!).

Although the distance constraints are most simply stored in a single $\#A \times \#A$ matrix whose lower and upper triangles contain the lower and upper bounds, this form of storage requires $O(\#A^2)$ memory and hence also $O(\#A^2)$ time to be scanned in each evaluation. In most cases of chemical interest, however, the total number of exact distance constraints (including bond lengths and geminal distances) is generally only of order #A, as is the number of additional upper bound constraints (at best). The number of lower bounds which must be stored explicitly is usually also quite limited. For this reason, most implementations designed for use on large problems store only a list of explicit distance constraints, containing the indices of those pairs of atoms for which explicit lower and/or upper bounds are experimentally available. This list can be stored as an array of records containing the indices of the atoms and the bounds, or this information can be stored in separate scalar arrays. For flexibility in adding or deleting distance constraints, we have also used a linked list [Havel1984].

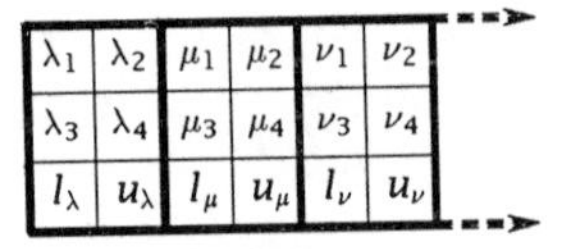

Figure 6.2

The records used to store nontrivial oriented volume constraints contain the indices of the atoms of the tetrahedron, together with lower and and upper bounds on its oriented volume. The complete list of such constraints is best stored as an array of these records, as shown.

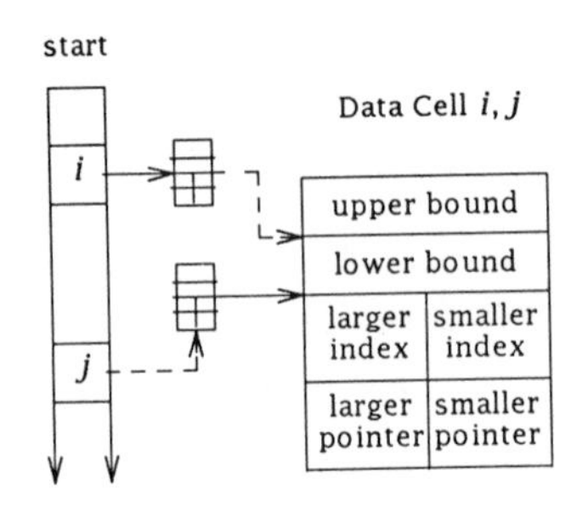

Figure 6.3

The intertwined linked list data structure used to maintain dynamic lists of distance constraints. Each data cell contains the lower and upper bounds on the distance whose smaller and larger atom indices are also contained in the cell. In addition, pointers to the next data cell on the smaller and larger index's lists are stored.

Another important feature common to all chemical problems is the presence of a large number of relatively small (i.e. 1 to 2 Å) lower bounds on the distances between the nonbonded, nongeminal pairs of atoms in the molecule. These are due to the fact that the van der Waals repulsion between atoms prevents these pairs from ever getting any closer together, and can accurately be reproduced as the sum of the *hard sphere radii* of the atoms. These radii in turn can be stored in a single array of #A elements. Checking a coordinate list $\mathbf{p} \in \mathbb{R}^{3\#A}$ for the presence of such lower bound violations by looking at all pairs of atoms, however, requires time $O(\#A^2)$. If (as is the case in molecular dynamics calculations) the coordinates $\mathbf{p}$ change by only a small amount between such checks, it is generally acceptable to prepare a list of van der Waals overlaps which is updated only every 10 evaluations or so [Gunsteren1984]. Unfortunately, in the early stages of error function minimization the changes in the coordinates between evaluations are sometimes substantial, so that this method is not applicable.

A better method of reducing the time spent checking for van der Waals overlaps places the embedded molecule inside a cube, which is then subdivided into subcubes large enough to ensure that any two atoms in nonadjacent subcubes are also not involved in van der Waals overlaps. Then by numbering the subcubes and sorting the atoms according to the number of the subcube in which each lies, it is possible to implement a check for van der Waals overlaps which (assuming the density of the atoms stays within reasonable limits) has an expected-time complexity of only $O(\#A\, ln(\#A))$ [Gunsteren1984, Braun1985]. This asymptotic performance, however, must be weighted against the relatively high constant factor inherent in the method, as well as the difficulties in implementing it. In molecules with $\#A \leq 100$, a simple quadratic check will actually be faster.

A good compromise between simplicity and asymptotic performance that we have used is based upon the *plane sweep method* for range queries [Bentley1980, Havel1984]. In this method, the coordinate axis with the greatest range in values is chosen, and the hard spheres which describe the atoms projected onto it to obtain a set of intervals on this line. The right and left end-points of these intervals are tagged by the atom they represent, and these end-points sorted by their values. This sorted list is then scanned from left to right. The set of atoms whose left but not right end-points have been encounted at any time in the scan is maintained, and each time a right end-point is encountered, the corresponding atom is checked for hard sphere contacts with all atoms in the set. This method is summarized as follows:

Algorithm 6.8:

INPUT: A list of points $[\mathbf{p}_i \,|\, i = 1, \ldots, N]$ in $\mathbb{R}^3$ together with the radii

$[r_i \mid i = 1, \ldots, N]$ of spheres centered upon them.
OUTPUT: The set of all pairs $\{i, j\}$ whose spheres overlap.
PROCEDURE:

Set C := {} for $i = 1, \ldots, N$.
Choose the coordinate axis x for which the maximum absolute difference $|x_i - x_j|$ in any pair of coordinates is maximum.
Let $w_i := x_i - r_i$ for $i = 1, \ldots, N$ and $w_i := x_i + r_i$ for $i = N+1, \ldots, 2N$ be the end-points of the intervals along the x-axis obtained by projection of the spheres.
Let $\pi : \{1, \ldots, 2N\} \rightarrow \{1, \ldots, 2N\}$ be the list of indices of the end-points sorted by increasing value, i.e. $w_{\pi(i)} < w_{\pi(i+1)}$ for $i = 1, \ldots, 2N-1$.
For each $i = 1, \ldots, 2N - 1$ do
  if $\pi(i) > N$ then:
    Set C := C \ i.
    For each $j \in$ C do
      if $\|\mathbf{p}_i - \mathbf{p}_j\| < r_i + r_j$ then
        print $\{i, j\}$.
  otherwise $(\pi(i) \leq N)$
    Set C := C $\cup$ i.

Another important component of our coordinate calculation procedure is triangle inequality bound smoothing. As shown in §5.3, the triangle inequality limits for any given set of distance bounds are equal to the shortest paths in a certain weighted digraph. Although general-purpose polynomial-time algorithms were given in §5.3 by which these shortest paths could be computed, more efficient algorithms which utilize the sparseness of the explicitly input distance bounds are available. Rather than directly computing a full matrix of shortest paths, these "sparse digraph" algorithms generally compute a shortest path tree whose branches are the shortest paths from the root to all other nodes in the digraph. Thus, to compute the shortest paths between all pairs of nodes in a digraph, it is necessary to compute a shortest path tree rooted at each node of the digraph. The two-fold redundancy which this procedure would usually entail can be largely eliminated by ignoring all nodes in the digraph whose indices are less than that of the root of the tree.

Several variations of these algorithms exist, but one of the simplest and most effective is the algorithm designated as "C2" in Dial *et al.*'s survey [Dial1979]. In order to implement this algorithm, it is necessary to use a data structure which enables one to scan all arcs leaving any given node in the digraph without any searching, such as the intertwined linked list described above. A few additional adaptations of the algorithm to the specific features of our particular digraph are also necessary [Havel1984]. Observe that the nodes of this digraph consist of two copies of the atoms in the molecule, and also two arcs for every explicitly given distance bound used to define its conformation (see §5.3). Rather than actually storing this redundant information, however, we effectively simulate the digraph from the information on the intertwined linked list. The final point to be aware of while reading the statement of the algorithm given below is that, in order to keep track of which "copy" of the molecule we are in, we use negative indices for the atoms in one copy and positive for those in the other. The following fills in the details.

Algorithm 6.9 (GEODESICS):

INPUT: A pair of symmetric matrices *LowerBound* and *UpperBound* containing the bounds on the distances between the atoms of the molecule, together with a list for each atom's index i of those indices k such that $UpperBound[i,k] < \infty$ (if $k > 0$) or $LowerBound[i,-k] > 0$ (if $k < 0$). (In order to keep the *LowerBound* matrix sparse, the sums of hard sphere radii are not stored in it.)

OUTPUT: The triangle inequality limits implied by the input distance bounds *LowerBound* and *UpperBound*. (If the sum of the hard sphere radii is larger than the lower triangle limit, the former is output instead.)

PROCEDURE:

```
Define function FIND_PATHS( root ):
  Initialize Queue := [root].
  While Queue not empty, do:
    Set j := HEAD(Queue) and delete j from Queue.
    For each arc [j,k] pointing out from j with Abs(k) > j, do:
      If k > 0 and PathLength[j]+UpperBound[j,k] < PathLength[k]:
        Set PathLength[k] := PathLength[j]+UpperBound[j,k].
        If k not in Queue, then:
          If ParentNode[k] <> j, add k to end of Queue;
          else add k to beginning of Queue.
        Set ParentNode[k] := j.
      If k < 0 and PathLength[j]-LowerBound[j,-k] > PathLength[k]:
        Set PathLength[k] := PathLength[j]-LowerBound[j,-k].
        If k not in Queue, then:
          If ParentNode[k] <> j, add k to end of Queue;
          else add k to beginning of Queue.
        Set ParentNode[k] := j.
End definition.

For i := 1 to #A, do:
  FIND_PATHS( i ).
  For m := i+1 to #A, do:
```

If *LowerBound*[*i*,*m*] > *UpperBound*[*i*,*m*] then:
Halt: bounds are erroneous.
Set *LowerLimit*[*i*,*m*] := Max(*Radius*[*i*]+*Radius*[*m*], *PathLength*[−*m*])
and *UpperLimit*[*i*,*m*] := *PathLength*[*m*]).

At the end of the calculation, the *ParentNode* array contains the immediate predecessor of each node in the tree, i.e. the first node on the path from that node to the root of the tree. Observe that when no restructuring of the tree occurs on updating the shortest path to a node, we add it to the head of the queue rather than the tail. In this way, we rapidly propagate changes through stable parts of the tree. On large problems, the above algorithm is easily an order of magnitude more efficient than the full matrix algorithm TRIANGLE given in §5.3.

Another nice feature of this approach to triangle inequality bound smoothing is that it is easy to update a shortest path tree after tightening a single arc length (i.e. lowering an upper or raising a lower bound) from the root to some other node of the tree, simply by applying the FIND_PATHS procedure to the root again. The ability to update shortest path trees after such modifications, in turn, is important in designing efficient algorithms for metrization (§5.3). Thus if we define FIND_PATHS as above, the following variation on the above algorithm computes a random metric space whose distances lie wholly within the bounds (if one exists).

Algorithm 6.10 (TRIAL_METRICS):
INPUT: As in Algorithm 6.9 above, including the definition of FIND_PATHS.
OUTPUT: A symmetric matrix *Distance* whose entries lie within their respective bounds and which obey the triangle inequality.
PROCEDURE:
For *i* := 1 to #A, do:
FIND_PATHS(*i*).
For *j* := *i*+1 to #A, do:
Set *LowerLimit* := Max(*Radius*[*i*]+*Radius*[*j*], *PathLength*[−*j*])
and *UpperLimit* := *PathLength*[*j*]).
Set *Distance*[*i*,*j*] := RANDOM(*LowerLimit*, *UpperLimit*).
Set *LowerBound*[*i*,*j*] := *UpperBound*[*i*,*j*] := *Distance*[*i*,*j*].
Make [*i*,*j*] and [*i*,−*j*] arcs in the digraph.
FIND_PATHS(*i*).
For *k* := *j*+1 to #A, do:
Set *LowerBound*[*i*,*k*] := *PathLength*[−*k*], *UpperBound*[*i*,*k*] := *PathLength*[*k*].
Reset digraph to its original state.

Here, RANDOM returns a random number between its two arguments. Since shortest path trees can be updated in time proportional to their size, the above is $O(\#A^3)$ in complexity.

6.3. The EMBED algorithm

For a typical molecule defined by distance and chirality constraints $(l, u, \tilde{\chi})$ and containing more than a dozen or so atoms, it is unlikely that the minimization of an error function which started from a random positioning of the atoms would converge to zero error. Instead it would become trapped in a local minimum, and it would be difficult to analyze the problem and fix it. In designing an algorithm to avoid this problem, our criteria were that all points must be treated equally, and the method should be so stable numerically that as long as the given constraints admit some embedding, representative coordinates will always be found. Fortunately, the basic theory of distance geometry provides us with the mathematical tools which are necessary to compute much better-than-random starting positionings efficiently, reliably and reasonably uniformly. The resulting procedure is called the *EMBED algorithm*. In this section we will present the basic theory on which this algorithm is founded, as well as some of the more heuristic tricks which have made it into an extremely useful tool for solving chemically relevant instances of the Fundamental Problem of Distance Geometry, and hence a wide variety of conformational problems.

The first step of the EMBED algorithm is to use the available geometric information to choose *trial distances* $\hat{d}(a,b)$ between the atoms of the molecule *a priori* which are, in some sense, "close" to the distances in an admissible conformation thereof. To this end we use the bound smoothing algorithms introduced in §5.3 and §5.4 to obtain the best available limits on the values of the individual distances. The trial distances $\hat{d}(a,b)$ are then chosen randomly between their respective lower and upper limits. Although reasonably good results have been obtained simply by choosing the $\hat{d}(a,b)$ to be uniformly distributed within the interval $[l(a,b), u(a,b)]$, it is preferable (although more time consuming) to use metrization (Algorithm 5.38 or 6.10) to compute a random metric space whose distances all lie within their respective limits.

Even though the distance function $\hat{d}$ which is obtained this way obeys both the triangle inequality and the given lower and upper bounds, it will not usually be three-dimensional Euclidean. Thus in the next, and most important, step of the EMBED algorithm we find a three-dimensional Euclidean distance function

$d : A^2 \rightarrow \mathbb{R}$ which is "close" to the trial distance function $\hat{d}$. In the process, we simultaneously find a positioning $\mathbf{p} : A \rightarrow \mathbb{R}^3$ of the atoms such that $d(a,b) = \|\mathbf{p}(a) - \mathbf{p}(b)\|$ for all $a,b \in A$. Since the Euclidean distance function d will not generally be equal to the function $\hat{d}$, the distances $d(a,b)$ will not generally satisfy the given lower and upper bounds. If the given distance constraints (l, u) are reasonably complete and precise, however, the corresponding Cartesian coordinates $\underline{\mathbf{p}} = \mathbf{p}(\underline{A})$ will often be close enough to admissibility to enable them to be refined by the minimization of an error function to fit *both* the distance and chirality constraints.

In order to fill in the details, we need a few new definitions.

Definition 6.11. The set of all $N \times N$ matrices over the real numbers can be regarded as an N^2-dimensional Euclidean vector space, which we call *matrix space*. The addition and multiplication operations in this space are matrix addition and scalar multiplication, respectively. The inner product of any two $N \times N$ matrices $\underline{\mathbf{X}}$ and $\underline{\mathbf{Y}}$ is given by $< \underline{\mathbf{X}}, \underline{\mathbf{Y}} > := tr(\underline{\mathbf{X}}^T \cdot \underline{\mathbf{Y}})$, and the matrix norm $\|\underline{\mathbf{X}}\| := < \underline{\mathbf{X}}, \underline{\mathbf{X}} >$ which is induced by this inner product is called the *Frobenius norm*. *Symmetric matrix space* is the $(N(N-1)/2)$-dimensional subspace of matrix space consisting of all symmetric matrices. *Zero-diagonal matrix space* is the subspace of symmetric matrix space consisting of all symmetric matrices with zeros down the diagonal.

It is interesting to note that the RMS distance deviation or *distance matrix error*

$$DME(d, d') := \left(\frac{\sum_{i<j} (d(a_i, a_j) - d'(a_i, a_j))^2}{N(N-1)/2} \right)^{1/2} = \left(\frac{\|\underline{\mathbf{M}}_d - \underline{\mathbf{M}}_{d'}\|^2}{N(N-1)/2} \right)^{1/2}, \tag{6.11}$$

which is often used as a measure of the dissimilarity between two conformations of a molecule (where $N := \#A$, and $\underline{\mathbf{M}}_d$ and $\underline{\mathbf{M}}_{d'}$ are the distance matrices versus some ordering of its atoms) is really just the distance between these matrices in zero-diagonal matrix space. This is, of course, only a lower bound on the length of the geodesic between the corresponding premetric spaces in the set

$$\mathbf{D}^{(1/2)}(l, u, \tilde{\chi}) := [\, \|\mathbf{p}(a_i) - \mathbf{p}(a_j)\| \mid 1 \leq i < j \leq N, \mathbf{p} \in \mathbf{P}(l, u, \tilde{\chi})] \tag{6.12}$$

which would be the preferable measure.†

The approximation of nonnegative matrices in zero-diagonal matrix space by Euclidean distance matrices plays an important role in an area of data analysis

† Note that the corresponding set $\mathbf{D}(l, u, \tilde{\chi})$ of squared distances is just a subset of the Cayley-Menger set $\mathbb{D}(N, 3)$), as defined in §3.5.

known as *multidimensional scaling* [Kruskal1978, Carroll1980], which seeks to produce geometric representations of abstract measures of dissimilarity between various sets of data in the social sciences. Given that we have found a trial distance function $\hat{d}$, the most obvious and perhaps also the best criterion for a "good" Euclidean approximation d is probably that which minimizes the *DME* between them. Another likely candidate would be that which minimizes the RMS difference between the *squared* distances $D := d^2$. Although both of these possibilities have been considered [Gower1984], as of the time of writing there exists no good method of finding these approximations.* Since the bordered matrix of squared distances $\underline{\tilde{\mathbf{M}}}_D$ must have rank five, our original approach to this problem was to find the rank five symmetric matrix $\underline{\mathbf{N}}$ which minimized the Frobenius norm $\|\underline{\tilde{\mathbf{M}}}_{\hat{D}} - \underline{\mathbf{N}}\|$. Although this matrix generally had neither zeros down its main diagonal nor ones along its borders, its other entries were often close enough to three-dimensional distances to enable coordinates to be found by triangulation [Crippen1978a]. Our current method, however, is substantially more robust.

In this method, we use the Euclidean distance function d which minimizes the Frobenius norm of the difference between the *metric matrices* obtained from the squared distances by the law of cosines relative to some fixed origin:

$$\underline{\mathbf{M}}_B := \left[\tfrac{1}{2}(D(\mathrm{a}_0,\mathrm{a}_i) + D(\mathrm{a}_0,\mathrm{a}_j) - D(\mathrm{a}_i,\mathrm{a}_j)) \mid i, j = 1, \ldots, N := \#\mathrm{A} \right], \tag{6.13}$$

where $\mathrm{a}_0 \notin \mathrm{A}$ denotes an atom at the origin. Since our results from Chapter 3 show that the premetric $D : \mathrm{A}^2 \rightarrow \mathbb{R}$ is three-dimensional Euclidean if and only if the corresponding metric matrix $\underline{\mathbf{M}}_B$ is positive semidefinite of rank three, this means we are seeking the positive semidefinite rank three matrix which best approximates a given symmetric matrix in terms of the Frobenius norm. Before going into how we actually compute this optimum approximation, however, let us show how the Cartesian coordinates can actually be obtained once we have it [Young1938]. This result provides us with a means of computing Cartesian coordinates from Euclidean distances which is numerically much more stable than triangulation (or, in matrix terms, the Cholesky decomposition of the metric matrix [MacKay1974, Faddeev1963]).

PROPOSITION 6.12. *Let* $\underline{\mathbf{M}}_B(\underline{\mathrm{A}})$ *be the positive semidefinite rank three matrix which corresponds to a Euclidean premetric space* $(\mathrm{a}_0 \cup \mathrm{A}, D)$ *via Equation (6.13), and let*

$$\underline{\mathbf{M}}_B = \underline{\mathbf{W}}^T \cdot \underline{\Omega} \cdot \underline{\mathbf{W}} \tag{6.14}$$

be its eigenvalue decomposition, where $\underline{\Omega}$ *is* 3×3 *diagonal and* $\mathbf{W}$ *is* $3 \times N$ *orthogonal. Then the mapping* $\mathbf{p} : \mathrm{a}_0 \cup \mathrm{A} \rightarrow \mathbb{R}^3$ *given by*

---

* Recent results suggest this *may* nevertheless be possible [Hayden1988].

$$\mathbf{p}(a_i) \ := \ [\omega_1^{½} w_{1i}, \omega_2^{½} w_{2i}, \omega_3^{½} w_{3i}] \quad (i = 1, \ldots, N) \tag{6.15}$$

and $\mathbf{p}(a_0) := \mathbf{0}$ *is an isometric embedding of* $(a_0 \cup A, D)$ *in* $\mathbb{R}^3$, *i.e.*

$$D(a_i, a_j) \ = \ \|\mathbf{p}(a_i) - \mathbf{p}(a_j)\|^2 \tag{6.16}$$

for all $i, j = 0, \ldots, N$.

Proof: Let $\underline{\mathbf{p}} = \mathbf{p}(a_0, \underline{A})$ as usual. The distances to the point $\mathbf{p}_0$ satisfy

$$\|\mathbf{p}_i - \mathbf{p}_0\|^2 \ = \ \|\mathbf{p}_i\|^2 \ = \ \sum_{k=1}^{3} \omega_k w_{ki}^2 \ = \ D(a_0, a_i) \tag{6.17}$$

by the definition of the metric matrix. For the remaining distances, we have

$$\begin{aligned} & \|\mathbf{p}_i - \mathbf{p}_j\|^2 \\ = \ & \|\mathbf{p}_i\|^2 + \|\mathbf{p}_j\|^2 - 2\mathbf{p}_i \cdot \mathbf{p}_j \\ = \ & D(a_0, a_i) + D(a_0, a_j) - 2 \sum_k \omega_k w_{ki} w_{kj} \\ = \ & D(a_0, a_i) + D(a_0, a_j) - 2(\tfrac{1}{2}(D(a_0, a_i) + D(a_0, a_j) - D(a_i, a_j))) \\ = \ & D(a_i, a_j) \, , \end{aligned} \tag{6.18}$$

as desired. QED

It follows from the orthogonality of the eigenvectors in $\underline{\mathbf{W}}$ that the coordinates $\mathbf{p}_i = [x_i, y_i, z_i]$ are principle axis coordinates, i.e. that they diagonalize the inertial tensor:

$$\begin{bmatrix} \Sigma(y_i^2+z_i^2) & -\Sigma x_i y_i & -\Sigma x_i z_i \\ -\Sigma x_i y_i & \Sigma(x_i^2+z_i^2) & -\Sigma y_i z_i \\ -\Sigma x_i z_i & -\Sigma y_i z_i & \Sigma(x_i^2+y_i^2) \end{bmatrix} \tag{6.19}$$

$$= \begin{bmatrix} \omega_2 \Sigma w_{2i}^2 + \omega_3 \Sigma w_{3i}^2 & -\omega_1^{½} \omega_2^{½} \Sigma w_{1i} w_{2i} & -\omega_1^{½} \omega_3^{½} \Sigma w_{1i} w_{3i} \\ -\omega_1^{½} \omega_2^{½} \Sigma w_{1i} w_{2i} & \omega_1 \Sigma w_{1i}^2 + \omega_3 \Sigma w_{3i}^2 & -\omega_2^{½} \omega_3^{½} \Sigma w_{2i} w_{3i} \\ -\omega_1^{½} \omega_3^{½} \Sigma w_{1i} w_{3i} & -\omega_2^{½} \omega_3^{½} \Sigma w_{2i} w_{3i} & \omega_1 \Sigma w_{1i}^2 + \omega_2 \Sigma w_{2i}^2 \end{bmatrix}$$

$$= \begin{bmatrix} \omega_2+\omega_3 & 0 & 0 \\ 0 & \omega_1+\omega_3 & 0 \\ 0 & 0 & \omega_1+\omega_2 \end{bmatrix}.$$

The sums of pairs of eigenvalues are therefore the moments of inertia relative to these axes and our choice of origin.

Thus once one has all the interatomic distances in a conformation of the molecule, Cartesian coordinates can be calculated simply by (partial) matrix diagonalization (as described below). It turns out that the positive semidefinite rank three matrix which optimally approximates a given symmetric matrix in the Frobenius norm can be computed by almost the same procedure. The fact that the best fixed-rank approximation can be obtained in this way was originally observed by [Eckart1936], and this result was first used by the present authors to calculate coordinates from trial distances in [Crippen1978b]. At about the same time, the simple extension of this procedure necessary to obtain optimum fixed-rank approximations which are also positive semidefinite was discovered by [Mardia1978]. Using the technique of *majorization* [Marshall1979], it was subsequently shown that this approximation is actually optimum in the l^p norm for all $p \geq 1$ [Mathar1985]. Here we prefer to give an elementary proof for the Frobenius or l^2 norm with which we are most concerned, which extends the proof of optimality given in [Havel1983, Havel1985] for the fixed-rank case.

One rather interesting feature of our proof is that it relys upon the following standard result from combinatorial optimization:

LEMMA 6.13. *Let* $\underline{\omega}$, $\underline{v}$ *be two vectors in* $\mathbb{R}^N$. *Then the optimization problem*

$$\text{maximize}: \ \textstyle\sum_{i,j} v_i \omega_j z_{ij} \tag{6.20}$$

$$\text{subject to}: \ \textstyle\sum_i z_{ij} = \sum_i z_{ji} = 1 \ \text{ for } j = 1, \ldots, N\,,$$

$$\text{and}: \ 0 \leq z_{ij} \ \text{ for } i, j = 1, \ldots, N$$

has as its optimum value

$$\textstyle\sum_i \omega_i v_{\pi(i)}\,, \tag{6.21}$$

where π *is a permutation of* $[1, \ldots, N]$.

Proof: This is an immediate consequence of the well-known fact that the basic solutions of the assignment problem are always integer-valued (see, for example, [Papadimitriou1982]). QED

We are now ready for our main result.

THEOREM 6.14. *Let* $\underline{\mathbf{M}}$ *be an* $N \times N$ *symmetric matrix with eigenvalue decomposition*

$$\underline{\mathbf{M}} = \underline{\mathbf{X}}^T \cdot \underline{\Omega} \cdot \underline{\mathbf{X}}\,, \tag{6.22}$$

where $\underline{\Omega} = \underline{\mathbf{diag}}[\omega_i \mid i = 1, \ldots, N]$ *with* $\omega_1 \geq \cdots \geq \omega_N$, *and let*

$$\underline{\Omega}' := \underline{\mathbf{diag}}[\max(0,\omega_1), \ldots, \max(0,\omega_n), 0, \ldots, 0] \in \mathbb{R}^{N \times N} \tag{6.23}$$

Then the $N \times N$ *positive semidefinite matrix of rank not exceeding* n *which best approximates* $\underline{\mathbf{M}}$ *in the Frobenius norm is given by*

$$\underline{\mathbf{M}}' := \underline{\mathbf{X}}^T \cdot \underline{\Omega}' \cdot \underline{\mathbf{X}}\,. \tag{6.24}$$

Proof: For any $N \times N$ matrix $\underline{\mathbf{N}} = \underline{\mathbf{Y}}^T \cdot \underline{\Upsilon} \cdot \underline{\mathbf{Y}}$, we have

$$\|\underline{\mathbf{M}} - \underline{\mathbf{N}}\|^2 = \|\underline{\mathbf{M}}\|^2 + \|\underline{\mathbf{N}}\|^2 - 2 < \underline{\mathbf{M}}, \underline{\mathbf{N}} > . \tag{6.25}$$

Thus whatever the optimum approximation to $\underline{\mathbf{M}}$ may be, it will maximize $< \underline{\mathbf{M}}, \underline{\mathbf{N}} >$ subject to the norm $\|\underline{\mathbf{N}}\|$ fixed. Now:

$$\begin{aligned} < \underline{\mathbf{M}}, \underline{\mathbf{N}} > &= tr(\underline{\mathbf{M}}^T \cdot \underline{\mathbf{N}}) = tr(\underline{\mathbf{M}} \cdot \underline{\mathbf{N}}) \quad \text{by symmetry} \\ &= tr(\underline{\mathbf{X}}^T \cdot \underline{\Omega} \cdot \underline{\mathbf{X}} \cdot \underline{\mathbf{Y}}^T \cdot \underline{\Upsilon} \cdot \underline{\mathbf{Y}}) \\ &= tr\left[\Sigma_k\left(\left(\Sigma_l x_{li} \omega_l x_{lk}\right) \cdot \left(\Sigma_m y_{mk} \upsilon_m y_{mj}\right)\right)\right] \\ &= tr\left[\Sigma_{l,m}\left(\omega_l \upsilon_m x_{li} y_{mj} \Sigma_k x_{lk} y_{mk}\right)\right] \\ &= \Sigma_{l,m} \omega_l \upsilon_m (\mathbf{x}_l \cdot \mathbf{y}_m)^2 , \end{aligned} \tag{6.26}$$

where $\mathbf{x}_l$ and $\mathbf{y}_m$ denote the l-th and m-th eigenvectors of $\underline{\mathbf{M}}$ and $\underline{\mathbf{N}}$, respectively. Since the dot products $\mathbf{x}_l \cdot \mathbf{y}_m$ are the directional cosines of the eigenvectors of $\underline{\mathbf{M}}$ with respect to those of $\underline{\mathbf{N}}$ and conversely, we have

$$\Sigma_l (\mathbf{x}_l \cdot \mathbf{y}_m)^2 = \Sigma_m (\mathbf{x}_l \cdot \mathbf{y}_m)^2 = 1 . \tag{6.27}$$

Since the summands themselves are independent quantities, it follows from Lemma 6.13 that the maximum value of $< \underline{\mathbf{M}}, \underline{\mathbf{N}} >$ is given by $\Sigma_i \omega_i \upsilon_{\pi(i)}$ for some permutation π of $[1, \ldots, N]$. In this case, for an appropriate ordering of the columns of $\underline{\mathbf{Y}}$ the vectors $\mathbf{x}_l$ and $\mathbf{y}_m$ are *biorthonormal*, i.e. $\mathbf{x}_l \cdot \mathbf{y}_m = \delta_{lm}$, so that the optimum approximation $\underline{\mathbf{N}} = \underline{\mathbf{M}}' =: \underline{\mathbf{X}}^T \cdot \underline{\Omega}' \cdot \underline{\mathbf{X}}$ is diagonalized by the same orthogonal transformation $\underline{\mathbf{X}}$ as is $\underline{\mathbf{M}}$. It follows that

$$\begin{aligned} tr((\underline{\mathbf{M}} - \underline{\mathbf{M}}')^2) &= tr(\underline{\mathbf{X}}^T \cdot \underline{\Omega} \cdot \underline{\mathbf{X}} - \underline{\mathbf{X}}^T \cdot \underline{\Omega}' \cdot \underline{\mathbf{X}}) \\ &= tr(\underline{\mathbf{X}}^T \cdot (\underline{\Omega} - \underline{\Omega}') \cdot \underline{\mathbf{X}}) \cdot (\underline{\mathbf{X}}^T \cdot (\underline{\Omega} - \underline{\Omega}') \cdot \underline{\mathbf{X}}) \\ &= tr(\underline{\mathbf{X}}^T \cdot (\underline{\Omega} - \underline{\Omega}')^2 \cdot \underline{\mathbf{X}}) \\ &= tr(\underline{\Omega} - \underline{\Omega}')^2 \\ &= \Sigma_i (\omega_i - \omega_i')^2 \end{aligned} \tag{6.28}$$

since the trace is invariant under orthogonal transformations. From this it is readily apparent that $\omega_i' = max(0, \omega_i)$ for $i = 1, \ldots, n$ and $\omega_i' = 0$ for $i = n+1, \ldots, N$, as claimed. QED

It should be mentioned that both [Sippl1985] and [Schlitter1987] have devised numerically stable pivoting schemes for the Choleski decomposition of indefinite metric matrices which require only a $4 \times N$ submatrix of the complete distance matrix (i.e. only the distances to and among the atoms of some fixed tetrahedron). The Cartesian coordinates that these procedures yield when applied to non-Euclidean distance matrices, however, are in no sense an optimum approximation to them.

The approximation $\underline{\mathbf{M}}'$ given by Theorem 6.14 is optimum for any fixed choice of origin for the metric matrix $\underline{\mathbf{M}}$. Although the signature of the metric matrix computed from any distance matrix does not depend upon our choice of origin, the relative magnitudes of the eigenvalues do. The approximation $\underline{\mathbf{M}}'$ will generally be better when the three eigenvalues of greatest absolute value are all positive and hence retained. This will usually be the case if the trial distance function $\hat{d}$ is reasonably close to being Euclidean, but the further our origin is translated in a negative direction of the bilinear form B_D in the amalgamation space (see §3.1), the larger the negative eigenvalues become in magnitude. The safest choice of origin seems to be at the centroid of the points, which also avoids singling out any one atom as special. It can be found directly in terms of the distances as follows.

PROPOSITION 6.15. *The squared distances D_{0i} from each point $\mathbf{p}_i \in \mathbb{R}^n$ to the centroid $\mathbf{p}_0$ of a collection of N points in an arbitrary bilinear affine space $(\mathbb{R}^n, \mathbb{R}^n, B)$ is given in terms of the mutual squared distances $D_{ij} := B(\mathbf{p}_i - \mathbf{p}_j, \mathbf{p}_i - \mathbf{p}_j)$ by:*

$$D_{0i} = \frac{1}{N}\sum_{j=1}^{N} D_{ij} - \frac{1}{N^2}\sum_{k>j=1}^{N} D_{jk} . \tag{6.29}$$

Proof: If we let $\mathbf{v}_{0i}$ be the vector from the i-th point to the centroid, we have

$$\sum_{j=1}^{N}\mathbf{v}_{0i} = \mathbf{0} = \sum_{j=1}^{N}(\mathbf{v}_{01}+\mathbf{v}_{1j}) , \tag{6.30}$$

where $\mathbf{v}_{ij} := \mathbf{p}_i - \mathbf{p}_j$ is the vector from the j-th to the i-th point. Thus $\mathbf{v}_{01} = -N^{-1}\sum_{j=2}^{N}\mathbf{v}_{1j}$, and:

$$\begin{aligned} D_{01} = B(\mathbf{v}_{01}; \mathbf{v}_{01}) &= \frac{1}{N^2}\sum_{j=2}^{N}\sum_{k=2}^{N} B(\mathbf{v}_{1j}; \mathbf{v}_{1k}) \\ &= \frac{1}{2N^2}\sum_{j=2}^{N}\sum_{k=2}^{N}(D_{1j}+D_{1k}-D_{jk}) \\ &= \frac{1}{2N^2}\left(2(N-1)\sum_{j=2}^{N} D_{1j} - 2\sum_{k>j=2}^{N} D_{jk}\right) \\ &= \frac{N-1}{N^2}\sum_{j=2}^{N} D_{1j} - \frac{1}{N^2}\sum_{k>j=2}^{N} D_{jk} \\ &= \frac{1}{N}\sum_{j=1}^{N} D_{1j} - \frac{1}{N^2}\sum_{k>j=1}^{N} D_{jk} . \end{aligned} \tag{6.31}$$

Since the numbering of the points is arbitrary, the proposition is established. QED

It was first shown by Lagrange in 1783 that in the Euclidean case the second term in Equation (6.29) is equal to the radius of gyration of the points, which (assuming each has unit mass) is also given by the trace of the inertial tensor. By adding an $(N+1)$-th atom a_0 to the set A and taking its distances to the remaining atoms to be equal to those given by Equation (6.29), we ensure that the row / column sums of the metric matrix $\underline{\mathbf{M}}_B^0$ obtained from Equation (6.13) are all zero, so that its rank is at most $N-1$. The coordinates computed from this metric matrix via Theorem 6.14 are then center of mass, principal axis coordinates. Observe also that the transformation between the matrix of squared distances $\underline{\mathbf{M}}_D$ and $\underline{\mathbf{M}}_B^0$ can be written in matrix form, as:

$$\underline{\mathbf{M}}_D \mapsto \underline{\mathbf{M}}_B^0 := \underline{\mathbf{T}}^T \cdot (-\tfrac{1}{2}\underline{\mathbf{M}}_D) \cdot \underline{\mathbf{T}}, \tag{6.32}$$

where $\underline{\mathbf{T}} = \underline{\mathbf{I}} - 1/N\,\mathbf{1} \cdot \mathbf{1}^T$ is the projection orthogonal to $\mathbf{1} \in \mathbb{R}^N$ (note that $\underline{\mathbf{T}}$ is a symmetric matrix). Thus in the Euclidean case, $\underline{\mathbf{M}}_B^0$ is really just the matrix of the bilinear form which is obtained by projecting the bilinear form B_D defined by the matrix $\underline{\mathbf{M}}_D$ onto the subspace $\{\mathbf{x} \in \mathbb{R}^N \mid \mathbf{x} \cdot \mathbf{1} = 0\}$ (i.e. the amalgamation space, as defined in §3.1). This fact provides us with an alternative proof of Theorem 3.6.

Fortunately, it is rarely if ever necessary to completely diagonalize the center of mass metric matrix in order to compute its three largest eigenvalues and their associated eigenvectors. Since the eigenvectors $\mathbf{w}_i$ of a symmetric matrix $\underline{\mathbf{M}}$ are (or can be chosen to be) orthonormal, we can expand any vector $\mathbf{v} \in \mathbb{R}^N$ in terms of them as $\mathbf{v} = \sum v_i \mathbf{w}_i$. Thus the k-fold product $\underline{\mathbf{M}}^k \cdot \mathbf{v}$ can be expressed as:

$$\underline{\mathbf{M}}^k \cdot \mathbf{v} = \sum_{i=1}^{N} \omega_i^k v_i \mathbf{w}_i, \tag{6.33}$$

where the ω_i are the eigenvalues of $\underline{\mathbf{M}}$, and hence $\underline{\mathbf{M}}^k \cdot \mathbf{v} \to \omega_1^k \mathbf{w}_1$ as $k \to \infty$, where $|\omega_1| = max(|\omega_i| \mid i = 1, \ldots, N)$. It follows that ω_1 and its associated eigenvector $\mathbf{w}_1$ can be computed with probability one simply by taking a random unit-length vector $\mathbf{v} \in \mathbb{R}^N$, and repeatedly multiplying it by $\underline{\mathbf{M}}$ and renormalizing until no further (significant) change is obtained. This method, sometimes known as *exhaustion*, converges in the l^2 norm at a rate proportional to $|\omega_2/\omega_1|$ (where ω_2 is the eigenvalue of second largest absolute value). Assuming a uniform distribution in the eigenvalues within some interval of length proportional to N, this makes it only $O(N^2)$ in expected-time complexity [Faddeev1963].

Once the eigenvalue ω_1 of largest absolute value and corresponding eigenvector $\mathbf{w}_1$ have been found, this eigenspace may be subtracted from the matrix by setting $\underline{\mathbf{M}} := \underline{\mathbf{M}} - \omega_1 \mathbf{w}_1 \cdot \mathbf{w}_1^T$. This process, known as *matrix deflation*, leaves all the eigenvalues of $\underline{\mathbf{M}}$ intact except for ω_1, which is set to zero, so that a second application of the method of exhaustion from a new random vector will yield the eigenvalue ω_2 of second greatest absolute value together with its eigenvector $\mathbf{w}_2$, etc. This process is continued until the three eigenvalues of greatest *signed* magnitude and their eigenvectors have been found.

Algorithm 6.16:

INPUT: A matrix $\underline{\mathbf{M}}$ and a precision $\epsilon > 0$.

OUTPUT: Its three largest eigenvalues $\omega_1, \omega_2, \omega_3$ and their associated eigenvectors $\mathbf{w}_1, \mathbf{w}_2, \mathbf{w}_3$.

PROCEDURE:

$k := 1$.
While $k \leq 3$ do:
  Set $\mathbf{v}$ to some random unit-length vector.
  Repeat for $i = 1,2,3,\cdots$
    Set $\omega := (\underline{\mathbf{M}}^i\mathbf{v})\cdot(\underline{\mathbf{M}}^i\mathbf{v})/((\underline{\mathbf{M}}^{i-1}\mathbf{v})\cdot(\underline{\mathbf{M}}^i\mathbf{v}))$
    and $\mathbf{w} := \underline{\mathbf{M}}^i\mathbf{v}/\|\underline{\mathbf{M}}^i\mathbf{v}\|$.
  Until successive values of ω differ by less than ϵ.
  If $\omega > 0$ then
    $\omega_k := \omega$, $\mathbf{w}_k := \mathbf{w}$, $k := k+1$;
  Set $\underline{\mathbf{M}} := \underline{\mathbf{M}} - \omega(\mathbf{w}\cdot\mathbf{w}^T)$.

Although this algorithm could fail if the metric matrix $\underline{\mathbf{M}}$ has less than three positive eigenvalues, this never happens in practice. In fact, when the trial distances obey the triangle inequality, the three eigenvalues of greatest absolute value are usually the three largest in signed value, so that it is rarely necessary to calculate more than three eigenvalues altogether.

While exhaustion has been found to be wholly adequate in practice, convergence may be unacceptably slow if by chance two eigenvalues have nearly the same absolute value. Should this ever prove a problem, the more robust method based on Chebychev matrix polynomials may be used [Braun1981]. First scale $\underline{\mathbf{M}}$

$$s := \frac{1}{N}\,\text{trace}\,\underline{\mathbf{M}} \tag{6.34}$$

$$\underline{\mathbf{M}}_s := \frac{1}{s}\underline{\mathbf{M}} \tag{6.35}$$

where $\underline{\mathbf{M}}$ is of order N. Then choose an arbitrary starting vector $\mathbf{u}$, and then find the nth Chebyshev polynomial of $\underline{\mathbf{M}}_s$ applied to $\mathbf{u}$ by the usual recursion relations, denoting $T_n(\underline{\mathbf{M}}_s)\mathbf{u}$ by $\mathbf{q}_n$:

$$T_0(\underline{\mathbf{M}}_s)\mathbf{u} = \mathbf{u} \tag{6.36}$$

$$T_1(\underline{\mathbf{M}}_s)\mathbf{u} = \underline{\mathbf{M}}_s\mathbf{u} \tag{6.37}$$

$$T_{n+1}(\underline{\mathbf{M}}_s)\mathbf{u} = 2\underline{\mathbf{M}}_s T_n(\underline{\mathbf{M}}_s)\mathbf{u} - T_{n-1}(\underline{\mathbf{M}}_s)\mathbf{u} \tag{6.38}$$

But since

$$\underline{\mathbf{M}}_s = \sum_{k=1}^{N} \lambda_k \mathbf{e}_k^T \mathbf{e}_k \tag{6.39}$$

we have

$$\mathbf{q}_n = T_n(\underline{\mathbf{M}}_s)\mathbf{u} = \sum_{k=1}^{N} T_n(\lambda_k)\mathbf{e}_k \cdot \mathbf{u} \mathbf{e}_k \tag{6.40}$$

Therefore as $n \rightarrow \infty$,

$$\frac{\mathbf{q}_n}{\|\mathbf{q}_n\|} \rightarrow \mathbf{e}_1 \tag{6.41}$$

and

$$\frac{\mathbf{q}_n^T \underline{\mathbf{M}}_s \mathbf{q}_n}{\mathbf{q}_n \cdot \mathbf{q}_n} \rightarrow \lambda_1 \tag{6.42}$$

Scaling with s ensures that $\lambda_1 \geq 1$, and outside the interval [-1,1] Chebyshev polynomials grow more efficiently than powers, so that convergence is faster. Then the largest eigenvalue of $\underline{\mathbf{M}}$ is just $s\lambda$, and $\underline{\mathbf{M}}$ can be deflated as before to find the second and third eigenvalues. Generally the first scaling is sufficient to locate all three.

Of course, the coordinates $\mathbf{p}(\underline{\mathrm{A}})$ calculated from the eigenvectors of $\underline{\mathbf{M}}$ may not satisfy the given distance constraints, even though the (squared) trial distances, $\hat{d}$(a,b), from which $\underline{\mathbf{M}}$ was derived, did satisfy them. The usual approach is to refine the coordinates by minimizing an error function with respect to them, as discussed above. An alternative initial "refinement" is to revise the trial distances by

$$\hat{d}(\mathrm{a,b})^{new} = \alpha \hat{d}(\mathrm{a,b}) + (1-\alpha)\|\mathbf{p}(\mathrm{a})-\mathbf{p}(\mathrm{b})\|^2 \quad \forall\ \mathrm{a,b} \in \mathrm{A} \tag{6.43}$$

Then the new trial distances produce a new $\underline{\mathbf{M}}$ and hence revised $\mathbf{p}(\underline{\mathrm{A}})$. Iterating some 5 to 10 times this way, letting α progress linearly from 1 to 0, can sometimes reduce the number and magnitude of distance violations in the coordinates, making final refinement against the error function an easier matter [Braun1981].

We summarize by presenting a complete statement of the EMBED algorithm.

Algorithm 6.17 (EMBED):

INPUT: A (possibly incomplete) matrix of lower and upper bounds on the distances between the N atoms of a molecule, together with an error function E for these distance constraints and any chirality constraints that may be available.

OUTPUT: The Cartesian coordinates of a conformation of the molecule which is admissible with respect to the given constraints, or "none" if none exists.

PROCEDURE:

Smooth the input distance bounds, obtaining a complete set of lower and upper limits l, u on all the interatomic distances in the molecule (see §5.3 & 5.4); if the bounds are found to be incorrect, exit with "none".

Choose a complete matrix $\underline{\mathbf{M}}_{\tilde{D}}$ of trial squared distances between all the atoms of the molecule, either uniformly from between their limits or else by metrization (§5.3).

Let $\underline{\mathbf{M}}_B := -\frac{1}{2}\underline{\mathbf{T}}\cdot\underline{\mathbf{M}}_{\tilde{D}}\cdot\underline{\mathbf{T}}$, where $\underline{\mathbf{T}} := \underline{\mathbf{I}} - 1/N\,\mathbf{1}\cdot\mathbf{1}^T$.

Compute the three largest eigenvalues ω_1, ω_2, ω_3 and their associated eigenvectors $\mathbf{w}_1$, $\mathbf{w}_2$, $\mathbf{w}_3$ by Algorithm 6.16 above.

Set $\mathbf{x} := \sqrt{\omega_1}\cdot\mathbf{w}_1$, $\mathbf{y} := \sqrt{\omega_2}\cdot\mathbf{w}_2$, $\mathbf{z} := \sqrt{\omega_3}\cdot\mathbf{w}_3$ and $\mathbf{p}_i := [x_i, y_i, z_i]$ for $i = 1, \ldots, N$.

Minimize $E(\underline{\mathbf{p}})$ via your favorite algorithm (Appendix F).

If poor convergence obtained, exit with "fail".

As previously stressed, in large problems convergence is more-than-likely to be less-than-perfect, and hence the whole process should be repeated as many times as feasible from a variety of trial distances.

As a simple example of the EMBED algorithm, we return to our cyclohexane framework (§4.5). We arrange the tetrangle limits in a square matrix with the lower limits below and the upper limits above the diagonal:

$$\begin{bmatrix} 0 & 1 & \sqrt{8/3} & \sqrt{(7+2\sqrt{6})/3} & \sqrt{8/3} & 1 \\ 1 & 0 & 1 & \sqrt{8/3} & \sqrt{(7+2\sqrt{6})/3} & \sqrt{8/3} \\ \sqrt{8/3} & 1 & 0 & 1 & \sqrt{8/3} & \sqrt{(7+2\sqrt{6})/3} \\ 5/3 & \sqrt{8/3} & 1 & 0 & 1 & \sqrt{8/3} \\ \sqrt{8/3} & 5/3 & \sqrt{8/3} & 1 & 0 & 1 \\ 1 & \sqrt{8/3} & 5/3 & \sqrt{8/3} & 1 & 0 \end{bmatrix}.$$

Thus $\underline{\mathbf{M}}_D$ has only three indeterminate squared distances D_{14}, D_{25} and D_{36}. If $\underline{\mathbf{M}}_B^0 = -\frac{1}{2}\underline{\mathbf{T}}\cdot\underline{\mathbf{M}}_D\cdot\underline{\mathbf{T}}$ is the center of mass metric matrix as above, the eigenvalues may be found by solving for the roots of the characteristic polynomial $p(D_{14}, D_{25}, D_{36};\,\omega)$:

$$\frac{-\omega}{7776}\Big(\big(3(D_{12}D_{13}+D_{12}D_{23}+D_{13}D_{23}) - 22(D_{12}+D_{13}+D_{23}) + 121\big) +$$

$$\big(12(D_{12}+D_{13}+D_{23}) - 132\big)\,\omega + 36\omega^2\Big)\cdot$$

$$\Big(\big(27D_{12}D_{13}D_{23} - 75(D_{12}+D_{13}+D_{23}) - 250\big) +$$

$$\big(-54(D_{12}D_{13}+D_{12}D_{23}+D_{13}D_{23}) - 450\big)\,\omega +$$

$$\left(108(D_{12}+D_{13}+D_{23}) \right) \omega^2 - 216\omega^3 \Big)$$

If we choose our values of the indeterminate squared distances to be those in the chair conformation, i.e. $D_{12} = D_{13} = D_{23} = 11/3$, this becomes:

$$\frac{\omega^3}{54} (6\omega - 1)(3\omega - 8)^2 ,$$

which has roots [8/3,8/3,1/6,0,0,0]. The eigenvectors of the three positive eigenvalues are

$$\mathbf{w}_1 = \tfrac{1}{2} \cdot [-1,-1,0,1,1,0], \quad \mathbf{w}_2 = \tfrac{1}{2} \cdot [1,0,-1,-1,0,1]$$

for $\omega = 8/3$ and

$$\mathbf{w}_3 = 1/\sqrt{6} \cdot [1,-1,1,-1,1,-1]$$

for $\omega = 1/6$, from which principal axis, center of mass coordinates may be obtained by multiplication with the square roots of the eigenvalues.

On the other hand, if we were to choose the squared distances which lead to a pseudo-Euclidean space, i.e. $D_{12} = D_{13} = D_{23} = (1/3)(19 - 2/3\sqrt{6})$, the roots of the characteristic polynomial are

$$[4-(2/3)\sqrt{6},\ 4-(2/3)\sqrt{6},\ (1/3)(-4+2\sqrt{6}),\ (1/3)(-4+2\sqrt{6}),\ 0,\ 2/3-3/2\sqrt{6}]$$

in decreasing order. It turns out that $\mathbf{w}_1$ and $\mathbf{w}_2$ above are eigenvectors of the eigenvalue $\omega = 4-(2/3)\sqrt{6}$, but $\mathbf{w}_3$ is the eigenvector of $\omega = 2/3-3/2\sqrt{6} < 0$. If we choose instead an eigenvector of $\omega = (1/3)(-4+2\sqrt{6}) > 0$, e.g. $\mathbf{w}'_3 = [0.24, 0.33, -0.58, 0.24, 0.33, -0.58]$, on multiplying by the roots of the eigenvalues we get the coordinates of a set of points in $\mathbb{R}^3$ which match the chosen distances to within 0.247Å *DME*.

Some applications of this algorithm to the determination of molecular structure from experimental data will be presented in the forthcoming chapter.

6.4. Other Methods

One consequence of the NP-hardness of the Fundamental Problem is that there is no single "best" way to solve it. Instead, there exist a large variety of methods, each of which has been specialized to handle a particular type of distance constraints well. The EMBED algorithm is probably the single most general method, but even so there are certain types of distance constraints for which its convergence is poor. For example, when a long flexible chain is present whose conformation is determined entirely by bounds on the vicinal distances, and no "long-range" constraints are present, the trial distances obtained by bound smoothing and metrization are likely to be far from Euclidean, in which case the best fit to them will not even come close to fitting the bounds. In this section we will therefore survey some alternative methods which have been pursued by other workers in the context of the various more specialized versions of the Fundamental Problem which are of greatest immediate concern to them.

The approach that first occurs to researchers familiar with molecular mechanics calculations is to add a distance constraint "penalty" function, similar to the error functions we have used, onto an ordinary intramolecular energy function, and to minimize the combined function as usual. When a sufficiently large weight is applied to the penalty function and the constraints are consistent, the global minimum of the combined function is assured of being a *feasible* energy minimum, i.e. one which satisfies the impose constraints. The addition of the energy to the penalty function, however, greatly exacerbates the local minimum problem, and the minimization of the combined function seldom converges to a global minimum in any but the most trivial of problems. The result is a difficult-to-understand compromise between the various constraint violations and the conformational energy. As a rule, the Lagrange multiplier based methods of constrained optimization which are discussed in Chapter 9 enforce the constraints much more rigorously, and should be used in preference to penalty function methods whenever possible.

One means of escaping from local minima in the combined energy and penalty functions which has been used with some success is a method of stochastic optimization based upon the *molecular dynamics* techniques which were developed to simulate the short-term dynamical behavior of molecules. This is usually known as *restrained molecular dynamics.* In this approach, the combined energy / penalty function is treated as a potential function, each atom is assigned a momentum as well as a position, and Newton's equations of motion are solved iteratively. Since the atoms are capable of moving uphill against the potential gradient a certain amount before losing their momenta, the conformation can escape from local minima in the course of this simulation. If one starts with a high temperature (i.e. mean square momentum) and *anneals* the system gradually, the conformation settles down in a minimum which is likely to be a rather low one, although it is by no means necessarily the global minimum.

For example, Zuiderweg *et al.* [Zuiderweg1985] have described their use of molecular dynamics with NMR derived distance constraints to produce allowed conformations of the *lac* repressor headpiece, a protein of 51 residues. In the computer model, the molecule was represented as 497 atoms, some of those being CH_2 groups, etc. The 169 constraints derived from NMR experiments happened to all be upper bounds, so the penalty function was

$$V_{ij} = \begin{cases} \frac{1}{2} W(d_{ij} - u_{ij})^2 & \text{if } d_{ij} > u_{ij} \\ 0 & \text{if otherwise} \end{cases} \qquad (6.44)$$

with a preferred weight of $W = 250$ kJ mol^{-1} nm^{-2}. Of course this approach requires a suggested starting conformation, which they found by constructing a physical model of the protein and adjusting it until most of the constraints were approximately satisfied. A molecular dynamics simulation of 10 picoseconds steadily reduced both the calculated energy and the number of constraint violations before locking into a conformation that resisted further improvement. This was manually altered via computer graphics so that an additional 30 ps of molecular dynamics substantially improved the energy, but only reduced the number of constraints violated by more than 1 Å from 14 to 10. Clearly the numerous local minima in the force field make satisfying the constraints difficult even for molecular dynamics, but having a low energy final result is desirable (see Chapter 9).

Subsequent work has shown that this method is capable of producing conformations which satisfy complex combinations of distance and dihedral angle constraints as well as any which have been obtained from the EMBED algorithm [Clore1987a]. The method has become especially popular for determining the conformation of proteins from nuclear magnetic resonance measurements [Vlieg1986, Clore1986a] (see also Chapter 7). Applied to such large molecules, however, the computer demands of this approach vastly exceed those of the EMBED algorithm, if only because the amount of time it takes to evaluate conformational energy functions and their gradients is considerably larger than the amount of time required for error functions. Fortunately, the computational demands of the method are substantially less if a good starting conformation is available, and the EMBED algorithm provides a good way to get one. Thus molecular dynamics is being used as a method of further reducing the constraint violations and simultaneously minimizing the energy of protein conformations which have been calculated by the EMBED algorithm [Clore1986b, Clore1987b, Clore1987c]. It would be interesting to see if such stochastic optimization methods could be applied to the error functions used by the EMBED algorithm to directly reduce the constraint violations without the high overhead of evaluating an energy function.

Leaving out the energy function and dealing strictly with geometric constraints leaves still an intractable problem [Goel1978, Goel1979, Goel1982,

Cariani1985]. With penalty terms similar to the one above, Goel and Ycas were able to reproduce the known crystal structure of bovine pancreatic trypsin inhibitor (BPTI) represented as 58 C^α points, given *all exact correct* interresidue distances. However, supplying only $d_{i-4,i}$, $d_{i-3,i}$, $d_{i-2,i}$, and $d_{i-1,i}$ exact correct distances for each residue i resulted in even Bremermann's minimizer locking into incorrect local minima, where the penalty remained greater than zero. Since Bremermann's local minimization algorithm is noted for its ability to leave small local minima in favor of nearby deeper valleys, this result is rather sobering. Wako and Scheraga [Wako1982] experienced similar difficulties in their repetition of Goel and Ycas's work, employing a number of sets of distance constraints. The conclusion is that satisfying constraints in a least squares sense by minimizing a penalty function is not very robust, if the starting conformation is poor, although it does enable one to formulate constraints of rather elaborate form, such as requiring certain average properties and given standard deviations.

A highly original approach to determining planar structures from sparse sets of exact distance constraints has been developed in the context of what is known as the *position location problem* by systems engineers [Yemini1979, Yemini1982]. This is the problem of determining the relative positions of an array of mobile units such as, for example, planes or ships, from dynamic measurements of their mutual distances, as derived from the time it takes sound, light or radio waves to travel between them. Spatial versions of this problem also occur in satellite geodesy. The method which Yemini *et al.* have used to solve the problem is called *incremental position location*. If (A, B, d) is the abstract bar and joint framework defined by the available exact distance constraints, the method starts by finding all occurrences of a set of small, generically rigid subgraphs in the graph (A, B), and determining coordinates for the joints in each trigonometrically, relative to its own frame of reference (i.e. origin and axes). Each such subset is called a "hyperedge".

At each step in the execution of the procedure, small subsets of hyperedges are compared against various "welders" which allow them to be merged into a single hyperedge. For example, if two hyperedges have three or more joints in common, the relative positions of all their joints may be uniquely determined relative to a common frame of reference. A similar result holds if there are three or more distance constraints between any pair of hyperedges. If a pair of hyperedges has exactly two joints in common, however, then there exist two possible planar arrangements related by reflection in the line through those joints. Trees of such pairs of hyperedges may be constructed and their possible planar arrangements enumerated to see which, if any, of these arrangements matches the distance constraints between their constituent joints. The resultant procedure is said to be both efficient and robust in the presence of small errors in the constraints. Unfortunately, its extension to computing three dimensional structures from unequal bounds on their distances appears to be rather difficult.

Klapper and DeBrota [Klapper1980] have used Cayley-Menger determinants to directly build up a full matrix of interatomic distances for a noncyclic molecule from its bond lengths, geminal distances and vicinal distances, which may then be readily converted into coordinates as described in the previous section. Although quite efficient, this method is unable to account for long-range information including, most notably, the absence of van der Waals overlaps between atoms which are separated by more than three covalent bonds. A more conventional, coordinate-based approach with much the same advantages and disadvantages may be found in [Essen1983]. Both these approaches have problems with the propagation of error as the chain is built up, but when the short-range information is sufficiently precise the resultant conformations are sufficiently close to admissibility that they can be corrected with a few cycles of error function minimization, using for example Newton's method [Essen1983].

Another approach specifically designed to construct polypeptide conformations which are consistent with NMR data has recently been proposed by Braun and Go [Braun1985]. This approach fixes the bond lengths and angles, and proceeds by minimization of the error function F (Equation 6.3) using the dihedral angles about single bonds as its variables. This choice of variables greatly reduces the dimensionality of the optimization problem, as well as the number of constraints which have to be explicitly included in the error function. Even so, the straightforward minimization of an error function for the complete list of constraints would rarely converge to zero if it were not for a clever heuristic they have found which allows them to avoid many local minima. Although the method is by no means infallible, it is probably the only direct approach to the optimization of protein conformation which has ever succeeded in finding a global minimum, and it would be interesting to see how it performs when applied to conformational energy functions.

The key to their success is that a protein is a linear polymer where one can expect experimental constraints that are "shortrange" (between atoms in residues near in *sequence*) and "longrange" (between residues distant in sequence but not necessarily distant in space). For a randomly chosen starting conformation, they minimized a sum of terms of the above form, initially including only shortrange terms and then gradually adding in the longer range terms. In this way, the secondary structure of the protein is built up before its tertiary structure, and in the early stages of the minimization it is possible for parts of the polypeptide chain which are widely separated in sequence to pass freely through one another as necessary to establish the long-range features of the conformation. As long as the shortrange constraints were plentiful (although not necessarily very precise), the local structure of the polypeptide chain became rather correct in the early stages, although the longrange conformation could be very wrong. Then as longer and longer range constraint terms were added, the overall conformation also became correct. In cases where the shortrange constraints failed to provide sufficient guidance, longrange constraints subsequently could not be satisfied.

The convergence which results is strongly dependent upon the precision of the short-range constraints, although much less so than in the methods of [Klapper1980, Essen1983]. In essence, the approach uses the relative ease of finding conformations consistent with short-range constraints to obtain a starting conformation for refinement with respect to longer range constraints. This is the opposite sequence of events from EMBED, since it relies on abundant long-range distance constraints to get a good starting conformation for its optimization. The currently available NMR data contains substantial quantities of both short and long-range information, and hence the quality of the convergence obtained by the two methods presently appears roughly comparable [Wagner1987].

Another method which uses the dihedral angles as its variables and attempts to globally minimize an error function is the *ellipsoid algorithm* [Billeter1986a]. This algorithm is a general method of constrained convex optimization which was developed by [Shor1977], and subsequently found to be applicable to more general nonlinear constrained optimization problems [Ecker1985]. As the name implies, it operates by generating a sequence of ellipsoids in the solution space, each of which contains a minimum (in convex problems), and whose volumes decrease geometrically towards zero while their centers home in on the minimum. Since the centers of the ellipsoids can move in large steps, the ellipsoid algorithm is capable of escaping from local minima which most other minimization algorithms would be trapped by, making it a potent method of finding globally optimal solutions. Although it is fundamentally a method of constrained optimization, it also functions well when only a feasible solution is sought. The ellipsoid algorithm is in fact the only known method capable of finding conformations wholly within the feasible region, so that they contain no violations whatsoever of the distance and chirality constraints, thereby proving that the given constraints were indeed consistent. Unfortunately, both the efficiency and reliability of the method decline rapidly as the number of variable dihedral angles becomes large, so that the method is not suitable for large molecules such as proteins. Nevertheless, a great many interesting problems which would be very difficult to solve exactly by other means are within its reach, most notably the problem of docking a drug molecule to its receptor protein [Billeter1986b].

The final method which we shall discuss is the brute-force solution of the system of equalities and inequalities in the squared distances which are imposed by the intrinsic characterizations of the Euclidean distance function given in Chapter 3, in particular Corollary 3.53 [Sippl1986]. This basic approach was first considered in [Havel1983], and discarded as being too unwieldy for practical use. Sippl and Scheraga, however, have simplified the numerical calculations involved somewhat, and in addition computers have since grown fast enough to allow such things to be attempted.

In their approach, a fixed, nondegenerate reference tetrahedron $\{a_1, \ldots, a_4\}$ is chosen, and the squared distances between its vertices and the remaining atoms are used to describe their positions relative to this tetrahedron. These so-called "Cayley-Menger coordinates" are not independent variables, however, since they must satisfy the nonlinear equations

$$D(a_1, \ldots, a_4, a_i) = 0 \qquad (6.45)$$

for all $i = 5, \ldots, \#A$. In the event that they do satisfy these equations, the remaining squared distances can readily be obtained from the vanishing of the Cayley-Menger bideterminant

$$\begin{aligned} 0 &= D(a_1, \ldots, a_4, a_i; a_1, \ldots, a_4, a_j) \qquad (6.46) \\ &= D(a_i, a_j) D(a_1, \ldots, a_4) + B_{56}(a_1, \ldots, a_4, a_i, a_j) , \end{aligned}$$

where $B_{56}(a_1, \ldots, a_4, a_i, a_j)$ is constant w.r.t. the unknown $D(a_i, a_j)$. Sippl and Scheraga then propose solving for these unknowns (which include the Cayley-Menger coordinates!) by standard nonlinear optimization methods.

In addition, when estimates $\hat{D}(a_i, a_j)$ of some of the squared distances are available from experimental sources, they have further proposed lumping the sum of the squares of the differences $(D(a_i, a_j) - \hat{D}(a_i, a_j))^2$ together with the violations of the Equations (6.45) and (6.46) by the D's into a single penalty function, which can then be minimized to yield what they apparently consider to be "the best solution to the incomplete embedding problem" [Sippl1986]. If convergence problems are encountered or the experimentally determined squared distances $\hat{D}(a_i, a_j)$ should not be three dimensionally realizable, however, the result will actually be an incomprehensible mixture of violations of the experimental data and the three-dimensionality of the conformation! As we have stressed earlier, in the calculation of molecular conformation from experimental distance information one should attempt to satisfy the constraints completely, and not seek to balance the constraint violations against one another. In our opinion and, we believe, the opinion of most experimentalists, it is an even worse idea to try to balance the violations of the distance constraints against those of three-dimensionality.

References

Bentley1980.

J.L. Bentley and D. Wood, "An Optimal Worst Case Algorithm for Reporting Intersections of Rectangles," *IEEE Trans. Comp., C29*, 571-577(1980).

Billeter1986a.

M. Billeter, T.F. Havel, and K. Wuthrich, "The Ellipsoid Algorithm as a Method for the Determination of Polypeptide Conformations from Experimental Distance Constraints and Energy Minimization," *J. Comp. Chem., 8*, 132-141(1986).

Billeter1986b.

M. Billeter, T.F. Havel, and I.D. Kuntz, "A New Approach to the Problem of Docking Two Molecules: The Ellipsoid Algorithm," *Biopolymers, 26*, 777-793(1986).

Braun1981.

W. Braun, C. Boesch, L.R. Brown, N. Go, and K. Wuthrich, "Combined Use of Proton-Proton Overhauser Enhancements and a Distance Geometry Algorithm for Determination of Polypeptide Conformations. Application to Micelle-Bound Glucagon," *Biochim. Biophys. Acta, 667*, 377-396(1981).

Braun1985.

W. Braun and N. Go, "Calculation of Protein Conformations by Proton-Proton Distance Constraints: A New Efficient Algorithm," *J. Mol. Biol., 186*, 611-626(1985).

Burkert1982.

V. Burkert and N.L. Allinger, *Molecular Mechanics*, ACS Monograph no. 177, American Chemical Society, Washington, D.C., 1982.

Cariani1985.

P. Cariani and N.S. Goel, "On the Computation of the Tertiary Structure of Globular Proteins. IV. Use of Secondary Structure Information," *Bull. Math. Biol., 47*, 367-407(1985).

Carroll1980.

J.D. Carroll and P. Arabie, "Multidimensional Scaling," *Ann. Rev. Psych., 31*, 607-649(1980).

Clore1986a.

G.M. Clore, A.T. Brunger, M. Karplus, and A.M. Gronenborn, "Application of

Molecular Dynamics with Interproton Distance Restraints to Three-Dimensional Protein Structure Determination: A Model Study of Crambin," *J. Mol. Biol., 191*, 523-551(1986).

Clore1986b.

G.M. Clore, M. Nilges, D.K. Sukumaran, A.T. Brunger, M. Karplus, and A.M. Gronenborn, "The Three-Dimensional Structure of α1-Purothionin in Solution: Combined Use of Nuclear Magnetic Resonance, Distance Geometry and Restrained Molecular Dynamics," *EMBO J.*, 2729-2735(1986).

Clore1987a.

G.M. Clore, M. Nilges, A.T. Brunger, M. Karplus, and A.M. Gronenborn, "A Comparison of the Restrained Molecular Dynamics and Distance Geometry Methods for Determining Three-Dimensional Structures of Proteins on the Basis of Interproton Distances," *FEBS Lett., 213*, 269-277(1987).

Clore1987b.

G.M. Clore, D.K. Sukumaran, M. Nilges, and A.M. Gronenborn, "Three-Dimensional Structure of Phoratoxin in Solution: Combined Use of Nuclear Magnetic Resonance, Distance Geometry and Restrained Molecular Dynamics," *Biochemistry, 26*, 1732-1745(1987).

Clore1987c.

G.M. Clore, D.K. Sukumaran, M. Nilges, J. Zarbock, and A.M. Gronenborn, "The Conformations of Hirudin in Solution: A Study Using Nuclear Magnetic Resonance, Distance Geometry and Restrained Molecular Dynamics," *EMBO J., 6*, 529-537(1987).

Crippen1978a.

G.M. Crippen, "Rapid Calculation of Coordinates from Distance Matrices," *J. Comp. Phys., 26*, 449-452(1978).

Crippen1978b.

G.M. Crippen and T.F. Havel, "Stable Calculation of Coordinates from Distance Information," *Acta Cryst., A34*, 282-284(1978).

Dial1979.

R. Dial, F. Glover, D. Karney, and D. Klingman, "A Computational Analysis of Alternative Shortest Path Algorithms and Labeling Techniques for Finding Shortest Path Trees," *Networks, 9*, 215-248(1979).

Eckart1936.

C. Eckart and G. Young, "The Approximation of one Matrix by Another of Lower Rank," *Psychometrica, 1*, 211-218(1936).

Ecker1985.

J.G. Ecker and M. Kupferschmidt, "A Computational Comparison of the

Ellipsoid Algorithm with Several Nonlinear Programming Algorithms," *SIAM J. Control and Optimization, 23*, 657-674(1985).

Essen1983.
H. Essen, "On the General Transformation from Molecular Geometric Parameters to Cartesian Coordinates," *J. Comp. Chem., 4*, 136-141(1983).

Faddeev1963.
D.K. Faddeev and V.N. Faddeeva, *Computational Methods of Linear Algebra*, W.H. Freeman, San Francisco, 1963.

Goel1978.
N.S. Goel, M. Ycas, and J.W. Jacobsen, "On the Computation of the Tertiary Structure of Globular Proteins," *J. Theor. Biol., 72*, 443-457(1978).

Goel1979.
N.S. Goel and M. Ycas, "On the Computation of the Tertiary Structure of Globular Proteins. II.," *J. Theor. Biol., 77*, 253-305(1979).

Goel1982.
N.S. Goel, B. Rouyanian, and M. Sanati, "On the Computation of the Tertiary Structure of Globular Proteins. III. Interresidue Distance and Computed Structures.," *J. Theor. Biol., 99*, 705-757(1982).

Gower1984.
J.C. Gower, "Distance Matrices and Their Euclidean Approximation," in *Data Analysis and Informatics, II*, ed. E. Diday *et al.*, pp. 3-21, Elsevier Science Publ., Amsterdam, 1984.

Gunsteren1984.
W.F. van Gunsteren, H.J.C. Berendsen, F. Colonna, D. Perahia, J.P. Hollenberg, and D. Lellouch, "On Searching for Neighbors in Computer Simulations of Macromolecular Systems," *J. Comp. Chem., 5*, 272-279(1984).

Havel1983.
T.F. Havel, I.D. Kuntz, and G.M. Crippen, "The Theory and Practice of Distance Geometry," *Bull. Math. Biol., 45*, 665-720(1983).

Havel1984.
T.F. Havel and K. Wuthrich, "A Distance Geometry Program for Determining the Structures of Small Proteins and Other Macromolecules from Nuclear Magnetic Resonance Measurements of 1H-1H Proximities in Solution," *Bull. Math. Biol., 46*, 281-294(1984).

Havel1985.
T.F. Havel, I.D. Kuntz, and G.M. Crippen, "Errata to "The Theory and Practice of Distance Geometry"," *Bull. Math. Biol., 47*, 157(1985).

Hayden1988.
T.L. Hayden and J. Wells, "Approximation by Matrices Positive Semidefinite on a Subspace," *Linear Algebra Appl.*(1988). To appear.

Klapper1980.
M.H. Klapper and D. DeBrota, "Use of Caley-Menger Determinants in the Calculation of Molecular Structures," *J. Comp. Phys.*, *37*, 56-69(1980).

Kruskal1978.
J.B. Kruskal and M. Wish, *Multidimensional Scaling*, Sage Publications, Beverley Hills, CA, 1978.

Luenberger1973.
D.G. Luenberger, *Introduction to Linear and Nonlinear Programming*, Addison-Wesley, Menlo Park, CA, 1973.

MacKay1974.
A.L. MacKay, "Generalized Structural Geometry," *Acta Cryst. A*, *30*, 440(1974).

Mardia1978.
K.V. Mardia, "Some Properties of Classical Multidimensional Scaling," *Comm. Statist.*, *A7*, 1233-1241(1978).

Marshall1979.
A.W. Marshall and I. Olkin, *Inequalities: Theory of Majorization and its Applications*, Academic Press, New York, NY, 1979.

Mathar1985.
R. Mathar, "The Best Euclidean Fit to a Given Distance Matrix in prescribed dimensions," *Linear Algebra Appl.*, *67*, 1-6(1985).

Papadimitriou1982.
C.H. Papadimitriou and K. Steiglitz, *Combinatorial Optimization: Algorithms and Complexity*, Prentice-Hall, Englewood Cliffs, NJ, 1982.

Schlitter1987.
J. Schlitter, "Calculation of Coordinates from Incomplete and Incorrect Distance Data," *J. Appl. Math. Phys.*, *38*, 1-9(1987).

Shor1977.
N.Z. Shor, "Cut-off Method with Space Extension in Convex Programming Problems," *Cybernetics*, *12*, 94-96(1977).

Sippl1985.
M.J. Sippl and H.A. Scheraga, "Solution of the Embedding Problem and Decomposition of Symmetric Matrices," *Proc. Natl. Acad. Sci. USA*, *82*, 2197-2201(1985).

Sippl1986.
M.J. Sippl and H.A. Scheraga, "Cayley-Menger Coordinates," *Proc. Natl. Acad. Sci. USA*, *83*, 2283-2287(1986).

Vlieg1986.
J. de Vlieg, R. Boelens, R.M. Scheek, R. Kaptein, and W.F. van Gunsteren, "Restrained Molecular Dynamics Procedure for Protein Tertiary Structure Determination from NMR Data: a Lac Repressor Headpiece Structure Based on Information on J-Coupling and from the Presence and Absence of NOE's," *Israel J. Chem.*, *27*, 181-188(1986).

Wagner1987.
G. Wagner, W. Braun, T.F. Havel, T. Schauman, N. Go, and K. Wuthrich, "Solution Structure of the Basic Pancreatic Trypsin Inhibitor Determined by Two Different Distance Geometry Algorithms from the Same Experimental Data," *J. Mol. Biol.*, *196*, 611-639(1987).

Wako1982.
H. Wako and H.A. Scheraga, "Distance-Constraint Approach to Protein Folding. II. Prediction of Three-Dimensional Structure of Bovine Pancreatic Trypsin Inhibitor," *J. Protein Chem.*, *1*, 85-117(1982).

Yemini1979.
Y. Yemini, "Some Theoretical Aspects of Position-Location Problems," *Proc. IEEE, 20th Conf.*, 1-8(1979).

Yemini1982.
Y. Yemini, A. Dupuy, and D.F. Bacon, "Distributed Position Location System," Preprint, Distributed Computing and Communications Group, Columbia Univ., New York, NY, 1982.

Young1938.
G. Young and A.S. Householder, "Discussion of a Set of Points in Terms of their Mutual Distances," *Psychometrika*, *3*, 19-22(1938).

Zuiderweg1985.
E.R.P. Zuiderweg, R.M. Scheek, R. Boelens, W.F. van Gunsteren, and R. Kaptein, "Determination of Protein Structures from Nuclear Magnetic Resonance Data Using a Restrained Molecular Dynamics Approach: The Lac Repressor DNA Binding Domain," *Biochimie*, *67*, 707-715(1985).

7. Applications of Embedding to Experimental Data

In order to use a program like EMBED to solve real chemical problems, it is very helpful to see examples of how other researchers have cast their physical constraints into distance and chirality constraints, and indeed what kinds of problems have been solved in the past. We begin this chapter with very practical commentary on how to convert real experimental data into terms EMBED can use, pointing out extensions and pitfalls. Since one of the most important uses of embedding calculations has been in the determination of conformations of biological macromolecules in solution by NMR, we complete the chapter with an extensive account of how such experimental data can be used, and what sort of results can be expected.

7.1. Relating Experimental Data to Distance Constraints

Distance geometry calculations require input in the form of lower and upper bounds on some of the interatomic distances in a molecule. In addition, lower and upper bounds on the oriented volumes of a limited number of the tetrahedra spanned by their quadruples of atoms may also be needed, in order to fix the chirality of these chemical groups (see §2.1 and §6.1). Although a great deal of experimental data can be directly translated into bounds on the distances and oriented volumes, many kinds of data are customarily expressed in terms of the angles between certain groups of atoms, or the overall shape and size of the molecule. The purpose of this section is to help the investigator to make the conversions from these other geometric parameters to the distance and oriented volume constraints used in distance geometry calculations. In so doing, we will also briefly illustrate how distance geometry has been applied to widely differing situations.

Bond lengths are usually known *a priori* from X-ray crystallography. If the length of the bond between atoms i and j is known to be $b \pm e$, then clearly

$u_{ij} := b+e$ and $l_{ij} := b-e$. Since most methods of minimizing error functions are not capable of enforcing the constraints with a precision better than the usual values of e, however, we usually just set $u_{ij} := l_{ij} := b$.

In order to translate a known *bond angle* θ_{ijk} into a geminal distance d_{ik}, we must have fixed ij and jk bond lengths. Then by the law of cosines,

$$d_{ik}^2 = d_{ij}^2+d_{jk}^2-2d_{ij}d_{jk}cos\theta_{ijk} \tag{7.1}$$

so that setting θ_{ijk} to the appropriate value fixes d_{ik}. In all but strained ring systems, one can generally use standard θ_{ijk} values from X-ray crystallography. In order to see that these three fixed distances then determine the bond angle uniquely, we look at the inverse relation

$$cos\theta_{ijk} = \frac{d_{ij}^2 + d_{jk}^2 - d_{ik}^2}{2d_{ij}d_{jk}} \tag{7.2}$$

Since the bond angle θ_{ijk} lies between 0 and 180 degrees by definition, it is uniquely determined by its cosine, and hence fixing the value of d_{ik} fixes the value of θ_{ijk} uniquely. It is also easy to see that the relation between the geminal distance d_{ik} and bond angle θ_{ijk} is monotone increasing, so that it is also possible to impose lower and upper bounds on θ_{ijk} by means of appropriate lower and upper bounds on d_{ik}. Moreoever, because bond angles are essentially always obtuse, it can further be shown that this is possible even when d_{ij} and d_{jk} are not constant: one simply uses u_{ij} and u_{jk} in place of d_{ij} and d_{jk} to calculate the value of u_{ik} which gives the desired upper bound on θ_{ijk}, and l_{ij} and l_{jk} to calculate l_{ik}.

In a similar fashion, the *dihedral angle,* τ_{hijk} (i.e., for the chain of atoms $a_h-a_i-a_j-a_k$, the torsional angle for rotation about the ij bond) imposes a restriction on d_{hk}, once all the other distances among the four atoms are fixed; or equivalently, when the hi, ij, and jk bond lengths, and the hij and ijk bond angles are fixed. If the *cis* conformation defines $\tau_{hijk}=0$, then

$$\begin{aligned} d_{hk}^2 = {} & (d_{ij}-d_{hi}cos\theta_{hij}-d_{jk}cos\theta_{ijk})^2 + (d_{hi}sin\theta_{hij}-d_{jk}sin\theta_{ijk}cos\tau_{hijk})^2 \\ & + (d_{jk}sin\theta_{ijk}sin\tau_{hijk})^2 \end{aligned} \tag{7.3}$$

Note that d_{hk} is unchanged when $-\tau_{hijk}$ is substituted for τ_{hijk}. Looking at it the other way around [Havel1983],

$$cos\tau_{hijk} = \frac{D(a_h,a_i,a_j;\, a_i,a_j,a_k)}{(D(a_h,a_i,a_j)\,D(a_i,a_j,a_k))^{1/2}}, \tag{7.4}$$

where the D's denote Cayley-Menger (bi)determinants in the squares of the distances $d_{hi}, \ldots, d_{jk}$ (see §3.6). Expanding these determinants and taking a careful look at the resultant expression shows that when all of these distances except

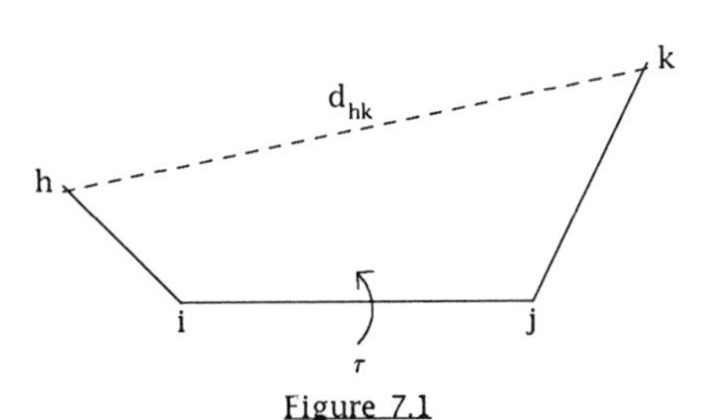

Figure 7.1

Relationship between the distances among four points and the dihedral angle τ defined by those points.

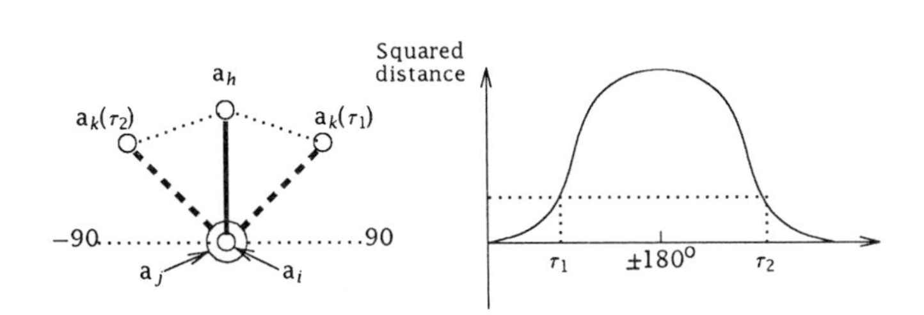

Figure 7.2

On the left is shown the Newman projection of a molecular fragment on four atoms labelled h through k; on the right the squared distance between h and k as a function of the torsion angle τ about the $i-j$ bond. Note that there are two possible conformers consistent with any nonextremal value of the squared distance, which differ only in chirality.

d_{hk} are held fixed, the relation between this vicinal distance and the cosine is once again monotone decreasing, so that τ_{hijk} can be fixed, up to its sign, by fixing d_{hk} (see Figure 7.2).

Fortunately, the sign of the dihedral angle is the same as the sign of the oriented volume of the tetrahedron spanned by the sequence of atoms $a_h, \ldots, a_k$, so that the angle can be unambiguously determined by setting both

The figure below shows the sinusoidal relation between the oriented volume and the torsion angle, from which we can derive bounds on the torsion angles from bounds on the oriented volume and vice versa.

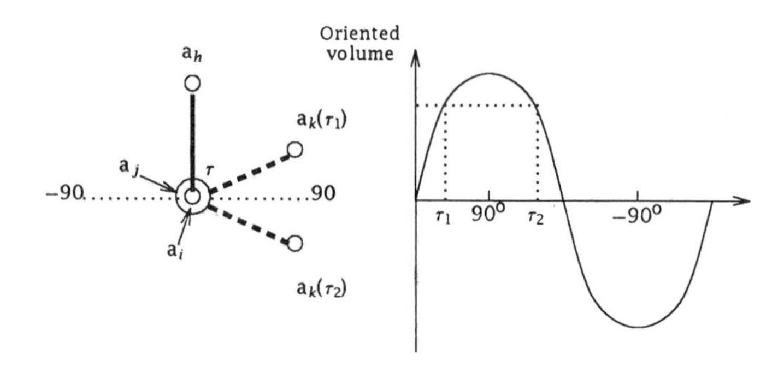

Figure 7.3

On the left is shown the Newman projection of a molecular fragment on four atoms labelled a_h through a_k, on the right the oriented volume that they span as a function of the torsion angle τ about the $a_i - a_j$ bond. Note that there are two possible conformers consistent with any nonextremal value of the oriented volume.

the vicinal distance and the sign of the oriented volume (see Figure 7.3).

Even though the oriented volume alone also does not uniquely determine the dihedral angle, since the rotational state of the i,j-bond can actually be specified by any of the dihedral angles among the substituents of the atoms a_i and a_j (for example, by any of $3 \times 3 = 9$ possible dihedral angles in the case of ethane), it turns out that the rotational state of the bond can be specified by setting the values of the oriented volumes of all bonded chains of four atoms whose central bond is $a_i - a_j$. By imposing bounds on the value of these oriented volumes equal to the extremes which can occur as these dihedral angles vary within given ranges, it has also proved possible to impose distinct lower and upper bounds on the dihedral angles. Although this could in principle also be done by a combination of bounds on the vicinal distance and oriented volume of a single quadruple, it turns out that the fit to the data after error function optimization is generally much better when the former method is used [Havel1985].

As already discussed in the previous chapter, a *chiral center* is treated in a rather *ad hoc* fashion as an extra penalty term during the optimization step. Whenever there are four atoms whose mutual distances are nearly fixed, and the input to the problem warrants fixing their chirality, then a chirality error term (Equation 6.9) must be added to the distance error function (Equation 6.7), which is expressed in terms of the oriented volume $vol_{\mathbf{p}}$, as calculated from coordinates for the four atoms using Equation (6.10). The precise magnitude of $vol_{\mathbf{p}}$ depends on the bond lengths and angles among the atoms, so in order to be consistent with the other distance constraints, the correct absolute value must be calculated from coordinates which may be either taken from X-ray crystal data or calculated using Equation (6.1). Since only four atoms are required, the asymmetric carbon itself or any one of its substituents is superfluous. Also note that each atom in a molecule is treated as unique in these calculations, so that for example methane could have chiral constraints even though *chemically* the hydrogens are indistinguishable.

It is often useful to impose nontrivial oriented volume constraints even on quadruples of atoms which are coplanar and so are not chiral (in the usual sense of the word). For example, although the coplanarity of sp^2-hybridized carbon atoms with their three covalent neighbors is a necessary consequence of the exact distance constraints obtained from the fixed bond lengths and angles among them, the corresponding bar and joint framework is *not* infinitesimally/statically rigid. This means that there exists an infinitesimal flexing (i.e. out-of-plane motion of the central carbon atom) which to a first-order approximation does not change any of the distances among these four atoms, and hence significant out-of-plane distortions can occur with only a small increase in the distance error. The oriented volume, however, changes rapidly with such distortions, so that by specifying a zero value for the oriented volume one can ensure that these atoms are almost perfectly coplanar in any set of coordinates with near-zero error. Experience has shown that adding precise distance constraints among the atoms immediately bonded to a planar group, such as a benzene ring, (and hence coplanar to the ring) is very helpful toward so strongly enforcing planarity that the resulting structures are pleasing to the eye. When significant convergence problems are encountered, we recommend including oriented volume constraints at any noncoplanar quadruple of atoms (which may not be all chemically distinguishable) whose mutual distances are all fixed by exact distance constraints. This has proven useful in order to prevent the bond angles from being bent to chemically unreasonable values by the "force" exerted by the error due to the other violated constraints. To do this, bounds equal to the oriented volumes spanned by these quadruples of atoms in their correct spatial arrangement must be used. However, as a rule one should attempt to satisfy

Some uses of constraints on the oriented volume to ensure chemical feasibility within the limits of practical convergence.

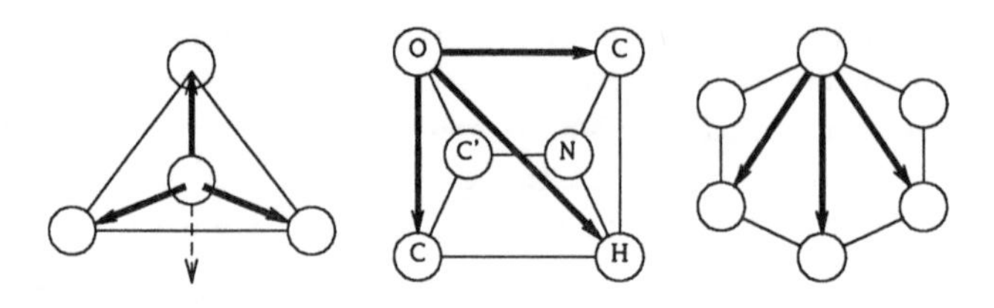

Figure 7.4

On the left is shown an sp^2-hybridized carbon atom, together with its three covalent neighbors. In the middle is shown a *trans* peptide bond. On the right is shown a benzene ring. In all cases, coplanarity can be assured by requiring that the oriented volume spanned by the three heavily drawn vectors (together with any symmetrically equivalent sets of vectors) is zero.

the geometric constraints completely, and not seek to balance one type of constraint violation against another.

In general, a *rigid group* of atoms contained within the molecule, such as a phenyl ring or even an α-helical segment in a protein, can be specified by setting interatomic upper and lower bounds equal to the distances found in model structures. The distance geometry algorithm in principle needs only barely enough constraints to completely specify the rigid group, but in practice convergence is more rapid if *all* distances among atoms in the group are fixed. This type of redundancy is particularly important because distance constraints are what guide the EMBED algorithm in its selection of trial distances and hence starting coordinates. As a general rule in setting up the input for a distance geometry calculation, one should always try to define a desired geometric condition in as many ways as possible, by imposing as many constraints characteristic of that geometric condition as possible, because the use of such redundant constraints often improves the fit of the computed conformations to the given condition.

So far we have considered constraints that are mostly known *a priori* from the chemical structure of the molecule under investigation. Of the kinds of data obtained from specific experiments on a (large biological) molecule, *crosslinking* constraints are particularly easy to incorporate into the calculation. For example, if a protein is represented at low resolution as only one point per residue, located at the C^{α}, then a disulfide bridge between residues i and j implies that u_{ij}=6.5 and l_{ij}=6.0 Å. Alternatively on an all atom basis, the constraints are that the S_i-S_j distance is between 2.0 and 2.1 Å, and that the $C_i^{\beta}-S_j$ and $C_j^{\beta}-S_i$ distances are between 3.0 and 3.1 Å. Knowing that a crosslinking reagent can join two particular parts of a molecule or molecular complex might be a helpful datum in addition to what is already known about the structure and would substantially reduce the upper bound on a particular distance. However, if the crosslink is long and flexible there would be still a large range between the upper and lower bounds. It has been estimated [Pillai1981] that one can obtain enough long range information from crosslinking on several antibody matrices to determine protein tertiary structure, but to our knowledge, no one has carried out such an extensive set of experiments.

Many different types of experiments yield useful information concerning distances between particular points in a molecule. We cite here only a sampling of the many examples from the literature. For example, a fluorescence study on aspartate transcarbamoylase yielded distances among active sites on subunits and some disulfides [Hahn1978]. Six distances in the glutamine synthetase active site were found by fluorescence and the NMR effects of a manganese ion probe [Villafranca1978]. By relating rate constants to electron-transfer distances for metalloprotein redox reactions, one can obtain protein surface to redox site distances [Mauk1980]. Anomolous small angle X-ray scattering from terbium labelled rabbit parvalbumin yields a terbium to center of mass distance of 13.2 Å [Miake-Lye1983].

These last two examples bring up the important point that distance geometry is ordinarily able to deal only with geometric constraints between *labelled* points in the molecule. At the opposite extreme, peaks in the Patterson diagram found by X-ray crystallography give the distance (and orientation) between pairs of atoms, but they contain very little information as to *which* pairs they represent. A useful exception to this condition is when the information applies to *all* pairs of atoms. For example, if one knows from hydrodynamic experiments or small angle X-ray diffraction what the diameter of the molecule must be, then this value can be simply used as the default upper bound for all pairs of atoms.

In the event there is some information available about distances of points to the center of mass of the molecule, the easiest approach is to add on an extra point to the problem which explicitly represents the center of mass. This trick works rigorously only when there are sufficiently many molecule points whose distances to the center of mass are constrained so that the extra point indeed finally lies near the center of mass calculated from the coordinates of all the atoms. As an extreme example, suppose we only knew the distance from center of mass to one point, out of a total of 100 molecule points. Then point 101, representing the center of mass, could well lie completely outside the molecule. In such a situation, Equation (6.29) would have to be used to constrain the choice of trial distances between molecule points, and the extra center of mass point could be dispensed with.

When more is known about the overall *shape* of the molecule than its spherical diameter, *outrigger points* are a convenient way of incorporating the information. These are extra points beyond those necessary to represent the molecule itself, which have fixed, large mutual distances. The shape of the molecule is expressed by confining all molecule points to lie within the intersection of appropriately chosen spherical shells centered at the outriggers. Although the outriggers could in principle be carried through all stages of the distance geometry algorithm, it has been our experience that convergence in the final error function minimization step can be substantially improved by fixing the outrigger coordinates so as to exactly satisfy their desired mutual distances and then minimize with respect to only the molecule point coordinates. As an example [Crippen1979] the coat protein monomer of tobacco mosaic virus is known to occupy in the intact virus a trapezoidal slab 25 Å thick along the virus particle's cylinder axis, 22° wide, running from 20 to 90 Å radially out from the axis. The TMV protein itself was represented by 79 points (one for every two amino acid residues) and four outrigger points were chosen, #80-83. These were positioned as shown in Fig. 7.5 to define the required shape. The distances among the outriggers,

Table 7.1

$\mathbf{D}_{outrig}$ =

	#80	#81	#82	#83
#80	0	188.272	188.272	205.760
#81		0	332.513	263.137
#82			0	263.137
#83				0

place #80 at the cylinder axis so that $u_{80,i}$=90 Å for i=1,...,79 determines the bottom of the wedge shaped allowed region of Fig. 7.5, and $l_{80,i}$=20 Å determines the top. Outrigger points #81 and #82 were located to the sides by means of appropriately chosen 80-81, 80-82, and 81-82 distances so that $u_{81,i} = u_{82,i}$ = 183.429 Å and $l_{81,i} = l_{82,i}$ = 149.084 Å confined the protein to the experimentally determined 22° wedge shape. Point #83 was then placed out of the plane of the previous three outriggers to confine the protein to the desired 25 Å thickness wedge by means of $u_{83,i}$=225 Å and $l_{83,i}$=200 Å.

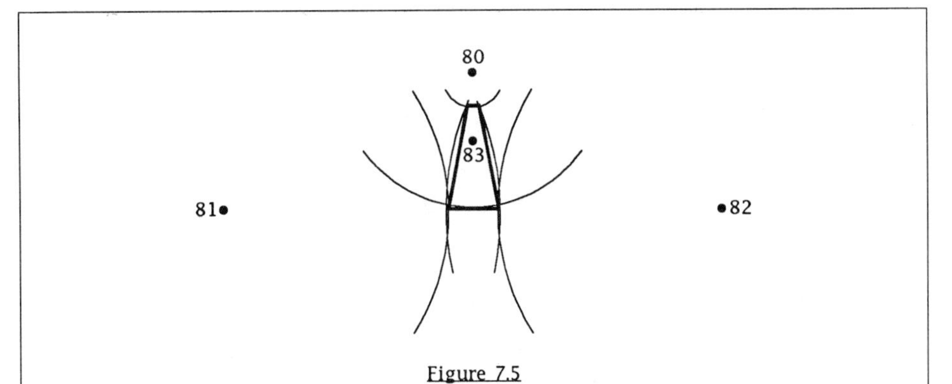

Figure 7.5

Plan view of the TMV protein wedge (trapezoid in heavy lines) showing the placement of outrigger points #80-83 to scale. Arcs indicate the upper and lower bound distances allowed between points of the polypeptide chain and the various outrigger points. Point #83 is 200 Å out of plane.

This TMV study is also noteworthy as an example of combining information from many different experimental sources. As of 1975 (before X-ray crystal structures became available) there was a great deal of conformational evidence on this cylindrical virus: X-ray diffraction studies on oriented gels of the virus (yielding radial distances for some residues), electron microscopy, hydrogen exchange, immunochemistry, various kinds of spectroscopy, and determination of accessibility to attacking reagents. Most of these amounted to constraints on the radial positions ($d_{80,i}$) for certain residues. In addition, chain connectivity provided some upper bounds, while the space filling property of the residues provided default lower bounds. Five random conformers fulfilling all constraints were generated and compared according to the root-mean-square deviation of their distance matrices, the distance matrix error (Equation 6.11), which is conveniently independent of translation and rigid rotation of one structure with respect to the other. The significance of this work with regard to distance geometry technique is that (i) minor inconsistencies in the input from different experimental techniques were detected as residual errors in the final coordinate optimization and as triangle inequality violations; (ii) otherwise all the constraints could be accommodated without difficulties in the calculations even though the data came from such diverse sources; and (iii) among the five generated structures there were some similar conformational features, but in overall rms deviation, they were grossly different in spite of the numerous experimental constraints. This last outcome is generally a desirable feature of distance geometry calculations, enabling the investigator to avoid overestimating the effectiveness of the evidence toward limiting the range of possible conformations.

A certain degree of *"unlabelling"* of molecule points can be tolerated by modifying the error function, Equation (6.7). Consider two disjoint sets of molecule points, $S_I = \{i_1,...,i_k\}$ and $S_J = \{j_1,...,j_l\}$, having the constraint that for at least one $i \in S_I$ and $j \in S_J$, $l_{IJ} \le d_{ij} \le u_{IJ}$. Then let the contribution from this constraint to the total error function, Equation (6.7), be given by

$$\min_{\substack{i \in S_I \\ j \in S_J}} \begin{cases} (d_{ij}^2 - u_{IJ}^2)^2 & , \; d_{ij} > u_{IJ} \\ 0 & , \; l_{IJ} \le d_{ij} \le u_{IJ} \\ (d_{ij}^2 - l_{IJ}^2)^2 & , \; d_{ij} < l_{IJ} \end{cases} \tag{7.5}$$

For example [Kuntz1980], some of the proteins in the 30S ribosome subunit are known to be near each other by chemical crosslinking, immuno-electron microscopy, and neutron diffraction evidence, but it is also apparent that the individual proteins are extended, rather than spherical. Therefore they were represented as three points each, and distance information between the protein indicated by points in S_I and the protein of S_J was introduced in the optimization of error function stage by a term of form Equation (7.5).

Suppose one needs to constrain some molecule points to lie within an arbitrarily complicated overall *shape*. This can be built into the error function if one is willing to devise a (differentiable) "shape function," $S(x,y,z)$, such that $S = 0$ on the desired boundary surface, $S > 0$ outside it, and $S < 0$ inside. Then for a constrained molecule point located at $[x,y,z]$, one adds to the error function the term

$$max(0, S(x,y,z)) \; . \tag{7.6}$$

For very complicated figures, it may be convenient to think of the allowed regions of space as a union of simpler, possible overlapping shapes, so that the S in the equation becomes the minimum of individual shape functions. In the ribosome study already mentioned above [Kuntz1980], the electron microscopy information came in the form of coordinates in an established reference frame, which was used for the shape function and positioning of some of the proteins. Three points per protein allowed us to represent each protein as appropriately extended, but required the semi-labelled penalty terms of Equation (7.5). The neutron scattering data gives not just upper and lower bounds between parts of two proteins, but the distribution of lengths of vectors connecting some part of one with some part of another. To a first approximation, the extremes of this distribution were used to determine the u_{IJ}s and l_{IJ}s used for the partially labelled constraints. In addition, the center of the distribution was taken as the desired distance between the central point of the one protein triple and the central point of the other protein triple. In order to do this, one of the three points representing a protein had to carry the extra label of "central." Of course it is a considerable simplification to represent an extended protein as only three points and to use only three parts of an entire distribution of distances, but it is an illustration of how to deal with such kinds of data within the distance geometry framework. Once again, small inconsistencies in the input data were easily detected and were resolved by essentially assigning somewhat larger error bars to the data. The optimization stage appeared to be extremely difficult, due mostly to the complicated shape function and due to Equation (7.5) being only piecewise differentiable. Unlike most distance geometry applications, local minima were frequently encountered and required small perturbations of the coordinates in order to continue optimizing.

A very rich source of experimental data suitable for distance geometry calculations is ordinary ("one-dimensional") NMR. To cite just a few examples, NMR lanthanide induced shift effect has been used to determine several metal-proton distances in carp parvalbumin [Lee1980]. Similarly, several Gd-proton distances were determined in a 13-residue peptide to show that its structure is consistent with that of the parvalbumin calcium binding loop, but NMR evidence was insufficient to exclude other possible conformations [Gariepy1985]. The cobalt-NAD distance in alcohol dehydrogenase has been measured by NMR [Drysdale1980]. A problem common to the above studies is that the very NMR experiments used to determine metal-proton distances must also be used to decide just where the metal ion binds, if indeed there is an unambiguous binding site.

A way to enrich the data gathering possibilities is to chemically modify the protein (or other large molecule) by covalently attaching a spin label to a reactive group on the protein, which can be verified by chemical means. The intricacies of protein chemistry permitting, there are many possible, well-defined, attachment sites and many different types of spin labels to perturb the NMR spectrum. Inasmuch as there will still be some conformational freedom in the linking group between the spin label and the protein even for a rather rigid protein, one must still determine the label's precise position by NMR. Just such a study [Schmidt1984] has located a spin label on hen egg lysozyme by NMR and distance geometry. Whereas spin labels allow one to estimate rather long distances (10 to 20 Å) between the label and assigned protons, the nuclear Overhauser enhancement (NOE) gives rough distance information between protons which are rather close ($\leq$ 5 Å) in space. Enough distances have been measured by NOE in tyrocidine A to approximately determine its conformation in solution [Kuo1980].

As an example of how to approach one of these rather detailed NMR determinations of conformation, consider the glycopeptide antibiotic, bleomycin [Crippen1981]. A minor part of the molecule was not included in the calculations in any way due to lack of data concerning it, but the rest was represented in great detail, using one point for each of the 75 non-hydrogen atoms. The default upper bound was 250 Å, and the default lower bound 2.8 Å (a conservative estimate of the van der Waals diameter of a CH_2 group). All bond lengths and bond angles were fixed at values derived from appropriate crystal structures, which also ensured planar carbonyl groups, tetrahedral sp^3 carbon atoms, etc. The NMR evidence indicated that the gulose and mannose residues contained in bleomycin have invariant conformations, so all distances among atoms in these groups were fixed at values obtained from appropriate crystal structures. The planarity of two conjugated rings was further enforced by including all distances among the atoms of each ring in the list of constraints. Three peptide bonds were fixed to be planar *trans* by setting appropriate 1-4 distances. The 19 asymmetric carbon atoms in the molecule were given their correct configuration by including the appropriate chiral terms in the error function. Six proposed ligand groups to a metal ion were expressed in the ordinary way, and octahedral ligation was encouraged by an appropriate choice of lower bounds among the ligand atoms. As it turned out, this constraint was not very strong. NMR experiments indicated that a certain (assigned) proton was close to the imidazole ring of bleomycin, which corresponds to small upper bounds on all the distances from the carbon attached to the proton to all the imidazole atoms. In addition to requiring the imidazole and pyrimidine ring nitrogens to coordinate to the metal, we insisted that they do so only in the direction of the nitrogen lone pair orbitals. This was done by fixing the distance between the metal and the atom on the opposite side of the ring from the nitrogen to the value obtained when metal, N, and opposite side of ring are colinear. The dihedral angles for rotation about two other bonds were restricted in accordance with the observed proton-proton coupling. All told there were approximately 300 constraints on the conformation. The major difficulty encountered was that sometimes the trial coordinates were near the correct full set of desired distances for one of the sugar rings, but in the mirror-image chirality. Then the distance constraints were fully satisfied and resisted the efforts of the chiral constraints to invert the entire ring. This could be avoided by weighting the chiral constraints more heavily. Furthermore, satisfying all the detailed constraints caused rather lengthy optimizations in the last stage of the calculation, so that structures required an average of 47 seconds apiece on a CDC 7600 computer. Precise bond lengths and angles were obtained only after optimizing until the error function had a very small value. Presumably appropriate weights on the error terms would have helped. Careful comparison of the few generated structures revealed a surprising variety of conformation, considering how difficult it is to build a space filling model of the molecule which incorporates all the above constraints. In the years since the original study, additional conformational constraints have been added from new experiments. In general, generating structures has become easier with more constraints, apparently because the embedding algorithm is more precisely directed toward embeddable conformations, and the old and new constraints are geometrically consistent. Furthermore, energy minimization from the generated structures leads more easily to lower minima. There is still unfortunately insufficient data to resolve major conformational alternatives.

7.2. Two-Dimensional NMR

When we first developed the EMBED algorithm (§6.3), we were at something of a loss to say where the large amount of distance data was to come from, in order to determine the solution conformation of something as large as a protein. (See the next section for how much data and what kinds are required.) Fortunately *two-dimensional NMR* (Nuclear Magnetic Resonance) soon arose as just such a source. It is not our purpose to explain the technique here or go into the details of performing two-dimensional NMR experiments. (See [Wider1984] for a review of 2-D NMR, with particular attention to the numerical data processing required before and after the fourier transform step. Other reviews of the technique [Bax1986, Wemmer1985] make special reference to protein conformational work. A particularly authoritative source is [Wuthrich1986].) Nevertheless, a brief account of what the technique can do, together with an indication of its strengths and weaknesses, is necessary for nonchemists to understand what follows.

Because hydrogen atoms are very common in organic molecules and relatively easy to detect by NMR, most NMR experiments are designed to measure proton resonances. There are two main kinds of two-dimensional proton NMR experiments which can yield conformational information suitable for distance geometry calculations. The first of these, known as two-dimensional *Correlated Spectroscopy* or *COSY* for short, yields information primarily about which pairs of protons are close together in the covalent structure of the molecule (i.e. are separated by at most three covalent bonds). In the event that they are separated by exactly three bonds, it can also provide information on the dihedral angles about the central bond. The second type of experiment, known as two-dimensional *Nuclear Overhauser Enhancement Spectroscopy* or *NOESY*, provides information about which pairs of protons are very close together in space.

The first problem to be solved in interpreting such two-dimensional NMR spectra is the determination of which atom(s) in the molecule each peak in the spectrum is due to. This problem, known as the *assignment problem*, must be solved before one can proceed with the distance geometry calculations, since they deal with distances between *labelled* points. Without going into any details, this problem is generally solved by first isolating the "spin systems" characteristic of the various kinds of amino acids in the COSY spectrum, and then linking each spin system together by means of the peaks between them in the NOESY spectrum, until the subsequence becomes unique in the protein [Wuthrich1982]. Although this procedure is rather laborious (sometimes requiring man-years of work), and not yet exactly a routine procedure [Neidig1984], it has successfully been applied to both proteins and nucleic acids (see above references). In particular, the *stereospecific* assignment of protons that are chemically equivalent but experience measurably different spatial environments is often difficult. We shall explain shortly how this ambiguity is accounted for in distance geometry calculations.

Although COSY is necessary for the assignment process, the NOESY experiment is the most important source of experimental data on the large scale conformation or *tertiary structure* of biopolymers. In order to better understand the limitations of NOESY, one must realize there is in general a dipole–dipole coupling between the spin populations of proton i and proton j, which are separated in space by some vector $\mathbf{r}_{ij} \in \mathbb{R}^3$, regardless of how many bonds separate them. Then the NOE is the fractional change in the intensity of one NMR peak when another peak is irradiated. The intensity, I, of the NOE is given by

$$I \propto \frac{f(\tau)}{\|\mathbf{r}_{ij}\|^6}\,, \tag{7.7}$$

where τ is the effective correlation time for rotation of $\mathbf{r}_{ij}$, and f is an (unknown) correlation function. As long as the protein is rigid, so that $\mathbf{r}_{ij}$ varies only due to the overall rotations of the whole molecule, $f(\tau)$ will be the same for all proton pairs throughout the protein, and there is a single constant of proportionality in Equation (7.7), which may be determined by comparison with internal references. In this favorable situation, one can usually detect NOEs for $\|\mathbf{r}_{ij}\| < 4$ Å, and the inverse sixth power dependence makes it essentially impossible to observe NOEs for protons separated by more than 5 Å. However, if a region of the molecule is rather flexible, $\mathbf{r}_{ij}$ may vary in length without significant reorientation, or it may reorient more rapidly, so that f will vary over the molecule and close protons may fail to show an NOE. Therefore in practice one can usually only distinguish a few classes of short interproton distances, for example $\|\mathbf{r}_{ij}\|<3$ Å ("strong NOE"), 3.0 Å$<\|\mathbf{r}_{ij}\|<3.5$ Å ("medium NOE"), and 3.5 Å$<\|\mathbf{r}_{ij}\|<5$ Å ("weak NOE"). Failure to observe an NOE between two hydrogens does not imply $\|\mathbf{r}_{ij}\|>5$ Å, because they might simply be in a highly flexible region of the protein. At least it is thought that the distance estimates are little affected by picosecond time scale motions [Olejniczak1984].

Some of the amino acid pseudostructures used to account for the lack of stereospecific assignments in proteins.

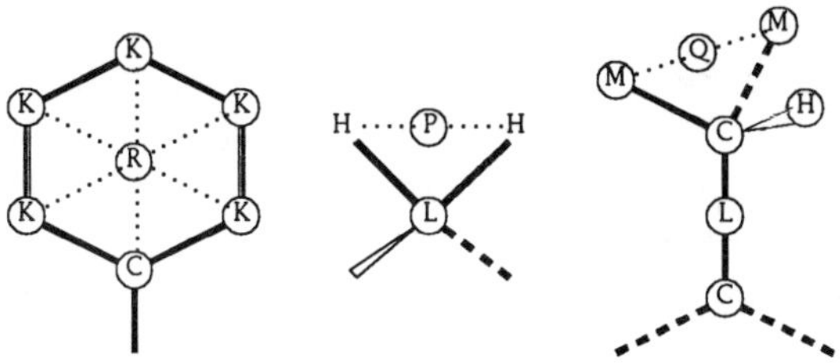

Figure 7.6

Pseudoatoms of types "L", "K" and "M" are used in place of methylene, methine and methyl groups, respectively. On the left is shown a phenyalanine ring and the dimensionless pseudoatom "R" in its center to which nonsterospecifically resolved peaks in the NMR spectra involving the ring protons are assigned. Similarly, peaks to pairs of methylene protons are assigned to dimensionless pseudoatoms of type "P" midway between them. On the right is shown the pseudostructure of leucine; nonstereospecifically resolved peaks involving the methyl protons are assigned to a pseudoatom of type "Q" midway between the methyls. The "corrections" to use are equal to the lengths of the dotted lines (see text).

The special features of the NMR data and protein structure have necessitated several extensions to our usual embedding procedure. To begin with, we have accounted for the lack of stereospecific assignments by using the "pseudostructure" representation proposed by [Wuthrich1983]. In this representation, groups bearing spectroscopically indistinguishable protons such as methyl groups are collapsed into single "pseudoatoms", which are given a radius slightly larger than their central carbon atom. Thus only distinguishable protons such as amide protons are used in these calculations, a fact which significantly reduces their size. Additional dimensionless (i.e. zero radius) pseudoatoms must also be added at the centroids of indistinguishable groups of protons, to which peaks involving any of those protons may be referred. In order to avoid imposing bounds which are unrealistically tight, when referring distance constraints to pseudoatoms it is further necessary to add a "correction" onto the estimated upper bounds equal to the distance from the pseudoatom to the protons which it substitutes. Figure 7.6 should be consulted for examples of this procedure.

An implementation of the EMBED algorithm known as the *DISGEO program*[*] specialized for the calculation of protein structure from NMR data has been developed [Havel1984]. The main change to the algorithm as presented in Chapter 6 which this program incorporates is a heuristic procedure to improve convergence known as "substructures". The idea behind this is as follows: the closer the trial distances which the EMBED algorithm uses are to actual three-dimensional distances, the better the fit of the resultant starting coordinates to them will be and hence the better the starting structure for the error function minimization will be (in the case that the trial distances are actually *equal* to three-dimensional distances, the starting coordinates will actually fit the distance constraints exactly and hence no error function minimization will be necessary at all). In order to get a better set of trial distances, the DISGEO program first embeds a "substructure", containing a subset of the atoms in the complete protein which are spread throughout its structure. In the substructure, the backbone is represented by all the carbonyl carbons, all amide nitrogens, and the α-protons of non-glycyl residues. Sidechains are represented by all β-carbons, non-terminal γ-carbons, and the ring center pseudoatoms of phenylalanyl and tyrosyl residues. This amounts to about a third of the total atoms of the full pseudoatom-atom representation, but it is enough to determine the C^{α} chiralities and otherwise make the step to the full embedding relatively easy.

The distance constraints for the substructure embedding are extracted from the triangle inequality limits on the distances in the complete molecule.[†] Since the substructure is so much smaller than the complete protein, coordinates for it can be computed with relative ease by a straightforward application of the EMBED algorithm followed by error function minimization, and the distances between the atoms of each computed substructure can then be included in the input for the calculation of each complete structure. This results in a very tight set of triangle inequality limits, so that the trial distances obtained by

* A version in VAX/VMS PASCAL is available for a nominal fee as program number 507 from the Quantum Chemistry Program Exchange, Chemistry Department, Room 204, Indiana University, Bloomington, Indiana 47401, U.S.A.

† Tetrangle inequality bound smoothing is too time consuming to be performed on molecules the size of proteins.

metrization from between those limits are substantially closer to three-dimensional distances than those which would be obtained using the experimental data alone. While this procedure has not been rigorously justified, it is a well-known "bootstrap" heuristic of a sort which has been found to work well on many computational problems. Our experience with the program shows it also works well in this case (see below).

As a general rule, in these calculations we have found it helpful to first minimize the distance error function, and then add on the chirality error function and further minimize the total error function. Although this can cause the optimizer to get trapped in local minima in certain idealized cases [Havel1983], in general the force of the chirality error function is large enough to ensure that all tetrahedra with the wrong chirality invert readily on minimization of the total error. The error with respect to the distance constraints alone, however, can be reduced more easily when the additional flexibility allowed by leaving out the error function C is present. In order to further increase the molecule's flexibility, vicinal (1,4) hard sphere overlaps are also not included in the error function.

In order to accomodate the complexity of the NMR data and the size of proteins, the DISGEO program also includes code which largely automates the process of generating the thousands of input distance constraints that are necessary. Given that BPTI, for instance, is represented by 666 atoms and pseudoatoms, there are on the order of 220000 default lower bounds which are not stored in advance, but rather produced on demand by adding the hard sphere radii of the two atoms involved. There are typically thousands of holonomic constraints required to enforce correct bond lengths, bond angles, asymmetric centers, and planar groups. Since these are the immediate (but somewhat complicated) consequences of the covalent structure, the program automatically generates them from the specified amino acid sequence. This is done by consulting a library of amino acid coordinates and computing bond lengths and geminal distances from them as needed, so that only the sequence of amino acids in the protein must be typed into a file in order to obtain all the constraints which follow from its covalent structure. Extensive checking for inconsistencies in the input is also done while reading in the files containing the distance and chirality constraints derived from the NMR data, since the probability of typographical errors in preparing such large amounts of input is high.

The marriage of 2-D NMR experiments with distance geometry calculations has been such a fruitful one that we cite here only a small sampling of the harvest. (See [Braun1987] for a recent review.) The earliest such calculations were performed with a version of the EMBED algorithm on micelle-bound fragments of the peptide hormone glucagon [Braun1981, Braun1983]. One study on a 22 residue peptide from myelin shows it is conformationally flexible in solution [Nygaard1984]. In another, the locations of helices in the sequence of the lac repressor DNA-binding domain were determined [Zuiderweg1983]. 2-D NMR applied to insectotoxin I_5A initially determined the secondary structure [Arseniev1983] and finally the full conformation in solution [Maiorov1984]. Similarly the solution conformation of Val-gramicidin A has been found [Arseniev1984], and the backbone conformation of gramicidin A in dioxane [Arseniev1985]. The first successful determination of the structure of a complete protein, the Bull Seminal Inhibitor, was made by [Williamson1985] using the DISGEO program described above [Havel1984], and a detailed account of this structure determination will be given in §7.4. Subsequently, the tertiary structure of several other proteins, tendamistat [Kline1986] and metallothionein [Braun1986], were made using the distance geometry algorithm of [Braun1985]. A number of further protein structure determinations from NMR data have also been made using restrained molecular dynamics in conjunction with distance geometry (see §6.4 for a brief account), and an implementation of the EMBED algorithm has recently been successfully applied to determine the conformation of a DNA duplex from NMR data [Hare1986].

7.3. Evaluation of Data by Simulation

Embedding calculations are easier to do than experiments, so we have tried to discover in advance how useful certain kinds of experiments would be toward the determination of the conformation of proteins. The trick is to abstract from the known crystal structure of some protein, such as pancreatic trypsin inhibitor (BPTI) [Deisenhofer1975], the kinds of data that one *could* obtain from the given experiment. By computing the coordinates of a number of distinct conformations which are all consistent with these *simulated* constraints, but which are otherwise random, and carefully examining the differences between them, it is possible to decide exactly what each kind of experiment can potentially tell us about protein conformations.

In our original paper of this type [Havel1979] BPTI (58 residues, little helix, and two strands of distorted β-sheet) and carp myogen (108 residues, mostly helical) were represented at the single point per residue level, placing the point at the C^α. In all, we tested 20 different sets of input information, generating 10 structures for each set. (see Table 7.2). These data sets were chosen to simulate the results of some of the wide variety of possible experiments and predictions which are possible for proteins, for example 10 Å cutoff contacts and noncontacts between residues (l_{ij} = 10 Å or u_{ij} = 10 Å as the corresponding distance in the crystal structure was greater than or less than 10 Å, respectively), secondary structure (all distances within regions of alpha helix or beta sheet were set to values observed in the crystal structures), center-of-mass distances (as in the crystal structures) and cross-links (selected exact interresidue distance constraints).

TABLE 7.2.
Mean *DME* among structures and with the crystal structure for each of the twenty data sets evaluated in the 1979 study.

Description	DME_s	DME_x	*SDB*
BPTI: contacts and noncontacts, cutoff 10 Å	1.0	1.0	5.6
CM: contacts and noncontacts, cutoff 10 Å	0.4	0.6	3.3
BPTI: contacts, cutoff 10 Å	2.5	2.8	25.0
BPTI: hydrophobic contacts, cutoff 10 Å	3.1	3.4	2.3
BPTI: hydrophilic contacts, cuttoff 10 Å	3.0	3.2	1.2
BPTI: noncontacts, cutoff 10 Å	4.3	10.3	0.2
BPTI: contacts and noncontacts, cutoff 25 Å	2.9	3.0	1.4
BPTI: contacts, cutoff 25 Å	4.9	5.7	0.4
BPTI: noncontacts, cutoff 25 Å	3.3	8.0	0.3
BPTI: Lys, Tyr and SS crosslinks	3.5	4.7	12.1
BPTI: Lys, Tyr, Asp, Glu and SS crosslinks	2.5	2.9	12.2
BPTI: center-of-mass distances	5.2	5.2	5.7
BPTI: center-of-mass and SS crosslinks	4.2	3.6	9.6
BPTI: distances within secondary structure	4.8	7.9	16.9
BPTI: secondary structure and SS crosslinks	3.1	5.6	13.8
CM: distances within secondary structure	6.2	7.3	16.8
BPTI: hydrophobic contacts and center-of-mass	3.6	2.5	31.6
BPTI: hydrophobic contacts and secondary structure	3.1	3.2	32.5
BPTI: center-of-mass and secondary structure	4.9	3.8	25.9
BPTI: center of mass, secondary structure, and hydrophobic contacts	4.1	2.3	50.1

Because of the limited computational facilities available to us at that time, only the alpha carbons of these proteins were included in the calculations, and the only comparisons made of the computed conformations were to calculate their distance matrix errors (*DME*; see §6.3), averaged over all 45 pairs of each set of ten conformations (DME_s) and over all ten pairs consisting of one of the computed conformations together with corresponding crystal structure (DME_x).

The *DME* is primarily sensitive to the overall size and shape of the molecule and to the way in which the chain is folded in space, i.e. to what is commonly known as its *tertiary structure*, but not to its overall handedness nor to the small scale details of its conformation. As a base line, one can expect about 10 Å rms deviation between BPTI conformers if they are generated as random freely jointed chains [McLachlan1984]. (The fact that the value of DME_x was sometimes much larger than DME_s is now known to be due to the fact that certain data sets allowed the computed structures to become much more expanded than the crystal structure.) In addition, the average sum of the distance bound violations, or *SDB*, was computed as a measure of the quality of convergence.

The main conclusions of this study were:

(1) Knowledge of only secondary structure is not very useful toward determining the overall tertiary folding, even for a protein which consists of mostly α-helices, such as carp myogen.

(2) It is better to know many nonoverlapping distances between distinct pairs of residues than it is to know an equal number of distances between any one residue and all the others. This very reasonable result shows that surface accessibility studies and shift-reagents do not, by themselves, tell one much about tertiary structure.

(3) It is better to know a few distances between residues widely separated in the sequence than it is to know many local distances.

(4) Of course, combining two sets of constraints gives a result better than either set alone.

(5) A simulated massive chemical crosslinking study, where we gave the exact distance between every residue in BPTI with a chemically reactive sidechain (66 such pairs in all), yields structures having substantial errors in the path of the polypeptide chain.

(6) It is better to know many distances imprecisely than it is to know a few distances well. In the extreme case that all the interresidue contacts and noncontacts within a BPTI-size protein are known at a 10 Å cutoff, the structure is actually determined with a precision comparable to that of many crystallographic analyses. This was perhaps the most startling result of the study, and speaks well for the efficacy of methods which yield such information (i.e. the NOESY experiment described in §7.2). Since the number of bits of information thus given increases with the square of the number of residues while the number of conformational degrees of freedom increases only linearly, the result for carp myogen was better than that for BPTI.

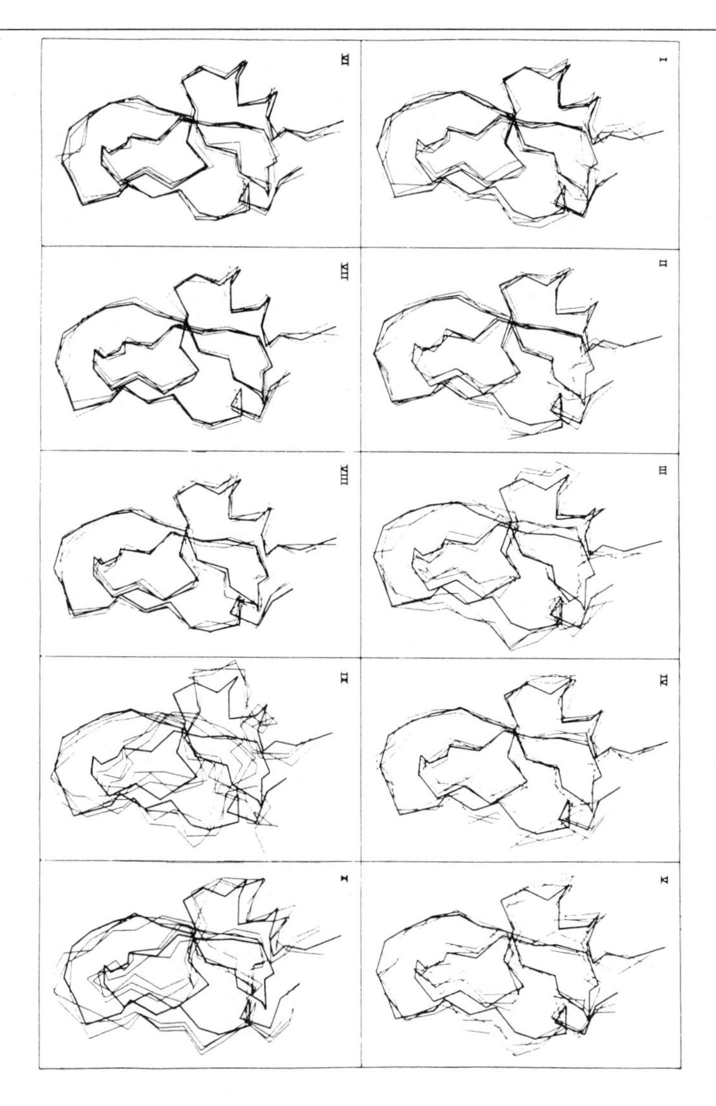

Figure 7.7

The BPTI α-carbons skeletons of the conformations computed for each of the ten data sets in the 1985 study, superimposed for best-fit on the crystal structure (heavier line). (Figure reproduced by permission [Havel1985].)

Several years later, a similar simulation study was performed, this time with special attention to the potential of the newly developed two-dimensional NMR methods [Havel1985]. This study was needed because, as described earlier, the NOESY experiment is only capable of yielding contact information at about a 5 Å cutoff, which is substantially smaller than the cutoffs used above, and information on the absence of contacts is not usually available. More important, however, is the fact that in order to obtain protein structures of quality comparable to those of crystallographic analyses, it is necessary to include far more than the alpha-carbons of proteins in the calculations. Thus these simulations were made using the same pseudostructures that are used to interpret the NMR data, as described in the previous section. In addition, since no single number is capable of adequately sumarizing the differences between two structures as complicated as complete protein conformations, the methods used to compare the structures had to be expanded upon substantially. The methods of comparison now included:

(A) The RMS difference in the coordinates minimized with respect to rotation of one conformation versus another about their superimposed centroids. This quantity correlates well with the *DME*, but in addition is sensitive to changes in the overall handedness of the chain fold. It was computed for the alpha carbon coordinates alone (the $RMSD^a$) as well as for all non-hydrogen atoms (the $RMSD^h$).

(B) The $RMSD^a$ for each consecutive segment of four amino acids in the sequence, averaged over the entire sequence. This is a measure of the "medium-range" differences in the two conformations, which we denote by $RMSD_4^a$.

(C) The RMS difference in the ϕ and ψ dihedral angles, which we abbreviate as $DHAD^b$ (the term "difference" here refers to the size in degrees of the smallest bond rotation necessary to make the rotational state of the bond in the two conformations being compared equal). This is a measure of the differences in the"short-range" structure, at the level of the Ramachandran maps.

(D) The RMS value of the coordinate vectors relative to a center of mass coordinate system, or what is commonly known as the *radius of gyration* R_G. This is a good measure of the overall size, which is not sensitive to the chain fold.

In addition, the computed structures were all examined by computer graphics techniques.

BPTI was represented as 666 points, corresponding to sufficient heavy atoms to define the full covalent structure and sufficient hydrogens to express any NOE constraints. Of course one knows *a priori* some thousands of interatomic lower distance bounds (taken to be the sum of the two corresponding van der Waals radii); 3290 fixed distances to get correct bond lengths, bond angles, and the three disulfide bonds in BPTI; and 450 chirality constraints to enforce planarity of certain groups and correct chirality of others. Beyond this, data sets were prepared by adding protons to the crystal structure of BPTI with standard geometries, and finding all 508 interproton, interresidue distances d_{ij} that were less than 4 Å in magnitude. As described below, these "NOEs" were sometimes classified as strong ($d_{ij} \leq 2.5$ Å), medium ($2.5 \leq d_{ij} < 3.0$ Å) and weak ($3.0 \leq d_{ij} < 4.0$ Å), in which case upper bounds equal to the upper limit defining the class were imposed on the corresponding distance (i.e. $u_{ij} := 2.5$ Å for strong NOEs etc.). These NOEs will be referred to as *quantitated.* Otherwise, a uniform upper bound of either 4.0 Å or 5.0 Å was imposed. In all cases the upper bound distance constraint was then referred to the appropriate psuedoatoms and the corresponding correction added to it, just as would be done with real NMR data.

TABLE 7.3
The data sets evaluated in the 1985 study. Right column is total number of NOE bounds (between backbone protons).

I	Complete quantitated set of all possible NOEs	508 (119)
II	Quantitated subset of all strong, aromatic & amide NOEs	356 (112)
III	As in II, but with 5 Å upper bound on nonsequential NOEs	356 (112)
IV	As in II, but with 4 Å upper bound on nonsequential NOEs	356 (112)
V	As in IV, but with all sequential NOEs deleted	234 (38)
VI	As in V, but with NOE bounds set to d_{ij}±0.1 Å	234 (38)
VII	As in VI, but with sequentials again included ±0.1 Å	356 (112)
VIII	As in VII, but with NOE bounds set to d_{ij}±0.5 Å	356 (112)
IX	As in IV, but with all weak NOE's eliminated	170 (59)
X	As in IX, plus supplementary constraints (see text)	170 (59)

The first data set (I) contained all 508 such NOEs, quantitated as just described. In the second (II) we kept only the strong aliphatic NOEs, together with those which involved an amide or aromatic proton, since these are the easiest to resolve in NMR spectra. In the next two data sets we quantitated only the "sequential" NOEs used in assignment (all of which are between amide, alpha and beta protons in adjacent residues), since these are usually the easiest to quantitate, with an upper bound on all the rest of 4.0 Å (in III) or 5.0 Å (in IV). In (V) we deleted all the sequential NOEs, to find out what they were telling us about the conformation, and in (VI) we imposed bounds on all the remaining NOEs equal to the crystal structure distances ±0.1 Å, to find out how much precision is really worth. In (VII) we added back the sequential NOEs, also with bounds equal to the crystal structure distances ±0.1 Å, and in (VIII) we relaxed all NOE bounds to the crystal structure distances ±0.5 Å. Finally, in (IX) we prepared a relatively poor data set, by eliminating all weak NOEs from data set (IV), and then in (X) tried to find out how much supplementary data (i.e. data not derived from NOESY) might help improve upon it. This supplementary data consisted of the distance constraints characteristic of the 21 hydrogen bonds observed in the crystal structure, together with the chirality constraints implied by the expected coupling constants on 27 of its ϕ angles.

TABLE 7.4
Summary of the results of the 1985 study. All numbers given are averages with the crystal structure (see text).

Data Set	$RMSD^a$	$RMSD^h$	$RMSD^a_4$	$DHAD^b$	R_G
I	1.2	2.1	0.37	50	10.6
II	1.2	2.2	0.35	48	10.5
III	1.9	3.1	0.61	63	10.6
IV	1.5	2.7	0.49	56	10.8
V	1.7	2.9	0.62	71	10.7
VI	0.8	2.1	0.30	62	10.6
VII	0.7	1.8	0.22	35	10.6
VIII	0.9	2.2	0.32	50	10.6
IX	2.5	3.3	0.74	75	11.1
X	2.2	3.1	0.54	53	11.0

For each of these ten data sets, a total of three conformations were computed. The numerical comparisions of these conformations are summarized in Table 7.4, wherein all differences reported are the averages of the differences of the three computed conformations with the crystal structure (the differences among the computed conformations were usually similar). Pictures of all of the three conformations computed for each of the data sets (I) through (X) are shown in Figure 7.7, superimposed for best fit upon the crystal structure (heavier line). There are a number of conclusions to be drawn from this study:

(1) If anything, the fit of the computed conformations to the crystal structure is even better than that expected on the basis of the 1979 study, probably because of the greater detail used in the representation of the polypeptide chain.

(2) DISGEO produced structures that agreed very well with the input constraints, which were, after all, certainly mutually consistent in these simulations. Planar groups were indeed planar within 1°, 90% of all distance constraints agreed within 0.1 Å, and the rare larger violations were always less than 1.0 Å. Apparently the random sampling of structures has a mild bias toward being more expanded than the native by some 5%, mostly for strands and loops exposed to the surface.

(3) In every case, the overall path of the backbone was correct, even when only the 170 strong NOEs were used.

(4) Using only 170 strong NOEs gave an rms deviation of 2.48 Å, whereas including all 356 easily detectable NOEs lowered that to 1.51 Å, even though in each case the bounds were the fairly broad ranges of the NOE classification. It is therefore advisable to detect as *many* NOEs as possible, with accuracy a secondary consideration.

(5) The sequential NOEs are rather unimportant to the determination of the global structure, since deleting them altogether only raised the rms deviation from 1.51 to 1.65 Å.

(6) On the other hand, the sequential NOEs are much more important than longer range NOEs in determining the fine details of local conformation, but even restraining these sequential distances to be within 0.1 Å of the native permitted average deviations in backbone dihedral angles of 35°. In other words, one might hope for a solution structure determination on the order of 1 Å accuracy in the sense of least squares superposition onto the native, but with regard to dihedral angles and detection of the hydrogen bonding network, the result is more comparable to an X-ray crystal structure with a nominal 3 Å resolution.

(7) Sidechain dihedral angles were even more poorly determined, although the sidechain positions were closer to the native than would be expected at random.

Overall, the most significant new result was the simple, undeniable fact that the information available from NMR data is quite sufficent to determine the overall conformation of a protein with a precision at least comparable to that of a low resolution crystal structure. This result was quite a surprise at the time, since prior to this study it was widely believed that because NMR only yields imprecise information on short distances, the cumulative effect of many small errors across the molecule would result in substantial uncertainty in the protein's large scale structure. In contrast to this expectation, however, we found that the large-scale structure was if anything even better defined than the local structure. In all of our computed conformations, the protein's secondary structure, the packing of these secondary structural elements in space, the polypeptide chain threading and the molecule's overall dimensions could be accurately reproduced from the data. The reasons behind this are, in a sense, what this book is all about, namely the fact that when one wishes to estimate the information content of a large number of distance measurements, one must add to it the information which is contained in the fact that they must all be mutually consistent.

7.4. Application to a Protein of Unknown Structure

In this section we present a case study of the determination of the conformation of a small (57 residue) protein, the Bull Seminal Inhibitor (BUSI), from NMR data [Williamson1985]. Our reason for choosing this particular example is that it was the first successful reported determination of the complete structure of a protein from NMR data, in which one of us (T. Havel) directly participated. It is, however, quite typical of several NMR structure determinations which have been made since that time. These calculations were made using the implementation of the EMBED algorithm known as the DISGEO program, which was briefly described in §7.2 (see also [Havel1984]).

For the details of the data's collection and interpretation in terms of distance constraints, we refer the reader to the original reference [Williamson1985]. All in all, a total of 202 NOE derived distance constraints were available. The sequential were classified as strong, medium and weak, as in the above simulations, while a uniform upper bound of 4.0 Å was imposed on those long-range NOEs involving at least one backbone amide or alpha proton, and an upper bound of 5.0 Å used for all the rest. The locations of the three disulfide bridges could also be determined from the NOESY spectra, and a *cis* peptide bond was found between Asp-12 and Pro-13. In addition to these NOESY data, a number of supplementary constraints were available. These included (the constraints necessary to define) the fourteen hydrogren bonds whose presence was inferred from amide proton exchange studies, together with the secondary structure of this protein, as determined by direct inspection of the NMR data [Williamson1984]. On the basis of the Karplus relation, it was also possible to confine those ϕ-angles whose amide-alpha coupling constants exceeded 8.0 Hz to the range between -160° and -80°, while those whose coupling constants were less than 5.5 Hz could be confined between -90° and -40°. This gave a total of 34 ϕ angle constraints, and finally the χ_1 angles of five residues with unique β protons were constrained to $180 \pm 30^\circ$ on the basis of the observed H^α to H^β proton couplings. These angle ranges were imposed by means constraints on the oriented volumes, as was described in §7.1.

Two sets of five conformations each were computed, one using only the NOE derived distance constraints (N1 – N5), and the other including the supplementary data as well (C1 – C5). The convergence in both cases was comparable to that obtained in the BPTI simulations discussed in the previous section. The worst distance constraint violations never exceeded 1 Å, and in some cases the worst violations were below 0.5 Å. Unfavorable atomic overlaps were negligable, covalent bond lengths within a few percent of their standard values, and the peptide bonds were planar to within 10°. This convergence is quite a tribute to the experimentalists who worked on this protein, for it strongly implies that they interpreted a large quantity of NMR data collected painstakingly over a span of several years without making any significant errors!

The differences between the computed conformations were also comparable to that obtained in the aforementioned BPTI simulations (see Figure 7.8). The mean $RMSD^a$ between conformations of 1.9 Å and 2.1 Å for the N and C structures, respectively, while the average $DHAD^b$ values were 77° and 51°. The protein itself consists of a central α-helix, packaged between a three stranded antiparallel β-sheet and a loop of 3_{10} helix, whose positions are similar in all of the ten random structures generated, even when NMR constraints not directly due to NOEs were excluded. Although there were relatively few direct constraints on the amino acid side chains, those which were buried in the interior of the protein were nevertheless quite well determined by the severe packing restrictions there. On the basis of the computed conformations, it was in fact possible to assign the salt bridge involving the His-24 side chain, whose existence was clear from the NMR data, to the proximal amino group of Lys-35. Even though no crystal structure was available for BUSI, the crystal structures of two close homologues were: the Porcine Pancreatic Secretory Inhibitor (PSTI; 26% homology), and the third domain of the Japanese Quail Ovomucoid (OMJPQ3; 45%). Since most of the mutations and deletions are confined to the N-terminal ends of the polypeptide chains, only the segment between residues 23 – 57 of BUSI were compared with the corresponding residues of the homologues. The average $RMSD^a$ with the PSTI was 2.2 Å for both the N and C data sets, respectively, while with OMJPQ3 these numbers were 1.9 Å and 1.8 Å, respectively. The $RMSD^a$ between these segments of the homologues themselves was 1.3 Å.

The best convergent structure from the complete data set (C1) was subjected to several cycles of further refinement by means of computer graphics manipulation of the side chains to eliminate undesirable staggered conformations, followed by restrained energy minimization (see Appendix of

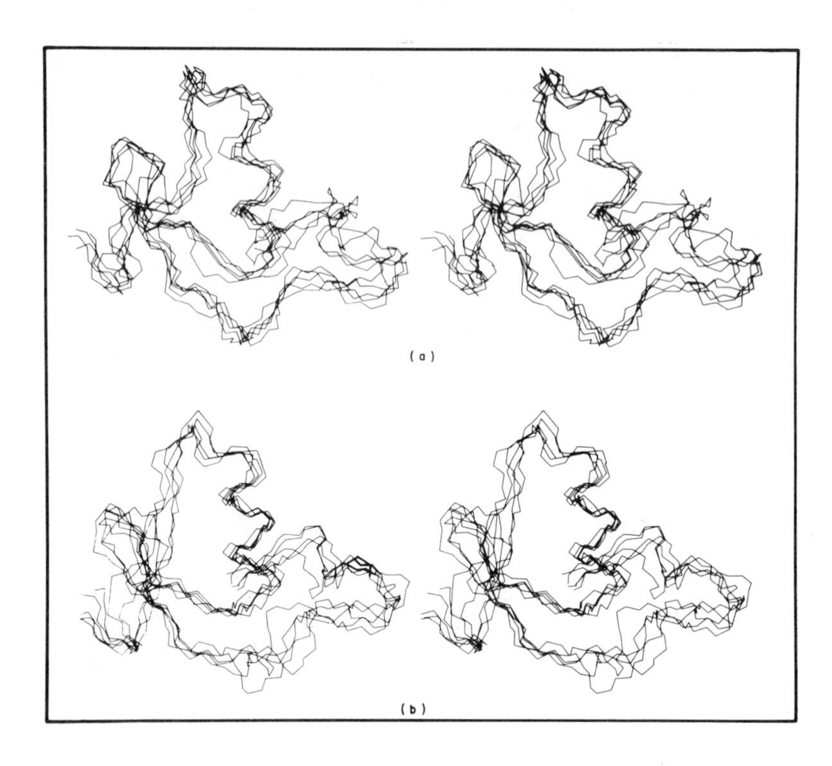

Figure 7.8

Stereo drawings of the polypeptide backbones of the five BUSI conformations computed for each of the two data sets, superimposed so as to minimize the *RMSD* with the first. Above, N1 – N5 (NOE derived constraints only); Below, C1 – C5 (NOE plus supplemental constraints). (Figure reproduced by permission [Williamson1985].)

[Williamson1985] for details). Although this did not substantially change the overall conformation in any way, it did eliminate energetically unfavorable local conformations and small van der Waals overlaps, and demonstrated the potential of such refinements to bring in those small details of the conformation which cannot adequately be accounted for by geometric considerations alone. In Figure 7.9, we show a picture of the central core of the refined structure after superposition on the corresponding residues of the two homologues; the unexpectedly excellent fit (which truly was found only *after* the refinement was complete!) proves conclusively that the combination of NMR and distance geometry is capable of producing structures which are in something very close to the right energy

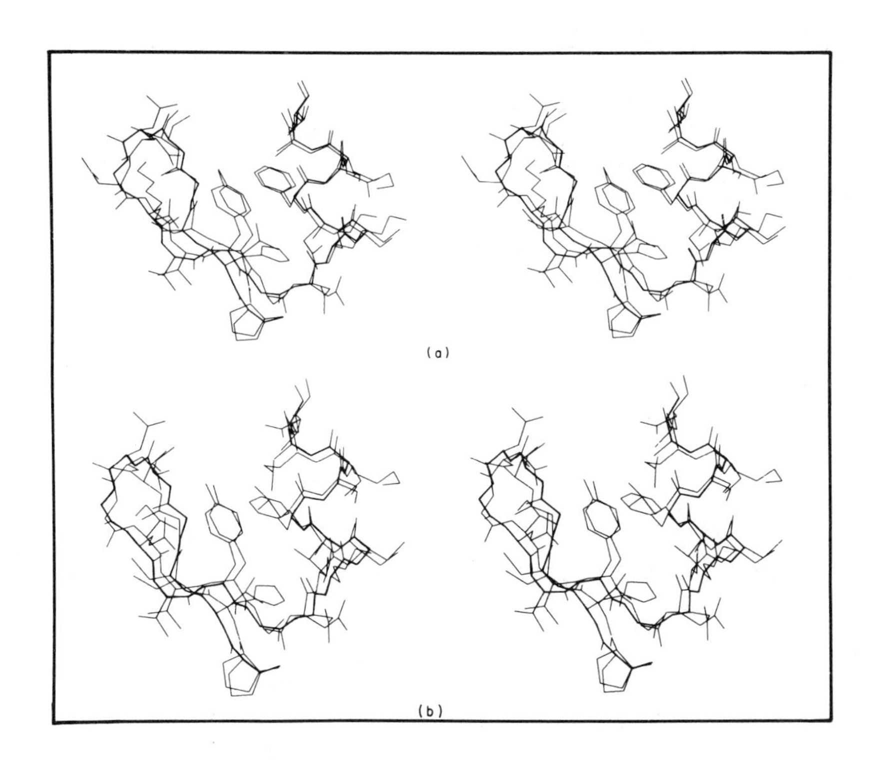

Figure 7.9

Stereo drawings of the the central core (residues 23 – 42) of the refined BUSI conformation (heavier lines) superimposed upon the corresponding residues of its homologues: OMJPQ3 (above); PSTI (below). (Figure reproduced by permission [Williamson1985].)

minimum.

It is clear from these examples that distance geometry permits one to combine results from a wide variety of experiments and thus determine the conformation of even small proteins in noncrystalline environments with an accuracy comparable to that of X-ray crystallography. Particularly with recent advances in NMR spectroscopy there is every reason to expect that sufficient data of adequate quality will be available for many interesting molecules.

References

Arseniev1983.

A.S. Arseniev, V.I. Kondakov, V.N. Maiorov, T.M. Volkova, E.V. Grishin, V.F. Bystrov, and Yu.A. Ovchinnikov, "Secondary Structure and Sequential Resonance Assignments in Two-Dimensional Proton Magnetic Resonance Spectra of Insectotoxin I_5A Buthus eupeus.," *Bioorg. Khim.*, *9*, 768-793(1983).

Arseniev1984.

A.S. Arseniev, V.F. Bystrov, V.T. Ivanov, and Y.A. Ovchinnikov, "NMR Solution Conformation of Gramicidin A Double Helix," *FEBS Lett.*, *165*, 51-56(1984).

Arseniev1985.

A.S. Arseniev, I.L. Barsukov, E.N. Shepel, V.F. Bystrov, and V.T. Ivanov, "2D ^1H NMR Conformational Analysis of Des-(Ala^3-D-Val^1) -(Val^1)Gramicidin A in Solution: The Right-Handed Parallel Double Helix," *Bioorg. Khim.*, *11*, 5-20(1985).

Bax1986.

A. Bax and L. Lerner, "Two-Dimensional Nuclear Magnetic Resonance Spectroscopy," *Science*, *232*, 960(1986).

Braun1981.

W. Braun, C. Boesch, L.R. Brown, N. Go, and K. Wuthrich, "Combined Use of Proton-Proton Overhauser Enhancements and a Distance Geometry Algorithm for Determination of Polypeptide Conformations. Application to Micelle-Bound Glucagon," *Biochim. Biophys. Acta*, *667*, 377-396(1981).

Braun1983.

W. Braun, G. Wider, K.H. Lee, and K. Wuthrich, "Conformation of Glucagon in a Lipid-Water Interphase by ^1H Nuclear Magnetic Resonance," *J. Mol. Biol.*, *169*, 921-948(1983).

Braun1985.

W. Braun and N. Go, "Calculation of Protein Conformations by Proton-Proton Distance Constraints: A New Efficient Algorithm," *J. Mol. Biol.*, *186*, 611-626(1985).

Braun1986.

W. Braun, G. Wagner, E. Woergoetter, M. Vasak, J.H.R. Kaegi, and K. Wuthrich, "Polypeptide Fold in the Two Metal Clusters of Metallothionein-2 by Nuclear Magnetic Resonance and Distance Geometry," *J. Mol. Biol.*, *187*, 125-129(1986).

Braun1987.

W. Braun, "Distance Geometry and Related Methods for Protein Structure

Determination from NMR Data," *Quart. Rev. Biophys.*, *19*, 115-157(1987).

Crippen1979.

G.M. Crippen, "The Effect of Distance Constraints on Macromolecular Conformation. I. The Effectiveness of the Experimental Studies on Tobacco Mosaic Virus Protein," *Int. J. Pept. Prot. Res.*, *13*, 320-326(1979).

Crippen1981.

G.M. Crippen, N.J. Oppenheimer, and M.L. Connolly, "Distance Geometry Analysis of the NMR Evidence on the Solution Conformation of Bleomycin," *Int. J. Pept. Prot. Res.*, *17*, 156-169(1981).

Deisenhofer1975.

J. Deisenhofer and W. Steigemann, "Crystallographic Refinement of the Structure of Bovine Pancreatic Trypsin Inhibitor at 1.5A Resolution," *Acta Cryst.*, *B31*, 238-250(1975).

Drysdale1980.

B.E. Drysdale and D.P. Hollis, "A Nuclear Magnetic Resonance Study of Cobalt II Alcohol Dehydrogenase: Substrate Analog-Metal Interactions," *Arch. Bioch. Biophys.*, *205*, 267-279(1980).

Gariepy1985.

J. Gariepy, L.E. Kay, I.D. Kuntz, B.D. Sykes, and R.S. Hodges, "Nuclear Magnetic Resonance Determination of Metal-Proton Distances in a Synthetic Calcium Binding Site of Rabbit Skeletal Troponin C," *Biochem.*, *24*, 544-550(1985).

Hahn1978.

L.-H. Hahn and G.G. Hammes, "Structural Mapping of Aspartate Transcarbamoylase by Fluorescence Energy-Transfer Measurements: Determination of the Distance Between Catalytic Sites of Different Subunits," *Biochem.*, *17*, 2423-2429(1978).

Hare1986.

D.R. Hare and B.R. Reid, "Three-Dimensional Structure of a DNA Hairpin in Solution: Two-Dimensional NMR Studies and Distance Geometry Calculations on d(CGCGTTTTCGCG)," *Biochemistry*, *25*, 5341-5350(1986).

Havel1979.

T.F. Havel, I.D. Kuntz, and G.M. Crippen, "The Effect of Distance Constraints on Macromolecular Conformation. II. Simulation of Experimental Results and Theoretical Predictions," *Biopolymers*, *18*, 73-82(1979).

Havel1983.

T.F. Havel, I.D. Kuntz, and G.M. Crippen, "The Theory and Practice of Distance Geometry," *Bull. Math. Biol.*, *45*, 665-720(1983).

Havel1984.

T.F. Havel and K. Wuthrich, "A Distance Geometry Program for Determining the Structures of Small Proteins and Other Macromolecules from Nuclear Magnetic Resonance Measurements of Intramolecular ^{1}H-^{1}H Proximities in Solution," *Bull. Math. Biol.*, *46*, 673-698(1984).

Havel1985.

T.F. Havel and K. Wuthrich, "An Evaluation of the Combined Use of Nuclear Magnetic Resonance and Distance Geometry for the Determination of Protein Conformations in Solution," *J. Mol. Biol.*, *182*, 281-294(1985).

Kline1986.

A.D. Kline, W. Braun, and W. Wuthrich, "Studies by ^{1}H Nuclear Magnetic Resonance and Distance Geometry of the Solution Conformation of Tendamistat, an α-Amylase Inhibitor," *J. Mol. Biol.*, *189*, 377-382(1986).

Kuntz1980.

I.D. Kuntz and G.M. Crippen, "A Computer Model for the 30S Ribosome Subunit," *Biophys. J.*, *32*, 677-696(1980).

Kuo1980.

M.C. Kuo and W.A. Gibbons, "Nuclear Overhauser Effect and Cross-Relaxation Rate Determinations of Dihedral and Transannular Interproton Distances in the Decapeptide Tyrocidine A," *Biophys. J.*, *32*, 807-836(1980).

Lee1980.

L. Lee and B.D. Sykes, "Nuclear Magnetic Resonance Determination of Metal-Proton Distances in the EF Site of Carp Parvalbumin Using the Susceptibility Contribution to the Line Broadening of Lanthanide-Shifted Resonances," *Biochem.*, *19*, 3208-3214(1980).

Maiorov1984.

V.N. Maiorov and A.S. Arseniev, "The Use of Distance Constraints in the Conformational NMR Analysis of Polypeptides. Insectotoxin I_5A Buthus eupeus," *J. Mol. Struct.*, *114*, 399-402(1984).

Mauk1980.

A.G. Mauk, R.A. Scott, and H.B. Gray, "Distances of electron transfer to and from metalloprotein redox sites in reactions with inorganic complexes.," *J. Am. Chem. soc.*, *102*, 4360-4363(1980).

McLachlan1984.

A.D. McLachlan, "How Alike are the Shapes of two random Chains," *Biopolymers*, *23*, 1325-1331(1984).

Miake-Lye1983.

R.C. Miake-Lye, S. Doniach, and K.O. Hodgson, "Anomalous X-Ray Scattering from Terbium-Labeled Parvalbumin in Solution," *Bioph. J.*, *41*, 287-292(1983).

Neidig1984.
K.P. Neidig, H. Bodenmueller, and H.R. Kalbitzer, "Computer Aided Evaluation of Two-Dimensional NMR Spectra of Proteins," *Bioch. Bioph. Res. Commun.*, *125*, 1143-1150(1984).

Nygaard1984.
E. Nygaard, G.L. Mendz, W.J. Moore, and R.E. Martenson, "NMR of a Peptide Spanning the Triprolyl Sequence in Myelin Basic Protein," *Biochem.*, *23*, 4003-4010(1984).

Olejniczak1984.
E.T. Olejniczak, C.M. Dobson, M. Karplus, and R.M. Levy, "Motional Averaging of Proton Nuclear Overhauser Effects in Proteins. Predictions from a Molecular Dynamics Simulation of Lysozyme," *J. Amer. Chem. Soc.*, *106*, 1923-1930(1984).

Pillai1981.
S. Pillai, "Can Multiple Low-Density Immunoadsorbents and "Negative" Distance Matrices Help Predict the Three-Dimensional Structure of Any Globular Protein," *Bioch. J.*, *199*, 277-280(1981).

Schmidt1984.
P.G. Schmidt and I.D. Kuntz, "Distance Measurement in Spin-Labeled Lysozyme," *Biochemistry*, *23*, 4261-4266(1984).

Villafranca1978.
J.J. Villafranca, S.G. Rhee, and P.B. Chock, "Topographical Analysis of Regulatory and Metal Ion Binding Sites on Glutamine Synthetase from Escherichia Coli: ^{13}C and ^{31}P Nuclear Magnetic Resonance and Fluorescence Energy Transfer Study," *Proc. Natl. Acad. Sci. USA*, *75*, 1255-1259(1978).

Wemmer1985.
D.E. Wemmer and B.R. Reid, "High Resolution NMR Studies of Nucleic Acids and Proteins," *Ann. Rev. Phys. Chem.*, *36*, 105-137(1985).

Wider1984.
G. Wider, S. Macura, A. Kumar, R.R. Ernst, and K. Wuthrich, "Homonuclear Two-Dimensional ^{1}H NMR of Proteins. Experimental Procedures," *J. Magn. Reson.*, *56*, 207-234(1984).

Williamson1984.
M. P. Williamson, D. Marion, and K. Wuthrich, "Secondary Structure in the Solution Conformation of the Proteinase Inhibitor IIA from Bull Seminal Plasma by Nuclear Magnetic Resonance," *J. Mol. Biol.*, *173*, 341-360(1984).

Williamson1985.

M.P. Williamson, T.F. Havel, and K. Wuthrich, "Solution Conformation of Proteinase Inhibitor IIA from Bull Seminal Plasma by ^{1}H Nuclear Magnetic Resonance and Distance Geometry," *J. Mol. Biol., 182*, 295-315(1985).

Wuthrich1982.

K. Wuthrich, G. Wider, G. Wagner, and W. Braun, "Sequential Resonance Assignments as a Basis for Determination of Spatial Protein Structures by High Resolution Proton Nuclear Magnetic Resonance," *J. Mol. Biol., 155*, 311-319(1982).

Wuthrich1983.

K. Wuthrich, M. Billeter, and W. Braun, "Pseudo Structures for the 20 Common Amino Acids for Use in Studies of Protein Conformations by Measurements of Intramolecular Proton-Proton Distance Constraints with Nuclear Magnetic Resonance," *J. Mol. Biol., 169*, 949-961(1983).

Wuthrich1986.

K. Wuthrich, *NMR of Proteins and Nucleic Acids,* J. Wiley & Sons, New York, NY, 1986.

Zuiderweg1983.

E.R.P. Zuiderweg, R. Kaptein, and K. Wuthrich, "Secondary Structure of the Lac Repressor DNA-Binding Domain by Two-Dimensional ^{1}H Nuclear Magnetic Resonance in Solution," *Proc. Natl. Acad. Sci. USA, 80*, 5837-5841(1983).

8. Ligand Binding

8.1. The Chemical Problem

Devising computational methods can be motivated by the inherent mathematical beauty of the problem or by the scientific pressure of numerous important applications. The drug-receptor mapping problem enjoys both these features. One of the many steps in designing new drugs, but a crucial one, is measuring the binding affinity of a series of compounds to a biologically important macromolecule, the receptor. From these measurements, one would like to deduce the chemical structure of new drug molecules which would bind even more strongly to the desired receptor while perhaps binding less tightly to other receptors [Ramsden1988, Martin1978, Hopfinger1985, Cohen1985]. It is quite a mathematical challenge to proceed from a collection of scalar binding constants to molecular structure, and the applications of such a method are enormous. The need for the method arises because firstly the binding experiments are easy to do: man-days instead of man-years for protein crystallography, for example. Therefore, there is a great fund of data available. Secondly, the commercial (and humanitarian) rewards for new drugs are enormous. Some pharmaceutical companies estimate it costs roughly $5000 for every new compound they synthesize, and to have one new drug in a hundred get through initial animal testing is considered a great success. Although several million organic compounds have been well characterized, this is only a sparse sampling of all the possibilities. Clearly, the Edisonian approach is impractical here.

Of course, we are focusing here on only one facet of total drug design. Our starting point is that biochemists and pharmacologists have pinpointed an isolated receptor, usually a protein, which must bind the drug to achieve the desired biological response. The conformation of the receptor is generally unknown. Then equilibrium constants for reversible binding have been determined experimentally for at least a few different compounds. If the receptor is an enzyme, and the desired response is enzyme inhibition, then we want each drug to compete for the same inhibition site, and the measured K_is (equilibrium constants for *dissociation)* can be converted to free energies of binding (*association*) by simply

$$\Delta G_{obsd} = RT ln K_i . \quad (8.1)$$

Often the experimental results are quoted in terms of I_{50}, the concentration of an inhibitor necessary to produce 50% inhibition. Assuming Michaelis-Menten kinetics:

$$\Delta G_{obsd} = RT ln \left[\frac{K_m [I_{50}]}{K_m + [S]} \right] \quad (8.2)$$

where K_m is the Michaelis constant, and $[S]$ is the substrate concentration used in the binding assay. If a drug is to be successful, it must interact strongly and specifically with its receptor, but there are of course many other factors to consider as well. For instance, the whole organism effectiveness of a drug is strongly correlated with its solubility, so using *in vivo* test data, one can explain and predict drug efficacy much better than chance by schemes relating molecular structure to solubility.

To explain *specific* binding to a single well-defined receptor, many workers in the field have sought common chemical structural features in the active compounds (the "pharmacophore") and important differences in the inactive compounds. This involves explicitly or implicitly superimposing molecules on one another. As long as the compounds form a homologous series, it is fairly unambiguous which atoms of one molecule are to correspond to which atoms of the other, but sometimes the data include molecules from radically different classes. The situation is worse still when dealing with conformationally flexible molecules, for then there may be whole ranges of superposition possibilities. In any case, the results of the study are necessarily expressed in terms of the drug molecular structures, e.g. bulky substituents at the 5-position are forbidden, etc. Thus it is hard to predict the binding of compounds for which there is no clear superposition onto the pharmacophore.

In order to be more generally applicable, a method should describe the receptor site in such a way that one could rationalize and predict the binding of molecules of diverse structures. Thus the analysis should concentrate on the receptor, not on the input molecules. For physical realism, drug molecules should be able to vary their orientation in the site and internal conformation to achieve an optimal fit. As we have seen from numerous X-ray crystal studies of enzyme-inhibitor complexes, the various regions constituting the site have individual energetic and steric preferences, and the site as a whole can make small movements to better accommodate a ligand.

8.2. The Mathematical Problem

Basically, one has to devise computer models of ligand molecules and receptor sites such that simulating the molecular motions and drug-receptor interactions is computationally feasible while being physically realistic enough to be useful. Ultimately, the ligand-receptor system should be described in terms of discrete quantum mechanical states, but with so many atoms involved, it is much more practical to have continuously variable atomic positions determining continuous inter- and intra-molecular energies. Even assuming for the moment that the receptor is fixed and that the ligand's bond lengths and angles are rigid, just modeling the docking of ligand in the site as a function of six or more continuous variables, consisting of translational, rotational, and conformational degrees of freedom, constitutes an infeasible global search problem [Crippen1975]. Consequently, we have considered coarse discrete models.

First, the ligands are represented as discrete atoms whose relative positions are determined by fixed bond lengths, fixed vicinal bond angles, and discretely variable torsion angles. Typical dihedral angle increments are from 10° to 120°. These ligands will often be referred to as "the molecules" in a study, since their chemical structures and conformational properties are taken as fully understood. The receptor is also ultimately a part of a macromolecule, but it is unknown, so we will refer to it only as the "site" or site model we are trying to construct. Like the molecules, the site is represented as a collection of discrete parts, either site "points" or "pockets" in our earlier work or site "regions" in §8.7. In either case, these parts are said to have fixed relative positions, and a part may represent either an empty portion of the real receptor which can be occupied by some atoms of the ligand, or it can be "repulsive" and represent volume excluded by the atoms of the real receptor. The possible ligand-receptor interactions can be summarized as a discrete set of binding "modes", each one of which tells which atom is in "contact" with which part of the site. When dealing with site points, we assume that each site point may be in contact with zero or one atom, and each atom of the molecule can be in contact with at most one site point. In the site region case, many atoms may occupy one region, each atom is located in exactly one region, and some regions may be empty. The total binding energy for a given mode is taken to be a sum over all contacts of the corresponding individual interaction energies. Note the important conceptual distinction between the usual view of a molecule's allowed conformation space as some complicated region in the product space of dihedral angles, and this idea of concentrating on a list of binding modes. We are saying that the only conformational property of interest in a molecule is what combinations of contacts can be made with the site. All other variability is irrelevant to a study of binding. Since we are describing a molecule's conformation space with respect to a given site, we now have some common ground with which to compare molecules. If two ligands both have a similar binding mode, then they have some conformational similarity, as far as that site is concerned. These ideas will be made more definite shortly.

Aside from the conceptual differences, there are strong computational advantages to the discretization of the ligands, their conformational search, their conformation space in terms of binding modes, and the energetics of ligand-receptor interactions. Whereas the global search over all ways to dock the ligand into the site is generally infeasible, producing the list of all sterically allowed binding modes is entirely possible, even for a molecule of as many as 40 atoms and a site consisting of 10 parts. In order to adjust the calculated energies to fit the observed binding, one only has to develop a table of interaction energies between each part of the site and each type of atom in the set of molecules. Furthermore, since the total interaction is just a sum over all contacts in a mode, we have a linear expression to fit.

The considerations just discussed have led to the following definitions for the ligand binding problem. Given a collection of molecules $m = 1,...,M$ and their experimentally determined free energies of binding $\Delta G_{m,obsd}$, we want to develop a model that simulates their binding to a site as accurately as is computationally feasible. Let $\mathbf{b}$ denote a vector encoding a particular binding mode for a particular molecule. Just how the combination of contacts is encoded in $\mathbf{b}$ will depend on the type of site model, and will be discussed later. Let

$$\mathbf{B}_m = \{\mathbf{b} \mid \text{geometrically and energetically allowed}\} . \tag{8.3}$$

Various schemes for checking geometric feasibility of an arbitrary mode $\mathbf{b}$ are discussed later. To explain the energetic constraint, let $\boldsymbol{\phi}_m$ be the vector of dihedral angles necessary to fix the conformation of molecule m once the bond lengths and bond angles are held fixed. The components of $\boldsymbol{\phi}_m$ can take on only discrete values arbitrarily chosen by the investigator. For any choice of $\boldsymbol{\phi}_m$, we can calculate the isolated molecule's internal energy, $E(\boldsymbol{\phi}_m)$, by some quantum mechanical or semiempirical method. Let the energetic global minimum be

$$E_m^* = \min_{\boldsymbol{\phi}_m} E(\boldsymbol{\phi}_m). \tag{8.4}$$

Then the (discretized) *allowed* conformation space, Φ_m, is given by

$$\Phi_m = \{\boldsymbol{\phi}_m \mid E(\boldsymbol{\phi}_m) \leq E_m^* + \Delta E\} \tag{8.5}$$

where the chosen $\Delta E > 0$ is on the order of the available thermal energy, $k_B T$, but less than $\Delta G_{m,obsd}$. In order for **b** in Equation (8.3) to be energetically allowed, it must correspond to a conformation in Φ_m. Incidentally, a very convenient summary of Φ_m can be produced in terms of interatomic distances, d_{mij}, by simply noting the maximum, u_{mij}, and the minimum, l_{mij}, values for all pairs of atoms i and j:

$$u_{mij} = \max_{\Phi_m} d_{mij} \tag{8.6}$$

$$l_{mij} = \min_{\Phi_m} d_{mij} \tag{8.7}$$

Clearly it is necessary that in every allowed conformation $l_{mij} \leq d_{mij} \leq u_{mij}$ for all i and j, but one cannot chose arbitrary values for d within its given range and always achieve embedability (see Chapter 3). An especially important case in practice is shown in Fig. 8.1, where $d_{R1,R2}$ is negatively correlated with $d_{R1,R3}$.

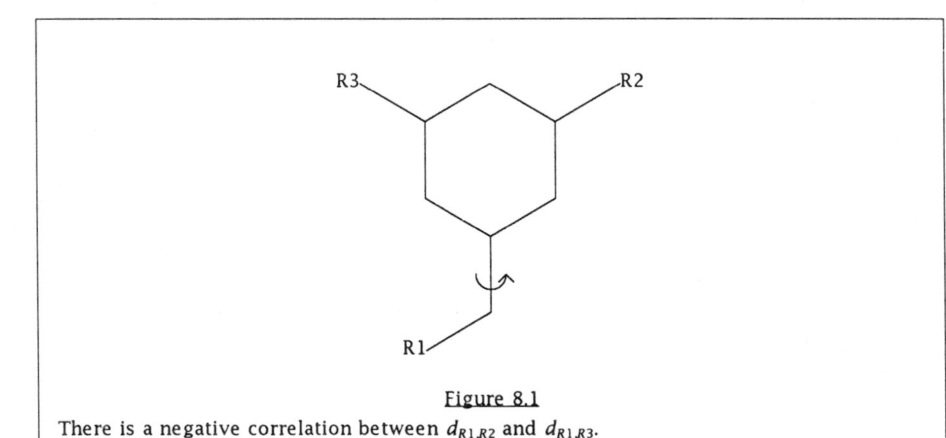

Figure 8.1
There is a negative correlation between $d_{R1,R2}$ and $d_{R1,R3}$.

If n_s = the number of site parts, and n_t = the number of atoms types in the whole set of molecules, then there will be an $n_s n_t$-dimensional energy parameter space, and for every mode, **b**, there corresponds a vector $\boldsymbol{\beta}$ in this space whose nonnegative integer components indicate how many times each energy parameter is used in calculating the total binding free energy for that mode. Thus we can just as well think of $\boldsymbol{\beta} \in \mathbf{B}_m$ as $\mathbf{b} \in \mathbf{B}_m$. One can map the set of **b**s *onto* the $\boldsymbol{\beta}$s, but not the reverse, since more than one **b** may correspond to one $\boldsymbol{\beta}$ when there is a choice of distinct groups of the same type to make contact with the same part of the site. The calculated free energy of binding for a particular mode is simply

$\boldsymbol{\beta}^T\mathbf{e}$, where $\mathbf{e}$ gives the current values of the set of energy parameters. Now the real molecule is free to translate and rotate as it approaches the site in repeated random collisions, eventually binding in one of the energetically best modes. We simulate this by letting *the* calculated binding of molecule m be

$$\Delta G_{m,calc} = \min_{\boldsymbol{\beta} \in B_m} (\boldsymbol{\beta}^T \mathbf{e}) \tag{8.8}$$

Note that this implies not only a particular value of $\Delta G_{m,calc}$, but one (or more) optimal mode(s). Finally, we adjust $\mathbf{e}$ until least squares agreement with the observed binding is obtained:

$$\min_{\mathbf{e}} \sum_m (\Delta G_{m,obsd} - \Delta G_{m,calc})^2 \tag{8.9}$$

Alternatively, one can be given an experimentally observed range of binding $[\Delta G_{m,obsd-}, \Delta G_{m,obsd+}]$ for each molecule, and then adjust $\mathbf{e}$ until

$$\Delta G_{m,obsd-} \leq \Delta G_{m,calc} \leq \Delta G_{m,obsd+} \quad \forall\, m \tag{8.10}$$

For a given optimal binding mode, $\Delta G_{m,calc}$ is linear in $\mathbf{e}$, so that Equation (8.9) is a linear least squares problem, and Equation (8.10) is a set of linear inequalities, but varying $\mathbf{e}$ can change the optimal binding mode discontinuously, thereby substantially complicating the problem. We will discuss means of solving Equations (8.9) and (8.10) in the next sections.

Before going through a full explanation of the algorithms for proceeding from binding data to a site model, we should consider more carefully the problem of how to represent the given ligands. Since the algorithms for the analysis of binding are combinatorial searches of one kind or another, it is important to keep the magnitude of the calculations to a minimum. In particular, the size of $\mathbf{B}_m$ tends to increase exponentially with the number of atoms in molecule m. Fortunately, it is neither necessary nor especially appropriate to consider every individual atom in each molecule, but rather only the groups of atoms which occur in the data set. Consider the fictitious set of binding data shown in Table 8.1.

$$
\begin{array}{c}
R_2 \\
| \\
CH_2 \\
| \\
NH_2 - C^* - COOH \\
| \\
R_1
\end{array}
$$

Table 8.1: Alanine derivatives binding to an imaginary site.

compound	ΔG_{obsd}	composition Ala	R_1	R_2
1	−5.0±0.3	L-	H	H
2	−7.0±0.3	L-	H	C_6H_5
3	−7.1±0.3	L-	H	p–C_6H_4Cl
4	−2.0±0.3	L-	CH_3	C_6H_5
5	−3.0±0.3	D-	H	H
6	−5.0±0.3	D-	H	C_6H_5
7	−4.9±0.3	D-	CH_3	C_6H_5

One way to think of a site model is as a device for recognizing which molecule is presented to it out of a given list. If two molecules have essentially the same observed binding affinity, within experimental error, then the model does not need to distinguish between them. Of course we would also like our model to be physically realistic and predictive, but consider for a moment simply the implications of being a discriminator. If in the set of molecules there is some structural aspect A, such as substitution at a particular position which might have various "values" A_i, $i=1,...,n_A$ (such as methyl, ethyl, propyl, etc.), then let $P(A_i)$ be the frequency of occurrence of A_i in the data set. Thus we treat P as a probability, since

$$0 \leq P(A_i) \leq 1 \quad \forall\ i \tag{8.11}$$

and

$$\sum_{i=1}^{n_A} P(A_i) = 1 . \tag{8.12}$$

The information theory entropy (see [Ash1965] and many other standard texts) η gives the average number of binary decisions necessary to determine which value feature A has in a molecule selected at random from the data set.

$$\eta(A) = -\sum_{i=1}^{n_A} P(A_i) log_2 P(A_i) \tag{8.13}$$

Thus in Table 8.1, $P(\text{L-})=\frac{4}{7}$ and $P(\text{D-})=\frac{3}{7}$, so $\eta(\text{chirality})=0.99$ bits, but $\eta(\text{R}_1)=0.86$ bits. The maximum value of $\eta(A) = log_2 n_A$, occurring when $P(A_i) = \frac{1}{n_A}$ for all i, as nearly happens for chirality. Since $1 \times log(1) = 0 \times log(0) = 0$, the presence of an amino group or a naphthyl group contributes nothing to η, i.e. they are foregone certainties the site model need not consider. Now suppose there is a second structural aspect, B, attaining values B_j, $j=1,...,n_b$. The total diversity the site model must recognize is

$$\eta(AB) \leq \eta(A)+\eta(B) \tag{8.14}$$

where equality holds only if A and B are statistically independent, that is, when

$$P(B_j) = P(B_j|A_j) \quad \forall \ i,j \tag{8.15}$$

where $P(B_j|A_j)$ is the probability of B_j occurring, given that A_i has occurred. In general,

$$\begin{aligned} \eta(AB) &= -\sum_i \sum_j P(A_i B_j) log_2 P(A_i B_j) \\ &= \eta(A)+\eta(B|A) \end{aligned} \tag{8.16}$$

where the conditional entropy is

$$\eta(B|A) = -\sum_{i=1}^{n_A} P(A_i) \left[\sum_{j=1}^{n_b} P(B_j|A_i) log_2 P(B_j|A_i) \right] . \tag{8.17}$$

In other words, $\eta(B|A)$ expresses the expected diversity in B given the outcome of A. Suppose A_1 = "R_2= H" and A_2 = "R_2= phenyl" in Table 8.1, and imagine for a moment B_1 = "presence of 2-position carbon in an R_2= phenyl" and B_2 = its absence. In this case, A and B are totally correlated, so that A_1 is true if and only if B_1 is true, and the same for A_2 and B_2. Thus $\eta(A) = \eta(B)$, $\eta(B|A) = 0$, and $\eta(AB) = \eta(A)$. The lesson is that it is pointless to describe the set of molecules in overly fine detail, since the complexity of the site model, and hence its structural discriminatory power, need be no greater than the structural entropy of the data set, which is independent of the structural descriptors we choose. In our example, letting A be chirality, B be the choice of R_1, C be whether R_2 is phenyl, and D

be chloro substitution of the phenyl, we have

$$\begin{aligned} \eta(ABCD) &= \eta(A)+\eta(B|A)+\eta(C|AB)+\eta(D|ABC) \\ &= 0.99 + 0.86 + 0.68 + 0.29 \\ &= 2.82 \text{ bits} \end{aligned}$$

By way of comparison, suppose we had to specify the location of each of the 23 atoms of compound **2** in a 10 Å cube to an accuracy of 1 Å in x, y, and z coordinates. That requires $23 \times 3 \times log_2 10 = 229$ bits, vastly more information than this compound represents as a member of the data set. Of course if the set of compounds is structurally very diverse, the information content per drug is higher.

8.3. Structural Composition

One of the first problems in any drug binding study is how to represent the molecules and how to suggest their modes of interaction with the receptor site. As we have seen, an economical representation is desirable for computing speed and for fundamental reasons from information theory. If we are dealing with a set of close analogues, one can usually assume they bind with the common group at the same place in the site, and it is immediately clear what are the structural similarities and differences. Suppose, however, we start with a set of unrelated compounds, such as those shown in Fig. 8.2.

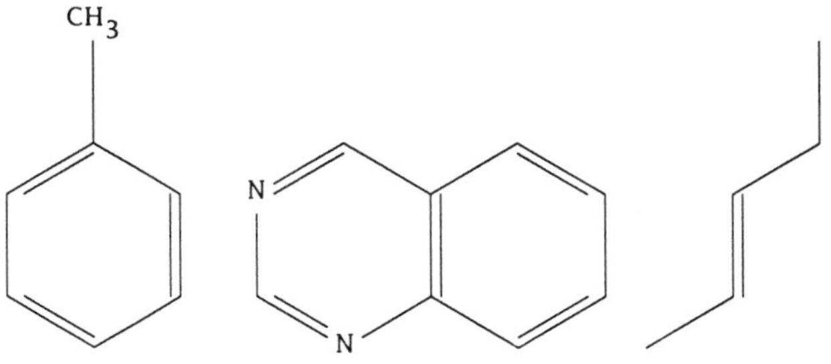

Figure 8.2
Molecular structural comparison example. **1** - toluene, **2** - quinazoline, **3** - trans-2-pentene.

It is not obvious how to represent these three molecules in an economical way. The following algorithm has proven to be a useful method for doing just that. First we must make our objective a little more precise.

Definition 8.1. The *base* is a set of atoms, their types, and upper and lower interatomic distance bound matrices. This is supposed to correspond to the common structural feature of the set of molecules, which is the "pharmacophore" for a set of *active* drugs. An *r-group* consists of a set of atoms, their types, and the upper and lower distance bounds from each atom to the atoms of the base. In other words, an r-group is positioned in space relative to the base, and this is why the base must consist of at least 3 non-collinear atoms.

What we seek is a base and a collection of various r-groups such that all of each molecule in the set can be expressed as the union of the (nonoverlapping) base and some of the r-groups.

Algorithm 8.2 Classification of Structure of a Set of Molecules:

INPUT: Given the set of molecules $m=1,\ldots,M$ and their upper and lower interatomic distance bound matrices $\mathbf{U}_m$ and $\mathbf{L}_m$ respectively, calculated according to Equations (8.6) and (8.7).

OUTPUT: A relatively small number of molecular fragments, and the decomposition of each molecule into some subset of them.

PROCEDURE:

```
| Mark every molecule as being in the current set of molecules,
|  and every atom as "not used".
| While there are molecules in the current set
| | Take the first current molecule as the configuration of atoms common
| |  to all these molecules, called the "base".
| | For all subsequent current molecules, find their "intersection" with
| |  the base.
| | | If it is less than 3 atoms, alter the molecule's mark to be
| | |  a member of a "future" set.
| | | Else the intersection becomes the new base.
| | For all current molecules
| | | Mark as "used" those atoms corresponding to the final base.
| | While there are unused atoms in the current molecule set
| | | Let the unused atoms of the first current molecule which has any,
| | |  constitute the latest "r-group".
| | | For all other current molecules with unused atoms,
| | | | Find the intersection of the unused atoms with the latest r-group.
| | | | If the intersection is not empty, make it the latest r-group.
| | | For all current molecules,
```

Mark the atoms corresponding to the r-group as used.
Mark molecules in the current set as used.
Mark "future" molecules as current.

The intersection of a base with a molecule amounts to a tree search over all combinations of matching the atoms in the base to those of the molecule, where not all atoms of either need be used. A branch of the tree of combinations is immediately pruned if the corresponding atom types are not the same. Furthermore, if atoms i_m and j_m of molecule m are to correspond to atoms i_b and j_b, respectively, of the base, then

$$\begin{aligned} u_{im,jm}+\delta &> l_{ib,jb} \quad \text{and} \\ u_{ib,jb} &> l_{im,jm}-\delta \end{aligned} \tag{8.18}$$

must hold, where $\delta > 0$ is a chosen distance match tolerance, typically on the order of 1 Å. Of all successful matchings, the first one of greatest cardinality is chosen. Then the intersection consists of all the atoms in the matching, their types, and interatomic distance bounds, where $u_{ij} = min(u_{im,jm}, u_{ib,jb})$ and $l_{ij} = max(l_{im,jm}, l_{ib,jb})$. In the event that $u_{ij} < l_{ij}$, set both to their mean. The intersection of unused atoms in a molecule and an r-group is found by an analogous process, except the distance bounds are all between atoms and the base group.

The general effect of Algorithm 8.2 is to either find a small base common to the entire data set, or else divide the set into subsets, each of which is treated as a separate problem. Within a molecule set, the tendency is toward a small number of small r-groups, each of which appears in most of the molecules. Finally, each molecule can be described as consisting of the base and some small collection of r-groups. This would be the economical representation described at the end of §8.2. Note that the algorithm is quite ignorant of chemistry, so the final component parts may not always correspond to recognized ring systems and functional groups. The distance tolerance, δ, is an important but arbitrary parameter in the process. A small value biases the algorithm toward finding more and smaller molecular fragments, and less in common throughout the data set. It is also worth noting when r-groups differ insignificantly in structure relative to the base, but differ only in atom type, since then relaxing the atom type classification would reduce the number of distinct r-groups. Consider, for example, the three molecules of Fig. 8.2. Initially, all 7 carbons of **1** are the base, and the matching of maximal cardinality with **2** involves all 7 atoms, as shown in Fig. 8.3. An equally good matching could be achieved by flipping **1** about an axis through its 1- and 4-positions (i.e. twirl the frying pan by its handle). All other matchings would either involve C-N type mismatches, or lower cardinality. The subsequent intersection of the base with **3** lowers the size of the new base to 5 atoms, but retains the old rigid conformation because **3** is able also to adopt that

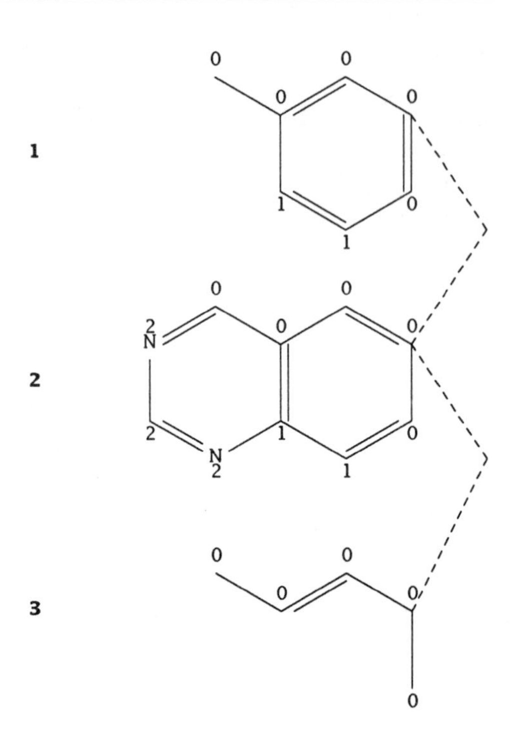

Figure 8.3

The molecules of Fig. 8.2 may be superimposed by vertically translating **2** and **3** onto **1** such that the atoms joined by the dashed line coincide. Atoms labelled 0 correspond to the base, and the others belong to r-groups 1 and 2.

form, as illustrated. Thus, there is a five carbon atom base common throughout, and these atoms have been marked with a "0". That leaves two unused atoms in **1**, which now constitute r-group 1. The equivalent two atoms can also be found in **2**, but all atoms in **3** are already used. At last the only unused atoms are two nitrogens and a carbon in **2**, which therefore amount to r-group 2. In this example, three molecules are described in terms of a base and two r-groups, but in a large set of chemical analogues, the breakdown would generally involve significantly fewer components than molecules.

The advantage of the algorithm is that it will suggest sometimes startling common features between structurally unrelated molecules, which is a great help in proposing a site model. By producing a list of base(s) and r-groups, it automatically provides a great data reduction for subsequent calculations, since one needs to retain only a representative point in each group for some applications. The disadvantages are that (i) the value chosen for δ strongly influences the outcome. A site model derived from these considerations also necessarily discriminates conformational differences to within δ. (ii) The ordering of the molecules influences the results because only the first largest match is retained. (iii) The summary of allowed conformation space in terms of $\underline{\mathbf{U}}_m = [u_{mij}]$ and $\underline{\mathbf{L}}_m = [l_{mij}]$ is a loss of information which may result in proposed molecular superpositions being unfavorable with regard to intramolecular energies, or outright geometrically impossible.

If one wishes to superimpose two molecules according to a *given* match, then the third shortcoming of the above method can be avoided [Blaney1985, Sheridan1986] even though the molecules are so conformationally flexible that rigid body optimal translation and rotation would be useless. The trick is to define an embedding problem in terms of all the atoms of the first and second molecules. The constraints are intramolecular fixed distances sufficient to leave only torsional degrees of conformational freedom, van der Waals minimal contact distances between all pairs of nonbonded atoms, and *inter*molecular upper bounds of δ between pairs of atoms involved in the match. The straightforward embedding procedure produces various ways to translate, rotate, and change the conformation of both molecules so as to achieve the required match, if possible at all. This is unfortunately such a time-consuming procedure (although not as time consuming as attempting it manually with interactive computer graphics!) that it can't be incorporated into Algorithm 8.2, but rather one must focus on a very few proposed matches.

8.4. The Matching Problem

We have seen in the preceding section how important it is to efficiently search for good matchings between molecules. Indeed locating the intersections is far and away the most time-consuming part of the classification algorithm. In §8.5 and §8.7 we will also see how crucial it is to determine an optimal matching between groups on the drug molecule and parts of the site. The kind of tree search described earlier is adequate as long as the two collections of objects to be matched are not very large, but for large molecules and detailed sites, the computing time mounts rapidly. We therefore will pause to consider matching algorithms in some generality.

The concern here is not *whether* we can calculate the binding mode of some molecule, m, with the site. All we have to do is examine all possible combinations of contacts between them and choose the best one. Rather we have to see how rapidly the required computer time rises with the size of the problem. In our case, the problem size is conveniently given by n_m, the number of atoms in the molecule, and n_s, the number of site points or regions in the given model. A useful classification of problems [Garey1979] puts in the "tractable" category any problem which can be solved in an amount of time expressible as a polynomial function of the size of the problem. The highest order exponent and the constant coefficients may be large, but the problem would still belong to this class P. At the opposite extreme are the "intractable" problems requiring time that increases exponentially with size. Not quite so bad are the NP problems, where the solution, once found, can be *verified* in polynomial time. A subset of NP is the NP-complete class, where no efficient (i.e. polynomial time) algorithm is known, but if such an algorithm were to be found for one member of the class, it could be used to solve all of them. (See §4.4 for further details.)

Clearly, docking a rigid molecule to a rigid site is in class P because all we have to do is try all triples of atoms making contact with all triples of site points, performing the required translations and rotations to see if these contacts are possible and if others are formed in addition, and then choosing the best combination. Running time goes as $O(n_m^3 n_s^3 min(n_m, n_s))$, but a lot of floating point arithmetic and trigonometry is required. A combinatorial algorithm might avoid this arithmetic overhead. Our original approach traverses the tree of possible binding modes in a relatively efficient manner, checking necessary (but not entirely sufficient) simple conditions for geometric feasibility. The main complication arises because some combinations of contacts imply other contacts must be also formed.

To make the mode search algorithm clear, suppose the site consists of only three points, and the molecule in question has only two points, called C1 and N2. Then there are $3^3 = 27$ possible mode vectors, since "0"= unoccupied is also a possibility for every site point. Figure 8.4 shows how the set of mode vectors can be organized into a tree hierarchy, where for example, (0,C1,.) stands for the set of all nodes having site point 1 free, molecule point C1 bound to site point 2, and any state of site point 3. The root of the tree is (.,.,.), and the fully specified binding modes in the fourth (bottommost) layer are the leaves. The search procedure is called a depth-first, leftmost tree traversal, with pruning at the highest levels possible. Thus the first mode to be located would be the trivial free mode, (0,0,0). Not all 27 leaves are examined. For instance, site point 1 may be attractive for N2, but suppose it is repulsive to C1. By "repulsive" we mean there would be an overwhelmingly unfavorable contribution to the total binding

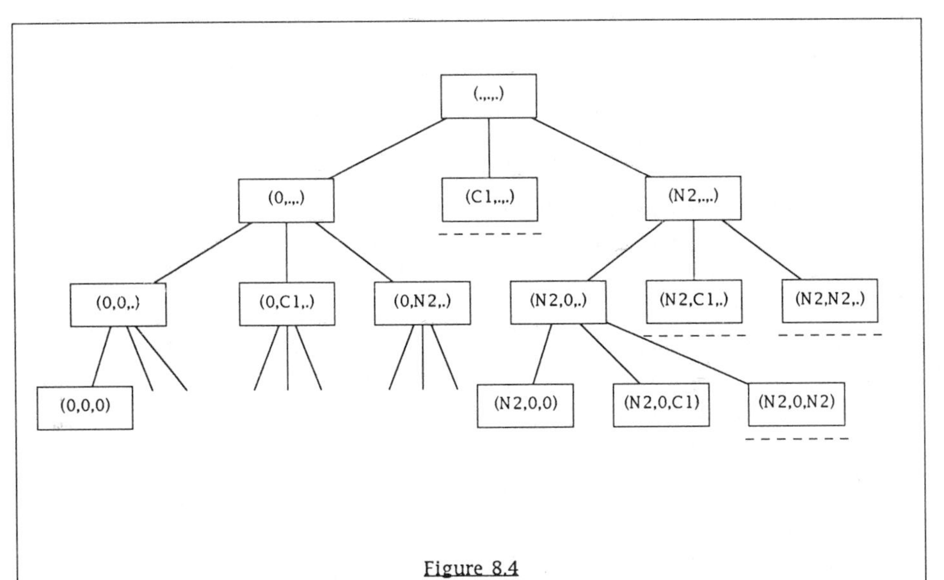

Figure 8.4

A partial illustration of the tree of all possible binding modes for a molecule consisting of only the two points, C1 and N2, and a site of three points. Not all nodes are shown. Dashed lines denote branches or leaves that are eliminated by the criteria discussed in the text.

energy due to this one contact. Then the (C1,.,.) branch can be pruned immediately, so that its 9 descendant leaves need not be checked at all. The second easiest thing to check is that no molecule point may be used twice in a mode vector, so for example, the (N2,N2,.) branch may be eliminated. The third check is that the distance range between a pair of used molecule points must overlap the distance between the corresponding site points up to a small flexibility, as in Equation 8.18. In the process of traversing the tree, satisfactory leaves (binding modes) will be found and their energy calculated. If some subsequent branch has already made such energetically poor choices that even making the energetically optimal contacts for the remaining unspecified parts of the vector would not result in a binding energy lower than the best already found, then this fourth check indicates the branch is not worth pursuing further. A fifth test occurs when all four molecule points of some asymmetric center are bound, such as the four different substituents of a tetrahedral carbon. Then their chirality must match that of their corresponding site points.

The last and most time consuming test is that of the forced contacts. Although a mode vector may indicate a number of unoccupied site points, perhaps some of these must be occupied as a consequence of the given contacts. These extra contacts must be included for geometric reality in the list of allowed mode vectors and for correct calculation of ΔG when searching for the mode with the most favorable binding energy. Take for example the binding of 1-bromo-3,4-dichlorobenzene to a site model such that the bromine, the *m*-Cl, and the 4-position carbon are all appropriately bound to site points. Then the entire molecule is rigidly fixed in the site, and if there is a site point near enough to the deduced location of the *p*-Cl, then they, too, must be in contact, whether or not the contact is energetically favorable. The basic notion is that the site points all have given Cartesian coordinates, so that the coordinates of an attached molecule point are identified with those of its site point. Then three molecule points can be employed to triangulate the position of a fourth molecule point. If this deduced position is close enough to another site point, a "forced contact" is added to the mode vector.

Before formally presenting what we call the Forced algorithm, let us illustrate some important cases it handles, by considering for a moment the two dimensional binding problem, where pairs of circles intersect, instead of triplets of spheres. Then we see in Fig. 8.5 an example where there are already contacts between molecule point N2 and site point s1, and between C1 and s2, but molecule point C3 is thought to be free. The molecule is fairly flexible due to rotatable bonds (which are no longer explicitly represented at this level of abstraction) so that the indicated upper and lower bound distances between C3 and C1 are not equal, and similarly for the C3-N2 distance. Site points s1 and s2 are of course occupied, but s3 and s4 are available to make contacts. Algorithm Forced makes the assumption that the region of the plane where C3 must be found is adequately outlined in terms of the intersection points between the pair of circles about s1 and the pair about s2. In this case, there are only three points of intersection, each marked by an "x". Note that the one lying within the steric radius, r3, of site point s3 is an example of a near intersection of the $l_{N2,C3}$ and $u_{C1,C3}$ circles. If s3 is attractive to C3 while s4 is repulsive, then that near intersection point would be the preferred contact, and would be made. Alternatively, if both s3 and s4 are repulsive, then C3 would opt for the third intersection point (at the lower left) which does not fall within the steric radius of any site point, and therefore no forced contact would be made. Now consider the situation shown in Fig. 8.6. Only one molecule point is bound (s1-N2), but C3 must lie nearby, and the molecule is apparently inflexible between atoms N2 and C3. Suppose that all the other site points, s2-s5, are repulsive. Then N2 has found a site point that is barely sterically permissible, but C3 must fall somewhere on the circle about s1 at a radius of $l_{N2,C3} = u_{N2,C3}$. Since there are not two bound molecule points with which to determine the position of C3 (three would be

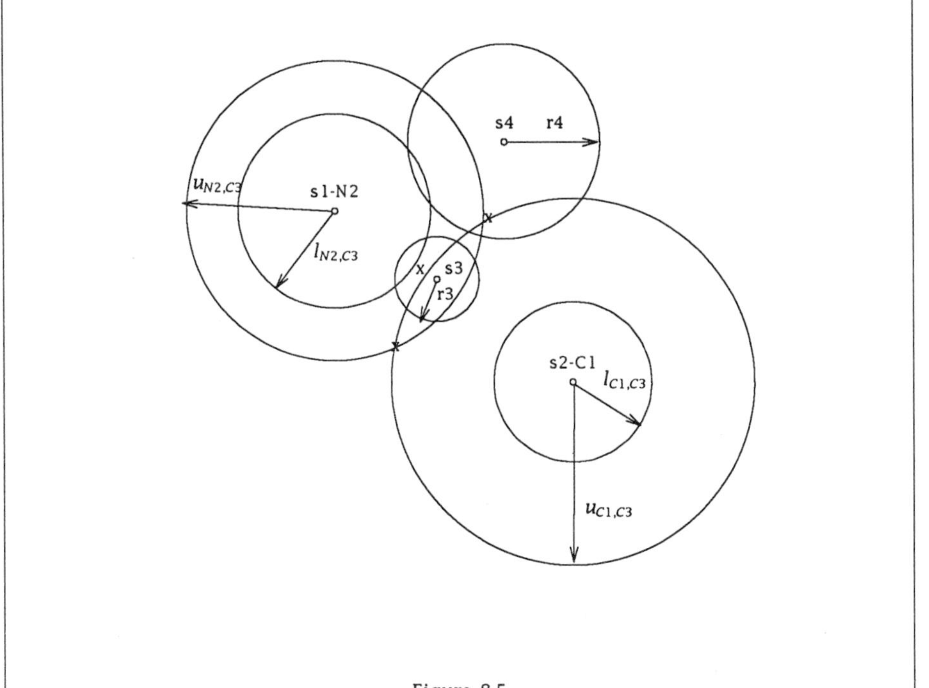

Figure 8.5
A two-dimensional illustration of forced contacts. Molecule point C3 must lie at one of the three points marked by an "x" because of the s1-N2 and s2-C1 contacts.

needed in three dimensions), the Forced algorithm considers also the intersections of the central circle with those of surrounding repulsive site points, as shown. Since the circles about s2-s5 overlap properly, each of the eight intersection points falls within the steric radius of another of the repulsive site points. Therefore the algorithm must declare a contact formed between C3 and one of s2, s3, s4, or s5. This would make the total energy of the mode so bad, that any mode having an s1-N2 contact will be rejected. Alternatively, if s2, s3, and s4 are all repulsive while s5 could have a favorable interaction with C3, the algorithm would conclude that C3 must seek the most favorable position and must lie at one of the two intersection points well within the s5 circle. It would then declare a forced s5-C3 contact.

Before describing the logic more precisely, we must present the pertinent equations for the calculations. Three spheres of arbitrary radii and positions may intersect at two or zero points. (A single point of intersection, or a circle, or

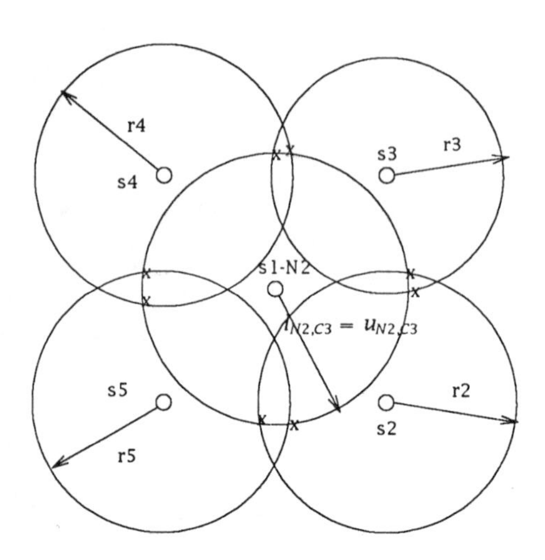

Figure 8.6

A two-dimensional illustration of forced contacts where the given s1-N2 contact constrains molecule point C3 to lie on a circle. Steric hindrance site points force C3 to be at one of the eight intersection points marked with an "x".

even a whole sphere of intersection are possible under special, highly unlikely circumstances.) Denote the spheres by a, b, and c, and let their radii be r_a, r_b, and r_c, and let the vector positions of their centers be $\mathbf{v}_a$, $\mathbf{v}_b$, and $\mathbf{v}_c$. The subsequent equations are much simpler if we translate and rotate to a local coordinate system where sphere a is centered at the origin (0,0,0), sphere b is on the positive x-axis at $(x_b,0,0)$, and sphere c is on the positive y half of the xy-plane at $(x_c,y_c,0)$. The rotation is specified by the 3×3 matrix $\underline{\mathbf{M}}$, whose rows $\mathbf{M}_1$, $\mathbf{M}_2$, and $\mathbf{M}_3$ are calculated by

$$
\begin{aligned}
\mathbf{s} &:= \mathbf{v}_b - \mathbf{v}_a \\
\mathbf{M}_1 &:= \frac{\mathbf{s}}{\|\mathbf{s}\|} \\
\mathbf{t} &:= \mathbf{v}_c - \mathbf{v}_a \\
\mathbf{u} &:= \mathbf{t} - \mathbf{M}_1 \cdot \mathbf{t} \\
\mathbf{M}_2 &:= \frac{\mathbf{u}}{\|\mathbf{u}\|}
\end{aligned}
\tag{8.19}
$$

$$\mathbf{M}_3 := \mathbf{M}_1 \times \mathbf{M}_2$$

Then the local sphere coordinates are

$$x_b := \|\mathbf{s}\| \tag{8.20}$$

$$x_c := \mathbf{M}_1 \cdot \mathbf{t}$$

$$y_c := \|\mathbf{u}\|$$

Local coordinates of a point, $\mathbf{p}_{loc}$, can always be converted to the original coordinate system by

$$\mathbf{p} := \mathbf{v}_a + \underline{\mathbf{M}}\mathbf{p}_{loc} \tag{8.21}$$

In the local coordinates, the intersection of the three spheres is (x_d, y_d, z_d), where

$$x_d := \frac{r_a^2 - r_b^2 + x_b^2}{2x_b} \tag{8.22}$$

$$y_d := \frac{r_b^2 - r_c^2 + x_c^2 + y_c^2 - x_b^2 + 2x_d(x_b - x_c)}{2y_c}$$

$$z_d := \pm\left(r_a^2 - x_d^2 - y_d^2\right)^{1/2}$$

The ± choice for z_d corresponds to the two different intersection points. If the argument to the square root is sufficiently small, say <0.01 Å, then there is effectively only a single point of intersection. If the argument is negative, there is no intersection. In practice we must be careful about the zero intersection case, since it may be a near miss of the three spheres where the single intersection point is supposed to lie in the plane of their centers, for example when forcing a contact to a fourth atom in a planar ring system because of contacts already made by three others in the same ring. After all, the coordinates of the used molecule points are consistent with the true molecular geometry only up to the site flexibility parameter, δ, as explained in the tree search third check. The approach to such a situation is to calculate the six points, $(x_1, y_1, 0)$, ..., $(x_6, y_6, 0)$, that are the in-plane intersection points of the three spheres taken two at a time, choose one point from each intersection pair such that the distances among the triplet are minimal, and then propose their center of mass as the "near intersection" point. The near intersection point is acceptable if it lies within δ of all three spherical surfaces. Specifically, points 1 and 2 resulting from the intersection of spheres a and b are located at

$$x_1 = x_2 = \frac{r_a^2 - r_b^2 + x_b^2}{2x_b} \tag{8.23}$$

$$y_1 = -y_2 = \left(r_a^2 - x_1^2\right)^{1/2}$$

If the argument to the square root is negative (indicating no intersection of even the two spheres), set $y_1 = y_2 = 0$, as this would be the point with equal errors in the squared distances:

$$x_1^2 - r_a^2 = (x_1 - x_b)^2 - r_b^2 \ .$$

Similarly, intersection points 3 and 4 between spheres a and c are found by solving

$$sx^2 + tx + u = 0 \tag{8.24}$$

taking only the real part of the solutions for x_3 and x_4, where

$$w := \frac{r_a^2 - r_c^2 + x_c^2 + y_c^2}{2y_c} \tag{8.25}$$

$$s := 1 + \left(\frac{x_c}{y_c}\right)^2$$

$$t := -\frac{2wx_c}{y_c}$$

$$u := w^2 - r_a^2$$

Then

$$y_3 := w - \frac{x_c x_3}{y_c} \tag{8.26}$$

$$y_4 := w - \frac{x_c x_4}{y_c}$$

For the intersection of spheres b and c, x_5 and x_6 are the real part of the roots of Equation (8.24) with the coefficients now given by

$$w := \frac{r_b^2 - r_c^2 + x_c^2 + y_c^2 - x_b^2}{2y_c} \tag{8.27}$$

$$s := 1 + \left(\frac{x_b - x_c}{y_c}\right)^2$$

$$t := 2\left(\frac{w(x_b - x_c)}{y_c} - x_b\right)$$

$$u := x_b^2 + w^2 - r_b^2$$

and the y-coordinates are found by

$$y_5 := w + \frac{(x_b - x_c)x_5}{y_c} \tag{8.28}$$

$$y_6 := w + \frac{(x_b - x_c)x_6}{y_c}$$

Now that the mathematical details of forced contacts have been presented, we can view the overall flow of logic. The situation consists of a given molecule with its inter-molecule point upper and lower bound distances, a proposed mode vector which is otherwise geometrically permissible and contains no gross energetic flaws, and the coordinates of the site points. There is an effective steric radius assigned to each site point, which is usually on the order of 1 Å but may be much greater for repulsive site points intended to make large portions of the site unattractive. There is also the given site flexibility parameter, on the order of 0.5 Å. The object is to add any extra contacts if they are a geometric consequence of those already there.

Algorithm 8.3 Forced:

INPUT: A molecule with upper and lower bounds on its inter-point distances, a proposed binding mode vector, site point coordinates, the radius for each site point, and the site flexibility, δ.

OUTPUT: The revised binding mode vector, perhaps containing additional contacts.

PROCEDURE:

Repeat until no more new contacts are forced
  For each unused molecule point
    Create a list of spheres: center coordinates and radii.
      Include every occupied site point using as radius the
        upper bound distance between its attached molecule point
        and the unused molecule point.
      If the upper and lower bounds differ by more than 0.5 Å
        Then include the site point coordinates again, using
          the lower bound distance as the radius.
      If the sphere list includes fewer than three unique centers
        Then include all (unused) repulsion sites, using their
          respective steric radii as the radii.
    For every triplet of spheres with distinct centers
      Find their intersection points with tolerance δ.
      If there are any intersection points,
        Reject those with distances to occupied site points
          inconsistent ($\pm\delta$) with the range between its
          attached molecule point and the unused molecule point.
        Reject those with chirality inconsistent with the
          unused molecule point and three other attached molecule points.
        Identify the remaining intersection points with the first
          site point lying less than its steric radius away.
        Build up a list of approved intersection points for the

unused molecule point in question, including site point assignment, if any.

If there are any approved intersection points, choose the one with the best interaction energy between the unused molecule point and the assigned site point, assuming zero energy if no site point is assigned.

If the best intersection point was assigned to an unused site point,

Then add to the mode vector the contact point between the unused molecule point and the site point.

Otherwise,

The unused molecule point remains unattached.

This forcing algorithm actually covers the whole spectrum of geometrically conceivable situations in a reasonably realistic way.

Subsequently we developed an alternative to searching the tree of possible binding modes with forced contact checking [Kuhl1984]. This alternative appears to be about as geometrically accurate for fairly rigid molecules while running faster, but at the conceptual expense of being more abstract. To see how this can be done, we need to formulate the docking problem in graph theoretic terms.

An undirected graph $\mathbb{G}$ = (V,E) where V = {nodes or vertices v_i} and E = $\{ \{v_i,v_j\} \}$ = the set of edges connecting some of the vertices, where we disregard the ordering of the vertex pairs. We are concerned largely with ***bipartite*** graphs (U,V;E), where the node set is U∪V, and every edge in E connects a node in U and a node in V. There are no connected pairs of nodes in U or in V.

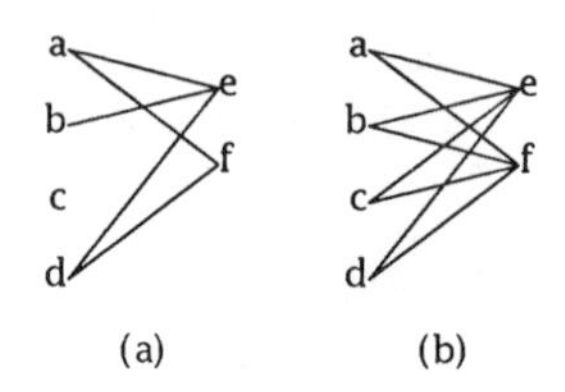

Figure 8.7

Examples of bipartite graphs with node sets U = {a,b,c,d} and V = {e,f}. Graph (a) is incomplete, but (b) is complete.

In Fig. 8.7, we see bipartite graphs representing the docking possibilities of a molecule having atoms a, b, c, and d to a site (or second molecule) consisting of points e and f. Some edges might be excluded as in Fig. 8.7(a) because of steric requirement or because of type mismatches, as in the structure classification algorithm. A bipartite graph is *complete* if $\{u,v\} \in E \ \forall \ u \in U$ and $v \in V$, as in Fig. 8.7(b). Weights might be assigned to the edges to indicate how good a contact is. The matching problem is to choose a subset of the edges such that each node is involved at most once and that the sum of the weights is maximized. There are efficient algorithms to do this, but unfortunately docking is significantly different, in that any single contact (graph edge) may be allowed, but certain *pairs* of edges are disallowed, exactly as in Equation (8.18). In other words, we are dealing with a modified assignment problem (MAP): given a complete bipartite graph (U,V,E) with weights $w(e) \geq 0 \ \forall \ e \in E$, and an arbitrary set of excluded pairs of edges $C = \{(e_i,e_j), \cdots\}$, and an integer $K \leq \sum_{e \in E} w(e)$, then we ask if there is a matching $S \subseteq E$ (where no more than one $e \in S$ is incident on every $u \in U$ and $v \in V$) with total weight $\sum_{e \in S} w(e) \geq K$ such that $\forall \ e,e' \in S, (e,e') \notin C$. Now there is a related combinatorial problem called the independent set problem (ISP): given graph (V,E) and integer $J \leq \#V$ (=cardinality of set V), does (V,E) contain an independent set $S \subseteq V$ such that $\#S \geq J$? "Independent set" means that $\forall \ s,s' \in S, (s,s') \notin E$, for example {b,c,f} in Fig. 8.7(a). For arbitrarily chosen C in MAP, ISP is equivalent to it [Kuhl1984], and the ISP is known to be NP-complete [Garey1979].

LEMMA 8.4. *(Kuhl) The independent set problem is equivalent to the modified assignment problem.*

Proof: Let the graph for the ISP be (V,E) and that for the MAP be (U',V',E'). Then we construct an instance of MAP from ISP as follows. Let $\#V =: n$, $U' := \{u_1, \ldots, u_n\}$, $V' := \{v_1, \ldots, v_n\}$, and $E' := \{(u,v) | \ \forall \ u \in U' \text{ and } v \in V'\}$. For $e := (u_i,v_j)$, let

$$w(e) := \begin{cases} 1, & \text{for } i=j \\ 0, & \text{otherwise.} \end{cases}$$

Let $C := \{(u_i,v_i),(u_j,v_j) | \ \forall \ (v_i,v_j) \in E\}$, and $K := J$. As shown in Fig. 8.8, individual nodes in the ISP become pairs in the MAP bipartite graph, and there is a weight 1 match (u_i,v_i) in MAP for each node i in ISP, while match pairs (u_i,v_i) and (u_j,v_j) are in the exclusion set C whenever i and j are connected in the ISP. Then the only way to achieve a total weight J is to find J of the (u_i,v_j) matches while not violating C. These matches correspond to J nodes of the ISP graph, none of which is connected. QED.

In the example, if $J=2$, the independent set is $\{1,3\}$, $C = \{((u_1,v_1),(u_2,v_2)),((u_2,v_2),(u_3,v_3))\}$, and the required matching consists of the two edges (u_1,v_1) and (u_3,v_3).

We are now left with an apparent contradiction: docking must be a class P problem when viewed as a geometric calculation, but class NP-complete when

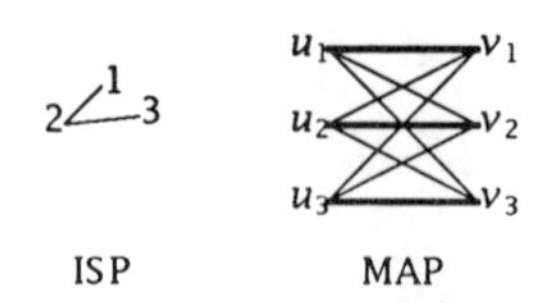

Figure 8.8
An example of the transformation of an instance of ISP to one of MAP. Heavy edges in the MAP have weight 1, while others have 0.

seen as MAP. The resolution is that geometry introduces strong, coordinated constraints involving more than contact pair exclusions and not just arbitrary sets of such exclusions. Just how to express these constraints depends in a complicated fashion on the conformation of site and molecule, and the internal degrees of freedom of the molecule. We have found it to be more practical to find all binding modes obeying the necessary contact pair exclusions, and then reject some of these modes by detailed geometric testing in the relatively rare instances when the exclusions were not sufficient to guarantee embeddability. Therefore, we will concentrate now on describing an efficient combinatorial docking algorithm taking into account contact pair exclusions.

Definition 8.5. A graph $\mathbb{G}' = (V',E')$ is a *subgraph* of $\mathbb{G}=(V,E)$ if $V' \subseteq V$ and $E' = \{(u,v) \in E \mid u,v \in V'\}$. $\mathbb{G}$ is *completely connected* if $E = \{(u,v) \mid u,v \in V\}$. $\mathbb{G}'$ is a *clique* of $\mathbb{G}$ if it is a maximal completely connected subgraph of $\mathbb{G}$; that is, adding an extra node to $\mathbb{G}'$ would cause it to be no longer completely connected. (See Appendix E for further background on graph theory.)

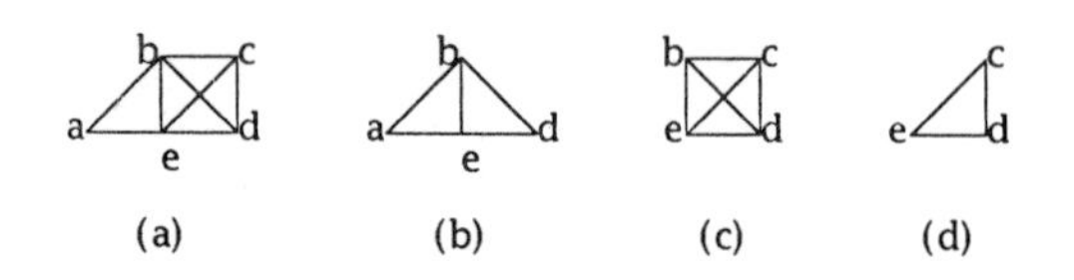

Figure 8.9
Graph definition examples. Graphs (b), (c) and (d) are all subgraphs of (a), but (c) is a clique by being completely connected, while adding point a, the only possibility, would destroy this. Graph (b) is not completely connected, and (d) is. However, (d) is not maximal (node b could be added) and therefore is not a clique.

Definition 8.6. If we have a molecule with atoms $i_m = 1, \cdots, n_m$ and a site with points $i_s = 1, \cdots, n_s$, define a *docking graph* $\mathbb{G}$, which has a node for every ordered pair (i_m, j_s). Let there be an edge in $\mathbb{G}$ between nodes (i_m, j_s) and (k_m, l_s) whenever

$$\begin{aligned} u_{i_m,k_m}+\delta &> d_{j_s,l_s} \quad \text{and} \\ d_{j_s,l_s} &> l_{i_m,k_m}-\delta \end{aligned} \tag{8.29}$$

Note Equation (8.29) is equivalent to Equation (8.18), except we are assuming the molecule might be flexible with distance bounds $u_{i_m,k_m} \geq l_{i_m,k_m}$, but the site has set distances d_{j_s,l_s} with a general given flexibility parameter $\delta > 0$. Fig. 8.10 shows a simple example of a docking problem and its docking graph.

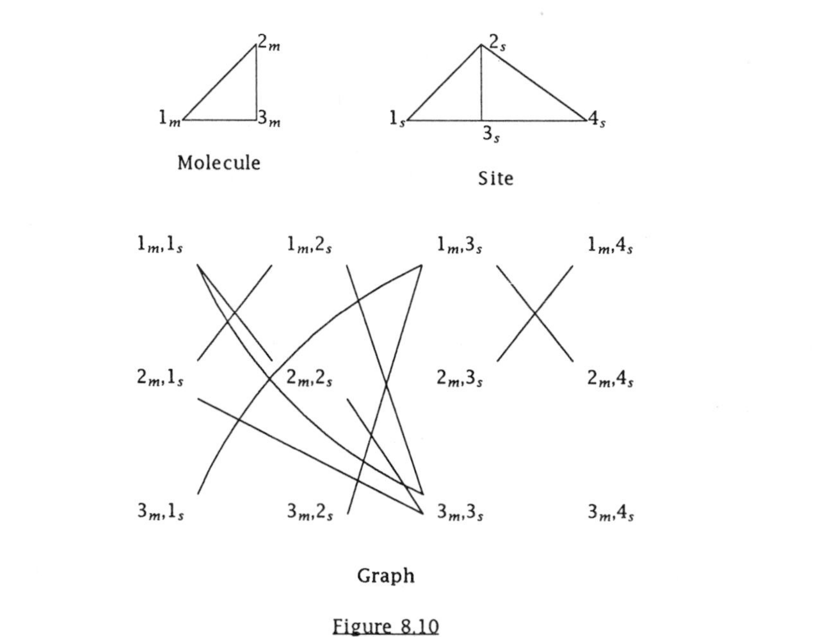

Figure 8.10

Docking graph example. The molecule has only three atoms arranged approximately rigidly in an isosceles right triangle, which is congruent with triangle $1_s2_s3_s$ in the site. Further, let $d_{1_m,2_m} = d_{3_s,4_s}$.

The connection between the docking problem and its docking graph is simply:

LEMMA 8.7. *(Kuhl) Given a molecule, a site, and the corresponding docking graph, $\mathbb{G}$, then every maximal matching between molecule and site is a clique of $\mathbb{G}$.*

Proof: Suppose (i_m,j_s) and (k_m,l_s) are two matches in a maximal matching V′, where $\mathbb{G} = (V,E)$ and $V' \subseteq V$. For both pairs to occur, Equation (8.29) must be satisfied, so the edge $((i_m,j_s),(k_m,l_s)) \in E$. Because this must be true for all pairs in V′, the subgraph induced by V′ must be completely connected. On the other hand, V′ being maximal means no additional matches can be added to V′, so the induced subgraph is a clique. QED.

Fortunately there are a number of general clique-finding algorithms available, of which we chose the one by Bron and Kerbosch [Bron1973] because it could be readily adapted to our needs. Namely, simply finding cliques in the docking graph is enough to properly take pairwise excluded contacts into consideration, but as in the Forced algorithm, one must sometimes consider more than pairwise exclusions. A binding mode might be very unfavorable because one of its contacts corresponds to a sterically forbidden interaction, like putting a bulky group into a small cavity. Yet simply excluding the contact is insufficient. Suppose for example a binding mode involves contacts a, b, and c in $\mathbb{R}^2$ (or in $\mathbb{R}^3$ with three collinear site points), and a is a repulsive contact. Then either b or c or both must also be avoided because having b and c forces a to occur. As long as the molecule is reasonably rigid, the docking graph also has this sort of information built into it. Each node on the graph corresponds to a contact; identify with that contact the coordinates of the site point involved. Then the match in distance between site points and corresponding molecule points represented by the connections b−a and c−a, means there must be also a connection b−c. Thus {a,b,c} is at least a subset of a clique that must be found once b and c have been selected. Similarly for four contacts *a, b, c,* and *d* in $\mathbb{R}^3$, where the chirality of the atom quartet agrees with that of the corresponding site points, if *a* is repulsive, then no more than two of *b, c,* and *d* can be simultaneously made without forcing *a* to occur. We have modified the algorithm of Bron and Kerbosch so that these repulsive cases are correctly handled, producing the following algorithm:

Algorithm 8.8 "Bronk":

INPUT: A site and a molecule.

OUTPUT: All geometrically (nearly) allowed binding modes.

PROCEDURE:

Convert the site and molecule descriptions to the equivalent docking graph as in Fig. 8.10.

Label the nodes of the docking graph arbitrarily as in Fig. 8.11, and note the repulsive nodes.

From here, search the tree of all combinations of contacts (=all subsets of nodes), keeping track of:

S := {set of nodes of completely connected subgraph}

C := {candidate set of nodes, each connected to all of s∈S}

N := {nodes also connected to all s∈S, but have already been members of a previously discovered clique, and therefore must *not* be added to S, in order to avoid repetition}

R := {repulsive set of nodes; these never are elements of N but must also not be added to S}

Initially, S = {}, R={all repulsive nodes}, N = {}, and C= {all nodes not in R}. This is the root vertex.

Traverse the tree, forming new vertices and backtracking until the procedure backtracks to the root vertex.

- At any vertex, if C = {},
  - S is a clique and is reported.
  - Then backtrack.
- Else if some n∈N or r∈R is connected to every c∈C
  - Can't possibly find a new clique.
  - Backtrack.
- Else form a new vertex.
  - Select a node c∈C to add to S.
  - Delete any c'∈C not connected to c.
  - Delete any n∈N not connected to c.
  - Delete any r∈R not connected to c.

Backtrack:

- Rise to the parent vertex.
- Put the c at that level into N.
- Continue on as at any vertex.

Fig. 8.12 shows the search tree examined by Bronk for the docking graph of Fig. 8.11. Traversal is in leftmost, depth-first order. Note how the clique S = {2,5,11} is avoided, because 2 is repulsive, and also {5,11} is rejected. The reason is that we are assuming the molecule and site in Fig. 8.10 are in $\mathbb{R}^2$, so that {5,11} => {2,5,11}, i.e. {2,5,11} is a completely connected subset that must be produced during the search for a maximal completely connected subset when starting from {5,11}. Note that because in the process of forming a new vertex, the new R is always contained in the old R, then the new S can never have a repulsive matching. Furthermore no binding mode (clique) found will imply a repulsive match. In our example, when S = {5}, C = {11}, and R = {2}, but 2 is connected to 11, the only member of C, so the algorithm backtracks. It is a general property of docking graphs that if contacts a, b, d, and e are collinear and {a,b} is (completely) connected while d and e are each connected to both a and b, then d and e are also connected. Thus if d∈R, S={a,b} and e∈C, no node e can be added to S, and it backtracks. Similarly in $\mathbb{R}^3$ when S = {a,b,c} and d∈R, any e∈C must also be connected to d by virtue of triangulation to a, b, and c.

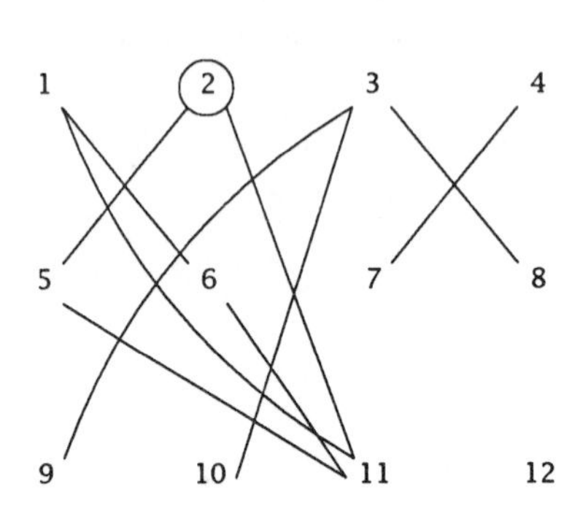

Figure 8.11
The docking graph example of Fig. 8.10 with simplified node labels. Node 2 is circled to indicate it is a repulsive contact.

For n_s site points and n_m atoms, Bronk has worst case performance of $O([n_s n_m]^4)$, but our average case experience [Kuhl1984] has computer time requirements of only $O([n_s n_m]^{2.8})$, which is even better than straightforward trigonometric docking.

8.5. Creating a Site Model from Binding Data

Now that the modelling problem has been stated, and some of the central algorithms have been explained, we can give an overview of how one develops a workable site model from experimental binding data. As usual, assume we have the chemical structures of some compounds and their experimentally determined ΔG_{obsd}, or equivalent measure of binding affinity. Then we will consider in turn the following basic steps: (i) find atomic coordinates for the molecules; (ii) determine the upper and lower interatomic distance bounds of each molecule; (iii) remove unnecessary atoms; (iv) suggest hypothetical binding modes; (v)

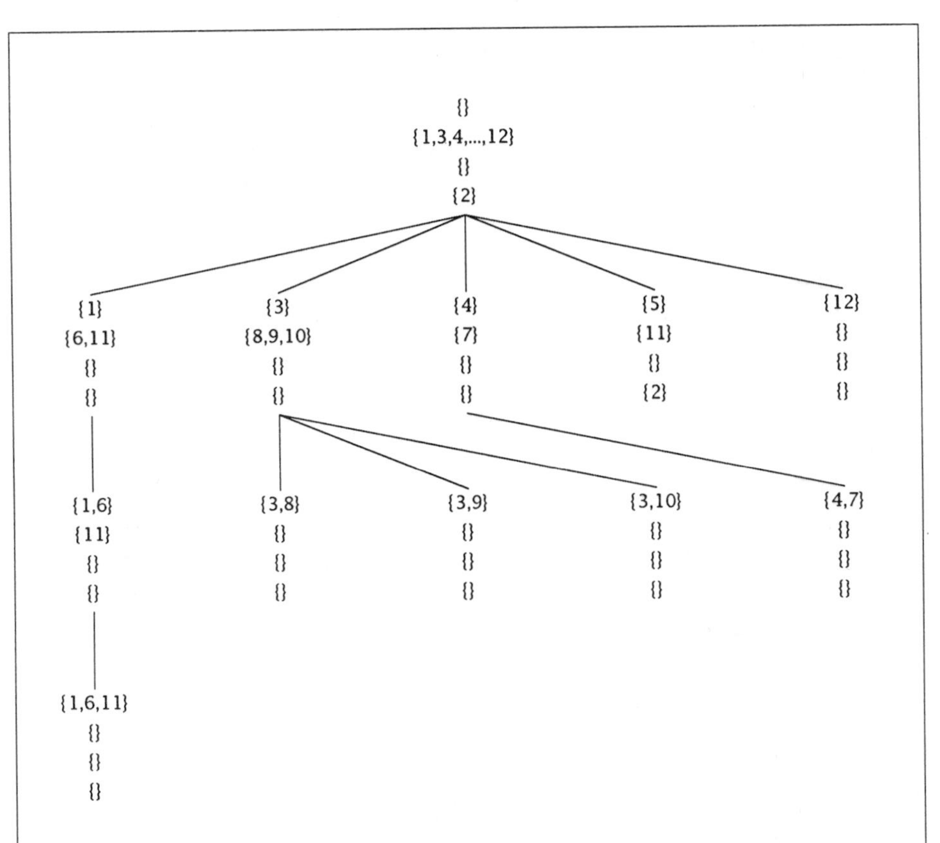

Figure 8.12

The tree traversed by Bronk searching for the cliques of the docking graph shown in Fig. 8.11. At each vertex are shown the S, C, N, and R sets written top to bottom as they appear when the vertex is first formed (as opposed to the backtracking state).

determine site point coordinates; (vi) generate lists of all geometrically allowed binding modes for each molecule; and (vii) determine interaction parameters such that the optimal binding modes have calculated binding energies in agreement with the observed values.

Step (i) atomic coordinates: Chemists draw structures on paper, but somehow one must convey to the computer the atomic coordinates, atom types, bond connectivity, selection of rotatable bonds, and chiral centers. There are many programs for inputing such information, usually as a front end to graphics display systems and/or conformational energy calculations. Since a crystal structure of exactly the drug molecule in question is generally not available, one usually resorts to estimating a reasonable starting conformation by molecular mechanics calculations. We have found it convenient to build up a library of standard ring systems, substituents and the like from the X-ray crystallographic literature. Then these fragments can be assembled as rigid bodies joined by rotatable bonds to form the desired molecule. Its conformation at this point is unimportant.

Step (ii) interatomic distance bounds: The molecular connectivity and knowledge of which bonds are rotatable is then used to explore all of conformation space, as in Equation (8.5). The results for each molecule are summarized as distance bound matrices, as shown in Equations (8.6) and (8.7). Note that all atoms, including hydrogens, are necessary for proper energy calculations here, or even for simply checking van der Waals contacts. The search over all combinations of all rotatable bonds can be done more or less cleverly, but basically computation time is on the order of the product over all rotatable bonds of the number of rotamers allowed to each. After this step, the molecular connectivity, atomic coordinates, and rotatable bonds are of no further interest.

Step (iii) edit: Subsequent steps are searches over combinations of atoms and site points, which can be extremely time consuming for large numbers of atoms. Algorithm 8.2 can act as a guide for deleting redundant atoms, but its execution time may be excessive if all atoms are included. Thus one may have to make subjective decisions at this stage about which atoms should be removed. Generally hydrogens can be removed here. Fine distinctions in atom types (e.g. aromatic vs. aliphatic sp^2 carbons) may be dispensed with here, at the investigator's discretion.

Step (iv) desired modes: An analysis of the chemical structural composition (Algorithm 8.2) of the data set is very helpful at this point. Constituent base group(s) and r-groups can henceforth be represented by either a single atom each or perhaps by a few atoms. The idea is to represent each molecule by the fewest number of points such that its atomic composition is still expressed and such that its special geometric properties are preserved. Thus if a phenyl group appears as an r-group, instead of representing it as six aromatic carbons and five aromatic hydrogens, just leaving the two *meta*-carbons would suffice to indicate the plane it occupies, its rotational possibilities relative to the rest of the molecule, and its steric protrusion from the point of attachment. Having reduced the structural information of the data set to its essentials, it is relatively easy to make the usual sorts of deductions about binding modes. Active compounds should have some kind of base in common, although it need not be a readily recognizable ring system or reactive group. Presumably this base is accommodated by the site. Active compounds having significantly different binding must interact differently with the site by virtue of presenting different r-groups to the same binding pocket or to different pockets. Having additional r-groups characteristic of inactive compounds but absent in active ones, is indicative of sterically forbidden regions in the site model.

Of course, there are often alternative explanations. For example, if adding a nitro group reduces activity, it may be due to steric repulsion at that position forcing the molecule out of the site, or it may be that the electron withdrawing power of the nitro group shifts the pK_a of some other group enough to change its ionization state and hence interaction energy with the site. Adding more bulk at some position may have no effect on binding because either the site interacts only weakly with that part of the molecule, or the molecule is forced to adopt a very different binding mode which coincidentally has much the same total interaction energy. Generally, the binding data alone are insufficient to distinguish between various alternative explanations, and the choice must be rather subjective. If additional evidence is available (NMR binding studies, titration curves of the different derivatives, quantum mechanical calculations, etc.), the decision can be less arbitrary. Simply having examined the available data to this level is often enough to suggest new experiments or compounds to synthesize, in hopes of resolving these questions.

Information theory can be of some help in deciding which chemical structural features are important for binding. The trivial case is where all compounds in the data set have essentially the same affinity. Then by Equation (8.13), the entropy in binding energy $\eta(E)=0$. As far as information theory is concerned, there is no site detail necessary whatsoever to explain the certain energy value. Now consider the nontrivial artificial data set of Table 8.1. The ΔG_{obsd} values fall into four distinguishable categories: –2, –3, –5, and –7. Then $\eta(E)= 1.8$ bits per compound, implying there is some variability in the energies that needs to be accounted for. Note that $\eta(E) < 2$ bits, the maximum value for four equally likely categories, because $P(\Delta G_{obsd} = -5) = \frac{3}{7}$, whereas $P(\Delta G_{obsd} = -3) = \frac{1}{7}$. Consider as before the structural characteristics A = chirality, B = choice of R_1, C = whether R_2 has a phenyl, and D = whether there is a chloro substitution on the R_2. Applying Equation (8.17), we find that $\eta(E|A) = \eta(E|D) = 1.5$ and $\eta(E|B) = \eta(E|C) = 1.4$, implying that knowing features B or C is more important than knowing A or D in reducing the uncertainty of the energy. If the site model can recognize both B and C then $\eta(E|BC) = 1.0$ indicates a further reduction in uncertainty, but knowing B and A is better since $\eta(E|AB) = 0.3$. The suggestion is then that it would be relatively important to have a site model that will recognize the handedness of the alanyl group and decide whether R_1 is H- or CH_3–. Although A and

B constitute only two structural features, more than two site points are necessary because detecting chirality requires four site points, one of which could serve to interact with R_1. About all one can say is that there must be at least as many site points as there are energetically significant chemical features.

From all these considerations, the investigator must finally settle upon a list of site points and what their roles are imagined to be. This amounts to drawing up a list of "desired" binding modes, one for each molecule. These subjective choices cannot be completely arbitrary, because they must pass two tests: there must be at least one set of site point coordinates consistent with the desired modes, and there must be interaction energy parameters such that the desired modes have the best ΔG_{calc} over all other possible modes, and such that these ΔG_{calc} values agree closely enough with the corresponding ΔG_{obsd}s. Fortunately, we can separate these two problems to some extent.

Step (v) site point coordinates: If for one of the molecules, the desired mode involves two contacts, (i_m, j_s) and (k_m, l_s), then the distance d_{j_s, l_s} is constrained by Equation (8.29), for some value of $\delta > 0$. Clearly if δ is made large enough, d_{j_s, l_s} could have any value. Therefore the method amounts to looking at each pair of contacts in the desired binding mode of each molecule, noting the smallest nonnegative value of δ such that Equation (8.29) can be satisfied for all pairs of site points. It may be that for some contact pair, this produces an unrealistically large value of δ, which means the investigator must change that desired mode. Otherwise, for the final value of δ, each d_{j_s, l_s} is constrained by the least upper bound and greatest lower bound implied by Equation (8.29) for all pairs of contacts (both within the same desired mode of one of the molecules) involving j_s and l_s. Finding site point coordinates in $\mathbb{R}^3$, if any, consistent with all these constraints is simply an embedding calculation. If that should fail, noting the mutually inconsistent constraints will help guide the investigator back to the desired binding mode(s) that must be revised. Otherwise the outcome will be a random selection of site point coordinate sets. If all are quite similar, the desired binding modes have constrained the site geometry to uniqueness, but otherwise there is little to choose between the alternate configurations.

In some cases, the desired mode may not really be the binding mode the investigator thinks a particular molecule will prefer. Suppose, for example, that adding a methyl group changes a compound from active to inactive. The binding hypothesis is that there must be some site point at that position representing a steric blocking bulge in the binding cavity. It is, however, important to know where to put this site point, so the interim "desired" mode looks like that of the demethyl derivative with the additional repulsive contact. Just what alternate mode the inactive methyl derivative finally assumes in the completed site model is of little interest as long as the calculated binding is adequately poor.

Step (vi) all modes: Now that the geometry of the site has been fixed, Algorithm 8.8 or the equivalent tree search will generate the list of all geometrically allowed binding modes for each molecule, $\mathbf{B}_m$. At this stage contacts are taken to be energetically not unfavorable, with the exception of repulsive contacts so designated in the site point roles. Finding several thousand allowed modes is not unusual. Since the mode checking only includes pairwise distance constraints and chiral quartets, some of these modes may in fact be geometric impossibilities, but certainly all correct modes are contained in the list. In practice, incorrect modes are relatively rare, so instead of eliminating all of them from the list, it is easier to check only the final energetically optimal binding modes. One way to do this is dock the molecule in the site model with interactive computer graphics, attempting to make the required contacts. Another way is described in §8.6.

Step (vii) energy refinement: The last hurdle is to determine the interaction energy parameters. As already explained in §8.2, now that we have found for each molecule m its set of binding modes $\mathbf{B}_m = \{\boldsymbol{\beta}\}$, where the $\boldsymbol{\beta}$s are vectors in the interaction energy space, we need to determine the vector of interaction energies, $\mathbf{e}$, such that the $\Delta G_{m,calc}$ values are least square fits to the $\Delta G_{m,obsd}$ data, as in Equation (8.9). The difficulty is that $\Delta G_{m,obsd}$ corresponds to the energetically most favorable mode, as in Equation (8.8), which in turn depends on the current value of $\mathbf{e}$. For molecules $m = 1, \cdots, M$ let $\boldsymbol{\beta}_m^*$ be the $\min_{\boldsymbol{\beta} \in \mathbf{B}_m} \boldsymbol{\beta}^T \mathbf{e}$, so that $\mathbf{B}^*(\mathbf{e}) = \{\boldsymbol{\beta}_1^*, \ldots, \boldsymbol{\beta}_M^*\}$ is the set of currently optimal binding modes, which clearly depends on $\mathbf{e}$. For a given set of binding modes, we seek to minimize the quadratic objective function

$$Z(\mathbf{B}^*,\mathbf{e}) = \sum_{m=1}^{M} (\boldsymbol{\beta}_m^{*T}\mathbf{e} - \Delta G_{m,obsd})^2 \tag{8.30}$$

with respect to $\mathbf{e}$. Now the requirement that $\boldsymbol{\beta}_m^*$ be (one of) the optimal binding modes of molecule m is simply the linear inequality

$$\boldsymbol{\beta}_m^{*T}\mathbf{e} \le \boldsymbol{\beta}'^T\mathbf{e} \quad \forall\ \boldsymbol{\beta}' \in \mathbf{B}_m \,. \tag{8.31}$$

Optimizing a quadratic function subject to linear inequality constraints is called quadratic programming, and there are good algorithms for this, e.g. Wolfe [Kuester1973] or Lemke [Ravindran1972]. Like ordinary least squares calculations, no initial guess for $\mathbf{e}$ is required. The basic approach is to assume $\mathbf{B}^*$ is the set of desired binding modes. Then if the full set of inequalities, Equation (8.31) for all m and all $\boldsymbol{\beta}' \in \mathbf{B}_m$, is inconsistent, one has to revise the list of desired binding modes and try again. Alternatively, the inequalities might be consistent, but the fit might be unacceptably poor, that is, Z is too large. Then the outliers are prime candidates for altering their desired binding modes. Of course, it is also possible that the calculation succeeds, and we are done. Generally one goes through a series of desired mode revisions, sometimes realizing there are additional site points necessary and thus backtracking to step (v). Assuming we have the correct number and placement of site points, the following systematic method for revising the desired modes has proven very effective: [Ghose1983a]

Algorithm 8.9 (Ghose) Energy refinement:

INPUT: Fixed site geometry, a set of molecules, and a list of their desired binding modes.

OUTPUT: A set of refined interaction energies and for each molecule a predicted binding mode which may differ from the originally desired one.

PROCEDURE:

Initially let $\mathbf{B}^*_{(0)}$ be the set of desired binding modes,
 and let there be no constraints.
Minimize $Z(\mathbf{B}^*_{(0)},\mathbf{e})$ with respect to $\mathbf{e}$, resulting in $\mathbf{e}_{(0)}$.
Determine the new set of energetically optimal binding modes, $\mathbf{B}^{*\prime}_{(0)}$,
 corresponding to $\mathbf{e}_{(0)}$.
Minimize $Z(\mathbf{B}^{*\prime}_{(0)},\mathbf{e})$, resulting in $\mathbf{e}'_{(0)}$.
Set the iteration count $k := 0$.
While $\mathbf{B}^{*\prime}_{(k)} \neq \mathbf{B}^*_{(k)}$
 If $Z(\mathbf{B}^*_{(k)},\mathbf{e}_{(k)}) < Z(\mathbf{B}^{*\prime}_{(k)},\mathbf{e}'_{(k)})$
  For each $\boldsymbol{\beta}^*_i \in \mathbf{B}^*_{(k)}$ and $\boldsymbol{\beta}^{*\prime}_i \in \mathbf{B}^{*\prime}_{(k)}$ such that $\boldsymbol{\beta}^*_i \neq \boldsymbol{\beta}^{*\prime}_i$
   Add $\boldsymbol{\beta}^{*T}_i\mathbf{e} \leq \boldsymbol{\beta}^{*\prime T}_i\mathbf{e}$ to the set of constraints.
  Let $\mathbf{B}^*_{(k+1)} = \mathbf{B}^*_{(k)}$.
 Else the quality of fit has improved.
  Delete all current constraints.
  Let $\mathbf{B}^*_{(k+1)}$ be $\mathbf{B}^{*\prime}_{(k)}$.
 Minimize $Z(\mathbf{B}^*_{(k+1)},\mathbf{e})$ subject to current constraints,
  resulting in $\mathbf{e}_{(k+1)}$.
  If constraints are inconsistent
   halt entire procedure and revise desired modes and/or site points.
 Determine the new optimal modes, $\mathbf{B}^{*\prime}_{(k+1)}$.
 Minimize $Z(\mathbf{B}^{*\prime}_{(k+1)},\mathbf{e})$, resulting in $\mathbf{e}'_{(k+1)}$.
 Increment the iteration counter, k.
Modes and interaction energies are now self-consistent.
Predicted binding modes are $\mathbf{B}^*_{(k)}$,
Optimal energies are $\mathbf{e}_{(k)}$, and
Accuracy of the fit is $Z(\mathbf{B}^*_{(k)},\mathbf{e}_{(k)})$.

Every time a constraint is added, Z can only increase. The advantage of the algorithm is that it will sometimes automatically switch to a better set of desired binding modes which *decreases* Z and improves the fit.

Fig. 8.13 shows schematically how the algorithm might proceed in the extremely simple case of only a single interaction energy, instead of a whole multidimensional vector **e**.

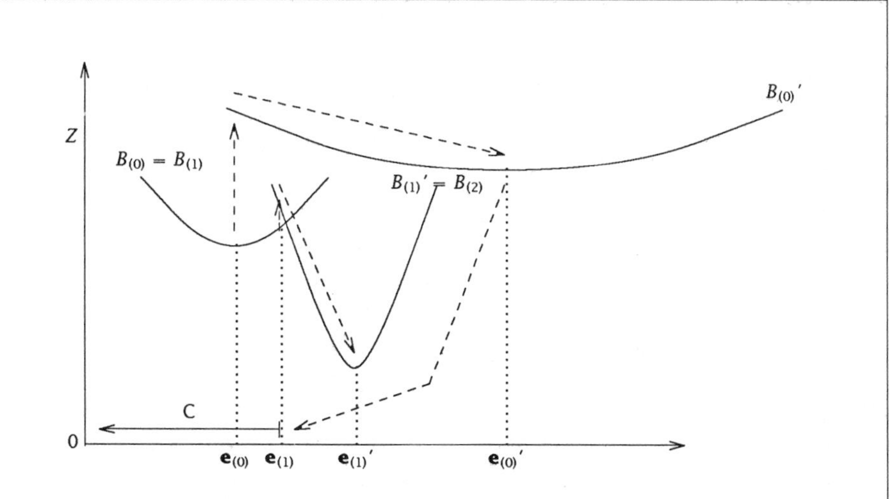

Figure 8.13

Illustration of the energy refinement algorithm for a one-dimensional energy space. Different sets of binding modes define the various quadratic surfaces. C indicates the values of **e** excluded by some linear constraint. See text for explanation.

Whatever the dimensionality of **e**, the $Z(\mathbf{B}^*,\mathbf{e})$ is a quadratic surface over the domain **e**, the shape of which depends on the modes $\mathbf{B}^*$. Suppose minimizing $Z(\mathbf{B}^*_{(0)},\mathbf{e})$ produces $\mathbf{e}_{(0)}$, but different binding modes, $\mathbf{B}^{*\prime}_{(0)}$, are then preferred. The fit with $\mathbf{B}^{*\prime}_{(0)}$ is in general worse for $\mathbf{e}_{(0)}$, but optimization leads to new $\mathbf{e}'_{(0)}$. Suppose the fit is still worse at $\mathbf{e}'_{(0)}$ as shown. Then some inequalities, indicated as the horizontal bar C, will restrict the optimization on the old $\mathbf{B}^*_{(0)}$ surface, now called $\mathbf{B}^*_{(1)}$. The constrained fit at $\mathbf{e}_{(1)}$ is worse than the unconstrained one, and these interaction energies may produce yet another set of preferred binding modes, $\mathbf{B}^{*\prime}_{(1)}$. Supposing that $\mathbf{B}^{*\prime}_{(1)}$ is so shaped that even constrained optimization improves the fit, new energies, $\mathbf{e}'_{(1)}$, result, and $\mathbf{B}^{*\prime}_{(1)}$ is made the new set of modes, $\mathbf{B}^*_{(2)}$. The constraints C are no longer needed, and the algorithm continues on from $\mathbf{B}^*_{(2)}$. In general, there is no guarantee that Z tends to decrease, and conceivably different paths in energy space would result from different choices of initial binding modes. In practice, however, the results tend to be satisfactory.

Remember that the components of **e** are the interactions between given types of atoms (or groups of atoms) and given site points. Although we have treated **e** as a vector for simplicity of notation, one generally thinks of them as a table with one row for every atom type and one column for every site point. It is sometimes reasonable to build a site model with an extended region represented by several site points, but with the intention that the environment be uniform within the region. Then really there is a column in the table for every site point *type*, and the site points of such a group would share the same type. Similarly, steric repulsion site points can always be thought of as having the same type, and correspond to a column with all entries fixed at large, positive values. All other entries are determined independently by the refinement algorithm. Often there is nevertheless remarkable internal consistency: superfluous site points have columns of small interaction energies; a given site point interacts favorably with every type of polar group but unfavorably with every hydrophobic group; or a site point has increasingly favorable interactions with atoms of increasingly positive partial charge. Fortuitous consistency certainly increases confidence in the correctness of the model, but the interaction matrix is seldom so consistent that one could add a row for, say, an iodo substituent by extrapolating from the chloro and bromo rows. There is also the nagging worry that with so many energy parameters, one could fit any arbitrary set of ΔG_{obsd}. Therefore, in order to enhance the predictive powers of these models while reducing the number of adjustable parameters, we label the rows not with atom types, but with physicochemical parameters. Suppose there is a fixed method of decomposition of molecular properties (partition coefficient between octanol and water, molar refractivity, electrical charge, etc.) into additive atomic contributions, such that the molecular value for any compound could be accurately predicted. Then for some binding mode **b** of some molecule where the n_a atoms $i = 1,\ldots,n_a$ each have n_p atomic properties, p_{ij}, $j=1,\ldots,n_p$, let the ΔG of the mode be

$$\Delta G(\mathbf{b}) = \sum_{\substack{\text{contacts} \in \mathbf{b} \\ \text{atom } i \text{ and} \\ \text{site point type } k}} \left[\sum_{j=1}^{n_p} p_{ij} e_{jk} \right] \qquad (8.32)$$

Since the p_{ij} are fixed, energy refinement involves adjusting the e_{jk}, $j=1,\ldots,n_p$ and $k=1,\ldots,$number of site point types. ΔG is still linear in the e_{jk}s, so Algorithm 8.9 proceeds exactly as before. Now e_{jk} can be interpreted as the coefficient for property j of the free energy contribution when a contact is made with any atom and a site point of type k. The advantage of this approach is that a novel compound having an atom of a type not found in the original data set, could still be tested with the site model to predict its ΔG_{calc} and binding mode. The disadvantage is needing a scheme for assigning atomic contributions to various molecular properties. Atomic partial charges can be fairly readily calculated by quantum chemistry programs, and predictive atomic hydrophobicity and molar refractivity assignments have been made [Ghose1985, Ghose1986, Ghose1987].

This finishes the outline of how one would develop a binding site model in terms of site points. §8.8 surveys the studies that already have been done on experimental data sets in order to further illustrate the procedure. We conclude this section with an artificial example demonstrating the complexities that can arise in a seemingly simple situation.

Suppose the data set consisted of only two molecules, chair cyclohexane and boat cyclohexane, each represented as six carbon atoms. Assume these conformers are rigid and non-interconverting. Then their approximate distance matrices are shown in Fig. 8.14.

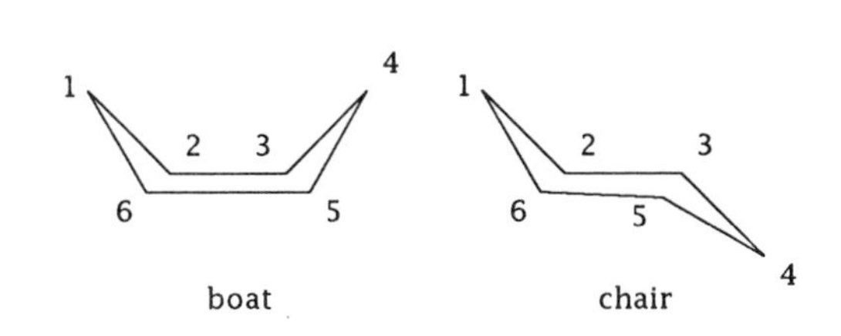

Figure 8.14

Cyclohexane inter-carbon distances in Å.
Upper triangle is the chair form, lower triangle the boat.

	1	2	3	4	5	6
1	0	1.5	2.5	3.0	2.5	1.5
2	1.5	0	1.5	2.5	3.0	2.5
3	2.5	1.5	0	1.5	2.5	3.0
4	2.6	2.5	1.5	0	1.5	2.5
5	2.5	3.1	2.5	1.5	0	1.5
6	1.5	2.5	3.1	2.5	1.5	0

Suppose the two forms have significantly different observed binding energies, and we must devise a model reflecting that. Clearly a site consisting of only one point is insufficient because both molecules consist of atoms all having the same type. There is only a single entry in the interaction energy table, $e_{C,1}$, namely that between carbon and site point 1. Both forms will have $\Delta G_{calc} = e_{C,1}$, which is too poor a fit to the observed values. Next try two site points. Fig. 8.14 shows there are only three significantly different interatomic distance categories: 1.5, 2.5, and 3.0 Å. If the two site points do not have a separation within δ of one of these three values, only one atom at a time can bind for either molecule, and we are back to the single site point case. If the separation is close to one of these values, then two carbons can bind simultaneously, giving a binding energy of $e_{C,1}+e_{C,2}$. Since examples of all three distance categories can be found in both molecules, they will both have optimal binding modes involving the same numbers of contacts, the same $\boldsymbol{\beta}$ energy space vectors, and hence the same ΔG_{calc}. Proceeding on, we are now forced to consider models having three site points. As before, unless the three site points have mutual distances chosen from the three categories, we will have effectively a one- or two-point site model. To form three contacts simultaneously, there are only three basic distance matrix possibilities for the chair form:

$$\begin{pmatrix} 0 & 1.5 & 2.5 \\ 1.5 & 0 & 1.5 \\ 2.5 & 1.5 & 0 \end{pmatrix}$$

as for atoms 1, 2, 3;

$$\begin{pmatrix} 0 & 1.5 & 3.0 \\ 1.5 & 0 & 2.5 \\ 3.0 & 2.5 & 0 \end{pmatrix}$$

as for atoms 1, 2, 4; and

$$\begin{pmatrix} 0 & 2.5 & 2.5 \\ 2.5 & 0 & 2.5 \\ 2.5 & 2.5 & 0 \end{pmatrix}$$

as for atoms 1, 3, 5. Unfortunately, atoms 1,2,3, 3,4,5, and 1,3,5 respectively of the boat form also have these distance matrices. Therefore, there is no arrangement (for δ= 0.2 Å, say) of three site points such that the chair will have three contacts while the boat has fewer. One the other hand, boat atoms 1,3,4 have distance matrix

$$\begin{pmatrix} 0 & 2.5 & 2.6 \\ 2.5 & 0 & 1.5 \\ 2.6 & 1.5 & 0 \end{pmatrix}$$

which cannot be matched by any triplet of chair atoms. If $\Delta G_{boat,obsd} = -6$ kcal and $\Delta G_{chair,obsd} = -4$ kcal, then arranging the three site points as in the last matrix and letting $e_{C,1} = e_{C,2} = e_{C,3} = -2$ fits the data. The boat form will have three contacts in its optimal binding mode, while the best the chair can do is make two contacts in various ways. On the other hand, suppose $\Delta G_{boat,obsd} =$

-6 kcal and $\Delta G_{chair,obsd} = -8$ kcal. The *e*s of a three point model cannot be adjusted to fit the data, and we are forced to go to *four* points. The only distinctive chair distance submatrix is

$$\begin{pmatrix} 0 & 1.5 & 3.0 & 1.5 \\ 1.5 & 0 & 2.5 & 2.5 \\ 3.0 & 2.5 & 0 & 2.5 \\ 1.5 & 2.5 & 2.5 & 0 \end{pmatrix}$$

as for atoms 1,2,4,6. If we position the four site points in this way, the chair can make four contacts simultaneously, but the boat can make at most three. Choosing $e_{C,1} = e_{C,2} = e_{C,3} = e_{C,4} = -2$ produces the desired site model. Note how fitting the binding energies for only two molecules has compelled us to employ four site points and four energy parameters. Viewed as a least squares fit of two observations to ten parameters (= 6 conformational degrees of freedom for the four site points + 4 energy parameters), this result is absolutely outrageous. The resolution of the conflict lies in remembering that the adjustable parameters of least squares fitting are continuously and independently variable from $-\infty$ to $+\infty$. In our site models, however, site point positions often cannot be varied without affecting the lists of allowed binding modes for the different molecules. Then the energies must be adjusted so that some one mode's total interaction fits the observed value *and* the ΔG for all other modes are no better. These implicit constraints and interdependencies are why the number of parameters is driven so high in our models.

8.6. Using a Site Model to Suggest New Drugs

At this point we have explained the binding affinities for some given drugs in terms of a site model consisting of site point coordinates and an interaction energy table. Simply inspecting the model on a computer graphics display may immediately suggest novel compounds capable of making several favorable contacts simultaneously. Quantitative binding predictions are always possible by carrying out steps *i, ii, iii,* and *vi* of the previous section for any chosen molecule. Since the interaction energies are already known, only the geometrically allowed mode(s) with optimal calculated binding are of interest. This mode would be its predicted binding mode, and the corresponding ΔG_{calc} would be the predicted binding energy. Note that the new molecule may be of an entirely different chemical class than those upon which the model was based. If the rows of the energy table correspond to given atom (or group) types, the new molecule must be composed of only those atom types, although the spatial arrangement could be quite different. Preferably, the rows of the energy table correspond to various atomic physico-chemical properties. Then any molecule whose atoms have known values of these properties can be tested in the site model. Unfortunately, there are so many possible small organic compounds that an exhaustive search for those best fitting the site is out of the question. A much less thorough, but systematic search would still be likely to uncover novel good compounds that might be missed by the intuitive inspection of the site. One such search algorithm is the subject of this section [Crippen1983].

The basic notion is that a chemist would interactively search for promising compounds that could be built up out of a library of given fragments: ring systems, functional groups, etc. The computer's task is to join pairs of fragments in "substitution reactions," test these molecules in the site, and present the chemist with the results, ordered by predicted binding strength. Then it is up to the chemist to delete compounds that are synthetically unreasonable or otherwise unpromising. He may also edit the library to guide the procedure toward more desirable derivatives. Thus the search tends to be more thorough than the unaided person would carry out, while focusing on more chemically correct and otherwise practical alternatives than the unaided computer would produce.

More formally, the top level procedure can be described as a tree search. As shown in Fig. 8.15, the tree consists of alternating levels of library and molecule nodes, starting at the top with a library node. The descendants of a library node, produced by procedure "newleads," consists of molecules taken directly from the library and derivatives formed by linking pairs of library molecules. Each molecule node has usually only a single descendant, a library consisting of the parent molecule and the parent library. The task of newleads is to present the best binding possibilities in order of preference, not mentioning the vastly more numerous collection of derivatives which are clearly inferior. The task of the chemist is to prune away molecule nodes (synthetically impossible, too costly, etc.) and to restrict the contents of library nodes so the search will focus on more promising compounds. The tree traversal is breadth-first in that the descendants of a library node are presented in the order of the newleads algorithm's preference which the chemist can revise as he sees fit. The search proceeds to lower levels only for the most promising molecule nodes. This is the way the chemist eventually terminates the tree traversal. In the example shown in Fig. 8.15, the search begins with a very small library consisting only of benzene and methane. Newleads may suggest that toluene fits the site best, followed by benzene, ethane, and methane. The chemist may decide to pursue only the toluene possibility, so he adds it to the current library. Newleads considers a large number of combinations of these molecules, with *p*-xylene and *m*-xylene being

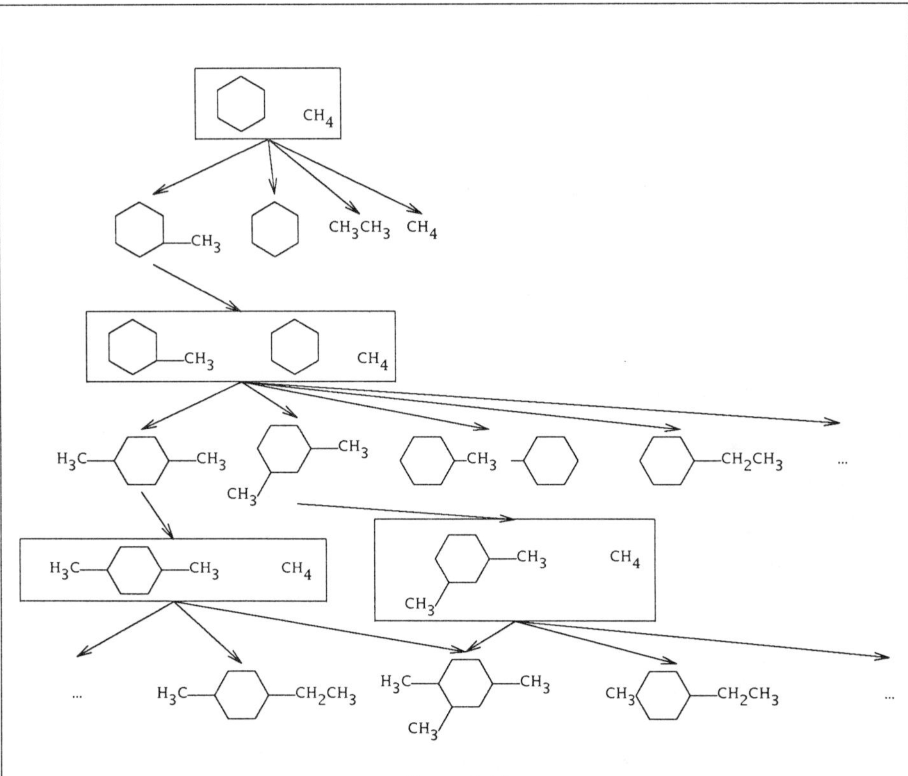

Figure 8.15

Using a very small library of molecular fragments to search for compounds that fit a given site. The tree nodes enclosed in boxes are fragment libraries from which descend the possibilities suggested by procedure newleads. From some of these descend in turn new trial libraries. Only a small fraction of the total possibilities is shown.

the best. The chemist decides to consider these two best possibilities, but drastically reduces the library so as to avoid derivatives of those molecules discarded at level 3. Newleads applied to these two libraries produces the illustrated compounds along with many others.

Even with the user drastically pruning the tree, the newleads procedure must be moderately clever about checking all library entries and pairwise combinations in such a way as to quickly pass over the numerous rather poor cases. For some specified binding energy tolerance, $\delta E = 2.0$ kcal, say, it will concentrate on those molecules that are likely to bind no more than δE worse than the best molecule found so far. A key trick in the screening procedure is ΔG_{est}, a fast lower (i.e. optimistic) bound on ΔG_{calc} made by disregarding all geometric considerations. Namely,

Algorithm ΔG_{est}:

Initially ΔG_{est}=0 and all site points and atoms
 are marked as unused.
While there are both unused site points and atoms
| Find $e_{ij} = \min_{k,l}\{e_{kl} \mid \text{atom } k \text{ site point } l \text{ unused}\}$
| If $e_{ij} < 0$
| | Mark i and j as used.
| | Increment ΔG_{est} by e_{ij}.

Then we have the algorithm

Algorithm 8.10. Newleads:

INPUT: The molecule library, site point coordinates, interaction energy table, and δE.

OUTPUT: A number of molecules and their derivatives which are predicted to bind well with the site.

PROCEDURE:
| Sort the molecules in the library according to their ΔG_{est}, and
| otherwise preferring the smaller molecules.
| Let the best binding so far be ΔG_{opt}=0 (i.e. not repulsive)
| For each molecule in the sorted list beginning with the best one
| | Skip to the next molecule if $\Delta G_{est} > \Delta G_{opt}+\delta E$.
| | Find its best mode, $\mathbf{b}_{calc}$, and corresponding ΔG_{calc}
| | by procedure "dock."
| | If $\Delta G_{calc} < \Delta G_{opt}+\delta E$
| | | Present the molecule to the user as a "molecule node"
| | | in the search tree.
| | If $\Delta G_{calc} < \Delta G_{opt}$, let $\Delta G_{opt} = \Delta G_{calc}$.
| | If ΔG_{calc}>0, and $\mathbf{b}_{calc}$ did not use all site points,
| | and this "parent" molecule has substitutable hydrogens,
| | | Make a second sorting of the library ("substituent") molecules
| | | by a second set of ΔG_{est} where those site points
| | | involved in $\mathbf{b}_{calc}$ are already used.
| | | For each substituent in the sorted list,
| | | | Reject it if ΔG_{calc}(parent)+ΔG_{est}(substituent) >
| | | | $\Delta G_{opt}+\delta E$ or >0.
| | | | Otherwise for all points of substitution on both substituent and parent
| | | | | Make the derivative.
| | | | | Find ΔG_{calc}(derivative) by "dock."

If $\Delta G_{calc} < \Delta G_{opt}+\delta E$,
Present the derivatives to the user as a molecule node in the search tree.
If $\Delta G_{calc} < \Delta G_{opt}$,
Let $\Delta G_{opt} = \Delta G_{calc}$.

The reason the lists are sorted in advance is that the reject steps above can confidently cut off the whole remainder of the list because certainly $\Delta G_{est} < \Delta G_{calc}$. In branch-and-bound terms, this amounts to a very strong fathoming criterion. As the algorithm proceeds, ΔG_{opt} decreases monotonically, and ΔG_{calc} of the successively reported molecules tends to decrease, but not monotonically for $\delta E>0$. For large δE, many molecules are presented to the user in rather rough order without omitting good candidates for further derivatives. At the opposite extreme, if $\delta E=0$, the presented molecules have monotonically decreasing ΔG_{calc} down to the very best molecule, but some which are nearly as good may have been skipped.

There remain to be explained just two important details: the docking procedure and joining of library molecules to form derivatives. For these, however, we must first understand the data structure used to represent molecules. Most of the algorithms in this chapter deal only with a molecule's atom types and interatomic distance bounds. Finding distance bounds (step *ii*, §8.5), joining molecules together, and docking a molecule in a site with necessary and sufficient geometric checking, all require a knowledge of the connectivity and rotatable bonds. Thus we represent a molecule as a list of atoms with unique labels, their types, a set of coordinates for some one conformation, bonds connecting atoms, and which bonds are rotatable. The bonding is given either as directed edges between the atom nodes of a tree graph, or as extra ring closure edges, as shown in Fig. 8.16. That way a program can traverse the tree, starting at the root atom, and find every atom without repetition, because a tree has no cycles. Any rotatable bond divides the tree into two sub-trees, allowing one to easily determine which atoms move with respect to the rest when that dihedral angle is changed. Van der Waals checking needs to be carried out between all pairs of atoms separated by more than two tree edges, except for atoms joined by a ring closure or one edge and a ring closure. Clearly, there is one ring closure per ring, and it is unimportant which molecular bond is chosen. If the root is always a non-tetravalent atom, then no node has more than three descendants because there are no pentavalent atoms. Assuming rigid bond lengths and bond angles, a molecule with n_r rotatable bonds generally has n_r+1 "rigid groups": overlapping subsets of atoms whose mutual distances are unaffected by rotatable bonds. Two atoms are members of the same rigid group if and only if either (1)

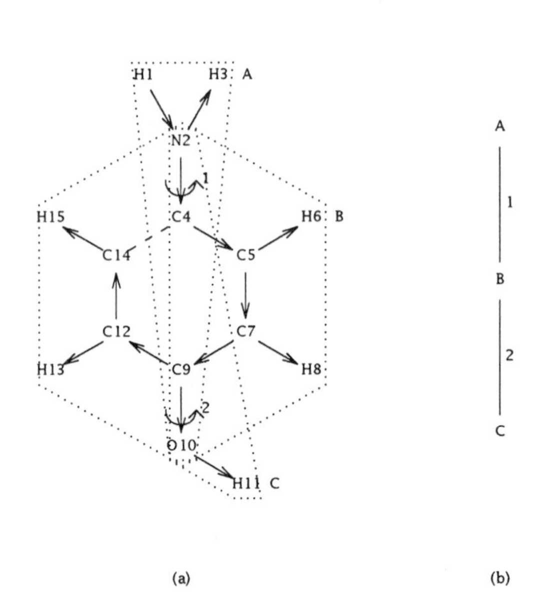

Figure 8.16
Encoding the molecular structure of 1-hydroxy-4-aminobenzene. (a) Atoms are labelled by type and by sequence in the connectivity tree traversal. Arrows are the bonds used to indicate tree connectivity, while the dashed line between C14 and C4 is a ring closure. Dotted loops enclose the three rigid groups, A, B, and C, that are the consequence of the two rotatable bonds, 1 and 2. (b) Corresponding rigid group graph.

they are linked by a series of non-rotatable bonds, or (2) at least one of them lies on the common axis of all rotatable bonds joining them. Thus in Fig. 8.16, H15 and H6 are in one rigid group by the first criterion, and N2 and H11 are in another group by the second. A procedure for determining all rigid groups of a molecule is the following:

Algorithm 8.11. Rigid Group Determination:

INPUT: A molecule in terms of its atomic coordinates for an arbitrary conformation, its bonds, and which bonds are rotatable.

OUTPUT: The list of rigid groups of atoms in the molecule, i.e. subsets of atoms which maintain their same relative positions in any conformation of the molecule having the same bond lengths and bond angles.

PROCEDURE:

Initially there are no rigid groups, and all atoms

are marked as unused.
While there is an unused atom,
This atom is used as the first member of a new rigid group.
Starting from this root, traverse the full connectivity graph,
including atom neighbors as used in the current group unless
(a) The neighbor is already a member of the group, or
(b) A rotatable bond links the neighbor.
This bond is the link to the next rigid group
(see Fig. 8.16(b)).
In this case, use the atom in the group but start a second
traversal from here which includes atoms in the group
only if they lie on the rotatable bond axis and all
further rotatable bonds crossed are collinear with the
first one.

Consider the graph in Fig. 8.16(b), where the nodes are the rigid groups, and the edges are the rotatable bonds linking them. We call two rigid groups adjacent in connectivity if they are adjacent nodes in this graph. A flexible cyclic molecule would even have cycles in such a graph. Dijkstra's algorithm [Dijkstra1959] is a handy way of determining the shortest path (i.e. least number of edges) between any two given nodes on a graph. Such a path would correspond to the set of rotatable bonds affecting the relative positions of two rigid groups.

Now given this sort of molecular description, each molecule in a library consists of one or more rigid groups. Joining two such molecules amounts to deleting one selected hydrogen on each, translating and rotating one of the fragments as a rigid body until the two broken bonds are collinear and the newly formed bond length is correct. The connectivity tree must be revised, and the new bond is presumed to be rotatable. Such a joining procedure will create (possibly branched) chains, but it will not form cycles. Thus all ring systems of interest must be included in the original library.

The last item is the docking procedure. Here we use BRONK, Algorithm 8.8, to generate a list of binding modes which includes all geometrically possible ones. Trying these in order of ΔG_{calc}, the algorithm soon comes to the most favorable one that can indeed be realized by rotating, translating, and bond-rotating the atomic coordinates onto the given site point coordinates. This is a special purpose algorithm that exploits the general feature of such molecules that they are constructed out of a small number of rigid groups each containing many atoms, and that rings are generally rigid. EMBED could be used instead to handle a wider range of molecules at the expense of more computer time. Another feature that helps one deal with large (e.g. >50 atoms) molecules with many (e.g. >6) important rotatable bonds, is that the interatomic distance bounds required by BRONK are found not by an exhaustive search over all conformation space, but by the triangle inequality and a knowledge of the rigid groups. These bounds are looser, but that does not result in excessive numbers of geometrically incorrect binding modes, in practice.

Algorithm 8.12. Dock:

INPUT: The site model and the molecule.

OUTPUT: The energetically optimal, geometrically allowed binding mode.

PROCEDURE:

- Determine the rigid groups of the molecule.
- For all pairs of atoms i, j in the same rigid group,
  - $u_{ij} = l_{ij} = d_{ij}$
- Otherwise
  - $u_{ij} = \infty$
  - l_{ij} = sum of the van der Waals radii of i and j.
- Smooth bounds by TRIANGLE, Algorithm 5.36.
- Generate the list of binding modes by BRONK.
- Sort the list in order of ΔG_{est}, assuming exactly the specified contacts will be made in each mode.
- While ΔG_{est} of the next mode < the best found so far by superposition,
  - Attempt to achieve these contacts:
  - Initially there are a translation, two rigid rotations, and n_r internal rotations as degrees of freedom.
  - Initially mark all rigid groups and desired contacts as untreated.
  - While degrees of freedom remain and there are untreated contacts,
    - Find the rigid group closest in molecular connectivity to a treated one, if any, and otherwise participating in the greatest number of contacts.
    - If translation is available,
      - Rigidly translate the molecule for a least-squares superposition of those atoms of the rigid group involved in contacts with the corresponding site points.
      - Translation is used.
      - If more than one contact is to be made,
        - Rotate the molecule for least-squares contact formation [McLachlan1982].
        - If two contacts had to be made,
          - One rotation is used.
        - Else if three or more noncollinear contacts were made,
          - Both rotations are used.
    - Else if both rotations are available,

Rotate as in the previous case, taking as origin the single contact involved in the previously treated rigid group.
Else if only one rotation is available,
Rotate as before, taking the axis of rotation defined by the last two contacts made.
Else only rotatable bonds are available.
Make a list of rotatable binds from this rigid group to the nearest treated group.
Cycling through the list up to three times,
Optimally rotate the rigid group so as to make the contacts, where the unnormalized vector in McLachlan's algorithm [McLachlan1982] indicating the direction and magnitude of the required rotation is modified by taking its component along the bond vector.
All rotatable bonds in the list are used.
All degrees of freedom are used.
The true ΔG is the sum of interaction energies over all contacts *actually* formed, whenever an atom and a site point are closer than a chosen $\delta > 0$, whether or not that contact was specified in the original binding mode.

This completes the discussion of the docking algorithm and the whole procedure of searching for new drugs given a site model. The computational details are rather elaborate compared to, say, §8.5, but it enables us to deal with larger, more flexible molecules than ever before, while ensuring complete geometric correctness of the binding modes.

8.7. Voronoi Polyhedra Site Models

Representing a receptor site as a collection of site points is computationally tractable, as has been shown in the last two sections. It is, however, not a very concise way to describe the site unless binding depends critically on the simultaneous interaction of a few key groups on the molecule having rather precisely defined relative positions. Consider, for example, the dopamine receptor model proposed by Humber *et al.* [Humber1979]. As shown in Fig. 8.17, there are binding pockets for a nitrogen (D), a hydrophobic group (H), a hydrogen bond donor (G), and a phenyl ring (E). Thus four site *points* would succinctly describe the site, and compounds unable to make contact with all four would correctly be predicted to bind poorly. Compounds having a nitrogen, a lipophilic group, a

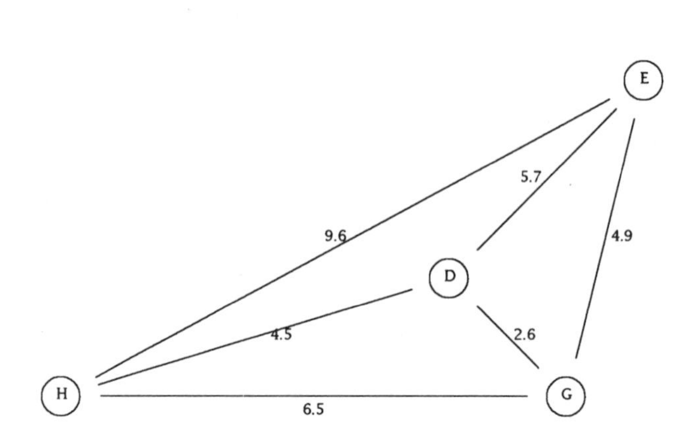

Figure 8.17

A model for the dopamine receptor [Humber1979] where site points D, E, G, and H economically describe its salient features. Interpoint distances are in Angstroms.

hydrogen bond donor, and a phenyl ring all in the correct relative positions, would be predicted to bind in a very specific mode making contacts with site points D, H, G, and E, respectively. On the other hand, suppose the site consisted of only a large hydrophobic slot into which a wide variety of fused aromatic ring systems could bind well. If naphthalene bound better than benzene, which bound better than cyclobutadiene, then one might have to propose the ten attractive site points in Fig. 8.18a, further claiming that benzene bound rather rigidly to s1, s2, s3, s4, s5, and s10. Also, one methyl of *m*-dimethylbenzene might bind to s6, but the other would fall beyond s1 without interacting. A much more non-committal and also physically more realistic approach would be to represent the slot as a flat slice of space having a few Angstroms thickness, as shown in Fig. 8.18b. Any atom lying in this *region,* r2, would interact with a homogeneous hydrophobic environment, while atoms protruding into r1 or r3 would experience steric repulsion. The proposed binding mode of benzene would not specify the two translational, one rotational, and one flipping degrees of freedom that keep the ring system in r2. Both methyls of o-dimethylbenzene would interact without having to specify more structural detail for the site.

Therefore, this section presents an alternative to §8.5 where we use regions instead of site points and absolute data fitting (Equation 8.10) instead of least squares (Equation 8.9). The regions are convenient to define in terms of Voronoi polyhedra [Crippen1984, Voronoi1908]. If we propose n_s site regions, we must

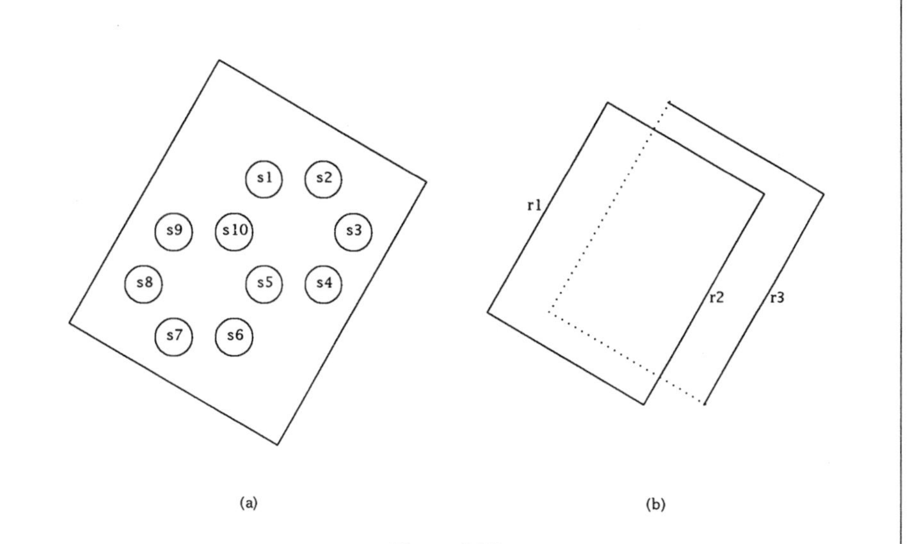

Figure 8.18

In (a) a hydrophobic binding slot is represented as ten site points, s1-s10, lying in a plane. The alternative representation in (b) is in terms of three *regions*, r1-r3, where the slot is r2 sandwiched between two parallel planes.

arbitrarily select coordinates for "generating" points $\mathbf{c}_i$, $i=1,\ldots,n_s$. Then the ith Voronoi polyhedron, which is our ith site region r_i, is the locus of points closer to its generating point than to any other:

$$r_i = \{\mathbf{x} \mid \|\mathbf{c}_i - \mathbf{x}\| < \|\mathbf{c}_j - \mathbf{x}\|, \ \forall \ j \neq i\} \tag{8.33}$$

Clearly, all space is accounted for, and the regions are non-overlapping. Each polyhedron is convex, and in three dimensions it is bounded by planes. A two-dimensional example is shown in Fig. 8.19. For a given translation, rotation, and conformation of a molecule, each atom must fall into one and only one region because for our purposes we insist on strict inequality in Equation (8.33), and in computer calculations we can assume an atom will never be exactly equidistant between two generating points. The convexity of the regions implies that if A-B-C are collinear atoms, and if both A and C are inside r_i, then B must also lie inside r_i. Similarly, if A, B, C, and D are coplanar atoms with D falling within triangle ABC, then if A, B, and C are in r_i, so must D. Lastly, if atom E falls within the tetrahedron formed by atoms A, B, C, and D, then if A, B, C, and D are within r_i, so must E. A (finite) region r_i can conveniently be described in terms of its set of vertices $\{\mathbf{v}\}_i$. For instance, $\{\mathbf{v}\}_1 = \{\mathbf{v}_1, \mathbf{v}_2, \mathbf{v}_3\}$ in Fig. 8.19. Then any point $\mathbf{p}$ in r_i

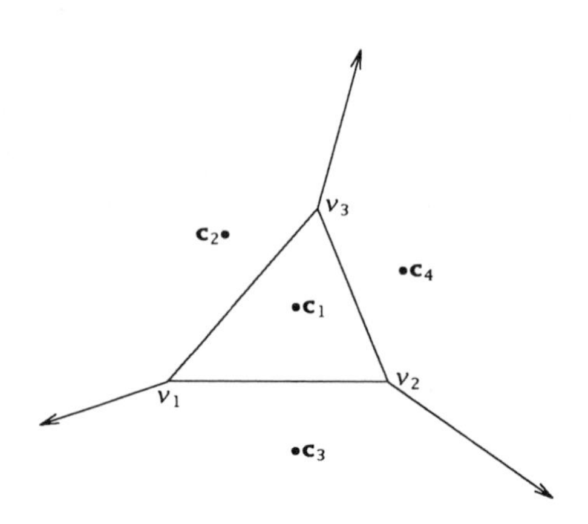

Figure 8.19

A two-dimensional example of Voronoi polyhedra forming a site model. Generating points $\mathbf{c}_1$-$\mathbf{c}_4$ determine the finite region r_1 and the three infinite ones r_2, r_3, and r_4, respectively. Regions meet at vertices v_1, v_2, and v_3. Perhaps r_1 would be a binding cavity, r_2 and r_3 the walls, and r_4 the open solvent.

must be a convex combination of the vertex set:

$$\mathbf{p} = \sum_{\mathbf{v}_j \in \{\mathbf{v}\}_i} \alpha_j \mathbf{v}_j \text{ and } \sum \alpha_j = 1 \tag{8.34}$$

Equation (8.34) can only be used for finite regions, so it is convenient to make all regions in the model finite by adding four outrigger points (or three in two dimensions) at (x,y,z) coordinates

$$\begin{array}{c} \left(-\frac{K}{2} , -\frac{K}{2\sqrt{3}} , -\frac{K}{2\sqrt{6}} \right) \\ \left(\frac{K}{2} , -\frac{K}{2\sqrt{3}} , -\frac{K}{2\sqrt{6}} \right) \\ \left(0 , \frac{K}{\sqrt{3}} , -\frac{K}{2\sqrt{6}} \right) \\ \left(0 , 0 , \frac{K}{\sqrt{2}\sqrt{3}} \right) \end{array} \tag{8.35}$$

where $K>0$ is some number large enough that the given generator points lie within the tetrahedron defined by the outriggers. All vertices can now be located by examining every quartet of generators $\mathbf{c}_i$, $\mathbf{c}_j$, $\mathbf{c}_k$, and $\mathbf{c}_l$, including outriggers. A vertex $\mathbf{v}$ would have equal squared distances to all four generating points. Solving the resulting linear system

$$\begin{pmatrix} 2(\mathbf{c}_j-\mathbf{c}_i)^T \\ 2(\mathbf{c}_k-\mathbf{c}_i)^T \\ 2(\mathbf{c}_l-\mathbf{c}_i)^T \end{pmatrix} \mathbf{v} = \begin{pmatrix} \|\mathbf{c}_j\|^2-\|\mathbf{c}_i\|^2 \\ \|\mathbf{c}_k\|^2-\|\mathbf{c}_i\|^2 \\ \|\mathbf{c}_l\|^2-\|\mathbf{c}_i\|^2 \end{pmatrix} \quad (8.36)$$

for $\mathbf{v}$ (all vectors are thought of as column vectors, so $(\cdots)^T$ indicates a row of the 3×3 matrix) results in a possible vertex. If $\mathbf{v}$ is no closer to any generator than it is to $\mathbf{c}_i$, then it is really a vertex and is a member of $\{\mathbf{v}\}_i$, $\{\mathbf{v}\}_j$, $\{\mathbf{v}\}_k$, and $\{\mathbf{v}\}_l$. The outriggers have no other purpose than to give each region a full complement of vertices with finite coordinates such that Equation (8.34) applies.

With site points, one could quickly check a possible binding mode by comparing the intersite distance with the corresponding interatomic distance range for every pair of contacts. The equivalent check here requires the greatest, u_{rij}, and the least, l_{rij}, distance between each pair of regions r_i and r_j. By this we mean

$$\begin{aligned} l_{rij} &= \min_{\substack{\mathbf{p}\in r_i \\ \mathbf{q}\in r_j}} \|\mathbf{p}-\mathbf{q}\| \\ u_{rij} &= \max_{\substack{\mathbf{p}\in r_i \\ \mathbf{q}\in r_j}} \|\mathbf{p}-\mathbf{q}\| \end{aligned} \quad (8.37)$$

Clearly $l_{rii}=0$ always, and $l_{rij}=0$ if r_i and r_j are adjacent regions. Using outriggers ensures that $u_{rij}<\infty$ for all i and j. For example, in Fig. 8.19, $u_{r11}=\|\mathbf{v}_1-\mathbf{v}_2\|$. The general algorithm is simply

Algorithm 8.13. Region distance bounds:

INPUT: The vertices of Voronoi polyhedra.

OUTPUT: The upper and lower bounds on distances between polyhedra.

PROCEDURE:

For all pairs of regions, r_i and r_j

For r_i, examine every vertex $\mathbf{v}_a$ one at a time, while for r_j examine all pairs of vertices $\mathbf{v}_b$ and $\mathbf{v}_c$; also consider single vertices, $\mathbf{v}_a$, of r_j and pairs, $\mathbf{v}_b$ and $\mathbf{v}_c$, of r_i.

The maximum distance, u, and the minimum, l, between $\mathbf{v}_a$ and the line segment $\mathbf{v}_b\mathbf{v}_c$ may be either some endpoint distances, $\|\mathbf{v}_a-\mathbf{v}_b\|$ or $\|\mathbf{v}_a-\mathbf{v}_c\|$, or perhaps $l=\|\mathbf{v}_a-\alpha\mathbf{v}_b-(1-\alpha)\mathbf{v}_c\|$ where $0<\alpha<1$ and

$$\alpha = \frac{(\mathbf{v}_a-\mathbf{v}_c)^T(\mathbf{v}_b-\mathbf{v}_c)}{\|\mathbf{v}_c-\mathbf{v}_b\|^2} \quad (8.38)$$

Over all vertex combinations, let $u_{rij}=\max u$ and

|| $l_{rij} = \min l$.

This algorithm is really only correct for two-dimensional Voronoi polyhedra. In three dimensions, the full approach would consider point-to-face distances, the face being a convex combination of triplets of vertices from one region. Then if a binding mode proposes that atom p lies in r_i while atom q lies in r_j, then a necessary (but not sufficient) condition is that

$$u_{mpq} \geq l_{rij} \text{ and } l_{mpq} \leq u_{rij} \tag{8.39}$$

for all atom pairs, where u_m and l_m are the usual interatomic upper and lower distance bounds.

Then developing a binding site model using regions instead of points follows very much the same outline set out at the beginning of §8.5: (i) find atomic coordinates for the molecules; (ii) determine all u_{mpq} and l_{mpq} for all atom pairs in all molecules; (iii) remove unnecessary atoms; (iv) choose the number of regions, n_s, and the coordinates of the generating points, $\mathbf{c}_i$, $i=1,...,n_s$; (v) calculate the interregion distance bounds by Algorithm 8.13; (vi) determine the geometrically allowed binding modes of each molecule (Equation 8.39); and (vii) determine the interaction parameters. Note that step iv involves subjective decisions and trial-and-error because we have not devised a deterministic algorithm for finding the required binding site geometry. However, we are no longer forced to explicitly propose the binding modes, due to a qualitative improvement in step vii, which we will now describe.

In the new determination of energy parameters, we seek $\mathbf{e}$ satisfying Equation (8.10) for all molecules by minimizing an error function $F(\mathbf{e})$ defined by

$$F(\mathbf{e}) = max\begin{Bmatrix} 0 \\ \max\limits_{m} \max\limits_{\boldsymbol{\beta}\in\mathbf{B}_m}(\Delta G_{obsd,m-} - \boldsymbol{\beta}^T\mathbf{e}) \\ \max\limits_{m} \min\limits_{\boldsymbol{\beta}\in\mathbf{B}_m}(\boldsymbol{\beta}^T\mathbf{e} - \Delta G_{obsd,m+}) \end{Bmatrix} \tag{8.40}$$

The first alternative for F applies if $\Delta G_{obsd,m-} \leq \Delta G_{calc,m} \leq \Delta G_{obsd,m+}$ for all m. In other words, $F(\mathbf{e}^*)=0$ at a solution $\mathbf{e}^*$. In the second alternative, the binding for at least one mode of at least one molecule is better (algebraically lower) than the observed limit, $\Delta G_{obsd,m-}$, and $F>0$ according to the best mode of the most offending molecule. In the third alternative, the binding of all modes for at least one molecule exceeds $\Delta G_{obsd,m+}$, and the value of $F>0$ shows how much the worst molecule would have to be corrected to bring at least one mode into the required range. If that can be achieved, then it doesn't matter that other modes will have energies higher than $\Delta G_{obsd,m+}$, since the molecule prefers the lowest mode. By construction, the global minimum of F is zero, but the corresponding $\mathbf{e}^*$ may not be unique. It is also possible for $F(\mathbf{e})>0 \ \forall \ \mathbf{e}$, given an unfortunate set of modes and ΔG_{obsd} values. F is continuous and piecewise linear in $\mathbf{e}$, but it may have local minima where $F>0$, while having a solution elsewhere. Given these properties, a generally successful algorithm for locating at least one solution is the following:

Algorithm 8.14. Inequality energy refinement:

INPUT: For each molecule, its observed range of binding energies and its set of binding modes.

OUTPUT: A table of interaction energies such that for each molecule the best calculated binding energy falls within its observed range.

PROCEDURE:

For each m, remove duplicate $\boldsymbol{\beta}'$, where
$\boldsymbol{\beta},\boldsymbol{\beta}' \in \mathbf{B}_m$ and $\boldsymbol{\beta} = \boldsymbol{\beta}'$.

For all pairs of molecules m and m'

For all respective pairs of modes, $\boldsymbol{\beta} \in \mathbf{B}_m$ and $\boldsymbol{\beta}' \in \mathbf{B}_{m'}$

If $\boldsymbol{\beta} = \boldsymbol{\beta}'$, then remove $\boldsymbol{\beta}'$ if
$\Delta G_{obsd,m'+} < \Delta G_{obsd,m-}$ because
then $\boldsymbol{\beta}'$ can never be the optimal mode for m'
without violating the constraints for m.

If $\Delta G_{obsd,m'-} < \Delta G_{obsd,m-}$ and $\Delta G_{obsd,m'+} \geq \Delta G_{obsd,m-}$,

Then $\boldsymbol{\beta}^T\mathbf{e} > \Delta G_{obsd,m'-}$ is a redundant constraint,
but $\boldsymbol{\beta}'$ might be the optimal mode for m'.

Take a random starting $\mathbf{e}^{(0)}$ with each component in the range
$[-R,+R]$, where $R = \max_m \Delta G_{obsd,m+}$.

Start from $\mathbf{e}^{(0)}$ and try to progress to a solution by
minimizing F in subgradient optimization iterations [Sandi1979].

$$\mathbf{e}^{(k+1)} = \mathbf{e}^{(k)} + t\nabla F(\mathbf{e}^{(k)}) \tag{8.41}$$

$$t = \lambda \frac{(F(\mathbf{e}^{(k)}) - F(\mathbf{e}^*))}{\|\nabla F(\mathbf{e}^{(k)})\|^2}$$

for the iteration counter k=0,...,2000, say, or until
$F(\mathbf{e}^{(k)}) < 10^{-5}$.
In Equation (8.41), $F(\mathbf{e}^*)=0$, and $\lambda = 0.5$ is a good choice.
In evaluating F, one needs to examine only the nonredundant
lower bounds.

If the iterations converged to a solution, halt.
Otherwise, try another random start for at most some given number of tries.

If the algorithm succeeds, there is clearly at least one solution, and possibly others besides. Failure after many random starts suggests — but does not prove — there is no solution. The only definitive proof of no solution is when all modes $\boldsymbol{\beta}'$ for one molecule are eliminated according to the inconsistency check above.

There are four main advantages to using Voronoi polyhedra models over the site point models of §8.5. First, as described at the outset of this section, regions may be a much more economical way to describe the site geometry. Often the data can be fit with an amazingly small number of regions, and in such cases it is entirely proper to refrain from overinterpreting the experimental results by leaving the picture of the site in terms of a few broad regions. Second, the predicted binding modes are left also quite indefinite, requiring for instance only that one end of the molecule lies in a large region without unwarranted specification of its conformation. The third point is that fitting ΔG_{obsd+} and ΔG_{obsd-} in an absolute inequality sense rather than as least-squares, amounts to an unambiguous test of whether the model fits the data. If not, the geometry of the site must be altered. A successful fit leaves no outliers, and success is not judged by some statistical measure like correlation coefficient. After all, the only random variables in the binding problem are variations in the experimental conditions that are not readily controlled. These manifest themselves in the quantity $\Delta G_{obsd+} - \Delta G_{obsd-}$ as supplied by the experimentalist; beyond that, there is no statistical component to the problem. If 99 compounds are well fit by a simple model, but the 100th compound would require substantial modifications, then so be it. That last compound should be regarded as a particularly informative one, rather than an aberrant result of an unusually large fluctuation of some random variable which should be overruled in the least-squares sense by the majority of the observations. The fourth point is that the determination of the interaction parameters according to Algorithm 8.14 does not depend on initial hypothesized binding modes. This represents a substantial step toward the fully objective deduction of the simplest binding site model consistent with the given data.

8.8. Review of Binding Study Applications

At this writing, binding site modelling in terms of Voronoi polyhedra is so new it has not yet been applied to a problem of practical interest, mostly waiting on efficient computer implementation. On the other hand, site models in terms of site points (§8.5) have been devised for a number of sets of genuine data. The seven main papers concerning such applications are somewhat confusing because the methodology was being developed simultaneously. Having explained the technique in the preceding sections, we can now give a relatively coherent summary of results. Those interested in more detail should consult the references cited; the purpose here is to give some indication of the scope of problems that can be treated and how they can be approached.

The first case considered was that of eight phenoxyacetone derivatives inhibiting the enzyme α-chymotrypsin [Crippen1979, Baker1967].

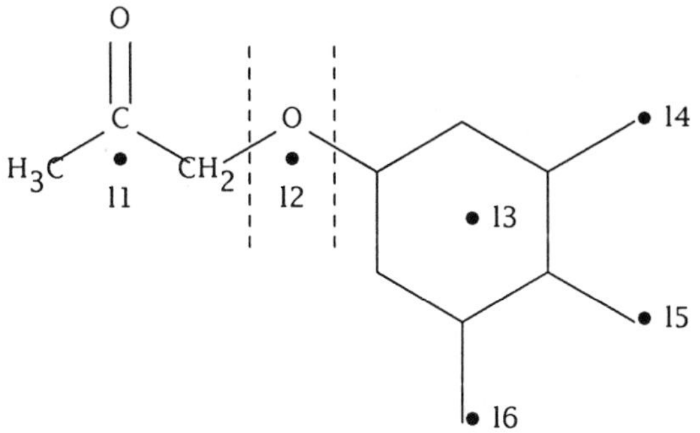

Table 8.2: Inhibitors of α-chymotrypsin.

ligand	12	14	15	16	ΔG_{obsd} (kcal)	ΔG_{calc} (kcal)
1	-	H-	H-	H-	−2.5	−2.5
2	-O-	H-	H-	H-	−2.8	−2.8
3	-O-	H-	CH_3-	H-	−2.8	−2.8
4	-O-	H-	Cl-	H-	−3.5	−3.7
5	-O-	Cl-	H-	H-	−4.2	−4.6
6	-O-	CH_3-	H-	H-	−2.8	−2.5
7	-O-	H-	CH_3O-	H-	−2.6	−2.5
8	-O-	CH_3-	H-	CH_3-	−0.2	−0.2

In this simple data set, the acetone and phenyl groups appeared in all ligands and were represented by two points, placed as shown in the above illustration. The other types of groups are -O-, $-CH_3$, and -Cl. Explaining the binding of just **1**-**7** is easy, but apparently the second m–CH_3 in **8** is sterically disallowed. Since there are three important rotatable bonds in most of these compounds, avoiding contact with one sterically repulsive site point is easy, and a whole unavoidable ring of them must be proposed. The result is shown in Fig. 8.20 with **8** bound in a mode that is strongly unfavorable, while **6** would have no trouble.

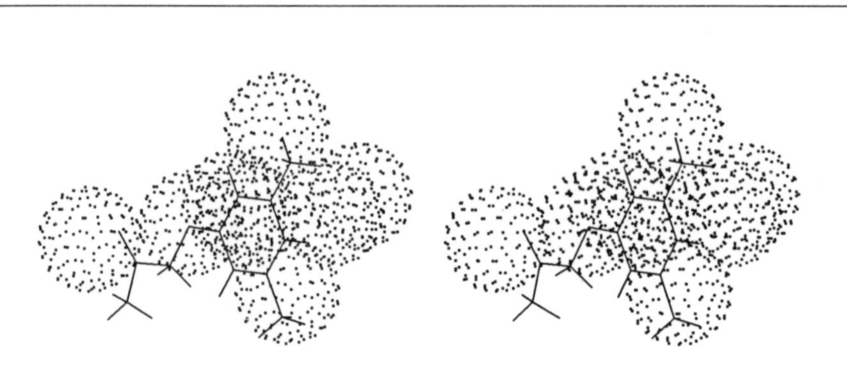

Figure 8.20

Stereo view of the site point model for phenoxyacetone inhibitors of α-chymotrypsin. Dotted spheres with radius $\delta d = 1$ Å represent the site points. Compound **8** is shown with interactions that are typical of the preferred modes of the other ligands. From left to right, the carbonyl group contacts s1, the ether oxygen contacts s2, and the benzene ring roughly centers on s3. Site points 4-7 form a ring with s4 above, s5 below, s6 (enlarged due to perspective) in front, and s7 behind. This compound can bring one m–CH_3 to interact favorably with s4 but the other interacts unfavorably with s5.

The energies of interaction that fit the observations are shown in Table 8.3. The large positive entries are not precisely determined by the least squares fit, but rather are arbitrarily set to a prohibitively large value.

Table 8.3: Interaction energies for the α-chymotrypsin site model (kcal/mol).

	s1	s2	s3	s4	s5-s8
CH_3COCH_2-	-0.01	0.1	10.0	1.0	1.0
-O-	1.0	-0.2	10.0	1.0	0.05
-C_6H_5	1.0	1.0	-2.6	1.0	1.0
-CH_3	1.0	1.0	-0.1	0.15	10.0
-Cl	1.0	1.0	10.0	-1.05	1.0

Although there have been numerous X-ray crystal structures of α-chymotrypsin complexed with various ligands, none is available with a phenoxyacetone, so one can only say that the model is geometrically plausible, but the agreement is not compelling.

There is, however, another case where the binding data are of extraordinarily high quality, and there is a comparable X-ray crystal structure, namely the binding of thyroxine analogues to human prealbumin [Crippen1981, Andrea1980, Blake1977]. These molecules tended to be rather large and conformationally flexible, as well as exhibiting stereospecific binding. Hence the initial determination of interatomic distance bounds was carried out with all atoms in order to correctly exclude sterically forbidden conformers, but for subsequent steps in this study, only those atoms were kept which were important in determining the binding, as in Fig. 8.21.

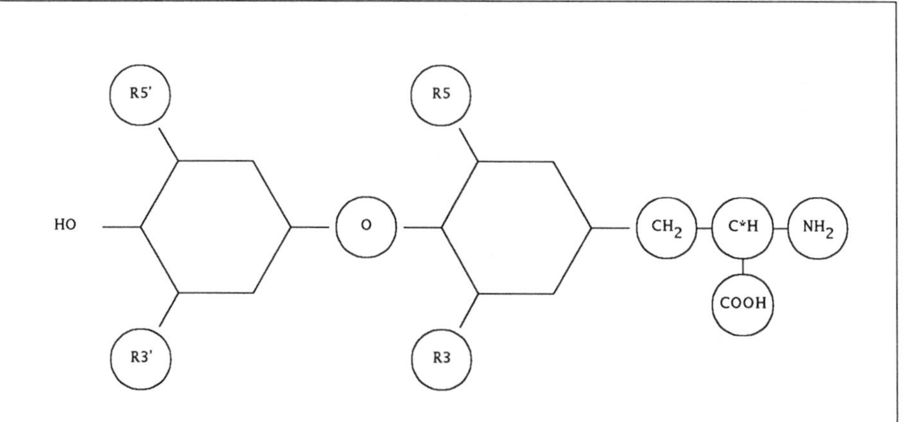

Figure 8.21

A typical thyroxine analogue structure. Circled atoms are significant in determining binding.

The asymmetric carbon required checking the chirality of the molecule against that of the site, in addition to the usual distance test (Equation 8.29). For any quartet of atoms i, j, k, and l, such that all interatomic distances are relatively invariant (e.g. $u_{ij} - l_{ij} < 0.1$ Å, etc.), consider the signed tetrahedral volume

$$T = \mathbf{v}_{ij} \cdot (\mathbf{v}_{ik} \times \mathbf{v}_{il}) \tag{8.42}$$

where $\mathbf{v}_{ij}$ is the normalized vector from atom i to atom j, etc., as calculated from the atomic coordinates of any arbitrary conformation. For $|T| > 0.15$, the quartet is chiral, and the chirality check consists of insisting that the sign of T must match that for the corresponding four site points with which atoms i, j, k, and l are in contact. This incidentally represents a useful generalization of asymmetric centers that includes the customary quartets, such as the C*, the methylene C, the amino N, and the carboxyl C shown in Fig. 8.21. In order to distinguish between L- and D- isomers of thyroxine, five site points were used: attachments for C*, CH_2, NH_2 in all isomers, and two alternate positions for the COOH. One could get by with having a site point for the L-isomer COOH and none in the D-case, for example. In addition, inspection of the data made it clear that the nature of the 3-, 5-, 3'-, and 5'-substituents is important, requiring another four site points. Some of the lengthy 3'-substituents had a significant influence on binding, so two more site points were added to accommodate them, making a grand total of 11 site points. The resulting site is shown in Fig. 8.22 superimposed upon the appropriate atoms of tetraiodothyronine in the crystal structure of its complex with prealbumin. The conformational flexibility and chemical similarity of the analogues keep the site point coordinates from being very precisely determined, but the rms deviation of intersite point distances from corresponding interatomic distances in the crystal structure can be as little as 1.0 Å. The model shows correct stereospecificity and an rms deviation between ΔG_{obsd} and ΔG_{calcd} of 0.5 kcal for the 27 analogues.

The advantage of the prealbumin-thyroxine study is that the data are of very high quality, and they provide a good test of building a stereospecific binding site model. If one is willing to go to less reliable data on a system where the receptor's purity is even under dispute, there is an excellent opportunity to test the method's ability to deal with chemically diverse drugs. Namely, in a paper by Crippen [Crippen1982] the assumption is that the *in vitro* assay used at the time primarily by Braestrup and Squires [Braestrup1978] measures ΔG_{obsd} for binding to the (unique) benzodiazepine receptor. There were 29 suitable compounds with published binding measured in this way: 13 benzodiazepines, 3 triazolobenzodiazepines, 9 carbolines, 1 pyrazolodiazepine, 1 benzoxazolone, and 1 phenothiazine. This is a case where chemical structural analysis by Algorithm 8.2 turned out to be extremely helpful in suggesting sets of equivalent atoms among these chemically diverse compounds. Computer time constraints dictated that each molecule be represented by about one half of its non-hydrogen atoms, but even so the algorithm was able to find a set of 5 atoms as a "base group" for the active compounds, given a 1.5 Å distance tolerance. No such commonality had been recognized by inspection. As it turned out, the binding modes proposed from this base group had to be modified considerably as steps iv-vii were repeated in the process of building the site, but it was at least a valuable starting point. One of the other novel features of this study was the elaborate steric walls required to force the inactive compounds to bind weakly while not interfering with the active compounds. In order to keep the number of site points small, points were given individual radii of interaction, not just the

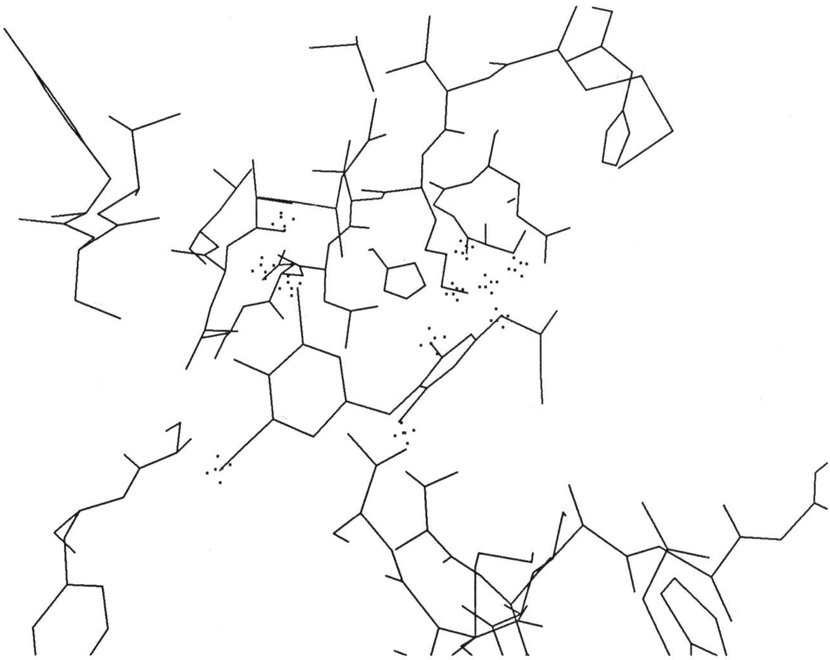

Figure 8.22

A rigid translation and rotation of the thyroxine binding model onto corresponding atoms in the prealbumin and 3,5,3',5'-tetraiodo-L-thyronine complex. Site points are shown as small clusters of dots with radius 0.36 Å=δd. Only the immediately neighboring amino acid residues of the protein are shown for the sake of clarity. Tetraiodothyronine runs horizontally across the middle of the picture roughly as shown in Fig. 8.21. The site points were positioned upon the crystal structure by a remarkably precise least squares superposition of the four site points intended to bind the corresponding four iodine atoms. With this thyroxine derivative, there are no atoms to contact the more distant site points intended for larger 3'-substituents, seen at the upper left of the ligand. At the right, the site point intended for the methylene C is well positioned, but those for the C*, CH_2, and NH_2 groups do not coincide well with the conformation chosen by the crystallographers.

common δd value. For instance, in Fig. 8.23, the attractive site points s1-s9 make favorable contributions to the binding energy with atoms coming within 0.35 Å = δd, but the other site points repel any atoms within their large indicated radii. In particular, the benzodiazepine diazepam can bind favorably as shown, but

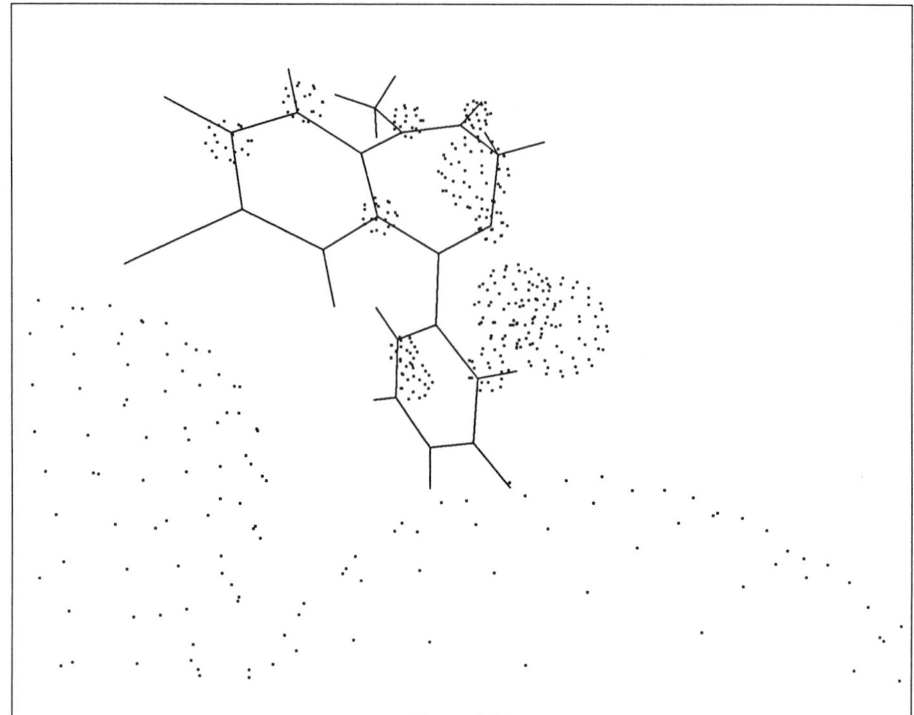

Figure 8.23
Predicted binding mode for diazepam in the benzodiazepine receptor. Site points are shown as dotted spheres with radii corresponding to their radii of influence. Site points 1-9 are all small spheres, and the large repulsive s11 is partly seen at the bottom of the picture.

virtually any *para*-substituent on the phenyl ring will cause disastrous steric overlap with s11. In that case, the next best binding mode is radically different and appropriately inferior in energy. As an example of a non-benzodiazepine binding to the same site, Fig. 8.24 shows the predicted binding mode for β-carboline-3-carboxylic acid ethyl ester. Site points s10 and s14 are arranged so that methyl substituents at the 1-position will produce strong steric overlaps, since the 1-methyl derivative is known to bind much worse. Note that the program suggests making the favorable contact between the ethyl group and s4, but this requires the nearly *cis* configuration of the ester linkage -CO-OCH$_2$-, which is energetically unfavorable due to the intrinsic torsional potential. This is simply a shortcoming of the initial survey of the molecule's conformation space, which only checked for van der Waals contacts. The energetic fit had an rms deviation

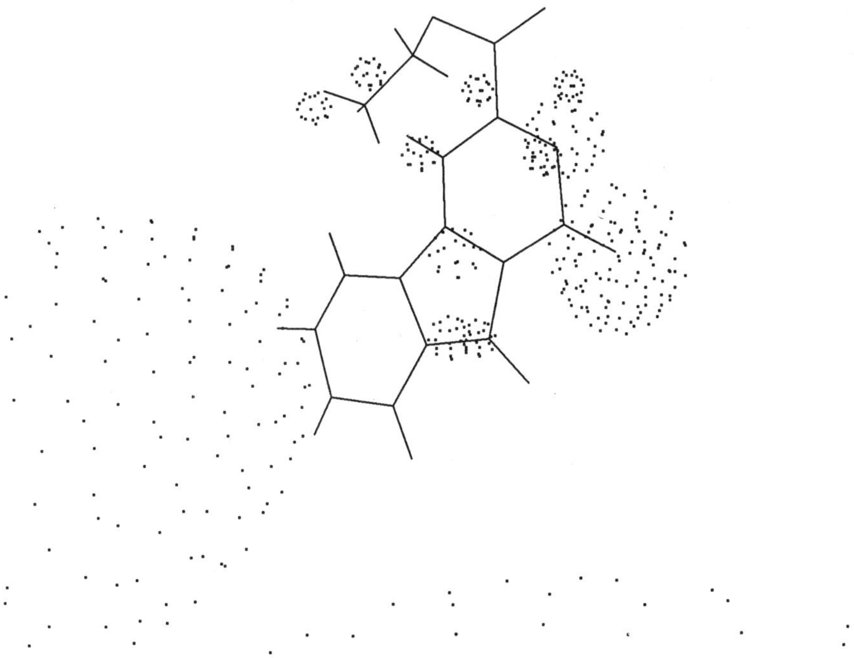

Figure 8.24
Predicted binding mode for β-carboline-3-carboxylic acid ethyl ester in the benzodiazepine receptor model. The site has the same orientation as in Fig. 8.23. Site points s10 and s14 are the dense dots on the far right; s4 is the small sphere on the upper left. The 1-methyl derivative is actually shown in order to illustrate the steric hindrance with s10 and s14.

of only 1.1 kcal/mol over the 29 compounds, due largely to the careful placement of the several repulsive site points. Fitting only the strongly bound compounds would have required many fewer site points, particularly repulsive ones.

The binding systems for which there are the most plentiful published data are the inhibition of various dihydrofolate reductases (DHFR). Aside from the considerable medicinal importance of such drugs, there is the opportunity to test the method on large data sets consisting of several chemical classes, and sometimes to compare against enzyme-inhibitor complex crystal structures. It is important to keep in mind that DHFRs from different organisms may show quite different binding preferences. Most of our work has been on *S. faecium* DHFR, for which there is a trustworthy data set of 68 quinazoline inhibitors [Hynes1974], but no crystal structure. Since these compounds are all closely homologous, one would expect them to bind in analogous modes, with differences in binding affinity caused by simple substituent effects. By and large, a Free-Wilson or Hansch analysis can rationalize the differences, but there are some striking exceptions.

Table 8.4: Binding of selected quinazoline derivatives to *S. faecium* dihydrofolate reductase [Hynes1974].

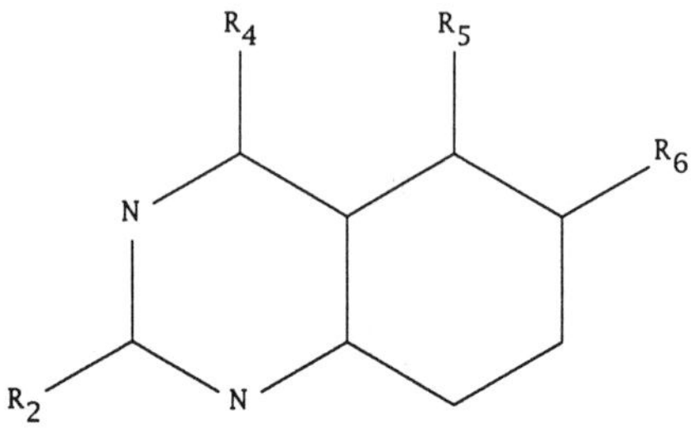

ligand	R_2	R_4	R_5	R_6	ΔG_{obsd} (kcal)
1	NH_2	OH	S-2-$C_{10}H_7$	H	−7.4
2	NH_2	NH_2	S-2-$C_{10}H_7$	H	−10.9
3	NH_2	NH_2	SO_2-2-$C_{10}H_7$	H	−6.5
4	NH_2	NH_2	H	S-2-$C_{10}H_7$	−12.1
5	NH_2	NH_2	H	SO_2-2-$C_{10}H_7$	−12.4

Ligands **1** and **2** are typical of the rule that 2,4-diamino derivatives bind better, apparently due to a shift in the pK_a of N1 and hence its protonation. However, this binding enhancement can be entirely negated by going from an -S- to an -SO_2- linkage, as in ligand **3**. The sensitivity as to linking group seems curiously confined to the 5-position, since it makes no difference for compounds **4** and **5**. There are two ways to rationalize all this. One way is to say that all the analogues bind in similar modes, but the increased electron withdrawing effect of -SO_2- over -S-, acting through the conjugated ring system, causes the ring nitrogens of **3** to be unprotonated. Resonance theory indicates this occurs from the 5-position, but not from the 6-position [Ghose1983a]. The alternative explanation, which is equally plausible without further experimental evidence, is that protonation at N1 leads to a preferred binding mode having no steric constraint near the 6-position but unable to accommodate a 5-SO_2- linkage. Such derivatives as **3** would be forced over into a less favorable ring orientation, perhaps such as that preferred by unprotonated rings (**1**). The number of site points can be kept low by further supposing the postulated ring flipping to take place about an axis runing through the 2- and 6-positions, so that 2- and 6-substituents bind to the same site points in either case. Note that both the electronic and steric explanations require having a separate atom type for the protonated N1. The steric and ring rotation explanation was first suggested by Crippen [Crippen1979], then applied to all 68 quinazolines [Crippen1980]. Once the improved energy refinement Algorithm 8.9 was available, we could identify unnecessary site points by their very weak interactions with all molecule points [Ghose1982, Ghose1983a]. About the simplest site (six site points) with a good fit to the data (ρ= correlation coefficient = 0.92, σ= standard deviation = 0.9 kcal/mol) is that shown in Figs. 8.25 and 8.26, showing how protonated and unprotonated derivatives are predicted to bind. (See [Ghose1983a] designated "study VII," with numbering of site points as in [Ghose1982].)

Being able to fit so many observations to only six site points (none of which were purely repulsive), is a tribute to the redundancy of the data. Dealing with a more diverse set of DHFR inhibitors (originally in reference [Ghose1983b] and subsequently generalized in reference [Ghose1984]) required 11 attractive site points and five repulsive ones. It also required shifting to the DHFR variety most extensively studied, namely that from rat liver, so the deduced site is expected to differ from the previously discussed studies. Instead of using the two-modes explanation, this work employed the uniform mode hypothesis discussed above in order to economize on computer time. The net result is a very good fit (ρ=0.95, σ=0.5) for 62 molecules of five chemical classes: triazines, quinazolines, pyrimidines, pyrroloquinazolines, and pyridopyrimidines. Figs. 8.27 and 8.28 illustrate some of the features of the model. Furthermore, it successfully predicts (ρ=0.84, σ=0.7) the binding affinities of 33 additional compounds: triazines, quinazolines, a pyrroloquinazoline, and a pteridine.

Figure 8.25

An example of a tight binding 2,4-diaminoquinazoline (**4**) in its predicted mode for the *S. faecium* DHFR model. Only six site points of those original 11 are necessary. Note that the protonated ring nitrogen in the 1-position is in contact with s2, located at the bottom of the quinazoline ring in this view.

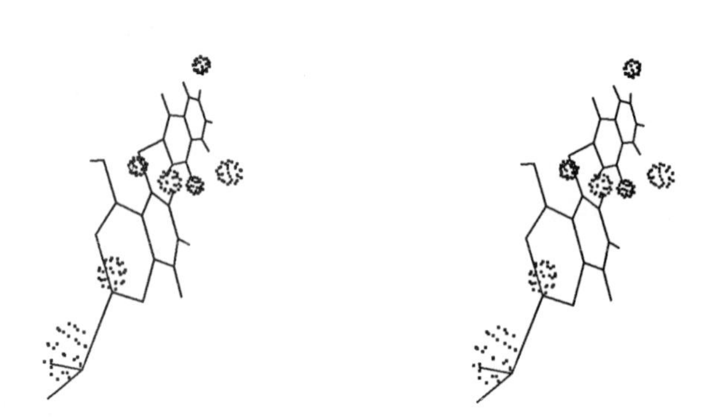

Figure 8.26

An example for the same DHFR model of the predicted binding mode for a 2-amino-4-hydroxy-quinazoline. Since the ring nitrogens are unprotonated, and the 5-substituent is not too bulky, N1 does not bind to s2 and the quinazoline ring system rocks so as to bring the 5-S in contact with s7.

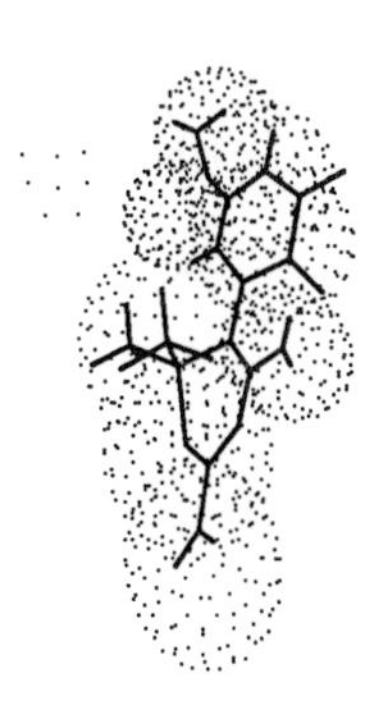
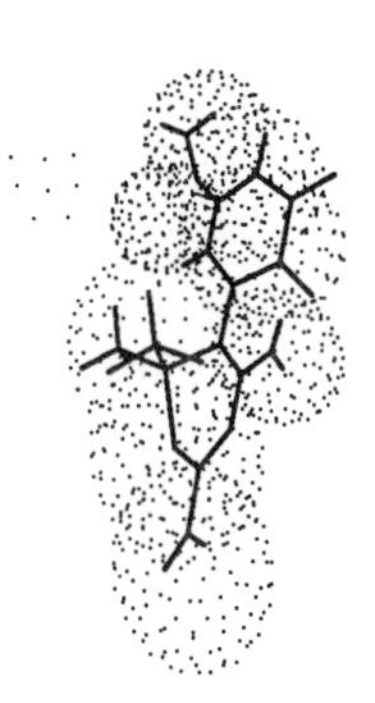

Figure 8.27

3'-Methoxytriazine is known to bind well to rat liver DHFR, explained largely by the favorable contacts with s8 and s9 in this model. Site points 12-16 are repulsive, but are not important in determining the preferred mode of this compound.

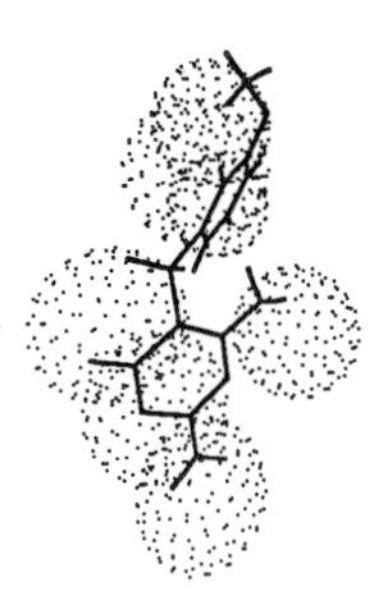
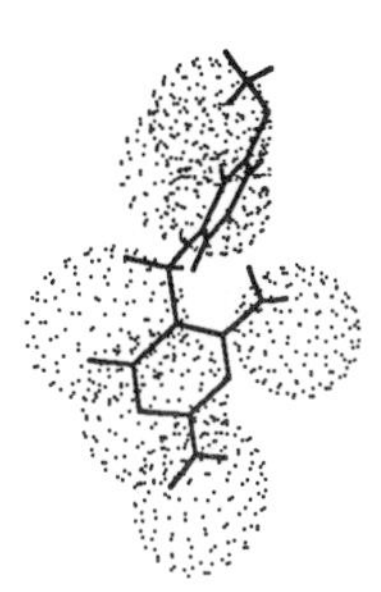

Figure 8.28

The predicted binding mode of 2,4-diamino-3'-methoxy-5-benzylpyrimidine at the same site as fig. 8.27, in approximately the same orientation. Steric repulsion from s12-14 prevents the favorable contacts enjoyed by triazines at s8, and s9. Note the hetero ring still experiences the same interactions with s1-3.

References

Andrea1980.

T.A. Andrea, R.R. Cavalieri, I.D. Goldfine, and E.C. Jorgensen, "Binding of Thyroid Hormones and Analogues to the Human Plasma Protein Prealbumin," *Biochemistry, 19,* 55-63(1980).

Ash1965.

R.B. Ash, *Information Theory,* Interscience Publishers, New York, 1965.

Baker1967.

B.R. Baker and J.A. Hurlbut, "Irreversible Enzyme Inhibitors. CVII. Proteolytic Enzymes. II. Bulk Tolerance within Chymotrypsin-Inhibitor Complexes," *J. Med. Chem., 10,* 1129-1133(1967).

Blake1977.

C.C.F. Blake and S.J. Oatley, *Nature, 268,* 115(1977).

Blaney1985.

J.M. Blaney, *private communication*(1985).

Braestrup1978.

C. Braestrup and R.F. Squires, "Pharmacological Characterization of Benzodiazepine Receptors in the Brain," *Eur. J. Pharmacol., 48,* 263-70(1978).

Bron1973.

C. Bron and J. Kerbosch, *Comm. ACM, 16,* 575(1973).

Cohen1985.

N.C. Cohen, "Drug Design in Three Dimension," *Adv. Drug Res., 14,* 41-145(1985).

Crippen1975.

G.M. Crippen, "Global Optimization and Polypeptide Conformation," *J. Comp. Phys., 18,* 224-231(1975).

Crippen1979.

G.M. Crippen, "Distance Geometry Approach to Rationalizing Binding Data," *J. Med. Chem., 22,* 988-997(1979).

Crippen1980.

G.M. Crippen, "QSAR by Distance Geometry: Systematic Analysis of Dihydrofolate Reductase Inhibitors," *J. Med. Chem., 23,* 599-606(1980).

Crippen1981.

G.M. Crippen, "QSAR by Distance Geometry: the Thyroxine Binding Site," *J. Med. Chem., 24,* 198-203(1981).

Crippen1982.

G.M. Crippen, "Distance Geometry Analysis of the Benzodiazepine Binding

Site," *Mol. Pharmacol.*, *22*, 11-19(1982).

Crippen1983.
G.M. Crippen, "Prediction of New Leads from a Distance Geometry Binding Site Model," *Quant. Struct.-Act. Relat.*, *2*, 95-100(1983).

Crippen1984.
G.M. Crippen, "Deduction of Binding Site Structure from Ligand Binding Data," *Ann. New York Acad. Sci.*, *439*, 1-11(1984).

Dijkstra1959.
E.W. Dijkstra, *Numer. Math.*, *1*, 269(1959).

Garey1979.
M.R. Garey and D.S. Johnson, *Computers and Intractability: A Guide to the Theory of NP-Completeness*, Freeman, San Francisco, 1979.

Ghose1982.
A.K. Ghose and G.M. Crippen, "Quantitative Structure Activity Relationship by Distance Geometry: Quinazolines as Dihydrofolate Reductase Inhibitors," *J. Med. Chem.*, *25*, 892-899(1982).

Ghose1983a.
A.K. Ghose and G.M. Crippen, "Essential Features of the Distance Geometry Model for the Inhibition of Dihydrofolate Reductase by Quinazolines," in *Proceedings of the 4th European Symposium on Chemical Structure - Biological Activity: Quantitative Approaches*, ed. J.C. Dearden, Quantitative Approaches to Drug Design, pp. 99-108, Elsevier, Amsterdam, 1983.

Ghose1983b.
A.K. Ghose and G.M. Crippen, "Combined Distance Geometry Analysis of Dihydrofolate Reductase Inhibition by Quinazolines and Triazines," *J. Med. Chem.*, *26*, 996-1010(1983).

Ghose1984.
A.K. Ghose and G.M. Crippen, "General Distance Geometry Three-Dimensional Receptor Model for Diverse Dihydrofolate Reductase Inhibitors," *J. Med. Chem.*, *27*, 901-914(1984).

Ghose1985.
A.K. Ghose and G.M. Crippen, "Use of Physicochemical Parameters in Distance Geometry and Related Three-Dimensional Quantitative Structure-Activity Relationships: A Demonstration using Escherichia coli Dihydrofolate Reductase Inhibitors," *J. Med. Chem.*, *28*, 333-346(1985).

Ghose1986.
A.K. Ghose and G.M. Crippen, "Atomic Physicochemical Parameters for Three-Dimensional Structure-Directed Quantitative Structure-Activity

Relationships I. Partition Coefficients as a Measure of Hydrophobicity," *J. Comp. Chem.*, *7*, 565-577(1986).

Ghose1987.

A.K. Ghose and G.M. Crippen, "Atomic Physicochemical Parameters for Three-Dimensional-Structure-Directed Quantitative Structure-Activity Relationships. 2. Modeling Dispersive and Hydrophobic Interactions," *J. Chem. Inf. Comp. Sci.*, *27*, 21-35(1987).

Hopfinger1985.

A.J. Hopfinger, "Computer Assisted Drug Design," *J. Med. Chem.*, *28*, 1133-1139(1985).

Humber1979.

L.G. Humber, F.T. Bruderlein, A.H. Philipp, M. Goetz, and K. Voith, "Mapping the Dopamine Receptor. 1. Features Derived from Modifications in Ring E of the Neuroleptic Butaclamol," *J. Med. Chem.*, *22*, 761-767(1979).

Hynes1974.

J.B. Hynes, W.T. Ashton, D. Bryansmith, and J.H. Friesheim, "Quinazolines as Inhibitors of Dihydrofolate Reductase. 2," *J. Med. Chem.*, *17*, 1023-1025(1974).

Kuester1973.

J.L. Kuester and J.H. Mize, *Optimization Techniques with Fortran*, McGraw-Hill, New York, 1973.

Kuhl1984.

F.S. Kuhl, G.M. Crippen, and D.K. Friesen, "A Combinatorial Algorithm for Calculating Ligand Binding," *J. Comp. Chem.*, *5*, 24-34(1984).

Martin1978.

Y.C. Martin, *Quantitative Drug Design*, Medicinal Research Series, 8, Macel Dekker, Inc., New York, 1978.

McLachlan1982.

A.D. McLachlan, *Acta Cryst.*, *A38*, 871(1982).

Ramsden1988.

C.A. Ramsden, *Comprehensive Medicinal Chemistry*, Pergamon Press, Oxford, England, 1988.

Ravindran1972.

A. Ravindran, "A Computer Routine for Quadratic and Linear Programming Problems," *Comm. ACM*, *15*, 818-820(1972).

Sandi1979.

C. Sandi, "Subgradient Optimization," in *Combinatorial Optimization*, ed. C. Sandi, pp. 73-91, Wiley, New York, 1979.

Sheridan1986.

R.P. Sheridan, R. Nilakantan, J. Scott Dixon, and R. Venkataraghavan, "The Ensemble Approach to Distance Geometry: Application to the Nicotinic Pharmacophore," *J. Med. Chem.*, *29*, 899-906(1986).

Voronoi1908.

G.F. Voronoi, "Nouvelles Applications des Paramètres Continus à la Théorie des Formes Quadratiques. Deuxième Mémoire. Recherches sur les Parallélloèdres Primitifs," *Reine Angew. Math.*, *134*, 198(1908).

9. Energy Minimization

So far, we have concentrated on finding conformations of a molecule which satisfy a list of *geometric* constraints. This is the appropriate response to the question, "Do these experimental data completely determine the molecule's conformation, and if not, what features remain undetermined?" Often, however, the question is a more heuristic search for energetically favorable conformations which also satisfy the necessary but insufficient constraints. These would be the physically most significant conformers because at equilibrium, the probability $P(\mathbf{x})$ of the (isolated) molecule being in state $\mathbf{x}$ is given by the Boltzmann distribution law,

$$P(\mathbf{x}) = \frac{e^{-[E(\mathbf{x})-E_o]/kT}}{\sum_{\mathbf{y}} e^{-[E(\mathbf{y})-E_o]/kT}} \tag{9.1}$$

where $E(\mathbf{x})$ is the internal energy of the molecule, E_o is the global minimum over all conformations, k = Boltzmann's constant, and T = the absolute temperature. As $T \rightarrow 0$, only the conformations with $E(\mathbf{x}) \approx E_o$ are of statistical importance, and one must locate all conformers having this global optimum energy value, but at ordinary temperatures, a state with $E(\mathbf{x}) = E_o + 3kT$ has a 5% probability, so the task is to locate *all* the relatively favorable conformations, $\Omega(\delta)$:

$$\Omega(\delta) = \{\mathbf{x} \mid E(\mathbf{x}) \leq E_o + \delta\} \tag{9.2}$$

where δ is on the order of kT. The difficulty is that for all but the simplest molecules, $\Omega(\delta)$ is not at all connected in conformation space (unless δ is absurdly large). Regardless of how conformations are parameterized (dihedral angles, Cartesian coordinates, generalized internal coordinates, etc.) and regardless of how the energy is calculated (quantum mechanical methods or empirical force fields), $E(\mathbf{x})$ has numerous local minima. For instance if we assume the molecule has fixed bond lengths and vicinal bond angles, a reasonable assumption for conformational studies on molecules without small flexible rings, then the most compact parameterization of conformation space is in terms of the n dihedral angles $\mathbf{x} := [x_1, \ldots, x_n]$ for the n rotatable bonds. The space is really an n-dimensional toroidal manifold because $x_i = 0$ is the same as $x_i = 2\pi$. There is generally a 3-fold rotation barrier around each such bond, so there are three local minima when $n-1$ torsional angles are held fixed. This would make 3^n minima regularly spaced in the conformation space, but since nearly every atom pair can interact strongly in some conformations but weakly in most others, the energy viewed as a function of one dihedral angle is extremely dependent on the values of the other $n-1$ variables. Experience indicates that 3^n is an underestimate, if anything, and the theoretical bounds from differential topology on the number of minima are too weak to be of any help [Csizmadia1983, Liotard1983, Mezey1981]. Denoting a local minimum as $\mathbf{x}^*$, let the "watershed" of $\mathbf{x}^*$ be $\{\mathbf{x}|$ there is a steepest descents path with respect to E from $\mathbf{x}$ to $\mathbf{x}^*\}$. Clearly each watershed is contiguous, and we can think of dividing all conformation space into such regions. Experience indicates that the watersheds are small and so irregularly spaced that subdividing the search problem is not feasible. Any numerical method for local optimization requires a starting conformation, makes rather gradual changes in this conformation so as to reduce the energy, and fairly quickly converges on the local minimum in whose watershed the process began. For small molecules, it is feasible to either map out the whole energy surface or to begin local minimization from so many closely spaced points in conformation space that one has confidence all local minima have been located. Then the desired Ω consists of the neighborhoods of all minima $\mathbf{x}^*$ such that $E(\mathbf{x}^*) \leq E_o+\delta$. For larger molecules, such exhaustive searches are not feasible. Random sampling of conformation space shows that Ω is quite rare, and most conformations (and therefore apparently most minima) lie well above reasonable values of $E_o+\delta$ [Crippen1971b].

Even though straightforward approaches to conformational energy minimization have not been successful for large molecules, this chapter covers many methods that are either intriguing or have actually been useful in realistic applications. So far in this book we have assumed basic familiarity with nonlinear optimization, but since many of the methods presented in this chapter are primarily global/local constrained/unconstrained minimization algorithms we first lay out the basic mathematics of optimization in §9.1. Then §9.2 outlines a number of methods for avoiding local minima and conducting global searches for conformations of molecules in $\mathbb{R}^3$. Finding very good local minima can be done more easily by the counterintuitive trick of initially working in higher dimensional spaces and then returning to $\mathbb{R}^3$ (§9.3). As another radical alternative, one can even discretize both the constraints and the energy function, and treat the whole thing as a combinatorial problem (§9.4).

9.1. Mathematical Background

Before we can discuss the problems connected with finding low energy conformations of molecules, we need to briefly survey the pertinent mathematical definitions and theorems. The intent is not to repeat the results of a book on optimization, but rather to make clear what kinds of situations can arise simply due to the optimization of an arbitrary function, and what features are special to molecular applications. For a more thorough discussion and proofs of what follows, see [Luenberger1973].

Consider a real-valued *objective function* $f: \mathbb{R}^n \rightarrow \mathbb{R}$ defined over some *feasible set* $\Omega \subseteq \mathbb{R}^n$.

Definition 9.1. The ***unconstrained minimization problem*** is

$$\text{minimize } f(\mathbf{x}) \qquad \text{subject to } \mathbf{x} \in \Omega \tag{9.3}$$

Generally we will be dealing with *completely unconstrained* minimization problems, where $\Omega = \mathbb{R}^n$.

Definition 9.2. A *local minimum point* of f over Ω is some $\mathbf{x}^* \in \Omega$ such that $f(\mathbf{x}) \geq f(\mathbf{x}^*)$ $\forall \mathbf{x} \in \Omega \backslash \mathbf{x}^*$ having $\|\mathbf{x}-\mathbf{x}^*\| < \epsilon$, for some $\epsilon > 0$. When $f(\mathbf{x}) > f(\mathbf{x}^*)$, we say $\mathbf{x}^*$ is a *strict local minimum point.*

Definition 9.3. A *global minimum point* of f over Ω is some $\mathbf{x}^* \in \Omega$ such that $f(\mathbf{x}) \geq f(\mathbf{x}^*)$ $\forall \mathbf{x} \in \Omega \backslash \mathbf{x}^*$. When $f(\mathbf{x}) > f(\mathbf{x}^*)$, we say $\mathbf{x}^*$ is a *strict global minimum point.*

Denote the gradient of f by ∇f, and the Hessian of f by $\nabla^2 f$, i.e. the matrix element $(\nabla^2 f)_{ij} = \dfrac{\partial^2 f}{\partial x_i \partial x_j}$ $\forall i,j = 1, \ldots, n$.

Definition 9.4. The $n \times n$ matrix $\underline{\mathbf{M}}$ is *positive (semi)definite* if $\mathbf{y}^T \underline{\mathbf{M}} \mathbf{y} >$ (or $\geq$) 0 $\forall \mathbf{y} \in \mathbb{R}^n$.

For real symmetric $\underline{\mathbf{M}}$, this can be shown to be equivalent to having all eigenvalues $\lambda >$ (or $\geq$) 0.

PROPOSITION 9.5. *Suppose $f \in C^2(\mathbb{R}^n)$, i.e. f is twice continuously differentiable. Let $\mathbf{x}^*$ be an interior point of Ω. Then $\mathbf{x}^*$ is a strict local minimum of f if and only if $\nabla f(\mathbf{x}^*) = \mathbf{0}$ and $\nabla^2 f(\mathbf{x}^*)$ is positive definite.*

When the Hessian is only positive semidefinite, $\mathbf{x}^*$ would be only a local minimum. The case of $\mathbf{x}^*$ being a boundary point of Ω is covered later under constrained minimization.

PROPOSITION 9.6. *Suppose* $f \in C^2(\mathbb{R}^n)$. *Then* f *is convex (see §6.1 for definitions) over a convex set* Ω *containing an interior point if and only if* $\nabla^2 f(\mathbf{x})$ *is positive semidefinite for all* $\mathbf{x} \in \Omega$.

PROPOSITION 9.7. *If* f *is a convex function defined on the convex set* Ω, *then the subset where* f *reaches its minimum is convex, and any local minimum of* f *is a global minimum.*

In molecular applications, this proposition is useful only when Ω has been narrowly restricted to the region around one of the numerous local energy minima.

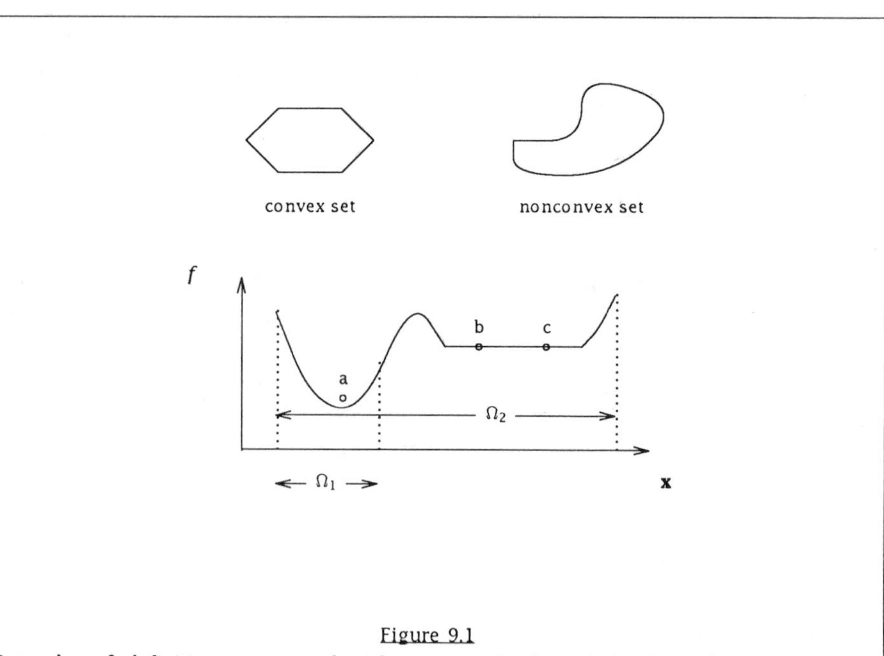

Figure 9.1

Examples of definitions connected with unconstrained minimization. f is not convex over Ω_2, but it is convex over the more restricted subset Ω_1. Then $\nabla^2 f$ is positive definite everywhere in Ω_1. The point a is the strict local and global minimum of f over Ω_1, and indeed it is the strict global minimum over all Ω_2. On the other hand, b and c are (not strict) local minima in a region where $\nabla^2 f$ is positive *semi*definite.

Thus we have outlined the main considerations in unconstrained minimization. There are a multitude of numerical algorithms for locating local minima, but the outlines of the more recommendable ones are relegated to Appendix F. For the moment, we need to consider only their most important property, namely how quickly they can locate the minimum. Algorithms sufficiently general to handle molecular energy minimization problems always need to be given an initial guess, $\mathbf{x}^{(0)}$, and then in an iterative process of some sort they produce a sequence $\mathbf{x}^{(i)}$, i=0,1,2,... which may or may not converge to a local minimum, $\mathbf{x}^*$. In retrospect, one can measure the degree of convergence in terms of some error function at each step by $e_i := \|\mathbf{x}^{(i)} - \mathbf{x}^*\|$ or $e_i := | f(\mathbf{x}^{(i)}) - f(\mathbf{x}^*) |$.

Definition 9.8. If the sequence $\{e_i\}_{i=0}^{\infty}$ converges to zero, then the *order* of convergence of the sequence is the supremum of the numbers $p \geq 0$ satisfying

$$0 \leq \lim_{i \to \infty} sup \frac{e_{i+1}}{(e_i)^p} < \infty \ . \tag{9.4}$$

Clearly the greater p is, the faster the convergence.

Definition 9.9. If the sequence $\{e_i\}$ converges to zero in such a way that

$$\lim_{i \to \infty} \frac{e_{i+1}}{e_i} = \alpha < 1 \tag{9.5}$$

then the sequence converges *linearly* with *convergence ratio* α. This corresponds to the case of $sup\ p = 1$ in the previous definition. When $sup\ p > 1$, we have *superlinear* convergence; in particular when $sup\ p = 2$, we speak of *quadratic* convergence.

The steepest descents algorithm is very "robust" in that it converges for almost any starting point with a wide variety of objective functions f, and the initial improvements of the error tend to be large, but eventually its convergence rate is linear. Newton's method for minimization may even diverge unless suitable precautions are made, as discussed in Appendix F, but when $\mathbf{x}^{(0)}$ is near enough to $\mathbf{x}^*$, its rate of convergence is quadratic. Roughly speaking, after sufficient iterations, the number of correct digits in $\mathbf{x}^{(i)}$ doubles with each iteration. There is no known minimization algorithm with a greater convergence rate.

In the applications that follow in this chapter, we often require $\mathbf{x}^*$ to be located very precisely, so that Newton's method is strongly recommended. The basic iteration is simply

$$\mathbf{x}^{(i+1)} = \mathbf{x}^{(i)} - \left[\nabla^2 f(\mathbf{x}^{(i)})\right]^{-1} \nabla f(\mathbf{x}^{(i)}) \ . \tag{9.6}$$

Unfortunately, if f is the internal energy of a molecule and $\mathbf{x}$ is the vector of all Cartesian coordinates of the atoms, then $\nabla^2 f$ is positive *semi*definite even in the vicinity of $\mathbf{x}^*$ because f is unchanged by overall translations and rotations $\in \mathbf{ST}$. Thus the Hessian cannot be inverted, or equivalently, the linear system implied in the Newton iteration cannot be solved. There are three ways around this problem. One is to augment f by penalty terms such that the first atom is held at the origin, the second remains on the first coordinate axis, the third stays on the plane of the first and second coordinate axes, etc. Thus if $\mathbf{y}_i \in \mathbb{R}^3$ is the vector of coordinates for atom $i=1,...,N$, let the objective function be

$$f'(\mathbf{y}_1, \ldots, \mathbf{y}_N) := f + \|\mathbf{y}_1\|^2 + y_{2,2}^2 + y_{2,3}^2 + y_{3,3}^2 \tag{9.7}$$

The second method is to replace the Hessian with a matrix $\underline{\mathbf{H}}$ defined by

$$\underline{\mathbf{H}} := \nabla^2 f + w\underline{\mathbf{I}} \tag{9.8}$$

where the weight $w > 0$ is chosen large enough to make $\underline{\mathbf{H}}$ positive definite but not so large as to effectively turn the algorithm into steepest descents. The third approach is to move only one atom at a time, which produces a much smaller linear system to solve each time and keeps the Hessian positive definite when near the minimum for most reasonable energy functions. Of course if $\mathbf{x}$ is a vector of internal coordinates, such as dihedral angles, the problem is also avoided.

Now that we have seen the kinds of situations that can arise in unconstrained optimization problems and in principle how one solves them, we can turn to *constrained* minimization. As we shall see, constrained problems are generally treated by transforming them into unconstrained ones.

Definition 9.10. The *constrained minimization problem* is

$$\begin{aligned} &\text{minimize } f(\mathbf{x}) \\ &\text{subject to } h_i(\mathbf{x}) = 0, \; i=1,...,m \\ &\qquad g_j(\mathbf{x}) \leq 0, \; j=1,...,p \\ &\qquad \mathbf{x} \in \Omega \subseteq \mathbb{R}^n \end{aligned} \tag{9.9}$$

where $m \leq n$, and usually $f, h_1, \ldots, h_m, g_1, \ldots, g_p \in C^2(\mathbb{R}^n)$.

It is notationally convenient to think of the constraints as vectors of functions, so that one can write $\mathbf{h}(\mathbf{x}) = \mathbf{0}$ and $\mathbf{g}(\mathbf{x}) \leq \mathbf{0}$. We have written the constraints always with zero as the right hand side and inequalities as $\leq$, but clearly any equality or inequality can be converted to such forms.

Definition 9.11. Any point $\mathbf{x} \in \mathbb{R}^n$ satisfying all the constraints is called *feasible.*

Definition 9.12. If $\mathbf{x}$ is a feasible point, then any inequality constraint having $g_i(\mathbf{x}) < 0$ is *inactive* at $\mathbf{x}$, and an inequality constraint having $g_j(\mathbf{x})=0$ is called *active.* Of course, equality constraints $h_i(\mathbf{x})=0$ must be active by the definition of feasibility.

Definition 9.13. Let $\mathbf{x}^*$ be a feasible point, where the set of indices for the active inequalities is $J := \{j \mid g_j(\mathbf{x}^*)=0\}$. Then we call $\mathbf{x}^*$ a *regular point* of the constraints if the set of gradient vectors $\{\nabla h_i(\mathbf{x}^*), \nabla g_j(\mathbf{x}^*) \mid 1 \leq i \leq m, \; j \in J\}$ is linearly independent.

PROPOSITION 9.14 Kuhn-Tucker Conditions. *Suppose* $\mathbf{x}^*$ *is a local minimum for the constrained minimization problem and also a regular point for the constraints. Then there exist vectors* $\lambda \in \mathbb{R}^m$ *and* $\mu \in \mathbb{R}^p$ *such that* $\mu \geq \mathbf{0}$ *and*

$$\nabla f(\mathbf{x}^*) + \lambda \nabla \mathbf{h}(\mathbf{x}^*) + \mu \nabla \mathbf{g}(\mathbf{x}^*) = \mathbf{0} \tag{9.10}$$

$$\mu \mathbf{g}(\mathbf{x}^*) = 0 \tag{9.11}$$

Note that since $\mu \geq \mathbf{0}$ and $\mathbf{g}(\mathbf{x}^*) \leq \mathbf{0}$, Equation (9.11) amounts to saying that a component $\mu_j > 0$ only when g_j is active.

Definition 9.15. An inequality constraint $g_j(\mathbf{x}) \leq 0$ is referred to as *degenerate* at some solution if it is active but its corresponding $\mu_j = 0$.

Definition 9.16. It is convenient to refer to the *Lagrangian* associated with a constrained minimization problem:

$$L(\mathbf{x},\lambda,\mu) := f(\mathbf{x}) + \lambda \mathbf{h}(\mathbf{x}) + \mu \mathbf{g}(\mathbf{x}) \ . \tag{9.12}$$

The vectors λ and μ are called *Lagrange multipliers.*

In these terms, the Kuhn-Tucker conditions along with the feasibility conditions can be summarized as $\nabla_{\mathbf{x}} L = \mathbf{0}$, $\nabla_{\lambda} L = \mathbf{0}$, and $\nabla_{\mu} L = \mathbf{0}$. Denote by $\underline{\mathbf{L}}$ the Hessian of L, that is, the matrix of second derivatives with respect to the components of $\mathbf{x}$.

PROPOSITION 9.17 Second Order Conditions. *Let all the functions* f, $\mathbf{h}$, *and* $\mathbf{g} \in C^2(\mathbb{R}^n)$, *and let* $\mathbf{x}^*$ *be a regular point of the constraints. If* $\mathbf{x}^*$ *is a local minimum point for the constrained minimization problem, then* $\exists\ \lambda \in \mathbb{R}^m$ *and* $\mu \in \mathbb{R}^p$ *such that the Kuhn-Tucker conditions hold and such that* $\underline{\mathbf{L}}$ *is positive semidefinite on the tangent space* $\mathbf{M}$

$$\mathbf{M} := \{\mathbf{y} \mid \nabla \mathbf{h}(\mathbf{x}^*)\mathbf{y} = \mathbf{0} \text{ and } \nabla g_j(\mathbf{x}^*)\mathbf{y} = 0 \ \ \forall\ j \in J\} \tag{9.13}$$

where

$$J := \{j \mid g_j(\mathbf{x}^*)=0\} \ . \tag{9.14}$$

Conversely, if there are λ *and* μ *satisfying the Kuhn-Tucker conditions and* $\underline{\mathbf{L}}$ *is positive definite on* $\mathbf{M}'$, *then* $\mathbf{x}^*$ *is a strict local minimum point of the constrained minimization problem. The possibly larger space* $\mathbf{M}'$ *is defined as above except that*

$$J' := \{j \mid g_j(\mathbf{x}^*)=0,\ \mu_j>0\} \tag{9.15}$$

where allowance is made for degenerate inequality constraints.

In other words, the second order conditions for optimality are much the same as in the unconstrained case, except that the Hessian of the Lagrangian takes the place of the Hessian of the objective function. Intuitively speaking, if the objective function represents potential energy, the Lagrange multipliers indicate the force with which the system presses against the corresponding constraints. This sensitivity interpretation is expressed more formally in the following proposition.

PROPOSITION 9.18. *For f, $\mathbf{h}$, and $\mathbf{g} \in C^2(\mathbb{R}^n)$, let there be a family of problems*

$$\begin{aligned} \text{minimize} \quad & f(\mathbf{x}) \\ \text{subject to } \mathbf{h}(\mathbf{x}) &= \mathbf{c} \\ \mathbf{g}(\mathbf{x}) &\leq \mathbf{d} \ . \end{aligned} \tag{9.16}$$

Suppose that when $\mathbf{c} = \mathbf{0}$ and $\mathbf{d} = \mathbf{0}$, there is a regular, local solution $\mathbf{x}^*$ satisfying the second-order sufficiency conditions for a strict local minimum. Assume for simplicity that there are no degenerate inequality constraints at $\mathbf{x}^*$. Treat $[\mathbf{c},\mathbf{d}]$ as a combined vector in $\mathbb{R}^{m+p}$. Then for every $[\mathbf{c},\mathbf{d}]$ in an open set containing $[\mathbf{0},\mathbf{0}]$ there is a solution $\mathbf{x}(\mathbf{c},\mathbf{d})$ which is a continuous function of $[\mathbf{c},\mathbf{d}]$ such that $\mathbf{x}(\mathbf{0},\mathbf{0}) = \mathbf{x}^*$, and such that $\mathbf{x}(\mathbf{c},\mathbf{d})$ is a local minimum of the above problem. Also,

$$\nabla_{\mathbf{c},\mathbf{d}} f(\mathbf{x}(\mathbf{c},\mathbf{d}))\Big]_{\mathbf{0},\mathbf{0}} = -(\lambda,\mu) \ . \tag{9.17}$$

This view of a family of constrained minimization problems will be important in §9.3.

Having now surveyed the basic mathematical considerations in constrained minimization, we are faced once again with actually solving such problems. As for unconstrained minimization, there is a host of algorithms, but we will discuss only the "augmented Lagrangian" approach, which has worked very well for us. Basically we want to transform the constrained problem into an unconstrained one so that we can use numerical minimization algorithms, such as Newton's method or steepest descent. The Kuhn-Tucker condition is very suggestive that we need to minimize the Lagrangian, and the second-order optimality condition even speaks of the positive definite Hessian of the Lagrangian. The catch is that the Hessian is required to be positive definite over only a tangent space, and not over an open set of Ω. When the Hessian is taken with respect to all components of $\mathbf{x}$, $\nabla^2_{\mathbf{x}} L$ may not at all be positive definite, and local unconstrained optimization methods may well fail to converge to a solution. Suppose in general we had (nonlinear) constraints:

$$\begin{aligned} & h_i(\mathbf{x})=0 \quad && , \ i=1,\dots,m \\ & g_i(\mathbf{x}) \leq 0 && , \ i=1,\dots,r \\ & \alpha_i \leq e_i(\mathbf{x}) \leq \beta_i && , \ i=1,\dots s \end{aligned} \tag{9.18}$$

and the usual objective function $f(\mathbf{x})$. The strategy is to force the Hessian to be positive definite near a solution by modifying it. For a more thorough explanation of what follows, see [Bertsekas1982].

Definition 9.19. For some weight $w > 0$, let the *augmented Lagrangian* be

$$L(\mathbf{x},\lambda,\mu,\kappa,w) := f(\mathbf{x}) + \sum_{i=1}^{m}\left(\lambda_i h_i + \frac{w}{2}h_i^2\right) + \sum_{i=1}^{r}\left(\mu_i g_i' + \frac{w}{2}g_i'^2\right) + \sum_{i=1}^{s} e_i' \tag{9.19}$$

where

$$g_i' = max\{g_i\,,\, -\frac{\mu_i}{w}\}$$

and

$$e_i' = \begin{cases} \kappa_i(e_i-\beta_i) + \frac{w}{2}(e_i-\beta_i)^2 & \text{if } \kappa_i+w(e_i-\beta_i) > 0 \\ \kappa_i(e_i-\alpha_i) + \frac{w}{2}(e_i-\alpha_i)^2 & \text{if } \kappa_i+w(e_i-\alpha_i) < 0 \\ -\frac{\kappa_i}{2w} & \text{otherwise} \end{cases}$$

If we neglect the terms involving w, we have essentially the usual Lagrangian except for neatly treating double sided inequalities with only a single multiplier each by exploiting the fact that each e_i can violate either its lower bound or its upper bound but not both simultaneously. On the other hand, neglecting the terms involving multipliers results in the sum of the objective function plus some penalty terms weighted by w. For $w < \infty$, local unconstrained minimization of such a function results in a more or less infeasible point which approaches feasibility only as $w \to \infty$. In practice this gives rise to rather slow convergence to $\mathbf{x}^*$. Equation (9.19) enjoys the best of both by choosing w large enough so that the augmented Lagrangian Hessian is positive definite but keeping w bounded above so that minimization of L leads rapidly to a solution $\mathbf{x}^*$, where the Lagrange multipliers have the values λ^*, μ^*, and κ^*.

Algorithm 9.20 Augmented Lagrangians (Bertsekas):
INPUT: A constrained minimization problem and initial guesses to the weight $w^{(0)}$, to the local feasible minimum point $\mathbf{x}^{(0)}$, and to the multipliers $\lambda^{(0)}$, $\mu^{(0)}$, and $\kappa^{(0)}$. There must also be a measure of constraint satisfaction, such as

$$\epsilon(\mathbf{x}) := \sum(h_i(\mathbf{x}))^2 + \sum_{g_i(\mathbf{x}) > 0}(g_i(\mathbf{x}))^2 + \sum_{e_i(\mathbf{x}) > \beta_i}(e_i(\mathbf{x})-\beta_i)^2 + \sum_{e_i(\mathbf{x}) < \alpha_i}(e_i(\mathbf{x})-\alpha_i)^2 \tag{9.20}$$

OUTPUT: The actual local feasible minimum $\mathbf{x}^*$ and the corresponding final multipliers λ^*, μ^*, and κ^*.

PROCEDURE:

For outer iteration k=0, 1,...

Locally (unconstrained) minimize $L(\mathbf{x}^{(k)},\lambda^{(k)},\mu^{(k)},\kappa^{(k)},w^{(k)})$ (Equation 9.19)

This is the inner iteration.

Update the multipliers and $w^{(k)}$ according to Equation (9.21).

Until $\epsilon(\mathbf{x}^{(k)})$ is adequately small.

The updating equations are

$$\lambda_i^{(k+1)} := \lambda_i^{(k)} + w^{(k)}h_i^{(k+1)} \tag{9.21}$$

$$\mu_i^{(k+1)} := \mu_i^{(k)} + w^{(k)}g_i'^{(k+1)}$$

$$\kappa_i^{(k+1)} := \begin{cases} \kappa_i^{(k)}+w^{(k)}(e_i-\beta_i) & \text{if } \kappa_i^{(k)}+c^{(k)}(e_i-\beta_i) > 0 \\ \kappa_i^{(k)}+w^{(k)}(e_i-\alpha_i) & \text{if } \kappa_i^{(k)}+c^{(k)}(e_i-\alpha_i) < 0 \\ 0 & \text{otherwise} \end{cases}$$

$$w^{(k+1)} := \begin{cases} \gamma w^{(k)} & \text{if } \epsilon^{(k)} > \delta\cdot\epsilon^{(k-1)} \\ w^{(k)} & \text{otherwise} \end{cases}$$

Given a particular set of constraints, γ and δ are chosen empirically to improve the performance of the algorithm. Suggested values are $4 \leq \gamma \leq 10$ and $\delta \approx 1/4$.

It remains to be shown that the algorithm actually converges well. For notational simplicity, consider only the case of equality constraints in what follows.

LEMMA 9.21. *Suppose* $\mathbf{x}^*$ *is a strict local minimum and a regular point of a constrained minimization problem, and* $f, \mathbf{h} \in C^2(\mathbb{R}^n)$ *on some open sphere around* $\mathbf{x}^*$. *Then there exists some minimal* $\overline{w} > 0$ *such that for all* $w > \overline{w}$ *the augmented Lagrangian* $L(\mathbf{x}^*,\lambda^*,w)$ *(Equation 9.19) has a positive definite Hessian.*

As usual in this section, the reader should consult [Bertsekas1982] for the proof. Qualitatively what is happening is that Proposition 9.17 assures positive definiteness in a tangent subspace, and the terms of the augmented Lagrangian weighted by w take care of the complementary subspace, which is spanned by the constraints.

PROPOSITION 9.22. *Make the same assumptions as in the previous lemma, and choose* $w > \overline{w}$. *Let* $\mathbf{S} := \{\mathbf{x} \mid \|\mathbf{x}-\mathbf{x}^*\| < \epsilon\}$. *Then there exist scalars* δ, ϵ, *and* $M > 0$ *such that:*

$$\forall\ [\lambda,w] \in D := \{[\lambda,w] \mid \|\lambda-\lambda^*\| < \delta w,\ w \geq \overline{w}\} \tag{9.22}$$

the problem

$$\underset{\mathbf{x}\in\mathbf{S}}{\text{minimize}}\ L(\mathbf{x},\lambda,w) \tag{9.23}$$

has a unique solution we will call $\mathbf{x}(\lambda,w)$. *Considering* $\mathbf{x}$ *as a function of* λ *and* w, *it is continuously differentiable in the interior of* D. *Also,* $\forall\ [\lambda,w] \in D$

$$\|\mathbf{x}(\lambda,w)-\mathbf{x}^*\| \leq \frac{M}{w}\|\lambda-\lambda^*\| \tag{9.24}$$

Furthermore, $\forall\ [\lambda,w] \in D$

$$\|[\lambda+w\mathbf{h}(\mathbf{x}(\lambda,w))]-\lambda^*\| \leq \frac{M}{w}\|\lambda-\lambda^*\| \tag{9.25}$$

And finally, $\forall\ [\lambda,w] \in D$*, the Hessian* $\nabla^2_{\mathbf{xx}}L(\mathbf{x}(\lambda,w),\lambda,w)$ *is positive definite and the matrix* $\nabla_{\mathbf{x}}\mathbf{h}(\mathbf{x}(\lambda,w))$ *has rank m.*

This is basically the justification of the augmented Lagrangians algorithm given above. Without knowing the multipliers initially, the weight w can eventually be raised high enough that the inner loop minimization will converge, and indeed the sequence $\mathbf{x}^{(k)}$ will converge to $\mathbf{x}^*$ while the multiplier update formula ensures that $\lambda^{(k)} \rightarrow \lambda^*$.

PROPOSITION 9.23. *As far as the outer loop of the augmented Lagrangians algorithm is concerned, convergence is at least linear if* $[w^{(k)}]$ *is bounded, and superlinear if the sequence is unbounded.*

For practical implementation of the algorithm, this is very important, because larger values of w tend to slow the convergence of the *inner* loop. Generally one hopes to adjust the parameters of the algorithm such that w grows sufficiently to achieve some sort of reasonable convergence in the outer loop but stops short of unduly impeding the inner loop minimizations.

9.2. Energy Minimization and Molecules

Perhaps a cleverer global search algorithm would succeed for larger molecules even though the brutal approach fails. Cleverness depends critically on what is known about the energy surface in advance. The conformational energy of any molecule can be defined as continuously differentiable almost everywhere, the exceptions being singularities where the distance between any two atoms $d_{ij} \rightarrow 0$. One can always define an arbitrary cutoff, E_{max}, beyond which the conformer cannot possibly have any physical importance, and then let the truncated energy be

$$E'(\mathbf{x}) = min(E(\mathbf{x}),E_{max})\ . \tag{9.26}$$

E' is continuous everywhere, bounded above by E_{max}, and below by some E_{min}. In the case of most empirical force fields,

$$E(\mathbf{x}) = \sum_{i<j}^{n_{atoms}} \epsilon_{ij}(d_{ij}) \tag{9.27}$$

where ϵ_{ij} is the contribution to the total from the pairwise interaction between atoms i and j. For all values of d_{ij}, $\epsilon_{ij} > -\infty$, and for some d^*_{ij}, ϵ_{ij} is optimal. Then a useful E_{min} is simply

$$E_{min} = \sum_{i<j}^{n_{atoms}} \epsilon_{ij}(d^*_{ij}) \tag{9.28}$$

When dihedral angles are the variables, the conformation space is certainly finite. Since the magnitude of the gradient ∇E is bounded, E' obeys a Lipschitz condition

$$|E'(\mathbf{x}) - E'(\mathbf{y})| \leq L\|\mathbf{x}-\mathbf{y}\| \tag{9.29}$$

where L is the Lipschitz constant, and $\|\cdot\|$ denotes some appropriate distance metric in conformation space. Clearly $L \leq \dfrac{(E_{max}-E_{min})}{\max\limits_{\mathbf{x},\mathbf{y}}\|\mathbf{x}-\mathbf{y}\|}$, and experience with the molecule in question can usually provide a better estimate. This is often all the *a priori* knowledge that can be brought to bear on the problem. Then the following algorithm [Crippen1975, Shubert1972a, Shubert1972b] requires a smaller number of energy evaluations (presumably more time consuming than any subsidiary calculations required by the global search algorithm) than any other sequential algorithm with the same starting point for *every* such function E'.

Algorithm 9.24 Lipschitz global optimization (Shubert):

INPUT: An energy function E' over a finite conformation space, a Lipschitz constant, and a required accuracy $\delta > 0$.

OUTPUT: All sampling points having energy within δ of the global minimum.

PROCEDURE:

Construct an adequately fine sampling grid over all conformation space.
Initially E' is unknown for every sampling point and the
 estimated energy of each is E_{min}.
Initially pick any point as the first one to evaluate.
While the lowest estimated $E' \leq \delta +$ lowest known E',
  Evaluate E' at the chosen point.
  Revise upwards the estimated E' of all nearby points by
   the Lipschitz condition, Equation (9.29).
  Sufficiently distant points will keep their old estimated E'.
  The next point to be evaluated is any one of the unevaluated points
   with the lowest estimated E'.

Applying this algorithm to oligopeptide global energy searches might be done with dihedral angles as variables, a step size of 20° between sampling points, $E_{max} = +15$ kcal/mol, $\delta = 3$ kcal/mol, $\|\cdot\|$ in Equation (9.29) being the usual Euclidean norm with differences in corresponding dihedral angles calculated *mod* 360°, and with the Lipschitz constant, L, estimated by $\max_{\mathbf{x},\mathbf{y}} \frac{|E'(\mathbf{x})-E'(\mathbf{y})|}{||\mathbf{x}-\mathbf{y}||}$ for many randomly chosen conformations **x** and **y**. If the energy function is such that the first two points evaluated differ in energy by their Lipschitz limit, and the points are maximally distant, and δ is smaller than E' could change between adjacent sampling points, then the search could take only two evaluations. At the opposite extreme, one can readily construct an example where every grid point must be evaluated, which means for n dihedral angles, some $\left(\frac{360}{20}\right)^n$ evaluations. Trials with oligopeptides have required roughly $(18^n)/3$ evaluations. At that rate, five dihedral angles would require on the order of 6×10^5 evaluations, and the prospects for handling yet more variables are poor indeed.

One way out of this "multiple minima" problem is to settle for less than a guarantee that all Ω has been located. At this point, we want a method that greatly increases the likelihood of finding a low energy conformation for a given investment in computation. Perhaps there are other, very different conformations with energy comparable to, or even better than that of the one already found, but at least we have found something useful. This is very much in the spirit of the embed algorithm in §5.4, where a sampling of geometrically constrained structures is produced without a definitive statement that the sampling is representative of all possibilities. This brings us first to general global optimization algorithms, a subject of active research today. For a review, see [McCormick1972, Pardalos1986]. Some methods depend on locating minima or stationary points, one after the other, and marking each one as it is found so the algorithm will not return to it [Crippen1969, Crippen1971a, Crippen1973]. Of course, these methods cannot be very useful on large molecules where there are a large number of local minima. Other approaches produce a trajectory in conformation space resembling the path a local minimizer takes from the starting point down to a local minimum, except that the trajectory tends to not converge on minor local minima while following the larger scale downward trends [Griewank1981]. At the opposite extreme in sophistication, merely jumping away in random directions will help explore conformation space and can be controlled enough to converge on new minima [Bremermann1970, Rao1981, Rao1981]. An appealing idea that continues to be rediscovered is that of finding some local minimum by the usual methods, and then leaving this by various means such that renewed minimization will converge on a new, nearby local minimum, which of course may not necessarily be any better [Crippen1971b, Simons1983, Levitt1976]. One strong factor in the appeal is that real molecules must undergo a similar process at ordinary temperatures in order to move over potential energy barriers. The direct simulation of conformational transitions, called molecular dynamics, unfortunately requires vast amounts of computer time for even the equivalent of 100 psec on a small protein [McCammon1980, Levitt1982, Levitt1983, McCammon1987]. Since real proteins exhibit overall refolding on the minutes time scale, molecular dynamics remains confined to the study of small molecules or rapid perturbations of large ones.

Another global minimization method which we have applied with considerable success to small constrained energy minimization problems is known as the *ellipsoid algorithm.* This algorithm was originally developed as a method of solving convex optimization problems subject to convex constraints [Shor1977], and was subsequently adapted to yield a polynomial time algorithm for linear programming [Khatchiyan1979] (although it is not competitive with the older simplex method on most linear programs of practical size). Subsequently the algorithm was found to be extraordinarily robust as a method of solving general constrained *non*linear optimization problems, especially in the face of multiple local minima, or ill-conditioned and redundant constraints [Ecker1983, Ecker1985].

The ellipsoid algorithm begins with an ellipsoid centered at a point $\mathbf{x}^{(0)} \in \mathbb{R}^n$ and chosen large enough to enclose a (constrained) minimum $\mathbf{x}^*$, and then calculates a sequence of ellipsoids of decreasing *hypervolume* centered at perhaps widely scattered points $\mathbf{x}^{(k)}$, but still (in principle) enclosing $\mathbf{x}^*$. The initial ellipsoid $\mathbf{E}^{(0)}$ has the starting point, $\mathbf{x}^{(0)}$, as its center and its shape is defined by a symmetric $n \times n$ matrix $\underline{\mathbf{Q}}^{(0)}$.† The initial guess of $\underline{\mathbf{Q}}^{(0)}$ must enclose the unknown solution, $\mathbf{x}^*$, which is the same as requiring

$$(\mathbf{x}^* - \mathbf{x}^{(0)})^T (\underline{\mathbf{Q}}^{(0)})^{-1} (\mathbf{x}^* - \mathbf{x}^{(0)}) \leq 1 \ . \tag{9.30}$$

At the kth iteration of the algorithm, we choose either some violated constraint or else the energy function if $\mathbf{x}^{(k)}$ is feasible, and calculate its gradient, $\mathbf{g}^{(k)}$. Then the ellipsoid center is moved according to

$$\mathbf{d} := -\frac{\underline{\mathbf{Q}}^{(k)} \mathbf{g}^{(k)}}{(\mathbf{g}^{(k)T} \underline{\mathbf{Q}}^{(k)} \mathbf{g}^{(k)})^{1/2}} \tag{9.31}$$

$$\mathbf{x}^{(k+1)} := \mathbf{x}^{(k)} + \frac{\mathbf{d}}{n+1} \tag{9.32}$$

and the shape and size are changed by

$$\underline{\mathbf{Q}}^{(k+1)} := \frac{n^2}{n^2-1} \left(\underline{\mathbf{Q}}^{(k)} - \frac{2}{n+1} \mathbf{d}\mathbf{d}^T \right) \ . \tag{9.33}$$

At any iteration k, the vector $\mathbf{x}^{(k)}$ and matrix $\underline{\mathbf{Q}}^{(k)}$ together define an ellipsoid $\mathbf{E}^{(k)}$. The gradient $\mathbf{g}^{(k)}$ defines a half-space $\mathbf{H}^{(k)} \subset \mathbb{R}^n$ by

† If the ellipsoid has axes parallel to the coordinate axes, then $\underline{\mathbf{Q}}^{(0)}$ is diagonal, and the diagonal elements are the squares of the ellipsoid radii.

$$\mathbf{H}^{(k)} := \{\mathbf{y} \in \mathbb{R}^n \mid (\mathbf{y}-\mathbf{x}^{(k)}) \cdot \mathbf{g}^{(k)} \leq 0\} \tag{9.34}$$

which (for convex problems at least) contains $\mathbf{x}^*$. Then the equations above give the smallest ellipsoid, in terms of $\mathbf{x}^{(k+1)}$ and $\underline{\mathbf{Q}}^{(k+1)}$, which contains $\mathbf{E}^{(k)} \cap \mathbf{H}^{(k)}$. At each iteration, the volume of the ellipsoid shrinks by a factor of

$$\left(\frac{n-1}{n+1}\right)\left(\frac{n}{(n^2-1)^{1/2}}\right)^n \tag{9.35}$$

independent of the energy function or constraints used. Although this factor rapidly approaches 1 as n increases, it is practical to solve problems with $n = 50$ or more. It is also possible that the $\mathbf{x}^{(k)}$ may approach a good solution long before the volume of the ellipsoid has become small.

In order to apply the ellipsoid algorithm to conformational problems, a number of critical implementation decisions had to be made [Billeter1986]. To begin with, the adverse dependence of the rate of convergence on the number of variables dictates that the dihedral angles be used instead of Cartesian coordinates. Thus an initial ellipsoid which contains the desired minimum can always be obtained simply by taking a sphere which encloses a cube of side length 360° centered on the origin (although smaller initial ellipsoids are preferable if the approximate location of the optimum solution is known in advance). Even though the energy function itself ensures the absence of unacceptable atomic overlaps, such overlaps can be checked for in much less time than it takes to evaluate an energy function, and hence we have often found it advantageous to include lower bounds derived from hard sphere radii among the constraints. In addition, we have used a list of upper bounds on the distances which are of such importance in the analysis of NMR data (see §7.3). Although the ellipsoid algorithm, as described above, leaves the order in which the constraints should be considered open, much better results are obtained if preference is given to the larger distance constraint violations. The gradients of these distance constraints with respect to the dihedral angles were calculated by Equation (4.51), and since hard sphere overlaps can usually be eliminated by means of small changes in the dihedral angles, the upper bound violations are always "preferred" to the hard sphere constraints. Finally, our choice of dihedral angles as the variables makes it easy to directly constrain their values, if desired. The gradients of these constraints are simply signed unit vectors, and we found it best to "prefer" them to both classes of distance constraints whenever possible. The following summarizes our procedure:

Algorithm 9.25 The Ellipsoid Algorithm:

INPUT: A list of dihedral angle constraints, upper bound distance constraints, and lower bound distance constraints (generally coming from hard sphere contact

checking); a penalty functional form $\mathbf{g} \geq \mathbf{0}$ for such constraints; thresholds for the violation of the three classes of constraints; an energy function; and a starting ellipsoid $\mathbf{E}^{(0)}$ specified by $\mathbf{x}^{(0)}$ and $\underline{\mathbf{Q}}^{(0)}$.
OUPUT: A point $\mathbf{x}^{(k)}$ for some $k \geq 0$ that approximates a solution to the constrained optimization problem.
PROCEDURE:

Iterate for $k = 0, 1, \cdots$

- If there is a violation of a dihedral angle constraint larger than a given threshold:
  - Let $\mathbf{g}^{(k)}$ be the gradient of its penalty term.
- Else if there is a violation of an upper bound constraint larger than its threshold:
  - Let $\mathbf{g}^{(k)}$ be the gradient of its penalty term.
- Else if there is a violation of a lower bound constraint larger than its threshold:
  - Let $\mathbf{g}^{(k)}$ be the gradient of its penalty term.
- Else let $\mathbf{g}^{(k)}$ be the energy gradient.
- If there were no constraint violations:
  - Save $\mathbf{x}^{(k)}$ as the best solution so far.
  - Decrease the three thresholds by a given factor.
- Compute the new ellipsoid $\mathbf{E}^{(k+1)}$.

Until either
- k exceeds the iteration limit,
- or there are no violations and the energy is adequately minimized,
- or the ellipsoid volume is adequately small,
- or the correction vector $\mathbf{d}$ is adequately small,
- or roundoff error results in a negative square root argument in the calculation of $\mathbf{d}$.

Note that the ellipsoid algorithm never looks at the energy until it has found a completely feasible solution. Because of this fact, equality constraints cannot be used since these are essentially always violated by at least some small amount. Nevertheless, recent results have indicated that with care even ring closure constraints can be imposed [Howard1987].

In one application of the ellipsoid algorithm to constrained energy minimization [Billeter1986], we have minimized the energy of a glucagon fragment (residues 17 – 27 containing 151 atoms and 48 rotatable bonds) subject to upper bounds on 18 of the interproton distances, as obtained by NMR [Braun1981]. The starting conformation had all angles set to 180°, and the starting ellipsoid was a sphere containing all conformations whose angles were within 180° of the starting conformation, so that we actually attempted to perform a 48 variable global optimization! The distance constraint violations were first eliminated after only 98 iterations, and the energy was reduced from 1295 to 306 kcal over the next 400 iterations. An additional 1500 iterations only succeeded in reducing the energy by another 10 kcal. Unconstrained minimization of the resultant conformation further reduced the energy to 277 kcal, but at the expense of introducing violations of the distance constraints of up to 1.4 Å. This is not surprising, since the NMR data was collected for miscelle-bound glucagon, while the ECEPP energy function employed [Nemethy1983] is essentially an *in vacuo* potential. It does emphasize, however, the importance of maintaining consistency with the constraints when using energy functions to refine conformations which have been determined from experimental data in solution.

In all of the above methods, there remains the question of how the molecule is to be represented and how this affects the success of searches for energy minima. Including all atoms, even hydrogens, makes possible a more precise potential function for simulating fine details of conformation [Levitt1982] at the expense of greater computer time for each energy evaluation and apparently a potential surface having more local minima due to the larger number of interacting particles. At the opposite extreme [Obatake1981, Levitt1976, Crippen1984, Crippen1985, Crippen1987a], potentials for proteins which involve only one or two points per amino acid residue can be useful in determining rather general conformational features much more economically for proteins. A related question is how the molecular conformation is to be parameterized. As we have discussed in earlier chapters, interparticle (interatomic) distances are most directly related to most potential functions, since they generally consist of pairwise interactions that depend on the interparticle separation. However there are the many geometric relations between the various distances which must be satisfied. Cartesian coordinates have generally been our preferred parameterization in that there are fewer of them than there are distances, distances are easy to calculate from them, and the only redundancies are the overall translational and rigid rotational degrees of freedom. We have never felt it necessary to remove these degrees of freedom because overall translation and rotation of a molecule is never observed in energy minimization. The most compact choice is dihedral angles, leaving bond lengths and vicinal bond angles fixed, a good assumption for large, unstrained molecules. Here there is a problem with numerical instability due to large atom displacements possible sometimes from small changes in dihedral angles. Nonetheless, with care this approach can also yield good results [Braun1983, Braun1985]. Of course, it is possible to use combinations of these various conformational parameters where suitable and then interconvert them [MacKay1974, Essen1983].

9.3. Energy Embedding

Another way to get around the global search problem is to know more about the energy function *a priori*. Empirical energy functions usually are made up of largely a sum of interatomic pairwise interaction terms. In most cases each term qualitatively resembles the curve shown in Fig. 9.2: a singularity of high energy when the atoms overlap, some distance r_{ij}^* of optimal separation, and then rising asymptotically to zero at large separations.

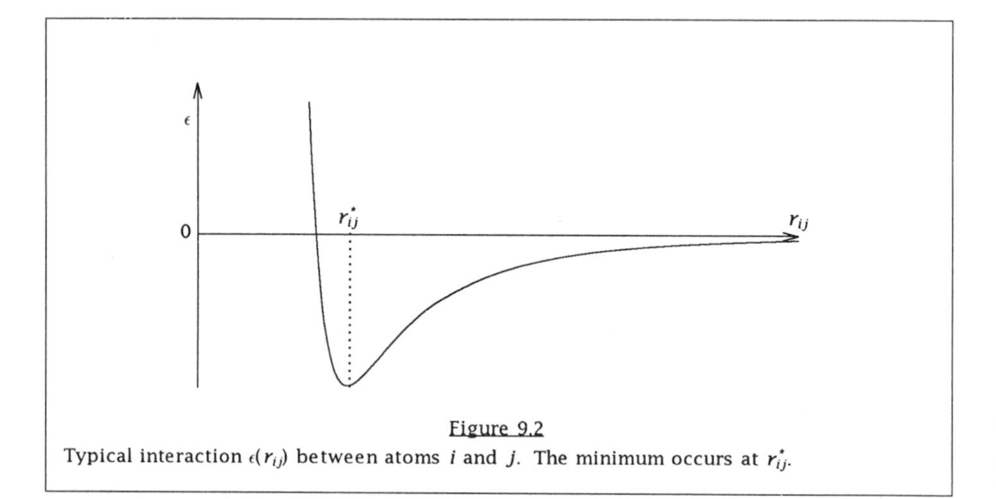

Figure 9.2
Typical interaction $\epsilon(r_{ij})$ between atoms i and j. The minimum occurs at r_{ij}^*.

Suppose the energy function consisted entirely of such unimodal terms:

$$E(\mathbf{x}) = \sum_{i<j}^{n_a}\sum \epsilon_{ij}(r_{ij}) \tag{9.36}$$

where n_a is the number of atoms. (Molecular mechanics potentials usually have 3-atom terms for vicinal bond angle bending and 4-atom terms for intrinsic torsional barriers, but a function can be constructed out of only pairwise terms.) Because the r_{ij}^*s are all roughly equal in value, they generally satisfy the conditions for embedding in n_a-1 dimensional space, since these are only inequality constraints involving the Cayley-Menger determinants of the r_{ij}^* values (see Proposition 3.28). Hence $E_{min} = \sum\sum \epsilon_{ij}(r_{ij}^*)$ could be realized in $\mathbb{R}^{n_a-1}$, there would be only one minimum, and it would be easy to find. Consider for example Fig. 9.3, where $n_a = 3$ and the energy function is

$$E = (r_{12}-3)^2+(r_{13}-4)^2+(r_{23}-5)^2 \tag{9.37}$$

Points **1**, **2**, and **3** can obviously be placed in $\mathbb{R}^2$ so as to achieve the global

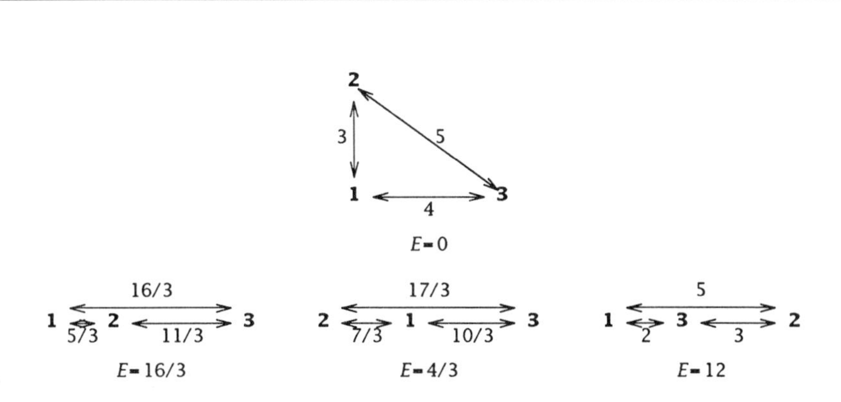

Figure 9.3

Three atoms with a unique energy minimum of eq. 9.37 in $\mathbb{R}^2$ have three local minima if forced to lie in $\mathbb{R}^1$.

minimum of energy, as shown at the top of Fig. 9.3. If, however, we require a minimal energy linear structure, ordinary embedding would tend to produce the middle structure because of its bias toward preserving long distances. In this example, that also corresponds to the lowest of the three minima, due to the equal weighting of terms in Equation (9.37). Unfortunately, in a practical intramolecular potential, the long distance terms are rather weak, as shown in Fig. 9.2, and embedding usually produces structures that cannot be minimized to exceptionally low energy. This led to the development of "energy embedding," where special projection directions were chosen in proceeding from $\mathbb{R}^{n_a-1}$ to $\mathbb{R}^3$ so as to preserve energetically important interactions, rather than overall shape [Crippen1982a, Crippen1982b].

There are many ways to project down to $\mathbb{R}^3$, and one would like to compress the (n_a-4)-dimensional subspace associated with the least energetic costs. This seems to be difficult to find, even for small compressions where we can approximate the the molecular energy by harmonic stretching terms:

$$E = \sum_{i<j}^{n_a}\sum k_{ij}(d_{ij}-r_{ij}^*)^2 \tag{9.38}$$

where the interatomic distances

$$d_{ij} = \left[\sum_{m=1}^{n_a-1}(c_{im}-c_{jm})^2\right]^{1/2} \tag{9.39}$$

and of course all the force constants $k_{ij} \geq 0$. In some cases, the directions of least initial resistance to projection are approximately given by the eigenvectors corresponding to the smallest eigenvalues of the $(n_a-1)\times(n_a-1)$ force constant weighted inertial tensor $\underline{\mathbf{T}}$ of the interatomic separation vectors.

$$t_{pq} := 2\sum_{i<j}^{n_a}\sum k_{ij}(c_{ip}-c_{jp})(c_{iq}-c_{jq})^T \tag{9.40}$$

Eventually, it was found more efficient to disregard the initial projection direction and use a continuous method to reduce the dimensionality from n_a-1 to 3, as described below. Then in the initial stages the molecule sometimes automatically undergoes large rotations in $\mathbb{R}^{n_a-1}$ to present the directions of least resistance to the compressive force.

The first step is to find coordinates, c_{ik}, $i=0,\ldots,n_a-1$, $k=0,\ldots,n_a-2$, in $\mathbb{R}^{n_a-1}$ for the n_a atoms. Initially, one can arbitrarily place the atoms at the corners of a regular hyper-tetrahedron with edge $r \approx 5.0$ Å.

$$\begin{aligned} c_{i,k} &:= 0,\ k \geq i \\ c_{i,k} &:= \frac{1}{i}\sum_{j=0}^{i-2} c_{j,k},\ k<i-1 \\ c_{i,i-1} &:= \left(r^2 - \sum_{k=0}^{i-2} c_{i,k}^2\right)^{1/2} \end{aligned} \tag{9.41}$$

This is a reasonable approach if all the optimal interatomic distances are close to r, or if the optimal distances cannot be embedded, but in the case they can be exactly embedded, analytic triangularization generalized to many dimensions will yield an excellent starting conformation if carried out at high numerical precision:

$$\begin{aligned} c_{i,k} &:= 0,\ k \geq i \\ c_{i,k} &:= \frac{r_{i,k+1}^{*2} - r_{i,k}^{*2} - c_{k+1,k}^2 - \sum_{j=0}^{k-1}(c_{i,j}-c_{k+1,j})^2-(c_{i,j}-c_{k,j})^2}{-2c_{k+1,k}},\ k<i-1 \\ c_{i,i-1} &:= \left(r_{i,0}^{*2} - \sum_{k=0}^{i-2} c_{i,k}^2\right)^{1/2} \end{aligned} \tag{9.42}$$

The initial guess is refined by Newton minimization, moving one atom at a time, with the step modified by the Armijo linesearch to ensure an energy decrease even when far from the minimum (see Appendix F). In the case of embeddable optimal distances, the triangulated initial coordinates are so good that this refinement halts immediately, and this energetically optimal conformation is unique, up to rigid rotations and translations. Otherwise, the refined structure tends to occupy some subspace of $\mathbb{R}^{n_a-1}$, and there are in general local minima even in such a high dimensional space.

Next we demand that the fourth and higher coordinates of each atom approach zero while otherwise trying to keep the energy minimal. To put it more precisely, let the coordinates matrix $\underline{\mathbf{C}} = [c_{ij}]$, for points $i=1,...,n_a$ and coordinates $j=1,...,n_a-1$. Then given the initial minimal energy conformation $\underline{\mathbf{C}}_{init}$ in $\mathbb{R}^{n_a-1}$,

$$\begin{aligned} &\underset{\underline{\mathbf{C}}}{\text{minimize}}\ E(\underline{\mathbf{C}}) \\ &\text{subject to} \quad c_{ij} = 0 \ \text{ for } i=1,...,n_a;\ j=4,...,n_a-1 \end{aligned} \tag{9.43}$$

Here we have singled out the first three coordinate axes for special treatment. If the initial coordinates are highly anisotropic, it would be advantageous to take as the preserved three dimensional subspace that space spanned by the eigenvectors of the three largest eigenvalues of the inertial tensor. However, the initial conformation is generally so spherical that there seems to be little reason to go to the trouble. This is simply a nonlinear constrained optimization problem which can be efficiently solved by augmented Lagrangians [Bertsekas1982], as already explained in §9.1. Let the vector of all unwanted coordinates be $\mathbf{g} := [c_{1,4}, c_{1,5}, \ldots, c_{n_a,n_a-1}]$. Then the augmented Lagrangian function is

$$L(\underline{\mathbf{C}},\lambda,w) = E(\underline{\mathbf{C}}) + \lambda\cdot\mathbf{g} + \frac{1}{2}wg^2 \tag{9.44}$$

where $g^2 := \|\mathbf{g}\|^2$, λ is the vector of Lagrange multipliers and w is a weight on the constraint terms. Initially in iteration 0, $w^{(0)}=0.01$ and $\lambda^{(0)} = \mathbf{0}$. We solve the inner loop of the augmented Lagrangians algorithm by the usual Newton method, moving one atom at a time. The outer loop is repeated until the undesired coordinates are reduced essentially to zero ($\|\mathbf{g}\| < 0.01$ Å). If the initial conformation in $\mathbb{R}^{n_a-1}$ was really well minimized, then E increases as $\|\mathbf{g}\|\rightarrow 0$ because the molecule cannot make as many favorable interactions in a lower dimensional space. Sometimes, the repeated cycles of minimization of L offer such additional opportunities for reducing E, that the energy actually slightly decreases. Meanwhile, the various components of λ increase from zero, reflecting the energetic cost of reducing the corresponding g_i to zero, but eventually $\lambda\rightarrow 0$ because there is no force component acting in the fourth and higher coordinate directions when all points lie in $\mathbb{R}^3$.

Equation (9.44) incorporates only the constraints $\mathbf{g} = \mathbf{0}$, but in general there would be also holonomic equality constraints for fixed bond lengths and bond angles, and inequality constraints $l_{ij} \leq d_{ij} \leq u_{ij}$ derived from experiment and/or theory. There might also be chirality inequality or equality constraints on quartets of atoms. As explained in §9.1, the general augmented Lagrangian algorithm can handle all such constraints, no matter how complicated or nonlinear.

The repetitions of the outer loop of the augmented Lagrangians produces a sequence $\{\mathbf{x}^{(k)}\}$ which can be viewed as steps along a continuous path parameterized by g^2 as g^2 proceeds from its initial value g^2_{init} in $\mathbb{R}^{n_a-1}$ monotonically toward zero in $\mathbb{R}^3$. Since there are sometimes substantial shifts in w and λ due to the updating equations, it is possible that the algorithm may jump from what was initially the best path to an alternate one, generally leading to an inferior result. This difficulty can be reduced by incrementing w only by a small factor, such as 1.5, albeit at the expense of extra computing time. A more serious difficulty with energy embedding is that it commits itself to such a path based on the characteristics of the energy surface in the neighborhood of the high-dimensional minimum. Thus it may be true that the initial energy gradient along the chosen path may be minimal compared to other directions, but the total integrated work over the whole path may be larger than along some other paths. Simple examples of this behavior can easily be constructed [Crippen1987b]. In other words, the method has no guarantee of locating the *global* minimum, but the average case performance in practice is that it reaches extremely good local minima as long as the curvature of each energy term at its minimum is representative of the depth of the well.

Given an initial set of coordinates in $\mathbb{R}^{n_a-1}$, one might ask if the return to $\mathbb{R}^3$ is even well defined.

LEMMA 9.26. *Number the atoms $1,\cdots,n_a$ and translate and rotate the minimal energy conformation $\mathbf{x}^* \in \mathbb{R}^{n_a-1}$ as a rigid body so that coordinates $i,\cdots,n_a-1$ of the ith atom are zero. Consider only the $N := n_a(n_a-1)/2$ remaining coordinates to be variable so that trivial rotations and translations of the molecule are excluded. Let these N coordinates be ordered in $\mathbf{x}$, such that the desired final coordinates are components x_i, $i=1,\ldots,3n_a-6$. Then there is some $\epsilon > 0$ such that the path proceeding from $\mathbb{R}^N$ to $\mathbb{R}^{3n_a-6}$ is unique from $g^2 = g^2_{init}$ to $g^2_{init}-\epsilon$.*

Proof: Since the energy E is at a minimum, we can approximate

$$E(\mathbf{x}+\mathbf{x}^*) \approx E(\mathbf{x}^*) + \frac{1}{2}\mathbf{x}^T\underline{\mathbf{H}}\mathbf{x} \tag{9.45}$$

for $\|\mathbf{x}\|$ small, where $\underline{\mathbf{H}}$ is the positive definite Hessian matrix. Therefore E is a convex function near $\mathbf{x}^*$, and hence

$$\mathbf{K} = \{\mathbf{x} \mid E(\mathbf{x}+\mathbf{x}^*) < E(\mathbf{x}^*)+\delta E\} \tag{9.46}$$

is a convex set for small $\delta E > 0$. The set of all conformations within $g^2 > 0$ of being three dimensional

$$\mathbf{L}(g^2) = \{\mathbf{x} \mid \sum_{i=3n_a-5}^{N} x_i^2 \le g^2\} \tag{9.47}$$

is also convex. To see this, suppose we choose any two conformations, $\mathbf{x}_1$ and $\mathbf{x}_2$, on the boundary of $\mathbf{L}(g^2)$, i.e. $\sum x_{1i}^2 = \sum x_{2i}^2 = g^2$. Then by the definition of a convex set, we must show $\mathbf{x}_3 = \alpha \mathbf{x}_1 + (1-\alpha)\mathbf{x}_2 \in \mathbf{L}(g^2)$ for any $0 \leq \alpha \leq 1$.

$$\sum_{i=3n_a-5}^{N} x_{3i}^2 = \sum (\alpha x_{1i} + (1-\alpha) x_{2i})^2 \tag{9.48}$$

$$= (1-2\alpha+2\alpha^2)g^2 + \alpha(1-\alpha)\sum x_{1i}x_{2i}$$

$$\leq (1-2\alpha+2\alpha^2)g^2 + \alpha(1-\alpha)[\sum x_{1i}^2]^{1/2}[\sum x_{2i}^2]^{1/2}$$

$$= (1-\alpha+\alpha^2)g^2 \leq g^2 \text{ for } 0 \leq \alpha \leq 1.$$

where the inequality in sums is from the Cauchy-Schwartz inequality. Since both **K** and **L** are convex, then $\mathbf{K} \cap \mathbf{L}(g_{init}^2 - \epsilon)$ is also convex, where $\epsilon > 0$ is chosen small enough so that the intersection is non-empty. Now the minimization of the convex function E restricted to the convex region $\mathbf{K} \cap \mathbf{L}$ always leads to a unique minimum. Therefore as g^2 decreases from its initial value g_{init}^2, there will be a *unique* path of minimal E. [Crippen1987b]. QED.

Of course eventually as $g^2 \rightarrow 0$ the argument breaks down as E ceases to be convex or $\mathbf{K} \cap \mathbf{L}$ becomes empty. Note that this argument depends on choosing a particular coordinate set for the high-dimensional minimal energy conformation, and then excluding further rigid rotations. In practice, rotating the high-dimensional conformation may result in different minima in $\mathbb{R}^3$.

Energy embedding has been applied to a number of small test cases and a few large examples [Crippen1982a, Crippen1982b]. In the small test cases where the results can be checked by an independent global search, the method always locates remarkably low energy structures, but not invariably the global minimum. In the absence of numerous geometric constraints, the results are determined strongly by the potential function, E. This has led to some surprising results when the empirical energy function used was tailored to agree very well with some experimentally determined conformations, because energy embedding's more global perspective allowed it to find minima for these same molecules which were very far from the experimental conformations, yet gave much lower energies! Apparently the method demands much more of the potential function than conventional local minimization approaches to conformational analysis. This has caused an extensive search for better potential functions at the one-point-per-residue resolution for predicting protein folding [Obatake1981, Crippen1984, Crippen1985, Crippen1987a]. Aside from the difficulties of choosing a good potential, energy embedding results seem remarkably uninfluenced by details of the algorithm. The work in 1982 by Crippen used a very different method for reducing the dimensionality of the molecule than the augmented Lagrangians algorithm given above. The quality of the results is the same, but the computational efficiency of the older work is worse. Other workers [Purisima1986], have followed our lead but have devised an apparently more difficult numerical method, involving minimizing penalty functions to achieve embeddability and dimensionality. Even so, they can locate unusually low energy minima for an all-atom representation of several isolated amino acid derivatives.

In all of the above discussion, the assumption has been that the energy consisted of strictly a sum of pairwise terms. Most accurate molecular mechanics potentials, however, also involve intrinsic bond angle bending and intrinsic torsional terms. To test whether it would work with more general, flexible geometry force fields, we [Crippen1987b] adapted to energy embedding AMBER version 3.0, a molecular potential function with the Cartesian coordinates of the atoms as variables [Weiner1984]. With this force field, N-acetyl-L-alanine-N'-methylamide, $CH_3CONHC^{\alpha}H(CH_3)CONHCH_3$, has only 12 united atoms because only the amide hydrogens are represented explicitly. An exhaustive search keeping peptide bonds in the *trans* conformation located only five local minima having $E = -42.41, -39.09, -38.34, -36.40$, and -41.40 kcal. Subsequent energy minimization in $\mathbb{R}^{11}$ starting from each of these lead to precisely the same high-dimensional minimum, where $E = -43.69$ kcal, but the five coordinate sets differed among themselves by rigid rotations and translations in $\mathbb{R}^{11}$. Returning the third and fifth coordinate sets to $\mathbb{R}^3$ led to the global minimum, but the other three returned to the second best minimum.

Of course one can easily generalize from $\mathbb{R}^3$ the simple pairwise distance dependent terms, such as bond stretching, nonbonded interactions, and electrostatic interactions. Bond angle bending can always be expressed in terms of $cos\theta$, where θ is the $i-j-k$ angle defined by three atoms, and

$$d_{ij}d_{kj}cos\theta = (\mathbf{c}_i-\mathbf{c}_j)\cdot(\mathbf{c}_k-\mathbf{c}_j) = \sum_{l=1}^{n}(c_{il}-c_{jl})(c_{kl}-c_{jl}) . \tag{9.49}$$

However, the dihedral angle ϕ for intrinsic torsional terms is generally calculated via cross products, which do not generalize very well to $\mathbb{R}^n$. Referring to Figure 9.5 for the dihedral angle ϕ defined by the four atoms $\mathbf{i}-\mathbf{j}-\mathbf{k}-\mathbf{l}$, we can calculate

$$\mathbf{a} = -\mathbf{v}_{kl}+\left(\frac{\mathbf{v}_{kl}\cdot\mathbf{v}_{kj}}{\mathbf{v}_{kj}\cdot\mathbf{v}_{kj}}\right)\mathbf{v}_{kj} \tag{9.50}$$

and

$$\mathbf{b} = \mathbf{v}_{ij}-\left(\frac{\mathbf{v}_{ij}\cdot\mathbf{v}_{kj}}{\mathbf{v}_{kj}\cdot\mathbf{v}_{kj}}\right)\mathbf{v}_{kj} \tag{9.51}$$

and then

$$cos\phi = \frac{\mathbf{a}\cdot\mathbf{b}}{(\mathbf{a}\cdot\mathbf{a})^{1/2}(\mathbf{b}\cdot\mathbf{b})^{1/2}} . \tag{9.52}$$

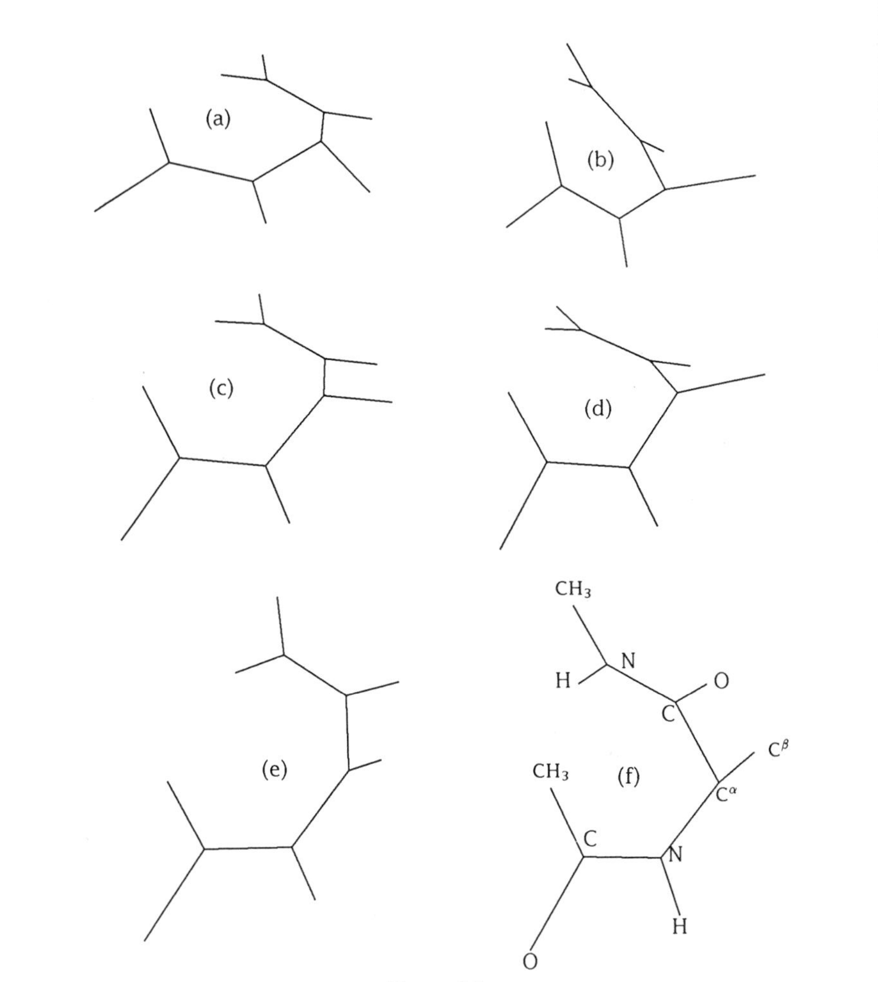

Figure 9.4

Stereographic projections onto the two-dimensional page of the first three dimensions of N-acetyl-L-alanine-N'-methylamide as it proceeds from the 11-dimensional minimum (a) up the path of least energetic resistance to the minimum in three dimensions (f). The lower left portion of each chain is closer to the viewer than the upper right end. Note how the molecule spreads out as it is flattened, and note how the C^{α} smoothly inverts chirality, since that is merely a local rotation in four dimensions.

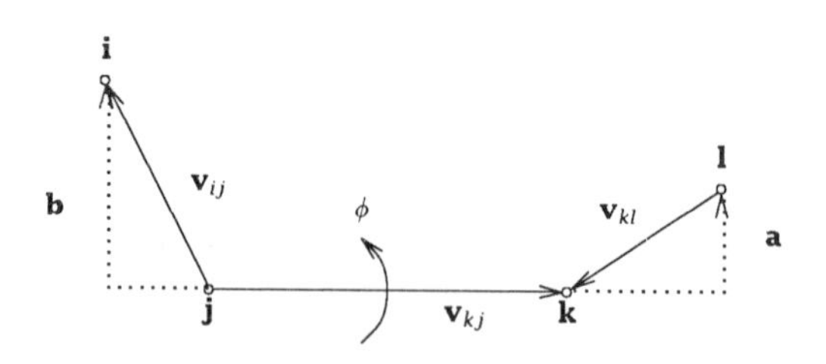

Figure 9.5
Definition of vectors used in calculating the dihedral angle ϕ using dot products.

The torsional terms in AMBER 3.0 can always be expressed as a function of $cos(n\phi+\psi)$, where the phase angle ψ is either 0 or π, and $n \in \{1,2,3,4,6\}$. With appropriate trigonometric identities, such expressions can always be written in terms of $cos\phi$, so although Equation (9.52) does not give ϕ unambiguously, it is sufficient for our purposes. In order to calculate analytical first and second derivatives with respect to some coordinate x of one of the four atoms, note that

$$\mathbf{a}\cdot\mathbf{b} = -\mathbf{v}_{ij}\cdot\mathbf{v}_{kl} + \frac{(\mathbf{v}_{ij}\cdot\mathbf{v}_{kj})(\mathbf{v}_{kl}\cdot\mathbf{v}_{kj})}{\mathbf{v}_{kj}\cdot\mathbf{v}_{kj}} \tag{9.53}$$

and similarly

$$\mathbf{a}\cdot\mathbf{a} = \mathbf{v}_{kl}\cdot\mathbf{v}_{kl} - \frac{(\mathbf{v}_{kl}\cdot\mathbf{v}_{kj})^2}{\mathbf{v}_{kj}\cdot\mathbf{v}_{kj}} \tag{9.54}$$

$$\mathbf{b}\cdot\mathbf{b} = \mathbf{v}_{ij}\cdot\mathbf{v}_{ij} - \frac{(\mathbf{v}_{ij}\cdot\mathbf{v}_{kj})^2}{\mathbf{v}_{kj}\cdot\mathbf{v}_{kj}} \quad . \tag{9.55}$$

Then

$$\frac{\partial cos\phi}{\partial x} = \frac{2\frac{\partial \mathbf{a}\cdot\mathbf{b}}{\partial x}(\mathbf{a}\cdot\mathbf{a})(\mathbf{b}\cdot\mathbf{b}) - (\mathbf{a}\cdot\mathbf{b})(\mathbf{b}\cdot\mathbf{b})\frac{\partial \mathbf{a}\cdot\mathbf{a}}{\partial x} - (\mathbf{a}\cdot\mathbf{b})(\mathbf{a}\cdot\mathbf{a})\frac{\partial \mathbf{b}\cdot\mathbf{b}}{\partial x}}{2(\mathbf{a}\cdot\mathbf{a})^{3/2}(\mathbf{b}\cdot\mathbf{b})^{3/2}} \tag{9.56}$$

and

$$\frac{\partial^2 cos\phi}{\partial x \partial y} = \frac{1}{4}\left[4\frac{\partial^2 \mathbf{a}\cdot\mathbf{b}}{\partial x \partial y}(\mathbf{a}\cdot\mathbf{a})^2(\mathbf{b}\cdot\mathbf{b})^2 - 2\frac{\partial \mathbf{a}\cdot\mathbf{b}}{\partial x}(\mathbf{a}\cdot\mathbf{a})(\mathbf{b}\cdot\mathbf{b})^2\frac{\partial \mathbf{a}\cdot\mathbf{a}}{\partial y}\right. \tag{9.57}$$

$$-2\frac{\partial \mathbf{a}\cdot\mathbf{b}}{\partial x}(\mathbf{a}\cdot\mathbf{a})^2(\mathbf{b}\cdot\mathbf{b})\frac{\partial \mathbf{b}\cdot\mathbf{b}}{\partial y} - 2\frac{\partial \mathbf{a}\cdot\mathbf{b}}{\partial y}(\mathbf{a}\cdot\mathbf{a})(\mathbf{b}\cdot\mathbf{b})^2\frac{\partial \mathbf{a}\cdot\mathbf{a}}{\partial x}$$

$$+(\mathbf{a}\cdot\mathbf{b})(\mathbf{b}\cdot\mathbf{b})\frac{\partial\mathbf{a}\cdot\mathbf{a}}{\partial x}(\mathbf{a}\cdot\mathbf{a})\frac{\partial\mathbf{b}\cdot\mathbf{b}}{\partial y}-2(\mathbf{a}\cdot\mathbf{b})(\mathbf{b}\cdot\mathbf{b})^2\frac{\partial^2\mathbf{a}\cdot\mathbf{a}}{\partial xy}(\mathbf{a}\cdot\mathbf{a})$$

$$-2\frac{\partial\mathbf{a}\cdot\mathbf{b}}{\partial y}(\mathbf{a}\cdot\mathbf{a})^2\frac{\partial\mathbf{b}\cdot\mathbf{b}}{\partial x}(\mathbf{b}\cdot\mathbf{b})+(\mathbf{a}\cdot\mathbf{b})(\mathbf{a}\cdot\mathbf{a})\frac{\partial\mathbf{b}\cdot\mathbf{b}}{\partial x}(\mathbf{b}\cdot\mathbf{b})\frac{\partial\mathbf{a}\cdot\mathbf{a}}{\partial y}$$

$$\left.-2(\mathbf{a}\cdot\mathbf{b})(\mathbf{a}\cdot\mathbf{a})^2\frac{\partial^2\mathbf{b}\cdot\mathbf{b}}{\partial x\partial y}(\mathbf{b}\cdot\mathbf{b})+3(\mathbf{a}\cdot\mathbf{b})(\mathbf{a}\cdot\mathbf{a})^2\frac{\partial\mathbf{b}\cdot\mathbf{b}}{\partial x}\frac{\partial\mathbf{b}\cdot\mathbf{b}}{\partial y}\right](\mathbf{a}\cdot\mathbf{a})^{-5/2}(\mathbf{b}\cdot\mathbf{b})^{-5/2}$$

There is, however, a persistent difficulty with strong intrinsic torsional barriers. Since the dihedral angle is calculated as a function of the positions of only four atoms, these can span at most a three-dimensional subspace of whatever space the molecule is embedded in. Clearly, the extra degrees of geometric freedom do not help the system to surmount a high torsional barrier. For example, in the case of N-acetyl-L-alanine-N'-methylamide, the two peptide bonds never switch between *cis* and *trans* either while locating the $\mathbb{R}^{11}$ minimum or in the process of returning to $\mathbb{R}^3$. Energy embedding is best suited to a potential function involving only pairwise interatomic distances.

9.4. A Combinatorial Approach

In this section we shall discuss another, *very* different approach to global energy minimization, which attempts to reduce the problem, as far as possible, to pure combinatorics. The advantage of this is that if it is done in the right way, it enables one to use combinatorial optimization techniques [Papadimetriou1982] to search conformation space for low energy structures in a way that is at least potentially far more efficient than grid searches. Even though the absolute running time of the version of the algorithm we present here is too large to be useful in most chemical applications, the principles which it demonstrates are applicable to a wide variety of problems in chemistry and elsewhere,* and the algorithm itself provides an excellent illustration of how many of the techniques presented in this book can be put into practice.

Since the *quantitatively* precise conformation of a molecule having more than about fifty atoms is determined by the *qualitative* information as to which atom pairs are or are not "in contact" relative to a uniform predefined cutoff (see §7.3, as well as [Havel1979]), an appealing way to discretize conformation space is to divide it up on the basis of which atoms are in contact. Moreover, since the

* For numerous examples, see Chapter 8.

energy of a molecule is closely related to the proximities of its atoms, this division of conformation space should conform closely to its natural subdivision into "watersheds" (see §9.1). In this work, we choose to make it conform exactly by using a "square-well" potential function to describe the energy. Specifically, suppose that for each pair of atoms indexed i and j we are given a lower bound $\bar{l}_{ij}$ on the interatomic distance d_{ij}, an upper bound $\bar{u}_{ij}$, a distance cutoff $\bar{c}_{ij}$ which determines when the atoms are "in contact", and the additive contribution $\bar{e}_{ij}$ (positive or negative) to the intramolecular energy when the atoms are in contact. Then the total energy is given by

$$E_{square\ well} := \sum_{i<j} \begin{cases} +\infty, & \text{if } d_{ij} < \bar{l}_{ij} \\ \bar{e}_{ij}, & \text{if } \bar{l}_{ij} \leq d_{ij} \leq \bar{c}_{ij} \\ 0, & \text{if } \bar{c}_{ij} < d_{ij} \leq \bar{u}_{ij} \\ +\infty, & \text{if } \bar{u}_{ij} < d_{ij} \end{cases}. \tag{9.58}$$

For some pairs of atoms (also referred to as *couples*), such as those which are covalently bonded or those which define vicinal bond angles, $d_{ij} = \bar{l}_{ij} = \bar{u}_{ij}$, and we will refer to such pairs as *holonomic*. These do not contribute to the energy. Other pairs are called *free*, and for these we may only know that $\bar{l}_{ij}$ is the sum of the van der Waals radii, and $\bar{u}_{ij}$ is the largest dimension of the molecule.

Definition 9.27. Let the atoms of a molecule correspond to the nodes of a graph, and let some of the couples be the graph edges (see Appendix E). Then the *graph* of all possible contacts (GOAPC) has edges corresponding to only the free couples, while the *contact graph* of a particular molecular conformation has edges corresponding to only those free couples which are in contact according to the given cutoffs. A missing edge implies that the corresponding couple is not in contact.

In these terms, the goal is to determine the minimum energy subgraph of the GOAPC that is also the contact graph of a realizable conformation of the molecule in question [Havel1983a].

Since the solution space consists of all subgraphs of the GOAPC and is therefore finite and discrete, the usual numerical methods for optimizing continuous functions cannot be applied here. Instead we use *branch and bound*, which is a practical compromise between two well-known ways of solving such combinatorial optimization problems, namely *relaxation methods* and *enumeration methods*. To explain what these methods are, we shall call those solutions (subgraphs) which are contact graphs of a conformation of the molecule the *feasible* solutions, and the restrictions on what constitutes a feasible solution the *constraints*. A feasible solution of minimum energy will be called an *optimum* solution. Figure 9.6 exhibits some solutions which are infeasible, thereby

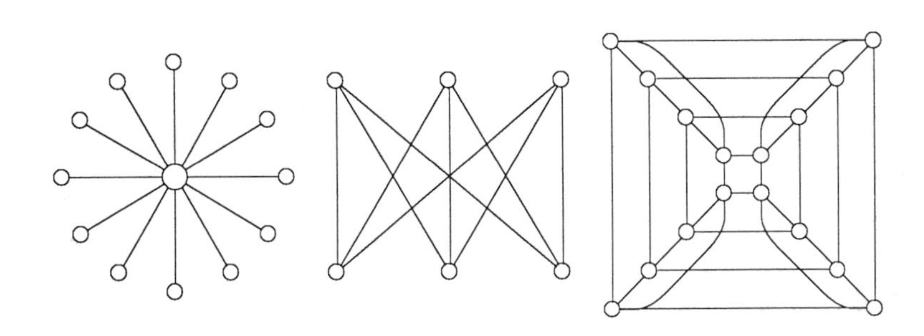

Figure 9.6

Some graphs are not the contact graphs of any configuration of points in $\mathbb{R}^3$ relative to a uniform cutoff.

showing that this distinction is nontrivial.

Sometimes, ignoring temporarily some of the constraints in a difficult combinatorial optimization problem results in a simpler problem which is easier to solve. The easier problem is termed a *relaxation* of the original. Since the set of solutions feasible in the relaxation is a (generally proper) superset of the set of all feasible solutions, the value of an optimum solution in the relaxation constitutes a *lower bound* on the value of an optimum solution to the original problem. In particular, should an optimum solution to the relaxation coincidentally also satisfy the constraints of the original problem, then it is also an optimum solution to the original problem. The easiest possible relaxation is obtained by ignoring *all* of the constraints, which in our case leads to a relaxation whose optimum solution consists simply of all couples in the GOAPC which have negative contact energies. We shall describe a better relaxation shortly.

Backtracking is an example of the second method which is involved in branch and bound, namely enumeration. Here, the collection of all possible subgraphs is successively bisected until a single feasible solution is reached, or a collection of subgraphs containing no feasible members is obtained. Each bisection is accomplished by choosing a couple from the GOAPC, and considering those subgraphs that do and those that do not contain this couple separately. Applying this same procedure to each of these two collections of subgraphs recursively, we eventually attain a feasible subgraph in which the presence and absence of all couples has been fixed, or find the contacts and non-contacts that we have fixed are by themselves infeasible. In the latter situation, no further bisection is required. Thus the time required for a backtrack enumeration grows with the number of feasible subgraphs instead of the total number of possible subgraphs.

To present the branch and bound method itself, we need to introduce some terms used in backtracking.

Definition 9.28. Backtracking may be visualized as a process of generating a tree whose *branches* correspond to fixing presence or absence of couples, and whose *nodes* correspond to the collections of subgraphs so obtained. Nodes are classified as *fathomed* or *active* as the enumeration of their collections is complete or not. The *search tree* is usually traversed in *depth-first* order. This means we descend branches of the tree until feasibility can be decided, and then *backtrack* by going up the tree towards its root until an active node is encountered, as shown in Figure 9.7. The *subproblem* at a node is the enumeration problem restricted to the collection of subgraphs at the node, and the *completions* of a node are the solutions of its subproblem. The *partial solution* at a node is the list of the couples which have been fixed in to or out of its completions, or *list of assumed (non)contacts* (LOANC).

In the problem at hand, the LOANC associated with the current node implies certain upper and lower distance bounds beyond those given *a priori* for the molecule. Bound smoothing (see Chapter 5) with the triangle and tetrangle inequalities may reveal inconsistencies in these bounds, thereby eliminating the current node and its descendents. If the LOANC at a node passes this test, it is subjected to the much more stringent and (on the small problems we have considered) time-consuming feasibility check, namely an attempt to embed the distance constraints which were obtained from the bound smoothing.

In order to convert a backtrack enumeration into a branch and bound search, we add to the feasibility tests an optimality test, as follows. The *subproblem* at a node of the tree is the problem of finding the minimum energy feasible solution which is consistent with the LOANC at that node. At every node of the tree which passes the above feasiblity tests, we now compute a lower bound on the energy of its completions by solving a relaxation of its subproblem. If this lower bound exceeds the energy of some conformation which has already been discovered by the search, then the optimum solution obviously cannot be a descendent of that node. Similarly, if the lower bound exceeds the energy of the N-th best structure found so far, then the solution of N-th lowest energy also cannot be a completion of the corresponding node. On the the hand, if the solution to this relaxation is found to be feasible with respect to the original problem, then it is an optimum completion of that branch of the tree. In fact, if we search the tree in a "best-first" manner, i.e. by always searching first a node

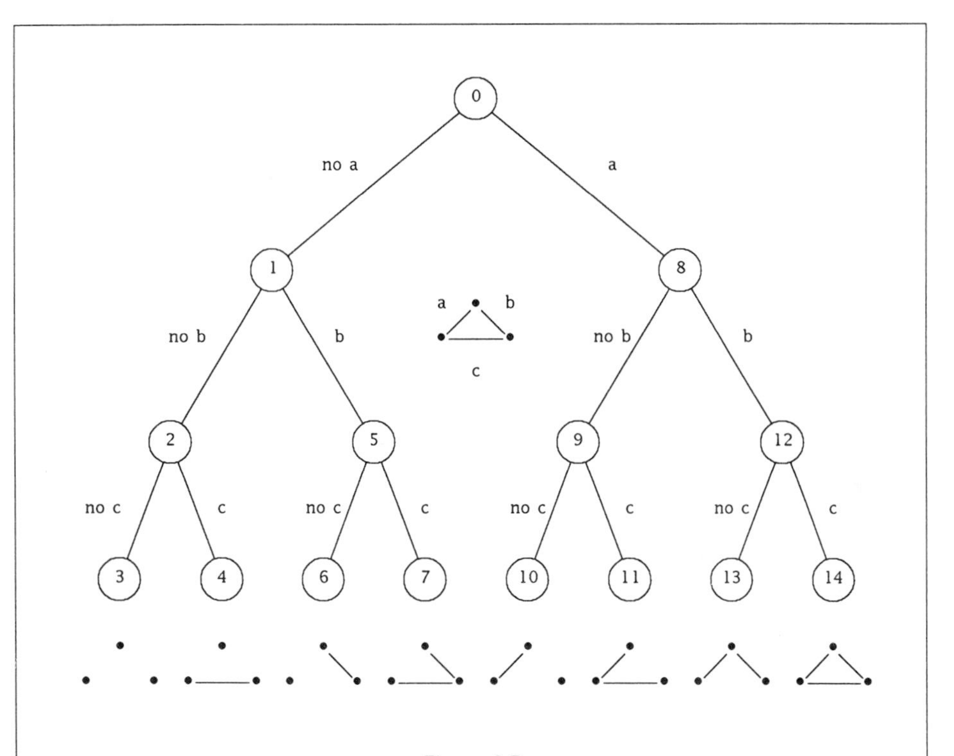

Figure 9.7

A search tree enumerating all possible graphs on three vertices. The possible edges of the graph are labelled a, b, and c, and the nodes of the treee are numbered in the order reached by a depth-first search.

whose energy lower bound is minimum over all unsearched nodes in the tree, we are guaranteed that the first feasible solution of the complete problem which is obtained by solving the relaxation of a subproblem is also an optimum solution to the complete problem. Since best-first seach can require an exponential amount of memory, however, we have used a compromise in which the children of each node are explored in a best-first manner, while the tree itself is traversed depth-first.

Obtaining good lower bounds on the energy of contact graphs is so important to the efficiency of the branch and bound search that it is worth employing fancier relaxations than to simply assume that all free couples with negative $\bar{e}$ are in contact. One obvious feature of molecular contact graphs is that the *degree* of each atom must lie within certain bounds. For example, the given experimental data may imply that some atoms must have a certain minimial number of contacts, and geometric packing considerations alone always imply a certain upper bound on the number of contacts any one atom can have. In other words, a lower bound on the energy of a class of solutions can be obtained by finding the subgraph of the current GOAPC which has minimum energy subject to constraints on the degrees of its atoms.

This problem, known as the *optimum degree constrained subgraph* (ODCSG) problem, belongs to a class of combinatorial optimization problems called *matching problems*, which can be solved in polynomial time. A general algorithm for solving matching problems has been implemented as a computer program, called BLOSSOM I, by [Edmonds1968]. The algorithm itself is well beyond the scope of this book (see [Havel1982] for a description), but for our present purposes it is sufficient to state that it reliably and efficiently solves ODCSG problems, completely untroubled by local minima. Although the degree constraints do ensure that the subgraphs of the GOAPC found are in some sense closer to *bona fide* contact graphs, they are insufficient to ensure embeddability, and drastically so for molecules with more than about ten atoms. Therefore BLOSSOM I must be built into a branch and bound search as a relaxation before it can accomplish anything useful.

The logic for the global combinatorial energy minimization of a molecule is presented in the following recursive algorithm. The space available to us unfortunately does not permit us to give a complete explanation of all the subtleties and tricks involved (see [Havel1982]), and the presentation given here can only attempt to make the overall stategy clear.

Algorithm 9.29 Combinatorial Energy Minimization:
INPUT: Lower and upper bounds on all distances $(\bar{l}, \bar{u})$, the contact energy function $(\bar{c}, \bar{e})$, degree constraints for each atom, and an integer $N_{min} > 0$.
OUTPUT: A list of *incumbent optima*, which contains the N_{min} lowest energy conformations, including the global minimum.
PROCEDURE:

Initialization: Smooth the bounds $(\bar{l}, \bar{u})$, obtaining (l, u).
The current LOANC consists of all contacts and noncontacts
which are *a priori* precluded by u and l;
the GOAPC contains all remaining nonholonomic couples.
The list of incumbent optima is empty and the energy
of the worst structure on it is infinite.
Mark the *root* "not searched".

Search(*root*).

```
Define procedure Search(node)
  If the node is marked "searched"
    If the node is the root
      Output the incumbent optima and halt.
    Else
      Return (ascending tree to parent node).
  Else
    Add the contacts and non-contacts on the LOANC to the initial
     distance bounds and smooth the resulting current bounds (l,u).
    If the current bounds are inconsistent with the triangle/tetrangle inequalities
      Return.
    If after of bound smoothing, we have l_ij > c̄_ij
      Transfer the ij couple from the current GOAPC to the LOANC
       as a noncontact.
    Similarly if u_ij was changed so that u_ij < c̄_ij
      Transfer the couple from the GOAPC to the LOANC as a contact.
    Degree bound preclusion: Repeatedly...
      Lower the current degree constraints by
       one for each new contact on the LOANC.
      If the degree upper bound of an atom reaches zero
        Add all couples in the current GOAPC containing that atom
         to the LOANC as noncontacts.
      If the degree of an atom in the GOAPC is equal to the degree lower
       bound of that atom
        Transfer all couples containing that atom to the LOANC as contacts.
    ...Until no more degree preclusions obtain.
    Attempt to embed using the current distance bounds.
    If the embedding failed
      Return.
    Else if embedded structure has energy better than worst incumbent optimum
      Selection: Add the structure to the list of incumbent optima,
       replacing the worst conformation there by it if the number
       of conformations on the list is equal to N_min.
    Bounding: Compute the ODCSG of the GOAPC w.r.t. the degree constraints.
    If the energy of the ODCSG plus energy due to contacts of the LOANC
     exceeds the worst energy on the list of incumbent optima
      Return.
    Completion: Add to the current distance bounds (l,u) all contacts
     corresponding to couples in the ODCSG, and noncontacts corresponding
```

to couples of the current GOAPC not in the ODCSG.
Smooth the current distance bounds (l, u)
If these distance bounds are inconsistent
Branching: Let the "branch set" consist of a minimal collection of couples
such that if they are added to the LOANC as contacts or noncontacts
according as they are present in or absent from the ODCSG,
the LOANC becomes inconsistent.
Else attempt to embed a structure which fits them
If unsuccessful:
Prepare a candidate "branch set" consisting of the couplesin the
current GOAPC sorted by decreasing violation of the constraints.
Else (successful):
The energy of this structure will be better than that of
the worst energy on the list of incumbent optima, so we
select it for the list of incumbent optima as above.
Return.
Mark the *node* as "searched".
Branching: Successively...
Add the couple of the branch set whose violation is largest to the LOANC,
and attempt to embed the bounds derived from it and (l, u).
If the embedding was successful
Select the structure for the list of incumbent optima if
its energy was an improvement on the worst, as above.
...Until embedding fails.
(Each combination of contacts and noncontacts involving the couples
of the branch set now represents a new branch to explore).
While current *node* has an unsearched branch whose energy lower bound
is less than the energy of worst structure on list of incumbent optima
Let *next_node* be the other end of a branch whose energy lower
bound does not exceed those of its siblings.
Search(*next_node*).
End definition.

The above algorithm has been implemented on a computer, and found to require exponentially increasing time as the size of the molecule increased over the range from cyclohexane to cyclodecane, with a growth rate of about $6^{N_{atoms}}$ [Havel1983b]. This was due at least in part to a decrease in the ability of the ODCSG to fathom partial solutions as the size of the molecule increased. A more ambitious test on N-acetyl-glycyl-N'-methylamide ($CH_3CONHCH_2CONHCH_3$) pressed the program to its current practical limits, but it did locate a conformation essentially equivalent to the global minimum as found by a $1°$ grid search over the two variable dihedral angles (peptide groups being held fixed). Unfortunately, this exhaustive search required much less computer time than the branch and bound search. Nonetheless, the branch and bound approach exhibited certain impressive efficiencies. There were 24 couples in the original GOAPC, 6 of which were excluded by the given distance constraints and the tetrangle inequality. Although that left $2^{18} \approx 10^5$ possible solutions to check, only 30 tree nodes had to be examined. The maximum tree depth was 5, and the maximum LOANC contained 15 couples. Solving the ODCSG involved a complicated algorithm, but less than 1% of the running time was consumed by it. Instead, about a third of the time was devoted to embedding the LOANC (once at each node tested), and another third went towards a necessary numerical energy minimization of the structures in $\mathbb{R}^3$. In order for the approach to be more widely useful, fundamental improvements must be made in our methods of solving the Fundamental Problem of Distance Geometry, as well as in developing an easily solved relaxation that is much more effective than ODCSG problem.

One possible direction which we have considered in an attempt to devise such improvements is to further "discretize" the problem by eliminating the use of explicit bounds l, u altogether! Because the cutoffs which occur in optimally realistic square well potentials are all equal to within a factor of two or so, if we make one more approximation and assume they are all equal, an ODCSG of the complete graph becomes simply a specification of which pairs of atoms are within a fixed distance of one another.

Definition 9.30. For a given graph $\mathbb{G} = (A,B)$, the *sphericity* $sph(\mathbb{G})$ is defined as the minimum dimension of a Euclidean space for which there exists a one-to-one correspondence between the points of a graph and points in the space such that pairs of graph points are adjacent if and only if the corresponding space points have distance less than or equal to some constant (without loss of generality, one).

The name sphericity stems from the evident fact that if we center unit radius balls or spheres on such a set of the points in space, then the distance between any pair is less than one if and only if the corresponding pair of balls or spheres has nonempty intersection. Graphs of sphericity one have been characterized in several ways [Fishburn1985], and can be recognized in linear time [Booth1976]. In higher dimensions, it has now been shown that all graphs have sphericity less than their orders, and that graphs of arbitrarily high sphericity exist [Maehara1984a, Maehara1984b, Maehara1986, Frankl1986] (see Figure 9.7 for some examples of graphs whose sphericity exceeds three). As of the time of writing, however, it is not known if this simplified problem can be solved, or to what extent solving it might yield useful insights into chemically relevant instances of the Fundamental Problem of Distance Geometry.

References

Bertsekas1982.

D.P. Bertsekas, *Constrained Optimization and Lagrange Multiplier Methods,* pp. 20, 96-156, Academic Press, New York, 1982.

Billeter1986.

M. Billeter, T.F. Havel, and K. Wuthrich, "The Ellipsoid Algorithm as a Method for the Determination of Polypeptide Conformations from Experimental Distance Constraints and Energy Minimization," *J. Comp. Chem., 8,* 132-141(1986).

Booth1976.

K. Booth and G. Lueker, "Testing for the Consecutive Ones Property, Interval Graphs and Graph Planarity Using PQ-Tree Algorithms," *J. Comp. Sys. Sci., 13,* 335-379(1976).

Braun1981.

W. Braun, C. Boesch, L.R. Brown, N. Go, and K. Wuthrich, "Combined Use of Proton-Proton Overhauser Enhancements and a Distance Geometry Algorithm for Determination of Polypeptide Conformations. Application to Micelle-Bound Glucagon," *Biochim. Biophys. Acta, 667,* 377-396(1981).

Braun1983.

W. Braun and N. Go, "Calculation of Protein Conformations which Satisfy a Given Set of Distance Constraints. A New Efficient Algorithm," *Biophysics (Japan), 23,* 88(1983).

Braun1985.

W. Braun and N. Go, "Calculation of Protein Conformations by Proton-Proton Distance Constraints: A New Efficient Algorithm," *J. Mol. Biol., 186,* 611-626(1985).

Bremermann1970.

H. Bremermann, "A Method of Unconstrained Global Optimization," *Math. Biosci., 9,* 1-13(1970).

Crippen1969.

G.M. Crippen and H.A. Scheraga, "Minimization of Polypeptide Energy VIII. Application of the Deflation Technique to a Dipeptide," *Proc. Natl. Acad. Sci. U.S.A., 64,* 42-49(1969).

Crippen1971a.

G.M. Crippen and H.A. Scheraga, "Minimization of Polypeptide Energy X. A Global Search Algorithm," *Arch. Bioch. & Biophys., 144,* 453-461(1971).

Crippen1971b.

G.M. Crippen and H.A. Scheraga, "Minimization of Polypeptide Energy XI. The Method of Gentlest Ascent," *Arch. Bioch. & Biophys., 144*, 462-466(1971).

Crippen1973.

G.M. Crippen and H.A. Scheraga, "Minimization of Polypeptide Energy. XII. The Methods of Partial Energies and Cubic Subdivision," *J. Comp. Phys., 12*, 491-495(1973).

Crippen1975.

G.M. Crippen, "Global Optimization and Polypeptide Conformation," *J. Comp. Phys., 18*, 224-231(1975).

Crippen1982a.

G.M. Crippen, "Conformational Analysis by Energy Embedding," *J. Comp. Chem., 3*, 471-476(1982).

Crippen1982b.

G.M. Crippen, "Energy Embedding of Trypsin Inhibitor," *Biopolymers, 21*, 1933-1943(1982).

Crippen1984.

G.M. Crippen and V.N. Viswanadhan, "A Potential Function for Conformational Analysis of Proteins," *Int. J. Pept. Prot. Res., 24*, 279-296(1984).

Crippen1985.

G.M. Crippen and V.N. Viswanadhan, "A Sidechain and Backbone Potential Function for Conformational Analysis of Proteins,," *Int. J. Pept. Prot. Res., 25*, 487-509(1985).

Crippen1987a.

G.M. Crippen and P.K. Ponnuswamy, "Determination of an Empirical Energy Function for Protein Conformational Analysis by Energy Embedding," *J. Comput. Chem., 8*, 972-981(1987).

Crippen1987b.

G.M. Crippen, "Why Energy Embedding Works," *J. Phys. Chem., 91*, 6341-6343(1987).

Csizmadia1983.

I.G. Csizmadia, "Topological Features of Conformational Potential Energy Surfaces," in *Symmetries and Properties of Non-Rigid Molecules*, ed. J. Serre, Studies in Physical and Theoretical Chem., vol. 23, pp. 315-321, Elsevier, 1983.

Ecker1983.

J.G. Ecker and M. Kupferschmidt, "An Ellipsoid Algorithm for Nonlinear Programming," *Math. Prog., 27*, 83-106(1983).

Ecker1985.

J.G. Ecker and M. Kupferschmidt, "A Computational Comparison of the Ellipsoid Algorithm with Several Nonlinear Programming Algorithms," *SIAM J. Control and Optimization, 23,* 657-674(1985).

Edmonds1968.

J. Edmonds, E. Johnson, and S. Lockhart, "BLOSSOM I: A Computer Code for the Matching Problem," Internal Memorandum, Thomas J. Watson IBM Research Center, 1968.

Essen1983.

H. Essen, "On the General Transformation from Molecular Geometric Parameters to Cartesian Coordinates," *J. Comp. Chem., 4,* 136-141(1983).

Fishburn1985.

P.C. Fishburn, *Interval Orders and Interval Graphs - A Study in Partially Ordered Sets,* J. Wiley & Sons, New York, NY, 1985.

Frankl1986.

P. Frankl and H. Maehara, "Embedding the N-Cube in Lower Dimensions," *Europ. J. Comb., 7,* 221-225(1986).

Griewank1981.

A.O. Griewank, *J. Optim. Theor. Applic., 34,* 11-39(1981).

Havel1979.

T.F. Havel, I.D. Kuntz, and G.M. Crippen, "The Effect of Distance Constraints on Macromolecular Conformation. II. Simulation of Experimental Results and Theoretical Predictions," *Biopolymers, 18,* 73-82(1979).

Havel1982.

T.F. Havel, *The Combinatorial Distance Geometry Approach to the Calculation of Molecular Conformation,* Ph.D. Dissertation, University of California, Berkeley, California, 1982.

Havel1983a.

T.F. Havel, I.D. Kuntz, and G.M. Crippen, "The Combinatorial Distance Geometry Method for the Calculation of Molecular Conformation. I. A New Approach to an Old Problem," *J. Theor. Biol., 104,* 359-381(1983).

Havel1983b.

T.F. Havel, I.D. Kuntz, G.M. Crippen, and J.M. Blaney, "The Combinatorial Distance Geometry Method for the Calculation of Molecular Conformation. II. Sample Problems and Computational Statistics," *J. Theor. Biol., 104,* 383-400(1983).

Howard1987.

A. Howard and P.A. Kollman, *Application of the Ellipsoid Algorithm to Cyclic*

Molecules, 1987. Personal communication.

Khatchiyan1979.
L.G. Khatchiyan, "A Polynomial Time Algorithm in Linear Programming," *Sov. Math. Dok., 20*, 191-194(1979).

Levitt1976.
M. Levitt, "A Simplified Representation of Protein Conformations for Rapid Simulation of Protein Folding," *J. Mol. Biol., 104*, 59-107(1976).

Levitt1982.
M. Levitt, "Protein Conformation, Dynamics, and Folding by Computer Simulation," *Ann. Rev. Biophys. Bioeng., 11*, 251-71(1982).

Levitt1983.
M. Levitt, *J. Mol. Biol., 168*, 595-620, 621-657(1983).

Liotard1983.
D. Liotard, "Global Analysis of Potential Surfaces: The Ground State of Methane," in *Symmetries and Properties of Non-Rigid Molecules*, ed. J. Serre, Studies in Physical and Theoretical Chemistry, vol. 23, pp. 323-333, Elsevier, 1983.

Luenberger1973.
D.G. Luenberger, *Introduction to Linear and Nonlinear Programming*, Addison-Wesley, Menlo Park, CA, 1973.

MacKay1974.
A.L. MacKay, "Generalized Structural Geometry," *Acta Cryst. A, 30*, 440(1974).

Maehara1984a.
H. Maehara, "Space Graphs and Sphericity," *Discrete Appl. Math., 7*, 55-64(1984).

Maehara1984b.
H. Maehara, "On the Sphericity of the Join of Many Graphs," *Discrete Math., 49*, 311-313(1984).

Maehara1986.
H. Maehara, "On the Sphericity for the Graphs of Semiregular Polyhedra," *Discrete Math., 58*, 311-315(1986).

McCammon1980.
J.A. McCammon and M. Karplus, "Simulation of Protein Dynamics," *Ann. Rev. Phys. Chem., 31*, 29-45(1980).

McCammon1987.
J.A. McCammon and S.C. Harvey, *Dynamics of Proteins and Nucleic Acids*, Cambridge University Press, Cambridge, England, 1987.

McCormick1972.

G.P. McCormick, "Attempts to Calculate Global Solutions of Problems that May Have Local Minima," in *Numerical Methods for Non-linear Optimization*, ed. F.A. Lootsma, pp. 209-221, Academic Press, 1972.

Mezey1981.

P.G. Mezey, "Lower and Upper Bounds for the Number of Critical Points on Energy Hypersurfaces," *Chem. Phys. Let., 82*, 100-104(1981).

Nemethy1983.

G. Nemethy, M. Pottle, and H.A. Scheraga, "Energy Parameters in Polypeptides. 9.," *J. Chem. Phys., 87*, 1883-1887(1983).

Obatake1981.

M. Obatake and G.M. Crippen, "A Residue-Residue Interaction Potential for Protein Conformational Calculations," *J. Phys. Chem., 85*, 1187-1195(1981).

Papadimetriou1982.

C.H. Papadimetriou and K. Steiglitz, *Combinatorial Optimization: Algorithms and Complexity*, Prentice Hall, Englewood Cliffs, New Jersey, 1982.

Pardalos1986.

P.M. Pardalos and J.B. Rosen, "Methods for Global Concave Minimization: A Bibliographic Survey," *SIAM Rev., 28*, 367-379(1986).

Purisima1986.

E.O. Purisima and H.A. Scheraga, "An approach to the multiple-minima problem by relaxing dimensionality," *Proc. Natl. Acad. Sci. USA, 83*, 2782-2786(1986).

Rao1981.

G.S. Rao, R.S. Tyagi, and R.K. Mishra, "Calculation of the Minimum Energy Conformation of Biomolecules using a Global Optimization Techinque. I. Methodology and Application to a Model Molecular Fragment (Normal Pentane)," *J. Theor. Biol., 90*, 377-89(1981).

Shor1977.

N.Z. Shor, "Cut-off Method with Space Extension in Convex Programming Problems," *Cybernetics, 12*, 94-96(1977).

Shubert1972a.

B.O. Shubert, "A Sequential Method Seeking the Global Maximum of a Function," *SIAM J. Numer. Anal., 9*, 379-388(1972).

Shubert1972b.

B.O. Shubert, "Sequential Optimization of Multimodal Discrete Function with Bounded Rate of Change," *Management Science, 18*, 687-693(1972).

Simons1983.

J. Simons, P. Jorgensen, H. Taylor, and J. Ozment, "Walking on Potential Energy Surfaces," *J. Phys. Chem.*, *87*, 2745-2753(1983).

Weiner1984.

S.J. Weiner, P.A. Kollman, D.A. Case, U.C. Singh, C. Ghio, G. Alagona, S. Profeta, and P. Weiner, "A New Force Field for Molecular Mechanical Simulation of Nucleic Acids and Proteins," *J. Amer. Chem. Soc.*, *106*, 765(1984).

APPENDICES

Mathematical Techniques

(G.M. CRIPPEN & T.F. HAVEL)

Appendices

A. Bilinear Algebra

A.1. The modern theory of Euclidean geometry is built upon the algebraic theory of bilinear forms defined on affine spaces. The purpose of this appendix is to bring together some basic results which we shall need from this field, and to fix the notation and terminology to be used elsewhere in the book. We assume that our readers already know the basics of vector and linear algebra (e.g. bases and dimension, linear transformations and their relation to matrices, determinants, rank, eigenvalue decompositions etc.). For the sake of brevity, no proofs have been included. Those readers interested in a detailed introduction to the subject are encouraged to read the excellent book [Snapper1971].

A.2. A *bilinear vector space*† $\mathbb{V}_B := (\mathbb{V}, B)$ consists of a vector space $\mathbb{V}$, which for our purposes may be taken to be real and finite dimensional, together with a function $B: \mathbb{V} \times \mathbb{V} \to \mathbb{R}$, called a *bilinear form*, which satisfies:

$$B(\mathbf{u};\mathbf{v}) \;=\; B(\mathbf{v};\mathbf{u}) \;\text{ for all } \mathbf{u},\mathbf{v} \in \mathbb{V}\,; \tag{A.1}$$

$$B(a\mathbf{u}+b\mathbf{v};\, \mathbf{w}) \;=\; a\,B(\mathbf{u};\,\mathbf{w}) + b\,B(\mathbf{v};\,\mathbf{w}) \;\text{ for all } a,b \in \mathbb{R} \text{ and } \mathbf{u},\mathbf{v},\mathbf{w} \in \mathbb{V}\,. \tag{A.2}$$

Condition (A.1) is described by saying the function B is *symmetric*, while condition (A.2) is described by saying that it is linear in the left argument. Condition (A.1) then implies that it is also linear in the right argument, or *bilinear*. Associated with each bilinear form is a unique *quadratic form* $Q: \mathbb{V} \to \mathbb{R}$, which is given by $Q(\mathbf{u}) := B(\mathbf{u};\mathbf{u})$. The quadratic form, in turn, uniquely determines the bilinear form according to the relations

$$2\,B(\mathbf{u};\,\mathbf{v}) \;=\; Q(\mathbf{u}) + Q(\mathbf{v}) - Q(\mathbf{u}-\mathbf{v}) \;=\; Q(\mathbf{u}+\mathbf{v}) - Q(\mathbf{u}) - Q(\mathbf{v})\,. \tag{A.3}$$

† Bilinear vector spaces are also called *metric vector spaces*, but we avoid that term here in order to avoid confusion with the classical metric spaces with which we deal elsewhere.

Thus which form we use is primarily a matter of convenience, and in what follows bilinear forms are usually more convenient.

A.3. A *subspace* $\mathbb{U}_B = (\mathbb{U}, B|_{\mathbb{U}})$ of $\mathbb{V}_B$ is a vector subspace $\mathbb{U}$ of $\mathbb{V}$ together with the restriction of B to $\mathbb{U}$ as its bilinear form; if $\mathbf{U} = \{\mathbf{u}_1, \ldots, \mathbf{u}_m\}$ is a set of generators for $\mathbb{U}$ we write $\mathbb{U} = <\mathbf{U}>$. Two vectors $\mathbf{u}, \mathbf{v}$ of a bilinear vector space are called *B-orthogonal* if $B(\mathbf{u}; \mathbf{v}) = 0$, and a vector which is B-orthogonal to itself is called *isotropic*. The *supplement* of a subset $\mathbf{W} \subseteq \mathbb{V}$ is the subspace consisting of all of vectors

$$\mathbf{W}^{\perp} \; := \; \{\, \mathbf{u} \in \mathbb{V} \mid B(\mathbf{u}; \mathbf{w}) = 0 \;\; \forall\, \mathbf{w} \in \mathbf{W} \,\} \tag{A.4}$$

which are B-orthogonal to all of $\mathbf{W}$, and the *radical* $\mathrm{rad}(\mathbb{V}_B)$ of a bilinear vector space is the supplement $\mathbb{V}^{\perp}$ of $\mathbb{V}$. For any bilinear vector space $(\mathbb{V}, B)$, there exists a B-orthogonal basis $[\mathbf{v}_1, \ldots, \mathbf{v}_n]$ of $\mathbb{V}$ with $\mathbb{V} = <\mathbf{v}_1> \oplus \cdots \oplus <\mathbf{v}_n>$ (where "$\oplus$" denotes the direct sum of subspaces) and $B(\mathbf{v}_i; \mathbf{v}_j) = 0$ for all $i \neq j$. Furthermore, in any B-orthogonal basis the isotropic vectors constitute a basis of the radical.

A.4. The dimension of the radical $dim(\mathrm{rad}(\mathbb{V}_B))$ itself is called the *nullity* of the space, and is denoted by $nul(\mathbb{V}_B)$. Bilinear vector spaces are called *nondegenerate* if their nullity is zero and *degenerate* otherwise. If $\mathbb{V}_B = (\mathbb{V}, B)$ is a nondegenerate bilinear vector space and $\mathbb{U}$ is a subspace of $\mathbb{V}$, then

$$dim(\mathbb{V}) \; = \; dim(\mathbb{U}) + dim(\mathbb{U}^{\perp})\,; \tag{A.5}$$

$$(\mathbb{U}^{\perp})^{\perp} \; = \; \mathbb{U}\,; \tag{A.6}$$

$$\mathrm{rad}(\mathbb{U}_B) \; = \; \mathbb{U} \cap \mathbb{U}^{\perp} \; = \; \mathrm{rad}(\mathbb{U}_B^{\perp})\,. \tag{A.7}$$

The quantity $rank(\mathbb{V}_B) := dim(\mathbb{V}) - nul(\mathbb{V}_B)$ is called the *rank* of the bilinear vector space, and the space is called *null* if its rank is zero. Bilinear vector spaces are further classified as *anisotropic* when they have no nontrivial null subspaces (i.e. $B(\mathbf{u}; \mathbf{v}) = 0 \;\; \forall\, \mathbf{v} \in \mathbb{V} \;\; \Rightarrow \;\; \mathbf{u} = 0$), and *semianisotropic* if every null subspace is a subspace of the radical (i.e. $B(\mathbf{u}; \mathbf{u}) = 0 \;\; \Rightarrow \;\; B(\mathbf{u}; \mathbf{v}) = 0 \;\; \forall\, \mathbf{v} \in \mathbb{V}$). The complete classification of bilinear vector spaces, however, rests on the concept of an *isometry*, which is a linear, bijective map $\Psi : \mathbb{V} \rightarrow \mathbb{V}'$ between two spaces $\mathbb{V}_B$ and $\mathbb{V}'_{B'}$ which preserves the bilinear form, i.e. such that for all $\mathbf{u}, \mathbf{v} \in \mathbb{V}$

$$B(\mathbf{u}; \mathbf{v}) \; = \; B'(\Psi(\mathbf{u}), \Psi(\mathbf{v}))\,. \tag{A.8}$$

Two bilinear vector spaces which are related in this way have essentially identical algebraic properties, and are called *isometric spaces.*

A.5. If $\mathbb{U}_B$ is a nondegenerate subspace of a bilinear vector space $\mathbb{V}_B$, then there exists a map $\mathbf{pr} : \mathbb{V} \rightarrow \mathbb{U}$ such that:

(i) $\mathbf{pr}$ is linear and onto;

(ii) the restriction $\mathbf{pr}|_{\mathbb{U}}$ of $\mathbf{pr}$ to $\mathbb{U}$ is the identity map on $\mathbb{U}$;

(iii) for all $\mathbf{u} \in \mathbb{U}$ and $\mathbf{v} \in \mathbf{V}$, $B(\mathbf{u};\mathbf{v}) = B(\mathbf{u};\mathbf{pr}(\mathbf{v}))$.

The map $\mathbf{pr}$ is called the *B-orthogonal projection* of $\mathbf{V}$ onto $\mathbb{U}$. If $\mathbb{W} = \mathrm{rad}(\mathbf{V}_B)$, the *quotient space*

$$\langle\!\langle \mathbf{V} \rangle\!\rangle \;:=\; \mathbf{V}/\mathbb{W} \;:=\; \{\mathbf{u}+\mathbb{W} \mid \mathbf{u} \in \mathbf{V}\} \;:=\; \{\{\mathbf{u}+\mathbf{w} \mid \mathbf{w} \in \mathbb{W}\} \mid \mathbf{u} \in \mathbf{V}\} \qquad \text{(A.9)}$$

can be made into a nondegenerate bilinear vector space by defining its bilinear form as $B(\mathbf{u}+\mathbb{W};\mathbf{v}+\mathbb{W}) := B(\mathbf{u};\mathbf{v})$. Then, for any maximal nondegenerate subspace $\mathbb{U}_B$ of $\mathbf{V}_B$, the B-orthogonal projection induces a well-defined isometry between $\mathbb{U}_B$ and the factor space $\mathbf{V}/\mathbb{W}$, which is given by $\mathbf{v}+\mathbb{W} \mapsto \mathbf{pr}(\mathbf{v})$.

A.6. A bilinear form can be uniquely *represented* relative to a given basis $[\mathbf{u}_i \mid i = 1, \ldots, n]$ of $\mathbf{V}$ by a (symmetric) matrix

$$\underline{\mathbf{M}}_B \;=\; \underline{\mathbf{M}}_B(\mathbf{u}_1, \ldots, \mathbf{u}_n) \;:=\; [B(\mathbf{u}_i;\mathbf{u}_j) \mid i,j = 1, \ldots, n]\,. \qquad \text{(A.10)}$$

Then if we also represent each $\mathbf{v},\mathbf{w} \in \mathbf{V}$ in terms of their coordinates versus these basis elements, i.e. as $\mathbf{v} = \Sigma x_i(\mathbf{v})\mathbf{u}_i$ and $\mathbf{w} = \Sigma x_i(\mathbf{w})\mathbf{u}_i$, it follows easily from bilinearity that:

$$\begin{aligned} B(\mathbf{v};\mathbf{w}) &= \sum_{i=1}^{n}\sum_{j=1}^{n} x_i(\mathbf{v})\,x_j(\mathbf{w})\,B(\mathbf{u}_i;\mathbf{u}_j) \qquad \text{(A.11)} \\ &= [x_1(\mathbf{v}), \ldots, x_n(\mathbf{v})] \begin{bmatrix} B(\mathbf{u}_1;\mathbf{u}_1) & \cdots & B(\mathbf{u}_1;\mathbf{u}_n) \\ \cdots & \cdots & \cdots \\ B(\mathbf{u}_n;\mathbf{u}_1) & \cdots & B(\mathbf{u}_n;\mathbf{u}_n) \end{bmatrix} \begin{bmatrix} x_1(\mathbf{w}) \\ \cdot \\ x_n(\mathbf{w}) \end{bmatrix}. \end{aligned}$$

Conversely, any $n \times n$ symmetric matrix defines a bilinear form on $\mathbb{R}^n$ in this way. The matrix $\underline{\mathbf{M}}_B$ is called the *metric matrix* of the space versus the basis $[\mathbf{u}_1, \ldots, \mathbf{u}_n]$. The rank of $\underline{\mathbf{M}}_B$ is the same as the rank of the space $\mathbf{V}_B$, and the matrix is either positive or negative (semi)definite whenever the space is (semi)anisotropic. The determinant $B(\mathbf{u}_1, \ldots, \mathbf{u}_n) := \mathit{det}(\underline{\mathbf{M}}_B)$ is obviously invariant under isometries, and the sign of this determinant, which is also independent of our choice of basis, is known as the *discriminant* of the space.

A.7. Note that if $[\mathbf{u}_1, \ldots, \mathbf{u}_n]$ is a B-orthogonal basis of $\mathbf{V}_B$, then the matrix $\underline{\mathbf{M}}_B$ will be diagonal. In this case, the matrix is said to be in *canonical form*. If $\mathbf{V}_B = (\mathbf{V},B)$ is a bilinear vector space, then the numbers of positive N_B^+, negative N_B^- and zero N_B^0 diagonal entries in the metric matrix for B versus a given basis is the same for all B-orthogonal bases of the space. Given any other bilinear vector space $(\mathbf{V}',B')$ together with a diagonal matrix $\underline{\mathbf{M}}_{B'}$ which represents B' versus a B'-orthogonal basis, the spaces $\mathbf{V}_B$ and $\mathbf{V}'_{B'}$ are isometric if and only if the numbers of positive, negative and zero diagonal entries of $\underline{\mathbf{M}}_B$ and $\underline{\mathbf{M}}_{B'}$ are the same. This fact, which is known as *Sylvester's law of inertia*, provides us with a complete description of all the isometry classes of (real) bilinear vector spaces. The three numbers (N_B^+, N_B^-, N_B^0) are sometimes referred to as the *signature* of the space.

A.8. Another important, though much more recent theorem in bilinear algebra, known as the *Witt extension theorem*, states that given two isometric bilinear vector spaces $\mathbf{V}_B$ and $\mathbf{V}'_{B'}$ together with subspaces $\mathbf{U}_B$ and $\mathbf{U}'_{B'}$ of $\mathbf{V}_B$ and $\mathbf{V}'_{B'}$, respectively, if there exists an isometry $\Psi: \mathbf{U}_B \rightarrow \mathbf{U}'_{B'}$, then there exists an isometry $\bar{\Psi}: \mathbf{V}_B \rightarrow \mathbf{V}'_{B'}$ such that for the restriction of $\bar{\Psi}$ to $\mathbf{U}$ we have $\bar{\Psi}|_{\mathbf{U}} = \Psi$. The isometry $\bar{\Psi}$ is called an *extension* of Ψ. Using this theorem, one can easily show that all B-orthogonal supplements of isometric subspaces of a bilinear vector space are likewise isometric, and that the maximal null subspaces all have the same dimension; this dimension is called the *Witt index* of the space. Although these facts are also trivial consequences of Sylvester's law of inertia, Witt's theorem shows that they are valid over arbitrary fields $\mathbb{K}$, where Sylvester's law fails to hold.

A.9. A *Euclidean vector space* $\mathbf{V}_B$ is simply an anisotropic bilinear vector space whose metric matrix $\underline{\mathbf{M}}_B$ is positive definite. Equivalently, for all $\mathbf{u} \in \mathbf{V}$ the associated quadratic form satisfies $Q(\mathbf{u}) \geq 0$, while $Q(\mathbf{u}) = 0$ implies that $\mathbf{u} = \mathbf{0}$. Since the signature is $(n, 0, 0)$ where n is the dimension of the space, it follows that all Euclidean vector spaces of equal dimension are isometric. Moreover, all subspaces of a Euclidean vector space are likewise Euclidean. The value of the bilinear form $B(\mathbf{u};\mathbf{v})$ on any two vectors $\mathbf{u},\mathbf{v} \in \mathbf{V}_B$ is called the *scalar product* of the vectors, and the positive square root of the value of the corresponding quadratic form $\sqrt{Q(\mathbf{u})}$ is called the *length* of $\mathbf{u} \in \mathbf{V}_B$. By dividing each element of a B-orthogonal basis by its length, we obtain a *B-orthonormal* basis $[\mathbf{u}_i \mid i = 1, \ldots, n]$ of $\mathbf{V}_B$, relative to which the metric matrix $\underline{\mathbf{M}}_B(\mathbf{u}_1, \ldots, \mathbf{u}_n) = \mathbf{I}$, the $n \times n$ unit matrix. If we represent the vectors $\mathbf{v},\mathbf{w} \in \mathbf{V}_B$ by their coordinates versus this basis, then their scalar product becomes simply $B(\mathbf{v};\mathbf{w}) = \Sigma x_i(\mathbf{v}) x_i(\mathbf{w})$, i.e. the classical vector dot product of the corresponding *coordinate vectors* $\mathbf{x}(\mathbf{v}), \mathbf{x}(\mathbf{w}) \in \mathbb{R}^n$. The B-orthonormal basis $[\mathbf{u}_1, \ldots, \mathbf{u}_n]$ is called an *orthonormal coordinate system* for $\mathbf{V}_B$.

B. Affine and Euclidean Geometry

B.1. In a vector space $\mathbf{V}$, the origin has properties which distinguish it from all other vectors; an affine space $\mathbb{A}$ is essentially a vector space from which this asymmetry has been eliminated. A Euclidean space, of course, is also an affine space, and the affine geometric properties of a Euclidean space are, roughly speaking, those which can be described without reference to its distance function. Mathematically, an n-dimensional *affine space* $\mathbb{A}_{\mathbf{V}} = (\mathrm{P}, \mathbf{V})$ consists of two parts: the first is a set P of elements called *points*; the second is an n-dimensional (real) vector space $\mathbf{V}$. The set of points and the vector space are connected by means of a function mapping $\mathrm{P}\times\mathbf{V}\rightarrow\mathrm{P}$, which for $\mathrm{p}\in\mathrm{P}$ and $\mathbf{v}\in\mathbf{V}$ is written as $[\mathrm{p},\mathbf{v}]\mapsto\tau_{\mathbf{v}}(\mathrm{p})$. This function is required to satisfy:

$$\forall\ \mathrm{p}\in\mathrm{P}:\ \tau_{\mathbf{v}}(\mathrm{p}) = \mathrm{p}\ \text{ if and only if } \mathbf{v} = \mathbf{0}\,; \tag{B.1}$$

$$\forall\ \mathrm{p}\in\mathrm{P},\ \mathbf{u},\mathbf{v}\in\mathbf{V}:\ \tau_{(\mathbf{u}+\mathbf{v})}(\mathrm{p}) = \tau_{\mathbf{u}}(\tau_{\mathbf{v}}(\mathrm{p}))\,; \tag{B.2}$$

$$\forall\ \mathbf{v}\in\mathbf{V}:\ \text{the map } \tau_{\mathbf{v}}\colon \mathrm{P}\rightarrow\mathrm{P}\ \text{ is a bijection}\,. \tag{B.3}$$

The map $\tau_{\mathbf{v}}$ itself is called the *translation* by $\mathbf{v}\in\mathbf{V}$.

B.2. Let us note some general properties of translations. Firstly, since by (B.2) $\tau_{\mathbf{v}}\circ\tau_{-\mathbf{v}} = \tau_{\mathbf{v}-\mathbf{v}}$, the inverse $\tau_{\mathbf{v}}^{-1} = \tau_{-\mathbf{v}}$ by (B.1). Secondly, for any two points $\mathrm{p},\mathrm{q}\in\mathrm{P}$, the vector $\vec{\mathrm{pq}} := \mathbf{v}\in\mathbf{V}$ such that $\tau_{\mathbf{v}}(\mathrm{p}) = \mathrm{q}$ is unique, for if $\tau_{\mathbf{v}}(\mathrm{p}) = \mathrm{q} = \tau_{\mathbf{w}}(\mathrm{p})$ then $\mathrm{p} = \tau_{\mathbf{v}}^{-1}(\tau_{\mathbf{w}}(\mathrm{p})) = \tau_{\mathbf{w}-\mathbf{v}}(\mathrm{p})$ by (B.2), whence $\mathbf{w}-\mathbf{v} = \mathbf{0}$ by (B.1). Finally, for any three points $\mathrm{p},\mathrm{q},\mathrm{r}\in\mathrm{P}$ with $\mathbf{u} := \vec{\mathrm{pq}}$, $\mathbf{v} := \vec{\mathrm{qr}}$, $\mathbf{w} := \vec{\mathrm{pr}}$ we have

$$\tau_{(\mathbf{u}+\mathbf{v})}(\mathrm{p}) = \tau_{\mathbf{v}}(\tau_{\mathbf{u}}(\mathrm{p})) = \tau_{\mathbf{v}}(\mathrm{q}) = \mathrm{r}\,, \tag{B.4}$$

i.e. $\mathbf{u}+\mathbf{v} = \mathbf{w}$.

B.3. The functions $\tau_{\mathbf{v}}$ enable us to "transplant" the linear structure of $\mathbf{V}$ onto the set of points P. To do this, we choose an arbitrary *origin* $\mathrm{o}\in\mathrm{P}$. Then we may associate to each point p the unique vector $\mathbf{v} := \vec{\mathrm{op}}$ and to each vector $\mathbf{v}$ the unique point $\mathrm{p} := \tau_{\mathbf{v}}(\mathrm{o})$. In terms of this representation of points by vectors, for any vector $\mathbf{v} := \vec{\mathrm{pq}}$ the action of the translation $\tau_{\mathbf{v}}$ upon a point p is given by $\vec{\mathrm{op}}+\vec{\mathrm{pq}} = \vec{\mathrm{oq}}$, i.e. by vector addition. Also, for any two points p and q the translation vector $\vec{\mathrm{pq}}$ is given by $\vec{\mathrm{oq}}-\vec{\mathrm{op}}$. Note that although the vectors $\vec{\mathrm{op}}$ and $\vec{\mathrm{oq}}$ which represent the points p and q depend upon our choice of origin, the translation vector $\vec{\mathrm{pq}}$ is independent of that choice.

B.4. More generally, given any collection of points $\mathrm{q}_1,\ldots,\mathrm{q}_m$ and two origins $\mathrm{o},\mathrm{p}\in\mathbb{A}$†, we have

† Following the usual practice, we shall frequently refer to our points as elements of the space **A**, rather than the set of points P thereof.

$$\sum_{i=1}^{m} a_i\vec{oq}_i - \sum_{i=1}^{m} a_i\vec{pq}_i = \sum_{i=1}^{m} a_i(\vec{oq}_i - \vec{pq}_i) = \sum_{i=1}^{m} a_i\vec{op} = \vec{op}\sum_{i=1}^{m} a_i . \qquad \text{(B.5)}$$

It follows that for all $q_1, \ldots, q_m \in \mathbb{A}$, if the coefficients $a_1, \ldots, a_m \in \mathbb{R}$ satisfy $\Sigma a_i = 0$, then the vector $\Sigma a_i \vec{oq}_i$ is *independent* of our choice of origin $o \in \mathbb{A}$. For a given set of points $q_1, \ldots, q_m$, the subspace of $\mathbf{V}$ which is defined by:

$$< q_1, \ldots, q_m > \;:=\; \{ \sum_{i=1}^{m} a_i\vec{oq}_i \,|\, \sum_{i=1}^{m} a_i = 0 \} \qquad \text{(B.6)}$$

is called the *direction space* of the q_i. If we let $o = q_k$ for any $k \in \{1, \ldots, m\}$, then for any coefficients $\{a_j \,|\, j \neq k\}$ we can satisfy the condition $\Sigma a_i = 0$ simply by setting $a_k = -\Sigma_{i \neq k} a_i$. It follows that $<q_1, \ldots, q_m>$ can also be expressed as the linear span of the vectors $< \vec{oq}_j \,|\, j \neq k >$ for all $o = q_k \in \{q_1, \ldots, q_m\}$.

B.5. An *affine subspace* with direction space $< q_1, \ldots, q_m >$ is defined as

$$((q_1, \ldots, q_m;\, p)) \;:=\; \{\tau_{\mathbf{v}}(p) \,|\, \mathbf{v} \in \, < q_1, \ldots, q_m > \} \qquad \text{(B.7)}$$

for any $p \in \mathbb{A}$. It is itself easily seen to be an affine space (containing the point p) whose associated vector space is just $< q_1, \ldots, q_m >$. If we replace p by any other point $r \in \mathbb{A}$ in (B.7), the result is another affine subspace, which is said to be *parallel* to the first. In particular, if $r = q_j$ for any $j = 1, \ldots, m$, we obtain an affine subspace $((q_1, \ldots, q_m))$ which is called the *affine span* of the points $q_1, \ldots, q_m$. For a given choice of origin $o \in \mathbb{A}$ and $j \in \{1, \ldots, m\}$, the vector which represents any point $p \in ((q_1, \ldots, q_m))$ of this subspace can be written as

$$\begin{aligned} \vec{op} &= \vec{oq}_j + \sum_{i=1}^{m} a_i\vec{oq}_i \quad \text{with } \sum_{i=1}^{m} a_i = 0 \qquad \text{(B.8)} \\ &= \sum_{i=1}^{m} b_i\vec{oq}_i \quad \text{with } \sum_{i=1}^{m} b_i = 1 \end{aligned}$$

where $b_i := a_i$ for all $i \neq j$ and $b_j := a_j + 1$. Note in particular that this equation holds independent of our choice of origin $o \in \mathbb{A}$, so that the point p such that $\vec{op} = \Sigma b_i \vec{oq}_i$ will likewise be independent of our choice of origin $o \in \mathbb{A}$ whenever $\Sigma b_i = 1$. This fact enables us to *symbolically* represent the point p by a *formal* linear combination $\Sigma b_i q_i$ with $\Sigma b_i = 1$ of the points q_i themselves. We call p the *barycentric sum* of the q_i.

B.6. Hence the affine subspace spanned by the points $q_1, \ldots, q_m$ can be written somewhat more elegantly as

$$((q_1, \ldots, q_m)) \;=\; \{ \sum_{i=1}^{m} b_i q_i \,|\, \sum_{i=1}^{m} b_i = 1 \} . \qquad \text{(B.9)}$$

If we further restrict the coefficients b_i to be nonnegative, the resultant set is called the *convex span* of the q_i, which we shall denote by $[(q_1, \ldots, q_m)]$. A set of points $q_1, \ldots, q_m \in \mathbb{A}$ is called *independent* if for any one and hence all choices of origin $\mathrm{o} \in \mathbb{A}$:

$$\sum_{i=1}^{n+1} a_i = 0 \quad \text{and} \quad \sum_{i=1}^{n} a_i \vec{\mathrm{oq}}_i = 0 \quad \Rightarrow \quad a_i = 0 \quad \text{for } i = 1, \ldots, n\,. \tag{B.10}$$

This is equivalent to saying that for $\mathrm{o} = \mathrm{q}_j$ the vectors $\{\vec{\mathrm{oq}}_i \mid i \neq j\}$ are linearly independent for one and hence all $j \in \{1, \ldots, m\}$. In this case, the coefficients occurring in the above representation of a point $\mathrm{p} = \Sigma b_i \mathrm{q}_i$ in terms of the q_i are *unique.* If in addition $m = n+1$, then all points of the space lie within the affine span $((\mathrm{q}_1, \ldots, \mathrm{q}_m))$, so that all points of the space can be uniquely represented by the vector of coefficients $[b_1, \ldots, b_{n+1}]$. The coefficients in such a representation are called the *barycentric coordinates* of the points p versus the (ordered) *barycentric basis* $[\mathrm{q}_1, \ldots, \mathrm{q}_{n+1}]$.

B.7. An *affine transformation* $T\colon \mathrm{P} \rightarrow \mathrm{P}$ is one which preserves the barycentric coordinates, i.e.

$$\mathrm{p} = \sum_{i=1}^{n+1} b_i \mathrm{q}_i \quad \Leftrightarrow \quad T(\mathrm{p}) = \sum_{i=1}^{n+1} b_i T(\mathrm{q}_i)\,. \tag{B.11}$$

As an immediate consequence of this definition, we see that affine transformations map affine subspaces onto affine subspaces of equal dimension, and that they preserve the parallelism of subspaces. In particular, affine transformations are *collineations* which map lines (i.e. 1-dimensional affine subspaces) onto lines. Note that translations are affine transformations, and that any nonsingular linear transformation, applied to the vectors representing the points relative to a common origin, induces an affine transformation. In fact, any affine transformation can be uniquely expressed as the composition of a translation and a linear transformation of the vectors representing the points from a constant origin.

B.8. We now choose an origin $\mathrm{o} \in \mathbb{A}$ together with a basis $[\mathbf{v}_i \mid i = 1, \ldots, n]$ of the associated bilinear vector space $\mathbf{V}_B$. This gives us a representation of points $\mathrm{p} \in \mathbb{A}$ by vectors $\vec{\mathrm{op}} \in \mathbf{V}$. At the same time, it allows us to represent each and every vector $\mathbf{x} \in \mathbf{V}$ by its coordinate vector $[x_1, \ldots, x_n] \in \mathbb{R}^n$. The result is that we now have a representation of each and every point $\mathrm{p} \in \mathbb{A}$ by such a coordinate vector. We call the pair $(\mathrm{o}, [\mathbf{v}_1, \ldots, \mathbf{v}_n])$ an *affine coordinate system* for $\mathbb{A}$, and given an affine coordinate system we write the *affine coordinates* of each $\mathrm{p} \in \mathbb{A}$ as $[x_1(\mathrm{p}), \ldots, x_n(\mathrm{p})]$. In terms of this representation, the translation $\vec{\mathrm{pq}}$ may be written as

$$\vec{\mathrm{pq}} = [x_1(\mathrm{q}) - x_1(\mathrm{p}), \ldots, x_n(\mathrm{q}) - x_n(\mathrm{p})]\,. \tag{B.12}$$

B.9. An *bilinear affine space* $\mathbb{A}_B = (\mathrm{P}, \mathbf{V}, B)$ is simply an affine space whose associated space of translations $\mathbf{V}$ has a bilinear form defined on it. Each such space is automatically endowed with a function

$$D(\mathrm{p,q;\,r,s}) \;:=\; B(\vec{\mathrm{pq}};\,\vec{\mathrm{rs}})\,, \tag{B.13}$$

which we call an *affine bilinear form.* Restricted to arguments $\mathrm{p}=\mathrm{r}$ and $\mathrm{q}=\mathrm{s}$, the function becomes an *affine quadratic form,* which we write as $D(\mathrm{p,q}) := D(\mathrm{p,q;\,p,q})$. An *isometry* is a map $\Psi:\mathbb{A}_B\rightarrow\mathbb{A}_B$ which satisfies $D(\Psi(\mathrm{p}),\Psi(\mathrm{q})) = D(\mathrm{p,q})$ for all $\mathrm{p,q}\in\mathbb{A}_B$. All the metrical properties of an affine bilinear space follow from its associated form, and in particular any two isometric spaces have identical metrical properties. At this point, it becomes almost trivial to define an *n-dimensional Euclidean space* $\mathbb{E} := (\mathrm{P}, \mathbf{V}, B)$: it is simply an n-dimensional bilinear affine space whose associated bilinear form B is positive definite.

B.10. We now define the *distance* between two points $\mathrm{p,q}\in\mathbb{E}$ to be $d(\mathrm{p,q}) := \sqrt{D(\mathrm{p,q})}$. If we choose an affine basis $(\mathrm{o}, [\mathbf{v}_1,\ldots,\mathbf{v}_n])$ such that $[\mathbf{v}_1,\ldots,\mathbf{v}_n]$ is orthonormal (as in section §9 of Appendix A), the distance can be written in a particularly simple form:

$$d(\mathrm{p,q}) \;=\; \left(\sum_{i=1}^{n}(x_i(\mathrm{q})-x_i(\mathrm{p}))^2\right)^{1/2}, \tag{B.14}$$

where $[x_i(\mathrm{p})\,|\,i=1,\ldots,n]$ are the corresponding affine coordinates of $\mathrm{p}\in\mathrm{P}$. This specific kind of affine coordinate system is known as a *Cartesian coordinate system,* and the corresponding *Cartesian coordinates* $[x_1(\mathrm{p}),\ldots,x_n(\mathrm{p})]$ are generally the coordinates of choice for Euclidean geometry. By using a Cartesian coordinate system, it in fact becomes possible to use all of the results of classical vector algebra [Wills1958] to analyze the geometric relations between points, lines and other subsets of a Euclidean space $\mathbb{E}$. In particular, we can describe the *isometries* of the space most simply by using Cartesian coordinates. For example, the translations of the space are clearly isometries, because given a translation $\mathbf{t}=\vec{\mathrm{pr}}=\vec{\mathrm{qs}}$ whose coordinates versus the basis $[\mathbf{v}_1,\ldots,\mathbf{v}_n]$ are $[t_1,\ldots,t_n]$, we have

$$D(\mathrm{p,q}) \;=\; \sum_{i=1}^{n}((x_i(\mathrm{q})+t_i)-(x_i(\mathrm{p})+t_i))^2 \;=\; \sum_{i=1}^{n}(x_i(\mathrm{r})-x_i(\mathrm{s}))^2 \;=\; D(\mathrm{r,s}). \tag{B.15}$$

In addition, we may ask when a matrix $\underline{\mathbf{R}}$ multiplying the Cartesian coordinates is an isometry. Very simply,

$$\begin{aligned}\sum_{i=1}^{n}\sum_{j=1}^{n}(r_{ij}(x_j(\mathrm{q})-x_j(\mathrm{p})))^2 \;&=\; \sum_{j=1}^{n}\sum_{k=1}^{n}(x_j(\mathrm{q})-x_j(\mathrm{p}))(x_k(\mathrm{q})-x_k(\mathrm{p}))\sum_{i=1}^{n}r_{ij}r_{ik} \qquad (\mathrm{B.16})\\ &=\; \sum_{j=1}^{n}(x_j(\mathrm{q})-x_j(\mathrm{p}))^2\end{aligned}$$

if and only if $\Sigma_i r_{ij}r_{ik}=1$ if $j=k$ and 0 otherwise. This can be expressed more simply in matrix terms as $\underline{\mathbf{R}}^T\underline{\mathbf{R}}=\underline{\mathbf{I}}$ (the unit matrix), and a matrix $\underline{\mathbf{R}}$ which fulfills this condition is called *orthogonal*. For a fixed origin p, the point r which is obtained from p by applying the translation with coordinates

$$[\Sigma_j r_{ij}(x_j(\mathrm{q})-x_j(\mathrm{p})) \mid i = 1, \ldots, n] \tag{B.17}$$

is said to be obtained by *rotation* of the point q about p by the orthogonal matrix $\underline{\mathbf{R}}$. The *fundamental theorem of Euclidean geometry* states that all the isometries of a Euclidean space $\mathbb{E}$ can be written uniquely as the composition of a translation and a rotation about a fixed (but arbitrary) origin $\mathrm{p} \in \mathbb{E}$. Since they can be written as the composition of a translation and a linear transformation in the associated vector space, we conclude that all isometries are affine transformations. *Euclidean geometry* may now be defined as the study of those properties of $\mathbb{R}^n$ which are invariant under translation and rotation.

B.11. A Euclidean space is still an affine space, however, and hence it should be possible to express the barycentric coordinates in terms of the Cartesian. To do this, we first express a barycentric sum p of points $\mathrm{q}_1, \ldots, \mathrm{q}_{n+1}$ in terms of the Cartesian coordinates:

$$[x_j(\mathrm{p}) \mid j = 1, \ldots, n] \;=\; [\sum_{i=1}^{n+1} b_i x_j(\mathrm{q}_i) \mid j = 1, \ldots, n] \tag{B.18}$$

This is a system of n equations in the $n+1$ unknown barycentric coordinates of the point p. The remaining condition is obtained by recalling that by definition the barycentric coordinates must satisfy $\Sigma b_i = 1$. Thus the barycentric coordinates of p are the solution of the system of linear equations:

$$\begin{bmatrix} 1 & \cdots & 1 \\ x_1(\mathrm{q}_1) & \cdots & x_1(\mathrm{q}_{n+1}) \\ \cdots & \cdots & \cdots \\ x_n(\mathrm{q}_1) & \cdots & x_n(\mathrm{q}_{n+1}) \end{bmatrix} \begin{bmatrix} b_1 \\ b_2 \\ \cdot \\ b_n \end{bmatrix} = \begin{bmatrix} 1 \\ x_1(\mathrm{p}) \\ \cdot \\ x_n(\mathrm{p}) \end{bmatrix}, \tag{B.19}$$

or $\underline{\mathbf{Q}}\mathbf{b} = \mathbf{p}$ in matrix notation. Using Cramer's rule for solving such systems, we obtain:

$$b_i = \frac{1}{det(\underline{\mathbf{Q}})} \, det \begin{bmatrix} 1 & \cdots & 1 & 1 & 1 & \cdots & 1 \\ x_1(\mathrm{q}_1) & \cdots & x_1(\mathrm{q}_{i-1}) & x_1(\mathrm{p}) & x_1(\mathrm{q}_{i+1}) & \cdots & x_1(\mathrm{q}_{n+1}) \\ \cdots & \cdots & \cdots & \cdots & \cdots & \cdots & \cdots \\ x_n(\mathrm{q}_1) & \cdots & x_n(\mathrm{q}_{i-1}) & x_n(\mathrm{p}) & x_n(\mathrm{q}_{i+1}) & \cdots & x_n(\mathrm{q}_{n+1}) \end{bmatrix}. \tag{B.20}$$

It is well known that the determinant $det(\underline{\mathbf{Q}})$, when divided by a normalization factor $1/n!$, is the *oriented hypervolume* of the simplex (higher dimensional tetrahedron) defined by the convex span of the basis $\mathrm{q}_1, \ldots, \mathrm{q}_{n+1}$. Thus the barycentric coordinates are equal to the hypervolumes of the simplices obtained by replacing each point q_i in turn by p, and dividing through by the volume of basis. Note in particular that if we place q_{n+1} at the origin of our Cartesian coordinate system, and each remaining q_i at unit distance along the i-th coordinate axis, we obtain $b_i = x_i(p)$ for $i = 1, \ldots, n$. Thus the Cartesian coordinates of p *are* its first n barycentric coordinates versus this particular basis.

B.12. In the above, we have shown how Euclidean geometry fits into the wider scheme of bilinear affine spaces. Because Euclidean spaces are bilinear affine spaces, *all of the facts given above for arbitrary bilinear affine spaces are also true of Euclidean spaces.* Although the above presentation was made using abstract vector and affine spaces, in most of the rest of the book we shall deal with concrete algebraic models of these spaces over $\mathbb{R}^n$. This is sometimes a little confusing, because then we have essentially two copies of $\mathbb{R}^n$ to deal with: the first is the coordinates of the points, and the second is the coordinates of the vectors. The context will generally make clear what our coordinates refer to.

C. The Analytic Projective Geometry of Three Dimensions

C.1. In Appendix B we saw that the n-fold Cartesian product of the real numbers $\mathbb{R}^n$ admits two essentially different geometric interpretations. The first is as a (Euclidean) vector space, while the second is as an affine space endowed with the function $\mathbf{x}, \mathbf{y} \mapsto \mathbf{x} - \mathbf{y}$ which maps pairs of points to vectors. In this appendix we present a brief account of yet another, more general interpretation, known as a *projective space*, which allows both affine and vector spaces to be treated together in a unified way. Because only the three dimensional version is needed in this book, we concentrate primarily on that case in what follows. More complete accounts may be found in many places, for example [Pedoe1970, Havel1987]. In contrast to the previous appendices, we shall formulate all our definitions directly in terms of coordinates.

C.2. A *projective point* is defined as a line through the origin in $\mathbb{R}^4$ with the origin itself removed, i.e. as a set of vectors of the form $\hat{\mathbf{x}} = \mathbb{R}^{\#} \cdot \mathbf{x} := \{a\mathbf{x} \mid a \neq 0\}$ for some nonzero $\mathbf{x} \in \mathbb{R}^4$, where $\mathbb{R}^{\#}$ denotes the set of all nonzero real numbers. The components of the vector $\mathbf{x}$ itself are called homogeneous coordinates for the projective point $\hat{\mathbf{x}}$; thus, the coordinates of a projective point are not unique even with respect to a fixed coordinate system. A projective *line* consists of a plane through the origin with the origin removed, i.e. a set of the form $\{a\mathbf{x} + b\mathbf{y} \mid \{a, b\} \neq \{0\}\}$ for linearly independent vectors $\mathbf{x}, \mathbf{y} \in \mathbb{R}^4$. Similarly, a projective *plane* is a set of the form $\{a\mathbf{x} + b\mathbf{y} + c\mathbf{z} \mid \{a, b, c\} \neq \{0\}\}$ for linearly independent vectors $\mathbf{x}, \mathbf{y}, \mathbf{z} \in \mathbb{R}^4$. The set of all projective points, taken together, is called a (three-dimensional) *projective space*, and is denoted by $\mathbb{P} = \mathbb{P}^3 = \mathbb{R}^4/\mathbb{R}^{\#}$.

Clearly, this space consists of all nontrivial linear combinations of any four linearly independent vectors in $\mathbb{R}^4$, modulo multiplication by $\mathbb{R}^{\#}$.

C.3. In order to see what the relationship between projective, affine and vector spaces is, we show how any three-dimensional affine space $\mathbb{A} = (\mathrm{P}, \mathbb{V})$ can be "completed", in a natural way, to obtain a copy of the projective space $\mathbb{P} = \mathbb{R}^4/\mathbb{R}^{\#}$. We do this by choosing a fixed coordinate system $[\mathrm{o},\mathbf{u}_1,\mathbf{u}_2,\mathbf{u}_3]$ for $\mathbb{A}$, and assigning to each affine point $\mathrm{p} \in \mathrm{P}$ with affine coordinates $[p_1,p_2,p_3]$ its *standard homogeneous coordinates* $\mathbf{p} := [1,p_1,p_2,p_3]$. At the same time, each vector $\mathbf{v}$ is given standard homogeneous coordinates $[0,v_1,v_2,v_3]$, where $[v_1,v_2,v_3]$ are its coordinates versus the vector coordinate system $[\mathbf{u}_1,\mathbf{u}_2,\mathbf{u}_3]$. Observe that the set of all standard homogeneous coordinates of points $\mathbb{H}_1 := \{\mathbf{p} \in \mathbb{R}^4 \mid p_1 = 1\}$ is an affine hyperplane of $\mathbb{R}^4$, while the standard homogeneous coordinates of vectors make up a vector hyperplane $\mathbb{V}_0 \subseteq \mathbb{R}^4$. If $\mathbb{P}_1 := \{\mathbb{R}^{\#}\cdot\mathbf{p} \mid \mathbf{p} \in \mathbb{H}_1\}$, then the mapping $\mathrm{P} \rightarrow \mathbb{P}_1 : \mathrm{p} \mapsto \mathbf{p}$ is an embedding of the points in the projective space $\mathbb{P}$, and $\mathbb{P}_0 := \{\mathbb{R}^{\#}\cdot\mathbf{v} \mid \mathbf{v} \in \mathbb{V}\} \subseteq \mathbb{P}$ is a projective plane therein, such that $\mathbb{P}_0 \cup \mathbb{P}_1 = \mathbb{P}$. Because they can be thought of as points "on the horizon" where parallel lines in $\mathbb{P}_1$ meet, the points of $\mathbb{P}_0$ are often called *points at infinity*, while the subspace $\mathbb{P}_0$ itself is called the *plane at infinity*.

C.4. *Projective geometry* is the study of those properties of sets of projective points which are preserved under arbitrary linear transformations $\underline{\mathbf{T}} = [t_{ij}]$ of the homogeneous coordinates of the points:

$$\underline{\mathbf{T}}\cdot\mathbf{p} \quad = \quad \left[\Sigma t_{1i}p_i, \ldots, \Sigma t_{4i}p_i\right]^T . \tag{C.1}$$

These are called *projective transformations*, and the corresponding group is $\mathbf{GL}(n; \mathbb{R})/\mathbb{R}^{\#}$. The projective properties of an affine space are those which are preserved under projective transformations of the standard homogeneous coordinates of the points and vectors therein. Observe that since $\Sigma t_{1i}p_i$ can be zero even if $p_1 = 1$, these transformations may map points at infinity to finite points and vice versa. Thus, just as affine geometry is the geometry of a vector space to which a new type of symmetry, the translations, has been added to eliminate the distinction which exists between the origin and any other vector of the space, projective geometry is essentially the geometry of an affine space whose symmetry group has been enlarged so that no further distinction exists between finite and infinite points.

C.5. The unique projective line

$$< \hat{\mathbf{p}}, \hat{\mathbf{q}} > \quad := \quad \{a\mathbf{p} + b\mathbf{q} \mid \mathbf{p} \in \hat{\mathbf{p}}, \mathbf{q} \in \hat{\mathbf{q}}, a,b \in \mathbb{R}^{\#}\} \tag{C.2}$$

which contains two given distinct projective points is called the *join* of $\hat{\mathbf{p}}$ and $\hat{\mathbf{q}}$. In the event that both points are finite, this corresponds to the unique line in the affine space $(\mathbb{P}_1, \mathbf{V}_0)$ which passes through the two points in question. If one of the two projective points is at infinity, say $\hat{\mathbf{q}}$, the result is again an affine line which passes through the finite point $\hat{\mathbf{p}} \in \mathbb{P}_1$ in the direction given by any vector $\mathbf{q} \in \mathbf{V}_0$ of homogeneous coordinates for $\hat{\mathbf{q}}$. If both $\hat{\mathbf{p}}, \hat{\mathbf{q}} \in \mathbb{P}_0$, however, the set of all their nontrivial linear combinations is the two-dimensional subspace $\mathbb{U} \subseteq \mathbf{V}_0$ (modulo $\mathbb{R}^{\#}$) which contains any two vectors $\mathbf{p}$ and $\mathbf{q}$ of homogeneous coordinates for $\hat{\mathbf{p}}$ and $\hat{\mathbf{q}}$, with the origin removed. Similarly, the join of any three non-collinear projective points is the unique projective plane which contains them. It may be interpreted as a plane in the affine space $\mathbb{P}_1$ unless all three points lie in the plane at infinity, in which case it is equal to that plane.

C.6. We now attempt to find an analytic description of these geometric operations, beginning with the join of three projective points whose homogeneous coordinates are $\mathbf{p}, \mathbf{q}, \mathbf{r} \in \mathbb{R}^4$. Since any four vectors in $\mathbb{R}^4$ are linearly dependent if and only if the determinant of the matrix whose columns are their components vanishes, the set of vectors $< \mathbf{p}, \mathbf{q}, \mathbf{r} >$ is equal to the locus of the homogeneous linear equation:

$$0 = det\begin{bmatrix} x_1 & p_1 & q_1 & r_1 \\ x_2 & p_2 & q_2 & r_2 \\ x_3 & p_3 & q_3 & r_3 \\ x_4 & p_4 & q_4 & r_4 \end{bmatrix} = x_1\, det\begin{bmatrix} p_2 & q_2 & r_2 \\ p_3 & q_3 & r_3 \\ p_4 & q_4 & r_4 \end{bmatrix} - x_2\, det\begin{bmatrix} p_1 & q_1 & r_1 \\ p_3 & q_3 & r_3 \\ p_4 & q_4 & r_4 \end{bmatrix} + \tag{C.3}$$

$$x_3\, det\begin{bmatrix} p_1 & q_1 & r_1 \\ p_2 & q_2 & r_2 \\ p_4 & q_4 & r_4 \end{bmatrix} - x_4\, det\begin{bmatrix} p_1 & q_1 & r_1 \\ p_2 & q_2 & r_2 \\ p_3 & q_3 & r_3 \end{bmatrix}.$$

Hence the values of these four 3×3 determinants in the components of $\mathbf{p}, \mathbf{q}, \mathbf{r}$ are all we need to know in order to determine the subspace $< \mathbf{p}, \mathbf{q}, \mathbf{r} >$. We now arrange these determinants in a vector ordered lexicographically by the indices of the components as:

$$\mathbf{p} \vee \mathbf{q} \vee \mathbf{r} \tag{C.4}$$

$$:= \left[det\begin{bmatrix} p_1 & q_1 & r_1 \\ p_2 & q_2 & r_2 \\ p_3 & q_3 & r_3 \end{bmatrix}, det\begin{bmatrix} p_1 & q_1 & r_1 \\ p_2 & q_2 & r_2 \\ p_4 & q_4 & r_4 \end{bmatrix}, det\begin{bmatrix} p_1 & q_1 & r_1 \\ p_3 & q_3 & r_3 \\ p_4 & q_4 & r_4 \end{bmatrix}, det\begin{bmatrix} p_2 & q_2 & r_2 \\ p_3 & q_3 & r_3 \\ p_4 & q_4 & r_4 \end{bmatrix} \right]^T.$$

Since any multiple of its components determines the same plane, we can regard the vector $\mathbf{p} \vee \mathbf{q} \vee \mathbf{r} \in \mathbb{R}^4$ as the homogeneous coordinates of the join of three projective points $< \hat{\mathbf{p}}, \hat{\mathbf{q}}, \hat{\mathbf{r}} > \subseteq \mathbb{R}^4/\mathbb{R}^{\#}$.

C.7. In order to obtain an analytic description of the join of two projective points with homogeneous coordinates $\mathbf{p}$ and $\mathbf{q}$, we observe that three given vectors $\mathbf{p}, \mathbf{q}, \mathbf{x} \in \mathbb{R}^4$ are dependent if and only if the four vectors $\mathbf{p}, \mathbf{q}, \mathbf{x}, \mathbf{y}$ are dependent for all $\mathbf{y} \in \mathbb{R}^4$. Letting $\mathbf{y}$ be each of the unit vectors of $\mathbb{R}^4$ in turn shows that this is the case if and only if $\mathbf{p}\vee\mathbf{q}\vee\mathbf{x} = \mathbf{0}$, i.e. $<\mathbf{p}, \mathbf{q}>$ consists of the solutions of the system of linear equations:

$$0 = -det\begin{bmatrix} p_1 & q_1 & x_1 \\ p_2 & q_2 & x_2 \\ p_3 & q_3 & x_3 \end{bmatrix} = -x_1\, det\begin{bmatrix} p_2 & q_2 \\ p_3 & q_3 \end{bmatrix} + x_2\, det\begin{bmatrix} p_1 & q_1 \\ p_3 & q_3 \end{bmatrix} - x_3\, det\begin{bmatrix} p_1 & q_1 \\ p_2 & q_2 \end{bmatrix};$$

$$0 = det\begin{bmatrix} p_1 & q_1 & x_1 \\ p_2 & q_2 & x_2 \\ p_4 & q_4 & x_4 \end{bmatrix} = x_1\, det\begin{bmatrix} p_2 & q_2 \\ p_4 & q_4 \end{bmatrix} - x_2\, det\begin{bmatrix} p_1 & q_1 \\ p_4 & q_4 \end{bmatrix} + x_4\, det\begin{bmatrix} p_1 & q_1 \\ p_2 & q_2 \end{bmatrix}; \qquad (C.5)$$

$$0 = -det\begin{bmatrix} p_1 & q_1 & x_1 \\ p_3 & q_3 & x_3 \\ p_4 & q_4 & x_4 \end{bmatrix} = -x_1\, det\begin{bmatrix} p_3 & q_3 \\ p_4 & q_4 \end{bmatrix} + x_3\, det\begin{bmatrix} p_1 & q_1 \\ p_4 & q_4 \end{bmatrix} - x_4\, det\begin{bmatrix} p_1 & q_1 \\ p_3 & q_3 \end{bmatrix};$$

$$0 = det\begin{bmatrix} p_2 & q_2 & x_2 \\ p_3 & q_3 & x_3 \\ p_4 & q_4 & x_4 \end{bmatrix} = x_2\, det\begin{bmatrix} p_3 & q_3 \\ p_4 & q_4 \end{bmatrix} - x_3\, det\begin{bmatrix} p_2 & q_2 \\ p_4 & q_4 \end{bmatrix} + x_4\, det\begin{bmatrix} p_2 & q_2 \\ p_3 & q_3 \end{bmatrix}.$$

These equations can be written in matrix form as $\mathbf{S}\cdot\mathbf{x} = \mathbf{0}$, where $\mathbf{S}$ is a skew-symmetric (i.e. $\mathbf{S}^T = -\mathbf{S}$) matrix whose absolute components are equal to the six 2×2 determinants occurring above. Since the matrix $\mathbf{S}$ determines the line completely and any nonzero multiple of $\mathbf{S}$ determines the same line, we can regard its components as the homogeneous coordinates of the projective line $<\mathbf{p}, \mathbf{q}>$, and as before we arrange these determinants lexicographically by index in a vector:

$$\mathbf{p}\vee\mathbf{q} := \left[det\begin{bmatrix} p_1 & q_1 \\ p_2 & q_2 \end{bmatrix}, \ldots, det\begin{bmatrix} p_3 & q_3 \\ p_4 & q_4 \end{bmatrix} \right]^T. \qquad (C.6)$$

C.8. The $\binom{4}{2} = 6$-dimensional vector $\mathbf{p}\vee\mathbf{q}$ is called the *outer product* of the vectors $\mathbf{p}, \mathbf{q} \in \mathbb{R}^4$, while a $\binom{4}{3} = 4$-dimensional vector of the form $\mathbf{p}\vee\mathbf{q}\vee\mathbf{r}$ is called the outer product of the three vectors $\mathbf{p}$, $\mathbf{q}$ and $\mathbf{r}$. These outer products may also be referred to as *k-vectors* for $k = 2, 3$, whereas 1-vectors are simply the original vectors of $\mathbb{R}^4$, and by convention 0-vectors are scalars. If we further define the outer product of any four vectors $\mathbf{p}$, $\mathbf{q}$, $\mathbf{r}$ and $\mathbf{s} \in \mathbb{R}^4$ to be the (scalar) 4-vector $det(\mathbf{p}, \mathbf{q}, \mathbf{r}, \mathbf{s})$, the outer product becomes an associative binary operation "$\vee$"

which maps a k-vector and an l-vector into a $(k+l)$-vector in $\mathbb{R}^{\binom{4}{k+l}}$ for all $k+l \leq 4$. Observe that since swapping any two columns of a determinant changes its sign, the outer product is *alternating*, meaning that $\mathbf{p} \vee \mathbf{q} = -\mathbf{q} \vee \mathbf{p}$, and in particular $\mathbf{x} \vee \mathbf{x} = \mathbf{0} \;\; \forall\, \mathbf{x} \in \mathbb{R}^4$. Also, since the determinant function is linear in each of its columns, the outer product of k and l-vectors $\mathbf{P}$ and $\mathbf{Q}$ extends in a well-defined way to an operation on the corresponding projective k and l-vectors $\hat{\mathbf{P}} := \{a \cdot \mathbf{P} \mid a \neq 0\}$ and $\hat{\mathbf{Q}}$. The components of these k-vectors are called the *Plücker coordinates* of the projective subspace obtained by taking the join of the corresponding projective points, and the vector space consisting of all linear combinations of k-vectors itself is denoted by $\bigvee_k(\mathbb{R}^4) \approx \mathbb{R}^{\binom{4}{k}}$.

C.9. Now for a word on the often neglected affine interpretation of outer products. If $\mathbf{p} = [1, p_1, p_2, p_3]$ and $\mathbf{q}$ are the standard homogeneous coordinates of the affine points $\hat{\mathbf{p}}, \hat{\mathbf{q}} \in \mathbb{P}_1$, their outer product has the form:

$$\mathbf{p} \vee \mathbf{q} = \begin{bmatrix} q_1 - p_1 \\ q_2 - p_2 \\ q_3 - p_3 \\ p_1 q_2 - p_2 q_1 \\ p_1 q_3 - p_3 q_1 \\ p_2 q_3 - p_3 q_2 \end{bmatrix}. \tag{C.7}$$

The first three components of this vector are those of the translation mapping $\hat{\mathbf{p}}$ to $\hat{\mathbf{q}}$, and as such are independent of our choice of origin in the affine space $(\mathbb{P}_1, \mathbf{V}_0)$, whereas the remaining components determine the position of the line in relation to the origin. Thus the outer product of affine points can be thought of as a vector of length $\sqrt{\Sigma(q_i - p_i)^2}$ contained in the line passing through $\mathbf{p}$ and $\mathbf{q}$, whose position along the line is not specified. This is sometimes known as a *line-bound vector*, to distinguish it from the *free vectors* in $\mathbf{V}_0$. If $\hat{\mathbf{p}}, \hat{\mathbf{q}} \in \mathbb{P}_0$, however, then the first three components of their outer product are zero, and the remaining components are, up to sign and order, those of the cross product of the three-dimensional vectors $[p_1, p_2, p_3]$ and $[q_1, q_2, q_3]$. This cross product, in turn, is a vector orthogonal to the subspace of $\mathbf{V}_0$ which is represented by $\mathbf{p} \vee \mathbf{q}$ whose magnitude is equal to the area spanned by these free vectors. Finally, we observe that if $\hat{\mathbf{r}}$ is any projective point and $\mathbf{r}$ are its standard homogeneous coordinates, the sum

$$\mathbf{p} \vee \mathbf{q} + \mathbf{q} \vee \mathbf{r} + \mathbf{r} \vee \mathbf{p} = (\mathbf{q} - \mathbf{p}) \vee (\mathbf{r} - \mathbf{p}) \tag{C.8}$$

is always the outer product of two free vectors in $\mathbf{V}_0$, which span an area twice that of the triangle $[(\mathbf{p}, \mathbf{q}, \mathbf{r})]$. This is analogous to the fact that the difference $\mathbf{q} - \mathbf{p}$ of the standard homogeneous coordinates of two affine points always represents a point at infinity.

C.10. For the standard homogeneous coordinates of 3-vectors, we have

$$\mathbf{p} \vee \mathbf{q} \vee \mathbf{r} \tag{C.9}$$

$$= \left[det\begin{bmatrix} 1 & 1 & 1 \\ p_1 & q_1 & r_1 \\ p_2 & q_2 & r_2 \end{bmatrix}, det\begin{bmatrix} 1 & 1 & 1 \\ p_1 & q_1 & r_1 \\ p_3 & q_3 & r_3 \end{bmatrix}, det\begin{bmatrix} 1 & 1 & 1 \\ p_2 & q_2 & r_2 \\ p_3 & q_3 & r_3 \end{bmatrix}, det\begin{bmatrix} p_1 & q_1 & r_1 \\ p_2 & q_2 & r_2 \\ p_3 & q_3 & r_3 \end{bmatrix} \right]^T .$$

In this case, the first three components are translation independent and vanish if $\hat{\mathbf{p}}, \hat{\mathbf{q}}, \hat{\mathbf{r}} \in \mathbb{P}_0$. If $\hat{\mathbf{s}}$ is any other projective point, then

$$-\mathbf{p} \vee \mathbf{q} \vee \mathbf{r} + \mathbf{p} \vee \mathbf{q} \vee \mathbf{s} - \mathbf{p} \vee \mathbf{r} \vee \mathbf{s} + \mathbf{q} \vee \mathbf{r} \vee \mathbf{s} = (\mathbf{q} - \mathbf{p}) \vee (\mathbf{r} - \mathbf{p}) \vee (\mathbf{s} - \mathbf{p}) \tag{C.10}$$

is again the outer product of free vectors, and the magnitude of its nonzero components is exactly six times the volume of the tetrahedron spanned by the affine points $\hat{\mathbf{p}}, \hat{\mathbf{q}}, \hat{\mathbf{r}}, \hat{\mathbf{s}}$. This is effectively the discrete version of Stoke's theorem.

C.11. There exist 2-vectors in $\bigvee_2(\mathbb{R}^4)$ which are not the outer products of any two vectors in $\mathbb{R}^4$. The reason is that the components of any such 2-vector $\mathbf{p} \vee \mathbf{q}$ must satisfy the equation

$$0 = det\begin{bmatrix} p_1 & q_1 & p_1 & q_1 \\ p_2 & q_2 & p_2 & q_2 \\ p_3 & q_3 & p_3 & q_3 \\ p_4 & q_4 & p_4 & q_4 \end{bmatrix} \tag{C.11}$$

$$= 2\left(det\begin{bmatrix} p_1 & q_1 \\ p_2 & q_2 \end{bmatrix} \begin{bmatrix} p_3 & q_3 \\ p_4 & q_4 \end{bmatrix} - det\begin{bmatrix} p_1 & q_1 \\ p_3 & q_3 \end{bmatrix} \begin{bmatrix} p_2 & q_2 \\ p_4 & q_4 \end{bmatrix} + det\begin{bmatrix} p_1 & q_1 \\ p_4 & q_4 \end{bmatrix} \begin{bmatrix} p_2 & q_2 \\ p_3 & q_3 \end{bmatrix} \right)$$

(where we have expanded the 4×4 determinant, whose value is obviously zero, along its first two columns to obtain the expression on the r.h.s.). Hence if $S_1 S_6 - S_2 S_5 + S_3 S_4 \neq 0$ for some $\mathbf{S} \in \bigvee_2(\mathbb{R}^4)$, there exists no $\mathbf{p}, \mathbf{q} \in \mathbb{R}^4$ such that $\mathbf{S} = \mathbf{p} \vee \mathbf{q}$, even though $\mathbf{S}$ *can* always be obtained by taking an appropriate linear combination of such outer products. Those 2-vectors in $\bigvee_2(\mathbb{R}^4)$ which are equal to the outer products of pairs of vectors in $\mathbb{R}^4$ will be called *simple 2-vectors*, whereas those which are not simple are called *compound*. We will prove in Chapter 2 that Equation (C.11) is also sufficient for a 2-vector to be simple. These equations themselves are known as the *Graßmann-Plücker relations*, and play much the same role in projective geometry that Schrodinger's equation plays in quantum mechanics or the Einstein field equations play in relativity, in that they provide a basic set of identities from which all the theorems of projective geometry can be derived.

C.12. We are now ready to introduce a new analytic operation into our calculus of projective geometry. This is called the *duality mapping*, and carries 1-vectors $\mathbf{s} \in \mathbb{R}^4$ to 3-vectors $\mathbf{s}^* \in \bigvee_3(\mathbb{R}^4)$. The 3-vector $\mathbf{s}^*$ itself is called the *dual* of $\mathbf{s}$, and is defined by the following identity:

$$\mathbf{x} \cdot \mathbf{s} = \mathbf{x} \vee \mathbf{s}^* \quad \forall\, \mathbf{x} \in \mathbb{R}^4, \tag{C.12}$$

where "$\cdot$" denotes the vector dot product. Since any 3-vector can be written as the outer product of three 1-vectors in $\mathbb{R}^4$, e.g. $\mathbf{s}^* = \mathbf{p} \vee \mathbf{q} \vee \mathbf{r}$, we can write this out in coordinates as

$$\sum_{i=1}^{4} x_i s_i = \mathbf{x} \vee \mathbf{p} \vee \mathbf{q} \vee \mathbf{r}. \tag{C.13}$$

Comparing this with Equation (C.3), we conclude that:

$$\begin{aligned}
s_1^* &= \det\begin{bmatrix} p_1 & q_1 & r_1 \\ p_2 & q_2 & r_2 \\ p_3 & q_3 & r_3 \end{bmatrix} = -s_4; \\
s_2^* &= \det\begin{bmatrix} p_1 & q_1 & r_1 \\ p_2 & q_2 & r_2 \\ p_4 & q_4 & r_4 \end{bmatrix} = s_3; \\
s_3^* &= \det\begin{bmatrix} p_1 & q_1 & r_1 \\ p_3 & q_3 & r_3 \\ p_4 & q_4 & r_4 \end{bmatrix} = -s_2; \\
s_4^* &= \det\begin{bmatrix} p_2 & q_2 & r_2 \\ p_3 & q_3 & r_3 \\ p_4 & q_4 & r_4 \end{bmatrix} = s_1;
\end{aligned} \tag{C.14}$$

which shows us how to write down the coordinates of $\mathbf{s}^*$ from those of $\mathbf{s}$ directly.

C.13. Thus the duality mapping is a linear transformation $\mathbb{R}^4 \rightarrow \bigvee_3(\mathbb{R}^4)$ with matrix:

$$\underline{\mathbf{M}}_* = \begin{bmatrix} 0 & 0 & 0 & -1 \\ 0 & 0 & 1 & 0 \\ 0 & -1 & 0 & 0 \\ 1 & 0 & 0 & 0 \end{bmatrix}. \tag{C.15}$$

In particular, the duality mapping is a homogeneous function, and since $\underline{\mathbf{M}}_* \cdot \underline{\mathbf{M}}_* = -\mathbf{I}$ this function is, up to sign, its own inverse. Thus the duality mapping extends in a well-defined way to an involution (i.e. self-inverse function) of $\mathbb{P}$ onto itself which takes each projective point $\hat{\mathbf{p}}$ to the projective plane with

Plücker coordinates $\hat{\mathbf{p}}^*$ and vice versa. In terms of the underlying vector space $\mathbb{R}^4$, the duality mapping takes each triple of vectors to a vector which is orthogonal to them, and which thus lies in the *orthogonal complement* of their join. It should be observed, however, that the concept of orthogonality has no meaning in projective geometry, since linear transformations of the homogeneous coordinates can map orthogonal sets of vectors to sets of vectors which are no longer orthogonal. Nevertheless, the relation between a projective point and its dual is preserved under projective transformations, so that duality is an intrinsic projective geometric relation.

C.14. It is also possible to define the dual of a 2-vector $\mathbf{U} \in \bigvee_2(\mathbb{R}^4)$. In this case, the result is another 2-vector $\mathbf{U}^*$, which is given in terms of its coordinates by:

$$U_1^* = U_6\,; \quad U_2^* = -U_5\,; \quad U_3^* = U_4\,; \qquad \text{(C.16)}$$
$$U_4^* = U_3\,; \quad U_5^* = -U_2\,; \quad U_6^* = U_1\,.$$

From this it is easily seen that $(\mathbf{U}^*)^* = \mathbf{U}$ and that $(a\cdot\mathbf{U}+b\cdot\mathbf{V})^* = a\cdot\mathbf{U}^*+b\cdot\mathbf{V}^*$ for all $a,b \in \mathbb{R}^{\#}$, so that once again the dual is an involutory projective relation between projective lines. If $\mathbf{U} = \mathbf{p}\vee\mathbf{q}$ is a simple 2-vector, examination of Equation (C.11) reveals that $\mathbf{U}^*$ also satisfies the Graßmann-Plücker relations, so that the dual is again a simple 2-vector. In terms of the underlying vector space $\mathbb{R}^4$, the duality mapping takes a nontrivial simple 2-vector $\mathbf{p}\vee\mathbf{q}$ to the outer product of a pair of vectors $\mathbf{r}\vee\mathbf{s}$ which span the orthogonal complement of $<\mathbf{p},\mathbf{q}>$. Finally, for the sake of completeness, we define the dual of a scalar to be the 4-vector whose single component has this same value.

C.15. In order to make the above seem more familiar to the reader, we point out that in $\mathbb{R}^3$ the outer product of two vectors $\mathbf{x} = [x_1, x_2, x_3]$ and $\mathbf{y} = [y_1, y_2, y_3]$ is the three-dimensional vector

$$\mathbf{x}\vee\mathbf{y} = \left[det\begin{bmatrix} x_1 & y_1 \\ x_2 & y_2 \end{bmatrix}, det\begin{bmatrix} x_1 & y_1 \\ x_3 & y_3 \end{bmatrix}, det\begin{bmatrix} x_2 & y_2 \\ x_3 & y_3 \end{bmatrix} \right]^T, \qquad \text{(C.17)}$$

and the dual thereof is

$$(\mathbf{x}\vee\mathbf{y})^* = \left[det\begin{bmatrix} x_2 & y_2 \\ x_3 & y_3 \end{bmatrix}, -det\begin{bmatrix} x_1 & y_1 \\ x_3 & y_3 \end{bmatrix}, det\begin{bmatrix} x_1 & y_1 \\ x_2 & y_2 \end{bmatrix} \right]^T, \qquad \text{(C.18)}$$

which is the usual cross-product $\mathbf{x}\times\mathbf{y}$.† Thus, the above definitions constitute a

† The reader should observe that although the magnitude of k-vectors has no meaning in projective geometry, when vectors are used to represent physical quantities the magnitudes of their outer products do have significance, as for example when angular momentum is expressed as the cross product of a radius vector and momentum.

generalization of the classical vector algebra of J. W. Gibbs and O. Heaviside to a four-dimensional setting. The n-dimensional version is known as the *calculus of extension* [Forder1940], and was in fact developed well before vector algebra, primarily by H. Graßmann [Grassmann1844, Grassmann1862].

C.16. We now introduce one last operation into our calculus of projective geometry, known as the *regressive product*,‡ which takes k and l-vectors $\mathbf{P}$ and $\mathbf{Q}$ to a $(4-k-l)$-vector:

$$\mathbf{P}\wedge\mathbf{Q} \;:=\; (\mathbf{P}^*\vee\mathbf{Q}^*)^* . \tag{C.19}$$

Since all the operations used to define it are projectively invariant, this extends to a well-defined operation on projective k-vectors. In order to determine the projective geometric meaning of this definition, let us consider first the very simple case in which $\mathbf{P} = \mathbf{u}_1\vee\mathbf{u}_3\vee\mathbf{u}_4$ and $\mathbf{Q} = \mathbf{u}_2\vee\mathbf{u}_3\vee\mathbf{u}_4$, where $\mathbf{u}_1, \ldots ,\mathbf{u}_4$ denote the unit vectors of $\mathbb{R}^4$. A straightforward application of the definitions gives:

$$\mathbf{P}^* = \begin{bmatrix}0\\1\\0\\0\end{bmatrix}, \quad \mathbf{Q}^* = \begin{bmatrix}-1\\0\\0\\0\end{bmatrix}, \tag{C.20}$$

so that

$$(\mathbf{P}^*\vee\mathbf{Q}^*)^* = \begin{bmatrix}1\\0\\0\\0\\0\\0\end{bmatrix}^* = \begin{bmatrix}0\\0\\0\\0\\0\\1\end{bmatrix} = \mathbf{u}_3\vee\mathbf{u}_4 , \tag{C.21}$$

which is a 2-vector representing intersection or *meet* $<\mathbf{u}_3, \mathbf{u}_4>$ of the subspaces $<\mathbf{u}_1, \mathbf{u}_3, \mathbf{u}_4>$ and $<\mathbf{u}_2, \mathbf{u}_3, \mathbf{u}_4>$. Since any four linearly independent vectors in $\mathbb{R}^4$ can be mapped onto any other four by a unique nonsingular linear transformation, it follows from the projective invariance of the regressive product that:

$$(\hat{\mathbf{p}}\vee\hat{\mathbf{r}}\vee\hat{\mathbf{s}})\wedge(\hat{\mathbf{q}}\vee\hat{\mathbf{r}}\vee\hat{\mathbf{s}}) \;=\; \hat{\mathbf{r}}\vee\hat{\mathbf{s}} \tag{C.22}$$

for any four projective points $\hat{\mathbf{p}}, \hat{\mathbf{q}}, \hat{\mathbf{r}}, \hat{\mathbf{s}}$ whose join is the entire projective space $\mathbb{P}$. Thus the regressive product of the 3-vectors which represent any two distinct projective planes is a 2-vector which represents their line of intersection. Similarly, the regressive product of a simple 2-vector and a 3-vector is zero if the projective line represented by the 2-vector lies in the projective plane

‡ If one views the duals of 1-vectors as (constant) 1-forms, this is essentially the exterior product which is used in the calculus of differential forms.

represented by the 3-vector, and otherwise the result is a 1-vector which represents their unique point of intersection, i.e.

$$(\hat{\mathbf{p}} \vee \hat{\mathbf{q}} \vee \hat{\mathbf{s}}) \wedge (\hat{\mathbf{r}} \vee \hat{\mathbf{s}}) = \hat{\mathbf{s}} . \tag{C.23}$$

C.17. All the theorems of projective geometry can be expressed as algebraic identities involving the products and duals of projective k-vectors, as above. As a simple but important example, we take the dual of both sides of (the projective version of) Equation (C.19) to obtain:

$$(\hat{\mathbf{P}} \wedge \hat{\mathbf{Q}})^* = \hat{\mathbf{P}}^* \vee \hat{\mathbf{Q}}^* . \tag{C.24}$$

Dualizing both sides, we have also:

$$(\hat{\mathbf{P}} \vee \hat{\mathbf{Q}})^* = \hat{\mathbf{P}}^* \wedge \hat{\mathbf{Q}}^* . \tag{C.25}$$

By applying this identity to any expression involving products and duals of projective k-vectors, it is possible to convert it into a *dual* expression in which all outer products have been replaced by regressive products, all regressive products have been replaced by outer products and all k-vectors have been replaced by their duals. Thus all the theorems of projective geometry come in pairs, in which the roles played by projective points and planes have been switched. This fact is known as the *principle of duality.*

D. Set Systems and Matroids

D.1. Set systems are the fundamental objects of interest in *combinatorics.* Matroids are a special type of set system which have particularly nice properties. The concept of a matroid was originally introduced in a geometric context by [Whitney1935], but they did not attract widespread interest until the 1960's, when it was gradually realized that a great many seemingly diverse combinatorial structures are actually special types of matroids. It has therefore been suggested that matroids may come to assume much the same unifying role in combinatorics that point set topology now plays in analysis [Crapo1970]. In this book, however, we shall be primarily interested in the geometric applications of matroids. This appendix has been written to provide readers unfamiliar with the theory with its most basic definitions and results. The classical reference on the subject (though unfortunately rather out-of-date) is [Welsh1976]. The first of a proposed three volume treatise has recently been released [White1986], while another recent introductory account with emphasis on algorithms may also be found in [Williamson1985].

D.2. Let E be a finite set, and $\mathbb{I} \subseteq 2^E$ be an arbitrary collection of subsets of E which includes the empty set {}. Then $\mathbb{I}$ is called a *set system* on E, and the members of $\mathbb{I}$ are traditionally called the *independent* subsets of E, while those not in $\mathbb{I}$ are called *dependent*. The function $\rho_{\mathbb{I}}: 2^E \to \mathbf{Z}$ defined by $\rho_{\mathbb{I}}(F) := max(\#I \mid I \subseteq F, I \in \mathbb{I})$ for all $F \subseteq E$ is called the rank function of $\mathbb{I}$, and its value $\rho_{\mathbb{I}}(F)$ on any subset $F \subseteq E$ is called the *rank* of F in $\mathbb{I}$. A set system is completely determined by its rank function, for given $\rho_{\mathbb{I}}$ as above, we have $\mathbb{I} = \{I \subseteq E \mid \rho_{\mathbb{I}}(I) = \#I\}$. Not every function $\rho: 2^E \to \mathbf{Z}$ is the rank function of a set system, however. In fact, it is an easy exercise to show that there exists a set system $\mathbb{I}$ such that $\rho = \rho_{\mathbb{I}}$ iff

$$\text{(S0)} \quad 0 \le \rho(F) \le \#F \quad \forall\ F \subseteq E\ ;$$

$$\text{(S1)} \quad G \subseteq F \subseteq E \Rightarrow \rho(G) \le \rho(F)\ ; \tag{D.1}$$

$$\text{(S2)} \quad \forall\ F \subseteq E,\ \exists\ I \subseteq F \text{ such that } \rho(I) = \#I = \rho(F).$$

Condition (S0) is actually a consequence of the other two, but is worth stating explicitly because we shall use it later on.

D.3. A set system (E, $\mathbb{I}$) is called *n-stable* if $I \in \mathbb{I}$, $\#I \le n+1$ and $J \subseteq I$ implies $J \in \mathbb{I}$; it is called *stable* if it is n-stable for $n = \#E-1$. In terms of the associated rank function $\rho_{\mathbb{I}}$, a set system is stable if and only if

$$\text{(M1)} \quad \rho_{\mathbb{I}}(F \cup e) \le \rho(F)+1 \quad \forall\ F \subseteq E \text{ and } e \in E\,. \tag{D.2}$$

An independent subset $I \subseteq F$ is called *maximal* in $F \subseteq E$ if $I \subseteq J \subseteq F$ and $J \in \mathbb{I}$ implies $J = I$. A set system (E, $\mathbb{I}$) is called *balanced* if $\#I = \#J$ for all maximal independent subsets $I, J \subseteq F \subseteq E$. This means that, given any two independent subsets $I, J \subseteq E$ with $\#J < \#I$, it is possible to *augment* J by the addition of some $H \subseteq I \setminus J$ with $H \neq \{\}$ so that $H \cup J \in \mathbb{I}$. Alternatively, in terms of its rank function $\rho_{\mathbb{I}}$ we can say that a set system is balanced if and only if

$$\text{(M2)} \quad G, F \subseteq E \text{ and } \rho_{\mathbb{I}}(G) < \rho_{\mathbb{I}}(F) \Rightarrow \tag{D.3}$$

$$\exists\ \{\} \subset H \subseteq F \setminus G \text{ with } \rho_{\mathbb{I}}(G \cup H) \ge \rho_{\mathbb{I}}(G) + \#H\,.$$

D.4. A *matroid* (E, $\mathbb{I}$) is a stable and balanced set system. Matroids may also be characterized in terms of abstract rank functions $\rho: 2^E \to \mathbf{Z}$ by requiring the function to satisfy (S0), (S1), (M1) and (M2) with $\#H = 1$. Axiom (M2) can also be replaced by

$$\text{(M2')} \quad \rho(F \cup e) = \rho(F \cup e') = \rho(F) \Rightarrow \rho(F \cup \{e, e'\}) = \rho(F) \tag{D.4}$$

for all $F \subseteq E$, $e, e' \in E$. Alternatively, we may simply require the rank function to obey (S0), (S1) and the *submodularity axiom*

$$\text{(M3)} \quad \rho(G \cup F) + \rho(G \cap F) \le \rho(G) + \rho(F) \tag{D.5}$$

for all $F, G \subseteq E$. The simplest examples of matroids are the *trivial matroid* $(E, \{\})$, whose rank is zero, and the *uniform matroid* $(E, \binom{E}{n})$, whose rank is n.

D.5. A *base* of a matroid is a maximal independent subset of E, and the set of all bases of a matroid is be denoted by $\mathbb{B}$. In terms of their bases, matroids can be defined by the single axiom

$$\text{(B0)} \quad A, B \in \mathbb{B} \;\Rightarrow\; \forall\, e \in A\backslash B,\ \exists\, e' \in B\backslash A \text{ with } (A\backslash e)\cup e' \in \mathbb{B}\,, \qquad \text{(D.6)}$$

which is usually called the *basis exchange axiom.* It is astonishing, yet not hard to prove, that all of the above axiom systems are equivalent.

D.6. We now consider some operations on matroids. Given a matroid $(E, \mathbb{I})$ whose bases are $\mathbb{B}$, the n-truncation $\mathbb{I}_n$ is the matroid whose bases are

$$\mathbb{B}_n \;:=\; \{I \in \mathbb{I} \mid \#I = n\} \qquad \text{(D.7)}$$

and whose rank function is $max(n, \rho)$. The *dual* $(E, \hat{\mathbb{I}})$ is the matroid whose bases are

$$\hat{\mathbb{B}} \;:=\; \{E\backslash B \mid B \in \mathbb{B}\} \qquad \text{(D.8)}$$

and whose rank function is given by

$$\hat{\rho}(F) \;:=\; \#F + \rho(E\backslash F) - \rho(E) \qquad \text{(D.9)}$$

for all $F \subseteq E$. The *restriction* $\mathbb{I}_F$ of $\mathbb{I}$ to $F \subseteq E$ is the matroid whose bases are

$$\mathbb{B}_F \;:=\; \{B\cap F \mid B \in \mathbb{B} \text{ and } B\cap F \text{ is maximal in } F\} \qquad \text{(D.10)}$$

and whose rank function is $\rho_F := \rho|_F$. We may also say that $\mathbb{I}_F$ is obtained from $\mathbb{I}$ by *deletion* of $E\backslash F$. Similarly, we may define the *contraction* $\mathbb{I}^F$ of $\mathbb{I}$ by F in terms of its bases as

$$\mathbb{B}^F \;:=\; \{B\backslash F \mid B \in \mathbb{B} \text{ and } B\cap F \text{ is maximal in } F\} \qquad \text{(D.11)}$$

or, in terms of its rank function, as $\rho^F(G) = \rho(G\cup F) - \rho(F)$ for all $G \subseteq E$. It can be shown that $\widehat{(\mathbb{B}^F)} = (\hat{\mathbb{B}})_F$ and $\widehat{(\mathbb{B}_F)} = (\hat{\mathbb{B}})^F$, i.e. restriction and contraction are *dual operations.*

D.7. The *closure operator* $\mathrm{cl}: 2^E \to 2^E$ of a matroid $(E, \mathbb{I})$ is defined by

$$\mathrm{cl}(F) \;:=\; \{e \in E \mid \rho(F\cup e) = \rho(F)\}. \qquad \text{(D.12)}$$

It can be shown that an arbitrary function $\mathrm{cl}: 2^E \to 2^E$ is the closure operator of a matroid on E if and only if for all $G \subseteq F \subseteq E$

$$\begin{aligned} &\text{(K1)} \quad F \subseteq \mathrm{cl}(F)\;; \\ &\text{(K2)} \quad \mathrm{cl}(G) \subseteq \mathrm{cl}(F)\;; \qquad \text{(D.13)} \\ &\text{(K3)} \quad \mathrm{cl}(F) = \mathrm{cl}(\mathrm{cl}(F))\;; \end{aligned}$$

(K4) $\quad e, e' \in E,\ e' \notin cl(F)$ and $e' \in cl(F \cup e) \quad \Rightarrow \quad e \in cl(F \cup e')$.

Thus, the closure operator provides us with yet another set of axioms for matroids.

D.8. The *hyperplanes* of the matroid are those subsets $H \subseteq E$ such that $H = cl(H)$. The *circuits* of a matroid are its minimal dependent subsets, i.e. those subsets $C \subseteq E$ such that $C \notin \mathbb{I}$ but $C \setminus c \in \mathbb{I}$ for all $c \in C$. These can also be used to characterize matroids via the axioms

(C1) $\quad C_1, C_2 \in \mathbf{C}$ and $C_1 \neq C_2 \quad \Rightarrow \quad C_1 \not\supseteq C_2$;

(C2) $\quad C_1, C_2 \in \mathbf{C},\ C_1 \neq C_2$ and $c \in C_1 \cap C_2$ (D.14)

$\quad \Rightarrow \quad \exists\ C_3 \in \mathbf{C}$ with $C_3 \subset (C_1 \cup C_2) \setminus c$.

The *cocircuits* of a matroid are the circuits of its dual. These are easily shown to be the complements of its hyperplanes. Finally, a *loop* is a one element circuit, while a *coloop* is a one element cocircuit.

D.9. The final characterization of matroids that we shall give is somewhat remarkable because of its algorithmic nature. If $\mathbb{I}$ denotes the independent subsets of a matroid on E, and $w : E \to \mathbb{R}$ is a set of nonnegative weights, the independent subset of maximal weight can be found by the following *greedy algorithm*:

Algorithm D.9:
INPUT: A matroid $(E, \mathbb{I})$ with nonnegative weights $w : E \to \mathbb{R}$.
OUTPUT: A subset $I \in \mathbb{I}$ such that $\Sigma_I w(e_i)$ is maximum.
PROCEDURE:

```
| Index E by decreasing weight as E := [e_1, . . . ,e_N]
|  with w(e_i) ≥ w(e_{i+1}) for i = 1,...,N−1.
| Set I := {}.
| For i from 1 to N:
| | If I∪e_i ∈ II:
| | | Set I := I∪e_i.
```

Moreover, it can be shown that a set system $(E, \mathbb{I})$ is a matroid if and only if the greedy algorithm yields the maximum weight $I \in \mathbb{I}$ for *all* weights $w : E \to \mathbb{R}$.

D.10. One classical example of a matroid is obtained by taking an $n \times m$ matrix $\underline{\mathbf{M}}$ over some field $\mathbb{K}$, letting $E := \{1, \ldots, m\}$, and defining

$$\mathbb{I}_{\underline{\mathbf{M}}} \ := \ \{ I \subseteq E \mid \{\mathbf{m}_i \mid i \in I\} \text{ is linearly independent} \}\ , \tag{D.15}$$

where $\mathbf{m}_i$ denotes the i-th column of $\underline{\mathbf{M}}$. Thus the bases of this matroid are the bases in $\underline{\mathbf{M}} = \{\mathbf{m}_i \mid i = 1, \ldots, m\}$ (of the subspace $\langle \mathbf{M} \rangle$) of $\mathbb{K}^n$, and the fact that the basis exchange axiom holds is the well known *Steinitz exchange theorem*. Alternatively, we can define this matroid in terms of its rank function by setting

$$\rho(\mathrm{F}) \;:=\; rank(\underline{\mathbf{M}}(\mathrm{F})) \tag{D.16}$$

where $\underline{\mathbf{M}}(\mathrm{F})$ denotes the submatrix of $\underline{\mathbf{M}}$ whose columns are those indexed by $\mathrm{F} \subseteq \{1, \ldots, m\}$. Such matroids are called *matric* matroids, and this example, of course, is the reason for the terms *hyperplane*, *rank*, etc. which were introduced above.

D.11. The terms *circuit*, *loop* etc. are explained by the following, equally classical example. If $\mathbb{G} = (\mathrm{V}, \mathrm{E})$ is a graph (see Appendix E), then

$$\mathbb{B}_G \;:=\; \{\mathrm{E}' \subseteq \mathrm{E} \,|\, \mathrm{E} \text{ is a spanning tree of } \mathbb{G}\} \tag{D.17}$$

is the set of bases of a matroid. The circuits of this matroid are the cycles of the graph, and in particular its loops (if any) are loops in the graph. These matroids are known as *graphic* matroids, and their duals are called *cographic*. They are actually a special case of the matric matroids introduced above, which are obtained from the columns of the vertex/edge incidence matrix of the graph $\mathbb{G}$, regarded as a matrix over the field $GF(2)$.

D.12. A matroid is called *representable* over a field $\mathbb{K}$ when there exists one to one correspondence between its elements and the columns of a matrix over $\mathbb{K}$ such that the independent subsets of the matroid correspond to the linearly independent subsets of the columns of the matrix (as above). In this case, the matrix together with this correspondence is called a *representation* of the matroid. There exist matroids which are representable over every field, over no field, or only over certain fields. For example, the Fano matroid with $\mathrm{E} := \{1, \ldots, 6\}$ and

$$\mathbb{B} \;=\; \binom{\mathrm{E}}{3} \setminus \left\{\{1,2,4\},\{1,3,6\},\{1,5,7\},\{2,3,5\},\{2,6,7\},\{3,4,7\},\{4,5,6\}\right\} \tag{D.18}$$

is representable only over the field $GF(2)$. In this book, we shall be mainly concerned with matroids which are representable over the real numbers $\mathbb{R}$. Note that it is also possible to represent matroids affinely. In this case, the independent subsets of the matroid correspond to independent subsets of points in an affine space.

E. Graph Theory

E.1. *Graph theory* is a branch of mathematics invented in 1736 by Leonhard Euler in conjunction with the famous *Koenigsberg bridge problem* [Euler1953]. Although graphs are also a specific class of topological space, their study is more properly considered a branch of *combinatorics*. Combinatorics may be briefly defined [Berge1971] as the specification, classification and enumeration of families of subsets of an arbitrary finite set. Although combinatorics was referred to disparagingly by von Neumann in his work on abstract automata [Neumann1951] as the most intractable of all fields of mathematics, the advent of real automata, i.e., electronic computers, has opened up new vistas in its scope and applications. Graph theory is the branch of combinatorics that results when one restricts the subsets in the families to contain at most two elements. Mathematically, it is of interest as the simplest structure one can impose upon a finite set. It is also a field that is rich in applications, to physics [Essam1971, Uhlenbeck1962], to chemistry [Balaban1976], and to the social sciences [Roberts1976], not to mention computer science and engineering [Hillier1980]. Its applications to biology have been relatively few and far between, a fact which is surprising given James Watson's critical observation that "biologically important things come in pairs" [Watson1968]. Graph theory has, however, been used in the analysis of multicellular development [Culik1976], ecological food webs [Patton1980], neural networks [McCulloch1965], and in the regulation of gene expression [Martinez1974]. The explosive increase in the results of graph theory, together with the rapidly expanding library of efficient algorithms for solving graph theoretic problems, makes graph theoretic models seem to be one of the most promising approaches available for dealing with many types of complex systems.

E.2. The notation we use will be for the most part that of Harary [Harary1972], altered only as necessary to make it consistent with the notation used in other chapters.

A *graph* is defined as a dual entity $\mathbb{G} := (P,Q)$, where P is an abstract finite set of indistinguishable elements, and $Q \subseteq \binom{P}{2}$ is a set of subsets of P such that each subset has cardinality exactly two. The elements of P will be referred to as the *points*, and the elements of Q as the *lines*, of the graph $\mathbb{G}$. The points of a graph are more commonly called vertices or nodes, and the lines are more commonly called edges or arcs. (The terms points and lines are Harary's.) In many applications, we will refer to points as atoms and lines as couples.

This is only one of many equivalent definitions of a graph that may be given. It is, however, the simplest and most direct. It must be made clear to the reader with no previous acquaintance with graph theory that graphs as defined above have absolutely no relation whatsoever with the two dimensional plots of a function vs. its coordinate with which he or she may be more familiar. It should also be observed that the mathematical meaning of the word *set* in the above precludes the presence of more than one line containing exactly the same points. The above definition may be extended by allowing the presence of one element sets in the set of lines Q; such lines are called *loops*. Unless otherwise stated, all our graphs will be loopless.

E.3. Given a graph $\mathbb{G} = (P,Q)$, a line $q \in Q$ is said to be *incident* to a point $p \in P$ if $q = \{p,p'\}$ for some $p' \in P$. A line q is said to be *incident* to another line q' if $q = \{p,p'\}$ and $q' = \{p',p''\}$ for p,p' and $p'' \in P$. Two points p and p' are said to be *adjacent* if there exists a line $q \in Q$ such that $q = \{p,p'\}$. Adjacent points are denoted by $p \sim p'$. The *order* of a graph is the number of elements in its point set, which we shall usually denote N. The number of lines in a graph will here be denoted by L.

The *support* of a set of lines $Q' \subseteq Q$ in a graph $\mathbb{G} = (P,Q)$ is the set of all points incident to Q', and is denoted by P(Q'). Similarly, the set of all lines each of which is contained in a set of points $P' \subseteq P$ is denoted by Q(P'). The set of all points in $P \setminus P'$ which are adjacent to points in P', on the other hand, is denoted by Adj(P'). The set of points adjacent to a single point $p \in P$ is called the *neighborhood* of p, and is denoted by $P_p := \mathrm{Adj}(p)$, whereas the set of all lines which are incident to p will be denoted by Q_p. Finally, the *valence* or *degree* of a point p in a graph, $val(p) = deg(p)$, is the number of points in its neighborhood.

The use of the words *points* and *lines* in the above has no immediate geometric meaning. It is possible to redefine the "points" and "lines" of our graphs to be entities homeomorphic, that is, topologically equivalent, to the corresponding geometric entities. In this case one has a *geometric graph*, as opposed to the *abstract graph* defined above. Unless otherwise stated, all our graphs will be abstract. By considering graphs as geometric entities, however, one can construct a very useful pictorial description of them. One simply puts down a point on a sheet of paper for every point in the point set of the abstract graph, and draws a line between a pair of points on paper whenever the corresponding points of the abstract graph are adjacent.

E.4. We will now demonstrate the utility of this descriptive tool by using it to introduce some common varieties of graphs. The *trivial graph* $\mathbb{K}_1$ is the graph consisting of but a single point and no lines. An *empty graph* is a graph on N points without lines, which we denote by $\bar{\mathbb{K}}_N$. A *complete graph* consists of N points and all $N(N-1)/2$ possible lines; it is denoted by $\mathbb{K}_N$. A *path* is a graph consisting of N points and $N-1$ lines whose points may be linearly ordered so that $P = [p_i]_{i=1}^{N}$ and $Q = [p_i,p_{i+1}]_{i=1}^{N-1}$; it is denoted by $\mathbb{P}_N$. A *cycle* is a graph for which any subset of $N-1$ points, together with those lines with both points included therein, forms a path; it is denoted by $\mathbb{C}_N$. A *tree* is a connected graph with exactly $N-1$ lines; this is denoted by $\mathbb{T}_N$. A *complete bipartite graph* $\mathbb{K}_{M,N}$ is a graph whose point set may be partitioned into two sets of M and N points each such that the line set consists of all MN possible lines with one point in one of the two sets and the other point in the other set. The *complete n-partite graph* $\mathbb{K}_{I,J,\ldots,N}$ is similarly defined for partitions of the point set into n sets.

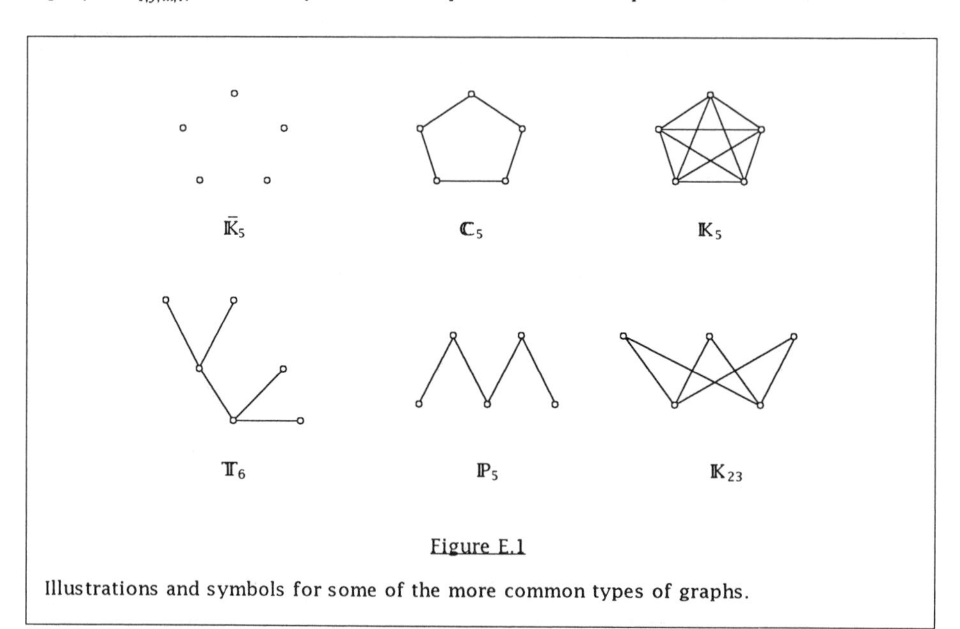

Figure E.1

Illustrations and symbols for some of the more common types of graphs.

These graphs are shown in figure E.1.

E.5. A graph $\mathbb{G}' = (P',Q')$ is said to be a *subgraph* of a graph $\mathbb{G} = (P,Q)$ if $P' \subseteq P$ and $Q' \subseteq Q$; this relation is denoted by $\mathbb{G}' \subset \mathbb{G}$. Given a graph $\mathbb{G} = (P,Q)$ and a set $P' \subseteq P$, the graph $\mathbb{G}(P') = (P',Q')$ is called the subgraph of $\mathbb{G}$ *induced* by P' if $Q' = \{q' = \{p,p'\} \mid p,p' \in P' \text{and} \, q' \in Q\}$. We say that a graph $\mathbb{G}'$ is *embeddable* in a graph $\mathbb{G}$ if it is isomorphic to an induced subgraph of $\mathbb{G}$; this relation is denoted by $\mathbb{G}' \subset \mathbb{G}$.

Common subgraphs of graphs are $\mathbb{G} \setminus p$, which is the graph induced by $P \setminus \{p\}$, and $\mathbb{G} \setminus q$, which is the subgraph $(P,Q \setminus \{q\})$. Subgraphs are called paths, trees, cycles etc., according to the same criteria as used above. In terms of their subgraphs, trees are characterized by the absence of cycles, and bipartite graphs, by the absence of *odd cycles*, that is, cycles of odd order. A subgraph whose point set equals that of its *supergraph* is called a *spanning subgraph*; thus a tree whose point set equals that of its supergraph is called a *spanning tree.* A spanning subgraph that is a cycle is, for historical reasons, called a *Hamiltonian cycle.* A subgraph that is a complete graph and is not properly contained in any other complete subgraph of its supergraph has yet another name: it is called a *clique.* The point set of an induced empty subgraph is called an *independent* set of points.

We now give some thought to the relations that can exist between graphs.

E.6. Two graphs $\mathbb{G}$ and $\mathbb{H}$ are said to be *isomorphic* when there exists a bijection between their point sets that preserves adjacency; in less mathematical terms, this is a one to one correspondence between the points of the two graphs such that if any two points are adjacent in one graph then the corresponding points are adjacent in the other. We denote this equivalence relation by $\mathbb{G} \approx \mathbb{H}$.

Two graphs are *homeomorphic* if both can be obtained from the same graph by a sequence of subdivisions of lines. A simple example of this is that $\mathbb{C}_4$ and $\mathbb{C}_3$ are homeomorphic because the former can be produced from the latter by changing one line $q = \{p_1, p_2\}$ into two lines $q' = \{p_1, p'\}$ and $q'' = \{p', p_2\}$ incident on a new point p′. The *complement* of a graph $\mathbb{G} = (P,Q)$ is defined as the graph $\bar{\mathbb{G}} = (P,\bar{Q})$, where $\bar{Q}$ is the set-theoretic complement of Q in the set of all pairs of elements of P, that is, $\bar{Q} = \{ \{p,p'\} \mid p,p' \in P \text{ and } \{p,p'\} \notin Q \}$.

Note that our notation for $\mathbb{K}_N$ and $\bar{\mathbb{K}}_N$ is consistent with that used in this definition. Complementation is an example of a *graph-valued function,* which is in general any operation that yields one graph from another. Note also that the complement of a complement is the original graph. There are graphs that are isomorphic to their complements, for example $\mathbb{C}_5$. A common means of describing a graph is as the complement of a more familiar graph. A *cotree* is a graph whose complement is a tree; a *cocycle* is a graph whose complement is a cycle. $\mathbb{C}_5$ is thus both a cycle and a cocycle. Other examples of graph-valued functions include the *union* of two graphs:

$$\mathbb{G} \cup \mathbb{G}' = (P \cup P', Q \cup Q'); \tag{E.1}$$

and the *join* of two graphs:

$$\mathbb{G} + \mathbb{G}' = (P \cup P', Q \cup Q' \cup \{\{p,p'\} \mid p \in P, p' \in P'\}). \tag{E.2}$$

E.7. The *invariants* of graphs are those properties which are preserved under isomorphisms. The degrees of the points of a graph turn out to be one of the simplest and most useful invariants of graphs. The maximum degree of any point in a graph will be denoted by $\Delta(\mathbb{G})$, the minimum degree will be denoted by $\delta(\mathbb{G})$. Graphs for which $\Delta(\mathbb{G}) = \delta(\mathbb{G})$ will be called *regular graphs.* A complete list of the degrees of the points of a graph will be known as its *degree* *sequence.* It is well-known that the sum of the entries of the degree sequence of a graph is equal to twice its number of lines.

E.8. A graph is referred to as *embedded* in a surface when it is drawn on that surface so that no two edges intersect. In particular, a graph is *planar* if it can be emdedded in the plane. There are several ways to identify and characterize planar and nonplanar graphs, the first and most elegant of which is due to [Kuratowski1930].

THEOREM E.8 (Kuratowski). *A graph is planar if and only if it has no subgraph homeomorphic to* $\mathbb{K}_5$ *or* $\mathbb{K}_{3,3}$.

The proof, which is rather involved, may be found in [Harary1972].

E.9. A graph is *n-colorable* if there exists an assignment of n colors (actually arbitrary labels) to its vertices such that no two adjacent vertices share the same color (label). The *chromatic number*, $\chi(\mathbb{G})$, of a graph $\mathbb{G}$ is the minimum n for which it is n-colorable.

The famous *four color problem* is an example of a graph coloring problem. It may be shown that, for any graph $\mathbb{G}$, $\chi(\mathbb{G}) \leq \Delta(\mathbb{G})+1$ [Harary1972]. If we define $\omega(\mathbb{G})$ to be the order of the largest clique in $\mathbb{G}$, then it is also plain that $\chi(\mathbb{G}) \geq \omega(\mathbb{G})$. Graphs for which $\chi(\mathbb{G}) = \omega(\mathbb{G})$ are called *perfect graphs.* Perfect graphs are important for algorithmic reasons because it is often possible to design efficient algorithms to solve problems on them that would be intractable on arbitrary graphs. The as yet unproven *strong perfect graph conjecture* states that all nonperfect graphs contain either an odd cycle or the complement of such as an induced subgraph. The reader is referred to Golumbic's recent book [Golumbic1980] for an in-depth account of these results, as well as of many of the results mentioned in this appendix.

E.10. A *walk* on a graph is an ordered list of points such that consecutive members of the list are adjacent in the graph. A walk is a *trail* provided no pair of points occurs twice as consecutive members of the list, that is, provided no line is *traversed* more than once. A trail is a *circuit* if the first and last members of its list are equal. If no point occurs more than once on the list, a trail is called a *path* and a circuit a *cycle.*

This means that for a path the subgraph obtained by taking the points of the list together with the lines between consecutive members of the list will be a path subgraph; a similar statement also holds for cycles. The path is said to *connect* the first and last members of its list. A graph in which every pair of points has a path connecting them is called a *connected graph.* A graph in which every pair of points has at least n disjoint paths connecting them is called an *n-connected graph.* The maximal connected induced subgraphs of a graph are called the *connected components* of the graph; these correspond to what we naturally think of as the pieces of the graph. The maximal 2-connected induced subgraphs are called the *blocks*.

A *cutset* of a graph is a set of points P^c such that the graph induced by $P \setminus P^c$ is not connected. A *cutpoint* is a one point cutset. The cutset is said to *separate* any two points which lie in different components of the subgraph induced by $P \setminus P^c$. The *connectivity* of a graph, $\kappa(\mathbb{G})$, is the size of its smallest cutset. *Menger's theorem* states that the minimum size of a cutset separating any two points of a graph is equal to the maximum number of disjoint paths between them. Thus the connectivity of an n-connected graph is exactly n. Trees may now be further characterized as the minimal graphs whose connectivity is one.

E.11. In practice, the elements of the point set of one's graph are rarely indistinguishable as demanded by the definition above. When they are not, we will call our graph a *labelled graph*. When stored on a computer, for example, the points of a graph are always distinguished, if arbitrarily, by their storage locations, which we shall consider to be the integers from one to N. There are three main data structures that may be used for the computer storage and manipulation of graphs [Aho1974, Golumbic1980]. The first of these is known as the *adjacency matrix* of the graph. This is a symmetric matrix whose ij'th element is one or zero as the points labelled i and j are adjacent or nonadjacent, respectively. This data structure has a $O(N^2)$ storage requirement. Then there is the *line-point incidence* list. This is a list of the pair of points incident to each line, and requires $O(2L)$ storage space. Finally, there is the *list of point-point adjacency lists*. This is a list of N lists, each of these lists being a list of the points adjacent to one of the points. The storage requirement here is also $O(2L)$. We shall see that in most cases it is the use of the list of point-point lists that results in the most efficient graph manipulation algorithms.

E.12. Although it is our intention to introduce algorithms for solving relevant graph theory problems in the context in which those problems arise, the following graph search technique, known as *depth-first search*, is important in solving so many graph theory problems that it is best presented in this introductory appendix. (Graph theory algorithms are covered in more detail in [Aho1974, Golumbic1980].) The data structure it uses is the list of point-point adjacency lists introduced above. The procedure terminates only after it has examined every point and line in the graph. It does so with a worst-case computation time that is $O(max\{N,L\})$, which justifies its being called a *linear time algorithm*, since the number of bits necessary to store the graph is also of this order. The algorithm is recursive in nature: the logic which it follows at each point of the graph is wholly independent of what it has done at the previous points. This logic is simple. If a point is found on the adjacency list of the current point that has not yet been visited, then that point becomes the new current point, and the search continues from there. We say that the line connecting the two points is *traversed* when this happens. Otherwise, the search returns to the previous point, and looks for another point on its adjacency list that has not yet been visited. The search partitions the lines of the graph into two sets, known as *tree lines* and *back lines*, depending on if they were traversed in the course of the search or not, respectively. The points of the graph, together with the tree lines, constitute a spanning tree of each connected component of the original graph; this is called a *spanning forest*. Each tree has the property that every pair of adjacent points of the original graph lie together on the unique path through the tree that connects them to the *root*, which is the first point visited in that component. Given any two points in a tree, we call the point nearest the root the *ancestor*, and the point furthest from the root, the *descendent*, of the other in our *depth-first tree*. Immediate neighbors in a *rooted tree* like this are usually referred to as *parent* and *child*. The points may be numbered 1 through *N* according to the order in which they are first reached; this numbering is called a *depth-first numbering* of the points of the graph. This numbering has many interesting properties that allow it to be used in recognizing features of graphs. In particular, depth-first search plays an essential role in the best algorithms available for finding the connected components of a graph, decomposing it into its blocks, and in testing it for planarity. A formal statement of the algorithm, without proof [Tarjan1972], follows:

Algorithm E.12:

INPUT: A graph represented by its point-point adjacency lists.

OUTPUT: A depth-first spanning forest and the corresponding depth-first numbering of its points.

PROCEDURE:

```
Define procedure DFSEARCH(p):
comment: DEPTH, COUNT and DFSNO are initially zero; FATHER
 is initially empty; ADJLST are the adjacency lists of each point;
  COUNT := COUNT+1;
  DFSNO(p) := COUNT;
  if FATHER(p) = {} then
  | DEPTH(p) := 1;
  else
  | DEPTH(p) := DEPTH(FATHER(p))+1;
  for each point p' ∈ ADJLST(p):
  | if DFSNO(p') = 0 then
  | | FATHER(p') := p;
```

```
DFSEARCH(p');
End definition.

while there exists p ∈ P with DEPTH(p) = 0:
    DFSEARCH(p);
```

E.13. The next graph search procedure we shall discuss is called *breadth-first search*. Like depth-first search, this procedure examines all the points and lines of the graph at least once, and has an overall complexity of $O(max\{N,L\})$. It is not a recursive procedure. Instead of immediately calling itself on the first unvisited point found on the adjacency list of the current point, it adds all unvisited points adjacent to the current point to the end of a queue, and then goes on to visit the first point at the beginning of the queue. A line is said to be *traversed* when one of its points is added to the queue while the other is being visited. The set of traversed lines again forms a rooted tree, or a forest if the graph is not connected, and the points may be numbered in the order first visited to obtain a *breadth-first numbering* of the points of the graph. In the case of breadth-first search, however, the *cross lines* of the original graph that are not part of the tree never connect a point to an ancestor or descendant. In fact, if one defines the *depth* of a point of the tree by one plus the number of points between it and the root, it may be shown that the points the cross lines connect differ in depth by at most one. We now present the algorithm, again without proof [Moore1957].

Algorithm E.13:

INPUT: A graph represented by its point-point adjacency lists.

OUTPUT: A breadth-first spanning forest and the corresponding breadth-first numbering of the points.

PROCEDURE:

```
comment: COUNT, DEPTH and BFSNO are initially zero; FATHER
  and QUEUE are initially empty; ADJLST are the adjacency lists.
while there exists a point p with BFSNO(p) = 0:
    QUEUE := p;
    while QUEUE is nonempty
        p := first point on the QUEUE;
        Delete first point on the QUEUE;
        COUNT := COUNT+1;
        BFSNO(p) := COUNT;
        if FATHER(p) = {} then
            DEPTH(p) := 1;
        else
            DEPTH(p) := DEPTH(FATHER(p))+1;
```

```
for each p' ∈ ADJLST(p):
  if BFSNO(p') = 0 then
    Add p' to the end of the QUEUE;
    FATHER(p') := p;
```

E.14. The most interesting application of breadth-first trees lies in that fact that all the paths from the points to the root through the tree are shortest paths in the original graph if each line of the graph is assigned unit length. The shortest paths that result from unit length lines are known as *geodesics*; the subgraphs induced by the points of the geodesics are known as *intervals*. Intervals have the property of being *chordless*, that is, there are no lines connecting nonconsecutive points of the path (else there exists a shorter path). The length of a geodesic is known as the *distance* between its two *endpoints*, and the greatest distance between any two points of a graph is called its *diameter*. The N by N matrix of all the distances between the points is the *distance matrix* of the graph. The use of the term "distance" is slightly misleading in this context, for the distances between the points of a graph are not the Euclidean distances, but simply integer-valued metrics.

F. Numerical Methods

In order to implement many of the algorithms outlined in this book, the reader would have to have some familiarity with numerical methods. We have glossed over many of the numerical details that are crucial for a successful calculation in the interests of focussing on the underlying concepts. Although we could refer the reader to a number of good books on numerical analysis, particularly [Press1986], we have included the following brief outlines of methods required in the earlier chapters, just to keep this book self-contained.

F.1. There are several closely related calculations required on linear systems. The first is *Gaussian elimination*, which uses a sequence of elementary row operations to reduce a given matrix $\underline{\mathbf{A}}$ to upper triangular form. The process involves dividing by selected "pivot" elements of $\underline{\mathbf{A}}$, and in order to be as numerically stable as possible, the algorithm permutes rows and columns (a process referred to as "full pivoting") such that the pivots are large in absolute value.

Algorithm F.1 Gaussian Elimination:

INPUT: An $n \times n$ matrix $\underline{\mathbf{A}}$.

OUTPUT: $\underline{\mathbf{A}}$ rewritten in upper triangular form along with the permutation vectors for rows and columns, $\mathbf{r}$ and $\mathbf{c}$ respectively.

PROCEDURE:

```
Initially all columns are unused.
For elimination on columns i from 1 to n
  Scan all columns j=1...n that have not been used once
    Let a_JK = element such that |a_JK| = max_k |a_jk|.
    If the column j has been used more than once
      A is singular.
      All remaining elements in the upper triangle are zero.
      Elimination is finished.
  Column K has been used again.
  If J ≠ K
    Interchange rows J and K of A.
  Set r_i = J and c_i = K.
  If |a_KK| is very small
    A is singular: cease elimination as before.
  Divide row K of A by a_KK.
  For each row l = 1 ··· n except row K
    Subtract a_l,K × row K from row l.
    Set a_l,K = 0.
 Having completed elimination, undo permutation of A.
  For l = n ··· 1
    If r_l ≠ c_l
      Exchange columns r_l and c_l of A.
```

The number of nonzero rows of the triangularized matrix immediately gives its *rank*. For such purposes, Gaussian elimination is extremely robust, the only questionable part being the criterion for the pivot element reaching essentially zero. That of course depends on the computer word length.

For actually solving linear systems, the method of choice is *LU decomposition*, where we solve

$$\underline{\mathbf{A}}\mathbf{x} = \mathbf{b} \tag{F.1}$$

for the unknown vector $\mathbf{x}$ by first finding a lower triangular matrix $\underline{\mathbf{L}}$ and an upper triangular matrix $\underline{\mathbf{U}}$ such that

$$\underline{\mathbf{LU}} = \underline{\mathbf{A}}\,. \tag{F.2}$$

Then the linear system is easily solved by first solving

$$\underline{\mathbf{L}}\mathbf{y} = \mathbf{b} \tag{F.3}$$

according to the forward substitution formula

$$y_1 = \frac{b_1}{l_{11}} \tag{F.4}$$

$$y_i = \frac{1}{l_{ii}}\left[b_i - \sum_{j=1}^{i-1} l_{ij} y_j\right] \qquad i=2,3,\cdots,n\ .$$

Subsequently one solves

$$\underline{\mathbf{U}}\mathbf{x} = \mathbf{y} \tag{F.5}$$

by the back substitution formula

$$x_n = \frac{y_n}{u_{nn}} \tag{F.6}$$

$$x_i = \frac{1}{u_{ii}}\left[y_i - \sum_{j=i+1}^{n} u_{ij} x_j\right] \qquad i = n-1, n-2, \cdots, 1\ .$$

There remains only to do the LU decomposition of $\underline{\mathbf{A}}$. The equations for this are simple, but in order to be numerically stable, at least partial pivoting (of the rows) is required. The diagonal of $\underline{\mathbf{L}}$ turns out to be unity, so the computer program generally overwrites $\underline{\mathbf{A}}$ with u_{ij} for $i \le j$ and l_{ij} for $i > j$. Since $det(\underline{\mathbf{A}}) = det(\underline{\mathbf{LU}}) = det(\underline{\mathbf{L}})det(\underline{\mathbf{U}}) = det(\underline{\mathbf{U}}) = \prod_{i=1}^{n} u_{ii}$, we have to keep track of whether the row permutation was even or odd and multiply the product by -1 in the latter case, in order to correctly determine the *determinant.*

Algorithm F.2 LU Decomposition:

INPUT: An $n \times n$ matrix $\underline{\mathbf{A}}$.

OUTPUT: $\underline{\mathbf{A}}$ overwritten with the packed, row permuted $\underline{\mathbf{L}}$ and $\underline{\mathbf{U}}$ matrices, along with the row permutation vector $\mathbf{r}$, and $p = \pm 1$, indicating even or odd permutation.

PROCEDURE:

- Initially $p=1$ (even).
- For each row i from 1 to n
  - Set scaling $v_i = 1/max_j |a_{ij}|$.
  - A zero row implies a singular matrix.
- For each column j from 1 to n
  - For i from 1 to $j-1$
    - $a_{ij} = a_{ij} - \sum_{k=1}^{i-1} a_{ik} a_{kj}$
  - For i from j to n
    - $a_{ij} = [a_{ij} - \sum_{k=1}^{j-1} a_{ik} a_{kj}]/a_{jj}$
  - Note I such that $|a_{Ij} v_I| = max_i |a_{ij} v_i|$.
  - If $j \ne I$
    - Interchange rows I and j.

Set $p = -p$.
Interchange v_l and v_j.
Set $r_j = l$.
If $|a_{jj}|$ very small
A is singular. Quit.
Else
For i from j+1 to n
Divide a_{ij} by a_{jj}.

F.2. The main nonlinear problem in this book is minimization of a scalar function $E(\mathbf{x})$ starting from some given vector starting point $\mathbf{x}^{(0)}$. E is generally twice continuously differentiable, and first and second derivatives can be calculated from the analytical formulae, rather than being approximated by differences. It is outside the scope of this book to fully consider and evaluate all of the hundreds of published algorithms, and we shall content ourselves with a brief survey of those with which good results have been obtained in the past (for more complete accounts of the possibilities, the reader is referred to standard texts, e.g. [Himmelblau1972, Luenberger1973, Fletcher1980, Gill1981, Dennis1983]). There are an enormous number of algorithms available for this unconstrained local optimization problem, but as long as good first and second derivatives are available, the clear choice is *Newton's method* for unexcelled convergence near the minimum:

$$\underline{\mathbf{H}}(\mathbf{x}^{(k)})\Delta\mathbf{x} = -\mathbf{g}(\mathbf{x}^{(k)}) \qquad \text{(F.7)}$$
$$\mathbf{x}^{(k+1)} = \mathbf{x}^{(k)} + \Delta\mathbf{x}, \quad k=0,1,\cdots$$

where $\underline{\mathbf{H}}$ is the Hessian matrix of second derivatives and $\mathbf{g}$ is the gradient vector. This of course amounts to solving a linear system, using the methods described above. Equation (F.7) is iterated until $\|\Delta\mathbf{x}\|$ and $\|\mathbf{g}\|$ are adequately small. If E is the energy of a molecule and $\mathbf{x}$ is the vector of all coordinates of all its atoms, then $\underline{\mathbf{H}}$ will be singular, but this can readily be avoided by either excluding sufficient coordinates so as to fix rigid translations and rotations, or by cycling repeatedly through the list of atoms applying Newton minimization to each atom alone. In the latter procedure, the size of the linear system is greatly reduced, and many inconsequential off-diagonal elements of $\underline{\mathbf{H}}$ need not be calculated, but there are some rare circumstances where the conformation may converge on a point that is not a minimum of E. Nevertheless, it has been found to be efficient and reliable in a wide variety of chemical minimizations (and is used, for example, in the MM2 program [Burkert1982]). The other difficulty is that Equation (F.7) describes a process that will converge very well when adequately near any kind of a stationary point (in particular, a minimum), but far from a minimum, $\Delta\mathbf{x}$ may seriously overstep the minimum, fall short of it, or indeed not be a descent direction at all, i.e. $\Delta\mathbf{x}\cdot\mathbf{g} > 0$.† Moreover it requires $O(n^2)$ of memory and $O(n^3)$ time to solve the linear system of equations in Equation (F.7), where n is the number of variables in the problem.

Recently a number of workers have developed *truncated* Newton methods, where a large number of variables may be successfully treated, even starting far from a local minimum, by concentrating on the diagonal elements of $\underline{\mathbf{H}}$ and approximately solving Equation (F.7) via a linear conjugate gradient method to an accuracy proportional the gradient norm, thereby obtaining good asymptotic convergence without the (usually unnecessary) expense of solving them exactly when far from a minimum. The efficiency of this method depends critically upon proper preconditioning of the Hessian to ensure a rapid solution of Newton's equations. Further improvements in efficiency can be obtained by using sparse approximations to the complete Hessian matrix [Ponder1987], or by approximating the directional derivatives it defines by finite difference schemes [Schlick1987]. For many variables, the implementation is complex, but trials of the approach exhibit good initial progress and unexcelled final convergence.

When far from the minimum, the simplest, most robust, and often the fastest converging method is *steepest descents*. Here the technique is to iterate

$$\mathbf{x}^{(k+1)} = \mathbf{x}^{(k)} - \frac{\alpha}{\|\mathbf{g}\|}\mathbf{g}(\mathbf{x}^{(k)}), \quad k=0,1,\cdots \tag{F.8}$$

where $\alpha > 0$ is chosen so as to minimize E along the search direction indicated by $-\mathbf{g}$. Although the initial improvement in E may be very good, the final convergence is often quite poor (see §9.1).

The method of *conjugate gradients* [Hestenes1980] initially does as well as steepest descents, outperforms it in the "mid-game", but often has poor final convergence. If B is the symmetric bilinear form defined by the Hessian $\nabla^2 E(\mathbf{x}^{(k)})$ at the current point $\mathbf{x}^{(k)}$, the direction in which to perform each line search is chosen to be B-orthogonal to the previous direction, and to lie in the two-dimensional subspace spanned by the previous direction and the current gradient. Applied to a positive definite quadratic function $Q: \mathbb{R}^n \rightarrow \mathbb{R}$ (whose Hessian, of course, is everywhere constant), the search direction generated by this procedure can also be shown to be B-orthogonal to all previous directions, and since the maximum size of a set of B-orthogonal vectors is equal to the dimension of the search space n, it follows that convergence to zero gradient is achieved in at most n iterations. Thus when one is close enough to a minimum

† For solving a single nonlinear equation in one unknown, Newton's method can be proven to converge *when adequately close to the solution.* Otherwise, examples can be constructed where the method converges from far away or diverges or oscillates endlessly.

of the error function for a quadratic approximation to be valid, one expects rapid asymptotic convergence. Because of this fact, its relative simplicity, and storage requirements of only $O(n)$, the conjugate gradients method has become the most widely used method of minimizing functions of large numbers of variables.

The small storage requirements arise because it is *not* actually necessary to compute the Hessian $\nabla^2 E(\mathbf{x}^{(k)})$, a fact which greatly decreases its computational and memory requirements. Suppose that in each one-dimensional minimization we set $\mathbf{x}^{(k+1)} := \mathbf{x}^{(k)} + \alpha_i \mathbf{v}^{(k)}$, and that the conjugate directions are given by $\mathbf{v}^{(k+1)} := \beta_i \mathbf{v}^{(k)} - \nabla E(\mathbf{x}^{(k)})$. If we determine the coefficient α_i by a line search along the direction $\mathbf{v}^{(k)}$, and we approximate the second derivative of E along the direction $\mathbf{v}^{(k)}$ by the formula

$$\left.\frac{\partial^2 E(\mathbf{x}^{(k)} + \alpha \mathbf{v}^{(k)})}{\partial \alpha^2}\right|_{\alpha = \alpha_i} \approx \frac{(\nabla E(\mathbf{x}^{(k+1)}) - \nabla E(\mathbf{x}^{(k)})) \cdot \mathbf{v}^{(k)}}{\alpha_i}, \qquad \text{(F.9)}$$

the coefficient β_i can be shown to be

$$\beta_i \approx \frac{\nabla E(\mathbf{x}^{(k+1)}) \cdot (\nabla E(\mathbf{x}^{(k+1)}) - \nabla E(\mathbf{x}^{(k)}))}{\mathbf{v}^{(k)} \cdot \nabla E(\mathbf{x}^{(k)})}. \qquad \text{(F.10)}$$

In the limit of a perfectly accurate line search, this can be further simplified to $\beta_i \approx \|\nabla E(\mathbf{x}^{(k+1)})\| / \|\nabla E(\mathbf{x}^{(k)})\|$. This leads us to the *Fletcher-Reeves conjugate gradient algorithm* for nonlinear minimization:

Algorithm F.3:

INPUT: A function $E : \mathbb{R}^n \to \mathbb{R}$ together with a starting point $\mathbf{x}^{(0)} \in \mathbb{R}^n$ and a target value $\epsilon > 0$ for the RMS gradient norm.

OUTPUT: A point $\mathbf{x}^* \in \mathbb{R}^n$ such that $\|\nabla E(\mathbf{x}^*)\| < \epsilon\sqrt{n}$.

PROCEDURE:

> Set $\mathbf{v}^{(0)} := \nabla E(\mathbf{x}^{(0)})$ and $i := 0$.
>
> repeat:
>
> > Find the minimum $\mathbf{x}^{(i+1)}$ along $\mathbf{v}^{(i)}$ by a line search.
> >
> > Set $\beta_i := \|\nabla E(\mathbf{x}^{(i+1)})\| / \|\nabla E(\mathbf{x}^{(i)})\|$, $\mathbf{v}^{(i+1)} := \beta_i \mathbf{v}^{(i)} - \nabla E(\mathbf{x}^{(i+1)})$ and $i := i+1$.
>
> until $\|\nabla E(\mathbf{x}^{(i)})\| < \epsilon\sqrt{n}$;
>
> $\mathbf{x}^* := \mathbf{x}^{(i)}$.

There exists a wide variety of variations on the basic idea behind the conjugate gradients method, most of which attempt to correct for the inexactness which is inherent in any finite line search (see [Nazareth1986, Dennis1986] for recent reviews).

In any minimization algorithm using first but not second derivatives, the "line search" to choose α can be quite time consuming and is generally an uncomfortable compromise between making some progress in the chosen direction but not wasting an undue effort at solving such a subproblem when the search direction will doubtless change in the next iteration. Among the numerous algorithms in the literature, we have found the Armijo linesearch to be quite satisfactory [Bertsekas1982].

Algorithm F.4 Armijo:

INPUT: The energy function E to be minimized, a maximum allowed step size m, $10^{-5} \leq \sigma \leq 10^{-1}$, $0.1 \leq \beta \leq 0.5$, the current gradient $\mathbf{g}$, the Newton method search direction $\Delta\mathbf{x}$, and the current position $\mathbf{x}^{(k)}$.

OUTPUT: The new position $\mathbf{x}^{(k+1)}$ such that $E(\mathbf{x}^{(k+1)}) < E(\mathbf{x}^{(k)})$.

PROCEDURE:

If $\Delta\mathbf{x}\cdot\mathbf{g} > 0$

  Search direction $\mathbf{s} := -\mathbf{g}$.

Else

  $\mathbf{s} := \Delta\mathbf{x}$.

If $\|\mathbf{s}\| > m$

  $\alpha := m/\|\mathbf{s}\|$.

Else

  $\alpha := 1$.

Repeat...

  $\mathbf{x}^{(k+1)} := \mathbf{x}^{(k)} + \alpha\mathbf{s}$

  $\alpha := \alpha\beta$

...Until $E(\mathbf{x}^{(k)}) - E(\mathbf{x}^{(k+1)}) \geq \sigma\alpha\mathbf{g}\cdot\mathbf{s}/\beta$

Although the Newton method calculates a search direction and a step size, cautiously submitting these two results to the Armijo linesearch can enforce convergence in situations when the unaided Newton method would diverge.

The most popular linesearch algorithm for steepest descents or conjugate gradients is the linear extrapolation / cubic interpolation procedure:

Algorithm F.5:

INPUT: A function $f: \mathbb{R} \to \mathbb{R}$ together with a starting point $x \in \mathbb{R}$ and a target value $\epsilon > 0$ for the absolute value of the derivative f'.

OUPUT: A point x' in the minimum nearest to x with $|f'(x')| < \epsilon$.

PROCEDURE:

$\delta :=$ *initial value*, $\xi := sign(f'(x))$, $y := x$.

While $\xi \cdot f'(y) > 0$ do

  $x := y$, $y := y - \xi\cdot\delta$, $\delta := 2\delta$.

If $|f'(y)| < |f'(x)|$ then $z := y$ else $z := x$.

While $|f'(z)| > \epsilon$ do:

  Let z be the minimum of the unique cubic polynomial which is determined by $f(x)$, $f(y)$, $f'(x)$, $f'(y)$.

If $sign(f'(z)) = sign(f'(x))$ then
  $z := x$;
else $sign(f'(z)) = sign(f'(y))$ then
  $z := y$.
Set $x' := x$.

G. Notation

G.0. In this book we have made a considerable effort to use a consistent system of mathematical notation throughout. The following "type" convention has been used for the individual symbols.

(1) *Italic* letters and names have been used for integer and real variables and functions.

(2) Small **bold** letters have been used for vectors, and large **bold** letters for subsets of vectors or matrices (for reasons given below, matrices are also underlined).

(3) Large roman letters are used for sequences and sets, and small roman letters stand for members thereof.

All of our vectors should be considered to be column vectors unless otherwise indicated. Compound entities such as groups, graphs, vector spaces, etc. will usually be indicated by overstriking, e.g. "$\mathbb{V}$".

It is often necessary to distinguish between a set and the same elements arranged in a *sequence* with a given order. The former is written as $\mathrm{X} = \{\mathrm{x},\mathrm{y}, \cdots\}$, whereas the latter is written as $\underline{\mathrm{X}} = [\mathrm{x},\mathrm{y}, \cdots]$. We say that X is the *underlying set* of $\underline{\mathrm{X}}$. Although coordinate vectors in $\mathbb{R}^n$ are sequences of n real numbers $\underline{X} = [x_1, \ldots, x_n]$, in this case the underlying set is rarely of any interest, so we have chosen to conform to the conventional notation and denote both abstract and coordinate vectors by lower case bold letters $\mathbf{x}$. Thus a matrix $\underline{\mathbf{X}} = [\mathbf{x}_1, \ldots, \mathbf{x}_n]$ is regarded as a sequence of (column) vectors whose underlying set is $\mathbf{X}$. Note also that the members of a sequence need not all be distinct, i.e. it is possible to have $\mathrm{x} = \mathrm{y}$ at two distinct locations in $\underline{\mathrm{X}}$. Hence $[\mathrm{x},\mathrm{y}] \neq [\mathrm{y},\mathrm{x}]$ unless $\mathrm{x} = \mathrm{y}$, whereas in the case of sets, $\mathrm{x} = \mathrm{y}$ implies that $\{\mathrm{y},\mathrm{x}\} = \{\mathrm{x},\mathrm{y}\} = \{\mathrm{x}\} = \{\mathrm{y}\}$. We shall often abbreviate *singletons* $\{\mathrm{x}\}$ by dropping the brackets, and denote the empty set by $\{\}$. The prefix "#", applied to a set, denotes its size, whereas the same symbol, applied to a sequence, denotes its length.

G.1. In the course of this work, it will frequently be necessary to impose an ordering on a set X. This will be done by means of an *indexing* of its elements, i.e. a one to one function mapping the set of integers $\{1, \ldots, \#X\}$ onto X, which is written in subscript notation. If $m := \#X$, the corresponding sequence with underlying set X is then denoted by $\underline{X} := [x_1, \ldots, x_m]$. It is also necessary to distinguish between indexings of sets and simple *numberings* thereof, which are not necessarily one to one: If $X = \{y_1, \ldots, y_m\}$ is a numbering of X, then it is possible to have $y_i = y_j$ for some $i \neq j$. We shall try to keep this important distinction clear by always using the corresponding lower case letter whenever an indexing is intended, i.e. $X = \{x_i \mid i = 1, \ldots, m\}$ will indicate an indexing of the set X, whereas $X = \{y_i \mid i = 1, \ldots, m\}$ is any numbering. We may apply some set theoretic operators directly to sequences, for example "$x \in \underline{Y}$" means $x \in Y$, "$\underline{X} \setminus y$" means the sequence $\underline{X}$ with y deleted, etc. The members of a sequence are always subscripted by their position, so that $x_i, x_j \in \underline{X}$ and $i \neq j$ implies that x_i and x_j do not occur in the same position of the sequence $\underline{X}$ (even if $x_i = x_j$), and $i < j$ implies x_i comes before (i.e. to the left of) x_j in the sequence.

G.2. Sometimes it will be necessary to specify a *subsequence* of a given sequence $\underline{X}$, which we shall do as follows. The symbol $\Lambda(m,n)$ denotes the set of all increasing subsequences of the integers $\{1, \ldots, m\}$ of length n:

$$\Lambda(m,n) \quad := \quad \{[\lambda_1, \ldots, \lambda_n] \mid 1 \leq \lambda_1 < \cdots < \lambda_n \leq m\} . \tag{G.1}$$

Then for a sequence $\underline{X} = [x_1, \ldots, x_m]$ and a given $\underline{\lambda} \in \Lambda(m,n)$, we define $\underline{X}_\lambda := [x_{\lambda_1}, \ldots, x_{\lambda_n}]$. The corresponding underlying set, in turn, is denoted by $X_\lambda := \{x_{\lambda_1}, \ldots, x_{\lambda_n}\}$. We shall indicate that $\underline{Y} := \underline{X}_\lambda$ is a subsequence of $\underline{X}$ by writing $\underline{Y} \subseteq \underline{X}$; note this is a stronger condition than $Y \subseteq X$. Also, for $\underline{\lambda} \in \Lambda(m,n)$ we use the symbol $\hat{\underline{\lambda}} \in \Lambda(m,m-n)$ to indicate the increasing sequence of integers with $\lambda \cup \hat{\lambda} = \{1, \ldots, m\}$. We call the sequence $X_{\hat{\lambda}}$ the *complement* of the X_λ in X.

G.3. The symbol 2^X denotes the *power set* of X consisting of all subsets of X, i.e. $2^X := \{Y \mid Y \subseteq X\}$. If X is infinite, $\hat{2}^X$ is the set of all finite subsets of X. Similarly, the symbol $2^{\underline{X}}$ indicates the set of all subsequences of $\underline{X}$, whereas $\underline{2}^X$ denotes the set of all sequences formed from the elements of X. The symbol $\binom{X}{m}$, on the other hand, denotes the set of all *m-element subsets* of X, i.e. $\{Y \subseteq X \mid \#Y = m\}$. An m-tuple is a sequence such that $\#\underline{X} = \#X = m$. Finally, X^n denotes the n-fold *Cartesian product* of the set X, which is the set of all sequences of length n composed of elements in X: $X^n := \{\underline{Y} \mid Y \subseteq X, \#\underline{Y} = n\}$.

G.4. Symbols List

$a, \ldots, z$	integer and real variables and constants
$A, \ldots, Z$	integer and real variables and constants

$\mathbf{a}, \ldots, \mathbf{z}$	vectors
$\vec{pq}$	vector from point p to point q
$\lVert\mathbf{v}\rVert$	norm of vector $((\sum v_i^2)^{1/2})$
$\mathbf{a}\times\mathbf{b}$	vector cross product
$\mathbf{a}\cdot\mathbf{b}$	vector dot product
$\hat{\mathbf{p}}$	projective point, or projective coordinates $[1,\mathbf{p}]$
$\dot{g}$	time derivative of some function g at $t = 0$
$\vee$	outer product (Appendix C)
$\bigvee_k(\mathbb{R}^n)$	vector space of all linear combinations of k-vectors (§C.8)
$\mathbf{A}, \ldots, \mathbf{Z}$	sets of vectors, or k-vectors
$\underline{\mathbf{A}}, \ldots, \underline{\mathbf{Z}}$	sequences of vectors, matrices
$\underline{\mathbf{A}}^T$	transpose of matrix
$det(\underline{\mathbf{A}})$	determinant of matrix $\underline{\mathbf{A}}$
$vol(\underline{\mathbf{P}})$	oriented volume, i.e. $det(\hat{\mathbf{p}}_1, \ldots, \hat{\mathbf{p}}_N)$
$\rho(A)$	rank of A (§D.1)
a, . . . , z	members of sets
$\{x,y, \cdots\}$	set composed of $x,y, \cdots$
$\{a \mid B\}$	set of all a with property B
$\{\}$	empty set
A, . . . , Z	abstract sets
$x \in X$	x is a member of set X
$x \notin X$	x is not a member of set X
$\cup$	set union
$\cap$	set intersection
$\bigcup_i A_i$	set union: $A_1 \cup \cdots \cup A_n$
$\bigcap_i A_i$	set intersection: $A_1 \cap \cdots \cap A_n$
$\dot{\cup}$	union of disjoint sets
$A \times B$	Cartesian product of sets: $\{[a,b] \mid a \in A, b \in B\}$
$A \subset B$	A is a proper subset of B
$A \subseteq B$	A is a not necessarily proper subset of B
$A \supset B$	A is a proper superset of B
$A \supseteq B$	A is a not necessarily proper superset of B
$A \setminus B$	set subtraction: all elements of A not in B
$[x,y, \cdots]$	sequence consisting of $x,y, \cdots$ in that order
$\underline{A}, \ldots, \underline{Z}$	sequences
#	cardinality (i.e. set size), sequence length
$\mathbb{A}, \ldots, \mathbb{Z}$	groups, graphs, affine and vector spaces
$\mathbb{A}[A]$	amalgamation space of A (Equation 3.10)
Å	Ångstrom unit = 10^{-10} meter
B	bilinear form

$C^2(\mathbb{R}^n)$	twice differentiable functions on $\mathbb{R}^n$
$C^\infty(\mathbb{R}^n)$	infinitely differentiable functions on $\mathbb{R}^n$
D	squared distance
d	distance
$\mathbb{G}$	a graph (Appendix E)
$\mathbb{K}$	an algebraic field, complete graph
P_p	set of vertices of a graph with vertex set P adjacent to a vertex p
P(Q)	set of vertices of a graph with vertex set P incident to edges in Q
Q_p	set of edges of a graph with edge set Q adjacent to a vertex p
Q(P)	set of edges of a graph with edge set Q in a set of vertices P
Q	quadratic form
$\mathbb{R}$	the real numbers
$\mathbb{R}/2\pi$	real numbers modulo 2π
$\mathbb{R}^n$	n-fold Cartesian product of real numbers
$rad(\mathbf{V}_B)$	radical of bilinear vector space $\mathbf{V}_B$ (§A.2)
$dim(\mathbf{V}_B)$	dimension of bilinear vector space $\mathbf{V}_B$
$nul(\mathbf{V}_B)$	nullity of bilinear vector space, i.e. $dim(rad(\mathbf{V}_B))$
$\mathbf{V}/\mathbf{W}$	quotient space (§A.4)
$\ll\mathbf{V}\gg$	nondegenerate factor space: $\mathbf{V}/\mathrm{rad}(\mathbf{V})$
$((\mathbf{V}))$	affine span of **V**
$[(\mathbf{V})]$	convex span of **V**
$< \mathbf{V} >$	space linearly spanned by set of vectors **V**
$\mathbf{V}^\perp$	orthogonal complement of set of vectors **V** (§A.2)
$\oplus$	direct sum
$A \Rightarrow B$	A implies B
$A \Leftarrow B$	B implies A
$A \Leftrightarrow B$	A if and only if B
$A \neq B$	A not equal to B
$A = B$	A equals B
$A := B$	A is defined to be B
$A =: B$	B is defined to be A
$\approx$	ϵ-equivalent (def. 2.3); isomorphism of spaces; isomorphism of graphs, congruence
$\dot{\approx}$	proper congruence
$\equiv$	equivalence of frameworks
$\succ$	dominance of frameworks
$\prec$	subdominance of frameworks
cl	closure operator of a matroid (§D)
inf	infimum
$\underline{\mathbf{diag}}$	diagonal matrix

$\underline{\mathbf{I}}$	identity matrix
$\underline{\mathbf{M}}_D$	matrix of squared distances
$\underline{\tilde{\mathbf{M}}}_D$	bordered matrix of squared distances
$\underline{\mathbf{M}}_{\mathbf{p}}$	rigidity matrix (Equation 4.26)
$sign(x)$	sign of: equals 1,0,–1 for $x > 0$, $x = 0$, $x < 0$, respectively
$sign(\pi)$	sign of: equals 1 if permutation π is even and -1 if odd
$\mathbf{ST}(n,\mathbb{R})$	group of screw translations in $\mathbb{R}^n$ (Equation 1.1)
sup	supremum
l.h.s.	left hand side
r.h.s.	right hand side
t.g.t.b.t.	too good to be true
w.r.t.	with respect to
w.l.o.g.	without loss of generality
$\forall$	for all
$\exists$	there exists
δ_{ij}	Kronecker delta: 1 if $i = j$, and 0 otherwise
$\boldsymbol{\beta}$	binding mode vector (Chapter 8)
$\boldsymbol{\phi}$	vector of dihedral angles (Chapter 8)
Δ	antisymmetric function (Chapter 2), difference of two reals
χ	chirality function
$\tilde{\chi}$	set of possible chiralities, aggregate chirality
$\overline{\chi}$	chirality function versus given indexing of atoms (Equation 2.10)
$\hat{\overline{\chi}}$	dual orientation versus an indexing (Equation 2.22)
σ	relative orientation (Equation 2.12)
$\hat{\sigma}$	dual relative orientation
$\Lambda(m,n)$	set of increasing subsequences of length n out of the integers $1, \ldots, m$ (see §G.2)
$\underline{\lambda}, \underline{\mu}, \underline{\nu}$	ordered ascending sequences
$\binom{A}{m}$	set of all m-element subsets of A
$\hat{\underline{\lambda}}$	complement tuple $\underline{\lambda} \in \Lambda(m,n)$
2^A	set of all subsets of A (see §G.3)
$\hat{2}^A$	set of all finite subsets of A (see §G.3)
$2^{\underline{A}}$	set of all subsequences of $\underline{A}$
$\underline{2}^A$	set of all sequences of elements of A
A^B	the set of all functions mapping a set B into a set A
$\sigma\|_B$	restriction of σ to B
$\sigma\|^B$	contraction of σ to B (Equation 2.24)
$a \rightarrow b$	a approaches the limit b
$a : B \rightarrow C$	a is a mapping or function from the set B to the set C
$a \mapsto b$	map defined by rule "a maps to b"

$a \leftrightarrow b$	interchange a and b, or a corresponds to b
$a \circ b$	composition of functions a and b
$\sqrt{A}$	positive square root
$\sum_{i=1}^{n} a_i$	summation: $a_1 + \cdots + a_n$
$\prod_{i=1}^{n} a_i$	product: $a_1 \times \cdots \times a_n$
$\int$	integral
∞	infinity
$n!$	factorial; $= 1 \times 2 \times \cdots \times n$
$\partial x / \partial y$	partial derivative of x with respect to y
$\underline{\partial \mathbf{D}}(\mathbf{f})$	Jacobian matrix: $[\partial f_i / \partial x_j]_{ij}$
∇f	gradient of scalar function $f(\mathbf{x})$
$\nabla^2 f$	Hessian of f, i.e. $(\nabla^2 f)_{ij} = \frac{\partial^2 f}{\partial x_i \partial x_j}$

References

Aho1974.

A. Aho, J.E. Hopcroft, and J. Ullman, *The Design and Analysis of Computer Algorithms,* Addison-Wesley Publishing Co., Menlo Park, Calif., 1974.

Balaban1976.

A.T. Balaban, *Chemical Applications of Graph Theory,* Academic Press, San Francisco, 1976.

Berge1971.

C. Berge, *Principles of Combinatorics,* Academic Press, San Francisco, 1971.

Bertsekas1982.

D.P. Bertsekas, *Constrained Optimization and Lagrange Multiplier Methods,* p. 20, Academic Press, New York, 1982.

Burkert1982.

V. Burkert and N.L. Allinger, *Molecular Mechanics,* ACS Monograph no. 177, American Chemical Society, Washington, D.C., 1982.

Crapo1970.

H. Crapo and G.C. Rota, "On the Foundations of Combinatorial Theory: Combinatorial Geometries," *The M.I.T. Press,* iv+289(1970). Preliminary edition.

Culik1976.

K. Culik and A. Lindenmayer, "Parallel Graph Generating and Graph Recurrence Systems for Multicellular Development," *Intnl. J. Gen. Sys., 3,* 53-66(1976).

Dennis1983.

J.E. Dennis, Jr. and R.B. Schnabel, *Numerical Methods for Unconstrained Optimization and Nonlinear Equations,* Prentice-Hall, Englewood Cliffs, NJ, 1983.

Dennis1986.

J.E. Dennis and K. Turner, "Generalized Conjugate Directions," Tech. Report 85-11 (revised May, 1986), Cornell University, 1986.

Essam1971.

J. Essam, "Graph Theory and Statistical Physics," *Discrete Math., 1,* 83-112(1971).

Euler1953.

L. Euler, "Solutio Problematis ad Geometriam Situs Pertinantis," *Sci. Amer.,* 66-70(July 1953).

Fletcher1980.

R. Fletcher, *Practical Methods of Optimization, vols. 1 & 2,* J. Wiley & Sons, New York, NY, 1980.

Forder1940.

H.G. Forder, *The Calculus of Extension,* Cambridge University Press, Cambridge, U.K., 1940. Reprint by Chelsea Publishing Co., Bronx, New York (1960).

Gill1981.

P.E. Gill, W. Murray, and M.H. Wright, *Practical Optimization,* Academic Press, Orlando, FL, 1981.

Golumbic1980.

M.C. Golumbic, *Algorithmic Graph Theory and Perfect Graphs,* Academic Press, New York, NY, 1980.

Grassmann1844.

H.G. Grassmann, *Die lineal Ausdehnungslehre,* Privately published., Stettin, Germany, 1844. Reprinted in *Hermann Grassmanns Gesammelte mathematische und physikalische Werke,* edited by F. Engels (Leipzig, 1984-1911).

Grassmann1862.

H.G. Grassmann, *Die vollstandige Ausdehnungslehre,* Privately published., Stettin, Germany, 1862. Reprinted in *Hermann Grassmanns Gesammelte mathematische und physikalische Werke,* edited by F. Engels (Leipzig, 1984-1911).

Harary1972.

F.A. Harary, *Graph Theory,* Addison-Wesley Publishing Co., Menlo Park, Calif., 1972.

Havel1987.

T.F. Havel and B. Sturmfels, *A Geometric Approach to the Graßmann Algebra*, 1987. In preparation.

Hestenes1980.

M. Hestenes, *Conjugate Direction Methods in Optimization*, Applications of Mathematics, 12, Springer-Verlag, New York, NY, 1980.

Hillier1980.

F. Hillier and G. Lieberman, *Operations Research, 3rd edn.*, Holdenday, Inc., New York, 1980.

Himmelblau1972.

D.M. Himmelblau, in *Numerical Methods for Nonlinear Optimization*, ed. F. A. Lootsma, pp. 69-97, Academic Press, 1972.

Kuratowski1930.

K. Kuratowski, "Sur le Probleme des Courbes Gauches en Topologie," *Fund. Math., 15*, 271-283(1930).

Luenberger1973.

D.G. Luenberger, *Introduction to Linear and Nonlinear Programming*, Addison-Wesley, Menlo Park, CA, 1973.

Martinez1974.

H. Martinez and G. Carlsson, "Genetic Nets and Dissipative Structures: An Algebraic Approach," *Bull. Math. Biol., 36*, 183-196(1974).

McCulloch1965.

W. McCulloch, *Embodiments of Mind*, M.I.T. Press, Cambridge, Mass., 1965.

Moore1957.

E. Moore, "The Shortest Path through a Maze," in *Proc. Intnl. Symp. Th. Switching*, pp. 285-292, Harvard University Press, Boston, Mass., 1957.

Nazareth1986.

J.L. Nazareth, "Conjugate Gradient Methods Less Dependent on Conjugacy," *SIAM Rev., 28*, 501-511(1986).

Neumann1951.

J. von Neumann, "The General and Logical Theory of Automata," in *Cerebral Mechanisms in Behavior: The Hixon Symposium*, J. Wiley & Sons, New York, 1951.

Patton1980.

B. Patton, "Some Odd (and Even) Relationships in Food Webs," *Ecol. Model., 9*, 171-177(1980).

Pedoe1970.

D. Pedoe, *A Course of Geometry for Colleges and Universities,* Cambridge University Press, Cambridge, U.K., 1970.

Ponder1987.

J.W. Ponder and F.M. Richards, "An Efficient Newton-like Method for Molecular Mechanics Energy Minimization of Large Molecules," *J. Comp. Chem., 8,* 1016-1024(1987).

Press1986.

W.H. Press, B.P. Flannery, S.A. Teukolsky, and W.T. Vetterling, *Numerical Recipes: the Art of Scientific Computing,* Cambridge University Press, Cambridge, U.K., 1986.

Roberts1976.

F.S. Roberts, *Discrete Mathematical Models,* Prentice-Hall, Englewood Cliffs, New Jersey, 1976.

Schlick1987.

T. Schlick and M. Overton, "A Powerful Truncated Newton Method for Potential Energy Minimization," *J. Comp. Chem., 8,* 1025-1039(1987).

Snapper1971.

E. Snapper and R.J. Troyer, *Metric Affine Geometry,* Academic Press, New York, 1971.

Tarjan1972.

R. Tarjan, "Depth-First Search and Linear Graph Algorithms," *SIAM J. Comput., 1,* 146-160(1972).

Uhlenbeck1962.

G. Uhlenbeck and G. Ford, *The Theory of Linear Graphs,* Studies in Statistical Physics, 1, North Holland, Amsterdam, 1962.

Watson1968.

J.D. Watson, *The Double Helix,* Mentor, New York, 1968.

Welsh1976.

D.J.A. Welsh, *Matroid Theory,* Academic Press, New York & London, 1976.

White1986.

N.L. White, editor, *Theory of Matroids,* Encyclopedia of Mathematics, 26, Cambridge Univ. Pr., Cambridge, U.K., 1986.

Whitney1935.

H. Whitney, "On the Abstract Properties of Linear Independence," *Amer. J. Math., 57,* 509-535(1935).

Williamson1985.

S.G. Williamson, *Combinatorics for Computer Scientists*, Computer Science Press, Rockville, MD, 1985.

Wills1958.

A.P. Wills, *Vector Analysis with an Introduction to Tensor Analysis*, Dover Publications, Inc., New York, NY, 1958. Reprint of 1931 edition.

Index